Mathematical & Computational Physics

Mathematical & Computational Physics

Editor:

Olga Moreira, Ph.D.

ArclerPress

www.arclerpress.com

Mathematical & Computational Physics

Editor: Olga Moreira, Ph.D.

708 3rd Avenue, 6th Floor
New York
NY 10017
United States of America
www.arclerpress.com

ISBN: 978-1-68094-432-7

Library of Congress Control Number: 2016951124

About the Editor

Olga Moreira, Ph.D.

Olga Moreira obtained her Ph.D. in Astrophysics from the University of Liege (Belgium) in 2010, her BSc. in Physics and Applied Mathematics from the University of Porto (Portugal). Her post-graduate travels and international collaborations with the European Space Agency (ESA) and European Southern Observatory (ESO) led to great personal and professional growth as a scientist. Currently, she is working as an independent researcher, technical writer, and editor in the fields of Mathematics, Physics, Astronomy and Astrophysics.

Contents

List of Contributors

Tom-Robin Teschner
Centre for Fluid Mechanics and Computational Science*Cranfield University*

László Könözsy
Centre for Fluid Mechanics and Computational Science*Cranfield University*

Karl W. Jenkins
Centre for Fluid Mechanics and Computational Science*Cranfield University*

Wade A. Walker
Austin, Texas, United States of America

Alison Ramage
Department of Mathematics and Statistics*University of Strathclyde*

André M. Sonnet
Department of Mathematics and Statistics*University of Strathclyde*

K. Murawski
Faculty of Mathematics, Physics, and Informatics*UMCS*

A. K. Srivastava
Aryabhatta Research Institute of Observational Sciences (ARIES)

J. A. McLaughlin
School of Computing, Engineering and Information Sciences*Northumbria University*

R. Oliver
Departament de Física*Universitat de les Illes Balears*

Satoshi Inoue
Max-Planck Institute for Solar System Research, Justus-von-Liebig-Weg 3, 37077 Göttingen, Germany
Institute for Space-Earth Environmental Research, Nagoya University, Chikusa-ku, Furo-Cho, 446-701 Nagoya, Japan

Julia Contreras-García
Université Pierre et Marie Curie, Univ Paris 06, UMR 7616, Laboratoire de Chimie Théorique, F-75005 Paris, France
Centre National de la Recherche Scientifique, Unité Mixte de Recherche 7616, Laboratoire de Chimie Théorique, F-75005 Paris, France
MALTA-Consolider Team Universidad de Oviedo, E-33006 Oviedo, Spain

Miriam Marqués
MALTA-Consolider Team Universidad de Oviedo, E-33006 Oviedo, Spain
Departamento de Química Física y Analítica, Universidad de Oviedo, E-33006 Oviedo, Spain
Departamento de Física Teórica, Universidad de Valladolid, E-47011 Valladolid, Spain

José Manuel Menéndez
MALTA-Consolider Team Universidad de Oviedo, E-33006 Oviedo, Spain
Departamento de Química Física y Analítica, Universidad de Oviedo, E-33006 Oviedo, Spain

José Manuel Recio
MALTA-Consolider Team Universidad de Oviedo, E-33006 Oviedo, Spain
Departamento de Química Física y Analítica, Universidad de Oviedo, E-33006 Oviedo, Spain

P.C. Bollada
University of Leeds, United Kingdom

C.E. Goodyer
University of Leeds, United Kingdom
Numerical Algorithms Group, United Kingdom

P.K. Jimack
University of Leeds, United Kingdom

A.M.Mullis
University of Leeds, United Kingdom

F.W. Yang
University of Leeds, United Kingdom
Numerical Algorithms Group, United Kingdom

Jari Hämäläinen
Centre of Computational Engineering and Integrated Design (CEID) and Department of Mathematics and Physics, Lappeenranta University of Technology (LUT), Lappeenranta, 53851, Finland.

Taija Hämäläinen
Metso Automation Oy, Vanha Porvoontie 229, Vantaa, 01380, Finland.

Teemu Leppänen
LUT Savo Sustainable Technologies (Varkaus), Lappeenranta University of Technology, Osmajoentie 75, Varkaus, 78210, Finland.

Heidi Niskanen
ELOMATIC, Itäinen Rantakatu 72, Turku, 20810, Finland.

Joonas Sorvari
Centre of Computational Engineering and Integrated Design (CEID) and Department of Mathematics and Physics, Lappeenranta University of Technology (LUT), Lappeenranta, 53851, Finland.

Marta Galanti
Dipartimento di Fisica e Astronomia and CSDC, Università degli Studi di Firenze, Firenze, Italia
Centre de Biophysique Moléculaire, Centre National de la Recherche Scientifique-UPR4301, Université d'Orléans, Orléans, France
Istituto Nazionale di Fisica Nucleare (INFN), Sezione di Firenze, Florence, Italy

Duccio Fanelli
Dipartimento di Fisica e Astronomia and CSDC, Università degli Studi di Firenze, Firenze, Italia
Istituto Nazionale di Fisica Nucleare (INFN), Sezione di Firenze, Florence, Italy

Francesco Piazza
Centre de Biophysique Moléculaire, Centre National de la Recherche Scientifique-UPR4301, Université d'Orléans, Orléans, France

Pietro Ballone
Department of Physics, Università di Roma "La Sapienza", Roma 00185, Italy; Tel.: +39-06-4991-4248; Fax: +39-06-4991-7697

Martin O. Steinhauser
Research Group of Shock Waves in Soft and Biological Matter, Fraunhofer Ernst-Mach-Institute for High-Speed Dynamics (EMI), Eckerstrasse 4, 79104 Freiburg, Germany

Stefan Hiermaier
Research Group of Shock Waves in Soft and Biological Matter, Fraunhofer Ernst-Mach-Institute for High-Speed Dynamics (EMI), Eckerstrasse 4, 79104 Freiburg, Germany

Robert Penfold
Institute of Food Research*Norwich Research Park*

Peter J. Wilde
Institute of Food Research*Norwich Research Park*

Jack A Tuszynski
Department of Physics and Department of Oncology, University of Alberta, Edmonton, Canada.

Philip Winter
Department of Oncology, University of Alberta, Edmonton, Canada.

Diana White
Department of Oncology, University of Alberta, Edmonton, Canada.

Chih-Yuan Tseng
Department of Oncology, University of Alberta, Edmonton, Canada.

Kamlesh K Sahu
Department of Physics, University of Alberta, Edmonton, Canada.

Francesco Gentile
Department of Mechanical and Aerospace Engineering, Politecnico di Torino, Torino, Italy.

Ivana Spasevska
Department of Biology, Ecole Normale Supérieure de Lyon, Lyon, France.

Sara Ibrahim Omar
Department of Oncology, University of Alberta, Edmonton, Canada.

Niloofar Nayebi
Department of Physics, University of Alberta, Edmonton, Canada.

Cassandra DM Churchill
Department of Chemistry, University of Alberta, Edmonton, Canada.

Mariusz Klobukowski
Department of Chemistry, University of Alberta, Edmonton, Canada.

Rabab M Abou El-Magd
Department of Biological Sciences, University of Alberta, Edmonton, Canada.

Xiaojun Xu
Department of Physics, University of Missouri, Columbia, MO 65211, USA;
Department of Biochemistry, University of Missouri, Columbia, MO 65211, USA;
Informatics Institute, University of Missouri, Columbia, MO 65211, USA

Shi-Jie Chen
Department of Physics, University of Missouri, Columbia, MO 65211, USA;
Department of Biochemistry, University of Missouri, Columbia, MO 65211, USA;
Informatics Institute, University of Missouri, Columbia, MO 65211, USA

Feng Luo
Environmental Sciences Division, Oak Ridge National Laboratory, Oak Ridge, Tennessee 37831, USA,
School of Computing, Clemson University, Clemson, SC, 29634, USA,

Yunfeng Yang
Environmental Sciences Division, Oak Ridge National Laboratory, Oak Ridge, Tennessee 37831, USA,

Jianxin Zhong
Computer Science & Mathematics Division, Oak Ridge National Laboratory, Oak Ridge, Tennessee 37831, USA,
Department of Physics, Xiangtan University, Hunan 411105, PR China,

Haichun Gao
Environmental Sciences Division, Oak Ridge National Laboratory, Oak Ridge, Tennessee 37831, USA,
Insitute for Environmental Genomics, and Department of Botany and Microbiology, University of Oklahoma, Norman, OK, 73019, USA

Latifur Khan
Department of Computer Science, University of Texas at Dallas, Richardson, TX 75083, USA,

Dorothea K Thompson
Environmental Sciences Division, Oak Ridge National Laboratory, Oak Ridge, Tennessee 37831, USA,
Department of Biological Sciences, Purdue University, West Lafayette, IN, 47907, USA

Jizhong Zhou
Insitute for Environmental Genomics, and Department of Botany and Microbiology, University of Oklahoma, Norman, OK, 73019, USA
Environmental Sciences Division, Oak Ridge National Laboratory, Oak Ridge, Tennessee 37831, USA,

G. MacDonald
Department of Mathematics and Statistics, University of Strathclyde, Glasgow, G1 1XH, United Kingdom

J.A. Mackenzie
Department of Mathematics and Statistics, University of Strathclyde, Glasgow, G1 1XH, United Kingdom

M. Nolan
Department of Mathematics and Statistics, University of Strathclyde, Glasgow, G1 1XH, United Kingdom

R.H. Insall
The Beatson Institute for Cancer Research, Garscube Estate, Switchback Road, Glasgow, G61 1BD, United Kingdom
J.A. Mackenzie:

Anotida Madzvamuse,
School of Mathematical and Physical Sciences, Department of Mathematics, University of Sussex, Brighton BN19QH, UK

Andy H. W. Chung,
School of Mathematical and Physical Sciences, Department of Mathematics, University of Sussex, Brighton BN19QH, UK

Chandrasekhar Venkataraman
School of Mathematical and Physical Sciences, Department of Mathematics, University of Sussex, Brighton BN19QH, UK

Preface

This book aims to provide an overview of the recent research devoted to numerical simulations of physical systems and processes by discussing the tools used in the analyses. It contains 17 chapters, organized into 3 sections:

1. Section 1 - Computational Fluid Dynamics, Magnetohydrodynamics, Solid State.
2. Section 2 - Computational Many-body problem, Condensed Matter, and Molecular Dynamics.
3. Section 3 - Computational Molecular Biophysics, Cellular Structure Prediction, and Reaction–Diffusion Systems.

Section 1 starts with a review of particle-based multiscale and hybrid methods applied to fluid mechanics, e.g. molecular dynamics, direct simulation Monte Carlo, lattice Boltzmann method, dissipative particle dynamics and smoothed-particle hydrodynamics. Thereupon, it will focus on numerical methods in the field of computational fluid dynamics, including the new Repeated Replacement Method (RRM) and iterative techniques for finite element Navier–Stokes approximations in incompressible fluid dynamics. Several methods and mathematical models are discussed in the encompassing context of magnetohydrodynamics modeling of the solar atmosphere, non-isothermal solidification of a metal alloy, electronic nature of compressibility in solids. This section ends with an illustration of solid state modeling applied to a papermaking process. Section 2 starts by reviewing of macroscopic transport equations in many-body systems which pose fundamental theoretical challenges in many domains ranging from inter and intra-cellular transport to diffusion in porous media. This section addresses various aspects of the many-body problem in molecular dynamics simulations and condensed matter. Section 3, starts by reviewing computational modeling in biology, covering a wide range of scales from organisms to electrons in atoms, also including the utility of molecular dynamics and quantum mechanical methods in biophysical and biochemical modeling. Next, overviews physics-based cellular structure prediction, focusing on a recently developed RNA statistical mechanical model (the Vfold model) and an application of Random Matrix Theory in the identification of gene expression. The last two chapters is devoted to new models for coupled systems of bulk-surface reaction–diffusion equations which have many applications in fluid dynamics and biological processes.

Editor
Olga Moreira, Ph.D.

INTRODUCTION

Computational mathematics involves mathematical research in areas of science where computing plays a central and essential role, emphasizing algorithms, numerical methods, and symbolic computations. Computation in research is prominent. Computational physics is a field that utilizes both existing physics formulas and numerical algorithms to make large-scale calculations with computers that would be extremely time consuming and tedious to perform by hand. Essentially, it is a branch of physics that is concerned with making mathematical models and solutions by using computers and programming. The mathematical models that physicists aim to create often involve vast amounts of information that require very powerful computers to process. The exact classification of computational physics in the overall field of physics is often debated. Some consider it to be a branch of theoretical physics, as it tends to be involved with areas of physics that are still deeply theoretical with little solid experimental support. Others believe that it should be considered a branch of experimental physics, as the data used typically comes from experiments. For the most part, though, scientists agree that it falls somewhere in between the two disciplines, and has both theoretical and experimental components.

Modern physics relies heavily on computers to work out much of the complex mathematical aspects of experiments and theories. Physics fields such as astrophysics, fluid mechanics, and accelerator physics both depend on programming and computation. In accelerator physics, for example, computers must monitor, record, and analyze vast quantities of information each time that particles are collided in a particle accelerator. Computational solid-state physics attempts to discover the link between the atomic properties of solids and their large-scale properties by analyzing large amounts of information about solids at the molecular level.

There are many other tasks solved through computation that can be loosely grouped under the field of computational physics. Often, such tasks as solving differential and integral equations or evaluating very large matrices are used to make calculations about physical systems. These tasks could easily be classified either as pure mathematics, which is mathematics performed purely for the sake of mathematics. However, when performed to discern information related to physics, they can just as easily fall into the category of computational physics.

Many colleges offer courses in computational physics, though any pre-collegiate instruction in the field is rare. Introductory college courses tend to teach basic programming principles and how to apply them to problems relating to physics. Later courses, often taught at the graduate level, teach how to manipulate and solve large problems made up of large quantities of data through the use of algorithms and advanced programming practices.

Challenges in computational physics

Physics problems are in general very difficult to solve exactly. This is due to several (mathematical) reasons: lack of algebraic and/or analytic solubility, complexity, and chaos. For example, - even apparently simple problems, such as calculating the wave function of an electron orbiting an atom in a strong electric field (Stark effect), may require great effort to formulate a practical algorithm (if one can be found); other cruder or brute-force techniques, such as graphical methods or root finding, may be required. On the more advanced side, mathematical perturbation theory is also sometimes used.

Divisions

It is possible to find a corresponding computational branch for every major field in physics, for example computational mechanics and computational electrodynamics. Computational mechanics consists of computational fluid dynamics (CFD), computational solid mechanics and computational contact mechanics. One subfield at the confluence between CFD and electromagnetic modelling is computational magneto hydrodynamics. The quantum many-body problem leads naturally to the large and rapidly growing field of computational chemistry.

Computational solid state physics is a very important division of computational physics dealing directly with material science. A field related to computational condensed matter is computational statistical mechanics, which deals with the simulation of models and theories that are difficult to solve otherwise. Computational statistical physics makes heavy use of Monte Carlo-like methods. More broadly, (particularly through the use of agent based modeling and cellular automata) it also concerns itself with (and finds application in, through the use of its techniques) in the social sciences, network theory, and mathematical models for the propagation of and the spread of forest fires. On the more esoteric side, numerical relativity is a (relatively) new field interested in finding numeric solutions to the field equations of general (and special) relativity, and computational particle physics deals with problems motivated by particle physics. Computational astrophysics is the application of these techniques and methods to astrophysical problems and phenomena. Some of the most important applications of computation in astrophysics are given below.

Stellar structure and evolution

Calculating multidimensional stellar models of rapidly rotating stars, modeling the effects of hydro dynamical processes such as convection from first principles, and understanding how stars generate magnetic fields through dynamo processes.

Radiation transfer and stellar atmospheres

Computational methods are required to calculate the propagation of light through the outer layers of a star, including its interaction with matter through absorption, emission, and scattering of photons. The calculation of cross sections for the interaction of light with matter for astrophysical relevant ions is itself a challenging computational problem.

Astrophysical fluid dynamics

The dynamics of most of the visible matter in the universe can be treated as a compressible fluid. Time-dependent and multidimensional solutions to the fluid equations, including the effects of gravitational, magnetic, and radiation fields, require numerical methods. A vast range of problems are addressed in this way, from convection and dynamo action in stellar and planetary interiors, to the formation of galaxies and the large scale structure of the universe

Planetary, stellar, and galactic dynamics

The most challenging problems today include accurate integration of the orbits of the planets over the age of the solar system, studying the dynamics of globular clusters including the effect of stellar evolution and the formation of binaries, studying galaxy mergers and interaction, and computing structure formation in the universe through the gravitational clustering of collision less dark matter.

Computational physics: "physics of the third kind"

Computational physics has emerged as the new third branch of physics besides the traditional branches of experimental and theoretical physics. The purpose of computational physics is not to crunch numbers, but to gain insight. This is particularly true if scientific workstations and supercomputers are coupled to sophisticated tools of visualization. During the last decade we have witnessed an almost exponential growth in computing power. This unparalled growth has redefined the classes of physics problems we are able to solve. What seemed at the forefront of research ten years ago can now be done on a high-speed scientific workstation? Today's massively parallel supercomputers allow us to address, at a fundamental rather than phenomenological level, some of the most challenging theoretical problems of modern-day physics.

The physical world: interacting quantum many-particle systems

One basic problem that is common to many areas of physics -- and other natural sciences such as chemistry -- is the quantum many-particle problem. Theorists working in atomic, condensed matter, nuclear and astrophysics (and some areas of elementary particle physics) face a very similar challenge: how to describe, usually at the quantum level, the features of many particle systems in terms of more basic interacting constituent particles. There are essentially two different approaches to the quantum many-particle problem which might be termed phenomenological and fundamental. In the first case, one tries to simplify the physical system to such an extent that the arising physical "model" can be solved analytically or with little computational effort. This approach is often the first stage in the development of a theory. As the field begins to mature, attention shifts from simple intuitive models to a "fundamental'" understanding, i.e. one attempts to describe physical systems "ab initio" starting from the most basic equations and physical principles. Computational physicists prefer the second approach. Almost all interacting quantum many-particle systems cannot be formulated perturbative; in fact, the interesting physical phenomenon (e.g. the ground state energy of the quantum system) is usually an infinite sum of perturbative diagrams. This means that the perturbative machinery of quantum field theory (Feynman diagrams etc.) is essentially useless, and the quantum field equations must be solved by new approximation schemes without invoking perturbation theory. Because of the complexity of quantum mechanical many-body problems this implies a numerical implementation on massively parallel architectures and requires substantial advances in both science and computational algorithms.

Computational Physics: Interdisciplinary Research

The essence of computational physics lies in the observation that many of the fundamental equations of physics have a similar mathematical structure: for example, the many-particle Schrodinger and Dirac equations as well as the nonrelativistic or relativistic hydrodynamics equations are all partial differential equations in space and time. All of the above-mentioned basic equations of physics can be implemented on spatial coordinate lattices; there will still be differences between many of the physics problems to be studied, for example short-range vs. long-range interactions, uniform vs. non-uniform lattice spacing etc. Nevertheless, with careful planning, computational physicists are capable of developing numerical methods and algorithms that can be utilized in a transparent manner in many subfields of physics. Computational physics is truly interdisciplinary.

Computational Complexity Theory

Computational complexity theory is an area of mathematics and computer science that is concerned with the resources necessary to solve problems on a computer system. A number of techniques are available to determine the resource requirements of a problem. Some problems might not be feasible on existing computer systems because of their resource demands. Researchers classify problems by difficulty and can divide computations into polynomial (P) versus nonterministic polynomial (NP) problems.

Solving a computation requires resources such as time, storage space and hardware. A computer system might have limitations that make a problem functionally impossible to solve because it does not have the available resources. As computer technology improves, a previously unsolvable problem might become solvable with the help of new technology and research in the field of computational complexity theory. The solvability of a problem is not necessarily determined by its complexity but on the algorithms used to solve it.

In computational complexity theory, a P problem is one that can be solved in polynomial time with a straightforward algorithm. It might still require substantial resources, but it is both solvable and checkable by computer. Such problems could be thought of as quickly solvable as long as a computer has the available resources to handle the necessary computations.

NP problems are more complex. It is not possible to apply a single algorithm, and it might be necessary to use more advanced options, such as parallel Turing machines that can explore several options. The problem might be solvable this way, but it will require substantially more resources. Such problems might be easier for human operators who are capable of advanced logical thinking, because the tipping point is often one of logic rather than sheer computation difficulty. The traveling salesman problem, in which the goal is to find the most efficient route between a numbers of cities along a route, is a classic example of an NP problem in computational complexity theory.

Classification of P versus NP problems through computational complexity theory can be a complex task, and problems might shift back and forth across the divide. A small set of computational problems do not fit neatly in either category and are sometimes classified as neither in order to reflect this. It might eventually be possible to develop an algorithm to solve an NP problem, and in some cases, it might apply to other problems that have a similar structure. In others, however, it might be problem-specific. The process of exploring such programs and developing approaches to solve them is an important area of mathematics and computer science that contributes to the development of advanced, high-powered computer systems.

Computational Fluid Dynamics

Computational Fluid Dynamics (CFD) is the study of the behavior of fluids, liquid and gaseous, by the use of powerful computers running numerical methods software. Knowledge of the interaction of solids with surrounding flowing fluids is of key interest in the design of many mechanical devices. CFD has expanded the topics to which fluid dynamic studies and experiments may be applied.

Traditionally, computational fluid dynamics studies were conducted in wind tunnels or flowing water tanks with real or model planes, cars, and boats. With the use of CFD, the mechanisms of such diverse events as volcanic eruptions, hurricanes, standing vortexes in water or in the air, ocean currents, the course of wildfires and more are potential targets. A limit to these studies is knowledge of the variables that must be defined for each system. Minimum variables include temperature, pressure, and compositions for systems undergoing chemical reactions at a defined boundary.

CFD software is based on the solution of the Navier-Stokes equations, or simplifications of them. The variables of interest are defined for one known boundary in the system. A virtual grid of either two or three dimensions is placed over the system, and the equations are solved for the properties of the incoming and outgoing fluid at each virtual boundary. The development of CFD software paralleled the availability of computational power, as the algorithms require repeated calculation and optimization until solutions are found.

Vehicle design is a frequent goal of fluid dynamics experiments. Air flows around cars effect performance, fuel consumption and noise level. Airplanes, boats, and especially space vehicles rely on these studies for predicting heat or ice buildup as well as streamlining to reduce frictional losses.

Heat dissipation is a major topic in computational fluid dynamics. All electronic components are susceptible to heat buildup and are often enclosed in small boxes with limited airflow. By the use of CFD models, designers can reroute components to better airflow and cooling.

The study of boundary layer conditions are tackled by computational fluid dynamics. The boundary layer refers to the very thin layer of fluid that is static along the surface of a solid that is in the path of a moving fluid. In this microenvironment is where corrosion, heat transfer, and component concentration levels are most critical.

The acquisition of skills to work in the field of computational fluid dynamics usually requires education in chemical engineering or similar pursuits. A thorough understanding of mass transfer, heat transfer, kinetics and fluid dynamics is necessary. The use of commercial CFD application packages is often taught by the software company or the skills are developed on the job.

Quantitative Physics

Quantitative physics is the branch of physics that involves research by repeated measurement and mathematical analysis of experimental results. It differs from some branches of theoretical physics, for instance, such as quantum mechanics or string theory research, where much of the underlying theory cannot be tested in the real world, or in a laboratory on Earth with current technology as of 2011. Any field of quantitative research like quantitative physics derives its conclusions from a statistical analysis of large amounts of experimental data. This data is often so vast and complex, however, that computers are used to do mathematical modeling of the data to interpret it better. An example of the use of quantitative physics would include that of climate studies run on supercomputers to predict climatological changes from various natural thermodynamic forces at play on, in, or near the Earth, as well as from changes in solar activity over long periods of time.

The study of physics at its core is the measurement of changes in matter and energy, and this makes most physics research quantitative physics in one form or another. Quantitative study is also important in physics because many of the physical laws, such as the speed of light or gravitational pull of the Earth, cannot be quantitatively defined just by human observation with the five senses. It is possible to observe a falling body, but, without precisely measuring its rate of descent, no clear picture is arrived at for how strong gravity actually is. Quantitative research physics, therefore, uses mathematics as an abstract way of understanding forces at work in the universe.

Processes that involve quantitative study, however, are not always intended to represent everyday reality. Physics determines the ideal conditions under which matter, energy, space, and time interact through repeated measurement and observation, and then determines the probability of events occurring. Physics equations used for this are based on abstract mathematical concepts that are only proven true with large numbers of repeated experiments. Quantitative physics, for instance, can predict the surface area of a spherical planet in space, but there is no such thing as a perfect sphere or any other perfect geometric shape in the natural world, so the process is, to some degree, an approximation.

Ideal representations in physics, such as the ballistic trajectory of a bullet through the air, are based on quantitative physics principles of gravitational pull and air resistance, but they can only predict a general trajectory for a bullet, not the actual, precise spot upon which it will land. Using equations and formula in quantitative physics often involves averaging out some of the variables that come into play or using mathematical shortcuts to negate their effect on the equation. This is because the goal is to understand the laws of nature in principle over that of specific, random applications.

Computational physics often complements quantitative physics in the laboratory, where equations cannot be formally or adequately tested in real world experiments. Often algorithms are used to streamline such calculations. Algorithms are a set of mathematical rules that the computer uses to reduce the number of calculations needed to solve a problem down to a finite series of steps. Computer assistance for quantitative physics is usually employed in areas where very complex interactions take place, such as in materials science, nuclear accelerator research, and molecular dynamics in biology.

SECTION I: COMPUTATIONAL FLUID DYNAMICS, MAGNETOHYDRODYNAMICS, SOLID STATE

CHAPTER 1

PROGRESS IN PARTICLE-BASED MULTISCALE AND HYBRID METHODS FOR FLOW APPLICATIONS

Tom-Robin Teschner, László Könözsy, and Karl W. Jenkins

Keywords: Multiscale simulations, Molecular dynamics, Direct simulation, Monte Carlo Lattice, Boltzmann dissipative particle dynamics, Smoothed-particle hydrodynamics

INTRODUCTION

Multiscale and hybrid methods have been successfully applied to fluid mechanics problems in recent years. The reason for combining different length and timescales is to overcome limitations of numerical approaches and increase computational efficiency. The Navier–Stokes equations are well suited for continuum flows but are limited at micro- or nanoscales. The Knudsen number is the ratio of the mean free path over a characteristic length scale and indicates whether the continuum model or the molecular approach should be applied. Karniadakis et al. (2006) stated that the continuum model is not valid for Kn>0.001 for gaseous flows. Gad-el Hak (2001) divided the flow regimes in which the no-slip boundary condition fails for $10^{-3}\leq Kn\leq 10^{-1}$, and the slip condition is expected at solid boundaries. The transition regime can be given between $10^{-1}\leq Kn\leq 10^{1}$, and flows are classified as free molecular flows when the Knudsen numbers are higher than ten. One of the most well-known space applications subject to all of these Knudsen number regimes is the Earth re-entry vehicle. The Stardust sample return capsule, a NASA project operating from 1999 to 2006, was specifically designed for collecting cosmic dust samples. Mitcheltree et al. (1997) reviewed numerical and experimental studies on this capsule to investigate the differences between continuum and particle-based approaches for various Knudsen numbers. They also reported that above the Earth at 130 km, the molecular effects become dominant (Kn> 10), and when the capsule is passing through the dense atmosphere, the medium starts to behave as a continuum. They observed discrepancies between the Navier–Stokes solvers and the particle-based direct simulation Monte Carlo (DSMC) method for the axial force coefficient at the altitude of 83 km. This difference can be explained as the flow is subject to different Knudsen numbers; therefore, the continuum approach breaks down.

Particle-based methods, such as molecular dynamics (MD), are derived from atomistic observations and therefore valid on much smaller physical length scales. It is, however, not feasible to employ these approaches on the continuum level as the computational cost becomes prohibitively expensive. Therefore, particle-based hybrid and multiscale methods were actively developed over the last two decades. MD is capable of simulating the assumed correct behaviour of the slip boundary conditions, and it starts to behave as a continuous medium at around 10–20 atomic diameter taken from the wall (Asproulis et al. 2012). For channels exceeding a couple of hundred atomic diameters in their height (Asproulis et al. 2012; Xu and Li 2007), computational efficiency is expected to be high by using multiscale methods. In addition to this, experimental capabilities such as micron-resolution particle image velocimetry (μ-PIV) (Santiago et al. 1998; Tretheway and Meinhart 2002) can still only operate at microscales. Advances towards smaller scales are hampered by the diffraction limit, noise in the particle image, interaction between fluid and seed particles, and the effects of the Brownian motion. Multiscale approaches are situated between pure particle simulations and experiments and able to obtain results in an efficient and confident manner; however, relatively few studies have been published concerning this overlapping region.

Micro- and nanofluidics became important in recent years as can be seen by the advent of MEMS, μ-TAS and lab-on-a-Chip devices (Gad-el Hak 2001; Ho and Tai 1998; Manz et al.1993). All of them are operating at length scales in which the continuum approach may not be valid. Squires and Quake

Citation: Teschner, T., Könözsy, L. & Jenkins, K.W. Microfluid Nanofluid (2016), Progress in particle-based multiscale and hybrid methods for flow applications 20: 68, doi:10.1007/s10404-016-1729-y.

(2005) investigated the flow physics at these small scales in terms of dimensionless numbers. They investigated the inertial effects (Reynolds number), advection and diffusion (Péclet number), interfacial tension (Capillary number), elastic effects occurring in deformable objects such as polymers (Deborah, Weissenberg and Elasticity number), density-driven flows (Grashof and Rayleigh number) and the validity of the continuum model (Knudsen number) concluding that the behaviour of the fluid flow at microscales differs due to the increased surface-to-volume ratio. More information can be found in the papers of Bayraktar and Pidugu (2006) focusing on flow physics in microchannels, and Gravesen et al. (1993) investigated micropumps, microvalves and microsensors. Koo and Kleinstreuer (2003) categorised the literature on microchannels into "flow instabilities", "viscous changes" and "no changes compared to macroscale". Understanding the fundamental different behaviours of the fluid flow at microscales is mandatory to appreciate the need for multiscale methods. The combination of two or more fluid dynamic approaches may accurately describe the fluid flow behaviour at the atomistic level with a particle-based approach while simulating the bulk of the fluid domain with an efficient Navier–Stokes solver.

In this paper, we present practical examples that set the scene for multiscale methods and we discuss the main findings in recent years related to several particle-based multiscale and hybrid methods including molecular dynamics (MD), direct simulation Monte Carlo (DSMC), lattice Boltzmann method (LBM), dissipative particle dynamics (DPD) and smoothed-particle hydrodynamics (SPH). Each method has been discussed separately in its own section where a brief overview has been given on the method and its governing equations. Each section presents the current state-of-the-art research followed by an interim conclusion. We summarise our findings on particle-based multiscale and hybrid methods in the last section.

MOLECULAR DYNAMICS METHOD

Molecular dynamics models the movement of atoms at the atomistic scale. We invoke Newton's second law for every atom i directly as

$$m_i \frac{\partial^2 \mathbf{r_i}}{\partial t^2} = \mathbf{F_i} = -\sum_{i \neq j} \frac{\partial V(r_{ij})}{\partial \mathbf{r}}. \tag{1}$$

The inter-atomic forces are replaced by the derivative of the inter-atomic potential. The most used one is the 12–6 Lennard–Jones (LJ) potential

$$V(r_{ij}) = 4\epsilon \left[\left(\frac{\sigma}{r_{ij}} \right)^{12} - \left(\frac{\sigma}{r_{ij}} \right)^{6} \right], \tag{2}$$

where ϵ is the depth of the potential well, σ is the distance at which the potential between two atoms is zero and r_{ij} is the distance between two atoms. In theory, we have to sum over all atoms for each atom to obtain V_{ij} which is computationally expensive and proportional to $\mathcal{O}(N^2)$. We define a cut-off distance r_{cut} after which we neglect the inter-atomic interactions. This is possible because the potential [Eq. (2)] asymptotically approaches zero for large values of r_{ij} and therefore, the computational time reduces to $\mathcal{O}(N \log N)$. From kinetic gas theory, we obtain the equipartition theorem for single atoms as

$$\frac{1}{2} m_i \mathbf{v_i}^2 = \frac{3}{2} k_b T, \tag{3}$$

where v_i is the velocity of the atom, k_B is the Boltzmann constant and T is the temperature. While we can conserve the overall momentum of a system by using Eq. (1), we cannot control the temperature. Therefore, it is common practice to rescale the velocity according to Eq. (3) or to use a thermostat. One widely used thermostat is that of Berendsen et al. (1984) where the equation of motion, Eq. (1), is coupled to a heat bath as

$$m_i \frac{\partial^2 \mathbf{r_i}}{\partial t^2} = -\sum_{i \neq j} \frac{\partial V(r_{ij})}{\partial \mathbf{r}} + m_i \gamma \left(\frac{T_0}{T} - 1 \right) \mathbf{v_i}, \tag{4}$$

where T_0 is the target temperature, T is the current (computed) temperature and γ is the strength at which the equation should relax towards T_0. The atoms are initialised with a random velocity at a temperature T obeying the Maxwell–Boltzmann distribution as

$$p(\mathbf{v_i}) = \left(\frac{m_i}{2\pi k_B T} \right)^{1/2} \exp \left[-\frac{1}{2} \frac{m_i \mathbf{v_i}^2}{k_B T} \right]. \tag{5}$$

Boundary conditions are not straight forward in MD simulations and require special attention. Where possible, periodic boundary conditions are applied. Particles can leave and enter the domain seamlessly. The problem arises when dealing with open boundary conditions. Particles may be inserted or removed easily; however, the force evaluation near the boundary results in an incorrect interaction potential due to the missing particles beyond the boundaries which may propagate into the domain. One remedy may be to sample the force near the boundary via a second simulation where periodic boundaries are applied, see, for example, Steijl and Barakos (2012). This, however, requires a second set of simulation data for describing correct boundary conditions. Solid boundaries are less problematic to impose. Atoms may be placed rigidly in a lattice structure to model a wall. Asproulis and Drikakis (2011) pointed out the danger of using non-physical, high mass values to effectively freeze atoms into position in that it produces incorrect slip length behaviour. Asproulis and Drikakis (2010) showed that using spring potentials for atoms on solid boundaries produces more realistic values for the slip length. See also Thompson and Troian (1997) for a description on solid boundaries and Delgado-Buscalioni et al. (2015) for an overview on open boundary conditions.

Review on hybrid and multiscale molecular dynamics methods

In 1995, O'Connell and Thompson (1995) conducted the first coupled computation of MD and continuum mechanics. Ever since there has been a broad interest in this field as indicated by the growing number of publications. They investigated the flow in a channel and split it into a continuum (Navier–Stokes) and atomistic (MD) region, separated by a hybrid solution interface (HSI) running parallel to the solid boundaries across the channel. The HSI is a buffer layer in which both descriptions are valid and it provides boundary conditions for each other. In this way, information from one description is passed to the other and an information exchange can take place. The buffer region introduced by the HSI is needed to obtain smooth profiles for the primitive variables across the interface. The MD and continuum region were coupled by using constrained dynamics, and the exchange of variables was done by state, i.e. velocity and density were imposed directly at the HSI. Their test case was the start-up Couette flow, for which good agreement between MD and continuum could be observed. Hadjiconstantinou and Patera (1997) investigated the flow around a square cylinder inside a channel and coupled non-equilibrium molecular dynamics (NEMD) with a spectral element solver in the wake of the cylinder. The solution at the interface was iteratively obtained by the Schwarz alternating method. Since the coupling region was chosen away from the wall, both NEMD and Navier–Stokes solutions were valid and could be compared against a full Navier–Stokes simulation. Good agreement was achieved between the full continuum and hybrid computation although the accuracy was limited due to statistical fluctuations, boundary condition imposition by NEMD and mismatch of the transport coefficient in the two models. Following up on their research, Hadjiconstantinou (1999) investigated the moving contact-line problem (two immiscible but otherwise identical fluids) in a microchannel for low Reynolds numbers. The continuum domain was placed at the channel centre, and the walls were resolved by MD. The initial velocity distribution was obtained via a full continuum solution, and therefore, only a few Schwartz iterations were necessary to converge the multiscale approach to its final solution. Comparison between a full MD solution showed a similar behaviour as in their previous study (Hadjiconstantinou and Patera 1997), where the overall trend was matched with fluctuations noticeably present.

Flekkøy et al. (2000) introduced a different approach to couple the atomistic and continuum region by imposing fluxes in the HSI. The Navier–Stokes equations are written in the following form

$$\frac{\partial \rho}{\partial t} + \nabla \cdot (\rho \mathbf{u}) = 0, \tag{6}$$

$$\frac{\partial \rho \mathbf{u}}{\partial t} + \nabla \cdot \mathbf{\Pi} = 0, \tag{7}$$

$$\frac{\partial \rho e}{\partial t} + \nabla \left(\rho e \mathbf{u} + \mathbf{\Pi} \cdot \mathbf{u} + q\right) = 0, \tag{8}$$

where we define Π and q as follows

$$\mathbf{\Pi} = \rho \mathbf{u}\mathbf{u} + P - \lambda \nabla \cdot \mathbf{u} - \mu \left[(\nabla \mathbf{u}) + (\nabla \mathbf{u})^T - \nabla \cdot \mathbf{u} \right], \tag{9}$$

$$q = -k_c \nabla T, \tag{10}$$

where μ and λ are the dynamic and bulk viscosities, T is the temperature and k_c is the thermal conductivity. We can compare the differences of coupling by states and fluxes in Tables 1 and 2, where we have defined ρ,u and e as the macroscopic density, velocity and total energy, A as the cross-sectional area across which the flux is evaluated, V_c as the volume of the cell inside the HSI, n the normal vector on A, m as the mass of an atom, s as the number of added/removed particles, v'_i and ε' as the velocity and energy of the added/removed particles, F_{ij} as the force between two atoms i and j (and F_i as the force acting on a singular atom i), ⟨JQ⟩ as the molecular energy flux, $E_{p,i}$ as the potential energy of atom i and ⟨ ⟩ as a time average. Π and q have been defined in Eqs. (9) and (10), respectively. We can see from Table 1 that density, momentum and energy are directly imposed on the atoms from the continuum when coupling by states is used. The macroscopic quantities for density, momentum and energy are obtained for each computational cell individually in the HSI by averaging over the atoms inside the cell c.

Fluxes have to be calculated first before they can be imposed on the continuum or atomistic side, as shown in Table 2. To conserve the overall mass, momentum and energy, it is necessary to remove or add atoms in the HSI. Since particles are free to move and interact inside computational cells, their energy level will change locally inside the cell. The energy level is associated with the temperature of that cell and so by randomly inserting and deleting atoms, the temperature may change and the overall conservation of energy is not guaranteed. Delgado-Buscalioni and Coveney (2003a, b) introduced the USHER algorithm to remove this shortcoming. In their algorithm, particles are introduced into a cell at a predefined energy level which has been determined by a steepest-descent approach so as to conserve the total energy. The USHER algorithm is equally applicable to coupling by states and fluxes. Both approaches were successfully applied, see O'Connell and Thompson (1995), Hadjiconstantinou and Patera (1997), Hadjiconstantinou (1999), Nie et al. (2004), Wang and He (2007) for state coupling and Flekkøy et al. (2000), Delgado-Buscalioni and Coveney (2003a, 2004) and Delgado-Buscalioni et al. (2005a, b) for flux coupling.

Table 1: Coupling by states

	Continuum → Atomistic	Atomistic → Continuum
Density	ρ	$\frac{1}{V_c}\left\langle \sum_{i \epsilon c} m_i \right\rangle$
Momentum	$\rho \mathbf{u}$	$\frac{1}{V_c}\left\langle \sum_{i \epsilon c} m_i \mathbf{v_i} \right\rangle$
Energy	e	$\frac{1}{V_c}\left\langle \sum_{i \epsilon c} E_{p,i} + \frac{1}{2} m_i \mathbf{v_i}^2 \right\rangle$

Table 2: Coupling by fluxes

	Continuum → Atomistic	Atomistic → Continuum
Mass flux	$A \rho \mathbf{u} \cdot \mathbf{n} \to ms$	$\frac{1}{V_c}\left\langle \sum_{i \epsilon c} m_i \mathbf{v_i} \right\rangle \cdot \mathbf{n} \to \rho \mathbf{u} \cdot \mathbf{n}$
Momentum flux	$A(\rho \mathbf{u}\mathbf{u} + \mathbf{\Pi}) \cdot \mathbf{n} \to$ $ms\langle \mathbf{v}'_i \rangle + \left\langle \sum_i \mathbf{F_i} \right\rangle$	$\frac{1}{V_c}\left\langle \left(\sum_{i \epsilon c} m_i \mathbf{v_i}\mathbf{v_i} - \frac{1}{2} \sum_{j \neq i} \mathbf{F_{ij}}\mathbf{r_{ij}} \right) \right\rangle \cdot \mathbf{n} \to$ $\mathbf{\Pi} \cdot \mathbf{n}$
Energy flux	$A(\rho e \mathbf{u} + \mathbf{\Pi} \cdot \mathbf{u} + q) \cdot \mathbf{n} \to$ $ms\langle \varepsilon' \rangle + \left\langle \sum_i \mathbf{F_i} \cdot \mathbf{v_i} \right\rangle - \langle J_Q \rangle \cdot \mathbf{n}$	$\frac{1}{V_c}\left\langle \left(\sum_{i \epsilon c} m_i \varepsilon_i \mathbf{v_i} - \frac{1}{2} \sum_{j \neq i} \mathbf{F_{ij}}\mathbf{r_{ij}}\mathbf{v_{ij}} \right) \right\rangle \cdot \mathbf{n} \to$ $q \cdot \mathbf{n}$

Delgado-Buscalioni et al. (2005b) demonstrated the need to incorporate the fluctuating component of the fluxes as they can impact the overall solution of the coupling scheme. They modified the boundary conditions for the continuum which did not influenced the flux conservation but allowed for fluctuations during the exchange. Applied to the oscillatory shear flow, the influence of the fluctuations was clearly seen and qualitatively good results achieved. Markesteijn et al. (2014) coupled the Landau-Lifhitz fluctuating hydrodynamic (LL-FH) equations with MD motivated by the fact that a peptide immersed in water showed a strong correlation between its deformation (dihedral angle) and the density fluctuation. Tested in two simulations (rectangular domain with separated LL-FH and MD domain), one with zero mean and one with a drift velocity on the continuum side, while in both cases the MD domain was initialised with no net momentum, the MD solution was approaching the continuum solution while fluctuations were preserved.

O'Connell and Thompson used constrained dynamics to exchange data at the HSI. They introduced a coupling strength parameter ξ which in their study was set to $\xi=0.01$. They observed that higher values negatively influenced the results although increasing ξ would cause lower computational times. Nie et al. (2004) were able to use a value of unity for the coupling strength in conjunction with an external force applied to the particles in the HSI to prevent them from escaping. Wang and He (2007) derived an equation for ξ allowing them to dynamically update the value of ξ, while Kamali and Kharazmi (2013) used yet another approach and imposed an arithmetic relation on ξ driven by the molecular time step. The simulation was initialised with $\xi(t_{MD}=0)=0.1$ and increased ξ via a third-order polynomial until $\xi=1$ was reached at a user-defined time step. They argued that fluctuations at the beginning of the exchange were responsible for the divergence. Using a low coupling strength parameter at the beginning suppressed possible divergence and the results converged as the coupling parameter was increased.

One of the challenges in coupling the continuum with an atomistic description is partially due to the imposition of boundary conditions from the continuum onto the atomistic level. While molecular data are easily averaged and imposed on the continuum, the reverse is not as evident due to the disparate degree of freedoms. Praprotnik et al. (2005) developed the adaptive resolution scheme (AdResS) in which MD is coupled with a coarse-grained version of MD which is able to blend between the two descriptions, changing the degrees of freedom on the fly. The inter-molecular force is blended between the two descriptions as

$$\mathbf{f}_{\alpha\beta} = w(x_\alpha)w(x_\beta)\mathbf{f}_{\alpha\beta}^{\mathbf{ex}} + [1 - w(x_\alpha)w(x_\beta)]\mathbf{f}_{\alpha\beta}^{\mathbf{cg}}, \tag{11}$$

where $w(x)$ is a blending function taking values from zero to unity, and α and β are the centre of masses of the two interacting molecules. The superscript ex denotes the explicit treatment (MD) and cg the coarse-grained version. While removing degrees of freedom and effectively increasing the molecular size, the approach is blended from a microscopic to a mesoscopic level at which boundary conditions may be imposed easier than on a pure MD boundary.

Ren (2007) investigated the stability of the state and flux coupling scheme as well as a combination of the two. Specifically, the *VV*, *FV*, *VF* and *FF* coupling schemes were investigated, where we have *V* = velocity (state), *F* = flux and the first letter indicates the direction from continuum to atomistic and the second from atomistic to continuum. *V* and *FV* were found to be stable for both small and large sampling intervals. *VF* performed well for small sampling intervals, while the purely flux-based *FF* scheme was claimed to be weakly stable. A stability analysis was carried out which confirmed the above findings and showed an amplification factor of unity for the *FF* scheme which, in combination with statistical noise, caused the weakly stable nature.

When dealing with multiscale methods, there are several ways to link the atomistic level with the continuum. The one used and discussed so far is known as the domain decomposition (DD), where the computational domain is split into sub-domains for which either an atomistic or continuum description is used while the coupling happens at the HSI. An alternative description is the heterogeneous multiscale method (HMM) (Weinan et al. 2003) where the entire domain is covered by a continuum solver and the microscopic part enters the computation locally at nodes where the continuum description is invalid. Asproulis et al. (2012) developed the point-wise coupling (PWC) based on the HMM and showed that this approach delivered good results for the velocity profiles in a start-up Couette flow. In their investigation, they placed the local atomistic regions at the wall nodes and analysed a range of parameter in the Lennard–Jones potential, as well as different channel heights and wall geometries. As evident, the approach lends itself to create bespoke multiscale flow domains and use the atomistic description only where it is necessary.

Borg et al. (2013) extended the idea of PWC to field-wise coupling (FWC). It operates in a similar fashion to the PWC approach and combines its strength with the domain decomposition. Sub-domains (fields) are placed continuously in the continuum domain but do not need to coincide with the continuum nodes as in the PWC. In a series of 6 steps, the continuum solver projects its solution on the atomistic level for which a new solution is obtained and imposed back onto the continuum. The stresses are corrected afterwards. As with the PWC, the atomistic domain can be tailored to the flow field and freely placed inside the macroscopic domain. Applied to a 1D Poiseuille flow with Newtonian and non-Newtonian fluids, they showed that by using only one microelement (field) near the wall, the resolution was enhanced while further elements needed to be placed inside the fluid domain to reduce the error compared to a full MD solution.

The Couette flow is usually used to test hybrid schemes as it is easy to set up and has an analytical solution. It does not require a complex computational domain, and hence, unstructured meshes, as are widely used in continuum CFD solvers, are not usually used. Borg and Reese (2008) set out to develop a framework that incorporates unstructured meshes and described their implementation in the open-source CFD solver openFOAM (Open Field Operation and Manipulation, http://www.openfoam.com/) towards a general purpose coupling approach. Although the implementation has been described in detail, no actual simulations were presented. Macpherson and Reese (2008) introduced the arbitrary interacting cell algorithm (AICA) which was designed to obtain the particles inside the cut-off radius r_{cut}, i.e. for building neighbour lists, within an unstructured framework. Geometrical constraints as well as parallel implementation issues have been addressed and discussed in detail. Using an unstructured mesh for a domain decomposition-based multiscale approach would require the HSI to handle complex interfaces. Borg et al. (2010) used a state controller to impose macroscopic conditions on the atomistic level. It has an *actuator* which is controlling the atoms, e.g. changing their velocity, and a *sensor* which is measuring the atomistic properties which are then used to drive the actuator in an iterative procedure. They successfully applied this technique to handle the exchange of boundary data inside the HSI for a Couette flow with an obstacle attached to the lower wall (inside the atomistic region). This method is, however, not only applicable for (complex) multiscale boundary data exchange but could also be used to impose non-periodic boundary conditions in pure MD simulations.

Steijl and Barakos (2012) reviewed and improved the treatment of non-periodic boundaries. Here, the cut-off distance of particles close to the boundary may exceed the boundary itself and hence will exhibit a lower inter-molecular force. They stated that neglecting this force could contribute 30–40 % in density fluctuation. To remove this shortcoming, a similar MD simulation is usually conducted for which the normal force of the particle component can be evaluated which is sampled and then imposed on particles in the simulation of interest. Furthermore, the authors also treated the tangential force component which has its equivalent in the continuum stress. They applied their methodology to a channel flow with the continuum domain in the centre and MD domains at the walls and obtained good agreement with the analytical solution for the Poiseuille flow. Holland et al. (2014) used MD to simulate a small portion of a channel and obtained the boundary conditions at the wall from it. They extended the channel to a high aspect ratio and imposed the boundary conditions from their MD data. The results showed that without the MD data, the continuum solver did not produce the correct physical behaviour, while with the MD data, a cheap pre-simulation technique was found to reproduce the expected results for geometrical similar, high aspect ratio channels. Even when placing an obstacle at the channel centre, the profile for velocity, density and pressure were accurately captured compared to a full MD simulation.

Alexiadis et al. (2013) developed a novel Laplacian-based HMM in which the shear stresses are directly obtained from MD computations (an alternative is to use the Irving–Kirkwood equation which is commonly employed). The idea behind their approach is to solve for the Laplacian term, i.e. the momentum equation of the Navier–Stokes equations becomes

$$\nabla^2\phi = \Omega(\mathbf{R}). \tag{12}$$

We have simply solved for the Laplacian term and collected the rest of the equation in $\Omega(R)$, where R is the macroscopic position vector. We can do the same on the atomistic level with r being the atomistic position vector as and thus have found a way to approximate the right-hand side of the Navier–Stokes equation, i.e. $\Omega(R)$ and, therefore, have solved for the stresses implicitly. Alexiadis et al. (2014) introduced the hybrid taxonomy to classify HMMs based on their order of approximation going from the atomistic to the continuum level. The order of approximation is equal to the order of the gradient that is being approximated. The Laplacian-based scheme is second-order in their taxonomy as a second-order gradient is approximated. HMMs approximating first-order derivatives are therefore first order, while zeroth-order schemes only approximated scalar transport coefficients. With increasing order, the noise may become dominant and they have stated that second-order schemes are currently the limit. To reduce noise, a filter may be applied as done in their study using the Savitzky–Golay filter which could allow higher-order approximation. Strictly speaking, hybrid–hybrid schemes are possible where, for example, velocity is obtained with the second-order Laplacian scheme, while the temperature could be supplied by a full continuum simulation without any coupling. This hybrid–hybrid scheme was then tested in a gravity driven microchannel, and good results were achieved after fine tuning the MD cell dimensions. It was found, similar to Borg et al. (2013), that the MD cells configuration played a crucial part in the accuracy of the overall solution and could deteriorate if inadequate dimensions were chosen.

$$\langle \nabla^2\phi \rangle \big|_{\mathbf{r}=\mathbf{R}} \cong \Omega(\mathbf{R}), \tag{13}$$

So far we have limited our discussion to monoatomic flows, but extension to multispecies flow may be simply achieved by accounting for each atomic mass as done by Kim et al. (2012). Instead of treating each atom equally, summing over its mass and force due to the Lennard–Jones potential allows for several species to be considered. They validated their approach for an incompressible flow with argon in a Couette flow and then used the same set-up but different weight ratios for the two types of atoms which otherwise satisfied the properties of argon. Incompressible flows have not been widely used for multiscale simulations, mainly due to high statistical scatter at low-speed flows. Ko et al. (2014) proposed an array of solutions to overcome this shortcoming. The first approach mentioned simulated the same geometry with different initial conditions for a number of times and then averaged the results. Clearly, for a large number of simulations it is expected that the noise will substantially reduce, but this gain comes with a high computationally cost. The second approach made use of a spatial regression which took data from neighbouring cells to improve the results in the cell at which information was exchanged. The third and final approach made use of temporal regression where data from the same cell, at which information was exchanged, were collected over the sampling time. The solution with the lowest statistical noise was obtained using several simulations and averaging the results; however, this approach can be repeated arbitrarily often to arrive at a defined noise threshold. The spatial and temporal regression techniques are more difficult to converge as sampling data are not abundantly available. However, encouraging results were obtained for the Couette flow and could pave the way for incompressible, hybrid schemes. Cosden and Lukes (2013) combined several aforementioned methods and developed an open-source MD-CFD solver based on openFOAM for the continuum computation and LAMMPS (Large-scale Atomic/Molecular Massively Parallel Simulator, http://lammps.sandia.gov/, see also Plimpton 1995) for the MD part. Coupling by state (constrained dynamics) was used, and the particle insertion method of Delgado-Buscalioni and Coveney (2003b) was employed in the HSI. They validate their open-source solver against the start-up Couette flow.

In the introduction to this section, we have mentioned the work of Hadjiconstantinou (1999) in which the moving contact-line problem was studied using a multiscale approach. The contact line was created between two immiscible fluid phases which has been achieved, on the MD side, by removing the attractive force in Eq. (2), i.e. by flipping the minus to a positive sign. This was only done for the potential between the two phases, while inter-phase potentials followed the standard form of Eq. (2). The topic of immiscible fluids is little studied in the context of multiscale simulations but has recently gained some interest in the field of polymer blends. Fermeglia and Pricl (2007) investigated the miscibility of such compounds found in industrial applications where MD was used to obtain coefficients which were mapped onto a mesoscale model. A widely found polymer blend in the automotive industry is formed of polymethylmethacrylate and polycarbonate, also referred to as PC-PMMA. They investigated a 70/30 mix of PC-PMMA for different shear rates and found that the miscibility was invariant with respect to shear. The blend remained immiscible and only changed its morphology, where spherical-like structures were found for low shear rates which elongated under higher shear. Only using a compatibiliser may the two phases be mixed. Similar results were found for polycarbonate–acrylonitrile-butadiene-styrene (PC–ABS) blends. Furthermore, the inhomogeneous density field distribution was mapped onto a FEM solver which then produced homogeneously distributed variables on the macroscale. Posocco et al. (2012), who investigated self-assembled monolayers composed of hydrocarbon and perfluorocarbon as a surfactant on gold nanoparticles, mentioned the importance of multiscale modelling in the context of immiscible flow domains. In their multiscale study, MD was utilised to obtain parameters for a mesoscale model (in this case dissipative particle dynamics, see also Sect. 5). These parameters had direct influence on the morphology on the mesoscale. The authors concluded that by influencing the parameters on the microscale, which can be easily done, 3D patterns can be controlled and created on the mesoscale. Furthermore, they stressed that tailored nanoparticles of complex nature may be studied and created for medical applications. Hence, a multiscale approach may offer new insight into the morphology of polymer blends, but its application may be far reaching and enabling the creating of new materials.

Further reviews dealing exclusively with multiscale MD schemes can be found in Kalweit and Drikakis (2008), Delgado-Buscalioni (2010) and Mohamed and Mohamad (2010).

Interim conclusion on hybrid and multiscale molecular dynamics methods

Multiscale MD computations have allowed to accurately capture the flow physics at the microscale while extending the computable domain into the mesoscale (Xu and Li 2007), where experimental measurements are feasible (Santiago et al. 1998; Tretheway and Meinhart 2002) and could provide validation data. Experiments may be performed at the microscale for low-speed flows, and hence, incompressible multiscale methods need to be further developed to cope with statistical scatter.

Most of the coupling schemes presented were either tested for the Couette flow (steady and unsteady), the oscillatory shear flow or the Poiseuille flow. Efforts towards more complex geometries have been presented, but simulations making use of complex HSIs and unstructured meshes are not known to us. This may be circumvented by using PWC or FWC as the underlying mesh can be structured or unstructured; however, the domain decomposition approach would benefit from a more general description for flows where the domain can be split into different domains, as, for example, in the case of a wall-bounded flow. The imposition of boundary data between atomistic and continuum domains remains challenging. We will discuss in Sects. 5and 6 how the inclusion of a description between MD and the Navier–Stokes equations can remove some of the noise while simplifying the overall coupling procedure.

DIRECT SIMULATION MONTE CARLO METHOD

The direct simulation Monte Carlo (DSMC) method is consistent with the Boltzmann equation but not derived directly from it (Palharini 2014; Shen 2005). It is a method developed for rarefied gas conditions, i.e. where the mean molecular diameter is much smaller than the mean molecular path. In DSMC, real atoms are grouped together and represented by a single particle which is then used for the simulation. The procedure of a DSMC simulation is as follows:

Populate mesh with particles, initialise simulation

- Advect particles
- Index particles
- Perform particle collisions
- Sample flow properties
- If $t=t_{max}$ proceed, else go to step 2
- Output solution

In step 1, particles are randomly seeded in the mesh obeying computational constraints. Each cell should have about 20–30 particles for low statistical noise (Bird 1994; Palharini 2014), and the time step Δt needs to be small in comparison with the local mean collision time (Shen2005). In step 2, particles are advected according to the temporal scheme used and could be as simple as $\mathbf{r}_p^{n+1} = \mathbf{r}_p^n + \mathbf{v}_p \Delta t$ with r_p being the particle coordinates at time *n* or n+1 and v_p its corresponding velocity. Step 3 determines the cell in which each particle currently resides. The collisions between particles are carried out in step 4, and the macroscopic flow properties are obtained in step 5. This process is repeated until the maximum time step is achieved or, if a steady solution is sought, after a time asymptotic solution is achieved.

While the particles are advected deterministically, the collisions occur statistically. The most common collision technique used nowadays is the no time counter (NTC) method introduced by Bird (1989) which removed unrealistic collisions rates found in highly non-equilibrium flows in its predecessor, the time counter (TC) method, previously introduced by the same author. First, the total number of collisions per cell is calculated as

$$N_{\text{coll}} = \frac{1}{2} \frac{N \overline{N} F_N (\sigma_{\text{T}} c_{\text{r}})_{\max} \Delta t}{V_{\text{c}}}, \tag{14}$$

where N is the number of simulated particles inside the cell, $\overline{N}$ is the average number of particles during the sampling time,

F_N is the number of real atoms represented by each simulated particle, σ_T is the total collision cross section and obtained from an appropriate molecular model, c_r is the relative speed and V_c is the cell volume. The simplest of molecular models to obtain σ_T is the Hard Sphere (HS) model where we have $\sigma_T = \pi d_{12}^2$ and $d_{12} = (d_1 + d_2)/2$, d_1 and d_2 being the diameters of the colliding particles. Its simplicity comes at the cost of lacking physical soundness, and various models have been introduced in the past to overcome these shortcomings, a detailed description of which can be found in Shen (2005) and Bird (1994).

For each cell, we calculate N_{coll} and loop over it. At each instance, we select a particle pair and test whether collision occurs via

$$P_{\text{coll}} = \frac{\sigma_T c_r}{(\sigma_T c_r)_{\max}}. \tag{15}$$

Equation (15) gives the collision probability. It is compared against a random number $R_n \in [0;1]$ and if $P_{coll} > R_n$ is true, the selected particles are accepted for collision. We should note that $(\sigma_T c_r)_{max}$ is difficult to estimate a priori and so is subject to be updated if $(\sigma_T c_r) > (\sigma_T c_r)_{\max}$. This does not tamper with the validity of the solution as it appears in the nominator of Eq. (14) and denominator of Eq. (15).

The collision occurs elastically, and energy is conserved. The velocities of the particles prior to collision are

$$\mathbf{c}_1 = \mathbf{c}_m + \frac{m_2}{m_1 + m_2}\mathbf{c}_r, \tag{16}$$

$$\mathbf{c}_2 = \mathbf{c}_m - \frac{m_2}{m_1 + m_2}\mathbf{c}_r, \tag{17}$$

where we have the centre of mass velocity as

$$\mathbf{c}_m = \frac{m_1\mathbf{c}_1 + m_2\mathbf{c}_2}{m_1 + m_2}, \tag{18}$$

and the relative velocity is

$$\mathbf{c}_r = \mathbf{c}_1 - \mathbf{c}_2. \tag{19}$$

The velocities after the collision for each particle can be expressed as

$$\mathbf{c}_1^* = \mathbf{c}_m^* + \frac{m_2}{m_1 + m_2}\mathbf{c}_r^* \tag{20}$$

$$\mathbf{c}_2^* = \mathbf{c}_m^* - \frac{m_2}{m_1 + m_2}\mathbf{c}_r^*. \tag{21}$$

Since an elastic collision is assumed and energy is conserved, the magnitude of relative and centre of mass velocity has to be conserved, i.e. we have $|\mathbf{c}_r| = |\mathbf{c}_r^*|$ and $|\mathbf{c}_m| = |\mathbf{c}_m^*|$ where post-collision values denote with an asterisk. In the HS model, the scattering occurs isotropically, and therefore, each particle's outbound direction is equally likely. We can therefore randomly select collision directions from an appropriate distribution which yields $\mathbf{c}_r^*$ and $\mathbf{c}_m^*$ and Eqs. (20) and (21) can thus be solved.

The macroscopic flow properties are then obtained with the following relations

$$\mathbf{u} = \frac{1}{N_c}\sum_{i=1}^{N_c}\mathbf{c} \tag{22}$$

$$\rho = nm \tag{23}$$

$$p = \rho\overline{(\mathbf{c} - \bar{\mathbf{c}})^2}, \tag{24}$$

where n represents the number density. Boundary conditions are easily imposed in the DSMC method. The simplest form is periodic boundaries in which particles re-enter the computational domain via a periodic interface. If a particle travels past a solid boundary during the advection step, its normal velocity component to the boundary is simply reversed and the position updated accordingly. At open boundaries, particles are removed when they exit the domain. At inflow boundaries, one needs to determine the velocity and number flux of particles. A popular choice to obtain the velocity of the particles is the acceptance–rejection method in which a random velocity is drawn from a suitable distribution function $f(v)$. This velocity will be accepted if $f(v)/f_{\max} > R_n$, where f_{max} is the maximum value of $f(v)$ and R_n a uniformly distributed random number. The number flux of particles across an open boundary can be calculated as

$$\dot{N} = \frac{n\beta K_t}{\sqrt{\pi}}, \tag{25}$$

where β is related to the most probable thermal speed as $\beta = (2RT)^{-1/2}$ with R being the specific gas constant and T the temperature and K_t is a normalisation constant and given by

$$2\beta^2 K_t = S\sqrt{\pi}(1 + erf(S)) + exp(-S^2), \tag{26}$$

where S is given by $S=\beta V$, V being the normal component of the velocity on the boundary and we have introduced *erf()* as the error function. For further information on boundary conditions, the reader may wish to consult Bird (1994), Lilley and Macrossan (2003) and Xiaohai (2005).

At this point, we wish to emphasise that despite a computational mesh is used, no information is exchanged with neighbouring cells, and hence, the choice of structured or unstructured meshes does not pose any higher implementation effort. When using unstructured grids, however, the neighbour cells need to be determined in order to change the cell ID once a particle passes from one cell to another. Fast neighbour search algorithms and implementation guidelines can be found in Löhner (2008). A further in-depth review of the DSMC method can be found in Oran et al. (1998).

Review on hybrid and multiscale direct simulation Monte Carlo methods

The DSMC method thrives for flow condition in the rarefied

gas regime. This is encountered at near space altitudes of 20–100 km (Xu et al. 2009) and hence lends itself for space applications. Modelling spacecrafts re-entering the atmosphere requires a more complex treatment as the full Knudsen number range is encountered, i.e from free molecular flow $(Kn > 10)$ in the outer atmosphere to the full continuum $(Kn < 0.001)$, while descending towards earth. La Torre et al. (2011) showed that at around K_n=0.1 the error from a Navier–Stokes solution was twice that of a full DSMC and confirmed that the continuum description is invalid in this region. Although the DSMC method could be extended into the low Knudsen number regime, it would be too computational expensive in comparison with a continuum-based Navier–Stokes solver. It is here that a multiscale DSMC solver benefits from the advantages of both worlds and computational efficiency is gained.

Boyd et al. (1995) introduced the gradient length local Knudsen number as

$$Kn_{\text{GLL}} = \frac{\lambda}{\phi}|\nabla\phi|, \tag{27}$$

where ϕ could be any flow variable of interest. They referred to Bird (1970) who showed that $Kn_{\text{GLL}} > 0.05$ implies a departure from the continuum behaviour. Lofthouse et al. (2007) investigated a Knudsen number range of $0.002 < Kn \leq 0.25$ for a Mach 10 flow around a cylinder using two independent solver: one based on the Navier–Stokes equations and the other on the DSMC method. They calculated the pressure, shear stresses and heat transfer on the cylinder surface and correlated the results with the local Kn_{GLL} to judge whether the region can be treated as a continuum or not. The shear stresses and heat transfer obtained from the Navier–Stokes solution showed considerable differences compared to the DSMC results in the $Kn_{\text{GLL}} > 0.05$ region. They explained this behaviour to be caused by the inaccurate boundary conditions imposed at the cylinder surface. The pressure was not as adversely affected and showed a better agreement. For $Kn = 0.002$, the total drag and peak heating was within 1 % of the predicted DSMC result, while for $Kn = 0.25$ the error increased to 26.2 and 32.1 %, respectively. The study showed the need for hybrid solvers where the fast Navier–Stokes solution should be applied in regions of $Kn_{\text{GLL}} < 0.05$ and the DSMC method elsewhere. This study crucially pointed out once more that an inaccurate boundary condition could cause a departure from the expected results despite the fact that the used method may be applicable away from the boundaries. We have shown results of Holland et al. (2014) in Sect. 2.1 that boundary conditions obtained and sampled from an appropriate molecular or particle approach can extend the Navier–Stokes equation further into the smaller length scales. Thus, multiscale method present a cure where the exact boundary conditions are uncertain on the macroscale, but not on the microscale.

Wu et al. (2006) developed a multiscale DSMC solver based on unstructured meshes. The breakdown parameter was based on Eq. (27), where □ was chosen to be the density, velocity and temperature so that Kn_{GLL} was obtained as $Kn_{\max} = \max\left[Kn_\rho, Kn_u, Kn_T\right]$, and a second breakdown parameter was used as

$$P = \left|\frac{T_{Tr} - T_R}{T_{Tr}}\right|, \tag{28}$$

where the indices T_r and R refer to the translational and rotational temperature. The initial solution was obtained via the Navier–Stokes equations, and if one of the two continuum breakdown parameters exceeded a threshold value, of which several values were tested, the domain was flagged for the DSMC solver. They tested their solver against a 2D flow over a wedge at 25° of inclination and a 3D near-continuum parallel orifice jet that expanded into near-vacuum. Good agreement between the full and multiscale DSMC results were attested and further experimental data for the 3D flow case confirmed this. Expanding on this work, Lian et al. (2011) used Eq. (27) with ϕ being the pressure. This has the advantage that the high-velocity gradient in the boundary layer, which may still be in thermal near equilibrium, is not excluded from the continuum domain, and hence, a costly DSMC calculation is avoided. They applied this to the same 2D, 25° inclined wedge flow at Mach 4 and to a Mach 12 square cylinder flow. The L2 norm of the density and temperature showed that the error was decreasing faster and achieved a slightly lower overall error compared to the breakdown parameter used before without diminishing the overall accuracy of the solver. Pantazis and Rusche (2014) coupled a DSMC and Navier–Stokes solver in openFOAM for complex, three-dimensional and unsteady flows. The authors emphasised that for unsteady flow coupling, numerical constraints become more restrictive and may deteriorate the solution and the efficiency of the multiscale scheme. Numerical noise, for example, may not as easily be reduced if the sampling interval is low and cannot be increased due to the time step requirement. The efficiency may suffer from a poorly designed parallel implementation. Any problems of this sort will build up over time and could cause the hybrid solver to perform worse than a monoscale approach (longer computational times) and diverge due to self-induced numerical instabilities. For geometric flexibility, unstructured meshes were supported. They tested their hybrid solver on the shock tube problem and the unsteady flow through an orifice. Their results closely matched those obtained with a pure DSMC solver. A comparison of the ideal and real speed-up further revealed good strong scaling capabilities.

So far in our discussion, we have only considered domain decomposition and heterogeneous multiscale methods and variations of these. These may not always be the best choices. If we consider a channel of height H that is of the order of the mean free path λ, as is the case in a Knudsen compressor, then either method (DD and HMM) would be difficult to implement. Using a continuum mesh and then refine it using the HMM approach, the mesh spacing inside the channel would be smaller than λ and, therefore, microelements placed on the nodes would overlap. One possible solution would be to make use of the field-wise coupling approach introduced in Sect. 2.1, which is what Docherty et al. (2014) have done. They used a hybrid Navier–Stokes DSMC code and considered the 1D micro-Fourier flow. As has been mentioned by Borg et al. (2013), the size of the microelement, which consists of a sampling and relaxation zone, needs to be appropriately defined in order to not adversely affect the results. Hence, a parametric study is needed a priori which has been done by the authors and the size of microelements had been adjusted to match a full DSMC simulation. Their approach worked well for the tested case but is currently limited to 1D flows. Therefore, the reported speed-up was only marginal as a wide scale separation could not be achieved. Another possibility is to join all microelements along transversal lines together, to form one single, cross-sectional microelement. This has been done by Patronis et al. (2013). They refer to this method as the internal multiscale method (IMM). They applied their new technique to a converging–diverging channel driven by an external acceleration, a curved, high aspect ratio channel driven by a pressure difference and the developing flow confined by two eccentric cylinders at different rotational velocities. Their approach agreed well against full DSMC simulations. The computational speed-up was reported to be 6, 50 and 300 times faster, respectively. The very high speed-up was due to the steady-state continuum solver for the multiscale method, whereas the full DSMC solver was fully unsteady. Patronis and Lockerby (2014) extended this method to the low-variance deviational simulation Monte Carlo (LVDSMC) as introduced by Homolle and Hadjiconstantinou (2007). It is based on the Boltzmann equation but retains the structure of the DSMC algorithm. The key idea is that the velocity distribution can be decomposed into an equilibrium part which can be solved analytically and a deviational part which is solved for by the simulated particles. Since the equilibrium is obtained in this way, small deviations from equilibrium can be simulated with higher signal-to-noise ratio and the overall noise decreases. For the case of the converging-diverging channel, good agreement was obtained compared to the DSMC results.

Statistical scatter is omnipresent and will remain a challenging problem due to the inherent stochastic treatment of the collisions. Averaged quantities based on the particles contain a high variance and the restriction imposed by current and foreseeable future generation of high-performance computing facilities means that researcher has to opt for noise reducing method to control the accuracy instead of increasing the particle count. Burt and Boyd (2008) introduced a low diffusion variant of the DSMC which replaces the collision by a set of explicit deterministically steps. Particles are seeded in cells and not allowed to freely travel from one cell to another, which is enforced by specular walls on the cell. This removes the necessity to index particles at every time step but requires to transfer and exchange momentum in-between neighbour cells. Furthermore, the cells are not rigid but move with a cell velocity obtained from the mass averaged particle velocities. Motivated by the successful application to the 1D shock tube, they coupled the LD variant with a conventional DSMC solver and simulated the flow around a cylinder at Mach 6 and a nozzle/plume expansion (Burt and Boyd 2009) as well as the flow around a cylinder at Mach 20 (Burt and Boyd 2010). The use of the LD-DSMC method to increase computational efficiency was justified by the fact that particle trajectories were correctly integrated while moving over cell-based length scales much larger than the mean free path. The results compared well to the reference solution and the speed-up for the Mach 20 cylinder flow was given to be about 2.6 higher compared to a full DSMC simulation. Jun et al. (2013) calculated the flow around a cylinder as well, this time at Mach 10. Compared to a full DSMC simulation, their speed-up was only about 1.2 using the same method as described above. They argued that a considerably large part of the calculation is spent finding the initial solution with the DSMC method which, for small Knudsen numbers, i.e. continuum regions, is rather time-consuming. Using a continuum solver instead to obtain the initial solution, improved their speed-up to a factor of 2.6.

Sun and Boyd (2005) introduced the sub-relaxation technique in which the history of a signal is removed during averaging once it starts to deteriorate the averaged result. A relaxation parameter θ can be freely set to influence the averaging procedure but may negatively affect the results and so a parametric study is advisable. Schwartzentruber and Boyd (2006) used the sub-relaxation technique in a hybrid Navier–Stokes DSMC solver and used it for boundary condition imposition at the interface between DSMC and continuum. They tested their approach on a one-dimensional shock tube for liquid argon and nitrogen gas. Their technique was able to provide smooth boundary conditions for the hybrid solver which compared well to full DSMC and experimental data. A pure Navier–Stokes solution was also tested which gave inaccurate results with the reciprocal shock thickness overestimated by almost a factor of 2. This is consistent with the observation of Carlson et al. (2004), who computed the flow inside a shock tube for Mach numbers of 1.55, 5 and 10. In their multiscale solver, the Navier–Stokes equations were coupled with the DSMC method based on the IP method (information preserving, Fan and Shen 2001). In this method, each particle is assigned a second, information preserving, velocity. This velocity is solely used to obtain the macroscopic quantities through particle averages but does not influence the movement of the particles and hence is less prone to statistical scatter. The IP velocity follows its own set of rules to update itself, a detailed list of which can be found in Shen (2005).

The results for the Mach 1.55 case yielded overall good agreement amongst the Navier–Stokes, DSMC and multiscale solver. Increasing the Mach number decreased the accuracy of the continuum solution, whereas DSMC and the hybrid version matched each other well. The reciprocal shock thickness confirmed a deviation by a factor of about 2 for the Navier–Stokes equations from the experiment, while DSMC and the multiscale solver matched the experiment well. The speed-up was also mentioned, and despite the multiscale solver being about 1.6 times faster than the pure DSMC version, the Navier–Stokes equations still yielded results 9 times faster than the multiscale solver. Thus, for those cases where the Navier–Stokes equations can be modified to incorporate appropriate boundary conditions, it is still cheaper while retaining a good degree of accuracy.

Boyd (2008) investigated the flow past a cylinder-flare and blunt planetary probe using a hybrid DSMC Navier–Stokes solver. Different mesh densities and time steps for the respective solver were used in order to speed up the calculation. The initial solution on which the domain has been separated into rarefied and continuum regime was obtained by the Navier–Stokes equations as they have previously proven to provide the initial solution in a shorter time (Jun et al. 2013). It has been mentioned that the mean free path varied over one order of magnitude for the cylinder-flare case (speed-up 1.4) and over two orders of magnitude for the blunt planetary probe (speed-up 12.5), where the speed-up in parenthesis are comparing to full DSMC simulations. This shows that multiscale solvers are becoming more efficient for a wide range of Knudsen number separation and it is under these conditions that they thrive.

Interim conclusion on hybrid and multiscale direct simulation Monte Carlo methods

Schwartzentruber and Boyd (2015) pointed out that experimental studies are vital and needed for the validation of current DSMC codes. Currently, most multiscale and hybrid DSMC codes are validated against pure DSMC solvers, see Scanlon et al. (2010) for a range of 1D, 2D and 3D test cases. They pointed out that MD could be used as a validation tool but is computational rather expensive for rarefied gases.

Another issue that arises when coupling DSMC with a second solver is the noise penalty that arises due to the stochastic treatment of the collision. A range of approaches have been presented herein (LD-DSMC, LVDSMC, IP, sub-relaxation) and have all been able to reduce the noise level. As seen in Ko et al. (2014) (see Sect. 2.1), using spatial or temporal regression reduced the noise with little extra computational effort while being independent of the method and hence could prove to be a viable alternative or even be used in conjunction with one of the above-mentioned methods.

The reported computational gains were highest for the cases with the highest Knudsen number separations, and hence, applications may be limited to these sorts of flow regimes. However, space applications have shown to exhibit the free molecular to continuum regime and thus rendered themselves as prime candidates for multiscale DSMC schemes.

LATTICE BOLTZMANN METHOD

The lattice Boltzmann method (LBM) is strictly speaking not a direct particle method. It is derived from the lattice gas cellular automata (LGCA) which in turn is based on physical particles. In the LGCA method, particles reside on so-called sites which are connected via links. It is essentially a computational mesh where sites refer to nodes and links to edges. The governing equation simply states that the number of outgoing and residing particles at a site is equivalent to the collision occurring at the same site, i.e.

$$N(\mathbf{r} + \mathbf{c}\mathrm{d}t, t + \mathrm{d}t) - N(\mathbf{r}, t) = \Omega. \tag{29}$$

Several collision models have been introduced, most notably the one of Hardy, Pomeau and de Pazzis (HPP model, Hardy et al. 1973, 1976) and Frisch, Hasslacher and Pomeau (FHP model, Frisch et al. 1986). Due to its binary nature (particles either do or do not exist at sites), the method is free of any numerical round-off errors. However, statistical noise is high and averaging over larger regions or several simulations is necessary to obtain macroscopic quantities. The model further lacks Galilean and rotational invariance. The interest in the method declined once the problems surfaced. In the LBM, the single particles are replaced by a density distribution function which removes some issues encountered in the LGCA. That, however, also means that the collision operator cannot be treated explicitly and has to be approximated numerically. The easiest and still one of the most widely used model is the one of Bhatnagar, Gross and Krook (BGK model, Bhatnagar et al. 1954). It is sometimes referred to in short as the LBGK (Lattice-BGK) model and given by

$$\Omega(f) = \frac{1}{\tau}\left(f_k^{\mathrm{eq}} - f_k\right), \tag{30}$$

where τ is the collision time, f_k is the density distribution at link k and f_k^{eq} is the corresponding equilibrium distribution. The collision time is given b0079

$$\frac{1}{\tau} = \frac{1}{(3\nu + 0.5)} = \omega, \tag{31}$$

where ν is the lattice viscosity and ω is the collision frequency. The full governing equation with the BGK operator is

$$\frac{\partial f}{\partial t} + \mathbf{c} \cdot \nabla f = \omega\left(f_k^{\mathrm{eq}} - f_k\right), \tag{32}$$

where Eq. (32) can be seen as a simple advection equation with a source term. If we ignore the unsteady term, we see the similarities to the LGCA method, i.e. Eq. (29), where we have streaming on the LHS and collision on the RHS. To solve the above equation, we need to define c and f_k^{eq}. The lattice speed cc is determined by the speed model. The speed model determines how the distribution function is streamed to its neighbours. Typically on a 2D structured mesh, there are 9 sites (nodes) to consider, located at $\mathbf{r}_{\text{neighbour}} = \mathbf{r}(r_x \pm \Delta x, r_y \pm \Delta y)$ which includes the site at $\Delta x = \Delta y = 0$. Therefore, the speed model would be termed D2Q9 (2D, 9 sites) and each site has a corresponding speed c_k and weight w_k. The speeds stream from one site to another and, therefore, take integer values of −1,0,1−1,0,1. They are obtained as

$$c_i = \begin{cases} 0 & i = 0 \\ (\cos[\alpha_1], \sin[\alpha_1]) & i = 1,2,3,4 \\ (\cos[\alpha_2], \sin[\alpha_2]) & i = 5,6,7,8 \end{cases} \tag{33}$$

where

$$\alpha_1 = (i-1)\,\pi/2$$
$$\alpha_2 = (i-5)\,\pi/2 + \pi/4$$

The corresponding weights are

$$w_i = \begin{cases} 4/9 & i = 0 \\ 1/9 & i = 1,2,3,4 \\ 1/36 & i = 5,6,7,8 \end{cases} \tag{34}$$

The weights are to some degree arbitrarily set but need to fulfil certain constraints, one of which is that $\sum w_k = 1$. More information on the weight constraints and how to develop higher-order speed models can be found in Succi (2001).

The density distribution function at equilibrium is given by

$$f_k^{\mathrm{eq}} = w_k \rho(\mathbf{r}, t)\left[1 + \frac{c_k \cdot \mathbf{u}}{c_s^2} + \frac{(c_k \cdot \mathbf{u})^2}{2c_s^4} - \frac{\mathbf{u}^2}{c_s^2}\right], \tag{35}$$

where we have the macroscopic density $\rho(\mathrm{r,t})$ and velocity u, while the lattice speed of sound is obtained from $c_s = 1/\sqrt{3}$.

The macroscopic quantities are then obtained from

$$\rho(\mathbf{r},t)=\sum f_k \tag{36}$$

$$\rho(\mathbf{r},t)\mathbf{u}=\sum f_k c_k. \tag{37}$$

Eq. (32) is now solved in the following way:

- Initialise domain
- Calculate f_k^{eq} from Eq. (35)
- Calculate Ω from Eq. (30)
- Stream the updated density distribution function to their neighbour sites
- Obtain macroscopic quantities for ρ and u via Eq. (36) and Eq. (37)
- Repeat from step 2 until $t=t_{max}$
- Output results

Due to the nature of the LBM in which macroscopic properties are determined via the density distribution function, we need to find boundary conditions for each contribution of the density distribution function f_i on the boundaries. For the D2Q9 model that we have discussed above, there will be 3 components of f_i outside the computational domain, 3 components on the boundary itself, i.e in-plane and 3 components pointing into the computational domain for which we need to formulate boundary conditions as they cannot stream from outside the boundary into the domain. There are several possibilities to find appropriate formulations for the inward pointing components of f_i and we will briefly describe the method of Zou and He (1997) which is one popular choice. Let us denote the inward facing components of f_i as f_-, f_c and f_+ where f_c is the component normal to the boundary and f_- and f_+ the component to its left and right, respectively. We first need to evaluate the density, which we can write for a Cartesian domain in compact form as

$$\rho=\frac{1}{1\pm u_\perp}\left[f_0+\sum_{i=1}^{inplane} f_i+2\left(\sum_{j=1}^{outside} f_j\right)\right], \tag{38}$$

where $\pm u_\perp$ is the perpendicular velocity component on the boundary and the sign is positive if the normal direction (pointing outside the domain) of the boundary is positively aligned with the coordinate system. We differentiate further between the components of f_i which are *inplane*, i.e. coincide with the boundary and *outside*, i.e which are outside of the computational domain. With these, we can calculate f_-, f_c and f_+ as

$$f_c=opp(f_c)\mp\frac{2}{3}\rho u_\perp, \tag{39}$$

$$f_+=opp(f_+)+\frac{1}{2}(f_{IP+}-f_{IP-})\mp\frac{1}{6}\rho u_\perp\pm\frac{1}{2}\rho u_\parallel, \tag{40}$$

$$f_-=opp(f_-)-\frac{1}{2}(f_{IP+}-f_{IP-})\mp\frac{1}{6}\rho u_\perp\pm\frac{1}{2}\rho u_\parallel. \tag{41}$$

We have introduced a function *opp*() which returns the component opposite to f_i, f_{IP+} and f_{IP-} which are the *inplane* component of f_i pointing in the positive and negative direction of the boundary with respect to the coordinate system and $u_\parallel$ as the velocity component parallel to the boundary. $\mp 2/3\rho u_\perp$ in Eq. (39) and $\mp 1/6\rho u_\perp$ in Eqs. (40) and (41) will be positive if the normal (pointing outside the domain) on the boundary is negative, and negative if the normal is positive. $1/2\pm\rho u_\parallel$ in Eq. (40) and Eq. (41) on the other hand will be positive for f_+ and f_-, if their direction with respect to the boundary is in positive coordinate direction and negative otherwise. The velocity component of $u_\perp$ is either prescribed via Dirichlet- or Neumann-type boundary conditions and $u_\parallel$ can be set for moving boundary problems. A comprehensive overview of various boundary conditions can be found in Latt et al. (2008) and Chen et al. (1996). Classical and recent applications of the LBM can be found in Chen and Doolen (1998) and Aidun and Clausen (2010).

Review on hybrid and multiscale lattice Boltzmann methods

The LBM has found wide acceptance and applicability among various research discipline, mainly due to its ease of implementation, handling of complex geometries and low computational cost along with its local behaviour which makes it a prime candidate for parallelisation. However, the LBM has also been found to work less well for flows involving heat transfer, compressibility and high Reynolds numbers. The main reason for the low-to-moderate Reynolds numbers that can be achieved with the classical LBGK model is due to the instability that arises with low lattice viscosities. One possible solution is to decrease the lattice spacing at the cost of increased computational cost. One could also use a multiple relaxation time (MRT) scheme (D'Humières et al. 2002) instead of the BGK operator for the collision. The main idea of the MRT scheme is that the collisions occur in moment space rather than velocity space and each moment is relaxed towards equilibrium at its own rate, while in the BGK model, all moments are relaxed at the same rate $1/\tau$. That allows for lower viscosities and hence higher Reynolds numbers. Karlin et al. (1998) introduced the notion of the entropic lattice Boltzmann method (ELBM), where the H-theorem is incorporated into finding the equilibrium distribution function f_k^{eq} in an attempt to adhere to the second law of thermodynamic, i.e. positive entropy production. In its limit, the ELBM will result in the classical LBM and subsequently allows lower viscosities as well. Compressible effects are difficult to model since the small number of discrete velocities in the speed model allows only small temperature variations. Successful applications have been presented, for example, by Guangwu et al. (1999), who simulated Sod's and Lax's shock tube problem. Despite following the trend of the reference solution, spurious oscillation, as well as numerical dispersion, was present. Compared to other published data obtained with the Navier–Stokes equations using different Riemann solvers and numerical schemes, its L1 norm in density, pressure, velocity and energy compared less favourable. Joshi et al. (2010) constructed a hybrid LBM in conjunction with a finite volume-based Euler solver where the primitive variables at the cell centres were obtained from the Euler equations, while the inter-cell fluxes were approximated by the LBM through a modified equilibrium distribution function to account for thermal effects. They applied their scheme to the 1D and 2D

Sod shock tube problem as done by Guangwu et al. (1999). Despite the absence of tabulated data in the work of Joshi et al. (2010), their scheme qualitatively exhibited the same behaviour as the pure LBM, see Guangwu et al. (1999). Compared to two solutions obtained from the pure Euler equation making use of the Godunov and Flux Vector Splitting (FVS) scheme, their solution compared well to the Euler results with spurious oscillation and numerical dispersion still present. To incorporate heat transfer and to compute the temperature field, as stated previously, the equilibrium distribution function needs to be modified and higher-order speed models should be used, i.e those that take their immediate neighbours and their neighbours into account. Various models have been introduced to account for the energy and thus for the temperature (Guangwu et al. 1999; Kataoka and Tsutahara 2004a, b; Shi et al. 2001) and are reviewed in Kun (2008). Mezrhab et al. (2004) discussed the thermal models available in the Lattice Boltzmann framework and concluded that these are still premature and need further development. They opted for a hybrid approach where the LBM was solved alongside an advection–diffusion heat equation to obtain the temperature field. They applied their coupled solver to a range of two- and three-dimensional cases and obtained results that compared well to reference data. The same approach was chosen by Wu et al. (2012) to simulated forced convection around a cylinder at various Reynolds numbers. The cylinder was resolved by the immersed boundary method (IBM) and its influence introduced into the advection–diffusion heat equation by a source term. Since the boundary itself did not coincide with the computational nodes as in body fitted meshes, evaluating gradients at the boundary was not straight forward. To calculate the Nusselt number, which contains the normal temperature gradient at the cylinder surface, it was recast using Fourier's law $-Q/k = \partial T/\partial n|_{\Omega}$ to eliminate the gradient and then finding an appropriate relation for the heat flux *Q*. Their results were compared against two other numerical simulations, and the discrepancies were at most 5 %. Chen et al. (2013) extended the advection–diffusion equation approach to handle species transport. Their test cases were species convection and diffusion with bulk reactions, species diffusion in a channel with surface reactions and natural convection inside a square cavity. The first two test cases were compared against analytic solutions and good agreement was attested while the same could be said about the third test case compared to the solution of a commercial Navier–Stokes solver. With their methodology tested and validated, they set out to simulate a more complex example of a wall-coated microreactor to simulate heat transfer, mass transport and chemical reactions. The velocity profiles were compared to a full-scale lattice Boltzmann simulation for which good agreement was obtained. To further accelerate the procedure, the grid spacing for the continuum domain was increased while keeping the spacing of the lattice Boltzmann domain the same. It has been found that at a grid spacing ratio of 10:1, results were still good while maximising the computational turnaround time. Yet another study has been done by Filippova and Hähnel (2000), using the same approach as chosen by the authors in the previous studies, and extended their approach to variable density flows in order to simulate chemical reacting flows. The LBM was further extended with adaptive mesh refinement capabilities and boundary fitting to account for the misalignment of the boundaries. They have applied their method to the flow around a porous burner, where the burner itself (cylindrical shaped) has been resolved with a refined grid. The ratio tested was 3 and 6 times finer than the far field grid. Since the time step is directly proportional to the lattice spacing, i.e. $\Delta x_k/\Delta t = c_k = const.$, a refined grid means that the time step decreases as well. For reactive flows, this becomes important as the timescales for reactions are rather small. Studies were conducted for a range of Reynolds and Damköhler numbers, matching the results of full Navier–Stokes solutions.

Li et al. (2014) simulated the melting and solidification process in a 2D rectangular domain where the west wall was above the melting temperature, the east wall below the melting temperature and the north and south wall adiabatic, i.e. $\partial T/\partial n|_{\Omega} = 0$. The velocities were obtained by the LBM, while the temperature was calculated from the energy equation which was solved by a finite volume-based SIMPLE (semi-implicit method for pressure linked equations, Patankar and Spalding 1972) scheme, see also Patankar (1980), Versteeg and Malalasekera (2007) for a detailed description of the SIMPLE scheme and its derivatives. The variables for the finite volume solver were stored at the cell centres while the LBM stored its variables at the vertices. In order to provide face-centred velocity components for the SIMPLE algorithm, the velocity needed to be interpolated to its required location. The interface was obtained via the interfacial tracking method. The method identifies the cells in which the interface currently resides and assumes that the cell temperature is at the melting condition, T_m. This is, however, only the case if the interface is exactly at the cell centre and, therefore, a correction was introduced to allow the interface to be off-centre. Simulations were then carried out for various Stefan numbers, $Ste = c_p\Delta T/L$ where *L* is the latent heat, and compared to experimental data. The bulk of the interface was agreeing well with the experiment. At the wall, discrepancies were observed and attested to the difficulties to maintain the adiabatic condition in experiments, as well as the assumption made in the numerical simulation, that the volume change during the phase transition can be neglected. Brent et al. (1988) developed an enthalpy–porosity technique for melting and solidification processes in which the Navier–Stokes equations remain valid throughout the flow for the liquid, transition and solid phase. The energy equation contains an additional source term in which the latent heat evolution is modelled for $T > T_m$; otherwise, the source term is zero. The momentum equation is equipped with a source term that mimics the Kozeny–Carman equation. It is zero for the liquid phase and increases in the transition region. For a fully solidified solution, its value is great enough to force all the other terms to vanish, enforcing a zero velocity field. The advantage of this approach is that the interface comes out as a product of the solution and does not need to be explicitly tracked. Chatterjee and Chakraborty (2006) reported results of a dendrite growth in an undercooled melt making use of the enthalpy–porosity technique. They used a modified thermal LBM from which hydrodynamic variables were computed and coupled it with an enthalpy density distribution function from which the thermodynamic variables were derived. Their results were reported to compare well against reference data although only qualitative figures were presented. However, their approach coupled the fast, modified (hybrid) LBM with the elegance of the enthalpy–porosity technique and hence lent itself to fast computational times for complex interfaces. Ladd (1994a, b) developed a general framework and solver (SUSP3D) to simulate suspended particles. Forces are imposed on the particles via fluctuating stresses rather than applying random forces directly to the particles. They are advected using classical Newtonian dynamics including angular momentum of the particles. The solver has been developed with an emphasis on good scalability for a large number of particles and hence has been subsequently used by various other researchers. Beetstra et al. (2006) made use of Ladd's solver and investigated the drag behaviour of various particle formations. They observed that particles in a cluster formation exhibited a lower average drag force per particle than a single particle. This was explained by the increased surface-to-volume ratio as each individual particle, on average, has less surface area to resist the oncoming flow. For a cluster of particles whose formation represented that of a sphere, at the closest packing possible (r/d=1), the average drag force per particle was only about 20 % of that of a single particle and only reached the same level of drag per particle for an inter-particle distance of r/d=7. Hence, the effect of clustering felt by each particle sustained several inter-particle distances away. Similar results were observed for other shaped clusters. They further investigated how the drag behaviour changed by increasing the Reynolds number and compared that against a single-particle drag law presented in Clift et al. (1978). They showed that for increasing Reynolds numbers, larger deviations from the drag law were observed. Following their work, Beetstra et al. (2007) investigated mono- and bidispersed particle systems, the latter with diameter ratios of 1:1.5–1:4. They derived a drag force relation for both cases and showed that failing to include the bidisperse nature in the drag law yielded significant differences. Their results were able to not only match the trend but also to accurately predict the particle drag for various Reynolds numbers and inter-particle distances. Furthermore, they showed that their model and simulation data could differ by as much as 3 times from current drag laws which, despite their shortcomings, are still widely used. Shah et al. (2013) investigated the drag behaviour of particle clusters using SUSP3D. Their cluster was surrounded by randomly seeded particles and, therefore, two voidages, that of the cluster and that of the surrounding cell, was defined. They explored an extensive range of various parameters, confirming that compared to the Ergun and Wen-Yu drag law, particle clusters exhibit a different drag behaviour. Furthermore for a constant cluster voidage of 0.7, they found that a minimum for the drag existed at an overall (cell) voidage of 0.96. Increasing the overall voidage further resulted in the drag to approach that of randomly seeded particles while decreasing the voidage resulted in a similar behaviour at a voidage of 0.85, which again demonstrated the importance of cluster formation in particle driven flows, such as the fluidised bed reactor. Usta et al. (2006) studied the migration of polymers in microchannels using SUSP3D, where polymers were modelled via particles connected by stiff linear Fraenkel springs. Their work was motivated by the discrepancies in the literature where polymers were reported to either migrate towards the channel centreline (including hydrodynamic interactions) or wall (without hydrodynamic interactions). From a thermodynamic perspective, no migration preference in a uniform shear flow is expected. Kinetic theory, on the other hand, predicts migration towards the centreline for uniform shear and Poiseuille flows. Their findings showed that for uniform shear flow the polymer migrated towards the centre of the channel as long as the channel height was large enough. Decreasing the size increased the influence of the walls, and for small enough channels, the polymer could migrate towards the walls as observed by previous studies. It is worthwhile to point out that this study presented an opportunity to study micro- and nanoscale phenomena using a mesoscale approach by simply incorporating a suitable polymer model into the solver making it a hybrid method. Thus, in comparison with a full molecular or multiscale approach, even more computational savings are to be expected while extending the capability of the solver to handle polymeric effects.

Ollila et al. (2011) mentioned that for thermal fluctuations to be present, the density needs to fluctuate as well. Using a squared speed of sound (as, for example, in Ladd 1994a) will zero out the viscosity which is relevant for fluctuating density flows. They incorporated their fluctuations via a forcing (source) term on the RHS of Eq. (32) and showed that the variance of the fluctuation is related to the fluctuation–dissipation in the stress tensor. They applied the new thermal fluctuating LBM (TFLBM) in a multiscale approach in Ollila et al. (2013c). Here they showed results, among other particle drag related studies, for MD particle in a cubic box where the solvent was modelled by their TFLBM. Tracking the mean square displacement over time of each particle, they were able to calculate the macroscopic diffusion coefficient without Langevin noise added to the system. Although the diffusion coefficient was observed to be smaller than anticipated, it has been attributed to the finite size of the simulation domain. Ollila et al. (2013b) further explored the effect of confined polymers using either the TFLBM or Langevin Dynamics (LD) in 2D and 3D. First, they studied the radius of gyration (a measure of how far the polymer is stretching) and the static structure factor (a measure of the equilibrium size of the polymer) in both planar and perpendicular direction with respect to the confinement and showed their different behaviours for various confinement levels. They showed that the TFLBM was able to capture the decrease in planar diffusion coefficient for increasing confinement levels, while LD was not able to accurately describe the polymers migration due to the absence of hydrodynamic interactions. In yet another study, Ollila et al. (2013a) investigated a microfluidic T-junction with one inlet at the south and two outlets at the east and west side. They showed that the separating streamline, i.e. the streamline which separates the streamlines going to the outlet to the east or west, can be controlled by adjusting the outlet pressure at both boundaries, which thus influenced the influx-to-outflux ratio (mass per unit time). Being able to control the separating streamline, they placed two particles in the channel downstream of the inlet with different geometrical parameters (initial separation of the two particles, distance of their centre of mass to the separating streamline, radius and channel geometry) and studied which outlet they would choose to exit. This has been validated against experimental data. In fact, they showed their findings in a phase diagram which made it possible to predict which outlet the particle preferred, based on their initial state. Their findings can thus be used to construct a microfluidic NOT gate (one particle) or NAND gate (two particles). Mackay and Denniston (2013) used the TFLBM to couple it via local forces, conserving momentum and energy, to MD particles. Here, the mass at the lattice nodes was used and coupled with the MD particles directly. MD particles were used to construct spheres so fluid could be captured inside. By changing the mass of the MD particles, either the mass of the fluid inside the spheres was much larger or smaller than the overall mass of the sphere (combined MD particles). Studying the velocity autocorrelation function (VACF), they showed that the long-time asymptotic behaviour was independent of the particles mass and agreed with theoretical findings. The particle mass only influenced the short-term behaviour of the VACF. Mackay et al. (2013) implemented the TFLBM into the open-source solver LAMMPS and validated their solver against drag on a single spherical particle, particle motion near a solid wall, hydrodynamic interactions between four particles without collision, two particles with collisions (hard sphere interactions) and a confined colloid undergoing a Couette flow. They supplemented their validated cases with a scalability analysis based on two high-performance computing clusters.

Cui et al. (2012) simulated the cavity formation in a particle covered domain, resembling a pipe leakage that initially was covered by soil particles. They used the discrete element method (DEM) for the particle motion, with and without surface energy for particle cohesion, and the LBM for the fluid surrounding the particles. The particle velocity and force were obtained from the fluid motion which in turn was influenced by the particle motion. They used a rectangular computational domain with an orifice at the bottom wall, located in the middle from which fluid was injected. The domain above was covered by particles. For relatively low exit velocities, the pore pressure was found to rise and stabilise over time. Increasing the exit velocity showed the formation of a cavity which was accompanied by a sudden drop in pore pressure. For even higher velocities, a blow out failure was observed where the fluid was flowing in a straight path to the particle surface, rupturing the layers of particles above. For moderate velocities, the cavity reached a stable size and stopped growing. The inter-particle cohesion was found to have no influence on the rate of orifice pressure build up. A greater pressure was, however, required to initiate the cavity formation as the inter-particle attraction forces needed to be overcome.

So far in our discussion, the particles suspended in the fluid have been modelled as rigid bodies. Dupin et al. (2007) coupled the LBM with deformable particle (DP) dynamics. The advantages of DP are that their membrane deforms as a direct response to the fluid motion. This approach is favourable in cases where the particle properties need to be resolved explicitly and influence the flow, however, that comes at an increased computational cost. Coarse meshes for a single DP have been reported to have approximately 500 nodes. Three types of particles were considered; capsules, spherical in shape, having a finite thickness and unbreakable membrane; vesicles, spherical in shape, having an infinite thickness and unbreakable membrane and red blood cells (RBCs), biconcave in shape. The motivation for this study was driven by the fact that studies up to date, using the computationally more expensive membrane potential function, only allowed small domains to be simulated. They validated their approach against several test cases and used their solver to simulate 200 RBC at low Reynolds numbers in a microchannel. RBC were initially placed 1 micron apart in a regular formation and then allowed to reach an equilibrium state. They found that the pressure field varies significantly in time and space. This fact is usually ignored in biological models and cells subjected to this pressure filed showed great deformation which could have a significant impact on biological processes such as gene expression, a process in which gene information are used to synthesis a functional gene such as a protein. Despite the focus on high scalability, with current limitation imposed by HPC facilities, simulations over the order of 100 cell diameters were not possible. However, it was argued that for such length scales the effect of the particles on the fluid can be modelled and used as a mesoscopic model to formulate boundary conditions for macroscopic simulations, as seen, for example, in Holland et al. (2014), Sect. 2.1.

Driven by the same motivation, to construct a highly efficient solver for general fluid dynamics applications, Zou et al. (2014) combined the efficient velocity, pressure and density computation of the LBM with the Navier–Stokes equations for the constitutive relations. They validated their solver against a Poiseuille flow, Taylor-Green vortex and sudden contraction with a ratio of 4:1 and obtained good agreement. Their focus then shifted to the computational efficiency of the hybrid solver. Comparisons for all three test cases showed that the solver coupling lattice Boltzmann and Navier–Stokes needed 20–25 % of the time compared to a full Navier–Stokes solution. The reduction was attributed to the removal of the pressure correction loop inside the PISO (pressure implicit with splitting of operator, Issa et al. 1986) algorithm that has been employed in the full Navier–Stokes solution. This is in accordance with the findings of Premnath et al. (2005) who commented that the pressure calculation in the Navier–Stokes equations usually takes about 80 % of the whole computational time. Salimi et al. (2015) presented a hybrid lattice Boltzmann/Navier–Stokes solver incorporating thermal effects directly into the LBM. Here the thermal LBM was coupled with the Navier–Stokes equations and the temperature, and the other hydrodynamic variables were exchanged at an interface. One of the goals was to elucidate the efficiency of the hybrid LBM solver, for which they first validated it against reference data for different Reynolds numbers, porosity levels, solid to fluid thermal conductivity and blockage ratios. They then simulated a fixed size domain with a pure LBM, pure Navier–Stokes and hybrid LBM for one lift cycle. When the time step in the hybrid and pure Navier–Stokes solver was set to that of the pure LBM, they needed 6.14 and 8.48 more computational time compared to the pure LBM solver. This is to be expected as the LBM itself performs less floating point operations per time step. At a ratio of approximately $\Delta t_{\mathrm{NS}}/\Delta t_{\mathrm{LBM}} = 50$, the hybrid and continuum Navier–Stokes solver performed as fast as the pure LBM solver, taking only 0.92 and 0.84 of its time, respectively. Pushing the time step ratio further to $\Delta t_{\mathrm{NS}}/\Delta t_{\mathrm{LBM}} = 200$, the computational time further decreased and took 0.55 for the hybrid and 0.33 for Navier–Stokes solver compared to the full LBM solver. This is not surprising as both methods had to perform fewer time steps to advance to the next time level. It is interesting to note that the pure continuum solver performed better for larger time steps than the hybrid solver. This indicates that for true macroscopic simulations the Navier–Stokes equations are still computationally efficient. In contrast, the hybrid solver was able to incorporate multiphysics phenomena on smaller scales at only slightly higher computational costs which may not be easily achieved with a conventional Navier–Stokes approach.

Thus far, we have seen applications in which the LBM was successfully coupled with particle and continuum domains. Due to its mesoscopic nature, however, it is a prime candidate for coupling with a microscale approach. Dupuis et al. (2007) coupled a LBM solver with MD and applied it for the flow past and through a carbon nano tube (CNT). The domain decomposition technique was used in conjunction with the Schwartz alternating method. State coupling was employed and the influence of imposing only velocities from MD to LBM and velocity gradients from MD to LBM investigated. Compared to a full MD solution, the error decreased for about 10 Schwartz iterations after which no considerable gain in accuracy could be achieved. The velocity gradient imposition was 3 times closer to the full MD solution than the solution in which only the velocities were exchanged. Using this approach, for the flow past and through a CNT, the average error (based on velocity differences) was 1.3 and 2.1 %, respectively. The speed-up for both cases was reported to be 7 and 2.4 times faster using the multiscale approach. This is due to the reduced computational demand away from the CNT where inter-molecular motions are adequately resolved by the LBM. Lobaskin and Dünweg (2004) used MD to simulate a charged colloidal system. Here the MD particles used the Lennard–Jones potential for the inter-particle forces and additionally finitely extendable nonlinear elastic (FENE) springs. Two test cases were presented, the first in which the colloid was given an initial horizontal velocity and the second with an initial angular velocity. For the first case, the colloid velocity autocorrelation function exhibited the expected long-term asymptotic behaviour in which momentum is transported away by diffusion. The colloid with initial rotation showed that its decay of angular velocity agreed well with Debye's law and again showed the expected long-term asymptotic behaviour. Results were then presented for a charged colloid during a self-diffusion process. Counterions were placed around the colloid to give a neutral overall charge. The velocity autocorrelation function of the centre of mass velocity showed good agreement with reference data for both neutral and charged colloid. Farahpour et al. (2013) investigate the translocation process of a single-stranded DNA (ssDNA) in which the ssDNA was passing through a pore, driven by an electrical field. The hydrodynamic interactions were captured with the LBM and the ssDNA was represented by a monomer through a MD simulation. The Laplace equation has been solved for the electric potential and counterions were added to overall balance the charge of the simulated system. By studying the velocity gradient tensor by its eigenvalues and eigenvectors, the principle axes of extension and compression could be obtained. It was found that the monomer stretched before approaching the pore, aligning itself with the electric field lines. It then compressed once passed through the pore. Further investigations revealed that replacing the counterions by a charged monomer alone produced incorrect approaching probabilities of the monomer to the pore. If the hydrodynamic interactions were further neglected, then field gradient effects were overestimated. Thus, the inclusion of both counterions and hydrodynamic interactions via a multiscale approach was deemed necessary to accurately predict the correct translocation behaviour of the ssDNA. Mackay et al. (2014) investigated the effect of shear flow on colloidal particles in a microchannel. Particles were initially placed in ordered layers into the channel, and each layer had the same volume fraction. An order parameter was defined as $\psi_6(r_{jk}) = 1/n \sum_{k=1}^{n} \exp[i6\theta(r_{jk})]$. Here, $\theta(r_{jk})$ is the angle between an arbitrary fixed vector and the vector from particle *j* to *k*. For $\langle|\psi 6|\rangle$=1.0, perfect order is achieved in the particle layer. A threshold was defined for $\langle|\psi 6|\rangle$=0.6 above which an ordered state was still encountered. Systems with a lower value of $\langle|\psi 6|\rangle$ were consequently classified as disordered states. Their results showed that $\langle|\psi 6|\rangle$ dropped rapidly at the start, reaching a steady state shortly after. The linear shear imposed by the opposite moving walls allowed particles close to the wall to move faster on average, allowing particles in their adjacent layers to jump into the layer due to the inter-particle forces. This behavior caused a change of volume fraction in each layer, and it was identified as a main cause for a drop in the order parameter. For a greater amount of particle exchanges, the order parameter eventually dropped below the critical value of 0.6 and disorder started to form. Particles close to the wall generally stayed in an ordered formation even though disorder was possible at the channel centre. Their findings have been summarized in a phase diagram showing the disordered, ordered and the coexistence layer for different shear rates and volume fractions.

Interim conclusion on hybrid and multiscale lattice Boltzmann methods

In our discussion, we have outlined that the thermal LBM is still immature and a great amount of hybrid solvers have been developed to use continuum mechanics to solve the temperature field on the mesoscopic scale. This is a valid approach which, however, is more concerned with overcoming the inherent shortcomings of the LBM rather than increasing the computational efficiency. Therefore, most studies discussed herein were concerned with hybrid solvers, while true multiscale approaches with bespoke flow domains in a domain decomposition fashion were less common. In a similar way, high Reynolds number flows are challenging due to the inherent numerical instabilities of the LBM. We have mentioned the MRT collision operator and the entropic LBM which currently extend the range of Reynolds number but do not remove the instabilities. For applications, however, in which solver efficiency is of importance and the temperature does not need to be computed, LBM enjoys a reputation of being easy to implement for complex geometries and handling multiphysics effects, while it outperforms comparable methods at the same resolution in terms of computational time.

DISSIPATIVE PARTICLE DYNAMICS METHOD

Dissipative particle dynamics (DPD) is a particle-based method and has high resemblance to MD which can be regarded as a coarse-grained version of it. By grouping atoms together into representative DPD particles, length and timescales are extended into the mesoscopic regime. DPD has evolved as an application driven need to simulate molecular systems for which the current limitations in space and time for MD are too restrictive. We can use DPD if we are only interested in the mesoscopic manifestation of the underlying molecular details. The governing equation is similar to that of MD as

$$\frac{\partial^2 \mathbf{r}}{\partial t^2} = \mathbf{F} = \mathbf{F}^{\text{int}} + \mathbf{F}^{\text{ext}}, \tag{42}$$

where we have assumed DPD particles of same mass m_i and

unity in magnitude. We see the difference to MD in the splitting of the force into internal and external forces. External forces are those imposed to particles by the system, such as gravitational forces. Internal forces arise due to the inter-particle interactions. We can further decompose the internal forces into

$$\mathbf{F}^{\text{int}} = \sum_{j \neq i} \left(\mathbf{F}_{ij}^{\text{C}} + \mathbf{F}_{ij}^{\text{D}} + \mathbf{F}_{ij}^{\text{R}} \right), \tag{43}$$

where the superscripts C, D and R stand for conservative, dissipative and random, respectively.

The conservative force models the particle interactions and is given by

$$\mathbf{F}_{ij}^{\text{C}} = a_{ij} w^{\text{C}}(r_{ij}), \tag{44}$$

where a_{ij} is the maximum repulsion force and the weight function $w^{C}(r)$ is defined as

$$w^{C}(r_{ij}) = \begin{cases} (1 - r_{ij})\hat{\mathbf{r}}_{ij} & r_{ij} < 1 \\ 0 & r_{ij} \geq 1 \end{cases} \tag{45}$$

Here, r_{ij} is the distance between two particle centre of masses and $\hat{\mathbf{r}}_{ij}$ the vector between particle *i* and *j* of unit length.

The dissipative force mimics the effect of viscosity at the atomistic level and is given by

$$\mathbf{F}_{ij}^{\text{D}} = -\gamma w^{\text{D}}(r_{ij})(\hat{\mathbf{r}}_{ij} \cdot \mathbf{v}_{ij})\hat{\mathbf{r}}_{ij}, \tag{46}$$

where v_{ij} is the particle velocity, γ a coefficient and again $w^{\text{D}}(r_{ij})$ a weight function. The random force is given by

$$\mathbf{F}_{ij}^{\text{R}} = \sigma w^{\text{R}}(r_{ij})\xi_{ij}\hat{\mathbf{r}}_{ij}, \tag{47}$$

where σ is a coefficient, ξ_{ij} is a random number satisfying a Gaussian distribution with a variance of unity, and $w^{\text{R}}(r_{ij})$ is a weight function. The coefficient γ and σ of Eqs. (46) and (47) are related by

$$\gamma = \frac{\sigma^2}{2k_B T}, \tag{48}$$

and their weight functions are

$$\begin{aligned} w^{\text{D}}(r_{ij}) &= \left[w^{\text{R}}(r_{ij})\right]^2 \\ &= \begin{cases} (1 - r_{ij}/r_c)^s & r_{ij} < r_c \\ 0 & r_{ij} \geq r_c \end{cases} \end{aligned} \tag{49}$$

Different values of *s* can lead to different dynamic behaviours of the system but for conventional DPD simulations, s=2 is chosen in general.

Much of what has been said about the boundary treatment for MD is applicable to DPD as well. Periodic boundary conditions are easily imposed, while open or solid boundaries require an appropriate treatment of the conservative, dissipative and random force. We refer the reader to Revenga et al. (1999), Mehboudi and Saidi (2014) and references therein for a detailed treatment of boundary conditions. We have outlined the similarities between DPD and MD and the coarse-grained nature of it. In addition to the conservative force, which essentially replaces the inter-particle potential as seen in MD, we have a dissipative and random force term. The random force adds heat to the system, while the dissipative force is reducing the particle velocities. Their effect together causes the temperature to be approximately constant accompanied by small fluctuations. Hence, DPD is an implicitly thermostated coarse-grained MD version.

An in-depth review of the DPD method and recent developments has been given by Liu et al. (2014).

Review on hybrid and multiscale dissipative particle dynamics methods

Although DPD and the LBM are both mesoscopic in nature, they have been used for different multiscale applications due to their inherent different descriptions. While the LBM is Eulerian based, DPD is a true Lagrangian method and due to its MD background, but upscaled mesoscopic regime, it has primarily found applications in medical and biological flows. Symeonidis et al. (2005) developed a new time integration technique and reported that DPD simulations can be 10^6 times faster than conventional MD simulations. In its limit, it adheres to both microscopic and macroscopic properties, making it a prime candidate for coupling in both directions.

Natural occurring molecules such as wax, fat and vitamins are referred to as lipids, which consist of a head and a tail. In water, lipids form bilayered membrane where the hydrophilic lipid heads form the surface of the membrane with their hydrophilic tails pointing inwards. Biological membranes are made up of different lipids, proteins and other molecules, contributing to the membrane properties. Mercker et al. (2012) studied the curvature modulated sorting of lipids in such a membrane. In the absence of chemical interactions, no sorting is to be expected, but they showed that sorting of lipids occur if one of the macroscopically defined elastic moduli for either lipid or proteins were different. Surface curvature and properties influenced the membrane's structure. Increasing the difference of lipid and protein elastic moduli increased the stability of the formed membrane patterns. The membranes themselves were simulated by a finite element-based membrane dynamics approach. The macroscopic parameters were directly derived from a DPD simulation that was coupled with the FE solver. They concluded that in the absence of a multiscale approach, results could deteriorate and produce incorrect physical systems. Thus, an intermediate DPD solver was needed to feed the macroscopic solver with micro/mesoscale information in order to capture the correct membrane dynamics. A similar approach was adopted by Ghoufi and Malfreyt (2012) who studied the effect of salt concentration in water on its surface tension. They used a method called many-body dissipative particle dynamics (MDPD) which at its core is essentially the DPD method, but the inter-particle force, i.e. Eq. (44), includes not only the inter-particle interactions but also a local particle density. This poses the difficulties that model parameters are not readily available. Therefore, atomistic Monte Carlo simulations were performed for a fixed parameter space for which a linear relationship could be established. This relation was then used in their MDPD simulation to derive the required parameters for which they simulated various salt concentration for two salts (*NaCl* and *NaF*) and showed that with an increase in salt concentration, the surface tension increased correspondingly. Their trends were matched by experimental data. The approach is similar to that of Holland et al. (2014), discussed in Sect. 2.1.

Fedosov and Karniadakis (2009) introduced their so-called triple-decker approach where they coupled MD, DPD and a Navier–Stokes solver. This allows a continuous description from microscopic through the mesoscopic regime up to the macroscopic layer. Not only is the length scale separation reduced via an intermediate, mesoscopic layer as seen in Sect. 2.1 where the microscale was directly coupled with the macroscale, but also this approach benefited from the fact that the mesoscopic solver, here using DPD, was particle based which made the coupling to the microscopic level simpler. On the other hand, the increase in time and length scale of the DPD method also meant that statistical scatter, in comparison with its MD counterpart, was reduced which made it easier to couple to a macroscopic solver. In their approach, MD particles were removed and inserted using the USHER algorithm (Delgado-Buscalioni and Coveney2003a, b) and DPD particles randomly inserted near the boundary. MD boundary conditions were enforced by specular walls, pressure forces to reduce density fluctuations and shear forces for the tangential velocity component. They applied their approach to low Reynolds number for Couette, Poiseuille and lid-driven cavity flows using a domain decomposition approach. Comparison against a full Navier–Stokes solution showed the applicability of the method. From a qualitative point of view, statistical scatter was low and profiles at the HSI were smooth for all cases. Reducing the HSI to zero, i.e. exchanging information directly at an interface without overlap, was also investigated for the Couette flow. They showed that their multiscale approach started to deviate from the analytical solution and showed a maximum discrepancy of about 15 % at the interface which subsequently reduced towards the channel walls.

Reducing the length scales even further, Kacar et al. (2010) also used a three-layer interaction model, here coupling quantum mechanics (QM) with MD and DPD. QM was employed to find the atomistic interaction energies which in turn were used on the molecular level to obtain the parameters for the DPD solver, specifically the solubility and cohesive energy densities. The results obtained on the mesoscopic level were then mapped back onto the microscopic level which completed one coupling cycle. They investigated styrene-co-fluorinated acrylate oligomers and found that an increase in oligomer concentration caused the system to self-assemble from spherical micelles to hexagonal cylinder-shaped structures. The use of a multiscale approach made it possible to reveal structures on the mesoscale that were influenced by the molecular motion on much smaller scales. Furthermore, the study showed that in certain cases a MD description might not be enough and information from even smaller scales are needed as quantum effects may have an impact on the solution at much larger scales, as seen here on the mesoscale.

Dzwinel et al. (2002) also investigated the self-assembly of micelles into hexagonal structures and further the Rayleigh-Taylor instabilities using a three-way coupling approach. Here, MD and DPD were used as well as the fluid particle model (FPM) which was introduced by Español (1998). The FPM is similar to DPD but further accounts for a non-central (shear) force. This is necessary as an orbiting particle around a central particle would not feel any drag, which is eliminated with the non-central force in the FPM and so conserves angular momentum. As the FPM is slightly more complex, it is best suited for length scales above those of DPD so that FPM particles are large enough to only influence their immediate neighbours. Similar to Kacar et al. (2010), Dzwinel et al. (2002) found that the self-assembly of colloidal particles into hexagonal, worm-like structures depends on its interaction parameters (here, the Lennard–Jones interaction potentials). Furthermore, they investigated the cluster growth rate and found a power law behaviour where the exponent initially took values of 1/2, whereas for longer times a transition towards an exponent of unity was observed which was consistent with theoretical predictions. In the case of the Rayleigh-Taylor instability, they showed that cluster fragmentation occurred over time where larger clusters separated into smaller ones while the average cluster size increased over time, which is characteristic for a mixing problem.

Li et al. (2013) used DPD and constructed a multiscale approach containing MD and a Monte Carlo-like polymerising model to study the effect of crosslink reactions and diffusion on the formation of carbon fibre and epoxy resin. They constructed a simple rectangular domain with a carbon surface at the bottom and periodic boundary conditions on the side. Initially, the hardener and epoxy resin were grouped together, away from the carbon surface and separated by a sizing agent. The simulation was then carried out at a constant temperature of 423 K and results presented at 60 min of simulation time. All three components, hardener, resin and sizing agent diffused into their neighbouring domain and combined into epoxy. The hardener was found to be chemically inadequate near the surface to form epoxy which was accompanied by a drop in crosslink density, compared to the bulk matrix density. The interface region, separated by the matrix and carbon fibres, was found to agree with experimental data. DPD presents a valid approach for studying larger atomistic groups but has shown to be inadequate for epoxy as greater molecular details were needed, and hence, MD was used in conjunction with a polymerising model to adequately represent such components.

A vesicle is a fluid-filled cell constructed from bilayered lipid membranes as described previously. They are involved in metabolism, transport processes or chemical reactions, which take place inside the vesicle. Sevink et al. (2013) used a DPD description to study vesicles with a radius of approximately 0.45 nm containing $\mathcal{O}(10^4)$ lipids for a duration of $\mathcal{O}(10^6)$ time steps. DPD was used to resolve the membrane structure and lipid interactions, conserving the necessary molecular details. The solvent surrounding the membrane was resolved by a combination of Brownian dynamics (BD) and dynamic density functional theory (DDFT) which allowed the solvent to be purely expressed by a continuous concentration variable. As a proof of concept, they tested their approach by initially placing lipids in a water solution with two lipid volume fractions. Lipids of approximately 5 v% transitioned into small aggregates which, due to the low percentage of lipids, remained at their initial small size. Increasing the volume fraction to approximately 23 v%, the results were substantially different. The lipids formed a discontinuous network, i.e. a membrane-like structure, but were prevented to form a fully closed membrane. Finally, they investigated a fully pre-assembled vesicle and removed a cylindrical portion of lipids at the membrane wall. After equilibration, the simulation showed that vesicle healing occurred by means of lipid reordering in the membrane to achieve a lower-energy state associated with a closed membrane. They concluded that the multiscale approach was needed to remove the unnecessary solvent details, which could be represented by a continuous field, in order to simulate realistic dimensions. Driven by the same motivation, Berezkin and Kudryavtsev (2013) studied reaction–diffusion flows. Their approach consisted of DPD which accounted for all the molecular reactions taking place, as a result of diffusion processes of initially separated particles or polymers. DPD was used and covered the entirety of the first particle/polymer domain and extended into the second particle/polymer domain, covering the interface. From there, a simple finite element-based diffusion equation was solved for the concentration in the form of $c_t = Dc_{xx}$ where D is the diffusion coefficient and c is the concentration field variable. The coupling, including an overlap region, was achieved by matching the continuum concentration field with the number density of the DPD particles. They validated their approach for a single reaction of different particles for which an analytical solution for the concentration/number density can be found. The FE solution showed a smooth profile throughout the domain and overlap region, while the DPD number density was subject to small fluctuations, however, overall matching the analytical solution. They then tested their approach on reactive coupling of immiscible polymer melts and interfacial polymerisation of two different monomers. In the first case, the number of copolymers per unit area n_c at the interface was observed to increase initially with time as $n_c \sim t$ and then slowed down to scale with the square root of time as $n_c \sim t^{1/2}$. The fast built up of n_c is explained in the reaction process taking place, while once the reaction had finished, diffusion became dominant and reduced the scaling behaviour. The same trends were observed for the second test case, in which monomer conversion into polymers exhibit the same time scaling law, as did the number-averaged and weight-averaged degree of polymerisation. Their simple coupling procedure allowed for greater computational efficiency by removing unnecessary molecular details in favour of a macroscopic concentration field description which is solely governed by diffusion.

A network of polymers can exhibit a loaded stage when external shear or normal forces act on the system. This behaviour can be observed in tissue engineering, paper manufacturing or biological systems. Masoud and Alexeev (2010) investigated such loaded polymer networks to study the permeation and hindered diffusion using DPD where polymer dynamics were obtained by a bond-bending lattice spring model. They validated their approach against experimental data for a non-loaded network for which they matched the data for permeability and diffusivity over a range of porosity values. For loaded networks, they found that the permeability and diffusion along the parallel direction to the normal forces experienced an increased rate compared to its non-loaded stage, while the components perpendicular to the force direction decreased. Under shear, greater permeability was observed for the components in the shear plane, i.e. for a shear force γ_{zx}, the components in x and z were increased, while it had no influence on the permeability normal to the shear plane, here the y direction. The diffusivity only increased for the x component, while both y and z direction exhibit a decrease in diffusion. To characterise the degree of network alignment, they introduced a second-order tensor for which an isotropic network (no loading) showed each main diagonal component to equal 1/3 of the tensor's trace, while the off-diagonal components were zero (and nonzero for an anisotropic network). Normal forces only altered the main diagonal components while shearing introduced nonzero, off-diagonal components. Aligning the main axis of deformation with the coordinate system by transformation revealed a quasi-linear master curve onto which all main diagonal components fell. Fedosov et al. (2012) considered only loaded networks under shear and investigated structural and rheological properties of star polymers for various Weissenberg numbers. These consist of f identical polymers which all connect to a centre node. The star polymers were modelled using a simple spring potential, and the excluded volume interactions were approximated by the shifted Lennard–Jones potential. The coupling occurred through forces in the DPD equations. They considered two hybrid approaches: one of which coupled DPD with the polymer chains and another using multiparticle collision dynamics (MCD). We will focus on the DPD results here although both MCD and DPD produced very similar results. The relative deformation of the star polymer showed to scale with Wi^2 up until $Wi \simeq 1$, after which it reached a saturated state and reduced. The alignment with the flow scaled initially with Wi^{-1} and then reduced to $Wi^{-0.43}$ for W_i greater than unity since these polymers, for high shear rates, were able to align with the flow. Interestingly, the shear rate only scaled as Wi^1 until $Wi \simeq 1$ after which a weaker scaling was observed, hinting that some shear energy at higher rates was transformed. All of the above findings were tested for five different concentration levels and two different number of nodes making up the polymer; however, they all fell onto one single master curve.

We saw in Eq. (44) that a maximum repulsion force needs to be provided in order to calculate the conservative force. This parameter is a microscopic dependent value and for complex systems best derived directly from MD simulations. This was done, for example, by Maly et al. (2008) who studied the morphology of self-assembling nanoparticles in copolymer mixtures, which are made up only of two different monomers. When mixing polymers, the energy associated with the mixing process is measured by the dimensionless Flory–Huggins interaction parameter χ_{AB}. For lamellar polymers, that is, layers of pure polymers stacked on top of each other, and the parameter a_{ij} of Eq. (44) can be assumed to be $a_{AA} = a_{BB}$, where we have two polymers of types A and B. The solubility parameters δ_i for both type A and B were obtained from MD simulations. With that knowledge, the Flory–Huggins parameter could be obtained as $\chi_{AB} = V_{DPD}(\delta_A - \delta_B)^2/(k_B T)$ where V_{DPD} is the volume of one segment of a DPD bead. This led to an empirical equation for a_{AB} as $a_{AB} = a_{AA} + 3.27(1 + (3.9/N_{DPD}^{0.51}))\chi_{AB}$. We refer the reader to Maly et al. (2008) for a detailed derivation on how to obtain the Flory–Huggins parameter and briefly discuss their findings (see also Fermeglia and Pricl 2007) for a general description on how to obtain the Flory–Huggins parameter from MD simulations). They constructed nanoparticles with a central, neutral DPD particle covered by 12 DPD particles of different combinations of A and B. They showed that the introduction of these nanoparticles initiated their self-assembly into distinct cylindrical shapes for various combinations of type A and B, covering the central DPD particle. Scocchi et al. (2007) investigated polymer–clay nanocomposites (PCNs) and used a similar multiscale approach by coupling MD to their DPD solver. PCN is a material that exhibit increased mechanical strength, heat resistance and reduced permeability for gases in comparison with a pure polymer matrix while having only a fraction of the weight. Using conventional macroscopic FE solvers, PCNs are of interest to material engineers but depend on their microscopic interactions, which are not fully understood. Critical parameters remain uncertain, making mesoscopic solvers a prime candidate for which the critical parameters are derived directly from MD simulations. They studied and compared the structure of a Nylon-6-MMT PCN to published data and found close agreement with expected results, paving the way for Toth et al. (2009) who studied different PCNs using MD and DPD and further added a macroscale FE solver. Microscopic details were preserved and passed from the lowest scale through the mesoscale up to the macroscale, and thus, no model constants needed to be defined. Careful validation of each stage showed the applicability of MD and DPD at the micro- and mesoscale. Results on the macroscale were then used to derive parameters of macroscopic interest, here the coefficients of thermal expansion and electrical conductivity. Four different polymer sizes in the PCNs were tested and showed that the thermal expansion coefficient decreased for all PCNs over the clay loading, while the rate of decrease was the steepest for the smallest polymer. The electrical conductivity, however, showed an increase over the clay loading, independent of the polymer size. The importance of this study was the elimination of model constants at the cost of a triple-decker approach. A trade-off in accuracy and computational speed has always to be found and assessed for each case individually. However, where accuracy is the sole motivation, this approach presents an excellent coupling procedure that involves all physical scales.

Biological and medical flow applications are often dominated by mesoscopic cell transport, such as red blood cells and vesicles. Filipovic et al. (2012) simulated different particle flow scenarios using a multiscale solver consisting of a macroscopic FE, and a DPD or LBM solver at the mesoscale. Their influence was introduced at the macroscale through forces at the mesh nodes which were then incorporated into the FE equations. They considered four different test cases: (1) one cylinder and (2) two cylinders depositing in a channel, (3) four elliptical particle subjected to shear flow and (4) particles in an arterial geometry of different shapes. For the first case, the drag force exerted on the cylinder agreed well with Oseen's approximation for both FE-DPD and FE-LBM. Comparison with experimental data of the cylinder displacement showed an initial, undisturbed settlement which was gradually disrupted by harmonic oscillation due to the vortex street forming at higher Reynolds numbers. Results for the second case were only presented qualitatively. For the shear flow, the FE-DPD results were compared to a pure FE solution and the trends were matched, and finally, the arterial geometry, essentially a Y-channel, was qualitatively simulated for cylindrical and elliptical particles. In their proof of concept study, they presented simplified flow geometries (often at the core of biomedical flows) and showed that mesoscopic quantities could successfully be projected onto the macroscopic domain. In fact, not in the spirit of multiscale methods, yet important to our discussion, Feng et al. (2012) investigated arterial (channel) flows with different values of stenosis (channel contraction) using independently a Navier–Stokes-based macroscopic solver and DPD. The transversal velocity profiles at different axial locations agreed well and both Navier–Stokes and DPD matched experimental data. The authors also investigated the recirculation zone just behind the stenosis, caused by the quasi sudden expansion. Smooth results were obtained from the Navier–Stokes solution while DPD suffered from larger statistical scatter, although still reproducing a distinct recirculation zone. The velocity magnitude was integrated in a plane covering the whole recirculation area for different stenosis values. For a stenosis of up to 40 %, the vorticity magnitude differed by a factor of two and only at higher values approached each other. The agreement at 75 % stenosis was within a few per cent. The standard deviation, due to the statistical scatter, was also higher for the DPD solution than for the Navier–Stokes part. The integrated backflow for 64 and 75 % stenosis showed a difference by a factor of 4 and 1.5, respectively, which was explained by the fact that the density immediately behind the stenosis varied while the incompressible Navier–Stokes solution assumed a constant density flow. Their study showed that DPD was able to capture the trend of the Navier–Stokes equation while being prone to statistical scatter. Therefore, it can be concluded that DPD possesses the ability to reproduce continuum effects while containing a larger level of noise compared to the Navier–Stokes equations. For a smooth solution, a hybrid or multiscale method is expected to work best, while details on the mesoscale are retained.

Platelets are small parts which are transported in the blood. They are smaller than red or white blood cells and are responsible to stop bleeding by coagulating (clotting) around wounds which on a macroscopic scale can be interpreted as changing from a liquid to a solid state. This process is referred to as haemostasis, and platelets are said to form a haemostatic plug. Zhang et al. (2014) investigated the platelet dynamics using DPD to solve for the blood and coarse-grained MD to study the platelets. The coupling was done by forces. Platelets needed to get*activated* before they attached to an opening in the endothelium (the inner most layer in the blood vessel) which was done via the platelet stresses. Therefore, the authors investigated rigid and deformable platelets and compared their shear stress distribution on the surface. The particles were subjected to a shear flow with moving walls applied to both upper and lower wall, equal in magnitude but opposite in direction. Initially, they compared the angular velocity and angular acceleration for which differences as high as a factor of 2 were observed. The average induced stresses on the membrane surface were 2.5 times higher for the rigid body than for the deformable one. Since the rigid body would not yield, it resisted the flow and hence higher values were possible. The stresses were also faster accumulated on the rigid platelet than on the deformable one. Their study showed the importance of accurately describing the platelets as the reduced shear stresses on the surfaces not only altered their flow behaviour subjected to shear flows but also suggested that rigid models, overestimating the shear stresses, yielded incorrect activation potentials which can have a significant impact on the correct modelling of the haemostasis. Tosenberger et al. (2013) investigated the mechanism of clot growth (coagulation) as the mechanism is still not fully understood and current theories were rendered inaccurate. In a first simulation, they only modelled the platelets in a blood flow where inter-platelet connections strengthened over time, while platelets already attached for a sufficient amount of time were not able to attach new ones. In this model, the essential process of clot growth was reproduced where a substance known as fibrin is implicitly included as the connections between platelets were growing over time. Fibrin is a protein that forms the haemostatic plug with the platelets. It can be regarded as the glue holding the platelets together. They showed that the obstacle formed by the clot disturbed the velocity distribution, while for large aggregates, the weak forces were not able to resist the drag caused by the platelet clot and ruptured, leaving a smaller platelet clot of only strong connections. In a second investigation, DPD was coupled with a reaction–diffusion–advection equation which was used to calculate the fibrin concentration explicitly. The second simulation was similar to the first one, in which platelets were not able to attach to other platelets in the clot if their fibrin concentration exceeded a threshold value, irrespective of the time they were in the clot. The mechanism of coagulation was essentially the same, while the added level of information available through the fibrin concentration field revealed a constant growing platelet clot. Ruptures of aggregates were still possible, however, a much finer distinction could be made between weak and strong connections. Furthermore, it was possible to show whether the platelets were covered by fibrin or not. Kojic et al. (2008) coupled DPD with a finite element representation of the Navier–Stokes equations. Here, DPD cells were locally introduced in the FE mesh and coincided in a way that the FE mesh edges were the boundary of the DPD region, which could span over several FE cells. This approach resembled the heterogeneous multiscale approach (HMM) as described in Sect. 2.1, where mesoscopic details are only locally computed. The FE solver covered the whole domain and the locally introduced DPD region was coupled to it by forces. This allowed for further computational savings by only considering mesoscopic details where necessary. They applied their approach for two simple benchmark cases: the channel and lid-driven cavity flow. For the channel, the DPD region was placed inside the centre of the channel, without contact to the wall. Compared to an unsteady channel flow, the multiscale approach matched the results of the analytical solution. For the lid-driven cavity, the top right corner (in this case the lid was moving from the right to the left) was presented by the DPD region and velocity profiles compared to the pure FE solution showed very good agreement. In fact, DPD provided very smooth velocity profiles for both cases, while no extra smoothing apart from averaging was done. It is interesting to note that the Reynolds numbers were of the order of 10^1–10^2 for which high statistical noise would be expected. Even spatial and temporal regression as discussed in Sect. 2.1 were not able to achieve similar level of noise reduction. This approach was recommended for use to study platelet dynamics as the clot region can effectively be modelled by DPD (and further coupled to a microscopic description if needed), while the vast majority of the blood can be modelled with an inexpensive continuum solver. In this approach, one would simplify the computations in favour of computational resources. The opposite approach would be to retain all molecular details in favour of high-fidelity results. For systems of realistic real-world sizes, however, simulations and high-performance computing (HPC) must go hand in hand. For that matter, Grinberg et al. (2011) investigated blood flow in a brain with an aneurysm (a sack-shaped expanded blood vessel). The geometry came from an MRI scan, and a coupled DPD/Navier–Stokes solver was developed from open-source tools. LAMMPS was used for the DPD simulation, calculating the blood plasma, red blood cells and platelets, while NekTar (http://www.nektar.info/, see also Cantwell et al. (2015) for NekTar++, its successor), a spectral element solver, was used on the continuum side for the blood flow. The mesoscopic side contained 10^{11} particles and the continuum solver 3×10^9 unknowns. In their study, they used up to 2×10^6 CPU cores on two different clusters, namely a Blue Gene/P and Cray XT5 cluster. For both systems, the strong scaling followed a linear trend up to about 10^6 CPUs. Only at 186,624 cores, run on the Cray XT5, the strong scaling dropped to 68 % efficiency. Their work was important towards efficient code design and coupling. Due to its general nature, it is equally applicable for a broader range of applications and is not limited to medical flows.

Mukhopadhyay and Abraham (2009) coupled a DPD and MD solver to study the flow over a square obstacle in a two-dimensional channel. The geometry was the same as studied by Nie et al. (2004). The domain was split in two parts: the bottom one containing the square obstacle was modelled using MD and the bulk flow above using DPD. While their validation showed good agreement to an analytical solution (Poiseuille flow without obstacle), they revealed further insight into the efficiency of multiscale methods (in this case for the Poiseuille flow). There were essentially two mechanisms that can increase the computational time: coarse graining, i.e. using different length scales and thus fewer particles (as is the case for DPD particles compared to MD particles) and considering timescale separation, for which fewer integrations need to be performed in order to advance to a given time level. They showed that the speed-up to be expected from only considering length scale separation scales with $\mathcal{O}\{[((N_m+\kappa)/(1+\kappa)N_m)]^2\}$ where the ratio of particle masses of each respective simulation is given by $N_m = m_{\mathrm{DPD}}/m_{\mathrm{MD}}$, and κ is the ratio of the volume of the DPD and MD region. Further introducing timescale separation yields $\mathcal{O}\{[((N_m\sqrt{\tau}+\kappa)/(1+\kappa)N_m\sqrt{\tau})]^2\}$, where τ is the relative timescale between DPD and MD. For different time steps of MD and DPD, i.e. $t_{\mathrm{DPD}} > t_{\mathrm{MD}}$, the numerator is growing faster than the denominator and further computational savings can be achieved.

Interim conclusion on hybrid and multiscale dissipative particle dynamics methods

We have shown some examples of hybrid and multiscale DPD applications. Due to its mesoscale nature, it can provide coarse-grained atomistic details for a macroscopic description or speed up a microscopic solver by extending the length scale. DPD is a versatile particle approach that can handle molecular details but provides less statistical noise as its MD counterpart. We have seen that coupling between DPD to either the microscopic or macroscopic level can be achieved by exchanging forces between the solvers. Furthermore, while coupling to the macroscale produced less fluctuations, coupling to the microscale was easily achieved due to the similarities of DPD and MD. Hence, the triple-decker approach used here by various authors coupled successfully the microscale with the macroscale, without experimental data input (i.e. the interaction parameter a_{ij} for the DPD solution) and provided a way to couple complex systems.

SMOOTHED-PARTICLE HYDRODYNAMICS METHOD

Smoothed-particle hydrodynamics (SPH), originally introduced and used for astrophysical applications (Gingold and Monaghan 1977; Lucy 1977), is a particle-based, meshless approach for the solution of the Navier–Stokes equations. The underlying approach is the *integral representation method* that in steps is approximated to yield the final form of the governing equations of the SPH framework. We will briefly explain the key equations and concepts, while full derivations can be found in Liu and Liu (2003).

We start with a trivial mathematical expression which is the underlying core of SPH. We can recast any function into an integral form as

$$f(\mathbf{x}) = \int_{\Omega} f(\mathbf{x}')\delta(\mathbf{x}-\mathbf{x}')\mathrm{d}\mathbf{x}', \tag{50}$$

where the Dirac delta function is defined as

$$\delta(\mathbf{x}-\mathbf{x}') = \begin{cases} 1 & \mathbf{x}=\mathbf{x}' \\ 0 & \mathbf{x}\neq\mathbf{x}' \end{cases} \tag{51}$$

As long as f(x) is continuous in Ω, Eq. (50) is an exact representation. In the SPH framework, the Dirac delta function is replaced by a *smoothing function*, also referred to as the *kernel*, which yields

$$\langle f(\mathbf{x})\rangle = \int_{\Omega} f(\mathbf{x}')W(\mathbf{x}-\mathbf{x}',h)\mathrm{d}\mathbf{x}'. \tag{52}$$

Here, $W(\mathbf{x}-\mathbf{x}',h)$ is the kernel with a support domain *h*, which is the region where the kernel is active and can be freely chosen. It needs to fulfil the following requirements:

1. Its integrated value needs to be unity, i.e.

$$\int_{\Omega} W(\mathbf{x}-\mathbf{x}',h)\mathrm{d}\mathbf{x}' = 1$$

2. It needs to be positively defined, i.e.

$$W(\mathbf{x}-\mathbf{x}') \geq 0 \quad \forall \mathbf{x} \in h$$

3. It needs to have compact support, i.e. $W(\mathbf{x}-\mathbf{x}') = 0$ if $|\mathbf{x}-\mathbf{x}'| > \kappa h$ where κ is a scaling factor

4. It should approach the Dirac delta function in its limit, i.e.

$$\lim_{h\to 0} W(\mathbf{x}-\mathbf{x}',h) = \delta(\mathbf{x}-\mathbf{x}')$$

5. It needs to be sufficiently smooth, be symmetric about its origin and decay away with increasing distance.

A Gaussian kernel may be used but only approaches zero at infinity. Research has been dedicated to find appropriate kernel functions; for example, the B-spline function is defined as

$$W(R,h) = \alpha \begin{cases} \frac{2}{3} - R^2 + \frac{1}{2}R^3 & 0 \leq R \leq 1 \\ \frac{1}{6}(2-R)^3 & 1 \leq R \leq 2 \\ 0 & R \geq 2 \end{cases} \tag{53}$$

where $R = |\mathbf{x}-\mathbf{x}'|/h$ and α depend on the computational dimension with values of 1/*h*, $15/7\pi h^2$ and $3/2\pi h^3$ for one, two and three dimensions, respectively.

Note that in Eq. (52) we follow the SPH convention and used the *kernel approximation operator* ⟨ ⟩ to indicate that we approximate Eq. (50) with a kernel. We make further use of a*particle approximation* in which we replace the integral by a summation over all particles within Ω as

$$\langle f(\mathbf{x_i})\rangle = \sum_{j=1}^{N} \frac{m_j}{\rho_j} f(\mathbf{x_j}) \cdot W_{ij}, \tag{54}$$

where *N* is the number of particles within *h* and a shorthand notation can be introduced as $W_{ij} = W(\mathbf{x_i}-\mathbf{x_j},h)$. By particle approximation, we mean that we fill the domain Ω with a sufficient amount of particles and solve Eq. (54) for each one of them. The integral $\int_{\Omega} \mathrm{d}\mathbf{x}'$ reduces to the volume of each particle V_j and is replaced by its mass and density in Eq. (54). The first derivative is found by differentiating both sides. Applying the Gauss theorem and dropping terms which are zero, one arrives at

$$\langle \nabla \cdot f(\mathbf{x_i})\rangle = -\sum_{j=1}^{N} \frac{m_j}{\rho_j} f(\mathbf{x_j}) \cdot \nabla_i W_{ij}. \tag{55}$$

We see from Eq. (55) that in order to find the derivative of $f(x_i)$, we need to find the derivative of W_{ij} which is analytically known and therefore readily available.

To this point, we have only approximated Eq. (50) by particles which is applicable to a general set of differential equations. We apply this methodology now to the Navier–Stokes equations for which we first write them in Lagrangian coordinates. Denoting the spatial coordinates by α and β, we have the continuity equation as

$$\frac{D\rho}{Dt} + \rho\frac{\partial \mathbf{v}^\beta}{\partial \mathbf{x}^\beta} = 0, \tag{56}$$

the momentum equation (without external forces) as

$$\frac{D\mathbf{v}^\alpha}{Dt} = \frac{1}{\rho}\frac{\partial \sigma^{\alpha\beta}}{\partial \mathbf{x}^\beta}, \tag{57}$$

and finally the energy equation as

$$\frac{De}{Dt} = \frac{\sigma^{\alpha\beta}}{\rho}\frac{\partial \mathbf{v}^\alpha}{\partial \mathbf{x}^\beta}. \tag{58}$$

The total stress tensor is decomposed into its pressure and viscous contribution as

$$\sigma^{\alpha\beta} = -p\delta^{\alpha\beta} + \tau^{\alpha\beta}. \tag{59}$$

We now combine Eqs. (56–58) with Eqs. (54) and (55) to arrive at the final equations for the SPH method. The full mathematical derivation is omitted but can be found in Liu and Liu (2003).

The continuity equation becomes

$$\frac{D\rho_i}{Dt} = \sum_{j=1}^{N} m_j \mathbf{v}_{ij}^{\beta} \cdot \frac{\partial W_{ij}}{\partial \mathbf{x}_i^{\beta}}, \tag{60}$$

for the momentum equation we have

$$\frac{D\mathbf{v}_i^{\alpha}}{Dt} = \sum_{j=1}^{N} m_j \frac{\sigma_i^{\alpha\beta} + \sigma_j^{\alpha\beta}}{\rho_i \rho_j} \frac{\partial W_{ij}}{\partial \mathbf{x}_i^{\beta}}, \tag{61}$$

and for the energy equation we obtain

$$\frac{De_i}{Dt} = \frac{1}{2}\sum_{j=1}^{N} m_j \left(\frac{p_i}{\rho_i^2} + \frac{p_j}{\rho_j^2}\right) \mathbf{v}_{ij}^{\beta} \frac{\partial W_{ij}}{\partial \mathbf{x}_i^{\beta}} + \frac{\mu_i}{2\rho_i}\epsilon_i^{\alpha\beta}\epsilon_i^{\alpha\beta}. \tag{62}$$

Boundary conditions are not as easily implemented in the SPH framework as in other particle-based methods. Similar as with MD in which particles interact with their neighbours, SPH particles interact with those in their support domain. There are no particles outside of the boundaries, and hence, the kernel will be truncated. The idea is similar to MD to place other particles outside of the boundaries which are also called *ghost* particles in the SPH framework. However, these *ghost* particles do not possess a well-defined physical property and need to be modelled via any reasonable physical assumption about their nature. Liu and Liu (2003) reviewed and presented a novel technique by combining existing treatments for solid boundaries. They introduced two new sets of *ghost* particles, the first of which is residing on the solid boundary itself. These particles repel particles within their vicinity back into the computational domain via a repulsive force. The second set of particles is residing outside of the boundaries and ensures that the support domain will not be truncated for particles near the solid boundary. They are mirrored particles of the real particles near the boundary and have the same physical properties except for the velocity component which is opposite in the boundary normal direction. Keat et al. (2015) reviewed various open boundary conditions and showed that there is still active research in this field due to the difficult nature of imposing correct open boundary conditions. Federico et al. (2012) provided an easy and intuitive approach in which two new sets of *ghost* particles were introduced, resulting in a total of four particle sets, i.e.*fluid*, *ghost*, *inflow* and *outflow* particles. Each of these *inflow* and *outflow* particles are placed at the open boundaries and have several layers, just as the *ghost* particles, to prevent a truncated support domain. The *inflow* particles are assigned with the desired pressure and velocity, while the physical properties for the *outflow* particles are frozen. The *inflow* particle are advected in the same manner as the *fluid* particles and change their set from *inflow* to *fluid*once they cross the open boundary. New particles are inserted into the *inflow* particle set to conserve mass. Similarly, the *outflow* particles are advected with their frozen velocity until they reach the end of the *outflow* domain and are then deleted. Each particle evolves with the equations given above for which standard CFD time stepping procedures can be used. Note that due to its Lagrangian particle nature, SPH represents a meshfree method for solving the Navier–Stokes equations and thus particles can easily be refined in regions of interest. This adaptive behaviour is easily done, in contrast to classical CFD techniques, where, for example, either structured meshes with hanging nodes or tetrahedral elements on unstructured grids need to be employed. This requires further solver modification which is not the case in the SPH method. Multi species transport is easily achieved by defining particles of different properties. However, it has been found that higher-order kernels may not conserve physical quantities as W_{ij} may locally have negative values. For the approximation of non-negative quantities such as the density, this could yield non-physical results. Stability, accuracy and convergence have been mainly tested for simple applications with equispaced particles in lower-dimensional space so that it is difficult to fully assess the strength and weaknesses of the method. We refer to Monaghan (2005) for a detailed discussion about the advantages and disadvantages as well as applications.

Review on hybrid and multiscale smoothed-particle hydrodynamics methods

The breakdown of the Navier–Stokes equations and flow phenomena at either very small length scales or rarefied gas conditions has fuelled research on particle methods. Inevitably, some applications can be described using several methods at comparable accuracy and computational cost. For a range of industrial applications, channel, pipe and flow through co-axial cylinders are often encountered. Filipovic et al. (2009) investigated those applications and compared DPD and SPH results independently to elucidate their strength and shortcomings in comparison. They investigated two laminar, low Reynolds number flow cases. The first was a simple Poiseuille flow and comparison with an analytical solution showed that there was little difference between the two methods, both showing accurate results. The second test case was the flow through a co-axial cylinder for three different conditions. The first set-up saw both cylinders rotate at the same angular velocity and the second only the outer cylinder, and for the third case, both cylinders were rotating with the same angular velocity but in opposite directions. The agreement with the analytical velocity profile was good for both methods and the tabulated error revealed that DPD, at best, was marginally more accurate than SPH. The density profile, on the other hand, showed variations between the cylinder walls for the DPD results which were absent in the SPH solution. They showed, however, that the computational time was about two orders of magnitude less for the SPH method which was due to the larger time step permissible as thermal fluctuations were not resolved as in the DPD method. Therefore, while maintaining a comparable level of accuracy, SPH is a fast method and may possess advantages features to be exploited by multiscale applications.

Unlike the continuous description of the Navier–Stokes equations, SPH models the same physical process as a particle which makes it easier to couple it to other particle methods.

Liu et al. (2002) coupled SPH directly to MD without the need for an HSI. This coupling, compared to direct Navier–Stokes/MD coupling, has several advantages. For instance, particles do not need to be inserted or deleted, and the initial SPH particle density can be matched at the interface and then smoothly decrease away from the interface. By further making SPH particles at the interface interact with MD particles through an interaction potential, MD particle does not feel a sudden loss of interaction potential at the interface, as on open boundaries, and hence, no correction needs to be done. Since SPH particles have an in-build smoothing ability by averaging over neighbouring particles, thermodynamic properties are constant across the interface. They tested a simple Poiseuille flow for which the wall regions were modelled by 1600 MD particles, while the centre part of the domain was covered with 600 SPH particles. The results matched the analytical solution, and indeed, a smooth profile across the contact interface could be observed. Jorn and Voth (2012) argued that MD is a very versatile approach but leads to disparate length scales for proton exchange membranes (PEM) which are found in fuel cells. Therefore, they developed a hybrid DPD-SPH approach in which DPD was modelling the water and polymers, while SPH accounted for the proton transport through the membrane. They validated their approach for a simple lamellar test case for which the proton concentration was fixed at both ends, i.e. high proton concentration at the anode (proton production) and low concentration at the cathode (proton consumption) and showed good agreement with an analytical solution for both concentration and proton flux profiles. They applied their hybrid model to a PEM geometry to study the conductivity behaviour with various water contents, which is the number of water molecules per sulphonate group. Their results revealed that explicitly accounting for the proton produced a linear increasing relationship between conductivity and water content, while excluding electrostatic effects yielded a nonlinear relationship. Furthermore, the conductivity was approximately two times greater for the highest water content when including the electrostatic effects, while for the lowest water content, no difference was noticeable which showed that electrostatic effects have an important influence on the conductivity. Experimental results for the conductivity were approximately a third higher and also showed a linear increase over the water content suggesting that proton exchange needs to be included for a correct physical representation. Murashima and Taniguchi (2010) used the modified smoothed-particle hydrodynamics (MSPH) method, introduced by Zhang and Batra (2004), and coupled it to a dumbbell model to obtain the stresses in a dilute polymeric fluid. MSPH is an extension of the conventional SPH method in which $\langle f(x_i)\rangle$ and its derivative is obtained via SPH and a set of equations, derived from a Taylor series expansion, are satisfied leading to more accurate results in the vicinity of boundaries compared to the classical SPH approach. The dumbbell model is a simple evolution equation of the end-to-end vector of a polymer chain, i.e. the distance between the end points of the polymer. The hybrid procedure was as follows: velocity, pressure and the new location of particles were calculated via MSPH and the divergence of the velocity then calculated. The dumbbell model was used to obtain the stresses and MSPH again to find its derivative. This procedure was repeated until tmaxtmax was reached. For a simple channel geometry, they showed that if the relaxation parameter is chosen too small, thermal fluctuations become dominant. Increasing the relaxation parameter reduced this behaviour and qualitative agreement with a pure macroscopic solution showed the expected velocity profile. Yuan et al. (2009) investigated several approaches to simulate the deposition behaviour of metal particles in low temperature, high-velocity air fuel spraying. The surface and impinging particles were modelled either continuously with finite elements or particle based with SPH, and thus for surface–particle interaction, the following three methods were applied: FE–FE, FE–SPH and SPH–SPH. They found for increasing particle temperature and velocity that the interface temperature and plastic strain were locally affected by the impinging particles. Due to the high particle energy, upon impact, particles deformed and thus heated/strained the surface over time. The effective plastic strain was further found to be a function of the impact velocity as particles below a critical velocity remained mostly intact, while for higher velocities, scattering was observed. Comparison between the different simulation approaches, i.e. SPH and FE for particles and surface, showed similar behaviour. Tartakovsky et al. (2008) used the core idea of SPH, i.e. Eqs. (54) and (55), and applied that to a reaction–diffusion equation (RDE), which solves for a concentration field, to simulate flow through porous media. Two mixing solutes A and B were used which formed a precipitate product C. It was assumed that C only accumulates at pores, and thus, a domain decomposition strategy could be adopted in which one RDE was solved at the pore level, while an averaged version of the RDE was solved on the macroscopic level. Since precipitation growth only occurred at the pores, reactions only needed to be captured with the RDE at the pores, while reactions were absent in the macroscopic counterpart. Different levels of resolutions were therefore used to reduce the computational time. The coupling was easily achieved in a non-iterative manner as both descriptions solved the same underlying equation. Ghost particles were introduced that could accumulate mass over time. They were transformed into solid particles if their mass exceeded a threshold value, and thus, precipitation growth could occur. They validated the approach against analytical solutions for planar and circular precipitation growth and tested the multiresolution scheme against a single resolution simulation of a passive scalar in porous media. Finally, they tested their approach in a porous medium using the hybrid scheme which showed that the porosity decreased over time due to precipitation accumulation in the pores. They tested it against a full-scale simulation which matched their hybrid approach. The advantage of their study was that the interface did not need explicit tracking but rather was a product of the solution.

Apart from molecular and particle-based applications, SPH has also been used to investigate geophysical problems. In rock removal for mines or tunnel excavation, rock blasting has become an important alternative to drilling. Fakhimi and Lanari (2014) used a hybrid solver comprising of SPH for gas dynamics and discrete element method (DEM) to model rock specimens. Rock particles were represented by circular discs and attached to each other. The domain was circular and housed DEM rock particles in its entire domain except for the centre, at which SPH particle resided. After a simulated explosion occurred, SPH particles expanded into the rock material. Collision via SPH and DEM particles were accounted for by conserving momentum through ideal plastic collisions.

Once the contact normal and shear stresses of DEM particles were exceeded, bond break ups and microcracks were formed. By further introducing a plane in a second simulation, located off-centre and slicing the circular domain into two parts at which the rock parameters were 1 % of that of the intact rock, simulations showed that the shockwave was not strong enough to cause further damage beyond the plane. Microcrack growth drastically reduced while the waves that had been reflected at the plane caused substantial damage to the rock material. Another geophysical example is that of a breaking dam. Hilton and Cleary (2012) constructed a multiscale method from SPH and the shallow water (SW) equations. Their initial study, to investigate each method on its own and compare it to experimental data and a multiscale method, was a simple dam break scenario. Here, water was initially held at a height *h* for a small portion of the domain, whereas the rest of the domain was left empty. Downstream of the water box was a pillar that interacted with the water. Velocity measurements were taking just upstream of the pillar and showed that all methods followed the experimental data. The pure SPH implementation was more computational expensive than the SW method so that the multiscale method was able to reduce the computational cost. Information exchanged between SPH and SW was the velocity and water height which were imposed at a single interface. With their multiscale approach validated, they applied it to a realistic scenario based on the Gordon dam in Australia spanning a region of approximately 71 ×× 62 km. They reported results for various intervals and showed that after a period of 80 min, a secondary lake formed due to flooding that spanned over 15 km. For such applications involving large length and timescales, a multiscale approach is necessary to capture the small and necessary details to accurately predict the outcome. The use of a multiscale method meant that, for a particle spacing of 15 m, the number of particles could be reduced by two orders of magnitudes (from 2×10^7 for a full SPH simulation to 1.5–2×10^5 for the multiscale simulation).

SPH has proven itself to be a very robust and fast method for visual applications and has been used in major movie productions (Monaghan 2005). Raveendran et al. (2011) noticed that due to the kernel approximation, sudden changes in pressure are not transferred instantaneously, as only the region within the kernel is updated. Hence, pressure changes manifest themselves as artificial acoustic waves which can be suppressed by increasing the stiffness or reducing the time step. They introduced a new hybrid approach where a coarse grid was superimposed onto the SPH domain and the pressure was solved via a Poison equation. In this way, the pressure could be obtained directly by interpolation, while a density correction procedure ensured that a constant density was kept throughout the domain. Their approach required 10 and 18 % more computational time for an incompressible and weakly compressible SPH solver; however, since numerical artefacts such as the spurious acoustic waves were reduced, much greater time steps could be employed, overall increasing the efficiency of the proposed algorithm. They applied their solver to several different water splashing cases using 10^5 to 5×10^5 particles and showed that their speed-up was 3–4 times compared to a standard SPH algorithm. Despite the visual effects orientated approach, the inclusion of a Poison solver on a coarser grid represented an elegant way to include the elliptic nature of pressure in the SPH method. Wang et al. (2013) followed a different approach for visual applications in which the fluid phase was solved by a fast LBM-based solver. The emphasis has been on the free surface dynamics, and hence, the use of LBM removed the problems encountered in Raveendran et al. (2011). The free surface front was tracked with massless particles where a coupling band existed that separated the fluid from the gas phase. Its centreline coincided with the free surface and expanded a distance *d* into both phases. Particles were inserted into the coupling band and were simulated via SPH. The velocity was imposed onto the particles via a weighted interpolation, gradually blending from a pure LBM velocity to a pure SPH velocity. The purpose of the added SPH particles was for visual effects only. Particles penetrating into the LBM domain from the coupling band were rendered as bubbles, while particles inside the coupling band were rendered as foam. Particles escaping the coupling band, outside the fluid domain, were subsequently rendered as spray. Their hybrid solver produced similar results compared to other visual approaches for various scenarios, while the computational speed, owing to the fast LBM solver, was among the fastest.

We now turn our attention to a new, hybrid method that has emerged in the last decade and has successfully been applied to various cases. In the introduction to this section, we have mentioned the work of Filipovic et al. (2009) and showed the close relation of SPH and DPD. Español and Revenga (2003) introduced a new method based on both SPH and DPD to harvest their respective advantages and termed it the smoothed dissipative particle dynamics (SDPD). This hybrid method reproduces the behaviour of the Navier–Stokes equations (SPH part), while thermal fluctuations on the mesoscopic level are retained (DPD part). Hence, the new method is able to be used for a broader range of scale separation, whereas no explicit coupling is needed in-between methods. The introduction of SPH into the DPD framework also means that SDPD particles have a physical dimension, unlike the arbitrary size of DPD particles. Vázquez-Quesada et al. (2009) investigated the scaling behaviour of the SDPD method for particles and polymers undergoing Brownian motion in suspension. In the case of a single colloidal particle (constant volume) surrounded by solvent particles that have been tested for different solvent particle volumes (by refinement), the thermal fluctuation of the colloidal particle has been shown to be scale invariant (over the solvent particle volume), while the thermal fluctuation of solvent particles showed a square root dependency over its volume. This behaviour is to be expected as the thermal fluctuation should only depend on the size of the particle and not on that of the surrounding ones. This mechanism is not observed in DPD as no scale information is available. Alternatively, either MD has to be employed to derive accurate microscopic parameter that can be imposed on the mesoscopic level or a trial and error approach with fine tuning parameters a posteriori. These are two common choices to circumvent this shortcoming of the DPD method.

Due to the DPD nature of the SDPD method, it can also be applied to microscale flows. Litvinov et al. (2008) investigated a suspended polymer chain in solution where the solvent and polymer molecules were modelled using SDPD, while polymer beads were bonded by a FENE spring model. First, they tested a two-dimensional polymer with zero imposed velocity and studied the stochastic motion of the polymer. The radius of gyration, which is the trace of the gyration tensor, increased linearly with the number of beads as expected from the theory. Here, the gyration tensor is a measure of how far the molecules of the polymer are apart, i.e. $\mathbf{G} = 1/2N^2 \sum_{i,j} \langle \mathbf{r_{ij}} \mathbf{r_{ij}} \rangle$ where N is the number of beads. In a second simulation, they applied their method to a polymer in a microchannel and altered the channel height to study the behaviour of confinement. Decomposing the radius of gyration to take only the x and y components of the polymer chain into account, i.e. the radius of gyration parallel and normal to the channel wall, showed anisotropic behaviour for channel heights H<10 where H was normalised by the unconfined radius of gyration. For H>10, both parallel and normal component collapsed onto one single curve, showing the diminishing effect of channel confinement and return to isotropic behaviour. In a similar study, Bian et al. (2012) investigated a single particle in suspension and applied it to several test cases ranging from typical examples found in micro- to macroscale flows which included flow through porous media, a particle with initially imposed velocity (kick started), particles under shear, interaction of two approaching spheres and a neutrally buoyant particle under Brownian motion close to and sufficiently far away from a solid boundary. Their test cases agreed well with reference data. In their last test case of a particle suspended near the solid surface, they also investigated the anisotropic behaviour of a single particle based on its diffusion coefficient. They decomposed it into diffusion parallel and normal to the wall, but unlike in Litvinov et al. (2008), they changed the position of the particle from the wall, while the second wall on the other side was placed sufficiently far away so that only the influence of the closest wall was felt by the particle. They observed that the diffusion parallel to the wall was always greater than normal to it, independent of the distance from the wall. When placed directly on the wall, the normal component was zero, while diffusion parallel to the wall was still permissible. When placed further away, the diffusion coefficient approached each other. Comparison with theoretical results showed that close to the wall (up to a couple of particle radii), SDPD and theory showed discrepancies of up to 30 % which vanished with increasing distance away from the wall. Hence despite the loss in accuracy close to the wall, the general behaviour of the microscale (anisotropic behaviour) was captured in both approaches. Accuracy close to the wall may be increased by using MD for near-wall regions and SDPD for the rest of the domain. Litvinov et al. (2009) derived an analytic expression for the diffusion coefficient to be used in conjunction with SDPD liquids. An accurate knowledge of the diffusion coefficient a priori is important for cases where the Schmidt number Sc influences the flow, as, for example, in the study of non-equilibrium polymer properties. They applied it to a periodic box and showed that flow properties were accurately captured. Furthermore, they showed that no secondary structures were formed during the simulation (crystallisation) by showing the mean square displacement to monotonically increase without plateauing (indicating solidification, decrease in diffusion). The radial distribution function followed the expected behaviour without secondary peaks, which would have indicated particle clustering.

SDPD in itself can be regarded as a hybrid approach combining SPH and DPD. However, it can be further coupled with a microscopic or continuum solver to further enhance details or gain computational efficiency. Moreno et al. (2013) presented a preliminary multiscale approach in which finite elements for the macroscopic Navier–Stokes equations were coupled with SDPD for biomedical applications. Fine scales were locally resolved with SDPD, while the majority of the domain could be represented by the FE method. Although a two-way coupling has been implemented, the coarse scale has been assumed to be steady as the time to reach equilibrium on the smallest scale was less than the macroscopic time step, i.e. $\Delta t_{\text{fine}} \ll t_{\text{eq}} \ll \Delta t_{\text{coarse}}$. Their test case was the lid-driven cavity, and they showed that the horizontal velocity profile of coarse and coupled velocity matched well. The extension to non-Newtonian fluids is necessary to study true biomedical applications such as blood flow which has been left for future work. Kulkarni et al. (2013) further investigated the scaling behaviour of the SDPD method and introduced a novel multiresolution scheme. First, they reported on the equilibrium properties of SDPD particles in a three-dimensional box. They showed that for different particle resolutions, the density stayed constant throughout the domain and the temperature stayed within 2 % of its imposed value. The velocity PDFs of the different particle systems showed that with increasing the particle resolution, the Maxwell–Boltzmann distribution showed narrower tails and higher peaks, approaching the Dirac delta function in its limit. This, again, showed the correct scaling behaviour of SDPD. They devised a multiresolution scheme in which particles were able to change their resolution from one domain to another. For example, if a particle was refined by a factor of 2, once it entered the refined domain, it would be deleted and two new particles inserted. To conserve mass and momentum, both would need to contain half the mass of the original particle while sharing the same, original velocity. One has to note that boundary conditions are sensitive to scale changes and, for such an approach, need to be considered carefully. They applied the multiresolution approach to an equilibrium test case showing that while the particle resolution was changed smoothly, the density was not adversely affected in the refinement region and showed a constant profile across the domain. The second test case was the Couette flow for which a fine resolution was chosen near the stationary, bottom wall, while a coarser representation was chosen at the moving, upper wall. The resolution was changed at the channel centreline in a region of finite thickness. They were able to match the analytical solution in this way. The ability to smoothly change between resolutions, while SDPD is applicable over a range of length scales, makes it a powerful and computational efficient method. Petsev et al. (2015) extended the idea of Kulkarni et al. to a multiscale, multiresolution SDPD-MD algorithm. First, they derived a multiscale approach based on the AdResS scheme of Praprotnik et al. (2005), see Sect. 2.1 for a detailed discussion, in which the pairwise force evaluation was blended between both description for a smooth change. Applied to an equilibrium test case, they showed that the temperature and density profile across the domain differed by 1.8 and 3.2 % at most, where the domain was divided into MD and SDPD with a buffer in-between. They applied it to the Couette flow where the moving wall was set along the MD domain in the first set-up and along all three domains (MD, SDPD and buffer layer) in the second set-up. For both cases, the correct start-up behaviour was observed with minor statistical scatter. In another simulation, they divided the SDPD region into a fine and coarse region where the fine region was located near the

MD interface to match its resolution. The moving wall was only applied to the MD domain, and again a good comparison with the analytical solution was observed with minor noise. Although the resolution was successfully changed within the SDPD domain, decreasing the computational demand, the same time step had to be used across all domains which removed some of the potential computational savings. A remedy could be the use of multiple timescale integrators which were not investigated. However, the ease of blending between MD and SDPD via the AdResS scheme and the multiresolution character of the SDPD method showed great potential of the SDPD method to be accurately used on the smallest scale, while upscaling showed the results to converge to the Navier–Stokes equations.

Interim conclusion on hybrid and multiscale smoothed-particle hydrodynamics methods

We have briefly discussed the progress in hybrid and multiscale SPH methods. Its drawback in accuracy has favoured other comparable methods; however, its ease of implementation and computational efficiency has led to applications such as visual effects. However, SPH remains a pure particle-based method that adheres to the Navier–Stokes equations [Eqs. (60–62)] and hence has shown superior coupling properties to other particle-based method, most prominently MD and DPD. The recent introduction of SDPD, a combination of the favour-able properties of SPH and DPD, provided new research interest into the SPH method, especially for the mesoscopic domain. It is, however, not limited to this regime as microscopic details are captured as well, although the inclusion of MD on the smallest scale has shown to be beneficial. In the study of Petsev et al. (2015), using the AdResS scheme, SDPD (and therefore SPH) showed that each simulated particle can be seen as a system carrying a set of equations which govern its dynamic behaviour. Changing from the micro- to a meso- or macroscale description and vice versa does not mean that the particles need to be destroyed or new particles inserted as in a Navier–Stokes-based coupling approach with MD particles. Instead, they may smoothly change from one description to another (changing the equations of motion via blending). This behaviour is truly advantages compared to traditional, grid-based Navier–Stokes algorithms, and further research into particle-based Navier–Stokes methods, such as the SDPD approach, could prove to simplify current challenges arising at the coupling interface.

SUMMARY AND CONCLUSIONS

In the previous five sections, we have looked at different hybrid and multiscale methods that were developed and used in recent years. Two features are commonly the driving motivation for opting to a multiscale or hybrid description over a monoscale approach, to increase the computational efficiency and to enhance details at smaller scales. We have focused our attention in this review on the most recent and most cited work so as to give an overview of the current state of research. In this spirit, we have not limited ourselves to just one single hybrid/multiscale approach but rather have tried to give a comprehensive overview in various particle-based methods. This also implies that it is not feasible to consider all published literature but rather to give an overview of what each method has achieved to date. It is hoped that the presented studies show an array of different approaches to tackle similar problems and that each method has advantages and disadvantages which may dictate the most suited approach for a particular problem. We have included the lattice Boltzmann and dissipative particle dynamics method in our discussion as two mesoscale representatives. We saw that coupling to other length scales, especially DPD to other particle-based methods, was easily achieved. Molecular dynamics, on the other hand, coupled directly to the Navier–Stokes equations presented difficulties due to disparate degrees of freedom arising from the length and timescale separation. Furthermore, due to the continuum–particle nature in multiscale MD simulations, particles need to be restricted or added/removed in the hybrid solution interface (HSI) in order to properly model the particle behaviour in the overlap region. This is not a trivial task, and various approaches have been discussed herein. We have discussed possibilities where more than two methods have been combined as seen in the triple-decker approach. MD to Navier–Stokes coupling may benefit from a mesoscopic middleman to reduce the effort to link both descriptions. For example, one could use MD at the atomistic level and couple it with the SDPD method. The SDPD method can then be blended from small to large scales (as done by Kulkarni et al. 2013) and then be coupled with the Navier–Stokes equations on the macroscale. Boundary conditions between MD and SDPD as well as SDPD and Navier–Stokes are easier to impose compared to a direct MD and Navier–Stokes coupling. At the macroscale, SDPD particles could easily exchange macroscopic quantities such as velocity, pressure and temperature, while on the smaller scales, they could make use of their particle nature to interact with MD particles. This approach presents a simpler coupling between descriptions, and indeed, a trend towards multiscale schemes that make use of more than two methods can be seen. The combination of micro, meso and macroscale together into one single approach may prove to yield more accurate and stable coupling methods, while physical phenomena are adequately modelled across all scales.

A different approach was proposed by Asproulis and Drikakis (2013) in which a microscale model was trained by using an artificial neural network (ANN). These are networks aiming to mimic the human's neural network. When presented with training input and knowing the corresponding output, an ANN can be trained to reproduce results for similar data (inside its confidence interval). ANNs are usually used for pattern recognition tasks, for example transforming handwriting into digital text or face recognition. However, Asproulis and Drikakis recognised that in MD to Navier–Stokes coupling, most of the MD simulations on the small scales are presented with similar input from the macroscale (Navier–Stokes solution) and thus produce similar outputs. They trained an ANN with MD data for a Couette flow and showed that it could speed up the computational time by disregarding computationally similar calculations and using the ANN to predict the microscopic properties. The benefit of an ANN is that it can take an arbitrary amount of input data and transform it into an arbitrary amount of output data. Latino et al. (2007) investigated the accuracy of ANNs for various parameters to predict the potential energy surface (PES) by training two different network architectures with MD simulations. They found that for a sufficient amount of training data, the error was less than 1 %. Studies in solid mechanics have also been presented that made successful use of ANNs, see, for example, Hambli (2011) for human bone geometry adaptation over time where long-time intervals prohibit microscale simulations and Unger and Könke (2009) for crack propagation in a reinforced concrete beam.

Returning to multiscale simulations, coupling two descriptions may prove to be a cumbersome task, especially deriving an efficient communication algorithm that works across processors. Tang et al. (2015) recently introduced the multiscale universal interface (MUI), which is a header-only library that takes most of the programming-related issues off the user's hands. A data sampler, derived from texture sampling concepts, is implemented which allows for spatial data interpolation between descriptions. This is needed in regions where there is a geometric mismatch between the two coupled schemes. The library operates in a multiple program, multiple data (MPMD) approach and uses point-to-point, non-blocking, asynchronous communication protocols for greater flexibility and efficiency. Furthermore, the concept of frames has been introduced which allows the storage of several time steps. This is needed to preserve time information, especially in cases where two solvers iterate at a different pace. Time information can be exchanged at different stages during the execution and the frames can also be used to calculate average quantities. These are sometimes needed where average flow quantities are imposed in the HSI. They applied it to a simple Couette flow and coupled two LAMMPS solver (SPH). Less than 100 lines of code were needed to establish the coupling. The computational efficiency showed good strong scaling over a range of up to 512 cores. In times where open-source fluid dynamic tools are readily available, their MUI library may provide a handy tool to couple different solvers at high efficiency.

In Fig. 1, we have summarised the length scale separations that have been achieved by the different particle-based multiscale methods discussed in the previous sections. Here we plot the smallest characteristic length scales against the size of simulation domain. The smallest characteristic length scales are: molecular diameter (for MD), mean free path λ (for DSMC), mesh spacing $\Delta x_{\min}$ (for LBM), particle diameter (for DPD) and average distance between particles (for SPH). These characteristic length scales are chosen arbitrarily but have been found to be a good indicator for the minimum resolution required for each respective method. Since simulations can be done using either dimensional or non-dimensional units, only those studies stating the physical length scales explicitly were considered. In some cases, the smallest characteristic length scales were approximated based on other geometrical information. A few observations can be made by studying Fig. 1. All the molecular dynamics-based multiscale simulations have the same smallest length scale $(0.34 \times 10^{-9}\,\mathrm{m})$ since all studies used liquid argon as their working fluid. This is a common choice and fixes the smallest scale, however, the domain size varied in orders of magnitude. A linear trend (in the log–log plot) is evident which shows that all presented studies operated at the same order of length scale separations of approximately $\mathcal{O}(10^2)$–$\mathcal{O}(10^3)$. This is also largely true for the non-dimensional simulations which were not considered in Fig. 1. This is explained by the fact that the channel flow is primarily used which in turn indicates that multiscale methods to date are still under development and primarily validated but less frequently applied to complex geometries. The channel flow offers a well understood example for which an analytic formula is available for both the Poiseuille and Couette flow. Hence, the expected scale separation does not change drastically for different particle-based method as the underlying geometry is not changing. However, to make the most use of multiscale simulations that range across scales, new methodologies need to be developed to couple the microscopic with the macroscopic world. It should be mentioned at this point that only those multiscale methods combining two approaches have been considered. We have addressed various studies combining more than two methods, see, for example, Fedosov and Karniadakis (2009), Kacar et al. (2010), Petsev et al. (2015), and it is expected that by combing more than two approaches, further length scale separation can be achieved. In terms of Fig. 1, multiscale approaches ranging over several length scales would be situated in the upper left part of the plot. There is only one example which can be considered a *true* multiscale approach which bridges the microscale $\mathcal{O}(10^{-9})$ with the macroscale $\mathcal{O}(10^{-1})$. This is the study of Grinberg et al. (2011) who elaborated on the difficulties of parallel computations that occur at these length scale separation. See Sect. 5.1 for a more detailed description. Another observation can be made from Fig. 1: if we define the micro- and macroscale to be approximately of the order of $\mathcal{O}(10^{-9})$ and $\mathcal{O}(10^{0})$, respectively, we can define the mesoscale to sit in-between these two scales. As seen from the plot, all methods, except for molecular dynamics, are located in this area. The DSMC and SPH methods are able to simulated flows on the macroscale as well but also seem to be successful on much smaller scales. In fact, the SPH method ranges from $\mathcal{O}(10^{-9})$ to $\mathcal{O}(10^{2})$ (here we include the SDPD method which makes it possible to be used on such small scales), the DSMC method from $\mathcal{O}(10^{-8})$ to $\mathcal{O}(10^{-1})$ and the DPD method from $\mathcal{O}(10^{-9})$ to $\mathcal{O}(10^{-4})$. These methods are valid over a wide range, but their capabilities are not leveraged due to HPC limitations and limited knowledge available on coupling approaches for complex geometries. We have summarised the advantages, disadvantages and applications of each presented particle method in Table 3. It may be consulted together with Fig. 1 to choose a suitable particle method to construct a multiscale scheme.

Table 3: Advantages, disadvantages and applications of the particle methods presented in this review

	Advantages	Disadvantages	Applications
MD	Explicit molecular structure	Computationally expensive ($\mathcal{O}(N^2)$)	Medical and biological systems
		Open boundaries difficult to treat	Coatings
			Surface-bounded flows
			Fundamental atomistic research
DSMC	Applicable to micro- and macroscales	High statistical scatter	Rarefied gas
	Easy treatment of complex geometries	Limited to high Knudsen numbers	Space related systems
LBM	Computational fast	classic LBM limited to low Mach numbers	Multiphase
	Applicable to meso- and macroscale	Thermal behaviour difficult to treat	Porous media
	Easy treatment of complex geometries		Multiphysics
DPD	Mesoscopic particle description, easy link to micro- and macroscale	Length scales of DPD particles not defined	Medical and biological systems
	Solvents explicitly treated (molecular structure)		Morphology of materials
			Coarse-grained molecular behaviour
SPH	Navier–Stokes in particle form	Still ongoing research on Boundary conditions	High speed flows
	Easy coupling due to particle nature	May not conserve properties for higher-order schemes	Explosions
			Navier–Stokes on the mesoscale

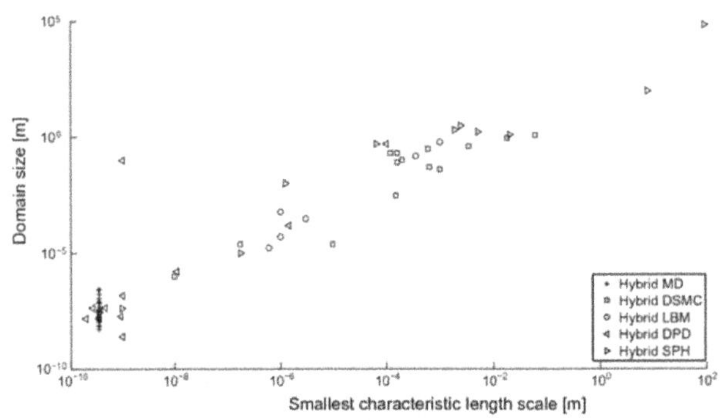

Figure 1: Length scale separations achieved with various particle-based multiscale methods. Only those methods that combine two different approaches are considered

Finally, we would like to mention the work of Martin Karplus, Michael Levitt and Arieh Warshel who in 2013 received the Noble prize in chemistry for having devised a physical sound manner to communicate information between quantum and classical mechanics in complex chemical systems. Multiscale schemes haven risen over the past decades and have demonstrated that current problems may be successfully simulated with such an approach. They have established themselves, and it is expected that their research interest will continue to grow, helping to demystify current and future challenges in fluid mechanics and beyond.

ACKNOWLEDGMENTS

This work has been supported by the Centre of Fluid Mechanics and Computational Science, EEB6001R.

REFERENCES

1. Aidun CK, Clausen JR (2010) Lattice-Boltzmann method for complex flows. Annu Rev Fluid Mech 42(1):439–472
2. Alexiadis A, Lockerby DA, Borg MK, Reese JM (2013) A Laplacian-based algorithm for non-isothermal atomistic–continuum hybrid simulation of micro and nano-flows. Comput Method Appl Mech Eng 264:81–94
3. Alexiadis A, Lockerby DA, Borg MK, Reese JM (2014) The atomistic–continuum hybrid taxonomy and the hybrid–hybrid approach. Int J Numer Methods Eng 98(7):534–546
4. Asproulis N, Drikakis D (2010) Boundary slip dependency on surface stiffness. Phys Rev 81:1–5
5. Asproulis N, Drikakis D (2011) Wall-mass effects on hydrodynamic boundary slip. Phys Rev 84:1–6
6. Asproulis N, Drikakis D (2013) An artificial neural network-based multiscale method for hybrid atomistic–continuum simulations. Microfluid Nanofluidics 15(4):559–574
7. Asproulis N, Kalweit M, Drikakis D (2012) A hybrid molecular continuum method using point wise coupling. Adv Eng Softw 46(1):85–92
8. Bayraktar T, Pidugu SB (2006) Characterization of liquid flows in microfluidic systems. Int J Heat Mass Transf 49(5–6):815–824
9. Beetstra R, van der Hoef MA, Kuipers JAM (2006) A Lattice-Boltzmann simulation study of the drag coefficient of clusters of spheres. Comput Fluids 35(8–9):966–970
10. Beetstra R, van der Hoef MA, Kuipers JAM (2007) Drag force of intermediate reynolds number flow past mono- and bidisperse arrays of spheres. AIChE J 53(2):489–501
11. Berendsen HJC, Postma JPM, van Gunsteren WF, DiNola A, Haak JR (1984) Molecular dynamics with coupling to an external bath. J Chem Phys 81(8):3684–3690
12. Berezkin AV, Kudryavtsev YV (2013) Hybrid approach combining dissipative particle dynamics and finite-difference diffusion model: simulation of reactive

polymer coupling and interfacial polymerization. J Chem Phys 139:154,102

13. Bhatnagar PL, Gross EP, Krook M (1954) A model for collision processes in gases. I. Small amplitude processes in charged and neutral one-component systems. Phys Rev 94(3):511–525
14. Bian X, Litvinov S, Qian R, Ellero M, Adams NA (2012) Multiscale modeling of particle in suspension with smoothed dissipative particle dynamics. Phys Fluids 24(1)
15. Bird GA (1970) Breakdown of translational and rotational equilibrium in gaseous expansions. AIAA J 8(11):1998–2003
16. Bird GA (1989) Perception of numerical methods in rarefied gas dynamics. In: Muntz EP, Weaver DP, Campbell DH (eds) Rarefied gas dynamics: theoretical and computational techniques. American Institute of Aeronautics and Astronautics, Washington, pp 211–226
17. Bird GA (1994) Molecular gas dynamics and the direct simulation of gas flows. Clarendon Press, Oxford
18. Borg M, Reese J (2008) A hybrid particle-continuum framework. In: ASME 2008 6th international conference on nanochannels, microchannels, and minichannels, Darmstadt, Germany, pp 995–1004
19. Borg MK, Macpherson G, Reese J (2010) Controllers for imposing continuum-to-molecular boundary conditions in arbitrary fluid flow geometries. Mol Simul 36:745–757
20. Borg MK, Lockerby DA, Reese JM (2013) Fluid simulations with atomistic resolution: a hybrid multiscale method with field-wise coupling. J Comput Phys 255:149–165
21. Boyd ID (2008) Hybrid continuum–particle computation of hypersonic flows. Technical report, Air Force Office of Scientific Research
22. Boyd ID, Chen G, Candler GV (1995) Predicting failure of the continuum fluid equations in transitional hypersonic flows. Phys Fluids 7(1):210MATH
23. Brent AD, Voller VR, Reid KJ (1988) Enthalpy-porosity technique for modelling convection-diffusion phase change: application to the melting of a pure metal. Numer Heat Transf 13(3):297–318
24. Burt JM, Boyd ID (2008) A low diffusion particle method for simulating compressible inviscid flows. J Comput Phys 227:4653–4670MATH
25. Burt JM, Boyd ID (2009) A hybrid particle approach for continuum and rarefied flow simulation. J Comput Phys 228(2):460–475
26. Burt JM, Boyd ID (2010) A hybrid particle scheme for simulating multiscale gas flows with internal energy nonequilibrium. In: 48th AIAA aerospace sciences meeting including the new horizons forum and aerospace exposition, Orlando, Florida, pp 2010–820
27. Cantwell C, Moxey D, Comerford A, Bolis A, Rocco G, Mengaldo G, De Grazia D, Yakovlev S, Lombard JE, Ekelschot D, Jordi B, Xu H, Mohamied Y, Eskilsson C, Nelson B, Vos P, Biotto C, Kirby R, Sherwin S (2015) Nektar++: an open-source spectral/hp element framework. Comput Phys Commun 192:205–219
28. Carlson HA, Roveda R, Boyd ID, Candler GV (2004) A hybrid CFD-DSMC method of modeling continuum-rarefied flows. In: 42nd AIAA aerospace sciences meeting and exhibit, Reno, Nevada, AIAA pp 2004–1180
29. Chatterjee D, Chakraborty S (2006) A hybrid lattice Boltzmann model for solid–liquid phase transition in presence of fluid flow. Phys Lett A 351(4–5):359–367
30. Chen L, He YL, Kang Q, Tao WQ (2013) Coupled numerical approach combining finite volume and lattice Boltzmann methods for multi-scale multi-physicochemical processes. J Comput Phys 255:83–105
31. Chen S, Doolen GD (1998) Lattice Boltzmann method for fluid flows. Annu Rev Fluid Mech 30(1):329–364
32. Chen S, Martínez D, Mei R (1996) On boundary conditions in lattice Boltzmann methods. Phys Fluids 8:2527–2536
33. Clift R, Grace J, Weber ME (1978) Bubbles, drops and particles. Academic Press, New York
34. Cosden IA, Lukes JR (2013) A hybrid atomistic–continuum model for fluid flow using LAMMPS and OpenFOAM. Comput Phys Commun 184(8):1958–1965
35. Cui X, Li J, Chan A, Chapman D (2012) A 2D DEM–LBM study on soil behaviour due to locally injected fluid. Particuology 10(2):242–252
36. Delgado-Buscalioni R (2010) Tools for multiscale simulations of liquids using open molecular dynamics. In: Engquist B, Runborg O, Tsai Y-HR (eds) Numerical analysis of multiscale computations. Springer, Heidelberg, pp 145–166
37. Delgado-Buscalioni R, Coveney PV (2003a) Continuum–particle hybrid coupling for mass, momentum, and energy transfers in unsteady fluid flow. Phys Rev 67:046,704
38. Delgado-Buscalioni R, Coveney PV (2003b) USHER: an algorithm for particle insertion in dense fluids. J Chem Phys 119(2):978–987
39. Delgado-Buscalioni R, Coveney PV (2004) Hybrid molecular–continuum fluid dynamics. Philos Trans R Soc A 362:1639–1654
40. Delgado-Buscalioni R, Coveney PV, Riley GD, Ford RW (2005a) Hybrid molecular–continuum fluid models: implementation within a general coupling framework. Philos Trans R Soc A 363:1975–1985
41. Delgado-Buscalioni R, Flekkøy EG, Coveney PV (2005b) Fluctuations and continuity in particle–continuum hybrid simulations of unsteady flows based on flux-exchange. Europhys Lett 69(6):959–965
42. Delgado-Buscalioni R, Sablić J, Praprotnik M (2015) Open boundary molecular dynamics. Eur Phys J Spec Top 2349:1–19
43. D'Humières D, Ginzburg I, Krafczyk M, Lallemand P, Luo LS (2002) Multiple–relaxation–time lattice Boltzmann models in three dimensions. Philos Trans R Soc A 360(1792):437–451
44. Docherty SY, Borg MK, Lockerby DA, Reese JM (2014) Multiscale simulation of heat transfer in a rarefied gas. Int J Heat Fluid Flow 50:114–125

45. Dupin MM, Halliday I, Care CM, Alboul L, Munn LL (2007) Modeling the flow of dense suspensions of deformable particles in three dimensions. Phys Rev 75(6):1–17

46. Dupuis A, Kotsalis EM, Koumoutsakos P (2007) Coupling lattice Boltzmann and molecular dynamics models for dense fluids. Phys Rev 75(4):046704

47. Dzwinel W, Da Yuen, Boryczko K (2002) Mesoscopic dynamics of colloids simulated with dissipative particle dynamics and fluid particle model. J Mol Model 8:33–43

48. Español P (1998) A fluid particle model. Phys Rev E 57(3):2930–2948

49. Español P, Revenga M (2003) Smoothed dissipative particle dynamics. Phys Rev 67(2):1–12

50. Fakhimi A, Lanari M (2014) DEM–SPH simulation of rock blasting. Comput Geotech 55:158–164

51. Fan J, Shen C (2001) Statistical simulation of low-speed rarefied gas flows. J Comput Phys 167(2):393–412

52. Farahpour F, Maleknejad A, Varnik F, Ejtehadi MR (2013) Chain deformation in translocation phenomena. Soft Matter 9(9):2750

53. Federico I, Marrone S, Colagrossi A, Aristodemo F, Antuono M (2012) Simulating 2D open-channel flows through an SPH model. Eur J Mech B Fluid 34:35–46

54. Fedosov DA, Karniadakis GE (2009) Triple-decker: interfacing atomistic–mesoscopic–continuum flow regimes. J Comput Phys 228(4):1157–1171

55. Fedosov DA, Singh SP, Chatterji A, Winkler RG, Gompper G (2012) Semidilute solutions of ultra-soft colloids under shear flow. Soft Matter 8(15):4109–4120

56. Feng R, Xenos M, Girdhar G, Kang W, Davenport JW, Deng Y, Bluestein D (2012) Viscous flow simulation in a stenosis model using discrete particle dynamics: a comparison between DPD and CFD. Biomech Model Mechanobiol 11:119–129

57. Fermeglia M, Pricl S (2007) Multiscale modeling for polymer systems of industrial interest. Prog Org Coat 58(2–3):187–199

58. Filipovic N, Ivanovic M, Kojic M (2009) A comparative numerical study between dissipative particle dynamics and smoothed particle hydrodynamics when applied to simple unsteady flows in microfluidics. Microfluid Nanofluidics 7(2):227–235

59. Filipovic N, Isailović V, Dukić T, Ferrari M, Kojic M (2012) Multiscale modeling of circular and elliptical particles in laminar shear flow. IEEE Trans Bio Med Eng 59(1):50–53

60. Filippova O, Hähnel D (2000) A novel lattice BGK approach for low Mach number combustion. J Comput Phys 158(2):139–160

61. Flekkøy EG, Wagner G, Feder J (2000) Hybrid model for combined particle and continuum dynamics. Europhys Lett 52(3):271–276

62. Frisch U, Hasslacher B, Pomeau Y (1986) Lattice-gas automata for the Navier–Stokes equation. Phys Rev 56(14):1505–1508

63. Ghoufi A, Malfreyt P (2012) Coarse grained simulations of the electrolytes at the water–air interface from many body dissipative particle dynamics. J Chem Theroy Comput 8(3):787–791

64. Gingold RA, Monaghan JJ (1977) Smoothed particle hydrodynamics—theory and application to non-spherical stars. Mon Not R Astron Soc 181:375–389

65. Gravesen P, Branebjerg J, Søndergård Jensen O (1993) Microfluidics—a review. J Micromech Microeng 3:168–182

66. Grinberg L, Insley JA, Morozov V, Papka ME, Karniadakis GEM, Fedosov D, Kumaran K (2011) A new computational paradigm in multiscale simulations: application to brain blood flow. In: 2011 international conference for high performance computing, networking, storage and analysis (SC), pp 1–12

67. Guangwu Y, Yaosong C, Shouxin H (1999) Simple lattice Boltzmann model for simulating flows with shock wave. Phys Rev 59(1):454–459

68. Hadjiconstantinou NG (1999) Hybrid atomistic–continuum formulations and the moving contact-line problem. J Comput Phys 265:245–265

69. Hadjiconstantinou NG, Patera AT (1997) Heterogeneous atomistic–continuum representations for dense fluid systems. Int J Mod Phys C 08:967–976

70. Gad-el Hak M (2001) Flow physics in MEMS. Mec Ind 2(4):313–341

71. Hambli R (2011) Numerical procedure for multiscale bone adaptation prediction based on neural networks and finite element simulation. Finite Elem Anal Des 47(7):835–842

72. Hardy J, Pomeau Y, de Pazzis O (1973) Time evolution of a two-dimensional model system. I. Invariant states and time correlation functions. J Phys 14(12):1746–1759

73. Hardy J, de Pazzis O, Pomeau Y (1976) Molecular dynamics of a classical lattice gas: transport properties and time correlation functions. Phys Rev 13(5):1949–1961

74. Hilton JE, Cleary PW (2012) A multiscale method for geophysical flow events. Int J Multiscale Comput 10(4):375–390

75. Ho CM, Tai YC (1998) Micro-electrical-mechanical-systems (MEMS) and fluid flows. Annu Rev Fluid Mech 30(1):579–612

76. Holland DM, Lockerby DA, Borg MK, Nicholls WD, Reese JM (2014) Molecular dynamics pre-simulations for nanoscale computational fluid dynamics. Microfluid Nanofluidics 18(3):461–474

77. Homolle TMM, Hadjiconstantinou NG (2007) A low-variance deviational simulation Monte Carlo for the Boltzmann equation. J Comput Phys 226(2):2341–2358

78. Issa RI, Gosman AD, Watkins AP (1986) The computation of compressible and incompressible recirculating flows by a non-iterative implicit scheme. J Comput Phys 62:66–82

79. Jorn R, Voth GA (2012) Mesoscale simulation of proton transport in proton exchange membranes. J Phys Chem C 116(19):10476–10489

80. Joshi H, Agarwal A, Puranik B, Shu C, Agrawal A (2010) A hybrid FVM–LBM method for single and multi-fluid compressible flow problems. Int J Numer Methods Fluids 62(4):403–427

81. Jun E, Boyd I, Burt J (2013) Assessment of an all-particle hybrid method for hypersonic rarefied flow. In: 51st AIAA aerospace sciences meeting including the new horizons forum and aerospace exposition, Grapevine (Dallas/Ft. Worth Region), pp 2013–1203

82. Kacar G, Atilgan C, Özen AS (2010) Mapping and reverse-mapping of the morphologies for a molecular understanding of the self-assembly of fluorinated block copolymers. J Phys Chem C 114(1):370–382

83. Kalweit M, Drikakis D (2008) Multiscale methods for micro/nano flows and materials. J Comput Theor Nanosci 5(9):1923–1938

84. Kamali R, Kharazmi A (2013) Investigation of multiscale fluid flow characteristics based on a hybrid atomistic–continuum method. Comput Phys Commun 184(10):2316–2320

85. Karlin I, Gorban A, Succi S, Boffi V (1998) Maximum entropy principle for lattice kinetic equations. Phys Rev 81(1):6–9

86. Karniadakis GE, Beskok A, Aluru N (2006) Microflows and nanoflows: fundamentals and simulation. Springer, New-York

87. Kataoka T, Tsutahara M (2004) Lattice Boltzmann method for the compressible Euler equations. Phys Rev 69(5):056,702

88. Kataoka T, Tsutahara M (2004) Lattice Boltzmann model for the compressible Navier–Stokes equations with flexible specific-heat ratio. Phys Rev 69(3):035,701

89. Keat S, Ns Cheng, Xie Y, Shao S (2015) Incompressible SPH simulation of open channel flow over smooth bed. J Hydro Environ Res 9(3):340–353

90. Kim N, Ko SH, Jha S, Novak B, Moldovan D, Nikitopoulos DE (2012) Multi-species fluid flow simulations using a hybrid computational fluid dynamics—molecular dynamics approach. In: 42nd AIAA fluid dynamics conference and exhibit, New Orleans, Louisiana, AIAA, pp 2012–3157

91. Ko SH, Kim N, Jha S, Nikitopoulos DE, Moldovan D (2014) Numerical methodologies for investigation of moderate-velocity flow using a hybrid computational fluid dynamics—molecular dynamics simulation approach. J Mech Sci Technol 28(1):245–253

92. Kojic M, Filipovic N, Tsuda A (2008) A mesoscopic bridging scale method for fluids and coupling dissipative particle dynamics with continuum finite element method. Comput Methods Appl Mech 197(6–8):821–833

93. Koo J, Kleinstreuer C (2003) Liquid flow in microchannels: experimental observations and computational analyses of microfluidics effects. J Micromech Microeng 13(5):568–579

94. Kulkarni PM, Fu CC, Shell MS, Leal LG (2013) Multiscale modeling with smoothed dissipative particle dynamics. J Chem Phys 138(23):234,105

95. Kun Q (2008) Development of lattice Boltzmann method for compressible flows. PhD thesis, National University of Singapore

96. La Torre F, Kenjereš S, Moerel JL, Kleijn CR (2011) Hybrid simulations of rarefied supersonic gas flows in micro-nozzles. Comput Fluids 49(1):312–322

97. Ladd AJC (1994a) Numerical simulations of particulate suspensions via a discretized Boltzmann equation. Part 1. Theoretical foundation. J Fluid Mech 271:285–309

98. Ladd AJC (1994b) Numerical simulations of particulate suspensions via a discretized Boltzmann equation. Part II. Numerical results. J Fluid Mech 271:311–339

99. Latino DARS, Freitas FFM, Aires-De-Sousa J, Fernandes FMSS (2007) Neural networks to approach potential energy surfaces: application to a molecular dynamics simulation. Int J Quantum Chem 107:2120–2132

100. Latt J, Chopard B, Malaspinas O, Deville M, Michler A (2008) Straight velocity boundaries in the lattice Boltzmann method. Phys Rev E 77(5):056703

101. Li M, Gu YZ, Liu H, Li YX, Wang SK, Wu Q, Zhang ZG (2013) Investigation the interphase formation process of carbon fiber/epoxy composites using a multiscale simulation method. Compos Sci Technol 86:117–121

102. Li Z, Yang M, Zhang Y (2014) A hybrid lattice Boltzmann and finite-volume method for melting with convection. Numer Heat Transf 66:307–325

103. Lian YY, Chen YS, Tseng KC, Wu JS, Wu B, Yang L (2011) Improved parallelized hybrid DSMC–NS method. Comput Fluids 45(1):254–260

104. Lilley CR, Macrossan MN (2003) Methods for implementing the stream boundary condition in DSMC computations. Int J Numer Methods Fluids 42(12):1363–1371

105. Litvinov S, Ellero M, Hu X, Na Adams (2008) Smoothed dissipative particle dynamics model for polymer molecules in suspension. Phys Rev 77(6):1–12

106. Litvinov S, Ellero M, Hu X, Adams NA (2009) Self-diffusion coefficient in smoothed dissipative particle dynamics. J Chem Phys 130(2):021101

107. Liu GR, Liu MB (2003) Smoothed particle hydrodynamics: a meshfree particle methods. World Scientific, New Jersey

108. Liu MB, Liu GR, Lam K (2002) Coupling meshfree particle method with molecular dynamics—novel approach for multiscale simulations. In: Advances in meshfree and X-FEM methods—proceedings of the 1st Asian workshop on meshfree methods, pp 211–217

109. Liu MB, Liu GR, Zhou LW, Chang JZ (2014) Dissipative particle dynamics (DPD): an overview and recent developments. Arch Comput Method Eng 22(4):529–556

110. Lobaskin V, Dünweg B (2004) A new model for simulating colloidal dynamics. New J Phys 6(54). doi:10.1088/1367-2630/6/1/054

111. Lofthouse AJ, Boyd ID, Wright MJ (2007) Effects of continuum breakdown on hypersonic aerothermodynamics. Phys Fluids 19(2):027105

112. Löhner R (2008) Applied computational fluid dynamics

techniques: an introduction based on finite element methods. Wiley, Chichester

113. Lucy LB (1977) A numerical approach to the testing of the fission hypothesis. Astron J 82:1013–1024
114. Mackay F, Pastor K, Karttunen M, Denniston C (2014) Modeling the behavior of confined colloidal particles under shear flow. Soft Matter 10(43):8724–8730
115. Mackay FE, Denniston C (2013) Coupling MD particles to a lattice-Boltzmann fluid through the use of conservative forces. J Comput Phys 237:289–298
116. Mackay FE, Ollila STT, Denniston C (2013) Hydrodynamic forces implemented into LAMMPS through a lattice-Boltzmann fluid. Comput Phys Commun 184(8):2021–2031
117. Macpherson GB, Reese JM (2008) Molecular dynamics in arbitrary geometries: parallel evaluation of pair forces. Mol Simul 34(1):97–115
118. Maly M, Posocco P, Pricl S, Fermeglia M (2008) Self-assembly of nanoparticle mixtures in diblock copolymers: multiscale molecular modeling. Ind Eng Chem Res 47(15):5023–5038
119. Manz A, Harrison DJ, Verpoorte E, Widmer H (1993) Planar chips technology for miniaturization of separation systems: a developingperspective in chemical monitoring. In: Brown PR, Grushka E (eds) Advances in chromatography, vol 33. CRC Press, New York, pp 2–66
120. Markesteijn A, Karabasov S, Scukins A, Nerukh D, Glotov V, Goloviznin V (2014) Concurrent multiscale modelling of atomistic and hydrodynamic processes in liquids. Philos Trans R Soc A 372:20130379
121. Masoud H, Alexeev A (2010) Permeability and diffusion through mechanically deformed random polymer networks. Macromolecules 43(23):10117–10122
122. Mehboudi A, Saidi M (2014) Physically based wall boundary condition for dissipative particle dynamics. Microfluid Nanofluidics 17(1):181–198
123. Mercker M, Ptashnyk M, Kühnle J, Hartmann D, Weiss M, Jäger W (2012) A multiscale approach to curvature modulated sorting in biological membranes. J Theor Biol 301:67–82
124. Mezrhab A, Bouzidi M, Lallemand P (2004) Hybrid lattice-Boltzmann finite-difference simulation of convective flows. Comput Fluids 33(4):623–641
125. Mitcheltree RA, Wilmoth RG, Cheatwood FM, Brauckmann GJ, Greene FA (1997) Aerodynamics of Stardust sample return capsule. Technical report, NASA Langley Research Center, Hampton, Virginia
126. Mohamed KM, Mohamad AA (2010) A review of the development of hybrid atomistic–continuum methods for dense fluids. Microfluid Nanofluidics 8:283–302
127. Monaghan JJ (2005) Smoothed particle hydrodynamics. Rep Prog Phys 68(8):1703–1759
128. Moreno N, Vignal P, Li J, Calo VM (2013) Multiscale modeling of blood flow: coupling finite elements with smoothed dissipative particle dynamics. Procedia Comput Sci 18:2565–2574
129. Mukhopadhyay S, Abraham J (2009) A particle-based multiscale model for submicron fluid flows. Phys Fluids 21:027,102
130. Murashima T, Taniguchi T (2010) Multiscale Lagrangian fluid dynamics simulation for polymeric fluid. J Polym Sci Part B Polym Phys 48(8):886–893
131. Nie XB, Chen SY, Robbins MO (2004) A continuum and molecular dynamics hybrid method for micro- and nano-fluid flow. J Fluid Mech 500:55–64
132. O'Connell ST, Thompson PA (1995) Molecular dynamics–continuum hybrid computations: a tool for studying complex fluid flows. Phys Rev 52(6):5792–5795
133. Ollila STT, Denniston C, Karttunen M, Ala-Nissila T (2011) Fluctuating lattice-Boltzmann model for complex fluids. J Chem Phys 134(6):064902
134. Ollila STT, Denniston C, Ala-Nissila T (2013) One- and two-particle dynamics in microfluidic T-junctions. Phys Rev E 87(5):050,302
135. Ollila STT, Denniston C, Karttunen M, Ala-Nissila T (2013b) Hydrodynamic effects on confined polymers. Soft Matter 9(13):3478
136. Ollila STT, Smith C, Ala-Nissila T, Denniston C (2013c) The hydrodynamic radius of particles in the hybrid lattice Boltzmann–molecular dynamics. Multiscale Model Simul 11(1):213–243
137. Oran ES, Oh CK, Cybyk BZ (1998) Direct simulation Monte Carlo: recent advances and applications. Annu Rev Fluid Mech 30(1):403–441
138. Palharini RC (2014) Atmospheric reentry modelling using an open-source DSMC code. PhD thesis, University of Strathclyde
139. Pantazis S, Rusche H (2014) A hybrid continuum–particle solver for unsteady rarefied gas flows. Vacuum 109:1–9
140. Patankar S, Spalding D (1972) A calculation procedure for heat, mass and momentum transfer in three-dimensional parabolic flows. Int J Heat Mass Transf 15(10):1787–1806
141. Patankar SV (1980) Numerical heat transfer and fluid flow. McGraw-Hill, New-York
142. Patronis A, Lockerby DA (2014) Multiscale simulation of non-isothermal microchannel gas flows. J Comput Phys 270:532–543
143. Patronis A, Lockerby DA, Borg MK, Reese JM (2013) Hybrid continuum–molecular modelling of multiscale internal gas flows. J Comput Phys 255:558–571
144. Petsev ND, Leal LG, Shell MS, Petsev ND, Leal LG, Shell MS (2015) Hybrid molecular–continuum simulations using smoothed dissipative particle dynamics. J Chem Phys 142:044101
145. Plimpton S (1995) Fast parallel algorithms for short range molecular dynamics. J Comput Phys 117(1):1–19
146. Posocco P, Gentilini C, Bidoggia S, Pace A, Franchi P, Lucarini M, Fermeglia M, Pricl S, Pasquato L (2012) Self-organization of mixtures of fluorocarbon and hydrocarbon amphiphilic thiolates on the surface of gold nanoparticles. ACS Nano 6(8):7243–7253

147. Praprotnik M, Delle Site L, Kremer K (2005) Adaptive resolution molecular-dynamics simulation: changing the degrees of freedom on the fly. J Chem Phys 123:1–14
148. Premnath KN, Nave JC, Banerjee S (2005) Computation of multiphase flows with lattice Boltzmann methods. Fluids Eng 2005:403–420
149. Raveendran K, Wojtan C, Turk G (2011) Hybrid smoothed particle hydrodynamics. In: Proceedings of the 2011 ACM SIGGRAPH/Eurographics symposium on computer animation—SCA'11, vol 1, pp 33–42
150. Ren W (2007) Analytical and numerical study of coupled atomistic–continuum methods for fluids. J Comput Phys 227:1353–1371
151. Revenga M, Zuniga I, Espanol P (1999) Boundary conditions in dissipative particle dynamics. Comput Phys Commun 121–122(122):309–311
152. Salimi M, Taeibi-Rahni M, Jam F (2015) Heat transfer analysis of a porously covered heated square cylinder, using a hybrid Navier–Stokes–lattice Boltzmann numerical method. Int J Therm Sci 91:59–75
153. Santiago JG, Wereley ST, Meinhart CD, Beebe DJ, Adrian RJ (1998) A particle image velocimetry system for microfluidics. Exp Fluids 25(4):316–319
154. Scanlon TJ, Roohi E, White C, Darbandi M, Reese JM (2010) An open source, parallel DSMC code for rarefied gas flows in arbitrary geometries. Comput Fluids 39(10):2078–2089
155. Schwartzentruber TE, Boyd ID (2006) A hybrid particle–continuum method applied to shock waves. J Comput Phys 215(2):402–416
156. Schwartzentruber TE, Boyd ID (2015) Progress and future prospects for particle-based simulation of hypersonic flow. Prog Aerosp Sci 72:66–79
157. Scocchi G, Posocco P, Fermeglia M, Pricl S (2007) Polymer-clay nanocomposites: a multiscale molecular modeling approach. J Phys Chem B 111(9):2143–2151
158. Sevink GJA, Charlaganov M, Fraaije JGEM (2013) Coarse-grained hybrid simulation of liposomes. Soft Matter 9(10):2816–2831
159. Shah MT, Utikar RP, Tade MO, Evans GM, Pareek VK (2013) Effect of a cluster on gas–solid drag from lattice Boltzmann simulations. Chem Eng Sci 102:365–372
160. Shen C (2005) Rarefied gas dynamics: fundamentals, simulations and microflows. Springer, Berlin
161. Shi W, Shyy W, Mei R (2001) Finite-difference based lattice Boltzmann method for inviscid compressible flows. Numer Heat Transf 40(1):1–21
162. Squires TM, Quake SR (2005) Microfluidics: fluid physics at the nanoliter scale. Rev Mod Phys 77(3):977–1026
163. Steijl R, Barakos G (2012) Coupled Navier–Stokes/molecular dynamics simulations in nonperiodic domains based on particle forcing. Int J Numer Methods Fluids 69:1326–1349
164. Succi S (2001) The lattice Boltzmann equation for fluid dynamics and beyond. Oxford University Press, Oxford
165. Sun Q, Boyd ID (2005) Evaluation of macroscopic properties in the direct simulation Monte Carlo method. J Thermophys Heat Transf 19(3):329–335
166. Symeonidis V, Karniadakis GE, Caswell B (2005) A seamless approach to multiscale complex fluid simulation. Comput Sci Eng 7(3):39–46
167. Tang YH, Kudo S, Bian X, Li Z, Karniadakis GE (2015) Multiscale universal interface: a concurrent framework for coupling heterogeneous solvers. J Comput Phys 297:13–31
168. Tartakovsky AM, Tartakovsky DM, Scheibe TD, Meakin P (2008) Hybrid simulations of reaction-diffusion systems in porous media. SIAM J Sci Comput 30(6):2799–2816
169. Thompson Pa, Troian SM (1997) A general boundary condition for liquid flow at solid surfaces. Nature 389(6649):360–362
170. Tosenberger A, Ataullakhanov F, Bessonov N, Panteleev M, Tokarev A, Volpert V (2013) Modelling of thrombus growth in flow with a DPD–PDE method. J Theor Biol 337:30–41
171. Toth R, Voorn DJ, Handgraaf JW, Fraaije JGEM, Fermeglia M, Pricl S, Posocco P (2009) Multiscale computer simulation studies of water-based montmorillonite/poly(ethylene oxide) nanocomposites. Macromolecules 42(21):8260–8270
172. Tretheway DC, Meinhart CD (2002) Apparent fluid slip at hydrophobic microchannel walls. Phys Fluids 14(3):12–15
173. Unger JF, Könke C (2009) Neural networks as material models within a multiscale approach. Comput Struct 87(19–20):1177–1186
174. Usta OB, Butler JE, Ladd AJC (2006) Flow-induced migration of polymers in dilute solution. Phys Fluids 18(3):14–18
175. Vázquez-Quesada A, Ellero M, Español P (2009) Consistent scaling of thermal fluctuations in smoothed dissipative particle dynamics. J Chem Phys 130(3):034901
176. Versteeg HK, Malalasekera W (2007) An introduction to computational fluid dynamics: the finite volume method, 2nd edn. Prentice Hall, Harlow
177. Wang CB, Zhang Q, Kong FL, Qin H (2013) Hybrid particle-grid fluid animation with enhanced details. Vis Comput 29(9):937–947
178. Wang YC, He GW (2007) A dynamic coupling model for hybrid atomistic–continuum computations. Chem Eng Sci 62:3574–3579
179. Weinan E, Engquist B, Huang Z (2003) Heterogeneous multiscale method: a general methodology for multiscale modeling. Phys Rev 67(9):2–5
180. Wu J, Lian Y, Cheng G, Koomullil R, Tseng K (2006) Development and verification of a coupled DSMC–NS scheme using unstructured mesh. J Comput Phys 219:579–607
181. Wu J, Shu C, Zhao N (2012) Simulation of thermal flow problems via a hybrid immersed boundary-lattice Boltzmann method. J Appl 2012:1–12

182. Xiaohai M (2005) Direct simulation Monte Carlo (DSMC) in micro flows. PhD thesis, National University of Singapore

183. Xu J, Li Y (2007) Boundary conditions at the solid–liquid surface over the multiscale channel size from nanometer to micron. Int J Heat Mass Transf 50:2571–2581

184. Xu SS, Wu ZN, Li Q, Hong YJ (2009) Hybrid continuum/DSMC computation of rocket mode lightcraft flow in near space with high temperature and rarefaction effect. Comput Fluids 38(7):1394–1404

185. Yuan X, Zha B, Hou G, Hou P, Jiang L, Wang H (2009) Multiscale model on deposition behavior of agglomerate metal particles in a low-temperature high-velocity air fuel spraying process. J Therm Spray Technol 18(3):411–420

186. Zhang GM, Batra RC (2004) Modified smoothed particle hydrodynamics method and its application to transient problems. Comput Mech 34(2):137–146

187. Zhang P, Gao C, Zhang N, Marvin SJ, Deng Y, Bluestein D (2014) Multiscale particle-based modeling of flowing platelets in blood plasma using dissipative particle dynamics and coarse grained molecular dynamics. Cell Mol Bioeng 7(4):552–574

188. Zou Q, He X (1997) On pressure and velocity boundary conditions for the lattice Boltzmann BGK model. Phys Fluids 9(6):1591–1598

189. Zou S, Yuan XF, Yang X, Yi W, Xu X (2014) An integrated lattice Boltzmann and finite volume method for the simulation of viscoelastic fluid flows. J Non Newton Fluid 211:99–113

CHAPTER 2

A PURE LAGRANGIAN MESHFREE METHOD FOR COMPUTATIONAL FLUID DYNAMICS

Wade A. Walker

INTRODUCTION

In this paper, we first present background material on CFD and discuss previous CFD methods which have informed this work. Then we motivate RRM and explain its workings in depth. Next, we show that RRM gives correct results for many standard test problems. We also demonstrate that RRM shows steadily decreasing error in its solutions as we increase the desired accuracy, and that RRM handles many common types of boundary conditions. Finally, we discuss the similarities and differences between RRM and other CFD methods.

Background

CFD is the use of numerical methods to model liquid and gas flow. CFD has many practical uses, from the analysis of the airflow over vehicles to the design of water turbines.

CFD covers a vast range of fluid compositions and flow types. For simplicity, we only consider a fluid that is:

- Continuous: Infinitely sub dividable, unlike real fluids which are made of discrete atoms and molecules.
- Simple: Completely described by density, velocity, and pressure at each point, which we call the "primitive variables", and write as ρ, u, and p. We do not consider other possible fluid properties like chemical reactivity. We also do not consider the action of non-pressure forces like gravity or electromagnetism on the fluid.
- Ideal: Described by the ideal gas law, in which the internal energy of a fluid is purely a function of ρ, p, and γ. The constant γ is called the ratio of specific heats, and has a value of about 1.4 for air.
- Single-phase: Consisting entirely of either liquid or gas, but not a mixture of the two. This means we need not model liquid-gas interfaces. We also do not consider the interaction of solid objects with the fluid.
- Inviscid: Having no resistance to deformation. This simplifies the equations of fluid motion.
- Adiabatic across contacts: Allowing no heat to flow from one side of a contact discontinuity to the other. This means that contact-adjacent regions will not tend towards the same temperature. We compare RRM's results to fluid flows that are adiabatic across contacts because of the availability of analytic solutions, but we show later that RRM is not adiabatic across contacts.
- One-dimensional: Having only one spatial dimension. This makes illustration and programming simpler.

Even though our fluid is infinitely subdividable, for illustration and analysis we divide it into finite-sized cells. Figure 1 shows a cell c_1 with its left edge at x_1 and its right edge at x_2. The density, velocity, and pressure components are shown on separate graphs.

Citation: Walker WA (2012), A Pure Lagrangian Meshfree Method for Computational Fluid Dynamics. PLoS ONE 7(7): e39999, doi:10.1371/journal.pone.0039999.

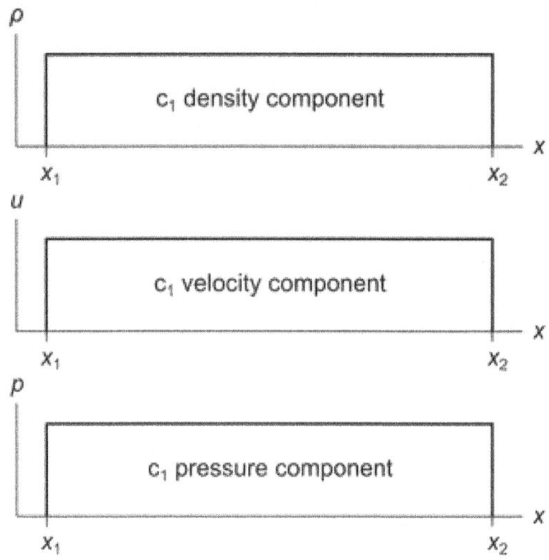

Figure 1: Fluid cell with three separate components. Fluid cell c_1 has density, velocity, and pressure components ρ, u, and p. The left and right coordinates of the cell are x_1 and x_2.

When we do not need to show all three components separately, we combine them onto one axis for simplicity as shown in figure 2, with the understanding that ρ, u, and p may have different values even though they are drawn with the same line.

We can describe fluid flow with cells in two main ways. The Eulerian description considers the cells to be stationary, and the fluid to flow across their edges and through them. The Lagrangian description considers the cells to move along with the fluid, so any given bit of fluid is always found in the same cell. We will initially use the Eulerian description since it is the most common. We will later switch to the Lagrangian description when we describe RRM in more detail.

Given the restrictions and cell definition above, we can model fluid flow with a set of equations called the Euler equations, which can be derived from the local conservation of mass, momentum, and energy. The Euler equations take on different forms depending on whether we write them for the Eulerian or Lagrangian description of fluid flow. For the Eulerian description, we write the Euler equations in English like this:

- Conservation of mass: The mass in a cell changes by the amount that flows across its edges.
- Conservation of momentum: The momentum in a cell changes by the amount that flows across its edges, and by the amount due to the pressure acting on its edges.
- Conservation of energy: The energy in a cell changes by the amount that flows across its edges, and by the amount due to work done by the pressure acting on its edges.

The Euler equations are typically written as partial differential equations, but we write them here as integral equations because it is more natural for our derivative-free numerical method. Here are the Euler equations for a cell, written for the Eulerian description of fluid flow, in conservation form:

$$\begin{aligned}
&\text{conservation of mass}\\
&\int_{x_1}^{x_2} (\rho(x,t_2)-\rho(x,t_1))dx\\
&=-\int_{t_1}^{t_2} (\rho(x_2,t)u(x_2,t)-\rho(x_1,t)u(x_1,t))dt\\
&\text{conservation of momentum}\\
&\int_{x_1}^{x_2} (\rho(x,t_2)u(x,t_2)-\rho(x,t_1)u(x,t_1))dx\\
&=-\int_{t_1}^{t_2} (\rho(x_2,t)u^2(x_2,t)-\rho(x_1,t)u^2(x_1,t))dt\\
&-\int_{t_1}^{t_2} (p(x_2,t)-p(x_1,t))dt\\
&\text{conservation of energy}\\
&\int_{x_1}^{x_2} (\rho(x,t_2)e_T(x,t_2)-\rho(x,t_1)e_T(x,t_1))dx\\
&=-\int_{t_1}^{t_2} (\rho(x_2,t)e_T(x_2,t)u(x_2,t)-\rho(x_1,t)e_T(x_1,t)u(x_1,t))dt\\
&-\int_{t_1}^{t_2} (p(x_2,t)u(x_2,t)-p(x_1,t)u(x_1,t))dt
\end{aligned} \tag{1}$$

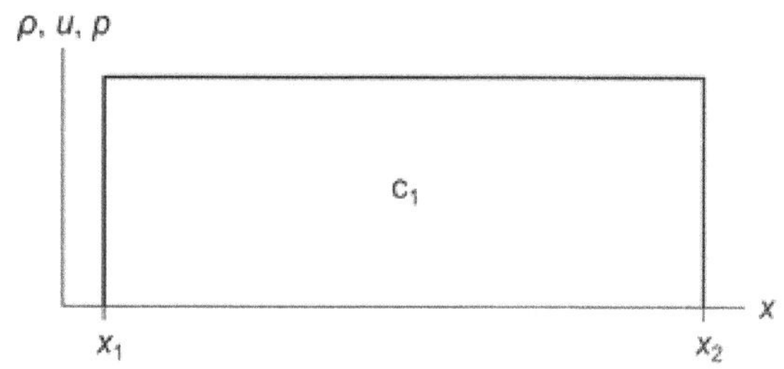

Figure 2: Fluid cell with three superimposed components. Fluid cell c_1 has density, velocity, and pressure components all superimposed on the same axis. The left and right coordinates of the cell are x_1 and x_2.

The coordinates x_1 and x_2 are the left and right edges of the cell. The times t_1 and t_2 are the starting and ending times of a period where fluid is flowing into and out of the cell, and pressure is acting on the cell edges.

This form is called the conservation form because it is written in terms of the conserved quantities per unit length. These conserved quantities are mass per unit length ρ, momentum per unit length ρu, and energy per unit length ρe_T.

The specific total energy e_T is the energy per unit mass due to both macroscopic and microscopic motion. The ideal gas law gives us equations for e_T and for the speed of sound a, which we will use later.

$$\text{specific total energy} \quad e_T=\frac{1}{2}u^2+\frac{p}{(\gamma-1)\rho} \tag{2}$$

$$\text{speed of sound} \quad a=\sqrt{\frac{\gamma p}{\rho}} \tag{3}$$

To write the Euler equations in a more compact form we define a vector of the conserved quantities

$$\mathbf{u}=\begin{bmatrix}\rho\\ \rho u\\ \rho e_T\end{bmatrix} \tag{4}$$

and a vector of the fluxes (plus the pressure and pressure work) at the edges

$$\mathbf{f}=\begin{bmatrix}\rho u\\ \rho u^2+p\\ \rho e_T u+pu\end{bmatrix} \tag{5}$$

Then the Euler equations can be written as a single vector equation

$$\int_{x_1}^{x_2} (\mathbf{u}(x,t_2)-\mathbf{u}(x,t_1))dx = -\int_{t_1}^{t_2} (\mathbf{f}(x_2,t)-\mathbf{f}(x_1,t))dt \quad \textbf{(6)}$$

For the general initial conditions $\rho(x,t_1)$, $u(x,t_1)$, and $p(x,t_1)$, the Euler equations have no known analytical solution. This is inconvenient when we wish to check the results of a numerical method. So in this paper we restrict ourselves to simple initial conditions known as the Riemann problem, where ρ, u, and p take on the constant values (ρ_l, u_l, p_l) and (ρ_r, u_r, p_r) on the left and right sides of an initial discontinuity, as shown in figure 3.

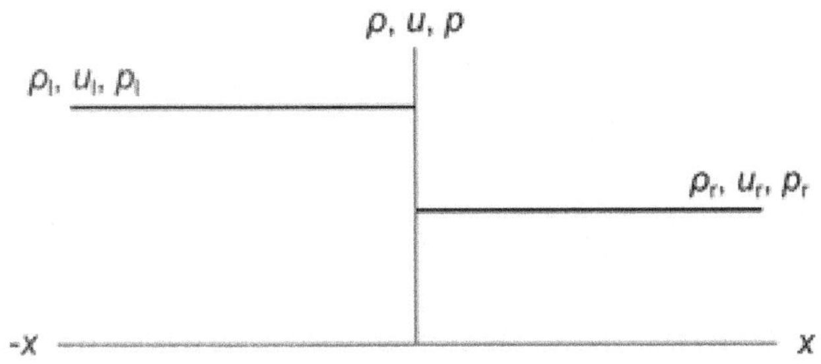

Figure 3: The Riemann problem. The Riemann problem specifies initial density, velocity, and pressure values of ρ_l, u_l, p_l on the left side of the origin and ρ_r, u_r, p_r on the right side of the origin.

Unlike the general initial conditions, the Riemann problem has an analytical solution, though this solution contains a nonlinear implicit equation and a number of special cases that we must treat carefully. In this paper, we use a Riemann solver due to Toro [1] as a standard to test RRM's results against. Many CFD methods, beginning with Godunov's method in 1959 [2], use an embedded Riemann solver as a part of their algorithms, though RRM does not.

Even for the Riemann problem, accurate numerical solutions to the Euler equations are challenging, mainly because the solutions can include discontinuities. At these discontinuities, the spatial derivatives in the differential form of the Euler equations are undefined, which spoils many simple numerical methods and requires special-case code in more advanced methods.

In the solutions to many other partial differential equations such as the heat equation, initial discontinuities will smear out and become increasingly smooth over time. But in the solutions to the Euler equations, initial discontinuities do not always smear out, and indeed new discontinuities may arise over time.

For example, consider Sod's shock tube problem [3], a special case of the Riemann problem. A shock tube is a gas-filled tube with a diaphragm in the center. The diaphragm is initially airtight, so the left and right sides of the tube can be separately charged to specific pressures and densities as shown in figure 4, which sets (ρ_l, u_l, p_l)=(1.0, 0.0, 1.0) on the left side, and (ρ_r, u_r,p_r)=(0.125, 0.0, 0.1) on the right side.

At time t=0.0, we instantly remove the diaphragm and let the fluid start flowing from left to right.Figure 5 shows the fluid at t=1.5 seconds. We can see both types of discontinuity that are possible in solutions to the Euler equations, as well as the "expansion fan" that joins the high-pressure left state to the flat area in the center.

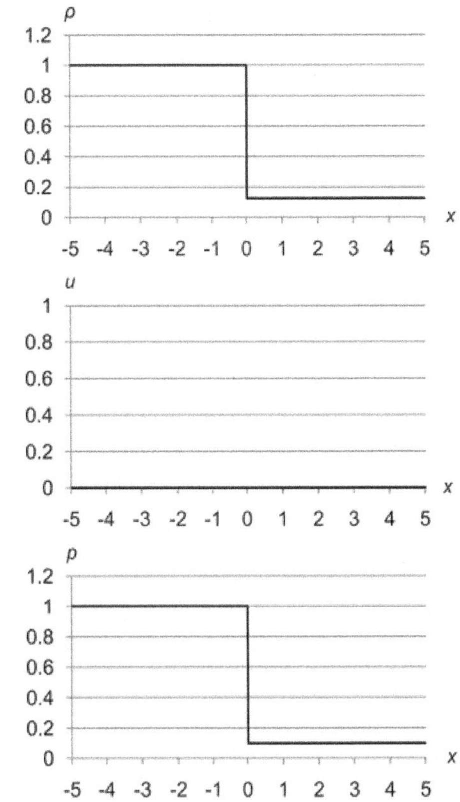

Figure 4: Sod's shock tube problem at t=0.0 seconds. Sod's shock tube problem showing initial density, velocity, and pressure values (ρ_l, u_l, p_l)=(1.0, 0.0, 1.0) and (ρ_r, u_r, p_r)=(0.125, 0.0, 0.1).

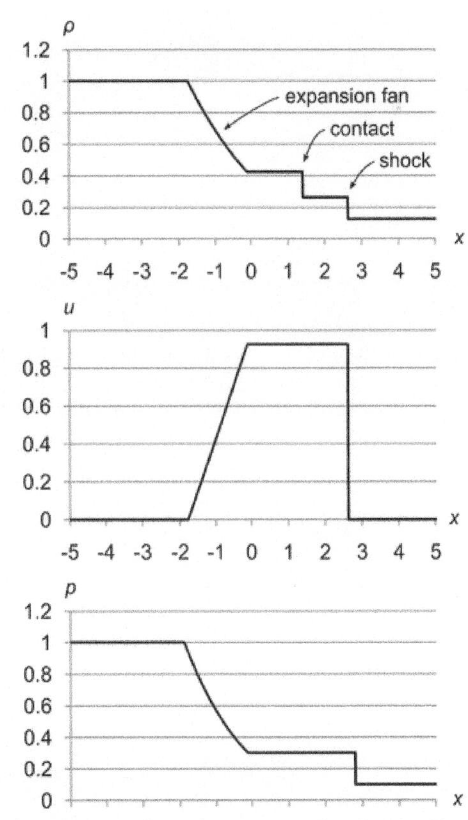

Figure 5: Sod's shock tube problem at t=1.5 seconds. Sod's shock tube problem showing density ρ, velocity u, and pressure p after 1.5 seconds of time evolution. We can see three flow features: an expansion fan, a contact, and a shock.

The first type of discontinuity, a contact, separates two areas that differ only in density. Contacts travel along with the fluid, and since velocity is constant across a contact, no fluid flows across them. Contacts cannot form spontaneously; they must either be present in the initial conditions as in Sod's problem, or they must be formed by the intersection of two shocks. As a real-world example, if stationary hot and cold water masses are carefully placed side by side, they will be separated by a contact discontinuity, at least until heat energy flows across the discontinuity and smears it out.

The second kind of discontinuity, a shock, can be formed by a pressure gradient steep enough to force the fluid to move faster than the local speed of sound a. Shocks can develop over time, and need not be present in the initial conditions. Density, velocity, and pressure can all change across a shock. As a real-world example, if you pilot a boat through the water faster than waves can travel through the water, the boat creates a shock at its bow.

The 2D graphs of Sod's shock tube problem above show flow features at specific times, but do not show how the fluid flow evolves over time. Figure 6 uses 3D to add a time dimension.

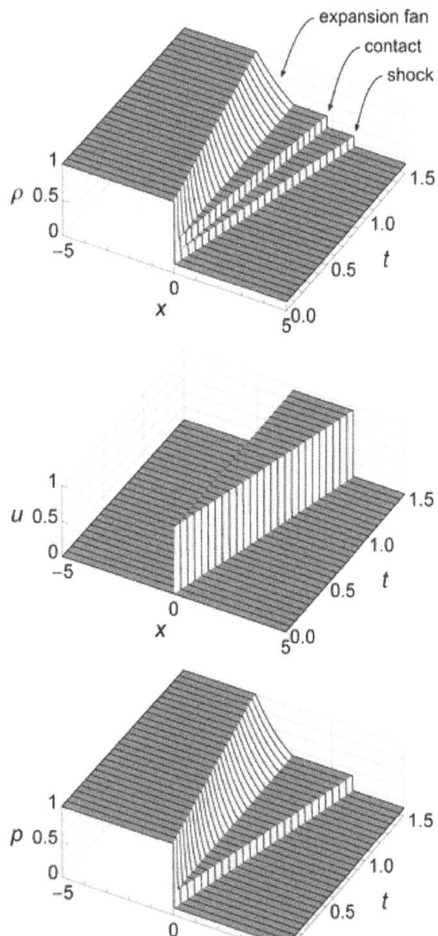

Figure 6: Sod's shock tube problem time evolution from t=0 to t=1.5 seconds. Sod's shock tube problem showing density ρ, velocity u, and pressure p from time t=0 to time t=1.5 seconds. We can see three flow features: an expansion fan, a contact, and a shock. The contact and the shock both start at the origin and move to the right, with the shock running ahead due to its higher speed. The expansion fan gradually slopes left as more and more fluid flows to the right to feed the shock.

These graphs show how the contact and the shock both start at the origin and move to the right, with the shock running ahead due to its higher speed. They also show how the expansion fan gradually slopes left as more and more fluid flows to the right to feed the travelling shock.

For subsequent figures we will mainly use 2D graphs, since they allow easier comparison of our results with those of a Riemann solver. We will use 3D only when the time evolution of the flow is of special interest, such as when we illustrate boundary conditions.

Previous Work

If you simply use the definition of the derivative to convert the Euler equations from differential equations to algebraic equations, you get the finite difference method (FDM). In conservation form, FDM models a fluid as a set of cells, each of which contains the values of the conserved quantities at a point within the cell. The explicit version of FDM calculates those values at the next time step from the values in nearby cells at the current time step. The set of nearby cells is called the stencil.

The finite volume method (FVM) also models a fluid as a set of cells, but it stores cell average values instead of point samples in the cells. In its explicit conservation form, FVM calculates the values at the next time step by adding and subtracting fluxes of the conserved quantities across each neighboring cell's edges during the time step.

The finite element method (FEM) was historically used for structural mechanics [4], but began to find use in fluid dynamics [5] as the method was generalized and applied to time-varying problems. FEM starts by creating a mesh of elements (cells in our terminology) which are shaped to fit the problem geometry. FEM then solves a system of equations at each time step to determine the unknown fluid values in each element. Fluid values in FEM are typically stored in piecewise polynomial form, as opposed to the point samples of FDM or the cell averages of FVM.

In FDM and FVM, the fluid is usually considered to move through stationary cells in a single, global coordinate system. This is the Eulerian description of fluid flow mentioned above.

Figure 7 shows an example with three stationary cells c_1, c_2, and c_3. The measuring pointsx_{m1}, x_{m2}, and x_{m3} are at the cell centers. The entire fluid has a rightward velocity u. In panel A at time t_1, we measure cell c_1's ρ and p values at x_{m1}, and c_2's values at x_{m2}. Cell c_3 is empty.

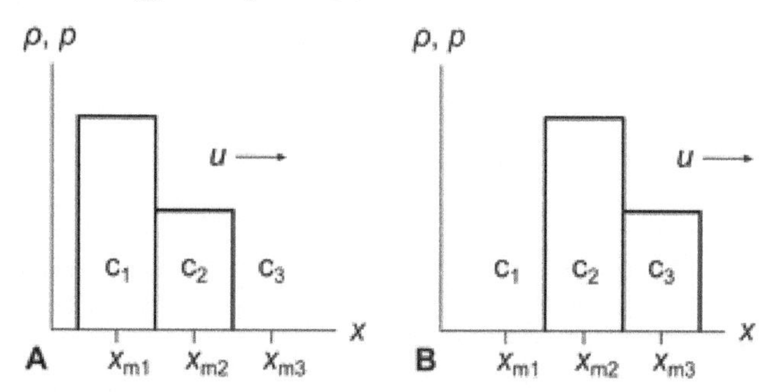

Figure 7: Eulerian description of fluid flow. In the Eulerian description of fluid flow, the fluid moves through stationary cells. Consider

a global coordinate system divided into three cells c_1, c_2, and c_3. The fluid is traveling rightwards with velocity u. In panel A at time t_1, a measurement at point x_{m1} will show the density ρ and pressure p of cell c_1. In panel B at a later time $t_2=(x_{m2}-x_{m1})/u$, we measure the same density and pressure at point x_{m2} because the fluid has moved to the right by one cell width.

In panel B at a later time $t_2=(x_{m2}-x_{m1})/u$, all the fluid from c_1 has moved into c_2, and all the fluid from c_2 has moved into c_3. Now we measure the same ρ and p values at x_{m2} that we previously measured at x_{m1}, and the same values at x_{m3} that we previously measured at x_{m2}. The fluid has moved one cell width to the right, but the cells themselves have stayed in place.

Eulerian methods are relatively simple to implement, but shocks, contacts and other steep gradients may smear out or gain unphysical oscillations as they cross cell edges, depending on the algorithm used. Researchers have proposed many refinements over the years to increase accuracy, such as Total Variation Diminishing (TVD) methods [6], Essentially Non-Oscillatory (ENO) methods [7], Monotone Upwind Schemes for Scalar Conservation Laws (MUSCL) [8], the Piecewise-Parabolic Method (PPM) [9], and many more.

Another approach to Eulerian fluid flow is the lattice Boltzmann method (LBM) [10]. Instead of a mesh of cells, LBM uses a lattice of connected sites, each of which can "stream" fluid to a fixed number of neighboring sites. Each site contains a distribution function that represents how much fluid is streaming in each direction. After each streaming step, LBM executes a "collision" step at each site to alter the distribution functions to maintain conservation. LBM has many attractive features, including ease of programming and simple handling of boundary conditions.

In contrast to FDM, FVM, and LBM, FEM often uses the alternative Lagrangian description of fluid flow, in which the cells travel along with the fluid.

Figure 8 shows an example, with two cells c_1 and c_2 moving to the right with a velocity u, similar to the Eulerian example above. However, in the Lagrangian description the fluid does not move across cell edges. Instead, the cells themselves move, carrying local coordinate systems along with them.

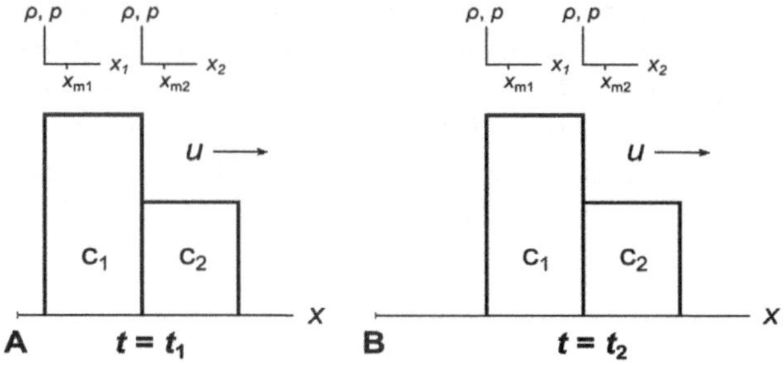

Figure 8: Lagrangian description of fluid flow. In the Lagrangian description of fluid flow, the cells are part of the fluid and move along with it. Consider fluid cells c_1 and c_2 traveling with their own local coordinate systems. The cells and their coordinate systems are both traveling rightwards at velocity u in the global coordinate system. In panel A at time t_1, and in panel B at any later time t_2, pointsx_{m1} and x_{m2} in the cells' coordinate systems remain at the same places in those cells. No fluid crosses cell edges.

Panel A shows us measuring the values at time t_1 of ρ and p at point x_{m1} in the local coordinate system of cell c_1. Panel B shows that we will measure the same values at any later time t_2, since cell c_1 and its coordinate system move together. The same holds true for cell c_2 in its local coordinate system at its own point x_{m2}.

Lagrangian methods handle shocks and contacts naturally, because those flow features travel with the fluid instead of smearing out as they cross cell edges. But pure Lagrangian methods are rare, because as the fluid flows, the cells can become excessively bunched up, stretched out, or deformed, which can reduce simulation accuracy and efficiency.

The cells of FDM, FVM, and FEM, and the lattice sites of LBM, are usually connected in a mesh. Each cell has a well-defined shape, and each cell or site has a fixed set of neighbors. In simple methods, these shapes and sets of neighbors are constant over the whole course of the simulation. But in Eulerian methods, a fluid may have complex flow features that move around over time, so we may want to create smaller cells in those complex areas and larger cells in other areas. Or in Lagrangian methods, some cells may become degenerate or singular in a complex flow, so that the method's equation solver no longer works correctly.

The process of changing the mesh to alleviate these problems is called remeshing. Figure 9panel A shows eight small fluid cells, and panel B shows those eight cells remeshed into two cells that cover the same area.

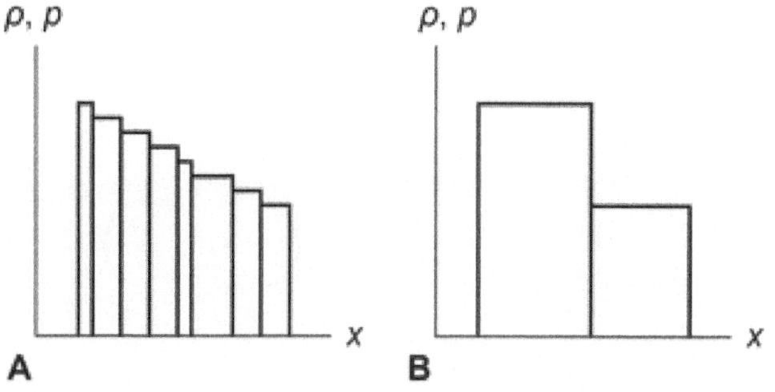

Figure 9: Remeshing. The eight cells in panel A can be remeshed into two cells in panel B that cover the same area and contain the same mass, momentum, and energy. Some CFD methods require remeshing to maintain accuracy or to prevent numerical difficulties.

To avoid this complication, the so-called meshfree methods do away with mesh connectivity entirely. One of the first meshfree methods was smoothed-particle hydrodynamics (SPH) [11],[12]. SPH is a purely Lagrangian method which models a fluid with a set of moving particles, and computes the fluid's properties at any point by summing the contributions of nearby particles using a kernel function which smooths out the particles' properties over some "smoothing length". SPH was originally motivated by the study of astrophysical problems such as galaxy formation, where the constituents were already discrete particles. SPH was later applied to other problems where the fluid was presumed to be continuous before being discretized.

The moving-particle semi-implicit method (MPS) [13] is a meshfree method similar to SPH, which was originally intended for simulation of incompressible fluids with interacting free surfaces. It also uses a kernel function (called a weight function in the MPS literature), but one which is specially designed to repel particles at short distance, thereby maintaining approximately constant density in the fluid. MPS has been applied to many situations, including simulations of coastal waves and dam breaks.

Much research in meshfree methods has been done in recent years, and there is now a great variety of such methods with different kernel functions, particle properties, and integration techniques. Li and Liu [14] and Huerta et al. [15] have both written excellent surveys of the field.

METHODS

Motivation

RRM was motivated by Chaikin's corner-cutting algorithm for curve generation, which evolved into the subdivision curves of computational geometry [16], [17]. A curve of this type starts as a set of lines joined end to end to form a roughly faceted curve, shown in figure 10 panel A. First we cut off each of the corners one-quarter of the way along each side, shown in panel B. Then we cut the corners off the new corners, shown in panel C, iteratively refining the curve into smaller and smaller line segments, until a desired level of smoothness is reached in panel D.

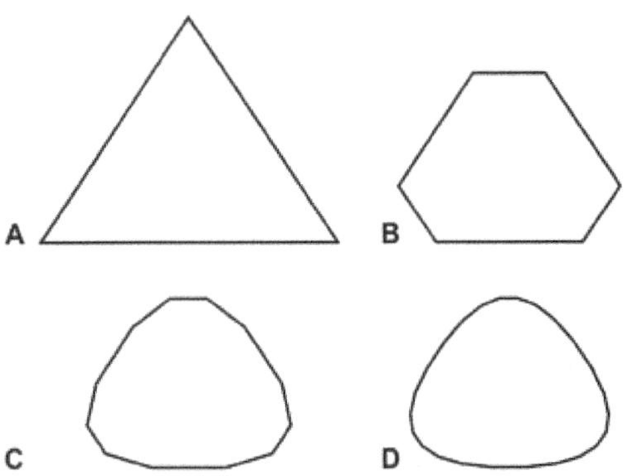

Figure 10: Chaikin's corner-cutting algorithm. Starting with the triangle in panel A, cutting the corners off one-quarter of the way along each side gives us panel B. Panels C and D show the process carried out two more times. We can repeat this process until the curve has any desired smoothness.

RRM does the same sort of iterative refinement, but on a moving fluid instead of a stationary curve, and with constraints on conservation of mass, momentum and energy rather than constraints on surface continuity and smoothness.

Overview

To begin, we divide a fluid into finite-sized cells. In one dimension, each cell is a line segment with an associated density, velocity, and pressure, all of which are constant across the cell.Figure 11 shows a fluid divided into three cells c_1, c_2, and c_3. For now we use periodic boundary conditions, so the left side of c_1 is adjacent to the right side of c_3. We indicate this with the dotted line on the right of c_3.

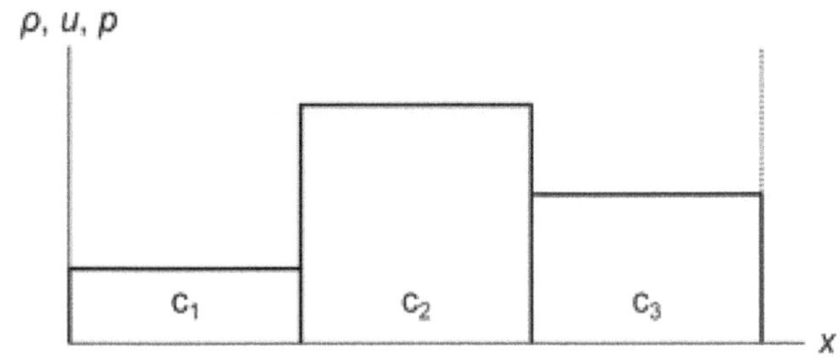

Figure 11: Fluid cells. A fluid divided into three cells c_1, c_2, and c_3. The dotted line at the right shows that there are periodic boundary conditions, so the right side of c_3 is adjacent to the left side of c_1.

At each cell edge, we send tracer particles left and right through the fluid at the local speed of sound a, as shown in figure 12. Each pair of tracer particles defines an expanding wavefront of change that originates at the cell edge. For example, in figure 12 we show w_{23}, the wavefront originating between c_2 and c_3, along with its left tracer particle p_l and its right tracer particle p_r.

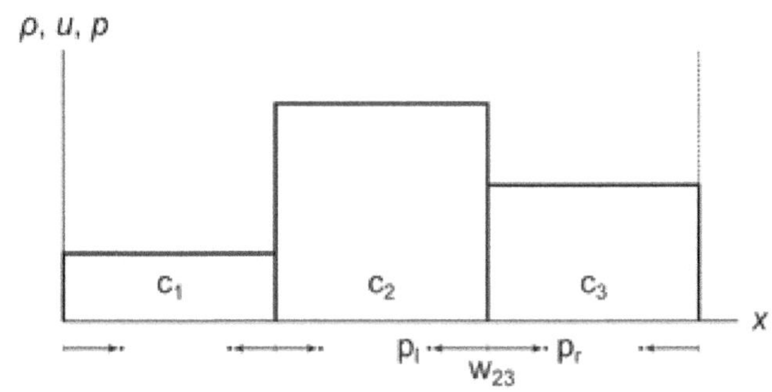

Figure 12: Fluid cells showing wavefronts and tracer particles. A fluid divided into three cells c_1, c_2, and c_3. The dotted line at the right shows that there are periodic boundary conditions. Wavefront w_{23} originates between c_2 and c_3, and contains tracer particles p_l and p_r that travel through the fluid at the local speed of sounda=sqrt(γp/ρ). The constant γ depends on the fluid (it has a value of 1.4 for air). The $\rho$and p values are those of the cell the particle is traveling through. Note that w_{31} (not labeled) extends into both c_3 and c_1 due to the periodic boundary conditions.

As each tracer particle travels, it accumulates an error metric that tracks how much each of the primitive variable values has changed, and over what distance. Figure 13 shows a tracer particle p, the right-hand particle of wavefront w, moving through the fluid. The particle's error metric Δ grows as the particle moves, with the slope of Δ changing when the particle crosses each cell edge.

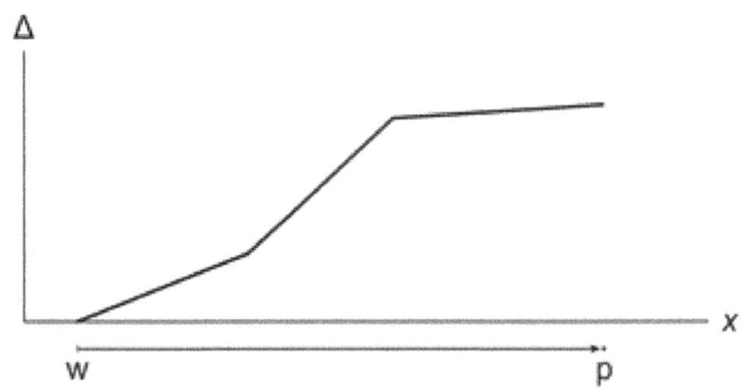

Figure 13: Error metric growing as a tracer particle travels. A particle p traveling right as part of wavefront w. Its error metric Δ increases as the particle travels, with the slope of Δ changing at cell edges.

When this error metric for either of the two tracer particles in any wavefront exceeds a preset maximum, we chop the wavefront area out of the fluid, flatten the chopped-out cell parts into a single new cell, and insert that new cell into the hole left by the chopping.

In areas of the fluid where primitive variable values differ greatly from cell to cell, tracer particles' error metrics will accumulate quickly, so new cells will be chopped out soon after wavefront creation. This leads to more, smaller cells in areas of the fluid with steep slopes. Conversely, in areas where values are very similar from one cell to the next, error metrics will accumulate slowly, so we will chop out fewer, larger cells in areas of the fluid with shallow slopes.

We illustrate this whole process in figure 14. In panel A, we chop wavefront w_{23} out of the fluid, removing the wavefront's tracer particles from the fluid at the same time. This leaves us with chopped cell parts c_{2c} and c_{3c}, shown in panel B. Panel C shows us flattening c_{2c} and c_{3c} into a new cell c_4 of the same mass, momentum and energy. Then in panel D we insert c_4 into the fluid and create new wavefronts w_{24} and w_{43} at the cell edges.

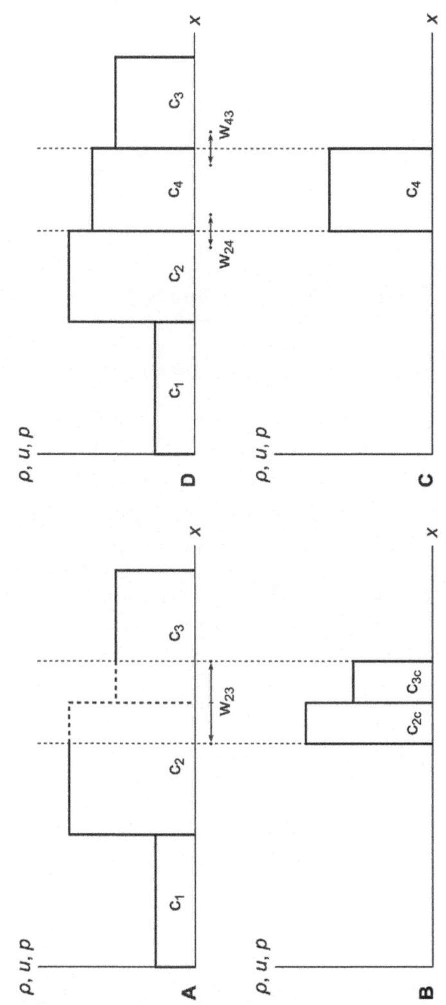

Figure 14: Chopping, flattening, and new cell creation. Panel A shows the chopping of wavefront w_{23} out of the fluid, which chops off the right side of c_2 and the left side of c_3. Panel B shows the resulting chopped parts c_{2c} and c_{3c}. Panel C shows the flattening of the two chopped parts into a new cell c_4 with the same mass, momentum, and energy. Panel D shows the insertion of the new cell c_4 into the fluid and the creation of the new wavefronts w_{24} and w_{43} on its edges.

The chop-flatten-create process always results in exactly one new cell, and always shrinks two other cells by chopping parts off of them. But this process can also remove any number of whole cells if the maximum error metric allows the wavefront to grow wide enough. For example, if wavefront w_{23} in figure 14 had grown wider, it could have chopped off the right side of c_1, entirely removed c_2, and chopped off the left side of c_3, resulting in no net change in the number of cells. An even wider wavefront which removes two whole cells would reduce the total number of cells in the fluid by one, and so on. This is how RRM increases and decreases the total number of cells over time to adapt to changing fluid conditions.

The last step in the RRM algorithm is to choose the next wavefront whose tracer particles have reached the maximum error metric and repeat the chop-flatten-create process detailed above. This repetition evolves the fluid simulation forward in time.

Stored Quantities

In each cell, we store three main types of data:

- The size, shape, and position of the cell. In one dimension, cells have only width, so we need only store the time-varying x coordinate $x_1(t)$ of the cell's left edge, and the cell's width w.
- The cell's three primitive variable values ρ, u, and p.
- Four extra vector quantities which help us ensure conservation.

Below we explain the relationships between these quantities and show how to derive other necessary values from them.

RRM is purely Lagrangian and represents the fluid as finite-sized cells, so we use the integral Lagrangian form of the Euler equations, written in terms of the primitive variables:

conservation of volume

$$\int_{m_1}^{m_2}(v(m,t_2)-v(m,t_1))dm = -\int_{t_1}^{t_2}(-u(m_2,t)+u(m_1,t))dt$$

conservation of momentum

$$\int_{m_1}^{m_2}(u(m,t_2)-u(m,t_1))dm = -\int_{t_1}^{t_2}(p(m_2,t)-p(m_1,t))dt$$

conservation of energy

$$\int_{m_1}^{m_2}(e_T(m,t_2)-e_T(m,t_1))dm$$

$$= -\int_{t_1}^{t_2}(p(m_2,t)u(m_2,t)-p(m_1,t)u(m_1,t))dt \quad \textbf{(7)}$$

The values m_1 and m_2 are the mass coordinates of the left and right side of the cell. The mass coordinates move with the fluid, unlike the fixed spatial coordinates x_1 and x_2 that we used in the Eulerian form of these equations in equation set 1. This means that the fluid between m_1 and m_2 stays between m_1 and m_2, with no fluid flow across the cell edges. We can get the mass coordinate m at a point in the fluid from the Eulerian coordinate x at that point by integrating all the mass up to that point:

$$m = \int_{x(0,t)}^{x(m,t)} \rho(x,t)dx \quad \textbf{(8)}$$

So the value of the mass coordinate at any point is the sum of all the mass to the left of that point in the fluid.

Note that the conservation of mass equation does not appear in equation set 7. That is because the mass between m_1 and m_2 is constant, so that equation would be trivial. Instead we show the conservation of volume equation, which merely says that a cell's volume v changes as its edges move towards or away from each other. In the Eulerian form of these equations, it was the conservation of volume equation that was trivially equal to a constant, so we omitted it from equation set 1.

Note also that the equations for the conservation of momentum and energy are simpler in the Lagrangian form than in the Eulerian form. This is because we do not need the flux terms that describe momentum and energy flowing across the cell edges, now that the cell edges move with the fluid.

As we saw with the Eulerian form, if we define a vector of the conserved quantities

$$\mathbf{u}=\begin{bmatrix} v \\ u \\ e_{\mathrm{T}} \end{bmatrix} \tag{9}$$

and a vector of the velocity, pressure and pressure work at the cell edges

$$\mathbf{f}=\begin{bmatrix} -u \\ p \\ pu \end{bmatrix} \tag{10}$$

then the Lagrangian form of the Euler equations can be written as a single vector equation

$$\int_{m_1}^{m_2}(\mathbf{u}(m,t_2)-\mathbf{u}(m,t_1))dm=-\int_{t_1}^{t_2}(\mathbf{f}(m_2,t)-\mathbf{f}(m_1,t))dt \tag{11}$$

We do not store the cells' conserved quantities directly, but we can calculate them by integrating over the primitive variables. Since our primitive variables are piecewise constant, the integrals are simply multiplications by w, the width of the cell.

$$\begin{aligned} &\text{cell right edge} && x_2(t)=x_1(t)+w \\ &\text{cell mass} && M=\rho w \\ &\text{cell momentum} && P=\rho wu \\ &\text{cell energy} && E=\rho we_{\mathrm{T}}=\frac{1}{2}\rho wu^2+\frac{pw}{(\gamma-1)} \end{aligned} \tag{12}$$

To allow our flattening process to exchange energy between kinetic and potential forms while remaining conservative, we store two extra vector quantities on each edge of each cell: pressure momentum and pressure energy.

Figure 15 shows these quantities for a single cell c_1. We define them as follows:

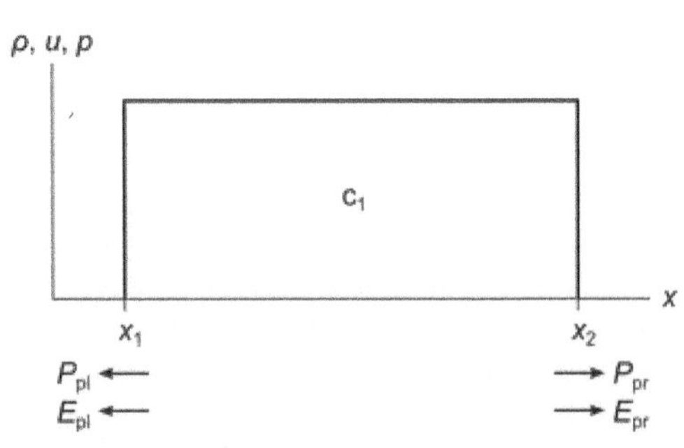

Figure 15: Pressure momentum and pressure energy vectors. A single cell c_1 showing left and right pressure momentum P_{pl} and P_{pr}, and left and right pressure energy E_{pl} and E_{pr}. Portions of these vectors are transferred to new cells during the chopping and flattening process, and this transfer is what causes the conversion between potential and kinetic energy and vice versa.

$$\begin{aligned} &\text{cell transit time} && \Delta t=wa \\ &\text{left pressure momentum} && P_{\mathrm{pl}}=-p\Delta t \\ &\text{right pressure momentum} && P_{\mathrm{pr}}=p\Delta t \\ &\text{left pressure energy} && E_{\mathrm{pl}}=-pu\Delta t \\ &\text{right pressure energy} && E_{\mathrm{pr}}=pu\Delta t \end{aligned} \tag{13}$$

The first quantity Δt is the time it takes a tracer particle to cross a cell, and also the time it takes for a cell to completely expend its store of pressure momentum and pressure momentum upon its neighbors. The next four quantities are the terms on the right-hand sides of the momentum and energy Euler equations from equation set 7. They represent the changes in momentum and energy due to pressure work that a cell has the potential to cause to its neighbors. We store equal quantities of P_p and E_p in each direction, so for each cell they sum to zero, leaving the overall momentum and energy of the fluid unchanged.

As the fluid evolves, the total fluid mass, momentum, and energy remain strictly constant when these extra P_p and E_p vectors are summed along with the cells' mass, momentum, and energy.

$$\begin{aligned} &\text{total fluid mass} && M_{\mathrm{f}}=\sum_{i=1}^{n} M_i \\ &\text{total fluid momentum} && P_{\mathrm{f}}=\sum_{i=1}^{n}(P_i+P_{\mathrm{pl}i}+P_{\mathrm{pr}i}) \\ &\text{total fluid energy} && E_{\mathrm{f}}=\sum_{i=1}^{n}(E_i+E_{\mathrm{pl}i}+E_{\mathrm{pr}i}) \end{aligned} \tag{14}$$

This insures strict conservation of mass, momentum, and energy over the course of the simulation.

Cell Chopping and Flattening

When we chop off one side of a cell, we are removing five quantities: mass, momentum and energy, plus part of the pressure momentum and pressure energy from the vectors on the chopped-off edge of the cell. Figure 16 shows this for a single cell c_1. Panel A shows the quantities remaining in c_1 after chopping, and panel B shows the quantities that are chopped off to form cell part c_{1c}.

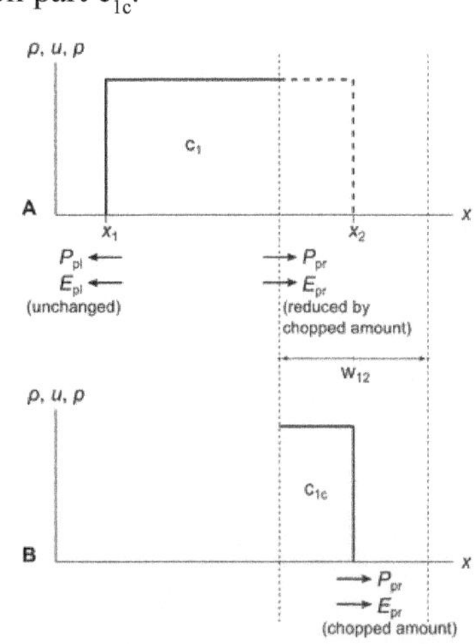

Figure 16: Chopping pressure momentum and pressure energy vectors. Panel A shows a single cell c_1 with the right side chopped off. Panel B shows the chopped mass, momentum, energy, pressure momentum P_p, and pressure energy E_p that are

now contained in the chopped cell part c_{1c}, which will be flattened into a new cell along with any other cells chopped out by the same wavefront. The amounts of mass, momentum, and energy transferred to c_{1c} are proportional to the width of c_{1c}, but the amounts of pressure momentum and pressure energy transferred to c_{1c} are proportional to the time since the creation of c_1.

Note that we chop off mass, momentum and energy in amounts proportional to the width of the chopped part, but we chop off pressure momentum and pressure energy in an amount proportional to the time since the chopped cell was created. This is because mass, momentum and energy are inherent properties of the fluid that must be integrated over space, whereas pressure momentum and pressure energy act over time to convert energy from potential to kinetic form when there is a gradient in the fluid pressure. We can see this in equation set 7, where the left-side integrals are spatial, and the right-side integrals are temporal. In RRM, we treat P_p and E_p as acting steadily over time, starting at time t_c when a cell is created, and ending at time $t_c+\Delta t$, the time at which both of the cell's tracer particles (the left wavefront's right particle, and the right wavefront's left particle) leave the cell.

When chopping out a new cell, we first determine its intersections with existing cells. Then we chop off those intersections and add up all the mass M, momentum P, and energy E the intersections contained. Then, using the width of the new cell w and the requirement that density, velocity, and pressure are constant across it, we can calculate the flattened values of the primitive variables for the new cell.

$$\begin{array}{ll} \text{flattened density} & \rho = \dfrac{M}{w} \\ \text{flattened velocity} & u = \dfrac{P}{M} \\ \text{kinetic energy} & KE = \dfrac{1}{2}Mu^2 \\ \text{potential energy} & PE = E - KE \\ \text{flattened pressure} & p = \dfrac{PE(\gamma-1)}{w} \end{array} \qquad \textbf{(15)}$$

Negative Pressure Fix

Very occasionally, the flattening process will produce a cell with negative pressure, either because of rounding or truncation error, or because a very small wavefront chops a large amount of pressure momentum and pressure energy, which would accelerate the newly created small cell more than its store of potential energy can support.

When this happens, we flatten the cell without adding in the chopped pressure momentum and pressure energy, which turns off pressure-to-momentum conversion for that cell and gives us a positive pressure after flattening. The unused pressure momentum and pressure energy are added to that of the newly created cell, which spreads the pressure-to-momentum conversion out over a slightly longer time.

Whitehurst's signal method [18] uses a similar fix for negative pressures, but averages over space instead of time. When negative pressure occurs in a cell, the signal method averages that cell's mass, momentum and energy with its three neighbors, in proportion to their volumes.

Tracer Particles and Their Error Metric

The movement of the tracer particles through the cells of the fluid models the movement of characteristics or acoustic wavefronts through the fluid. The tracer particles do not represent real physical particles, they are merely a computational device. They do not carry mass, momentum, or energy, they do not interact with each other, and they do not affect cells' properties. They always travel at the local speed of sound a in the cell that contains them.

As the tracer particles move through the fluid, we accumulate an error metric that tells us when to stop and chop out a new cell. The error metric $\mathbf{\Delta}_{1,n}$ is the error accumulated by a tracer particle as it travels from cell 1 to cell n.

$$\text{error metric} \qquad \Delta_{1,n} = \sum_{i=1}^{n} \max(|\mathbf{M}_i - \mathbf{M}_{i-1}|)d_i \qquad \textbf{(16)}$$

where d_i is the distance the tracer particle travels in cell i, and the metric vector $\mathbf{M}_i$ for cell i is

$$\mathbf{M}_i = \begin{bmatrix} \rho_i \\ u_i \\ p_i \end{bmatrix} \qquad \textbf{(17)}$$

When $\mathbf{\Delta}_{1,n}$ for either tracer particle exceeds a user-supplied Δ^{max}, we chop out a new cell.

This error metric needs a bit of explaining. First, the metric is a vector of all the primitive variables (instead of choosing just one or two) so that variation in any of them across the fluid can trigger the chopping of a new cell. So our maximum error metric $\mathbf{\Delta}_{max}$ is a vector $\mathbf{\Delta}_{max}=[\Delta_{max\ \rho}, \Delta_{max\ u}, \Delta_{max\ p}]^T$, with each value set separately by the user. When we say that $\mathbf{\Delta}_{1,n}$ exceeds $\mathbf{\Delta}_{max}$, we mean that some element of $\mathbf{\Delta}_{1,n}$ exceeds the corresponding element of $\mathbf{\Delta}_{max}$.

We take the absolute value of $\mathbf{M}_i - \mathbf{M}_{i-1}$ so the error metric will increase monotonically as the tracer particle travels. If we did not, the error metric might go up and down many times without exceeding $\mathbf{\Delta}_{max}$, which could lead to chopping out new cells that contain more total variation in the primitive variable values than we meant to allow.

We multiply the error metric by distance so that the error metric grows even as the tracer particles move across cells, not just as the particles cross cell edges. This prevents us from chopping out unduly large new cells in areas of shallow density, velocity, or pressure gradients.

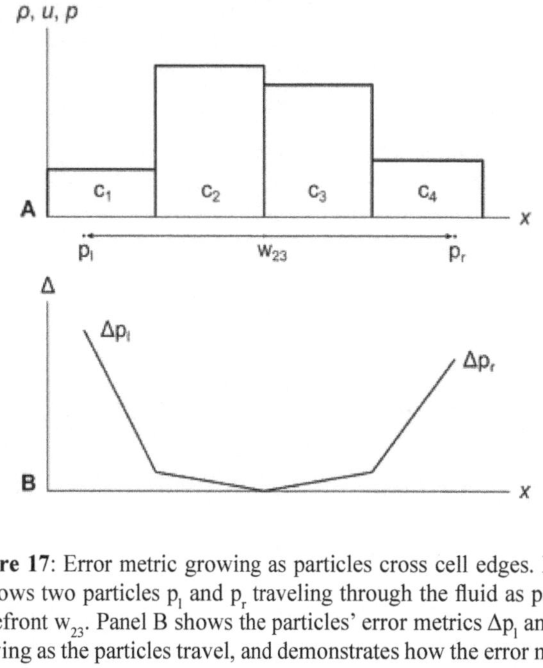

Figure 17: Error metric growing as particles cross cell edges. Panel A shows two particles p_l and p_r traveling through the fluid as part of wavefront w_{23}. Panel B shows the particles' error metrics Δp_l and Δp_r growing as the particles travel, and demonstrates how the error metric

of each particle in a wavefront is tracked separately. Note how the slope of the error metric across each cell is proportional to the difference in the cells' density ρ, velocity u, and pressure p at the edge the particle crossed to get into the cell.

There is also a special case in this formula. When i is 1, $\mathbf{M}_0$ is the metric vector of the cell on the other side of the edge from where the tracer particle was created. So the tracer particle does not travel through cell 0, but its metric vector contributes to the overall error metric.

Figure 17 panel A shows two tracer particles p_l and p_r traveling through a fluid as part of wavefront w_{23}. Panel B shows how the error metrics Δp_l and Δp_r of the two particles change as the particles travel.

Note that the slopes of Δp_l and Δp_r are shallow in the center of the graph, because the density, velocity, and pressure of c_2 and c_3 are similar. As the particles cross into c_1 and c_4, the slopes of Δp_l and Δp_r increase substantially, which means that w_{23} will reach $\mathbf{\Delta}_{max}$ sooner than it would have with a shallower gradient in the fluid.

Wavefront Unioning

When we choose a wavefront that we wish to chop the fluid with, we first must check for overlap with other wavefronts. The final area we chop out will be the union of the first wavefront with all the wavefronts that overlap it, and all the wavefronts that overlap them, and so on. Figure 18shows an example: if we choose w_{12}, we see that it overlaps w_{23}, which overlaps w_{34}, so the final area we would have to chop is w_{union}.

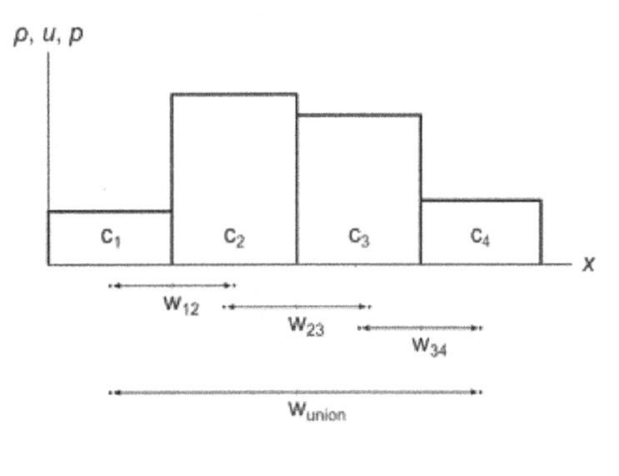

Figure 18: Wavefront unioning. A fluid divided into four cells c_1, c_2, c_3 and c_4. Wavefront w_{12} overlaps wavefront w_{23}, which overlaps wavefront w_{34}, so we must chop out the union wavefront w_{union} to properly account for the effects of each wavefront on the others.

Wavefront unioning was motivated by the observation that once two expanding wavefronts overlap, the fluid in each one has affected the fluid in the other, so they can no longer be treated separately.

Wavefront unioning turns out to be essential for the stability of the simulation. Without wavefront unioning, it is possible to chop out an area that contains unbalanced pressure momentum and pressure energy, even in a perfectly "flat" fluid that has no density or pressure gradient. This imbalance can cause a newly-created cell's velocity to be abnormally high, which causes a glitch in the simulation where fluid cells pile up or spread out in an unphysical way.

Consider figure 19 panel A, which shows three cells c_1, c_2, and c_3 with ρ=1, u=0, and p=1. We call this the "213 problem" because the widths of the cells are 2, 1, and 3 from left to right.

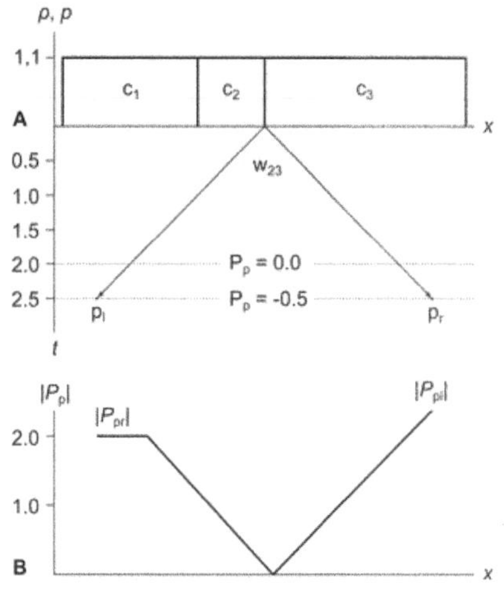

Figure 19: The "213 problem". Panel A shows a fluid divided into three cells c_1, c_2, c_3 of widths 2, 1, and 3 from left to right (hence the name "213 problem"). All three cells were created at time t=0, and all three have density ρ=1, pressure p=1, and velocity u=0. In a simulation without wavefront unioning, if wavefront w_{23} chopped out a new cell at time t=2.5, that new cell would have a net momentum of −0.5. Panel B shows that this is because the rightward momentum P_{pr} from c_1 levels off at t=2.0, while the leftward momentum from $c_3$continues to increase until t=3.0. This demonstrates that wavefront unioning is required to avoid unphysical changes in cell velocity during simulation.

Assume that all three cells were created at time t=0, and that the speed of sound a=1. Since there is no density or pressure gradient, chopping out a new cell anywhere in this fluid should result in a new cell with ρ=1, u=0, and p=1.

Let us consider wavefront w_{23} expanding from the right side of c_2 and see if this is true. At timet=2, w_{23} will contain equal and opposite amounts of pressure momentum from c_1 and c_3, since$P_{prl}=2$ and $P_{pl3}=-2$. The pressure momenta P_{pl2} and P_{pr2} from c_2 will cancel since the whole cell is covered, so the overall pressure momentum $P_p=P_{prl}+P_{pl2}+P_{pr2}+P_{pl3}$ contained in $w_{23}$is zero, as shown by the dotted line at t=2. So far, so good.

At time t=2.5, the pressure momentum P_{prl} from c_1 is still 2, since it ran out of pressure momentum to contribute at t=2. But the pressure momentum P_{pl3} from c_3 is −2.5, since it will not run out until time t=3.0. Figure 19 panel B shows how the wavefront's left and right pressure momenta $P_{pl}=P_{pl2}+P_{pl3}$ and $P_{pr}=P_{prl}+P_{pr2}$ increase as the wavefront expands, withP_{pr} leveling off at t=2 when P_{prl} stops increasing.

So if we chop out a new cell at time t=2.5, it will have an overall pressure momentum P_p of −0.5, as shown by the dotted line at t=2.5. This will make the new cell move to the left, even though there is no pressure gradient in the fluid! Unioning w_{23} with w_{12} (not shown) fixes this problem.

The analysis of the 213 problem shows that if we run a simulation without wavefront unioning, it will show occasional unphysical glitches. Since mass, momentum, and energy are all strictly conserved, the glitches sometimes smooth out over time, but if a glitch is big enough, it may create a large local gradient and significantly slow down simulation. Wavefront unioning avoids this problem.

Discrete Event Simulation

RRM uses a discrete event simulation flow. We keep a priority queue of events, sorted in order of increasing event time. There are two kinds of events: particle events, where particles intersect cells, and wavefront events, where one of the particles in the wavefront exceeds the maximum error metric.

Particle events merely transfer particles from their current cell to the intersected cell, which changes their speed and the rate at which they accumulate error. Wavefront events chop new cells out of the fluid.

Figure 20 shows what the event queue might look like for the previous example of the 213 problem at time t=0, if we assume that the wavefront w_{23} would chop out a new cell at time t=1.5. For simplicity, we show only the events associated with wavefront w_{23}. In a real simulation there would be a wavefront between each pair of cells, so the event queue would be much more cluttered.

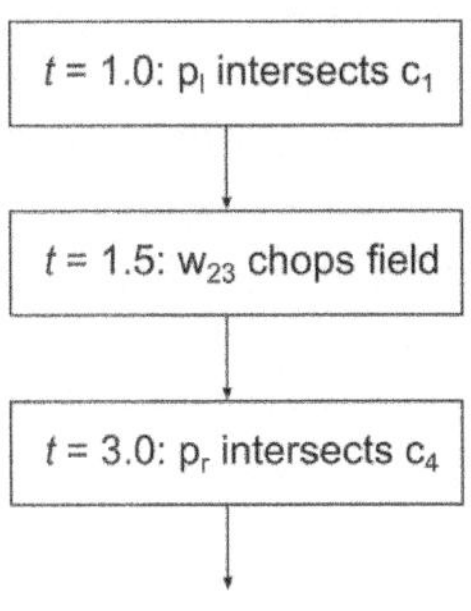

Figure 20: Simulation event queue. Queue of events associated with wavefront w_{23} in the previous example of the 213 problem. Events are ordered by increasing time, and the simulator always executes the event at the head of the queue.

The first event transfers particle p_l from c_2 to c_1. The second event uses wavefront w_{23} to chop a new cell out of the fluid. The third event is removed at the same time the second event is processed, since a wavefront's tracer particles are removed in the chopping process.

Here is how we determine the event times. When we create a wavefront with its pair of tracer particles, we find the intersection time $t_{intersection}$ of each particle with the nearest cell edge in its direction of travel, and the time $t_{max\ error}$ when the error metric of each particle will exceed the maximum error metric. Figure 21 shows all four of these times for particles p_l and p_r in wavefront w_{23}.

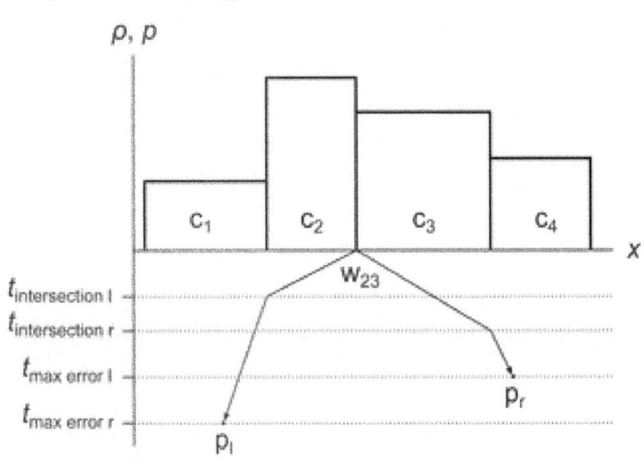

Figure 21: Particle intersection time and maximum error time. This figure shows all four possible event times for a single wavefront. Consider a fluid divided into four cells c_1, c_2, c_3 and c_4. Wavefront w_{23} contains particles p_l and p_r. Particle p_l intersects cell c_1 at time $t_{intersection\ l}$, and reaches the maximum error metric at $t_{max\ error\ l}$. Particle p_r intersects cell c_4 at time $t_{intersection\ r}$, and reaches the maximum error metric at $t_{max\ error\ r}$.

We enqueue the two tracer particles as events, using the intersection times as the event times. We also enqueue the wavefront as an event, using the soonest of the max error times as the event time.

$$\text{wavefront event time} \quad t_e = \min\left(t_{maxerrorl}, t_{maxerrorr}\right) \quad (18)$$

Each time we pull an event off the event queue, we check whether the event is a particle intersecting a cell, or a wavefront whose particle is reaching the maximum error metric. If a particle has intersected a cell, we transfer it into the intersected cell, recalculate the event time, and requeue the particle.

If either particle in a wavefront has reached the maximum error metric, we union the wavefront with any overlapping wavefronts, chop and flatten the area of the union into a new cell, and insert the new cell into the fluid. Then we create a new wavefront for each edge of the new cell and insert the corresponding events into the event queue. Finally, we transfer any particles caught in the chopped-out area into the new cell, which changes their speeds to the local speed of sound in the new cell, recalculates their event times, and requeues them.

RRM Algorithm Flowchart

For reference, figure 22 is a flowchart that outlines the entire RRM algorithm.

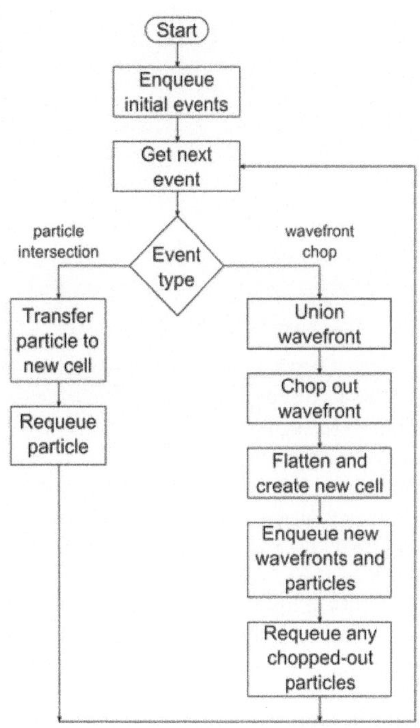

Figure 22: RRM flowchart. Flowchart showing the outline of the complete RRM algorithm.

RESULTS

We tested RRM on nine standard test problems, and plotted RRM's results (solid lines) against the output of Toro's Riemann solver (dashed lines). The two match closely in most cases, with some exceptions discussed below.

In the following tests, RRM typically uses a maximum of 200 to 400 cells during the simulation, depending on the maximum error metric we set. Most of those cells are concentrated in areas of high gradient, with only a few wide cells in flat areas. We set the maximum error metrics to obtain good results in a relatively short time. In the error analysis section that follows these test results, we will justify our choices of these maximum error metrics and show how the quality of the results varies as the maximum error metrics are varied.

Test 1

Figure 23 shows test 1, which is Sod's problem [3] with the initial conditions $(\rho_l, u_l, p_l)=(1.0, 0.0, 1.0)$ and $(\rho_r, u_r, p_r)=(0.125, 0.0, 0.1)$. The maximum error metrics for ρ, u, and p are $(\Delta_{max\ \rho}, \Delta_{max\ u}, \Delta_{max\ p})=(1.0e\text{-}5, 1.0e\text{-}3, 1.0e\text{-}3)$. The results are for time t=1.5.

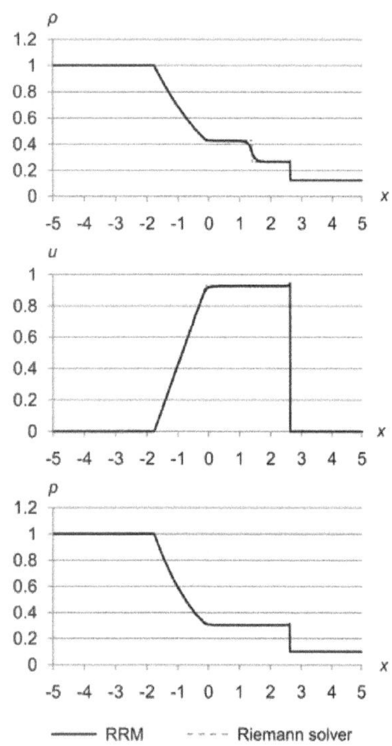

Figure 23: Test 1: Sod's test problem at high accuracy. Sod's problem with initial conditions $(\rho_l, u_l, p_l)=(1.0, 0.0, 1.0)$ and $(\rho_r, u_r, p_r)=(0.125, 0.0, 0.1)$, with maximum error metric $(\Delta_{max\ \rho}, \Delta_{max\ u}, \Delta_{max\ p})=(1.0e\text{-}5, 1.0e\text{-}3, 1.0e\text{-}3)$, at time t=1.5. This test shows typical RRM results: an s-shaped contact because RRM is not adiabatic across contacts, and a slight peak at the shock due to finite shock thickness.

These results are typical of RRM, and match the Riemann solver's results closely with two exceptions: the s-shaped contact, and the slight peak where the shock shows a finite thickness.

The s-shaped contact occurs because unlike a Riemann solver, RRM is not adiabatic across contacts, and models heat diffusion as a side effect of the algorithm. Wavefronts are created at contacts the same as at any other cell edges, so new cells are created across contacts, and gradual diffusion is the result.

We could easily make RRM adiabatic across contacts by adding a rule that when a tracer particle reaches a contact, its error metric is set to the maximum. This would insure that new cells are always created on one side of the contact or the other, keeping the contact sharp. We have not tried this yet, so RRM's current behavior is more like a real fluid than a Riemann solver in this respect.

Shocks in RRM have a finite thickness that manifests as a thin peak at the shock front. The shock thickness decreases as the accuracy is increased. This is because RRM creates new cells at the shock front at a rate proportional to the accuracy, and the more frequently cells are created there, the more quickly the change in density, velocity, and pressure is propagated to the area behind the shock. In the limit of infinite accuracy, the shock would be infinitely thin as it is in the Riemann solver's results.

Shocks in a real fluid also have a finite thickness of a few mean free paths, for a similar reason. It takes fluid atoms or molecules a few collisions to transition from their state in front of the shock to their state behind the shock. But because real fluids are not continuous, the shock thickness at a given set of conditions is essentially fixed by the fluid's physical properties.

Unlike contacts, shocks in RRM will always be sharply defined, because they are formed by the edge of a supersonic cell pushing into slower fluid. Our wavefronts always travel at the local speed of sound a, so shocks are naturally self-forming because nearby wavefronts cannot outrun them.

Test 2

Figure 24 shows test 2, which is a modified version of Sod's problem due to Laney [19]. This problem has a 100-to-1 pressure differential instead of the 10-to-1 differential of Sod's problem. The initial conditions are $(\rho_l, u_l, p_l)=(1.0, 0.0, 1.0)$ and $(\rho_r, u_r, p_r)=(0.01, 0.0, 0.01)$. The maximum error metric is $(\Delta_{max\ \rho}, \Delta_{max\ u}, \Delta_{max\ p})=(1.0e\text{-}4, 1.0e\text{-}4, 1.0e\text{-}3)$, and the results are for time t=1.5.

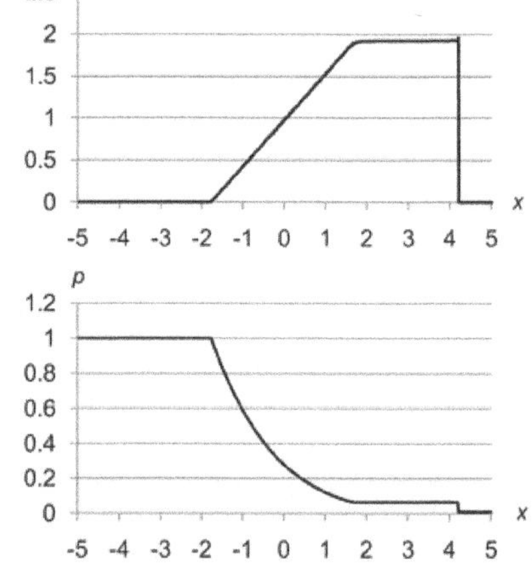

Figure 24: Test 2: Modified Sod's test problem with 100-to-1 pressure differential. A modified version of Sod's problem with initial conditions are $(\rho_l, u_l, p_l)=(1.0, 0.0, 1.0)$ and $(\rho_r, u_r, p_r)=(0.01, 0.0, 0.01)$, with maximum error metric $(\Delta_{max\ \rho}, \Delta_{max\ u}, \Delta_{max\ p})=(1.0e\text{-}4, 1.0e\text{-}4, 1.0e\text{-}3)$,

at time t=1.5. This test shows that RRM still gives good results on a problem that has a 100-to-1 pressure differential instead of the 10-to-1 differential of Sod's problem.

This test shows that RRM can handle strongly supersonic flows. We can see that the contact is s-shaped as usual, and there is just a hint of a u peak at the shock front, but otherwise the results are in agreement with the Riemann solver. The velocity at the shock front is higher than in the original Sod's problem, as we expect due to the greater pressure differential.

Test 3

Figure 25 shows test 3, which is a modified version of Sod's problem where the entire fluid moves right with u=1.0. The initial conditions are $(\rho_l, u_l, p_l)=(1.0, 1.0, 1.0)$ and $(\rho_r, u_r, p_r)=(0.125, 1.0, 0.1)$. The maximum error metric is $(\Delta_{max\,\rho}, \Delta_{max\,u}, \Delta_{max\,p})=(5.0e\text{-}5, 1.0e\text{-}3, 1.0e\text{-}3)$, and the results are for time t=1.5.

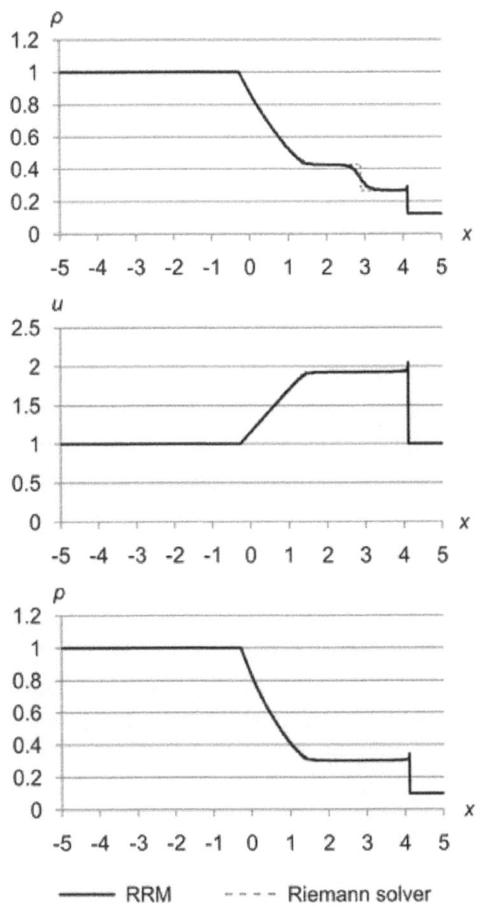

Figure 25: Test 3: Modified Sod's test problem with initial u=1.0. A modified version of Sod's problem where the entire fluid moves right with u=1.0, with initial conditions $(\rho_l, u_l, p_l)=(1.0, 1.0, 1.0)$ and $(\rho_r, u_r, p_r)=(0.125, 1.0, 0.1)$, with maximum error metric $(\Delta_{max\,\rho}, \Delta_{max\,u}, \Delta_{max\,p})=(5.0e\text{-}5, 1.0e\text{-}3, 1.0e\text{-}3)$, at time t=1.5. This test shows one of the benefits of the fully Lagrangian nature of RRM. Since the cells all move to the right with u=1.0, the shock front does not have to cross cell edges during the simulation, so the shock is just as sharp as in the u=0 case.

This test shows one of the benefits of the fully Lagrangian nature of RRM. Since the cells all move to the right with u=1.0, the shock front does not have to cross cell edges during the simulation, so the shock is just as sharp as in the u=0 case. The u curve is identical to the u=0 case, but shifted upwards by 1.0.

Test 4

Figure 26 shows test 4, which is test problem 1 from page 225 of Toro's book on Riemann solvers and numerical methods [1]. The initial conditions are $(\rho_l, u_l, p_l)=(1.0, 0.75, 1.0)$ and $(\rho_r, u_r, p_r)=(0.125, 0.0, 0.1)$. The maximum error metric is $(\Delta_{max\,\rho}, \Delta_{max\,u}, \Delta_{max\,p})=(1.0e\text{-}5, 1.0e\text{-}4, 1.0e\text{-}4)$, and the results are for time t=0.8.

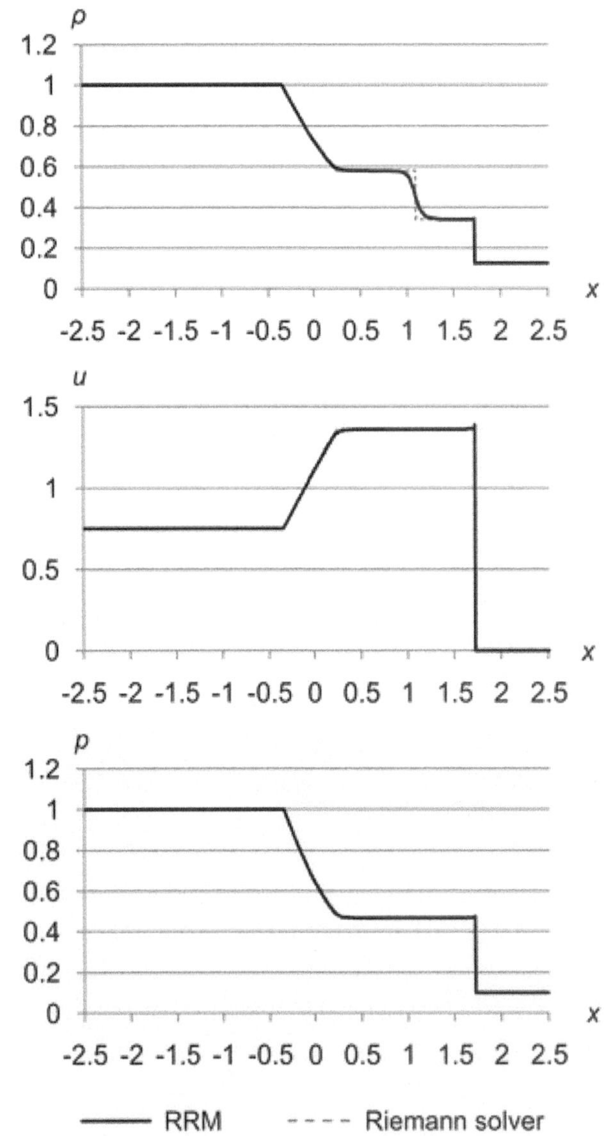

Figure 26: Test 4: Toro test 1. Toro's test problem 1, with initial conditions $(\rho_l, u_l, p_l)=(1.0, 0.75, 1.0)$ and $(\rho_r, u_r, p_r)=(0.125, 0.0, 0.1)$, with maximum error metric $(\Delta_{max\,\rho}, \Delta_{max\,u}, \Delta_{max\,p})=(1.0e\text{-}5, 1.0e\text{-}4, 1.0e\text{-}4)$, at time t=0.8. This test is similar to Sod's problem, but the left cell is initially ramming into the right cell, so the velocity at the shock front is somewhat higher.

This test is similar to Sod's problem, but the left cell is initially ramming into the right cell, so the velocity at the shock front is somewhat higher.

Test 5

Figure 27 shows test 5, which is test problem 2 from Toro's book [1]. The initial conditions are $(\rho_l, u_l, p_l)=(1.0, -2.0, 0.4)$ and $(\rho_r, u_r, p_r)=(1.0, 2.0, 0.4)$. The maximum error metric is $(\Delta_{max\,\rho}, \Delta_{max\,u}, \Delta_{max\,p})=(1.0e\text{-}5, 1.0e\text{-}5, 1.0e\text{-}5)$, and the results are for time t=0.6.

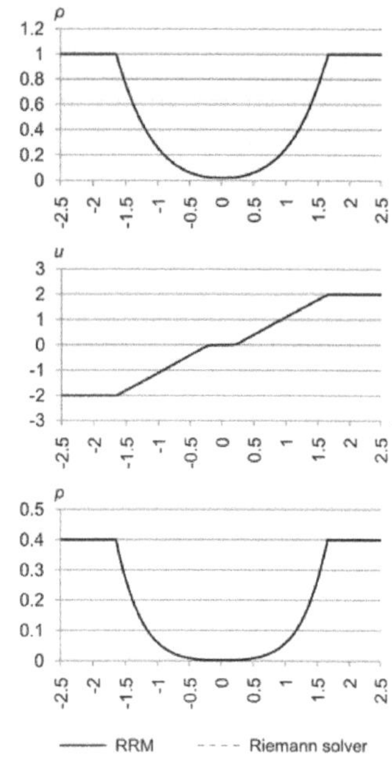

Figure 27: Test 5: Toro test 2. Toro's test problem 2, with initial conditions are $(\rho_l, u_l, p_l)=(1.0, -2.0, 0.4)$ and $(\rho_r, u_r, p_r)=(1.0, 2.0, 0.4)$, with maximum error metric $(\Delta_{max\ \rho}, \Delta_{max\ u}, \Delta_{max\ p})=(1.0e\text{-}5, 1.0e\text{-}5, 1.0e\text{-}5)$, at time t=0.6. This test shows that RRM can correctly handle the near-vacuum state created in the center.

This test creates a near-vacuum in the center, which can cause problems in the iteration schemes that some Riemann solvers use to find p. RRM does not have any special difficulty with vacuum areas, either as part of the initial conditions, or evolved during the simulation as we see here.

Note that in this test we set $\Delta_{max\ u}$ relatively low. This is to resolve the velocity features near the origin that are far from the large density and pressure gradients on either side.

Test 6

Figure 28 shows test 6, which is a modified "converging" version of test problem 2 from Toro's book [1]. The initial conditions are $(\rho_l, u_l, p_l)=(1.0, 3.0, 0.4)$ and $(\rho_r, u_r, p_r)=(1.0, -3.0, 0.4)$. The maximum error metric is $(\Delta_{max\ \rho}, \Delta_{max\ u}, \Delta_{max\ p})=(5.0e\text{-}4, 5.0e\text{-}4, 5.0e\text{-}4)$, and the results are for time t=1.1.

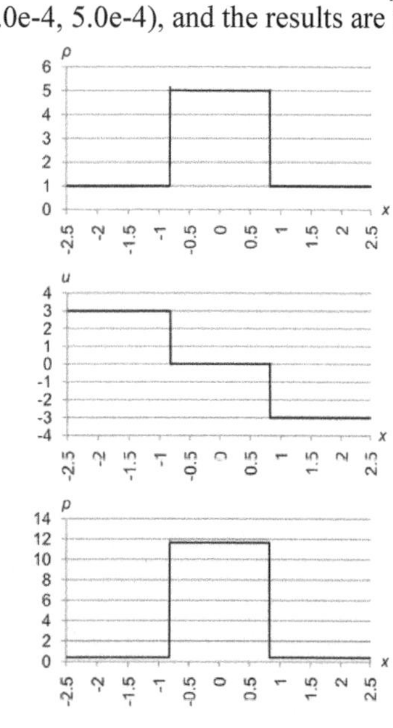

Figure 28: Test 6: Toro test 2 converging. A modified "converging" version of Toro's test problem 2, with initial conditions $(\rho_l, u_l, p_l)=(1.0, 3.0, 0.4)$ and $(\rho_r, u_r, p_r)=(1.0, -3.0, 0.4)$, with maximum error metric $(\Delta_{max\ \rho}, \Delta_{max\ u}, \Delta_{max\ p})=(5.0e\text{-}4, 5.0e\text{-}4, 5.0e\text{-}4)$, at time t=1.1. This is a test of symmetry and momentum conservation, to make sure that two colliding cells will pile up into one stationary mass with sharp edges.

This is a test of symmetry and momentum conservation, to make sure that two colliding cells will pile up into one stationary mass with sharp edges.

Test 7

Figure 29 shows test 7, which is test problem 3 from Toro's book [1]. The initial conditions are $(\rho_l, u_l, p_l)=(1.0, 0.0, 1000.0)$ and $(\rho_r, u_r, p_r)=(1.0, 0.0, 0.01)$. The maximum error metric is $(\Delta_{max\ \rho}, \Delta_{max\ u}, \Delta_{max\ p})=(1.0e\text{-}5, 5.0e\text{-}3, 1.0e\text{-}2)$, and the results are for time t=0.04.

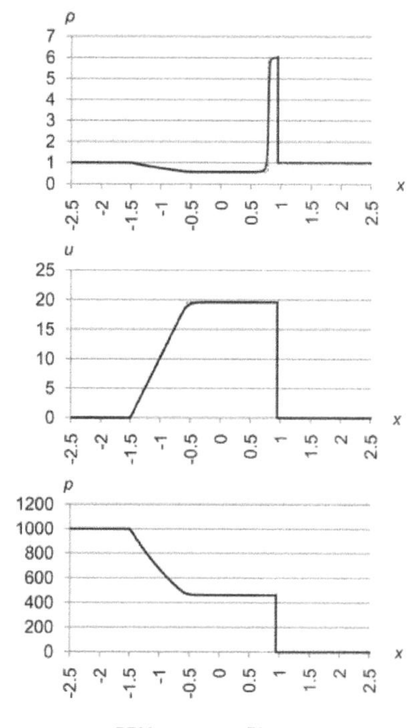

Figure 29: Test 7: Toro test 3. Toro's test problem 3, with initial conditions $(\rho_l, u_l, p_l)=(1.0, 0.0, 1000.0)$ and $(\rho_r, u_r, p_r)=(1.0, 0.0, 0.01)$, with maximum error metric $(\Delta_{max\ \rho}, \Delta_{max\ u}, \Delta_{max\ p})=(1.0e\text{-}5, 5.0e\text{-}3, 1.0e\text{-}2)$, at time t=0.04. This test's solution contains a strong shock very close to a contact. Since RRM is spatially adaptive, it simply creates many new cells between the shock and the contact to get the required accuracy.

The solution to this test requires a strong shock to be placed very close to a contact. Since RRM is spatially adaptive, it simply creates many new cells between the shock and the contact to get the required accuracy.

Test 8

Figure 30 shows test 8, which is test problem 4 from Toro's book [1]. The initial conditions are $(\rho_l, u_l, p_l)=(5.99924, 19.5975, 460.894)$ and $(\rho_r, u_r, p_r)=(5.99242, -6.19633, 46.0950)$. The maximum error metric is $(\Delta_{max\ \rho}, \Delta_{max\ u}, \Delta_{max\ p})=(5.0e\text{-}4, 1.0e\text{-}2, 1.0e\text{-}2)$, and the results are for time t=0.15.

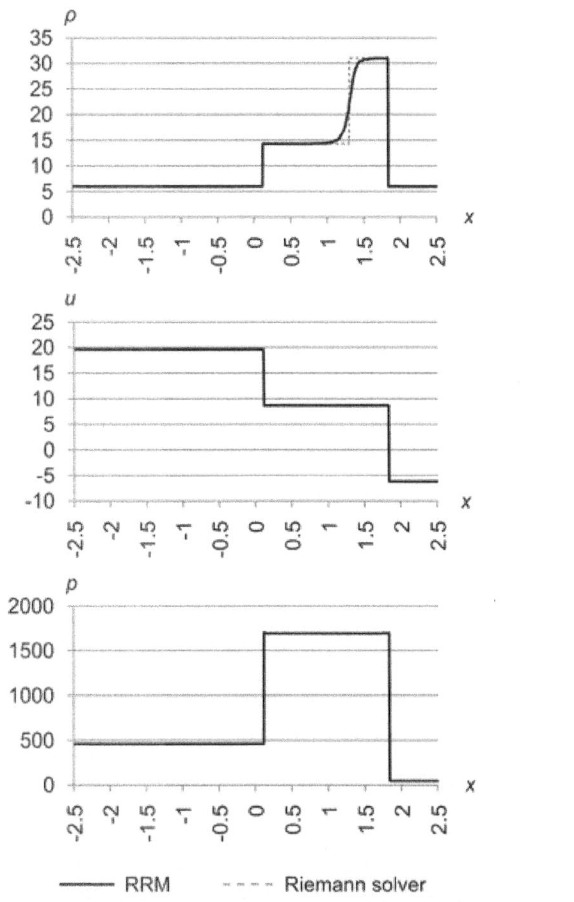

Figure 30: Test 8: Toro test 4. Toro's test problem 4, with initial conditions (ρ_l, u_l, p_l)=(5.99924, 19.5975, 460.894) and (ρ_r, u_r, p_r)=(5.99242, −6.19633, 46.0950), with maximum error metric $(\Delta_{max\ \rho}, \Delta_{max\ u}, \Delta_{max\ p})$=(5.0e-4, 1.0e-2, 1.0e-2), at time t=0.15. The solution to this test has two rightward-traveling shocks with a contact between them. As usual, the shocks are sharply resolved and the contact is s-shaped due to RRM's modeling of heat diffusion.

The solution to this test has two rightward-traveling shocks with a contact between them, which can be smeared out by some non-adaptive methods. As usual, the shocks are sharply resolved and the contact is s-shaped due to RRM's modeling of heat diffusion.

Test 9

Figure 31 shows test 9, which is test problem 5 from Toro's book [1]. The initial conditions are (ρ_l, u_l, p_l)=(1.0, −19.59745, 1000.0) and (ρ_r, u_r, p_r)=(1.0, −19.59745, 0.01). The maximum error metric is $(\Delta_{max\ \rho}, \Delta_{max\ u}, \Delta_{max\ p})$=(1.0e-5, 1.0e-2, 1.0e-2), and the results are for time t=0.03.

The initial values of this test were designed to give an almost stationary contact at the origin, which causes difficulties for some numerical methods. RRM handles stationary contacts the same as it does moving contacts, due to the Lagrangian nature of the simulation.

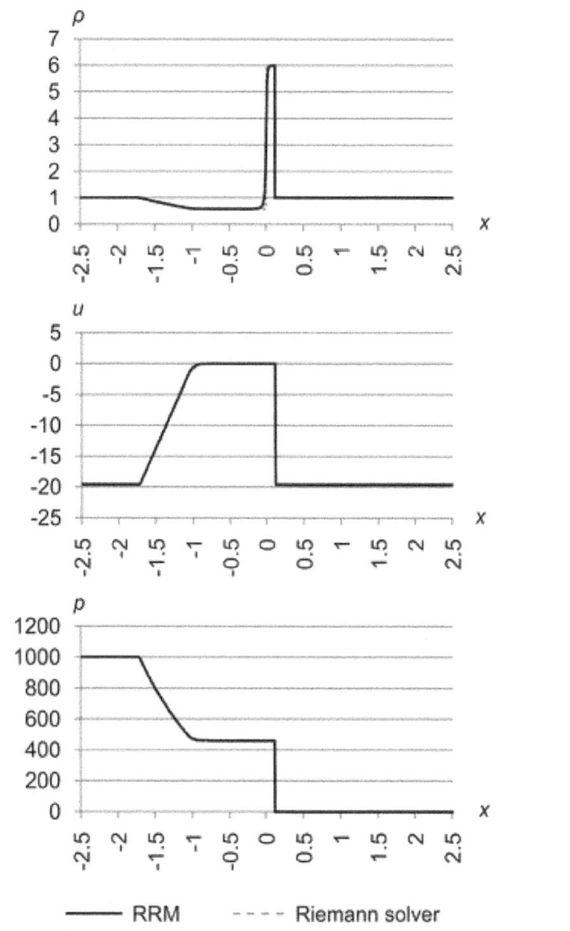

Figure 31: Test 9: Toro test 5. Toro's test problem 5, with initial conditions (ρ_l, u_l, p_l)=(1.0, −19.59745, 1000.0) and (ρ_r, u_r, p_r)=(1.0, −19.59745, 0.01), with maximum error metric $(\Delta_{max\ \rho}, \Delta_{max\ u}, \Delta_{max\ p})$=(1.0e-5, 1.0e-2, 1.0e-2), at time t=0.03. The initial values of this test were designed to give an almost stationary contact at the origin, which causes difficulties for some numerical methods. RRM handles stationary contacts the same as it does moving contacts, due to the Lagrangian nature of the simulation.

Absolute Error Analysis

To analyze RRM's error as compared to a Riemann solver, first we will show qualitatively how the accuracy of the simulation decreases as the maximum error metrics are increased. Then we will define a quantitative measure of the error between RRM's solution and that of the Riemann solver, and show how it decreases as each of the maximum error metrics is decreased. We will also show how the number of cells in the simulation increases as the error is reduced.

Figure 32 shows test 1 again, at the same accuracy as before, but this time only drawing one dot per cell (except for the edge cells, which have a dot on each side). The results are for time t=1.5.

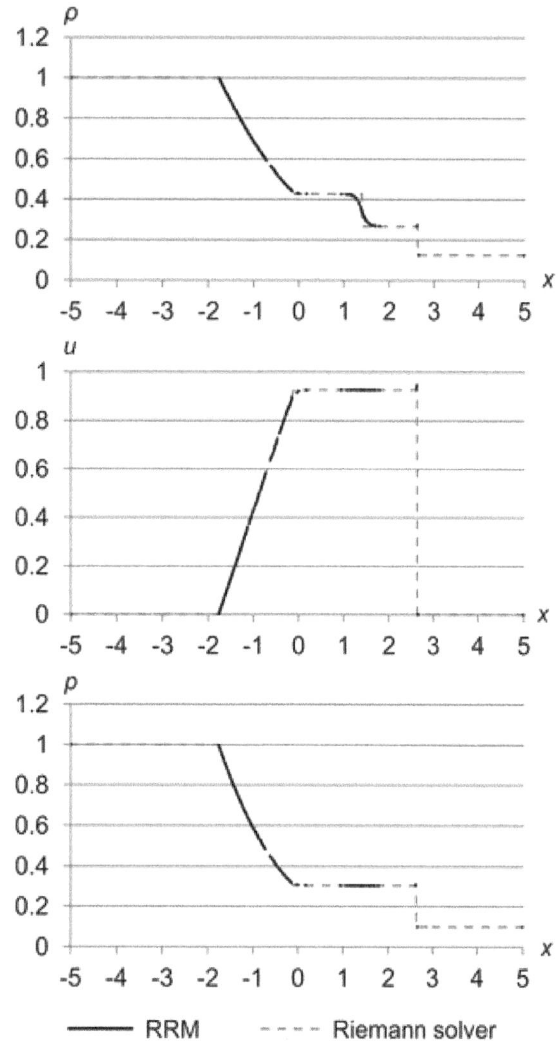

Figure 32: Sod's test problem at high accuracy, showing cell density. Sod's problem with initial conditions $(\rho_l, u_l, p_l)=(1.0, 0.0, 1.0)$ and $(\rho_r, u_r, p_r)=(0.125, 0.0, 0.1)$, with maximum error metric $(\Delta_{\max\,\rho}, \Delta_{\max\,u}, \Delta_{\max\,p})=(1.0\text{e-}5, 1.0\text{e-}3, 1.0\text{e-}3)$, at time t=1.5. Each of the approximately 800 cells is shown by a single dot, except the two edge cells which have two dots apiece. This figure shows that RRM is good at concentrating cells (and thereby computational effort) in areas of primitive variable gradient.

We can see that almost all of the approximately 800 cells are concentrated along the expansion fan and at the contact, with only one or two cells for each flat area. This illustrates how well RRM concentrates its computational effort on the active areas of the fluid.

Figure 33 shows test 1 again, but with the accuracy reduced by increasing the maximum error metric to $(\Delta_{\max\,\rho}, \Delta_{\max\,u}, \Delta_{\max\,p})=(1.0\text{e-}3, 1.0\text{e-}2, 1.0\text{e-}2)$ to show how the simulation begins to degrade. The results are for time t=1.5.

At this accuracy, the widths of the fluid cells are directly visible in the jagged curve of the expansion fan, and the contact is mostly smeared out. As expected, the shock thickness is greater due to the decreased accuracy. The shock front is also "blown back" so that it trails a constant distance behind the correct location. This is because the shock in RRM is a dynamic phenomenon with no special-case code. If the accuracy is not set high enough, fluid will pile up at the shock front where it cannot be redistributed fast enough to maintain the correct wave shape.

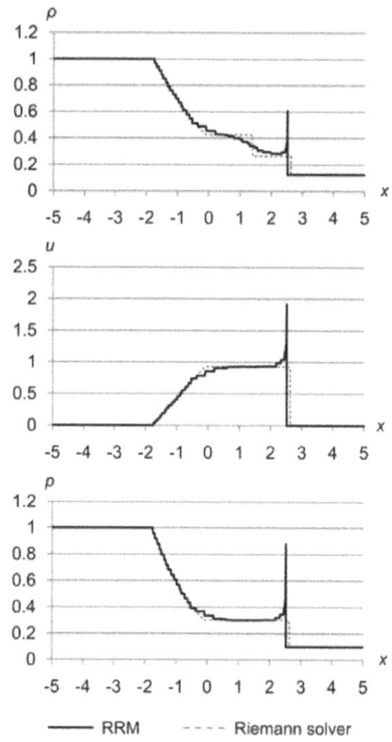

Figure 33: Sod's test problem at medium accuracy. Sod's problem with initial conditions $(\rho_l, u_l, p_l)=(1.0, 0.0, 1.0)$ and $(\rho_r, u_r, p_r)=(0.125, 0.0, 0.1)$, with the accuracy reduced by increasing the maximum error metric to $(\Delta_{\max\,\rho}, \Delta_{\max\,u}, \Delta_{\max\,p})=(1.0\text{e-}3, 1.0\text{e-}2, 1.0\text{e-}2)$ to show how the simulation begins to degrade, at time t=1.5. The shock is of increased thickness due to the lower accuracy, and is "blown back" from the correct location.

Figure 34 shows test 1 yet again, but with the accuracy further reduced by increasing the maximum error metric to $(\Delta_{\max\,\rho}, \Delta_{\max\,u}, \Delta_{\max\,p})=(1.0\text{e-}2, 1.0\text{e-}2, 1.0\text{e-}2)$ to show a more extreme failure. The results are for time t=1.5.

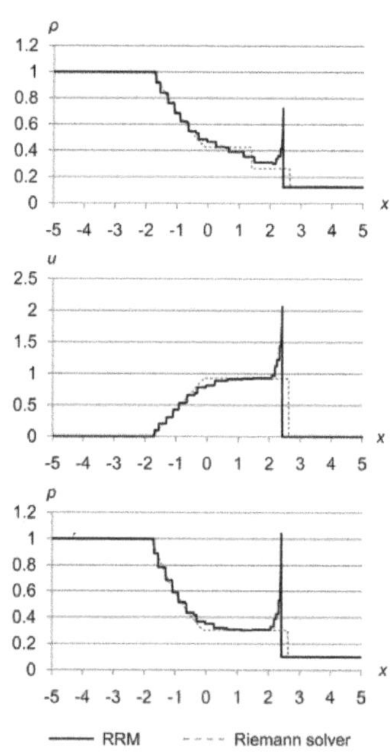

Figure 34: Sod's test problem at low accuracy. Sod's problem with initial conditions $(\rho_l, u_l, p_l)=(1.0, 0.0, 1.0)$ and $(\rho_r, u_r, p_r)=(0.125, 0.0, 0.1)$, with the accuracy further reduced by increasing the maximum error metric to $(\Delta_{\max\rho}, \Delta_{\max\,u}, \Delta_{\max\,p})=(1.0\text{e-}2, 1.0\text{e-}2, 1.0\text{e-}2)$ to show a more extreme failure, at time t=1.5. The shock is even thicker due to the lower accuracy, and is "blown back" even farther from the correct location.

The jagged expansion fan is even more pronounced here, since the cells are even larger. The contact is completely gone, the shock is blown back even further, and the spike at the shock front is even higher, since more fluid is piled up there.

Of course, we would never run a real simulation at such low accuracy. These figures are merely to show how the weaknesses of RRM differ from those of other methods. In particular, though RRM is a conservative method, that alone is does not guarantee correct shock placement as it does in FDM and FVM. But shocks in RRM remain very sharp even at very low accuracy, and there are no Gibbs oscillations near the shocks. This is because our cells are a piecewise-linear representation of the primitive variable values of the fluid.

Now we present a more quantitative analysis. We define.

$$\mathbf{e}(x,t) = \mathbf{u}_{RRM}(x,t) - \mathbf{u}_{Riemann}(x,t) \quad (19)$$

to be the vector error between the primitive variable values in RRM's solution and the primitive variable values in the Riemann solver's solution. Then we define a maximum integral error norm

$$e_{\text{maxinorm}} = \max\left(\int \|\mathbf{e}(x,t)\| dx\right) \quad (20)$$

to represent the maximum value, from the start time to some chosen end time, of the spatial integral of the norm of the error e(x,t) over the whole fluid.

Note that we choose the maximum integral error norm instead of the simpler maximum error norm.

$$e_{\text{maxnorm}} = \max\left(\|\mathbf{e}(x,t)\|\right) \quad (21)$$

because the maximum error norm for RRM is typically the thin peak right at the shock front, which is of almost constant height (though decreasing thickness) as simulation accuracy is increased.

Figure 35 shows $e_{maxinorm}$ and the maximum number of cells n_{max} vs. the maximum density error metric $\Delta_{max\,\rho}$ for Sod's problem. $\Delta_{max\,\rho}$ is swept from 1.0e-1 to 1.0e-5, while $\Delta_{max\,u}$ and $\Delta_{max\,p}$ are held constant at 1.0e-1. Simulation was from time t=0.0 to t=1.5.

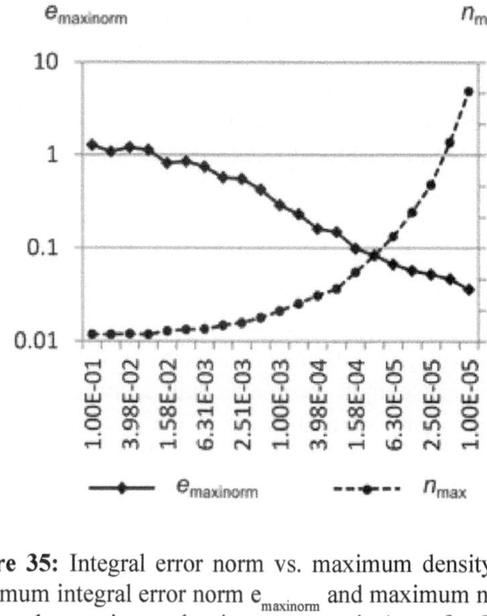

Figure 35: Integral error norm vs. maximum density error metric. Maximum integral error norm $e_{maxinorm}$ and maximum number of cells n_{max} vs. the maximum density error metric $\Delta_{max\,\rho}$ for Sod's problem. $\Delta_{max\,\rho}$ is swept from 1.0e-1 to 1.0e-5, while $\Delta_{max\,u}$ and $\Delta_{max\,p}$ are held constant at 1.0e-1. Simulation was from time t=0.0 to t=1.5. This figure shows that the error decreases logarithmically as the number of cells (and thus the computational effort) increases logarithmically.

We can see that once $\Delta_{max\,\rho}$ gets smaller than about 1.5e-4, the maximum number of cells in the simulation (which is a good proxy for the computational effort required) increases rapidly to maintain the approximately logarithmic decrease in maximum integral error norm. This computational effort goes into squaring off the contact, which is inherently diffusive in RRM.

Since $e_{maxinorm}$ is integrated over a width of 10, $e_{maxinorm}$=0.1 corresponds to an average absolute density error of 0.01, or about 1%. But the error is not evenly distributed. Most of the error is around the s-shaped contact, with a lesser amount near the shock front due to the transition region.

Figure 36 shows $e_{maxinorm}$ and the maximum number of cells n_{max} vs. the maximum velocity error metric $\Delta_{max\,u}$ for Sod's problem. $\Delta_{max\,u}$ is swept from 1.0e-1 to 1.0e-5, while $\Delta_{max\,\rho}$ and $\Delta_{max\,p}$ are held constant at 1.0e-1. Simulation was from time t=0.0 to t=1.5.

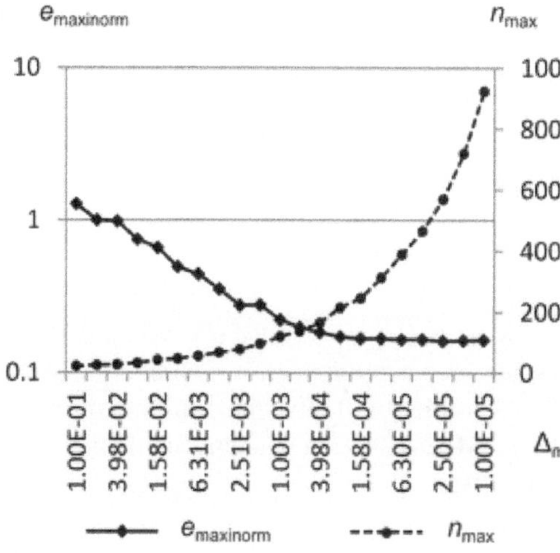

Figure 36: Integral error norm vs. maximum velocity error metric. Maximum integral error norm $e_{maxinorm}$ and the maximum number of cells n_{max} vs. the maximum velocity error metric $\Delta_{max\,u}$ for Sod's problem. $\Delta_{max\,u}$ is swept from 1.0e-1 to 1.0e-5, while $\Delta_{max\,\rho}$ and $\Delta_{max\,p}$ are held constant at 1.0e-1. Simulation was from time t=0.0 to t=1.5. This figure shows that the error cannot be decreased past a certain point solely by adjusting $\Delta_{max\,u}$, since there is little velocity gradient across the contact (where most of the error is concentrated in this test).

Note that we get less than a decade of decrease in maximum integral error norm as we decrease $\Delta_{max\,u}$, and for values lower than about 4.0e-4 we get very little additional benefit, though we increase computation effort by a factor of 5. This is because the velocity gradient across the contact is small, so decreasing $\Delta_{max\,u}$ will not cause more cells to be created there.

Figure 37 shows $e_{maxinorm}$ and the maximum number of cells n_{max} vs. the maximum pressure error metric $\Delta_{max\,p}$ for Sod's problem. $\Delta_{max\,p}$ is swept from 1.0e-1 to 1.0e-5, while $\Delta_{max\,\rho}$ and $\Delta_{max\,u}$ are held constant at 1.0e-1. Simulation was from time t=0.0 to t=1.5.

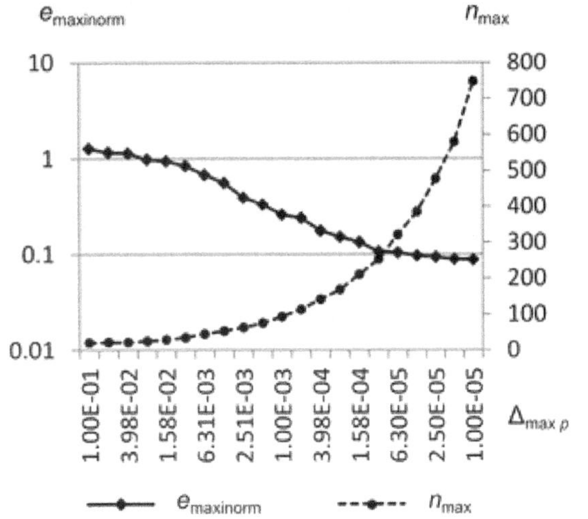

Figure 37: Integral error norm vs. maximum pressure error metric. Maximum integral error norm $e_{maxinorm}$ and the maximum number of cells n_{max} vs. the maximum pressure error metric $\Delta_{max\ p}$ for Sod's problem. $\Delta_{max\ p}$ is swept from 1.0e-1 to 1.0e-5, while $\Delta_{max\ \rho}$ and $\Delta_{max\ u}$ are held constant at 1.0e-1. Simulation was from timet=0.0 to t=1.5. This figure shows that the error cannot be decreased past a certain point solely by adjusting $\Delta_{max\ p}$, since there is little pressure gradient across the contact (where most of the error is concentrated in this test).

We see similar behavior to the $\Delta_{max\ u}$ sweep, where we get about one decade of decrease in the maximum integral error norm, with little further improvement as $\Delta_{max\ p}$ is reduced further. Again, this is because there is little pressure gradient across the contact, so reducing $\Delta_{max\ p}$ cannot improve the contact shape.

Figure 38 shows $e_{maxinorm}$ and the maximum number of cells n_{max} vs. all three maximum error metrics $\Delta_{max\ \rho}$, $\Delta_{max\ u}$, and $\Delta_{max\ p}$ for Sod's problem. All three maximum error metrics are swept from 1.0e-1 to 1.0e-5 in tandem. Simulation was from time t=0.0 to t=1.5.

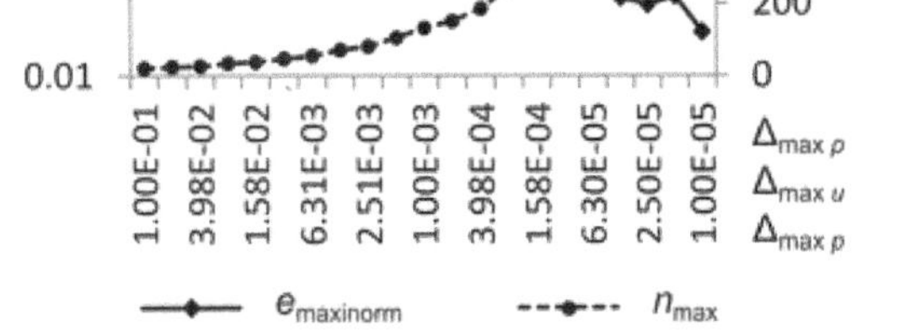

Figure 38: Integral error norm vs. all maximum error metrics. Maximum integral error norm $e_{maxinorm}$ and the maximum number of cells n_{max} vs. all three maximum error metrics $\Delta_{max\ \rho}$, $\Delta_{max\ u}$, and $\Delta_{max\ p}$ for Sod's problem. All three maximum error metrics are swept from 1.0e-1 to 1.0e-5 in tandem. Simulation was from time t=0.0 to t=1.5. This figure shows that there is a slight synergistic effect between the three maximum error metrics, since the minimum error achieved here is slightly lower than when each of the three is set to 1.0e-5 individually.

This last graph shows that if all three maximum error metrics are decreased together, we can get a bit less error than if they are decreased separately. This indicates that there is some interaction between their effects, though it is small for this test case.

Taken together, the previous four figures demonstrate that for this particular test case, RRM's error decreases almost logarithmically as $\Delta_{max\ \rho}$ is decreased logarithmically, whether by itself or in combination with the other maximum error metrics. However, other tests are sensitive to different maximum error metrics. For example, test 5 is sensitive to $\Delta_{max\ u}$, because its solution has velocity features near the origin that are far from the largest density or pressure gradients.

The maximum number of cells used by the simulation goes up almost logarithmically as the maximum error metrics are decreased logarithmically. And since the current RRM implementation performs O(n) operations per new cell created, where n is the number of cells in the fluid, this implies that the computational effort goes up logarithmically as well. This restricted our ability to run simulations with more than about 1000 cells in a reasonable time, which we define to be less than 5 minutes on one core of a 2.4 GHz Intel Core2 Quad CPU.

Future RRM implementations could perform as few as O(log n) or even O(1) operations per new cell created, if they used a more sophisticated data structure for cell intersection. We have so far avoided improving this data structure, since it makes the code much more difficult to maintain and alter for research purposes.

A note on the error-reducing efficiency of $\Delta_{max\ \rho}$, $\Delta_{max\ u}$, and $\Delta_{max\ p}$ in RRM is appropriate here. In the sweeps of $\Delta_{max\ u}$ and $\Delta_{max\ p}$ above, we see that the integral error norm decreases by fewer orders of magnitude than the number of cells increases by. This may simply show that decreasing the maximum error metrics of velocity and pressure does not efficiently reduce an error which is mostly in the density near the contact. Indeed, it appears that reducing $\Delta_{max\ \rho}$ reduces the integral error norm more efficiently, by approximately one order of magnitude as the number of cells increases from 100 to 1000. But for fewer than 100 cells, the error-reducing efficiency of $\Delta_{max\ \rho}$ is not as great. Overall, the error-reducing efficiency of the maximum error metrics in RRM is not yet fully understood.

These results indicate that RRM requires some further refinement if it is to efficiently produce results at any desired precision. If we assume the RRM implementation can be improved so that it performs only O(1) operations per cell, we would still like to insure that the number of cells always increases at a rate slower than the integral error norm decreases. One possibility is to adapt RRM to produce results more like those of a Riemann solver. As mentioned earlier, it would be straightforward to split new cells across contacts to maintain their sharpness, which would reduce a major source of error. Another possibility is to attempt to develop a new analytical solution to the Riemann problem that incorporates heat diffusion across contacts, and measure RRM's error against that instead.

Conservation Error Analysis

Conservation error is the difference between the conserved quantities currently present in all cells, and the original conserved quantities at the start of the simulation, assuming any boundary effects are properly accounted for. We define one conservation error for each conserved quantity:

$$\text{mass conservation error} \quad e_{mass}(t) = M_f(t) - M_f(0)$$

$$\text{momentum conservation error} \quad e_{momentum}(t) = P_f(t) - P_f(0)$$

$$\text{energy conservation error} \quad e_{energy}(t) = E_f(t) - E_f(0) \qquad \textbf{(22)}$$

where $M_f(t)$ is the total fluid mass, $P_f(t)$ is total fluid momentum, and $E_f(t)$ is total fluid energy, all functions of time.

Figure 39 shows the three conservation errors for Sod's problem over the first five seconds of flow time, at a time resolution of 0.01 seconds.

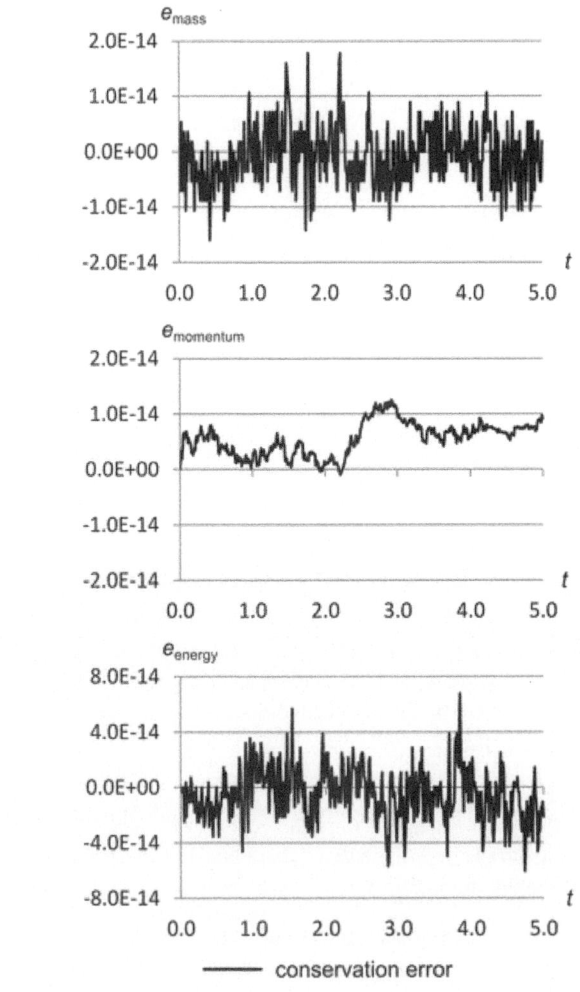

Figure 39: Conservation error vs. time on Sod's problem over 5 seconds. The three conservation errors for Sod's problem over the first five seconds of flow time, at a time resolution of 0.01 seconds. This graph shows that mass, momentum and energy are all conserved to within about ±6.0e-14. Total mass in the simulation is 11.25 kg, total momentum is 0.0 m·s, and total energy is 27.5 J.

This graph shows that mass, momentum and energy are all conserved to within about ±6.0e-14. Total mass in the simulation is 11.25 kg, total momentum is 0.0 m·s, and total energy is 27.5 J.

The first detail to note about this graph is that the conservation error is roughly two orders of magnitude larger than the floating-point precision ε_{fp}, which on our test machine is about 2.22e-16 for 64-bit IEEE floating point.

This is due to our use of the primitive variable form of the Euler equations in the current RRM implementation. The conserved quantities are derived from the primitive variable values of cell width, density, velocity, and pressure in a series of floating-point operations, each of which may be incorrect by roughly ε_{fp}. It takes only a few multiplicative operations for the error to grow to the observed value. Fortunately, since the signs of the individual errors are essentially random, the overall error does not tend to grow over time.

If we instead used the conservation form of the Euler equations in the RRM implementation, with a careful treatment we could get the error down to a smaller multiple of ε_{fp}. But since the error is already small in an absolute sense and does not grow over time, we chose to stay with the primitive variable form because it is simpler to code.

The second detail to note about the conservation error graph is the overall trends of the lines. The mass and energy lines are as expected, with floating-point truncation error causing random fluctuation about the horizontal axis. However, the momentum line differs, showing instead a fluctuation around approximately $e_{momentum}$=0.5e-14.

This is due to the initial conditions and time evolution of Sod's problem. At time t=0, the density and pressure are between 0.1 and 1.0, and calculating with these numbers to get mass and energy results in some nonzero error. However, the initial velocities are exactly zero, so initially $e_{momentum}$ will also be exactly zero. As the simulation proceeds and cell velocities increase, the effective baseline of $e_{momentum}$ is raised, since the calculations leading to momentum are no longer involve exact values of zero. Other test problems show variations on this behavior, but no problems tested so far show any time trend in conservation error.

Other Boundary Conditions

So far we have considered only periodic boundary conditions, which are simple to implement since they do not affect the conserved quantities of the fluid. But RRM can handle many other types of boundary conditions by adjusting the conserved quantities of each new cell just before it is flattened. We will illustrate three more types of boundary conditions: solid, Dirichlet, and free.

Solid boundaries are immovable and impermeable. To make a solid boundary, we check if each new cell touches or crosses the boundary. If so, we set its momentum to zero, and adjust its width so its edge just touches the boundary. Figure 40 shows Sod's problem with solid boundaries at x=−5 and x=5, from time t=0 to t=20. The initial conditions are (ρ_l, u_l, p_l)=(1.0, 0.0, 1.0) and (ρ_r, u_r, p_r)=(0.125, 0.0, 0.1). The maximum error metric is $(\Delta_{max\,\rho}, \Delta_{max\,u}, \Delta_{max\,p})$=(1.0e-4, 1.0e-2, 1.0e-2).

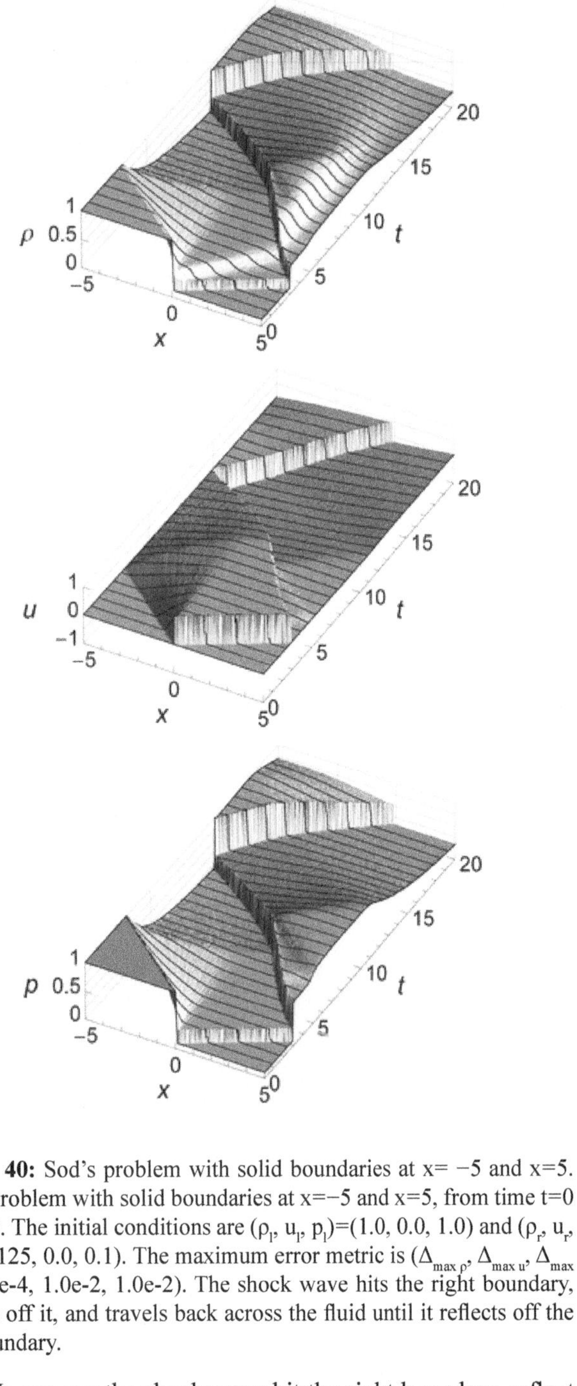

Figure 40: Sod's problem with solid boundaries at x= −5 and x=5. Sod's problem with solid boundaries at x=−5 and x=5, from time t=0 to t=20. The initial conditions are (ρ_l, u_l, p_l)=(1.0, 0.0, 1.0) and (ρ_r, u_r, p_r)=(0.125, 0.0, 0.1). The maximum error metric is ($\Delta_{\max \rho}$, $\Delta_{\max u}$, $\Delta_{\max p}$)=(1.0e-4, 1.0e-2, 1.0e-2). The shock wave hits the right boundary, reflects off it, and travels back across the fluid until it reflects off the left boundary.

We can see the shock wave hit the right boundary, reflect off it, and travel back across the fluid until it reflects off the left boundary. If we let the simulation run indefinitely, the shock will travel back and forth many times, until numerical dissipation finally smooths it out. Eventually, the density and pressure will be flat, and the velocity will be everywhere zero.

Dirichlet boundaries hold the primitive variable values of the fluid constant at the boundaries. To make a Dirichlet boundary, we check if each new cell touches or crosses a boundary. If so, we set its density, velocity, and pressure to some constant boundary values, and adjust its width so its edge just touches the boundary. Figure 41 shows an inrush problem with Dirichlet boundaries at x=−5 and x=5, from time t=0 to t=10. The initial conditions are (ρ_l, u_l, p_l)=(0.1, 0.0, 0.2) and (ρ_r, u_r, p_r)=(0.1, 0.0, 0.2). The boundary values are (ρ_l, u_l, p_l)=(0.3, 0.6, 0.4) and (ρ_r, u_r, p_r)=(0.6, −0.6, 0.5). The maximum error metric is ($\Delta_{\max \rho}$, $\Delta_{\max u}$, $\Delta_{\max p}$)=(1.0e-4, 1.0e-3, 1.0e-3).

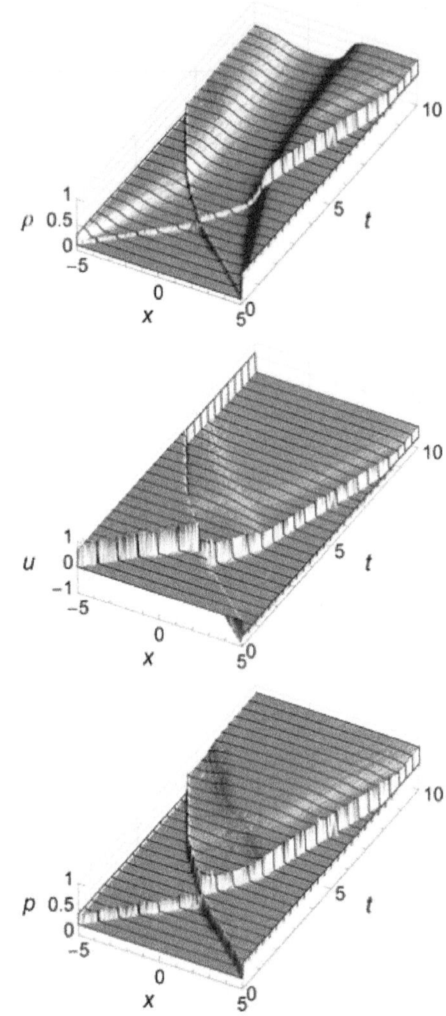

Figure 41: Inrush problem with Dirichlet boundaries at x= −5 and x=5. An inrush problem with Dirichlet boundaries at x=−5 and x=5, from time t=0 to t=10. The initial conditions are (ρ_l, u_l, p_l)=(0.1, 0.0, 0.2) and (ρ_r, u_r, p_r)=(0.1, 0.0, 0.2). The boundary values are (ρ_l, u_l, p_l)=(0.3, 0.6, 0.4) and (ρ_r, u_r, p_r)=(0.6, −0.6, 0.5). The maximum error metric is ($\Delta_{\max \rho}$, $\Delta_{\max u}$, $\Delta_{\max p}$)=(1.0e-4, 1.0e-3, 1.0e-3). Two asymmetrical shocks propagate in from the edges, cross near the center, and continue to the opposite edges, where they are squelched by the boundary conditions.

We can see the two asymmetrical shocks propagate in from the edges, cross near the center, and continue to the opposite edges, where they are squelched by the boundary conditions. If we let the simulation run longer, the continuous fluid inflow fills the area higher and higher, with velocity everywhere zero, and density and pressure eventually becoming flat due to diffusion.

Free boundaries let fluid flow in or out of the boundaries, without creating any disturbance that might propagate back into "interesting" parts of the fluid. To make a free boundary, we check if each new cell touches or crosses a boundary. If so, we set its density, velocity, and pressure to those of the intersected cell nearest the boundary (for inflow) or farthest from the boundary (for outflow), and adjust its width so its edge just touches the boundary. Figure 42 shows a rightward flow problem with free boundaries at x=−5 and x=5, from time t=0 to t=15. The initial conditions are (ρ_l, u_l, p_l)=(0.8, 0.8, 0.1) and (ρ_r, u_r, p_r)=(0.1, 0.7, 0.05). The maximum error metric is ($\Delta_{\max \rho}$, $\Delta_{\max u}$, $\Delta_{\max p}$)=(5.0e-5, 1.0e-4, 1.0e-4).

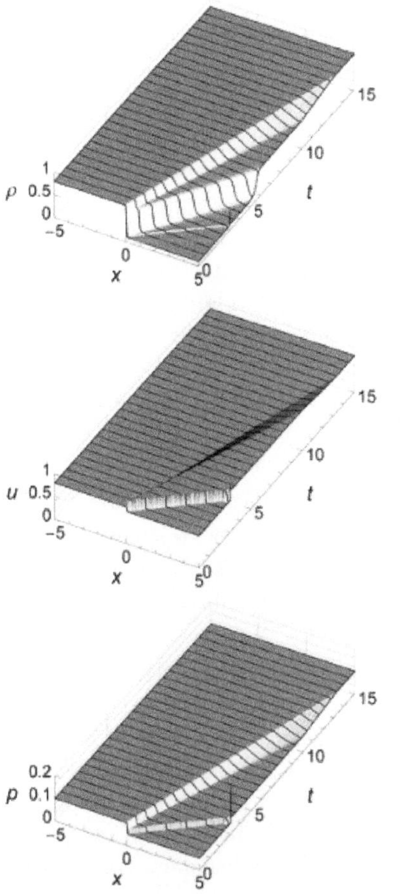

Figure 42: Rightward flow problem with far-field boundaries at x= −5 and x=5. A rightward flow problem with free boundaries at x=−5 and x=5, from time t=0 to t=15. The initial conditions are $(\rho_l, u_l, p_l)=(0.8, 0.8, 0.1)$ and $(\rho_r, u_r, p_r)=(0.1, 0.7, 0.05)$. The maximum error metric is $(\Delta_{\max\,\rho}, \Delta_{\max\,u}, \Delta_{\max\,p})=(5.0e\text{-}5, 1.0e\text{-}4, 1.0e\text{-}4)$. Fast-moving fluid flows in from the left boundary and pushes the slower fluid in front of it, forcing it out of the right boundary. After about t=14 all of the fluid is in the left state, since all the right fluid as been pushed out.

We can see that fast-moving fluid flows in from the left boundary and pushes the slower fluid in front of it, forcing it out of the right boundary. After about t=14 all of the fluid is in the left state, since all the right fluid as been pushed out. Note that when the shock hits the right boundary at about t=4, we can see a glitch. This is because if a cell is near the edge of the fluid, and there is no other cell between it and the edge, we extend it to touch the edge. In this case, it just happens to catch a narrow cell in the transition region and widen it so it is visible.

Finally, we mention two details that apply to all three boundary condition types discussed above. First, when producing 3D graphs of RRM simulations, we remove the very thin cells that can occur in the transition regions at shock fronts. This is simply to make the graphs more legible, since otherwise these very thin cells hide the details of the fluid behind them. You can see one of these cells at t=4 on the right side of the very last graph above.

Second, we note that these adjustments to the conserved quantities of new cells require us to change the stored initial conserved quantities of the entire fluid by a commensurate amount, so the simulation will not fail its ongoing per-event conservation checks. This models the mass, momentum, and energy that are being added and removed at the boundaries.

Source Terms

Our analysis so far has only treated the Euler equations in the homogeneous case, where no mass, momentum, or energy are added to or removed from the fluid during the simulation, except for a special case at the boundaries.

In the more general case, we augment the Euler equations with source terms thus:

$$\int_{m_1}^{m_2} (\mathbf{u}(m,t_2)-\mathbf{u}(m,t_1))dm = -\int_{t_1}^{t_2} (\mathbf{f}(m_2,t)-\mathbf{f}(m_1,t))dt + \int_{t_1}^{t_2}\int_{m_1}^{m_2} \mathbf{s}(m,t)\,dm\,dt \qquad \textbf{(23)}$$

using a new vector of source terms

$$\mathbf{s} = \begin{bmatrix} s_{mass} \\ s_{momentum} \\ s_{energy} \end{bmatrix} \qquad \textbf{(24)}$$

Presenting a scheme to simulate these equations is beyond the scope of this paper. However, we can make a few remarks about how it might be possible.

A few cases would be simple. For example, mass or energy source terms that are constant in space and time could easily be implemented by adding mass or energy to new cells during flattening.

A few more cases are somewhat difficult, but feasible using splitting schemes similar to those described in chapter 15 of Toro's book [1]. For example, momentum source terms that are constant in space and time could be implemented by changing the cells' equations of motion from constant-velocity to varying-velocity, at the cost of complicating the intersection calculation of particles with cells.

The general case becomes very difficult. If spatially- and temporally-varying source terms are allowed, cells' masses and energies could change in ways that would require both space and time integration to resolve at flattening time. The intersection calculation of particles with cells would also require solving ordinary differential equations, rather than simple algebraic equations.

DISCUSSION

So far we have shown that RRM gives correct results for many standard test problems, that RRM's error decreases steadily as we increase the desired accuracy, and that RRM handles many common types of boundary conditions. Now we explain the similarities and differences between RRM and other CFD methods in detail, list some of RRM's limitations, and suggest directions for future research.

Comparison with Adaptive Eulerian Methods

Simple CFD methods advance time across the whole fluid in lockstep. But this wastes effort in smooth areas of the fluid, and gives suboptimal resolution in steep areas. To solve this problem, Osher and Sanders proposed locally varying time steps [20]. Such methods can make simulation much more efficient, but they require that special care be taken at the interfaces between areas of differing time resolution.

Adaptive mesh refinement (AMR) methods such as the one proposed by Berger and Oliger[21] flag points in the fluid with high estimated error for possible refinement. Then every so often, the flagged points are clustered together to determine the size, shape, and orientation of a new, finer sub-mesh to cover them. Finer sub-meshes are integrated using proportionately shorter time steps, so AMR is adaptive in both space and time. AMR requires special care when integrating the sub-meshes, to insure that the boundary conditions with the rest of the fluid maintain conservation.

RRM does not divide the fluid into areas of different mesh fineness or time step size. Instead, each new cell is created at its own individually chosen time, which need not bear any relation to the creation times of other cells. This gives us a very fine-grained spatio-temporal adaptivity, with the disadvantage that such a simulation must use an event queue instead of a simple time-stepping loop. RRM does not require a clustering algorithm, since new cells are preferentially created in high-gradient areas. But RRM does require the unioning of wavefronts before creating a new cell, which is a similar operation.

RRM differs from FDM and FVM in that it does not use numerical derivatives in the cell chopping and flattening process, only integrals. This means that RRM does not need a flux limiter or slope limiter to smooth spurious oscillations that can be caused by the extremely large gradients near shocks.

Comparison with the Lattice Boltzmann Method

RRM differs from LBM firstly in that RRM is meshfree and Lagrangian, where LBM has a mesh (though it is called a lattice in LBM literature) and is Eulerian (since the fluid flows through fixed lattice sites). A more interesting difference between RRM and LBM is that in RRM, cells are free to move in any direction, where in LBM, fluid can only move in a fixed number of directions between adjacent sites. Fluid flows in LBM can therefore exhibit anisotropies, depending on the choice of lattice type and connectivity.

Cells in 2D and 3D RRM will have angular velocity, which will complicate the flattening process somewhat, where in LBM the collision process is much the same for all dimensionalities. Since collision takes place only at zero-size lattice sites, angular quantities do not arise, which keeps the programming simpler than RRM.

Comparison with Adaptive Meshed Lagrangian Methods

To adapt to the time-varying features of a fluid, a meshed Lagrangian method can move the mesh relative to the fluid, change the mesh connectivity, or both.

The Arbitrary Lagrangian-Eulerian method (ALE) [22] combines the Eulerian and Lagrangian forms by creating a third "referential" coordinate system that is independent of both the fixed world coordinates and the moving material coordinates. This allows cells to move independently of both the fixed coordinate system and the material.

As an ALE simulation progresses in Lagrangian mode, the cells can be "rezoned" by allowing the mesh to move relative to the fluid, while keeping the same mesh connectivity. This rezoning helps keep the cells from becoming tangled or degenerate, which would prevent further simulation. Rezoning is an Eulerian process, since it allows fluid to flow across cell edges, and it can smear out contacts and shocks unless one is careful when rezoning near them. For this reason, if the fluid motion is complex enough, it may be impossible to keep the original mesh connectivity without unacceptably degrading accuracy.

The "free Lagrange" methods such as FLAG [23] and Whitehurst's signal method [18] allow mesh points to be dynamically linked and unlinked over the course of the simulation, thereby changing the initial mesh connectivity as the fluid moves. These methods use a variety of heuristics to maintain a reasonable mesh, such as trying to keep nearest neighboring points connected, or trying to keep the angles of mesh triangles as equal as possible.

In RRM, there is no mesh connectivity, so there is no need to track or alter it over the course of simulation. RRM constantly creates new cells, which has an effect similar to rezoning in that it allows fluid to flow across the edges of chopped cells.

Moving finite element (MFE) methods [24]–[27] generalize the finite element method to better track moving fluid flow features using moving elements. MFE methods result in extremely stiff systems of ordinary differential equations (ODEs), and so require sophisticated implicit ODE solvers. They also require careful tuning with user-chosen parameters to keep the elements from becoming too small or bunching up at shocks. It is also possible to adaptively create and destroy nodes in an MFE method, as shown by Kuprat in 1992 [28].

Since each new RRM cell chops out and replaces what was underneath it, cells can never bunch up, and there is only one user-chosen parameter to set, the maximum error metric (though this metric does have three components). And since RRM does not use systems of equations, it does not need numerical solvers.

Comparison with Previous Meshfree Methods

Of all current CFD methods, the meshfree methods are the most similar to RRM. Indeed, we categorize RRM itself as a meshfree method, since it shares the characteristics of being purely Lagrangian and not using a mesh. But as we will see, there are many differences between previous meshfree methods and RRM in how cells or particles are formed, how their motion is calculated, and how long they persist during a simulation.

Previous meshfree methods and RRM both discretize a fluid into a set of particles or cells with no connectivity between them. In previous meshfree methods like SPH and MPS, the particles are acted upon by forces over time, and thus change their velocities. In contrast, once a cell is created in RRM, it moves at a constant velocity even as parts of it are chopped away by the creation of subsequent new cells. This means RRM does not require integration of the equations of motion of its cells.

A major difference between RRM and SPH or MPS lies in how the primitive variable values like density and pressure are determined at each point in the fluid. SPH and MPS store the conserved quantities in moving material particles. These particles have zero extent, but a smoothing function allows us to find the primitive variable values in the spaces between the particles. In contrast, RRM does not store the conserved quantities of the fluid directly. Instead, they are the result of integrating the stored primitive variable values over the cell areas. Since the cells have non-zero sizes, and new cells are constantly being created to fill any gaps, we do not need a smoothing function. The tracer particles in RRM are non-material particles that do not carry any conserved quantities, they simply trace out the expanding acoustic waves in the fluid.

A particle-based meshfree method like SPH can be made adaptive by allowing the smoothing length to vary inversely with density, as shown by Benz in 1990 [29]. This is refined by Owen et al. in 1998 [30] to give each particle a time-varying, anisotropic smoothing length that attempts to keep the number of neighboring particles the same in each direction.

The motion of the tracer particles in RRM gives an effect similar to the use of anisotropic smoothing length in adaptive SPH, since the tracer particles sweep out an area that varies with the local speed of sound and the local fluid motion.

SPH can also be made adaptive by splitting and merging particles during simulation [31]–[33]. Splitting is done in low-density areas, and merging in high-density areas, to insure that the number of particles is appropriate to accurately track the fluid motion.

The previous meshfree method most similar to RRM is the Finite Mass Method (FMM) [34],[35]. FMM divides the fluid into finite-sized cells (called mass packets in FMM papers) with an internal distribution typically described by third-order B-splines. These cells can move, deform, and interact during simulation. If the cells become too deformed, the simulation is stopped, the fluid is remeshed into new, undeformed cells and the simulation is restarted.

Cells in RRM do not change shape, except in that parts of them are chopped away by the creation of new cells. Therefore RRM does not need to remesh to fix excessive cell deformation. RRM constantly creates and destroys cells, so any excessive bunching or gapping due to cell movement is fixed incrementally rather than all at once in a remeshing operation.

Realm of Applicability

RRM should work for most systems of conservation equations that have a complete wave description, which are often described as hyperbolic. In one dimension, such a system typically has three types of waves: entropy waves, left-propagating acoustic waves, and right-propagating acoustic waves. In RRM, the motions of the cells model the entropy waves, and the moving tracer particles model the propagation of the acoustic waves.

RRM's tracer particle methodology is useful only for compressible fluids, since incompressible fluids have no acoustic waves. So for example, RRM should work for the compressible Navier-Stokes equations, but not the incompressible Navier-Stokes equations.

RRM's "chop out and flatten" methodology is most applicable to systems that are locally conservative. A system that is globally conservative but not locally conservative cannot easily be chopped out and flattened, since the flattening would require access to parts of the fluid other than those that were chopped out.

Limitations

One limitation of RRM is that its data structures are complex and difficult to code correctly. Allowing cells to move and be chopped up over time is straightforward, but involves extra bookkeeping that many other methods do not require. Extension to 2D and 3D is possible, but will be even more complex since cells must be allowed to rotate as well as translate.

Another limitation of RRM is its use of event-driven simulation. This gives good spatial and temporal adaptivity, but it makes the algorithm more difficult to parallelize, especially on GPGPUs (General-Purpose Graphics Processing Units) such as NVIDIA's Tesla where data-dependent branching is penalized.

A final limitation applies to von Neumann boundary conditions. Since RRM does not use spatial derivatives, it is difficult to hold them constant at the fluid boundaries. RRM could implement von Neumann boundary conditions with some difficulty by creating "ghost cells" just outside the boundaries to give the correct behavior, but it would not fit neatly into the "adjusting the conserved quantities of new cells" paradigm discussed in the above section on boundary conditions.

Further Research

The obvious future research directions for RRM are extension to 2D and 3D, replacement of the Euler equations with the compressible Navier-Stokes equations, and parallelization. A not-so-obvious direction is using RRM to simulate the behavior of non-fluid fields containing inherent discontinuities or intractable nonlinearity. Since RRM does not evaluate numerical derivatives or solve systems of equations, it might be applicable to fields whose traditional discretizations are numerically ill-behaved or difficult to formulate.

RRM could also benefit from further investigation into its rate of convergence. As noted at the end of the section on absolute error, increases in computational effort do not always result in proportional decreases in the integral error norm. More research into this area would be helpful to insure that RRM can efficiently achieve any desired level of error.

REFERENCES

1. Toro EF (1999) Riemann solvers and numerical methods for fluid dynamics: A practical introduction. Berlin: Springer Verlag. 624 p.
2. Godunov SK (1959) A difference scheme for numerical computation of discontinuous solutions of hydrodynamic equations. Matematicheskii Sbornik 47: 271–306.
3. Sod GA (1978) A survey of several finite difference methods for systems of nonlinear hyperbolic conservation laws. Journal of Computational Physics 27: 1–31.
4. Clough RW (1960) The finite element method in plane stress analysis. Proceedings of the 2nd ASCE Conference on Electronic Computation 345–378.
5. Martin HC (1968) Finite element analysis of fluid flows. Proceedings of the 2nd Conference on Matrix Methods in Structural Mechanics 517–535.
6. Harten A (1983) High resolution schemes for hyperbolic conservation laws. Journal of Computational Physics 49: 357–393.

7. Harten A, Engquist B, Osher S, Chakravarthy SR (1987) Uniformly high order essentially non-oscillatory schemes III. Journal of Computational Physics 71: 231–303.
8. van Leer B (1979) Towards the ultimate conservative difference scheme V: A second-order sequel to Godunov's method. Journal of Computational Physics 32: 101–136.
9. Colella P, Woodward PR (1984) The piecewise parabolic method (PPM) for gas-dynamical simulations. Journal of Computational Physics 54: 174–201.
10. Chen S, Doolen GD (1998) Lattice Boltzmann method for fluid flows. Annual Review of Fluid Mechanics 30: 329–364.
11. Gingold RA, Monaghan JJ (1977) Smoothed particle hydrodynamics: Theory and application to non-spherical stars. Monthly Notices of the Royal Astronomical Society 181: 375–389.
12. Lucy LB (1977) A numerical approach to the testing of the fission hypothesis. Astronomical Journal 82: 1013–1024.
13. Koshizuka S, Oka Y (1996) Moving-particle semi-implicit method for fragmentation of incompressible fluid. Nucl. Sci. Eng. 123: 421–434.
14. Li S, Liu WK (2002) Meshfree and particle methods and their applications. Applied Mechanics Review 55: 1–34.
15. Huerta A, Belytschko T, Fernández-Méndez S, Rabczuk T (2004) Meshfree methods. In: Stein E, de Borst R, Hughes TJR, editors. pp. 279–309. Chichester: John Wiley & Sons, Ltd.
16. Chaikin GM (1974) An algorithm for high-speed curve generation. Computer Graphics and Image Processing 3: 346–349.
17. Catmull E (1974) A subdivision algorithm for computer display of curved surfaces. Ph.D. dissertation. University of Utah, UT, USA. Technical Report UTEC-CSc-74–133.
18. Whitehurst R (1995) A free Lagrange method for gas dynamics. Monthly Notices of the Royal Astronomical Society 277: 655–680.
19. Laney CB (1998) Computational gasdynamics. Cambridge: Cambridge University Press. 613 p.
20. Osher S, Sanders R (1983) Numerical approximations to nonlinear conservation laws with locally varying time and space grids. Mathematics of Computation 41: 321–336.
21. Berger MJ, Oliger JE (1984) Adaptive mesh refinement for hyperbolic partial differential equations. Journal of Computational Physics 53: 484–512.
22. Hirt CW, Amsden AA, Cook JL (1974) An arbitrary Lagrangian-Eulerian computing method for all flow speeds. Journal of Computational Physics 14: 227–253.
23. Crowley WP (1971) FLAG: A free-Lagrange method for numerically simulating hydrodynamic flows in two dimensions. Proceedings of the Second International Conference on Numerical Methods in Fluid Dynamics, Lecture Notes in Physics 8: 37–43.
24. Miller K, Miller RN (1981) Moving finite elements I. SIAM Journal on Numerical Analysis 18: 1019–1032.
25. Miller K (1981) Moving finite elements II. SIAM Journal on Numerical Analysis 18: 1033–1057.
26. Carlson NN, Miller K (1998) Design and application of a gradient-weighted moving finite element code I: In one dimension. SIAM Journal on Scientific Computing 19: 728–765.
27. Carlson NN, Miller K (1998) Design and application of a gradient-weighted moving finite element code II: In two dimensions. SIAM Journal on Scientific Computing 19: 766–798.
28. Kuprat AP (1992) Creation and annihilation of nodes for the moving finite element method. Ph.D. dissertation. Department of Mathematics, University of California, Berkeley, CA, USA. UMI Order No. pp. GAX93–04972.
29. Benz W (1990) Smooth particle hydrodynamics: A review. In: Buchler JR, editor. pp. 269–288. Dordrecht: Kluwer Academic Publishers.
30. Owen JM, Villumsen JV, Shapiro PR, Martel H (1998) Adaptive smoothed particle hdrodynamics: Methodology II. The Astrophysical Journal Supplement Series 116: 155–209.
31. Meglicki Z, Wickramasinghe D, Bicknell GV (1993) 3D structure of truncated accretion discs in close binaries. Monthly Notices of the Royal Astronomical Society 264: 691–704.
32. Kitsionas S, Whitworth AP (2002) Smoothed particle hydrodynamics with particle splitting, applied to self-gravitating collapse. Monthly Notices of the Royal Astronomical Society 330: 129–136.
33. Lastiwka M, Quinlan N, Basa M (2005) Adaptive particle distribution for smoothed particle hydrodynamics. International Journal for Numerical Methods in Fluids 47: 1403–1409.
34. Gauger C, Leinen P, Yserentant H (2000) The finite mass method. SIAM Journal on Numerical Analysis 37: 1768–1799.
35. Klingler M, Leinen P, Yserentant H (2005) The finite mass method on domains with boundary. SIAM Journal on Scientific Computing 26: 1744–1759.

CHAPTER 3 COMPUTATIONAL FLUID DYNAMICS FOR NEMATIC LIQUID CRYSTALS

Alison Ramage and André M. Sonnet

Keywords: Computational fluid dynamics Nematic liquid crystals Alignment tensor Iterative solvers Communicated by Zhong-Zhi Bai.

INTRODUCTION

In recent years, significant advances have been made in the development of effective preconditioned iterative solvers for finite element models in incompressible fluid dynamics, such as solution of Stokes and Navier–Stokes equations [5]. The ready availability of these methods in public domain codes such as IFISS [3] and TRILINOS [10] has extended the range of possible applications by making it easier for the practitioner to apply these fast solvers to specific situations. In particular, the MATLAB package IFISS, as well as being a useful source of benchmark problems, also provides a convenient starting point for developing solvers dedicated to a particular application [4, §3.4]. In this paper, we describe one such situation where an IFISS-based fast solver is used as part of an algorithm designed to compute flow in a liquid crystal cell.

Computing flow in complex fluids such as liquid crystals and polymers is very challenging because of the altered structure of the flow equation: the underlying Navier–Stokes problem contains additional terms representing the interaction between the flow and the orientation of the molecules within the fluid. In liquid crystal applications, the usual form of the stress tensor is very complicated. Our method relies on reformulating the time derivative in the stress tensor in a way which effectively decouples the flow and orientation, with each appearing only on the right-hand side of the other equation. In this way, difficulties associated with solving the fully coupled problem are circumvented and a stand-alone solver can be used for the flow equation.

The flow of a nematic liquid crystal can be described in various ways. While the most common approach uses the Ericksen–Leslie theory for the nematic director, a more general description using the second rank alignment tensor is needed for problems that involve defects. Different constitutive theories for the alignment tensor have been derived [11, 16–18, 25, 26] , and numerical solutions for some special cases have been produced [6, 7, 28]. The creation of backflow and its influence on the annihilation of defects in two space dimensions has been examined in [27, 29]. In this example, the reorientation of the alignment is the driving force. Also the impact of flow on the orientation has received much attention. Possibly the earliest application was given by Leslie: the flow alignment of the director in a simple shear [13]. The behaviour of the alignment tensor under shear in a monodomain was also extensively studied in [8, 19], and other studies have considered lid-driven cavity flow [9, 14, 33].

Even in a homogeneous, simple shear flow, many different types of behaviour can be found, such as flow aligning, tumbling, and chaos. Furthermore, to obtain a complete picture, spatially inhomogeneous situations have to be considered. The evolution is determined by two equations: the flow is governed by a generalised Navier–Stokes equation, in which the divergence of the stress tensor also depends on the alignment tensor and its time derivative, and the evolution of the orientation is governed by a convection-diffusion type equation that contains non-linear terms that stem from a Landau-DeGennes potential [22].

In this paper we consider a specific model with three viscosity coefficients that allows us to write the contribution of the orientation to the viscous stress in the form of an orientation dependent force. As an example application, we consider the standard fluid flow test problem of Stokes flow in a lid-driven cavity. We propose a time-discretised strategy for solving the flow-orientation problem that involves two alternating steps. First, for a given flow field, one time step for the orientation equation is carried out according to the methods described in [20]. Then, the flow field of the Stokes flow is computed for the given orientation field. This is done using state-of-

Citation: Alison Ramage, and André M. Sonnet, Computational fluid dynamics for nematic liquid crystals, DOI: 10.1007/s10543-015-0586-5.

the-art Krylov subspace and multigrid iteration techniques implemented in IFISS [3].

GENERAL UNDERLYING EQUATIONS

We consider a nematic liquid crystal whose orientational order is described by the second rank alignment tensor Q. If u denotes a unit vector parallel to the symmetry axis of an effectively uniaxial molecule, Q can be defined as the local average

$$\mathbf{Q} := \langle \overline{\boldsymbol{u} \otimes \boldsymbol{u}} \rangle = \left\langle \boldsymbol{u} \otimes \boldsymbol{u} - \frac{1}{3}\mathbf{I} \right\rangle \tag{2.1}$$

where II is the identity tensor and $\overline{\ldots}$ denotes the symmetric traceless part of a tensor.

Equations of motion for incompressible flow and alignment can conveniently be formulated in terms of a frame-independent invariant rate of the alignment tensor [23]. Here we use the co-rotational time derivative

$$\overset{\circ}{\mathbf{Q}} = \dot{\mathbf{Q}} - 2\,\overline{\mathbf{WQ}} \tag{2.2}$$

where $\mathbf{W} = \frac{1}{2}(\nabla \boldsymbol{v} - (\nabla \boldsymbol{v})^T)$ is the skew part of the velocity gradient, with v satisfying the incompressibility constraint

$$\nabla \cdot \boldsymbol{v} = 0, \tag{2.3}$$

and $\dot{\mathbf{Q}} = \frac{\partial \mathbf{Q}}{\partial t} + (\nabla \mathbf{Q})\boldsymbol{v}$ is the material time derivative of Q. If the free energy connected with the alignment is given as a function $W = W(\mathbf{Q}, \nabla\mathbf{Q})$, the dissipation is specified as a function $R = R(\overset{\circ}{\mathbf{Q}}, \mathbf{Q}, \mathbf{D})$ that is a quadratic form in $\overset{\circ}{\mathbf{Q}}$, and the symmetric part of the velocity gradient is $\mathbf{D} = \frac{1}{2}(\nabla \boldsymbol{v} + (\nabla \boldsymbol{v})^T)$, then the equations for flow and alignment take the general form [22,25]

$$\rho\dot{\boldsymbol{v}} = \operatorname{div}\mathbf{T}$$

$$\frac{\partial W}{\partial \mathbf{Q}} - \operatorname{div}\frac{\partial W}{\partial \nabla \mathbf{Q}} + \frac{\partial R}{\partial \overset{\circ}{\mathbf{Q}}} = \mathbf{0} \tag{2.4}$$

where the stress tensor T is given by

$$\mathbf{T} = -p\mathbf{I} - \nabla\mathbf{Q} \odot \frac{\partial W}{\partial \nabla \mathbf{Q}} + \frac{\partial R}{\partial \mathbf{D}} + \mathbf{Q}\frac{\partial R}{\partial \overset{\circ}{\mathbf{Q}}} - \frac{\partial R}{\partial \overset{\circ}{\mathbf{Q}}}\mathbf{Q}. \tag{2.5}$$

This tensor contains an isotropic contribution from the hydrostatic pressure p, a viscous stress with symmetric part ∂R/∂D and skew part $\mathbf{Q}\,\partial R/\partial \overset{\circ}{\mathbf{Q}} - \partial R/\partial \overset{\circ}{\mathbf{Q}}\,\mathbf{Q}$, and an elastic stress

$$\left(\nabla\mathbf{Q} \odot \frac{\partial W}{\partial \nabla \mathbf{Q}}\right)_{ij} := Q_{kl,i}\frac{\partial W}{\partial Q_{kl,j}}$$

which is analogous to the Ericksen elastic stress in a director based description.

SPECIFIC MODEL

To obtain a specific model, we choose the free energy to be of the form

$$W = \phi + \frac{1}{2}L_1\|\nabla\mathbf{Q}\|^2, \tag{3.1}$$

where $\phi = \frac{1}{2}A(T)\mathrm{tr}\mathbf{Q}^2 - \frac{\sqrt{6}}{3}B\mathrm{tr}\mathbf{Q}^3 + \frac{1}{4}C(\mathrm{tr}\mathbf{Q}^2)^2$ is a Landau-deGennes potential, and a curvature elastic energy with one elastic constant L_1 is used. Although other models involving additional elastic constants exist (see, for example, [2]), we choose this commonly-used one-constant approximation for simplicity in the equations below. For an alignment tensor theory to be consistent with Ericksen–Leslie theory (in the case of uniaxial alignment with constant scalar order parameter), the dissipation function R needs to contain at least five terms. The choice

$$R = \frac{1}{2}\zeta_1\overset{\circ}{\mathbf{Q}} \cdot \overset{\circ}{\mathbf{Q}} + \zeta_2\mathbf{D} \cdot \overset{\circ}{\mathbf{Q}} + \frac{1}{2}\zeta_3\mathbf{D} \cdot \mathbf{D} + \frac{1}{2}\zeta_{31}\mathbf{D} \cdot (\mathbf{DQ}) + \frac{1}{2}\zeta_{32}(\mathbf{D} \cdot \mathbf{Q})^2 \tag{3.2}$$

with five phenomenological viscosity coefficients $\zeta_1, \zeta_2, \zeta_3, \zeta_{31}$, and ζ_{32} leads to the stress tensor proposed in [18]. Although more general forms for R are available (see [24], [25, eq. (4.23)]), omitting terms other than those in (3.2) simply amounts to neglecting higher-order corrections to the Ericksen–Leslie viscosity coefficients, so we retain the simpler form here. Using (3.1) and (3.2) in (2.4) yields the equation for the alignment

$$\zeta_1\overset{\circ}{\mathbf{Q}} = -\Phi - \zeta_2\mathbf{D} + L_1\Delta\mathbf{Q},$$

where Φ is the derivative ∂ϕ/∂Q of the Landau-deGennes potential ϕ. The different contributions to the stress tensor (2.5) then take the following explicit forms: the skew-symmetric part is

$$\mathbf{T}^{\mathrm{skew}} = \zeta_1(\mathbf{Q}\overset{\circ}{\mathbf{Q}} - \overset{\circ}{\mathbf{Q}}\mathbf{Q}) + \zeta_2(\mathbf{QD} - \mathbf{DQ}) \tag{3.3}$$

and the symmetric traceless part of the viscous stress is

$$\overline{\mathbf{T}^{(v)}} = \zeta_2\overset{\circ}{\mathbf{Q}} + \zeta_3\mathbf{D} + \zeta_{31}\overline{\mathbf{DQ}} + \zeta_{32}(\mathbf{Q} \cdot \mathbf{D})\mathbf{Q}. \tag{3.4}$$

In the one elastic constant approximation used here in (3.1), the elastic contribution to the stress is symmetric and given by

$$\mathbf{T}^{(e)} = -L_1\,\nabla\mathbf{Q} \odot \nabla\mathbf{Q}. \tag{3.5}$$

SOLUTION STRATEGY

We begin by writing the stress tensor in a more convenient form, namely, we remove its explicit dependence on the time derivative of the alignment tensor. To this end, we observe that on a solution

$$\overset{\circ}{\mathbf{Q}} = \frac{1}{\zeta_1}(-\Phi - \zeta_2\mathbf{D} + L_1\Delta\mathbf{Q}). \tag{4.1}$$

Using this in expression (3.3) for the skew part of the viscous stress, we find that

$$\begin{aligned}\mathbf{T}^{\mathrm{skew}} &= \zeta_1(\mathbf{Q}\overset{\circ}{\mathbf{Q}} - \overset{\circ}{\mathbf{Q}}\mathbf{Q}) + \zeta_2(\mathbf{QD} - \mathbf{DQ}) \\ &= \Phi\mathbf{Q} - \mathbf{Q}\Phi + L_1[\mathbf{Q}(\Delta\mathbf{Q}) - (\Delta\mathbf{Q})\mathbf{Q}] \\ &= L_1[\mathbf{Q}(\Delta\mathbf{Q}) - (\Delta\mathbf{Q})\mathbf{Q}],\end{aligned} \tag{4.2}$$

where the last equality holds because Φ is simply a polynomial in Q and hence commutes with Q. Applying the same procedure to the symmetric part of the viscous stress yields

$$\overline{\mathbf{T}^{(v)}} = \frac{\zeta_2}{\zeta_1}(L_1\Delta\mathbf{Q} - \Phi) + \zeta_4\mathbf{D} + \zeta_{31}\overline{\mathbf{DQ}} + \zeta_{32}(\mathbf{Q}\cdot\mathbf{D})\mathbf{Q} \tag{4.3}$$

where we have introduced a renormalised isotropic viscosity ζ_4 according to $\zeta_4 := \zeta_3 - \zeta_2^2/\zeta_1$.

From now on, we will neglect the last two terms in (4.3), that is, we will assume that $\zeta_{31} = \zeta_{32} = 0$. In terms of the Leslie viscosities, this amounts to making the assumptions $\alpha_1 = 0$ and $\alpha_5 = -\alpha_6$, see [22]. We note that while these assumptions are reasonable for small molecule liquid crystals, they will have to be modified for polymeric liquid crystals (see Sect. 7). The advantage of making these assumptions is that with $\zeta_{31} = \zeta_{32} = 0$, the divergence of the stress tensor takes a very convenient form, namely,

$$\operatorname{div}\mathbf{T} = -\nabla p + \frac{1}{2}\zeta_4\Delta\boldsymbol{v} + \operatorname{div}\mathbf{F} \tag{4.4}$$

with

$$\mathbf{F} = L_1\left(\mathbf{Q}(\Delta\mathbf{Q}) - (\Delta\mathbf{Q})\mathbf{Q} + \frac{\zeta_2}{\zeta_1}\Delta\mathbf{Q} - \nabla\mathbf{Q}\odot\nabla\mathbf{Q}\right) - \frac{\zeta_2}{\zeta_1}\Phi. \tag{4.5}$$

We non-dimensionalise by expressing all lengths in terms of the nematic coherence length $\xi = \sqrt{9CL_1/(2B^2)}$ and all times in terms of the relaxation time $\tau_1 = 9C\zeta_1/(2B^2)$. In addition, we rescale the alignment tensor according to $\tilde{\mathbf{Q}} = 3C/(2B)\,\mathbf{Q}$. This leads to the dimensionless Landau-deGennes potential

$$\tilde{\Phi} = (\vartheta + 2\operatorname{tr}\tilde{\mathbf{Q}}^2)\tilde{\mathbf{Q}} + 3\sqrt{6}\,\overline{\tilde{\mathbf{Q}}\tilde{\mathbf{Q}}}, \tag{4.6}$$

where $\vartheta = 9C/(2B^2)\,A(T)$ is a dimensionless temperature parameter. In these units, the clearing point T_c and the pseudo critical temperature T^* correspond to $\vartheta = 1$ and $\vartheta = 0$, respectively [12]. Note that, for convenience, the tildes are dropped in all subsequent formulae.

The final dimensionless equations then are

$$\overset{\circ}{\mathbf{Q}} = \Delta\mathbf{Q} - \Phi - \mathrm{Tu}\,\mathbf{D} \tag{4.7}$$

for the orientation and

$$\rho\dot{\boldsymbol{v}} = -\nabla p + \Delta\boldsymbol{v} + \operatorname{div}\mathbf{F} \tag{4.8}$$

with

$$\mathbf{F} = \mathrm{Bf}\left\{\frac{1}{\mathrm{Tu}}[\mathbf{Q}(\Delta\mathbf{Q}) - (\Delta\mathbf{Q})\mathbf{Q} - \nabla\mathbf{Q}\odot\nabla\mathbf{Q}] + \Delta\mathbf{Q} - \boldsymbol{\Phi}\right\} \tag{4.9}$$

for the flow, together with the incompressibility constraint (2.3). Note that here we have introduced two dimensionless parameters: the backflow parameter, $\mathrm{Bf} = 4B\zeta_2/(3C\zeta_4)$, measures the impact of the orientation on the flow, and the tumbling parameter, $\mathrm{Tu} = 3C\zeta_2/(2B\zeta_1)$, measures the relative strength of the viscosities ζ_2 and ζ_1. In a simple shear one can expect flow alignment for $\mathrm{Tu} > 1$, where the liquid crystal aligns at an angle of $\cos 2\phi_a = -1/\mathrm{Tu}$ to the direction of the flow gradient [13]. For values of $\mathrm{Tu} < 1$ some dynamic state such as tumbling should prevail.

IMPLEMENTATION AND SOLVER DETAILS

In what follows, we will assume flow at a low Reynolds number, that is, we assume that flow inertia can be neglected so $\dot{\boldsymbol{v}} = 0$. Equation (4.8) is then simply a Stokes equation with a force equal to the divergence of the tensor F in (4.9) that depends only on Q and its spatial derivatives. This suggests the following iterative solution strategy:

Algorithm: Coupled flow-orientation algorithm

1. Calculate an initial orientation field **Q**.
2. Solve the Stokes equation (4.8) and incompressibility constraint (2.3) with div **F** as a force [for **F** in (4.9)].
3. Use the obtained flow field $\boldsymbol{v}$ to compute one time step in a discretised version of the orientation equation (4.7).
4. With the new orientation field **Q**, go back to step 2.

Note that, within this framework, any two stand-alone solvers (one for the orientation equations and one for the Stokes equation) can be used. That is, the algorithm structure is independent of the discretisation methods used within each solver, and specific details of the underlying flow and orientation problems (such as shape of the domain and boundary conditions). Furthermore, changing the form of the elastic energy in (3.1) would change only the right-hand side of the flow problem (through F) and would not affect the Stokes iteration matrix (see below).

For the orientation solver, we note that a symmetric, traceless second-rank tensor has five independent degrees of freedom. Once Q is expressed in terms of a suitable basis [20,21], the orientation equation (4.7) takes the form of five coupled non-linear partial differential equations. In the numerical experiments which follow, these were discretised in space using finite differences on a uniform grid and in time using an explicit Euler method. Although stability considerations mean that the size of timestep which can be used is limited with an explicit method, that is not a concern here as small time-steps are already needed for accuracy in terms of modelling flow evolution. Also, the complexity and computational expense of implementing a matrix-based nonlinear iterative solver for the system of five coupled equations in (4.7) makes a fully implicit approach impractical.

As highlighted in the introduction, the Stokes equation (4.8) was solved at each time-step using a finite element based iterative solver adapted from the public domain MATLAB package IFISS [3,4]. The particular finite element discretisation used was a $Q_2 - Q1$ Taylor-Hood mixed approximation (that is, quadratic elements for velocity and linear elements for pressure). The resulting linear equations take the form of a saddle point system

$$\begin{bmatrix} A & B^T \\ B & 0 \end{bmatrix}\begin{bmatrix} \mathbf{u} \\ \mathbf{p} \end{bmatrix} = \begin{bmatrix} \mathbf{f} \\ \mathbf{g} \end{bmatrix} \equiv \mathcal{A}\bar{u} = \bar{f} \tag{5.1}$$

for the vectors of velocity and pressure unknowns, u and p respectively (see, for example, [5, §3.3] for more details). Here (5.1) was solved using one of the state-ofthe-art preconditioned MINRES solvers from IFISS. Specifically, a block diagonal preconditioner of the form

$$\mathcal{M} = \begin{bmatrix} P & 0 \\ 0 & S \end{bmatrix} \quad (5.2)$$

was used (with preconditioned coefficient matrix equivalent to $M^{-1}A$). With this form of preconditioner, choosing P = A and $S = B\,A^{-1}B^T$ (the so-called Schur complement of the saddle-point problem) results in a MINRES solver which converges in three iterations [15]. This is clearly not practical for realistic problems, because it involves explicitly inverting A and the Schur complement several times. However, it suggests that choosing P and S to be approximations to A and the Schur complement which are cheap to invert will result in an effective preconditioner M.

For P, any good preconditioner for the Laplacian A can be used. For the Schur complement approximation, we use $S = M_p$, where M_P is the finite element mass matrix corresponding to the pressure, which is spectrally equivalent to the Schur complement. Note that, although we invert M_P explicitly in the experiments below, if necessary the action of M_P^{-1} can be effectively approximated using a small number of steps of Chebyshev iteration (see [5, Remark 4.5], [31]).

In the numerical experiments of §6, we show results obtained using three different preconditioners of the form (5.2):

- *Diagonal preconditioning (DP):* P = diag(A), S = diag(M_P). This basic preconditioner should offer a modest reduction in iteration counts but is clearly very cheap to invert.
- *Ideal preconditioning (IP):* P = A, S = M_p. This represents the best possible preconditioner of this form, as A is inverted exactly. It can be shown that the eigenvalues of $M^{-1}A$ lie in small intervals that are uniformly bounded away from $\pm\infty$ and the origin, meaning that MINRES will converge rapidly and in a number of iterations which is independent of the discrete problem size [30].
- *MG preconditioning (MGP):* P = mg(A), S = M_p. Here we apply geometric multigrid (denoted by mg) to the Laplacian component. This involves one Vcycle with two directional sweeps (left→right, bottom→top) of line Gauss-Seidel iteration as smoother (see, for example, [32]).

Having access to good flow solvers is a key ingredient of our approach, as efficient solution of system (5.1) is critical to the overall practicality of the coupled floworientation algorithm.

NUMERICAL EXPERIMENTS

To illustrate how the coupled flow-orientation algorithm in §5 can be applied in practice, in this section we present the results of some numerical experiments on a lid-driven cavity problem [1]. The lid-driven cavity is a classic test problem in fluid dynamics where flow in a square cavity is driven by the lid moving from left to right, see Fig. 1. The flow boundary conditions are of Dirichlet type everywhere, with the velocity fixed at some positive rate in the x-direction on the lid and zero along all other cavity walls. Here we use a 'watertight' cavity, that is, the velocity is fixed to be zero at the top corner points on both left and right boundaries. The resulting discontinuous horizontal velocity generates a strong singularity in the pressure solution, but away from these corners the pressure is essentially constant.

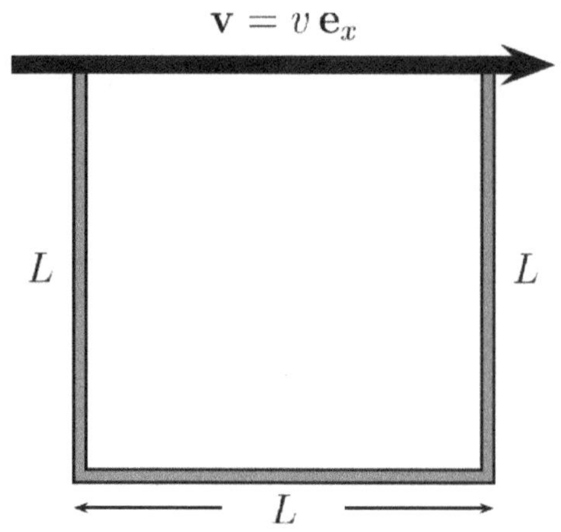

Figure 1: Specification of lid-driven cavity problem

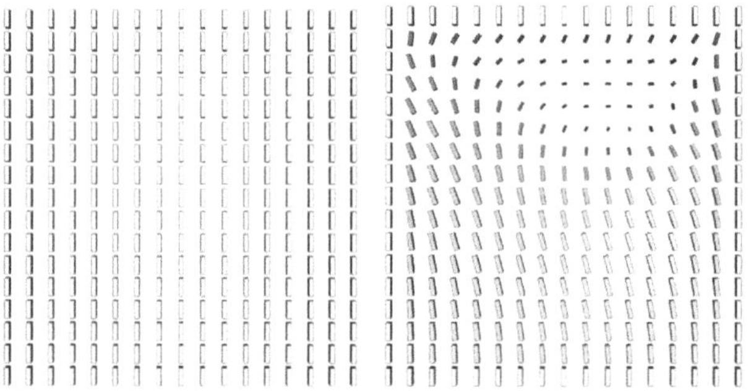

Figure 2: In-plane orientation. The initial homogeneous orientation with fixed boundary conditions is shown on the left. The non-homogeneous alignment field caused by the moving lid (on the right) shows regions of flow alignment in the lower part of the cavity and a periodic tumbling alignment close to the lid. There, the scalar order parameter is reduced significantly, as is visible from the smaller size of the boxes

Dirichlet boundary conditions are also used for the alignment tensor. The same uniaxial alignment with equilibrium order parameter is prescribed at all boundaries and also as an initial condition in the bulk. In this way, without driving flow a homogeneous uniaxial orientation, as given by the initial condition (see the left of Fig. 2), would result. As mentioned above, alternative problem domains and boundary conditions could be implemented directly in the flow and orientation solvers.

In-plane orientation

For a pure in-plane evolution, we used lid velocity v = 10 and cavity length L = 8. This corresponds to a Reynolds number of Re = V $L\rho/\zeta_4 \approx 10^{-5}$ for a typical small-molecule liquid crystal. The Ericksen number is then Er = $\zeta_1 V\, L/L_1 \approx 80$, and we chose Bf = 2/3 and Tu = 1/5. The temperature was chosen as □ = 0, corresponding to the pseudo-critical temperature T^*. The time-step used in the explicit Euler method (for the orientation equations) was $\Delta t = 0.0001$. This ensures stability of the method for the range of spatial discretisation parameters used (from h = 1/16 to h = 1/256 for the experiments reported on below). For orientation boundary conditions, we used homeotropic anchoring on the top and bottom of the cavity and planar anchoring on the lateral sides. The initial orientation is shown on the left of Fig. 2. The boxes shown lie parallel to the eigensystem of the alignment tensor, and the lengths of the edges correspond to the respective eigenvalues, see [20]. The shading of the box shows its degree of biaxiality: a white box corresponds to uniaxial alignment, where two eigenvalues are equal (such as in the initial configuration), and a black box corresponds to perfectly biaxial alignment, where one eigenvalue is zero. In general, the level of darkness of a particular box is proportional to the biaxiality measure $\beta^2 = 1 - 6(\mathrm{tr}\,\mathbf{Q}^3)^2/(\mathrm{tr}\,\mathbf{Q}^2)^3$ used in [12]. Note that the number of boxes plotted has been chosen for clear representation of the solutions, and does not correspond to the number of degrees of freedom used in the calculations.

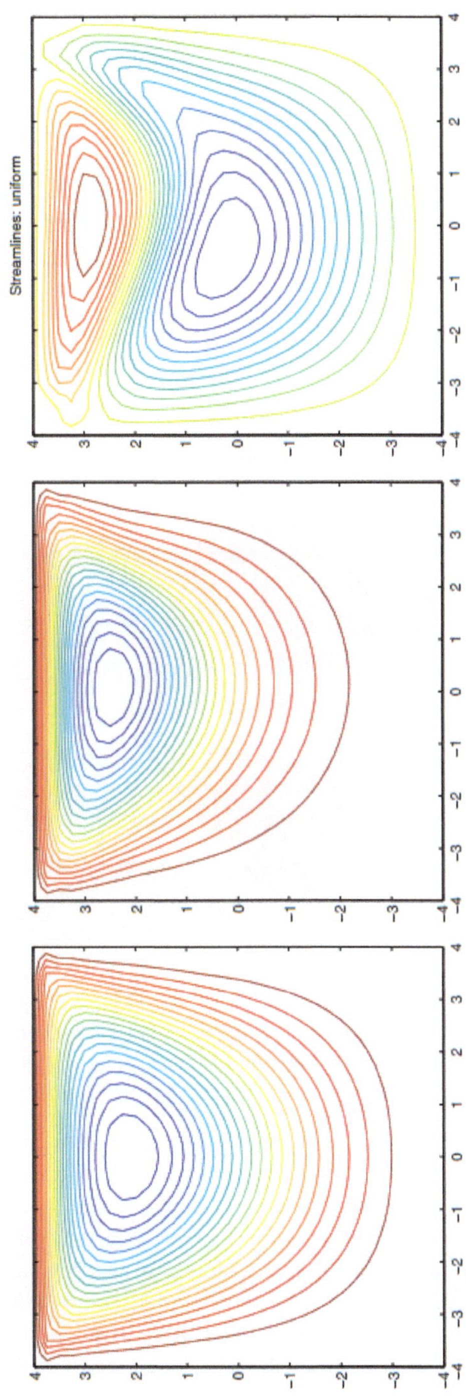

Figure 3: Flow field during the evolution. The left picture shows the streamlines of the flow field for the initial homogeneous configuration: they are symmetric with respect to a vertical axis through the centre of the cavity. The picture in the middle shows the streamlines at a later time, which are no longer symmetric but, due to the changes in the orientation field, are shifted slightly to the right. The right picture shows a contour plot of the difference between the two flow fields

The right picture in Fig. 2 shows a snapshot of the alignment field after the flow has developed. The evolution displayed shows two distinct types of orientation. On the one hand, in the lower part of the cavity, the orientation is dominated by the elastic forces and a stationary state of aligned flow is found. On the other hand, close to the lid where the velocity gradient is large, a periodic solution of in-plane tumbling orientation is found. Furthermore, because of the fixed boundary conditions, the orientation necessarily shows defects. In the alignment tensor description, these defects are characterised by a planar uniaxial orientation. They are generated close to the upper right corner of the lid and travel towards the centre of the cavity and from there to the upper left corner.

For the given choice of the parameters, the flow field is only slightly affected by the orientation (see Fig. 3). Initially, with a homogeneous orientation, the stream lines are symmetric about a vertical axis through the centre of the cavity. This reflects the time-reversal symmetry of the Stokes equation. When the orientation is no longer homogeneous, however, this symmetry is broken and the streamlines shift to the right. This is an effect similar to that found in isotropic fluids at high Reynolds numbers. It is found here in a linear flow equation because of the influence of the orientation on the flow.

To illustrate the efficiency of the flow solver, in Table 1 we present a summary of the performance of the three preconditioners discussed in §5 (as compared to results with unpreconditioned MINRES, which are in the column labelled 'none').

For each method, two quantities are tabulated: k is the average number of MINRES iterations required at each time-step to compute the flow field (with convergence tolerance 0.0001), and s is the amount of time associated with this computation (in seconds, as calculated using the MATLAB commands tic and toc). In both cases, the results have been averaged over the first 200 time-steps, as this initial phase poses the greatest challenge for the flow solver. As expected, the number of MINRES iterations required with no preconditioning grows with the problem size. Although using diagonal preconditioning (DP) reduces the iteration count slightly, it can be seen by comparing the values of s that the expense involved in constructing the preconditioner outweighs its benefits as h decreases. For ideal preconditioning, we see that k is, as predicted by the theory in [30], essentially independent of the discretisation parameter h, although the cost of explicitly inverting A grows rapidly. When this inversion is avoided by replacing the action of A^{-1} by using one multigrid V-cycle based on A, as in MGP, there is a slight growth in the number of iterations needed but, crucially, the method is still essentially grid-independent, and at a much reduced cost. In this framework, MGP is clearly extremely efficient, as is necessary for the overall coupled flow-orientation algorithm to be practical.

Table 1: A comparison of iteration counts and times for various preconditioners averaged over the first 200 time-steps

h	None		DP		IP		MGP	
	k	s	k	s	k	s	k	s
1/16	46.6	4.61e−3	15.4	3.49e−3	4.0	1.11e−2	6.1	8.50e−3
1/32	97.8	1.80e−2	33.9	1.16e−2	5.8	6.10e−2	7.2	1.43e−2
1/64	141.1	8.85e−2	60.5	5.50e−2	5.4	2.42e−1	6.9	3.28e−2
1/128	176.2	4.69e−1	101.0	3.68e−1	4.7	1.18e+0	5.6	9.37e−2
1/256	219.8	1.87e+0	161.9	2.08e+0	3.4	4.17e+0	4.2	2.70e−1

Out-of-plane orientation

To obtain an out-of-plane evolution, both the boundary and initial conditions were tilted by an angle of 15◦ out of the shear plane. The initial orientation is again homogeneous; with respect to the in-plane orientation on the left of Fig. 2, the top of the alignment tensor is simply tilted by 15◦ out of the plane towards the observer. Here we used v = 15 and L = 16, which corresponds to a Reynolds number of 3×10^{-5} and an Ericksen number of 240. As before, Bf = 2/3 and $\vartheta = 0$, but this time we chose Tu = 4/5 to facilitate the occurrence of out-of-plane periodic solutions (see, for example, the phase diagrams for monodomains in [8]). The time-step used for the orientation solver was t = 0.0001 as before.

The resulting evolution (as illustrated by the snapshot on the left of Fig. 4) again shows in the lower part of the cavity a region of flow alignment that here is out of the plane. Close to the lid, periodic kayaking is found. A close-up view as seen from the top right of the cavity is shown on the right of Fig. 4, where this periodic out-of-plane behaviour is clearly visible. This is again accompanied by the creation of defects in the upper right corner and their annihilation in the upper left corner. A notable difference from the in-plane evolution, however, is that the reduction of the scalar order parameter is far less pronounced. Here, it takes place mostly around the defects: the orientation can go out of the plane to avoid the frustration induced by the flow gradient.

When the orientation has components that lie out of the plane, the force density div F that is due to the orientation-related contributions to the stress can have a component that lies out of the plane even when both flow field and orientation field are assumed to be homogeneous in that direction. This out-of-plane component of div F is shown in Fig. 5. It is most noticeable at the corners where the pressure is divergent, but it is present throughout the cavity.

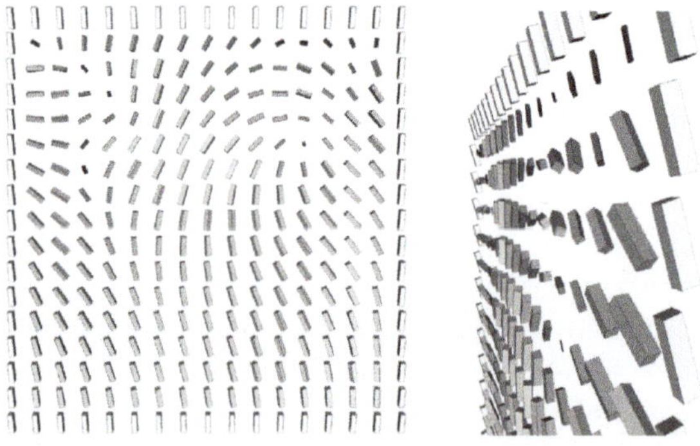

Figure 4: Out-of-plane orientation. On the left, in the lower part of the cavity the orientation is again one of flow alignment, but here it is out of the plane. Close to the lid, periodic kayaking is found and, as in the in-plane case, defects are created in the upper right corner and eventually annihilate in the upper left corner. The reduction of order is considerably less pronounced than for the in-plane case. A close-up view as seen from the top right of the cavity is shown on the right

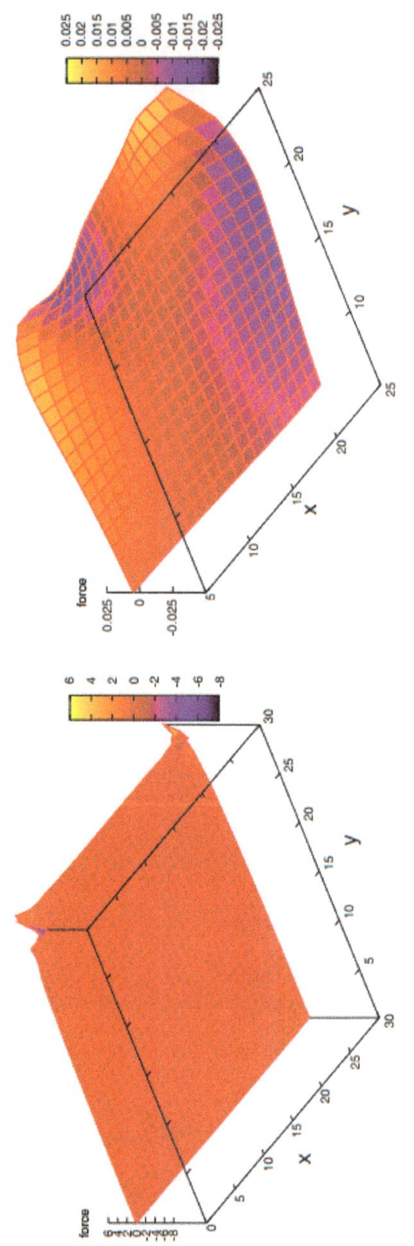

Figure 5: Component of the force div F perpendicular to the shear plane. This force component is particularly large at the corners where the pressure is divergent (left) but it is present throughout the cavity, as evidenced by the close-up of the central region (right)

This shows that, for out-of-plane evolutions, truly three-dimensional flow fields will arise and that two-dimensional computations are therefore only of limited value in this context. The relative performance of the various preconditioned Stokes solvers discussed in §5 is very similar to that for the in plane orientation example of §6.1, so iteration counts and timings have not been displayed here. The MGP preconditioner again significantly outperformed the other methods.

CONCLUSIONS

In this paper we have described a highly efficient algorithm for the computation of flow and orientation in nematic liquid crystals. Writing the influence of the orientation field on the flow in the form of a force density allows us to solve the flow equation by using well established fast solvers. This aspect of the modelling dominates the computational time required, so that the overhead added to the computational fluid dynamics by the anisotropic liquid is rather small. We note that, although here we have focussed on the use of a fast solver from the IFISS package, other existing software could also be used.

One disadvantage of the method that we have presented is that only three viscosity coefficients enter the viscous stress. However, when the co-rotational time derivative that we have used here is replaced by a general co-deformational time derivative the same numerical procedure as before can be employed. As long as only the terms proportional to ζ_1, ζ_2, and ζ_3 are considered in the dissipation function, the influence of the orientation on the flow still takes the form of a force density This makes the type of algorithm presented in this paper suitable for a more general class of materials, such as polymeric liquid crystals.

$$\mathring{\mathbf{Q}} = \dot{\mathbf{Q}} - 2\overline{\mathbf{WQ}} - 2\sigma\,\overline{\mathbf{DQ}},$$

Generalisation to high Reynolds numbers is also straightforward: it requires the retention of the inertial term $\rho v^{\cdot}$ on the left hand side of (4.8) and solution of the resulting Navier–Stokes equation with a specified force term. At each time-step, the latter leads to a saddle-point system of a form similar to (5.1) but with the diffusion component replaced by a discrete convection-diffusion operator. Such systems could be solved efficiently using advanced preconditioned iterative techniques for finite element Navier–Stokes approximations, such as the pressure convection-diffusion and least-squares commutator preconditioners described in [5] and implemented in IFISS.

REFERENCES

1. Burggraf, R.: Analytical and numerical studies of the structures of steady separated flows. J. Fluid Mech. 24, 113–151 (1966)
2. de Gennes, P.G., Prost, J.: The Physics of Liquid Crystals, 2nd edn. Clarendon Press, Oxford (1993)
3. Elman, H.C., Ramage, A., Silvester, D.J.: Algorithm 886: IFISS, Incompressible Flow & Iterative Solver Software. ACM T. Math. Softw. 33(2) (2007)
4. Elman, H.C., Ramage, A., Silvester, D.J.: IFISS: a computational laboratory for investigating incompressible flow problems. SIAM Rev. 56(2), 261–273 (2014)
5. Elman, H.C., Silvester, D.J., Wathen, A.J.: Finite Elements and Fast Iterative Solvers with applications in incompressible fluid dynamics, 2nd edn. Oxford University Press, Oxford (2014)
6. Farhoudi, Y., Rey, A.D.: Shear flows of nematic polymers. I. Orienting modes, bifurcations, and steady state rheological predictions. J. Rheol. 37, 289–305 (1993)
7. Grecov, D., Rey, A.D.: Texture control strategies for flow-aligning liquid crystal polymers. J. Non-Newton. Fluid Mech. 139, 197–208 (2006)
8. Grosso, M., Maffettone, P.L., Dupret, F.: A closure approximation for nematic liquid crystals based on the canonical distribution subspace theory. Rheol. Acta 39, 301–310 (2000)
9. Hernández-Ortiz, J.A., Gettelfinger, B.T., Moreno-Razo, J., de Pablo, J.J.: Modeling flows of confined nematic liquid crystals. J. Chem. Phys. 134 (2011)
10. Heroux, M.A., Bartlett, R.A., Howle, V.E., Hoekstra, R.J., Hu, J.J., Kolda, T.G., Lehoucq, R.B., Long, K.R., Pawlowski, R.P., Phipps, E.T., Salinger, A.G., Thornquist, H.K., Tuminaro, R.S., Willenbring, J.M., Williams, A., Stanley, K.S.: An overview of the trilinos project. ACM Trans. Math. Softw. 31(3), 397–423 (2005)
11. Hess, S.: Irreversible thermodynamics of nonequilibrium alignment phenomena in molecular liquids and in liquid crystals. Z. Naturforsch. 30a, 728–738 & 1224–1232 (1975)
12. Kaiser, P., Wiese, W., Hess, S.: Stability and instability of an uniaxial alignment against biaxial distortions in the isotropic and nematic phases of liquid crystals. J. Non-Equilib. Thermodyn. 17, 153–169 (1992)
13. Leslie, F.M.: Some constitutive equations for liquid crystals. Arch. Ration. Mech. Anal. 28, 265–283 (1968)
14. MathSciNetCrossRefMATH
15. Liu, P.: Simulations of nematic liquid crystals: Shear flow, driven-cavity and 2-phase (isotropic-nematic) step flow. Ph.D. thesis, Brown University, Providence (2012)
16. Murphy, M.F., Golub, G.H., Wathen, A.J.: A note on preconditioning for indefinite linear systems. SIAM J. Sci. Comput. 21(6), 1969–1972 (2000)
17. Olmsted, P.D., Goldbart, P.: Theory of the nonequilibrium phase transition for nematic liquid crystals under shear flow. Phys. Rev. A 41, 4578–4581 (1990)
18. Pereira Borgmeyer, C., Hess, S.: Unified description of the flow alignment and viscosity in the isotropic and nematic phases of liquid crystals. J. Non-Equilib. Thermodyn. 20, 359–384 (1995)
19. Qian, T., Sheng, P.: Generalized hydrodynamic equations for nematic liquid crystals. Phys. Rev. E 58, 7475–7485 (1998)
20. Rienäcker, G., Hess, S.: Orientational dynamics of nematic liquid crystals under shear flow. Phys. A 267, 294–321 (1999)
21. Sonnet, A., Kilian, A., Hess, S.: Alignment tensor versus director: Description of defects in nematic liquid crystals. Phys. Rev. E 52(1), 718–722 (1995)
22. Sonnet, A.M.: Numerical methods for calculating the orientational dynamics of chiral and ordinary nematic liquid crystals. Ph.D. thesis, Technische Universität Berlin, Berlin (1996). ISBN 3-89685-405-4 (in German)
23. Sonnet, A.M., Maffettone, P.L., Virga, E.G.: Continuum theory for nematic liquid crystals with tensorial order. J. Non-Newton. Fluid Mech. 119, 51–59 (2004)
24. Sonnet, A.M., Virga, E.G.: Dynamics of dissipative ordered fluids. Phys. Rev. E 64(3), 031,705 (2001)
25. Sonnet, A., Maffettone, P.L., Virga, E.G.: Continuum theory for nematic liquid crystals with tensorial order. J.

Non-Newton. Fluid mech. 119, 51–59 (2004)

26. Sonnet, A.M., Virga, E.G.: Dissipative Ordered Fluids. Springer, Berlin (2010)
27. Stark, H., Lubensky, T.: Poisson-bracket approach to the dynamics of nematic liquid crystals. Phys. Rev. E 67 (2003)
28. Svenšek, D., Žumer, S.: Hydrodynamics of pair-annihilating disclination lines in nematic liquid crystals. Phys. Rev. E 66, 021,712 (2002)
29. Tiribocchi, A., Henrich, O., Lintuvuoria, J.S., Marenduzzo, D.: Switching hydrodynamics in liquid crystal devices: a simulation perspective. Soft Matter 10, 4580–4592 (2014)
30. Tóth, G., Denniston, C., Yeomans, J.M.: Hydrodynamics of topological defects in nematic liquid crystals. Phys. Rev. Lett. **88**(10), 105–504 (2002)
31. Wathen, A.J., Silvester, D.: Fast iterative solution of stabilised stokes systems. part i: Using simple diagonal preconditioners. SIAM J. Numer. Anal. 30(3), 630–649 (1993)
32. MathSciNetCrossRefMATH
33. Wathen, A.J., Rees, T.: Chebyshev semi-iteration in preconditioning for problems including the mass matrix. Electron. Trans. Numer. Anal. 34, 125–135 (2009)
34. Wesseling, P.: An Introduction to Multigrid Methods. Wiley, New York (1992)
35. Yang, X., Forest, G., Mullins, W., Wang, Q.: 2-D lid-driven cavity flow of nematic polymers: an unsteady sea of defects. Soft Matter 6, 1138–1166 (2010)

CHAPTER 4

NUMERICAL SIMULATIONS OF MAGNETOACOUSTIC–GRAVITY WAVES IN THE SOLAR ATMOSPHERE

K. Murawski, A. K. Srivastava, J. A. McLaughlin, and R. Oliver

Keywords: Waves Magnetic fields Corona Granulation

INTRODUCTION

The complicated magnetic-field configuration of the Sun plays a key role in various types of dynamical plasma processes in its atmosphere, including all significant plasma dynamics of the lower solar atmosphere. The resulting magnetic structures channel energy from the photosphere into the upper atmosphere in the form of magnetohydrodynamic (MHD) waves. These waves experience mode conversion, resonances, trapping, and reflection, which result in the complicated dynamical processes in the lower solar atmosphere, the details of which depend on the plasma properties as well as on the strength of the magnetic field. The complex magnetic-field and plasma structuring in the lower solar atmosphere support the excitation of various kinds of MHD waves. Their propagation, reflection, and trapping have been extensively studied theoretically and observationally (e.g. McAteer et al.2003; Hasan et al.2005; Srivastavaet al.2008; Fedun, Erdélyi, and Shelyag 2009; Srivastava 2010; Murawski, Srivastava, and Zaqarashvili 2011; Gruszecki et al.2011; and references therein). The evolving magnetic fields of the lower solar atmosphere also lead to transient processes across a wide range of spatial–temporal scales in the form of eruption and associated phenomena. For example, various types of solar jets are formed at short spatial–temporal scales, which play a significant role in mass and energy transport and also couple the various layers of the solar atmosphere (Shibata et al.2007; Katsukawa et al.2007; Srivastava and Murawski 2011; and references therein). In addition, the magnetic activity and injections of helicity into the lower solar atmosphere result in large-scale eruptive phenomena, including solar flares and coronal mass ejections (CMEs) in the outer part of the magnetized solar atmosphere (Srivastava et al.2010; Shibata and Magara2011; Zhang, Cheng, and Ding 2012; and references therein). All this makes the coupling of the complex magnetic field in various layers of the Sun, caused by waves and transients, one of the most significant areas of contemporary solar research.

In the quiet-Sun magnetic networks, cavities are important locations where the magnetic fields are sufficiently inclined due to their well-evolved horizontal components. The cavities are formed over the granular cells in the form of field-free regions through the transport of plasma at their boundaries, and are overlaid by bipolar magnetic canopies. The vertical magnetic fields, however, reside mostly in the core of these magnetic networks (Schrijver and Title 2003; Centeno et al.2007). The magnetic cavity–canopy systems are thought to be ideal resonators of the various MHD waves that can be trapped in the cavity, and can also leak upward through the core of these magnetic networks in the form of magnetoacoustic–gravity waves. It is thought that the field-free cavity regions underlying the bipolar canopy can trap the high-frequency acoustic oscillations, and the low-frequency components may leak into the higher atmosphere in the form of magnetoacoustic–gravity waves (Kuridze et al.2008; Srivastava et al.2008; Srivastava 2010). Therefore, these magnetic structures in the lower solar atmosphere may play an important role in wave filtering (McIntosh and Judge 2001; Krijger et al.2001; McAteer et al.2002; Vecchio et al.2007; Vecchio, Cauzzi, and Reardon 2009; Srivastava 2010; and references therein).

In addition to recent high-resolution observations of MHD waves in the lower solar atmosphere, extensive efforts have been made in the area of analytical and numerical modeling of these waves: Fedun, Erdélyi, and Shelyag (2009) have

Citation: K. Murawski, A.K. Srivastava, J.A. McLaughlin, and R. Oliver, Numerical Simulations of Magnetoacoustic–Gravity Waves in the Solar Atmosphere, DOI 10.1007/s11207-012-0202-7.

investigated the 3D numerical modeling of the coupled slow and fast magnetoacoustic wave propagation in the lower solar atmosphere; and recently Fedun et al. (2011) have reported the first numerical results of the frequency filtering of torsional Alfvén waves in the chromosphere. In addition to general numerical modeling of the waves in the lower solar atmosphere, models have also investigated the acoustic-wave spectrum in the localized magnetic structures of the lower solar atmosphere, e.g. the magnetic cavity–canopy system (Kuridze et al.2008; Srivastava et al.2008; Kuridze et al.2009; and references cited therein).

It is also noteworthy that Bogdan et al. (2003), Fedun, Erdélyi, and Shelyag (2009), and Fedunet al. (2011), discussed in detail the excitation, propagation, and conversion of magnetoacoustic waves in a realistic 3D MHD simulation. However, in these references the waves driven by a periodic driver were discussed, whereas we numerically simulate the excitation of magnetoacoustic–gravity waves generated by pulses in the gas pressure and the vertical velocity component, which mimics an isolated solar granule. We aim to investigate and understand this simpler (but still complex enough) system before we move on to the more realistic, multiple-granule system. Our philosophy is to build up our models incrementally, with a clear focus on the underlying physical processes at each step.

We here investigate the excitation of magnetoacoustic–gravity waves generated from localized pulses in the gas pressure as well as in the vertical velocity component by modeling the effect of an isolated solar granule. These pulses are initially launched at the top of the solar photosphere, which is permeated by a weak magnetic field. We investigate three different configurations of the background magnetic-field lines: vertical, horizontal, and oblique to the gravitational force. We aim to show that small-amplitude perturbations that are associated with such a granule are able to trigger large-amplitude, complicated oscillations in the solar corona, which exhibit periodicities within the detected range of three to five minutes.

The structure of the article is as follows: In Section 2 we describe the numerical model. We report the numerical results in Section 3 and present the discussion and conclusions in Section 4.

NUMERICAL MODEL

We consider a gravitationally stratified solar atmosphere that is described by the ideal two-dimensional (2D) MHD equations:

$$\frac{\partial \varrho}{\partial t} + \nabla \cdot (\varrho \mathbf{V}) = 0, \tag{1}$$

$$\varrho \frac{\partial \mathbf{V}}{\partial t} + \varrho (\mathbf{V} \cdot \nabla)\mathbf{V} = -\nabla p + \frac{1}{\mu}(\nabla \times \mathbf{B}) \times \mathbf{B} + \varrho \mathbf{g}, \tag{2}$$

$$\frac{\partial p}{\partial t} + \nabla \cdot (p\mathbf{V}) = (1-\gamma) p \nabla \cdot \mathbf{V}, \tag{3}$$

$$\frac{\partial \mathbf{B}}{\partial t} = \nabla \times (\mathbf{V} \times \mathbf{B}), \qquad \nabla \cdot \mathbf{B} = 0. \tag{4}$$

Here ϱ is mass density, **V** is the flow velocity, **B** is the magnetic field, $p=k_B \varrho T/m$ is the gas pressure, T is the temperature, $\gamma=5/3$ is the adiabatic index, $\mathbf{g}=(0,-g,0)$ is the gravitational acceleration, where $g=274$ m s^{-2}, m is the mean particle mass, and k_B is Boltzmann's constant. Throughout this article, we use the Cartesian coordinate system with the vertical axis denoted by y and the horizontal axis x. Henceforth, we assume that the medium is invariant along the z-direction with $\partial/\partial z=0$ and set the z-components of plasma velocity and magnetic field equal to zero, i.e.$V_z=0$ and $B_z=0$. This assumption removes Alfvén waves from the system but still allows magnetoacoustic–gravity waves to propagate freely.

Equilibrium Configuration

We assume that the solar atmosphere is static [$\mathbf{V}_e=\mathbf{0}$] at its equilibrium, and that it is threaded by a straight magnetic field,

$$\mathbf{B} = B_0 \hat{\mathbf{s}}, \tag{5}$$

where $\hat{\mathbf{s}}$ is a unit vector that is either vertical, horizontal, or oblique. We choose B_0 by requiring that at the reference point (0,10) Mm the Alfvén speed

$$c_A(y) = \frac{B_0}{\sqrt{\mu \varrho_e(y)}} \tag{6}$$

and sound speed

$$c_s(y) = \sqrt{\frac{\gamma p_e(y)}{\varrho_e(y)}} \tag{7}$$

satisfy the constraint $c_A(y=10\ \mathrm{Mm})=10c_s(y=10\ \mathrm{Mm})$. This constraint reproduces typical conditions in the solar corona where, typically, $c_s=100$ km s^{-1} and $c_A=1000$ km s^{-1}. As a result, the solar corona is magnetically dominated with $B_0 \approx 14.5\times10^{-4}$ T. The plasma β,

$$\beta(y) = \frac{\gamma}{2}\frac{c_s^2(y)}{c_A^2(y)}, \tag{8}$$

in the solar corona attains a value of $\beta(y=10\ \mathrm{Mm})=0.012$. It slowly grows with depth within the chromosphere and abruptly reaches a value of $\beta(y=0\ \mathrm{Mm})\approx 4\times10^5$ below y=2 Mm at the bottom of the photosphere (Figure 1, bottom). This high value of β shows that in these low regions of the solar atmosphere the effect of the magnetic field is negligibly small.

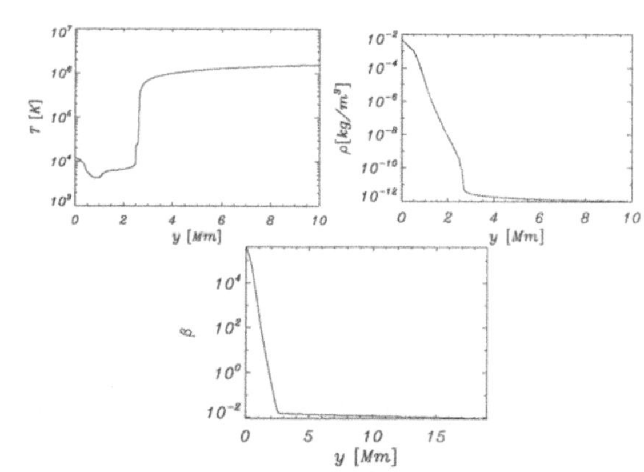

Figure 1: Equilibrium profile of the solar temperature (top-left), the mass density (top-right), and the plasma β (bottom).

The straight magnetic field of Equation (5) is a simplified model of curved magnetic-field lines that are located in the low regions of the solar atmosphere. However, because the plasma β grows rapidly below y=2.5 Mm, where it reaches high values (see Figure 1, bottom), we in fact model the very weakly magnetized low layers of the atmosphere, which do not correspond to flux-tubes and sunspots. Therefore, in this first approximation, the curved magnetic-field lines can be replaced by a straight magnetic field, and our models are justified.

In Equations (6) and (7), $\varrho_e(y)$ and $p_e(y)$ denote the equilibrium mass density and gas pressure, respectively. They are specified by the hydrostatic constraint, which in the context of Equation (5) states that the pressure gradient is balanced by the gravity force,

$$-\nabla p_e + \varrho_e \mathbf{g} = \mathbf{0}. \tag{9}$$

With the ideal gas law and the y-component of Equation (9), we arrive at

$$p_e(y) = p_0 \exp\left[-\int_{y_r}^{y} \frac{dy'}{\Lambda(y')}\right], \qquad \varrho_e(y) = \frac{p_e(y)}{g\Lambda(y)}, \tag{10}$$

where

$$\Lambda(y) = \frac{k_B T_e(y)}{mg} \tag{11}$$

is the pressure scale-height, and p_0 denotes the gas pressure at the reference level that we choose in the solar corona at y_r=10 Mm.

We adopt an equilibrium temperature profile [$T_e(z)$] for the solar atmosphere that is close to the VAL-C atmospheric model of Vernazza, Avrett, and Loeser (1981): see Figure 1, top-left panel.T_e attains a value of about 5700 K at the top of the photosphere, which corresponds toy=0.5 Mm. At higher altitudes, $T_e(y)$ falls off until it reaches its minimum of 4350 K at an altitude of y≈0.95 Mm. Higher up, $T_e(y)$ grows gradually with height up to the transition region, which is located at y≈2.7 Mm. Here $T_e(y)$ experiences a sudden increase to the coronal value of 1.5 MK at y=10 Mm. Then with Equation (10) we obtain the corresponding gas pressure and mass-density profiles. Both $p_e(y)$ (not shown) and $\varrho_e(y)$ (Figure 1, top-right panel) experience a sudden decrease from photospheric to coronal values at the transition region.

Initial Conditions

At t=0 seconds, we initially perturb the equilibrium impulsively by simultaneously using localized Gaussian pulses in the gas pressure and the vertical-velocity component, viz.

$$p(x,y,t=0) = p_e(y) + A_p f(x,y), \qquad V_y(x,y,t=0) = A_v f(x,y) \tag{12}$$

$$f(x,y) = \exp\left[-\frac{(x-x_0)^2}{w_x{}^2} - \frac{(y-y_0)^2}{w_y{}^2}\right]. \tag{13}$$

Here A_p and A_v are the amplitudes of the perturbations, (x_0,y_0) is their initial position, and (w_x,w_y) denotes their widths along the x- and y-directions. We set and hold fixed x_0=0 Mm,y_0=0.5 Mm, w_x=0.5 Mm, w_y=0.5 Mm, A_p=0.02$p_e(y_0)$, and A_v=0.2 km s^{-1}. These magnitudes ofA_v and A_p lie within the range of the detected flow and temperature in solar granulation (e.g.Rieutord et al.2002; Kostik, Khomenko, and Shchukina 2009; Baran 2011). We separately consider three orientations for our unidirectional equilibrium magnetic field lines: (a) horizontal, (b) vertical, (c) oblique to the gravitational force.

NUMERICAL RESULTS

Equations (1)–(4) are solved numerically using the FLASH code (Lee and Deane 2009). This code implements a second-order unsplit Godunov solver with various slope limiters and Riemann solvers, as well as adaptive mesh refinement (AMR). The main advantage of using the AMR technique is to refine a numerical grid at steep spatial profiles while keeping a coarse grid at the places where fine spatial resolution is not essential. This technique usually introduces interpolation errors at different-sized numerical cells. These errors can result in a vertical flow that, albeit initially small, can grow with height to an unacceptable magnitude. The remedy we adopted for this inherent phenomenon is to refine the whole region below the transition region. We used the Roe solver and minmod flux limiter, and set the simulation box for the horizontal and vertical equilibrium magnetic field cases in the x-direction to be −5≤x≤5 Mm. To trace plasma structures that extend more horizontally in the case of the oblique magnetic field, we used −2≤x≤8 Mm. Along the y-direction the numerical box was set to −1≤y≤19 Mm in all cases. At all boundaries, we fixed all plasma quantities to their equilibrium values, which led only to weak numerical reflections of the incident wave signals. Additionally, we increased the physical domain size and dampened incident waves by adopting a coarse numerical grid at the top boundary to avoid the influence of reflections during the simulation time range. As a result, these reflections do not exert any noticeable effect on the dynamics of the system. In all our studies, we used an AMR grid with a minimum/maximum level of refinement set to 3/7.

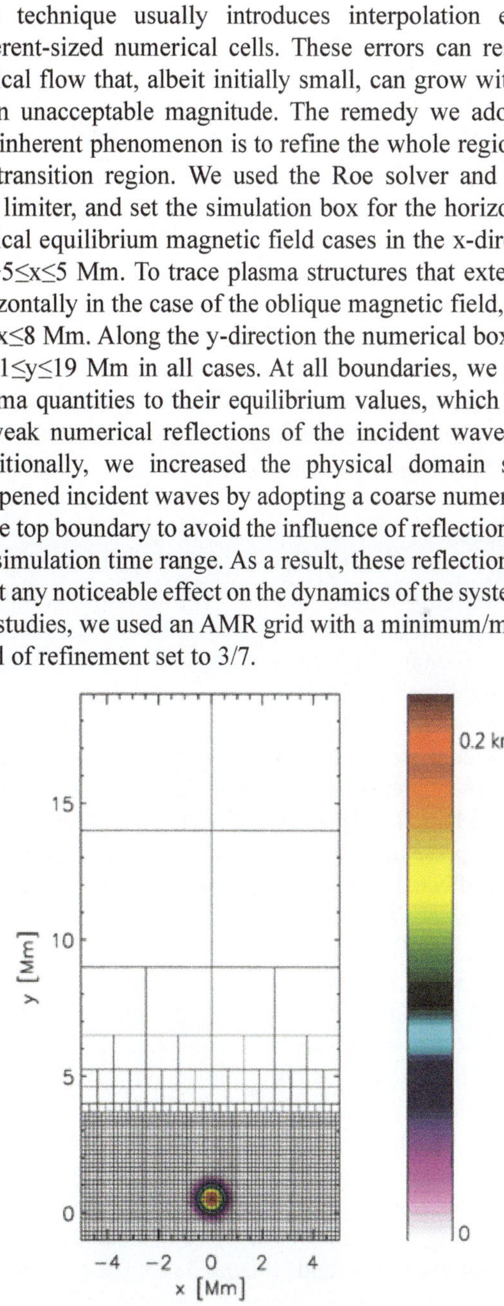

Figure 2: Numerical blocks with their boundaries (solid lines) and the pulse in velocity of Equation (12) (color maps) at t=0 seconds for the vertical and horizontal equilibrium magnetic fields. The maximum velocity (red color) corresponds to 0.2 km s^{-1}.

The initial system of blocks is shown in Figure 2. For the initial vertical and horizontal magnetic fields, at t=0 seconds the whole simulation region is covered by 2666 blocks, with a similar number of blocks for the oblique magnetic field. Because every block consists of 8×8 numerical cells, this number of blocks corresponds to 170 624 numerical cells. This results in the finest (poorest) resolution of $\Delta x = \Delta y \approx 19.5$ km ($\Delta x = \Delta y \approx 625$ km) in the region below y=3.7 Mm (abovey=9 Mm) at t=0 seconds (see Figure 2). The refinement strategy is based on controlling numerical errors in mass density, which results in an excellent resolution of steep spatial profiles and greatly reduces numerical diffusion at these locations. The duration of a typical numerical run was 1500 seconds. Although the numerical simulations were carried out for $-5 \leq x \leq 5$ Mm, $-1 \leq y \leq 19$ Mm, only the results in the regions of interest are displayed here.

Vertical Equilibrium Magnetic Field: $\hat{s} = \hat{y}$

First, we investigate the case of the vertical equilibrium magnetic field, which corresponds to $\hat{s} = \hat{y}$ in Equation (5). Here $\hat{y}$ is a unit vector along the y-direction. The initial pulse of Equation (12) triggers magnetoacoustic–gravity waves in the solar atmosphere. These waves are compressive, and so we can trace their evolution in both temperature and velocity. Figure 3illustrates logarithmic temperature profiles (color maps) and velocity vectors resulting from the perturbations. At t=200 seconds (top-left panel), the excited perturbations in the vertical component of velocity have already penetrated the solar corona, arriving at the point (0,5.5) Mm. The perturbation in the vertical-velocity component propagates by ≈ 5 Mm in the first 200 seconds of its evolution, which gives the speed of propagation as ≈ 25 km s^{-1}. The average sound speed in this region is significantly higher. This is also confirmed by Fedun, Shelyag, and Erdélyi (2011) (cf. their Figures 4–7), who reported that their waves propagated from the transition region up to 2 Mm within about 40 seconds of the simulation time. From this comparison we infer that the perturbation we initially imposed triggers flows in the solar atmosphere first, and thereafter these flows produce magnetoacoustic–gravity waves. At this moment in time, the transition region experiences a weak perturbation and the signal reflected from the transition region is represented by arrows around x=0 Mm, y=1.5 Mm. This reflected signal clashes with the slowly upwardly propagating wave that is triggered at the launching place, i.e.,x=0 Mm, y=0.5 Mm. As a result of this clash, the leading front is generated. Cold chromospheric plasma lags behind, and at t=400 seconds (Figure 3, top-right panel) it reaches an altitude of $y \approx 6.5$ Mm. The cold material is lifted because of the pressure gradient, which works against gravity and forces the chromospheric material to penetrate the solar corona (Sterling and Hollweg 1989). At t=800 seconds (Figure 3, bottom-left), the transition region experiences well-developed oscillations whose central regions reach $y \approx 7$ Mm at x=0 Mm, and also off-side propagating oscillations. These oscillations are of lower amplitudes at t=1000 seconds (Figure 3, bottom-right panel). At this time, the central plasma experiences gravitational deceleration while the region of cold plasma widens. The jet, which is associated with the shock, lasts for several hundred seconds. Its velocity reaches a maximum of ≈ 60 km s^{-1}.

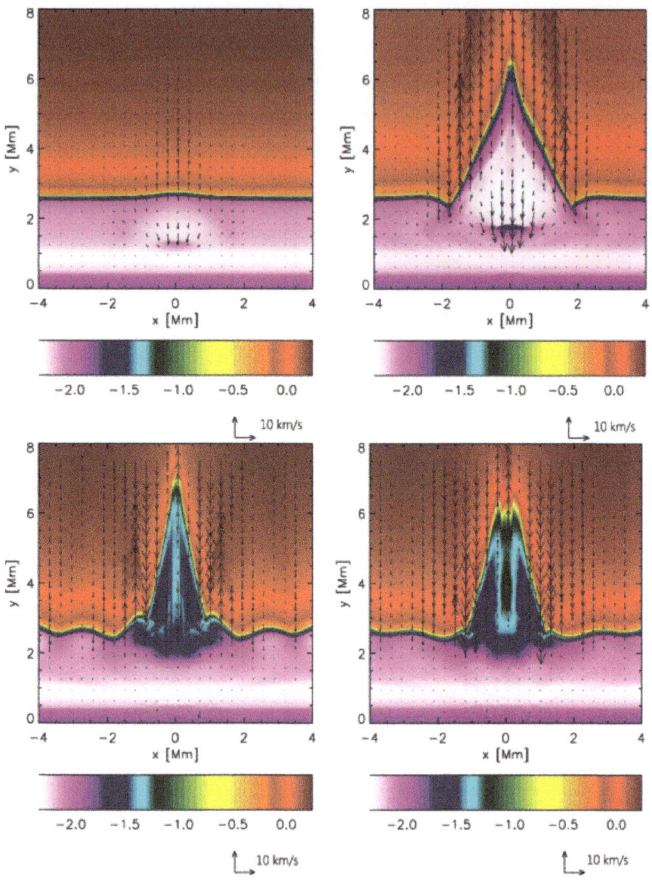

Figure 3: Spatial profiles of the logarithm of the temperature (color maps) and velocity (arrows) profiles att=200 seconds (top-left), t=400 seconds (top-right), t=800 seconds (bottom-left), t=1000 seconds (bottom-right), for the vertical equilibrium magnetic field. Temperature is expressed in units of 1 MK. The arrow below each panel represents the length of the velocity vector, expressed in units of 10 km s^{-1}. The corresponding movie can be found in fig3.avi in the Electronic Supplementary Materials.

Figure 4 illustrates the temporal variation of the vertical velocity component at the detection point (0,5) Mm for the case of Figure 3. As a result of the rapid mass density decrease with height, upwardly propagating waves grow in amplitude and steepen rapidly to form shocks. The arrival of the first shock-front at the detection point is clearly seen at $t \approx 290$ seconds. The second shock-front reaches the detection point at $t \approx 540$ seconds, i.e. approximately 250 seconds later. This secondary shock (and subsequent shocks that arrive after about 160 seconds and 240 seconds) results from the nonlinear wake, which lags behind the leading shock (Sterling and Hollweg 1989; Srivastava and Murawski 2011). These times can be compared with the acoustic cut-off period (Roberts 2006),

$$P_{ac}(y) = \frac{4\pi\Lambda(y)}{c_s(y)\sqrt{1+2\Lambda'(y)}}, \tag{14}$$

which for y=0.5 Mm has a value of $P_{ac}(y) \approx 180$ seconds. This value differs by 70 seconds from the time-span between arrivals of neighboring shocks, which is 250 seconds. However, the 2D model we discuss here is more complex than the 1D scenario described by the Klein–Gordon equation (Roberts 2006). The wave period we detected is altered by the interaction between up-going waves from the launching place of the initial pulse and the reflected wave from the transition-region signals. Because wave reflections originate at the large temperature gradients, the chromosphere sustains a cavity for these waves, which are represented by the oblique stripes located at altitudes of 0.7 Mm<y<1.4 Mm (Figure 5). In the neighborhood of the point (0.4,0.4) Mm we observe the formation of vortices (Figure 5, top panel) that experience an energy cascade into smaller scales. These vortices are present until the end of our simulation runs (Figure 5, bottom panel). The first vortex is triggered by the initial pulse in V_y, which is a characteristic feature of velocity perturbations. This first vortex causes convection in the convectively unstable plasma layers. According to Schwarzschild's instability condition, a medium is convectively unstable if the squared buoyancy (or Brunt–Väisälä) frequency,

$$\omega_g^2 \equiv -g\left(\frac{1}{\gamma\Lambda} - \frac{1}{\Lambda}\right) = \frac{g}{\Lambda} - \frac{g^2}{c_s^2}, \qquad (15)$$

is negative (e.g. Roberts 2006). Because this criterion is satisfied for y<0.6 Mm, convection sets in there. Such vortices were also described by Konkol, Murawski, and Zaqarashvili (2011), in a similar context.

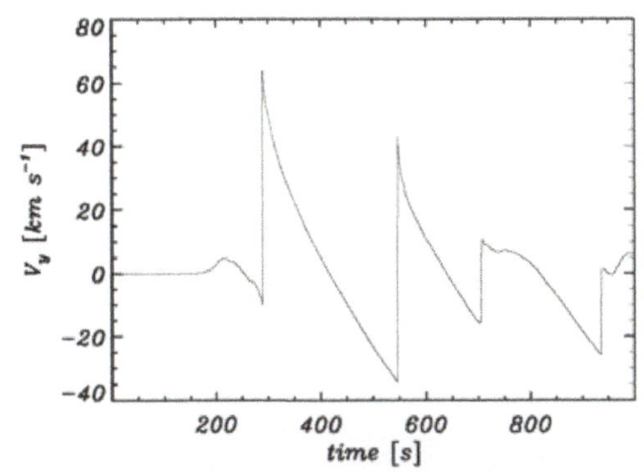

Figure 4: Temporal signature of V_y at the point (0,5) Mm for the vertical equilibrium magnetic field.

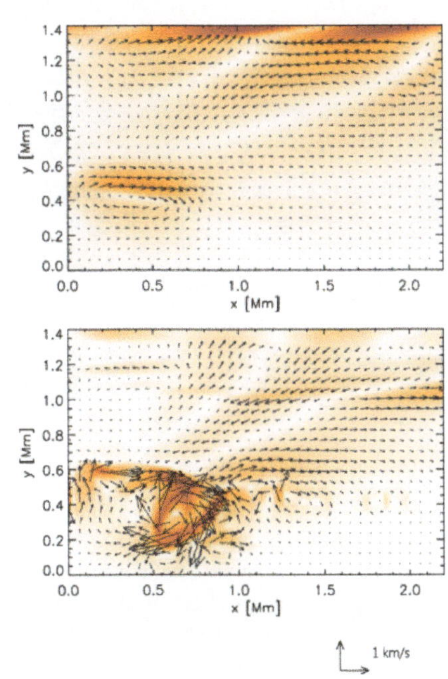

Figure 5: Total velocity [|**V**|] (color map) and velocity (arrows) profiles at t=400 seconds (top) and t=1000 seconds (bottom) for the vertical equilibrium magnetic field. The velocity vectors are expressed in units of 1 km s^{-1}. The corresponding movie can be found in fig5.avi in the Electronic Supplementary Materials.

Horizontal Equilibrium Magnetic Field: $\hat{\mathbf{s}} = \hat{\mathbf{x}}$

In this section, we investigate the horizontal equilibrium magnetic field, which corresponds to $\hat{\mathbf{s}} = \hat{\mathbf{x}}$ in Equation (5). In this case, the spatial profiles of T drawn at t=250 seconds and t=1000 seconds (Figure 6) reveal small-amplitude oscillations of the transition region without the jet that was observed for the vertical magnetic field. Comparing this with Figure 3, we see that the waves resulting from the initial pulses already arrived at the solar corona at t=250 seconds (Figure 6, left panel). The magnetic-field lines are horizontal, therefore these waves are essentially fast magnetoacoustic–gravity waves in the region of low plasma β that occurs for y>2.6 Mm (Figure 1, bottom panel). It is interesting that the orientation of the magnetic field plays a very crucial role in determining the amplitude of transition-region oscillations.

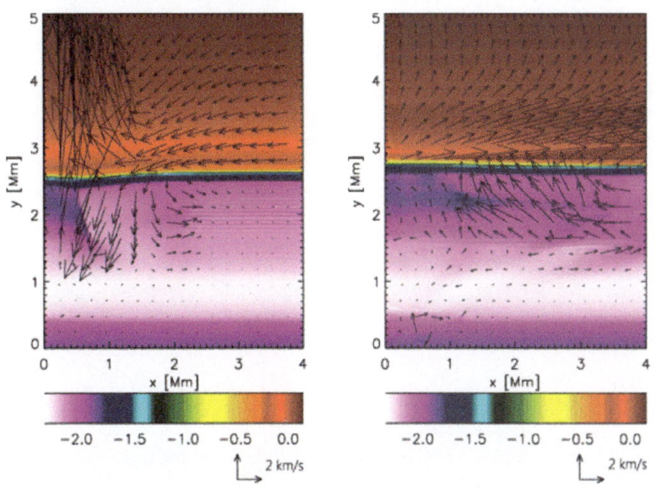

Figure 6: Spatial profiles of the logarithm of the temperature (color maps) and velocity (arrows) at t=250 seconds (top) and t=1000 seconds (bottom) for the horizontal equilibrium magnetic field. Temperature is in units of 1 MK. The arrow below each panel represents the length of the velocity vector, expressed in units of 2 km s^{-1}. The corresponding movie can be found in fig6.avi in theElectronic Supplementary Materials.

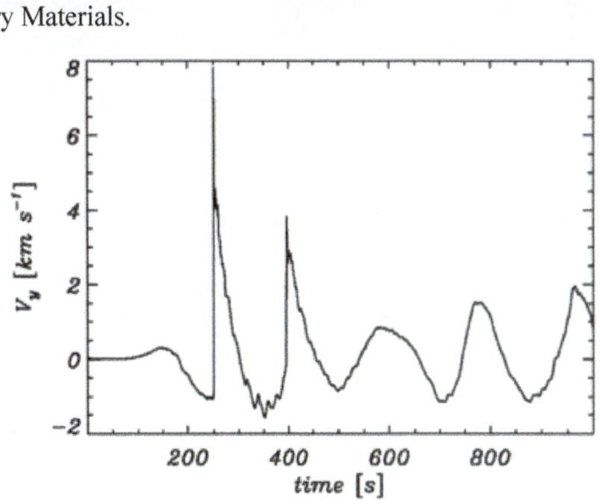

Figure 7: Time signature of V_y at the point (0,5) Mm for the horizontal equilibrium magnetic field.

Similar to the temporal-signatures of Figure 4, V_y at the detection point (0,5) Mm reveals the shocks. However, we now observe only two shocks for the horizontal magnetic field (as seen in Figure 7). The first shock arrives at the detection point at t=250 seconds and the second shock reaches this point at t=400 seconds. We note that the time-span of 150 seconds between these events is 100 seconds shorter than for the vertical magnetic field in Section 3.1. However, it is shorter by only 30 seconds than the acoustic cut-off period [P_{ac}] resulting from Equation (14). The subsequent three oscillations, which are of smaller amplitude, reveal periods on the order of P_{ac}.

The chromosphere exhibits essentially similar features to those seen for the vertical magnetic field in Section 3.1. The vortex motion at (x=0.4,y=0.4) Mm and the trapped waves in the solar chromosphere for 0.7 Mm<y<1.4 Mm are again discernible (Figure 8). At t=400 seconds (top panel), the vortex is clearly developed. It is similar to that of Figure 5 (top panel). The flow patterns at the line y=1.4 Mm in Figures 5 and 8 differ in some details, though. In the case of the vertical magnetic field, the resulting waves are essentially slow magnetoacoustic–gravity waves, while Figure 8 corresponds to the horizontal background magnetic field with much contribution from fast magnetoacoustic–gravity waves. At a later time, turbulence results from this vortex, and it differs in some details from its analog of Figure 5 (bottom panel). In both the vertical and horizontal magnetic fields, the turbulence originates from the impulsively triggered perturbations that are initially launched in the convectively unstable atmospheric layers.

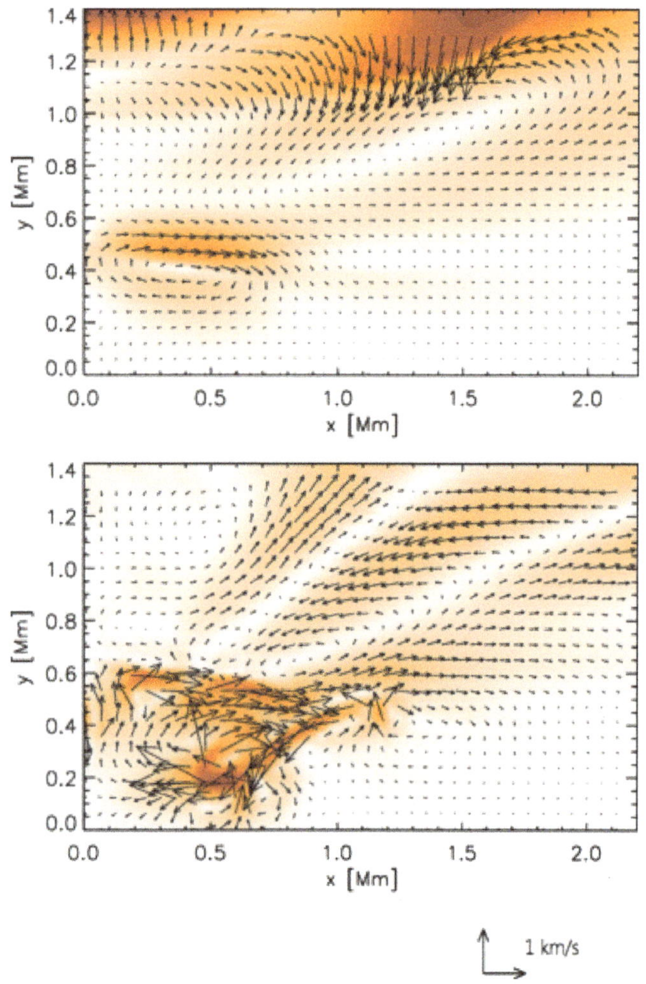

Figure 8: Total velocity [|**V**|] (color map) and velocity (arrows) profiles at t=400 seconds (top) and t=1000 seconds (bottom) for the horizontal equilibrium magnetic field. The velocity vectors are expressed in units of 1 km s^{-1}. The corresponding movie can be found in fig8.avi in the Electronic Supplementary Materials.

Oblique Equilibrium Magnetic Field: $\hat{\mathbf{s}} = (\hat{\mathbf{x}} + \hat{\mathbf{y}})/\sqrt{2}$

In this section, we discuss the case of the oblique magnetic field, for which $\hat{\mathbf{s}}$ in Equation (5) is at an angle of $\pi/4$ to the horizontal axis. Figure 9 displays temperature profiles at t=400 seconds (top-left panel), t=600 seconds (top-right panel), t=800 seconds (bottom-left panel), and t=1000 seconds (bottom-right panel). These profiles are more complex than those for either the horizontal or the vertical magnetic field of Sections 3.1 and 3.2. At t=400 seconds (Figure 9, top-left), we clearly see the rising transition region. At t=1000 seconds (Figure 9, bottom-right), we observe surface waves propagating along the transition region. However, for the oblique equilibrium magnetic field, we find that the temporal signatures of V_y reveal shocks, and the periods are about 250–300 seconds (Figure 10). The jet, which is associated with the shock, lasts for several hundred seconds and reaches a velocity maximum of ≈30 km s^{-1}. This is a result of complex interaction of the waves with the background plasma in the upper regions of the solar atmosphere. This complexity is seen at t=1000 seconds in the velocity profiles of Figure 11, which clearly illustrates the trapped and reflected waves in the solar chromosphere as well as clearly developed vortices.

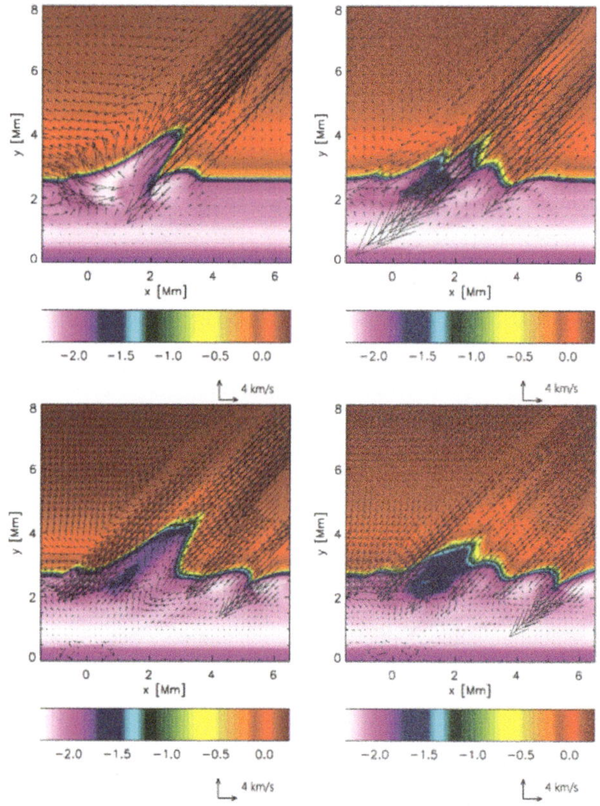

Figure 9: Spatial profiles of the logarithm of the temperature (color maps) and velocity (arrows) profile att=400 seconds (top-left), t=600 seconds (top-right), t=800 seconds (bottom-left), t=1000 seconds (bottom-right) for the oblique equilibrium magnetic field. The temperature is given in units of 1 MK. The arrow below each panel represents the length of the velocity vector, expressed in units of 4 km s^{-1}. The corresponding movie can be found in fig9.avi in the Electronic Supplementary Materials.

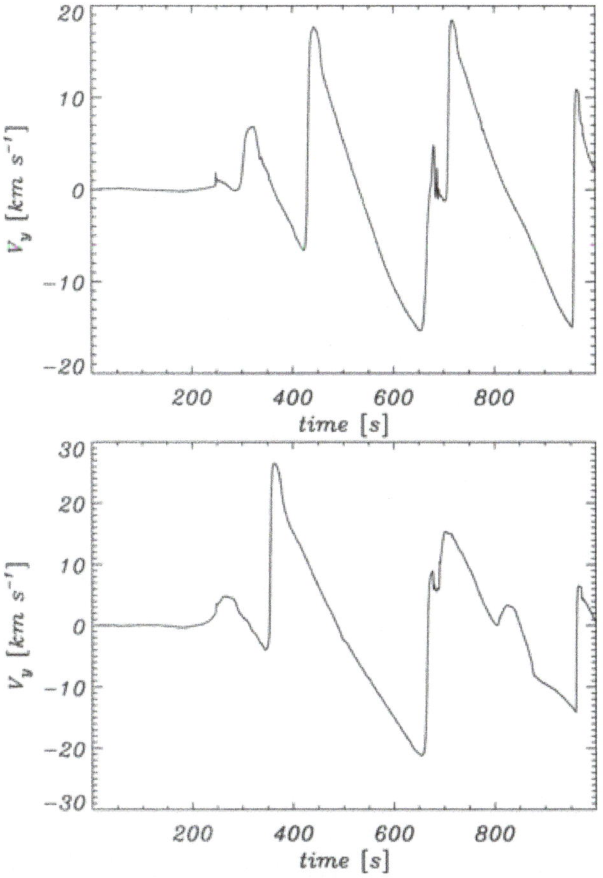

Figure 10: Time signature of V_y at the points (4.5,5) Mm (top) and (3.84,4.98) Mm (bottom) for the oblique equilibrium magnetic field.

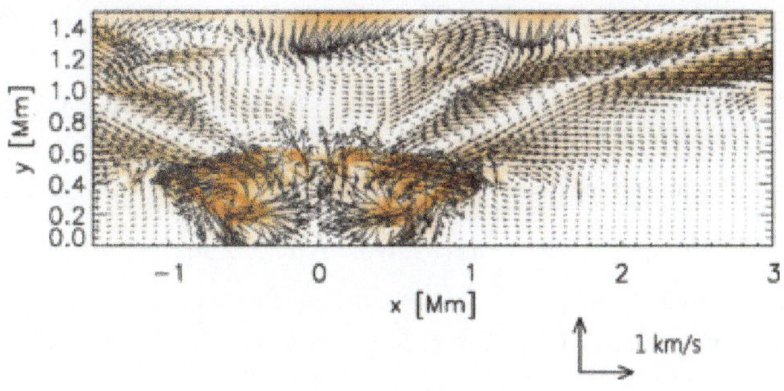

Figure 11: Total velocity [|**V**|] (color map) and velocity (arrows) profile at t=1000 seconds for the oblique equilibrium magnetic field. The velocity vectors are expressed in units of 1 km s^{-1}. The corresponding movie can be found in fig11.avi in the Electronic Supplementary Materials.

DISCUSSION AND CONCLUSION

We have investigated the impulsive excitation of magnetoacoustic–gravity oscillations and compared and contrasted the resulting propagation under different orientations of equilibrium magnetic fields. We performed 2D numerical simulation of the velocity and gas-pressure pulses, which mimic a solar granule. These pulses were initially launched at the top of the solar photosphere in a stratified solar atmosphere using the VAL-C temperature profile. We find that in the cases where the background magnetic field possesses a non-zero vertical component, the amplitude of the upwardly propagating perturbations rapidly grows with height due to the rapid decrease in the equilibrium mass density. Therefore, the perturbation quickly steepens into shocks in the upper regions of the solar chromosphere, which launches the cool material behind it.

We found that the excitation of magnetoacoustic–gravity waves in a non-horizontal equilibrium magnetic-field configuration is channeled upwards and able to cause large-amplitude transition-region oscillations. Meanwhile, the horizontal equilibrium magnetic field configuration results in relatively smaller-amplitude transition-region oscillations caused by fast magnetoacoustic–gravity waves. These spread their energy across magnetic-field lines, in contrast to slow magnetoacoustic–gravity waves, which in a strongly magnetized plasma region are guided along the magnetic-field lines. This indicates that the wave energy is transferred into the upper atmosphere in larger amounts when the magnetic field is more vertical. However, the complexity and therefore the evolution of a horizontal equilibrium magnetic field allows the reflection and trapping of the waves in the lower solar atmosphere and can influence the localized dynamics and heating of the atmosphere. The reflection of the waves in the lower solar atmosphere and its trapping may also cause chromospheric cavities.

By analyzing the temporal-signature of V_y at a fixed spatial point, we also found that the oscillation period is longer when comparing the vertical magnetic field (here, the characteristic oscillation period was ≈250–300 seconds) with the horizontal equilibrium magnetic system (in which the characteristic period of oscillation was ≈150–200 seconds). For the vertical magnetic-field system in low plasma-β regions, the magnetoacoustic–gravity waves are adequately described as slow magnetoacoustic–gravity waves (propagating along the magnetic-field lines at approximately the sound speed [$c_s(y)$], see Equation (7)). For the horizontal magnetic-field system in strongly magnetized plasma, the oscillations transverse to the magnetic field are appropriately described as fast magnetoacoustic–gravity waves (propagating across the magnetic-field lines at the fast speed, $c_f^2(y) = c_A^2(y) + c_s^2(y)$, see Equations (6) and (7)). For y=0.5 Mm we have $c_A \ll c_s$; the cut-off frequencies of the fast and slow magnetoacoustic–gravity waves are very close to each other, which results in very close values of the cut-off periods. Therefore, we might expect that the periods detected for the vertical and horizontal magnetic fields exhibit similar values. However, the cut-off periods are derived on the basis of the linear theory, which is valid for small-amplitude oscillations only. Such small oscillations are present in the case of the horizontal magnetic field, therefore the numerical data lie close to the analytical prediction. In the vertical magnetic field, the upwardly propagating waves interact with the waves that are reflected from the transition region. This gives rise to larger-amplitude oscillations, which significantly alter the background plasma. The waves that propagate through this strongly modified medium exhibit modified velocities, and they are reflected from the transition region, which is locally strongly curved. This results in periods within the range 250–300 seconds, which differ from P_{ac}.

It is known that radiation is an effective mechanism of wave damping in the low photosphere (Mihalas and Toomre 1982). It might not radically alter the system's behavior, but radiative damping is at least likely to reduce the amplitude of the waves that reach the transition region, which leads to shorter jets than those that we see in our models, and also to lower amplitudes of the coronal shocks. We did not invoke radiative cooling or thermal conduction in our model atmosphere, because we aimed to model the small-scale atmospheric regions above the solar photosphere where these effects are not believed to be dominant. However, we intend to include these effects in our future studies.

The solar atmosphere is structured by convective overshoot, which is absent from our model. Instead, we isolated a single granule-like perturbation with the aim to mimic a magnitude of flow and plasma temperature that is associated with the solar granulation (Baran 2011). We considered the complex scenario that results from this simple model to understand the physics of wave phenomena above such magnetic structures in the solar atmosphere. We intend to develop more advanced models in our future studies.

In conclusion, our numerical simulations clearly demonstrate that small-amplitude initial pulses in vertical velocity and gas pressure are able to trigger a plethora of dynamic phenomena in the upper regions of the solar atmosphere with periods in the range 150–300 seconds, a value that depends on the orientation of the background magnetic field. However, it should be noted that our 2D simulations are idealized in the sense that they do not include radiative transfer and thermal conduction along field lines. The magnetic-field configuration and the equilibrium stratification are simple and we modeled a single granule only. These limitations require additional studies, which we intend to carry out in the near future.

ACKNOWLEDGEMENTS

The authors express their thanks to the referee for the stimulating comments. This work has been supported by a Marie Curie International Research Staff Exchange Scheme Fellowship within the 7th European Community Framework Program (KM). The software used in this work was in part developed by the DOE-supported ASC/Alliance Center for Astrophysical Thermonuclear Flashes at the University of Chicago. JM acknowledges IDL support provided by STFC, UK. RO acknowledges financial support from MICINN/MINECO and FEDER funds through grant AYA2011-22846 and also CAIB through the "Grups Competitius" scheme and FEDER funds. KM expresses his thanks to Kamil Murawski for his assistance in drawing the numerical data.

REFERENCES

1. Baran, O.A.: 2011, In: Choliy, V., Ivashchenko, G., Ivaniuk, O. (eds.) 2010, Adv. Astron. Space Phys., Proc. the 17th Young Scientists' Conference on Astron. Space Phys. 53.
2. Bogdan, T.J., Carlsson, M., Hansteen, V.H., McMurry, A., Rosenthal, C.S., Johnson, M., Petty-Powell, S., Zita, E.J., Stein, R.F., McIntosh, S.W., Nordlund, A.: 2003, Astrophys. J. 599, 626.
3. Centeno, R., Socas-Navarro, H., Lites, B., Kubo, M., Frank, Z., Shine, R., Tarbell, T., Title, A., Ichimoto, K., Tsuneta, S., Katsukawa, Y., Suematsu, Y., Shimizu, T., Nagata, S.: 2007,Astrophys. J. Lett. 666, L137.
4. Fedun, V., Erdélyi, R., Shelyag, S.: 2009, Solar Phys. 258, 219. doi:10.1007/s11207-009-9407-9.
5. Fedun, V., Shelyag, S., Erdélyi, R.: 2011, Astrophys. J. 727, 17.
6. Fedun, V., Verth, G., Jess, D.B., Erdélyi, R.: 2011, Astrophys. J. Lett. 740, L46.
7. Gruszecki, M., Murawski, K., Kosovichev, A.G., Parchevsky, K.V., Zaqarashvili, T.: 2011, Acta Phys. Pol. 42, 1333.
8. Hasan, S.S., van Ballegooijen, A.A., Kalkofen, W., Steiner, O.: 2005, Astrophys. J. 631, 1270.
9. Katsukawa, Y., Berger, T.E., Ichimoto, K., Lites, B.W., Nagata, S., Shimizu, T., Shine, R.A., Suematsu, Y., Tarbell, T.D., Title, A.M., Tsuneta, S.: 2007, Science 318, 1594.
10. Konkol, P., Murawski, K., Zaqarashvili, T.V.: 2011, Astron. Astrophys. 537, A96.
11. Kostik, R., Khomenko, E., Shchukina, N.: 2009, Astron. Astrophys. 506, 1405.
12. Krijger, J.M., Rutten, R.J., Lites, B.W., Straus, Th., Shine, R.A., Tarbell, T.D.: 2001, Astron. Astrophys. 379, 1052.
13. Kuridze, D., Zaqarashvili, T.V., Shergelashvili, B.M., Poedts, S.: 2008, Ann. Geophys. 26, 2983.
14. Kuridze, D., Zaqarashvili, T.V., Shergelashvili, B.M., Poedts, S.: 2009, Astron. Astrophys. 505, 763.
15. Lee, D., Deane, A.E.: 2009, J. Comput. Phys. 228, 952.
16. McAteer, R.T.J., Gallagher, P.T., Williams, D.R., Mathioudakis, M., Phillips, K.J.H., Keenan, F.P.: 2002, Astrophys. J. Lett. 567, L165.
17. McAteer, R.T.J., Gallagher, P.T., Williams, D.R., Mathioudakis, M., Bloomfield, D.S., Phillips, K.J.H., Keenan, F.P.: 2003, Astrophys. J. 587, 806.
18. McIntosh, S.W., Judge, P.G.: 2001, Astrophys. J. 561, 420.
19. Mihalas, B.W., Toomre, J.: 1982, Astrophys. J. 263, 386.
20. Murawski, K., Srivastava, A.K., Zaqarashvili, T.V.: 2011, Astron. Astrophys. 535, A58.
21. Rieutord, M., Ludwig, H.G., Roudier, T., Nordlund, A., Stein, R.: 2002, Nuovo Cimento C 025, 523.
22. Roberts, B.: 2006, Phil. Trans. Roy. Soc. London A 364, 447.
23. Schrijver, C.J., Title, A.M.: 2003, Astrophys. J. Lett. 597, L165.
24. Shibata, K., Magara, T.: 2011, Living Rev. Solar Phys. 8, 6.http://solarphysics.livingreviews.org/open?pubNo=lrsp-2011-6&page=articlese8.html. ADS
25. Shibata, K., Nakamura, T., Matsumoto, T., Otsuji, K., Okamoto, T.J., Nishizuka, N., et al.: 2007,Science 318, 1591.
26. Srivastava, A.K.: 2010, New Astron. 15, 621.
27. Srivastava, A.K., Murawski, K.: 2011, Astron. Astrophys. 534, A62.
28. Srivastava, A.K., Kuridze, D., Zaqarashvili, T.V., Dwivedi, B.N.: 2008, Astron. Astrophys. Lett.481, L95.
29. Srivastava, A.K., Zaqarashvili, T.V., Kumar, P., Khodachenko, M.L.: 2010, Astrophys. J. 715, 292.
30. Sterling, A.C., Hollweg, J.V.: 1989, Astrophys. J. 343,

985.

31. Vecchio, A., Cauzzi, G., Reardon, K.P.: 2009, Astron. Astrophys. 494, 269.

32. Vecchio, A., Cauzzi, G., Reardon, K.P., Janssen, K., Rimmele, T.: 2007, Astron. Astrophys. Lett.461, L1.

33. Vernazza, J.E., Avrett, E.H., Loeser, R.: 1981, Astrophys. J. 45, 635.

34. Zhang, J., Cheng, X., Ding, M.D.: 2012, Nat. Commun. 3, 747.

CHAPTER 5

MAGNETOHYDRODYNAMICS MODELING OF CORONAL MAGNETIC FIELD AND SOLAR ERUPTIONS BASED ON THE PHOTOSPHERIC MAGNETIC FIELD

Satoshi Inoue

Keywords: Sun Magnetic field, Photosphere Corona, Magneto hydro dynamics (MHD), Solar active region, Solar flare, Coronal mass ejection (CME)

REVIEW

Introduction

Solar flares are explosive phenomena observed in the atmosphere of the Sun (the solar corona). These events are observed as sudden bursts of electromagnetic radiation, such as extreme ultraviolet radiation (EUV), X-rays, and even white light; some examples are shown in Fig. 1a–c. The scale is classified as soft X-rays, using the 1–8 Å band obtained by the GOES-5 satellite (one of the Geostationary Orbiting Environment Satellites), as shown in Fig. 1d. The Sun is known to be a magnetized star. Figure 2a shows the line-of-sight component of the magnetic field, and the positive and negative polarities cover the whole sun. Figure 2b shows the three-dimensional (3D) magnetic field lines traced from the positive to the negative polarities; these have been extrapolated under the assumption of the potential field approximation (this will be discussed below). Solar flares often occur above the sunspots corresponding to a cross section of strong magnetic flux. In addition, because the solar corona satisfies the low- β plasma condition (β= 0.01–0.1) (Gary 2001) in which the magnetic energy dominates that of the coronal plasma, solar flares are widely considered to be a manifestation of the conversion of the magnetic energy of the solar corona into kinetic and thermal energy, culminating in the release of high-energy particles and electromagnetic radiation.

Citation: Satoshi Inoue, Magnetohydrodynamics modeling of coronal magnetic field and solar eruptions based on the photospheric magnetic field, OI: 10.1186/s40645-016-0084-7.

Figure 2c is an enlarged view of the region that is marked by an arrow in Fig. 2b; here, the field lines are responsible for the current density accumulation, which initiates the flare. These field lines are extrapolated using the nonlinear force-free field (NLFFF) approximation; this is one of the main topics of this paper.

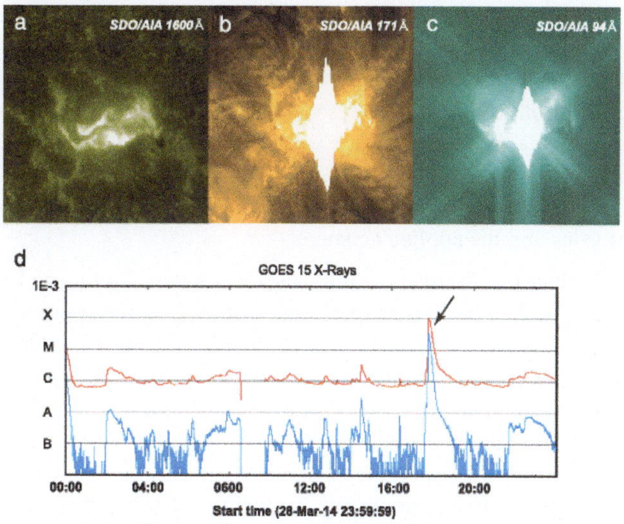

Figure 1: Observations of the solar flare. **a–c** The solar flares in the EUV images for different wavelengths observed on the solar surface or in the solar atmosphere. From left to right, the wavelengths are 1600, 171, and 94 Å. The flares were observed by SDO/AIA at around 18:00 UT on 29 March 2014. **d** Time profile of the X-ray flux measured by the GOES 12 satellite on 29 March 2014. The solar X-ray outputs in the 1–8 Å and 0.5–4.0 Å passbands are plotted

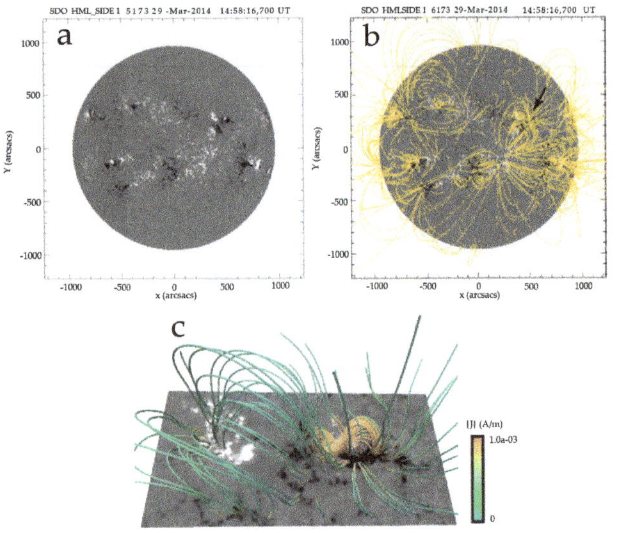

Figure 2: Magnetic fields of the sun. **a** Full-disk image of a line-of-sight component of the solar magnetic field observed by SDO/HMI at 15:00 UT on 29 March 2014, which corresponds to 2.5 h before an X1.0-class flare. **b** The magnetic field lines in yellow are superimposed on **a**. The field lines are extrapolated under the approximated potential field. This figure is courtesy of Dr. D. Shiota (Shiota et al. 2012). **c**The active region, corresponding to the region marked by an arrow in **b**, is the region in which a sunspot with a strong magnetic field is concentrated. The field lines are plotted according to the NLFFF approximation, in which they accumulate the strong current density

Furthermore, this causes a huge amount of coronal gas (a typical mass is 10^{15} g) with a velocity of 100–2000 kms $^{-1}$ to be released into interplanetary space; this is called a coronal mass ejection (CME; e.g., Forbes (2000)). The CMEs are sometimes associated with solar flares; however, the detailed understanding of the relationship between these two phenomena remains elusive (Chen 2011; Schmieder et al. 2015). It is important to understand these phenomena in order to better understand the nonlinear plasma dynamics of the processes involving the magnetic energy or helicity of the solar coronal plasma; this includes storage-and-release processes as well as the forecasting space weather (Tóth et al. 2005; Liu et al. 2008; Kataoka et al. 2014). Investigations of solar flares and CMEs are thus important in terms of both the elemental plasma physics and the applied science.

Since the discovery of the solar flares by Carrington (1859), many studies have been performed (including observational, theoretical, and numerical studies) for clarifying their dynamics (Benz 2008; Priest and Forbes 2002; Shibata and Magara 2011; Wang and Liu 2015). Many new insights on solar flares and related phenomena have been obtained by analyzing the data collected by satellites. For instance, the Yohkoh satellite obtained much data on dynamical features of the sun, some of which had not been predicted; this can be seen in Fig. 3a; this image, taken by a soft X-ray telescope, shows several important aspects that have helped our understanding of solar flares. For example, Tsuneta et al. (1992) discovered the cusp-shaped structure during the solar flare seen in the lower right panel in Fig. 3a. A detailed analysis (Tsuneta 1996) produced evidence of the reconnection, and this lent support to a theoretical flare model based on reconnection; this model is named for its developers, Carmichael, Surrock, Hirayama, Kopp, and Pneumann (CSHKP) and explains the observations at multiple wavelengths (Carmichael 1964; Hirayama 1974; Kopp and Pneuman 1976; Sturrock 1966). Masuda et al. (1994) confirmed the CSHKP model of the solar flare by analyzing the hard X-ray signals obtained during a solar flare. In addition, Sterling and Hudson (1997) found a characteristic pattern of X-rays that are released prior to a flare; this is shown in the upper right panel of Fig. 3a. This pattern is a sigmoid (an S- or inverse S-shaped structure) that changes into a cusp-shaped loop structure after the flare occurs. Su et al. (2007) and McKenzie and Canfield (2008) demonstrated the fine structure and topology of the field lines that were later observed by an X-ray telescope (Golub et al. 2007) on board the Hinode satellite (Kosugi et al. 2007). In addition, Yokoyama et al. (2001)found evidence of reconnection inflow in extreme ultraviolet observations of the Solar and Heliospheric Observatory (SOHO). The images of the coronal loop shown in Fig. 3b are reminiscent of reconnection. Because these observations were based on imaging of electromagnetic waves, the data were mapped onto a 2D plane. Thus, obtaining a 3D reconstruction of these events is extremely difficult.

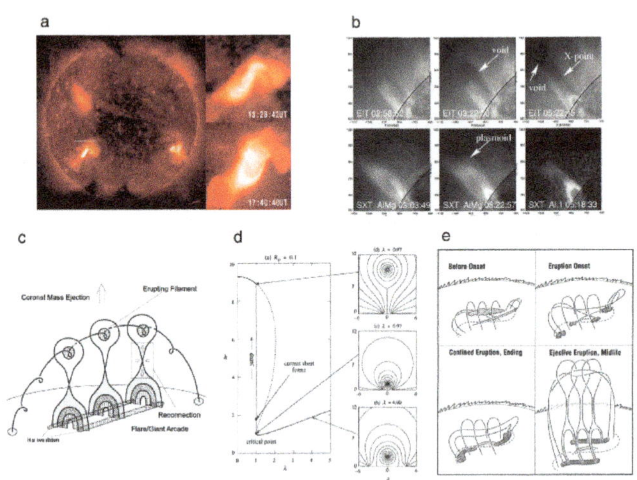

Figure 3: Observations and models of the solar flares. **a** The solar corona observed by soft X-ray from on board the Yohkoh satellite. The left panel shows the whole sun; the upper and lower right panels show the sigmoid and cusp-loop structures, observed before and after the flare, respectively. This figure is courtesy of ISAS/JAXA. **b** The reconnection process in the solar flare observed by SOHO satellite fromYokoyama et al. (2001). **c** 3D view of the magnetic field during the solar flare inferred from the observations from Shiota et al. (2005). **d** The loss-of-equilibrium model proposed by Forbes and Priest (1995). The flux tube loses the equilibrium by changing the boundary conditions; as a result, an eruption occurs. **e** The tether-cutting reconnection model proposed by Moore et al. (2001). The flux tube is created by the reconnection taking place between the two sheared field lines formed before onset; eventually, the flux tube can erupt away from the solar surface. The images in (**b**–**e**) are copyright AAS and reproduced by permission

Based on this observational evidence, there have been several attempts to construct the 3D magnetic structure (e.g., Shibata (1999)). Figure 3c is an image of a 3D magnetic structure inferred from observations during the onset of the solar eruption depicted in Shiota et al. (2005); the reconnection model can be used to explain various observed phenomena, e.g., the two H- αflare ribbons, and giant arcades. In addition, various models have been proposed that predict the onset of solar flares and CMEs. For instance, Forbes and Priest

(1995) proposed the catastrophic model shown in Fig. 3d; this shows that the flux tube in the solar corona does not remain at equilibrium when the boundary conditions are changed, and this results in a sudden eruption. The tether-cutting model, proposed by Moore et al. (2001), is shown in Fig. 3e. They assumed that two sheared field lines existed along the polarity inversion line (PIL) prior to the onset of the flare; this is shown in the upper left panel of Fig. 3e. Note that this has a somewhat sigmoidal structure. If there is reconnection between the sheared field lines, then long twisted lines are formed, and an eruption may occur. The final state shown in the right bottom panel of Fig. 3e is very similar to that shown in Fig. 3c.

The dramatic increase in computer power allows us to perform 3D magnetohydrodynamics (MHD) simulations and to estimate the 3D dynamics of magnetic fields during solar flares. Several studies have modeled sunspots to be asymmetric or as simple dipole fields and have analytically obtained the 3D coronal magnetic fields by fitting appropriate boundary conditions (e.g., Amari et al. (2000) and Amari et al. (2003a)). Figure 4a shows the results from Amari et al. (2003b); this shows the formation of a flux tube, which is initiated by the initial potential field, through twisted and converged motion on the photosphere. The twisted motion imposed on a dipole sunspot causes the accumulation of sheared field lines, and the motion converging toward the PIL creates a flux tube composed of highly twisted field lines due to the flux cancellation. Aulanier et al. (2012) and Janvier et al. (2013) constructed similar MHD models, and these generated a 3D view that extended the well-established 2D CSHKP model. This view produced a 3D feature that was not seen in the 2D model; their simulations produced strong-to-weak sheared post-flare loops, which are consistent with observations (Asai et al. 2003). On the other hand, Kusano et al. (2012) successfully reproduced an eruption in a different way, as shown in Fig. 4b. They created a linear force-free field that had shearing field lines as the initial condition; a small dipole emerging flux was imposed at a local area on the PIL. They found that only two types of emerging flux can produce a flux tube; this shows that the eruption is due to interactions with a pre-existing sheared magnetic field. Later, this scenario was confirmed in observations by Toriumi et al. (2013) and Bamba et al. (2013).

Other MHD models have been derived from an initialized flux tube. Solar filaments are often observed on the sun; these are composed of a denser plasma than that in the solar corona (Parenti 2014). It is widely agreed that the highly helical twisted lines in the filament sustain the dense plasma in the solar corona (Priest and Forbes 2002). Recent observations clearly show the helical structure of the magnetic field, i.e., the flux tube and the dynamics (e.g., Cheng et al. (2013); Nindos et al. (2015); Vemareddy and Zhang (2014)). In addition to this, the flux tube/filaments have often been observed to erupt away from the solar surface. Following these observations, extensive MHD modeling, focusing on the flux tube dynamics, has been performed. Inoue and Kusano (2006) investigated the dynamics of a flux tube that was initially embedded in the solar corona, as shown in Fig. 4c. This extended the studies of Forbes (1990)and Forbes and Priest (1995) showing the dynamics in a 2D space. This study found that the flux tube eruption was caused by a kink instability in 3D space, rather than by a loss of equilibrium in 2D space, as discussed by Forbes (1990). Recently, a higher-resolution simulation was performed by Nishida et al. (2013), who reported complex reconnections and plasmoid motions associated with flux tube eruption. Chen and Shibata (2000) numerically confirmed that a flux tube eruption is triggered by a small emerging flux that is the result of the reconnection with magnetic fields lines surrounding the flux tube, and it can reduce the downward tension force acting on the flux tube. Török et al. (2009) extended this into 3D space. As shown in Fig. 4d,Török and Kliem (2005) and Fan (2005) constructed more realistic MHD models by noting that the flux tube roots are tied to the solar surface (Titov and Démoulin 1999), rather than by assuming infinitely long flux tubes as in Inoue and Kusano (2006) and Nishida et al. (2013). Török and Kliem (2005) reported that the eruption depends on the decay rate of the external magnetic field, and later, this scenario was explained as torus instability (Kliem and Török 2006). To address this instability, detailed stability and equilibrium analyses of flux tubes in the solar corona were performed by Isenberg and Forbes (2007) and Démoulin and Aulanier (2010), and the dynamics were numerically confirmed by Török and Kliem (2007), Fan (2010), and Aulanier et al. (2010). Attempts are being made to meet the challenge of simulating a solar eruption through the emergence of highly twisted flux tube embedded in the convection zone (e.g., An and Magara (2013); Archontis et al. (2014); Leake et al. (2014)).

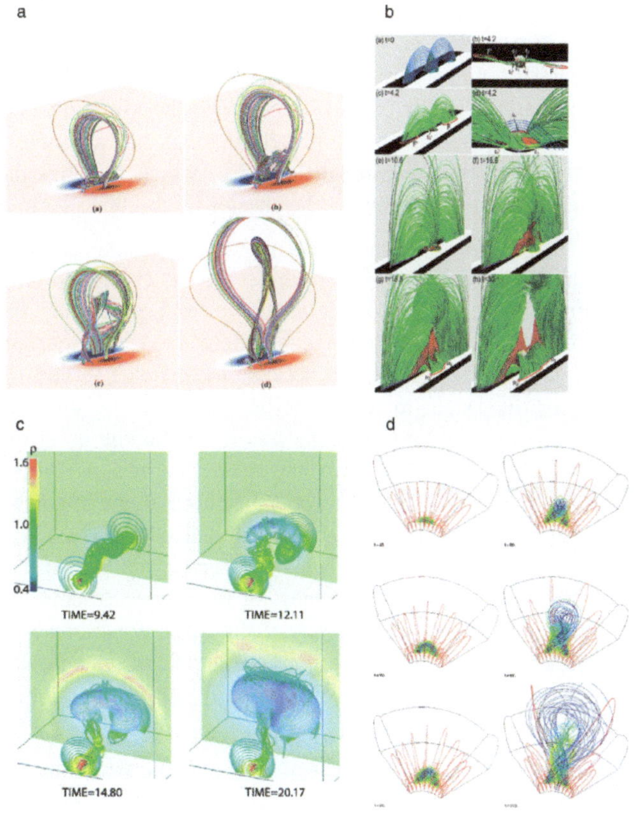

Figure 4: 3D MHD simulation of solar flares by pioneers in the field. **a** MHD modeling of the solar flare byAmari et al. (2003a). The potential field was reconstructed from the given simple dipole fields, which were imposed on the twisted and converged motion. Consequently, the potential field was converted into a non-potential field, leading to the eruption. **b** MHD modeling by Kusano et al. (2012) shows that the emergence of small flux can destroy the initial equilibrium condition of the linear force-free field, leading to the formation of a large flux tube and an eruption. **c**Inoue and Kusano (2006) investigated the flux tube dynamics associated with the solar flares and causing a CME. The flux tube was assumed to be infinitely

long and was driven by kink instability, leading to a CME for a certain supra-threshold height. **d**Fan (2005) employed a more realistic flux tube (Titov and Démoulin 1999) with footpoints tied to the solar surface. The eruption was first driven by kink instability and later by torus instability (Fan 2010). All images are copyright AAS and reproduced by permission

Several studies have shown the formation and dynamics of a large-scale CME in the range of a few solar radii. Antiochos et al. (1999) proposed a breakout model in which a moving magnetic field surrounding the core fields triggers the CME; those dynamics were later confirmed in a high-resolution simulation (e.g., Lynch et al. (2008) and Karpen et al. (2012)). Shiota et al. (2010)reported that an interaction between the core field (modeled as a spheromak) and the ambient field is important for determining whether an ejection will occur.

However, most of the studies presented above assumed hypothetical and ideal situations. Although these studies clarified many elementary physical processes related to the onset and dynamics of solar flares, they did not incorporate the data collected by solar satellites (in particular, they did not incorporate magnetic field data). One of the reasons for this is that only the photospheric magnetic field can be measured, and this implies that the coronal magnetic field cannot be observed directly. Nevertheless, several models have been proposed in which the photospheric magnetic field is treated as a boundary surface (.e.g., Török et al. (2011); van Driel-Gesztelyi et al. (2014); Zuccarello et al. (2012)). Challenging simulations considered a wide domain that extended from the Sun to the Earth; their major objectives included the initiation of a CME, its propagation in interplanetary space, and ultimately its interaction with the magnetosphere, which governs the dynamics of the ionosphere (Manchester et al. 2004; Tóth et al. 2005).

On the other hand, most of these models employed only the normal components of the magnetic field, neglecting the horizontal fields. Horizontal magnetic fields are very important for explaining the solar flares because these fields serve as a proxy for the extent to which the field lines are twisted and sheared, i.e., for determining the free magnetic energy at the solar surface. The MHD modeling of solar eruptions, which accounts for the three components of the photospheric magnetic field, has only recently been demonstrated, thanks to a state-of-art solar physics satellite. However, several problems remain open; these include the uniqueness of the numerical solution and the mathematical consistency of the MHD equations on a specified boundary (these questions will be discussed below).

In this paper, we present state-of-the-art MHD modeling, which accounts for the photospheric magnetic field, and we will focus on applying this to solar eruptions. In particular, we introduce the modeling of the coronal magnetic field and solar eruptions, based on the three components of the photospheric magnetic field. This area of research has been recently revived, beginning with a study by Jiang et al. (2013), and followed by Inoue et al. (2014a), Amari et al. (2014), and Inoue et al. (2015). The structure of this article is as follows. We first introduce a method for 3D reconstruction of the coronal magnetic field, based on the photospheric magnetic field; this includes a potential field that is easily reconstructed from one of the components of these fields and a nonlinear force-free field that is based on all of the components. Next, we describe recent MHD models that use a magnetic field that is reconstructed from the measured photospheric field. Finally, we draw some important conclusions.

Extrapolation of the coronal magnetic fields

Because we can obtain observations of the magnetic field components only for the photosphere, it is necessary to extrapolate to obtain information about the 3D coronal magnetic fields in 3D. The solar corona is considered to be in low- β plasma state, where $\beta=P/2B^2$ is defined to be the ratio of the plasma gas pressure (P) to the magnetic pressure (B^2). From this, we have that the force-free state

$$\boldsymbol{J} \times \boldsymbol{B} = 0 \tag{1}$$

is a good approximation for describing the state of the coronal magnetic field, where $\boldsymbol{B}$ is the magnetic field satisfying the solenoidal condition,

$$\nabla \cdot \boldsymbol{B} = 0, \tag{2}$$

and $\boldsymbol{J}$ is the current density,

$$\boldsymbol{J} = \nabla \times \boldsymbol{B}. \tag{3}$$

In this section, we introduce a method for extrapolating the solar coronal magnetic field given only the photospheric magnetic fields in the force-free approximation.

Potential field

The potential field is the simplest force-free field approximation:

$$\nabla \times \boldsymbol{B} = 0, \tag{4}$$

where the current density vanishes everywhere. In this formulation, the magnetic field can be replaced with the scalar function ψ, as follows:

$$\boldsymbol{B} = -\nabla\psi. \tag{5}$$

If we use the solenoidal condition of Eq. (2), then Eq. (5) can be rewritten as

$$\nabla^2\psi = 0. \tag{6}$$

This corresponds to the Poisson equation, for which a unique solution is guaranteed for a boundary valueproblem. In this way, we can calculate the solar coronal magnetic field, given the normal component of the magnetic field (B_n) and its Neumann condition,

$$B_n = \frac{\partial\psi}{\partial n}, \tag{7}$$

on each boundary. Although the photospheric magnetic field can be considered to be the bottom surface, conditions are required on the other boundaries in order to solve Eq. (6). Several such methods have been proposed, some of which are described below.

One approach is to use Green's functions (Sakurai 1982; 1989). In this approach, the potential field is created by monopoles that are located at different points on the bottom boundary (x′,y′,0), at which the magnetic flux $B_z dx'dy'$ exists. The scalar potential ψ is

$$\psi = \frac{1}{2\pi} \int B_n\left(x',y'\right) G\left(x,y,z,x',y'\right) dx'dy', \tag{8}$$

where $G = 1/\sqrt{|r' - r|}$. The scalar function is determined automatically by the normal component of the observed magnetic field, whereas ***B***=0 is assumed as ***r*** approaches ∞. This method can be applied to an isolated active region that is not influenced by the magnetic fields of other regions. On the other hand, if the magnetic field lines in the active region extend into another active region, the boundary conditions at the sides and top are no longer appropriate.

The Fourier expansion can be used for deriving the solution of Eqs. (5) and (6). The solution was presented by Priest (2014), as follows:

$$B_x = -\sum_k \frac{k_x}{|k|} B_k e^{ik_x + ik_y - kz},$$
$$B_y = -\sum_k \frac{k_y}{|k|} B_k e^{ik_x + ik_y - kz},$$
$$B_z = B_0 + \sum_k B_k e^{ik_x + ik_y - kz}, \tag{9}$$

where the bottom boundary values are expanded into Fourier components k_x and k_y. This formulation implies that all of the components decay exponentially, implying ***B***=0 at z=∞. However, the side boundaries automatically obey periodic boundary conditions, so this method is useful only for describing areas far from the side boundaries.

We can easily extend Eq. (6) in spherical coordinates (r, θ, ϕ) and thus obtain a solution for the whole sun, as shown in Fig. 2b. This overcomes the problem mentioned above regarding the connectivity of the field lines. In spherical coordinates, the solution to Eq. (6) can be written using Legendre polynomials (Altschuler and Newkirk 1969), as follows:

$$\psi = \sum_{n=1}\sum_{m=0}\left[\left(\frac{1}{r}\right)^{n+1} (g_n^m \cos\phi + h_n^m \sin\phi)\, P_n^m(\cos\theta)\right], \tag{10}$$

where $P_n^m(\cos\theta)$ are Legendre polynomials, and g_n^m and h_n^m are coefficients obtained from spherical harmonics analysis. The boundary condition is based on the normal component of the photospheric magnetic field, and the Neumann condition of ψ is the same as that in Eq. (7). Using the above calculations, the potential fields can be expressed as follows:

$$B_r = -\sum_{n=1}^{N}\sum_{m=0}^{n}\left\{(n+1)\left(\frac{1}{r}\right)^{n+2} (g_n^m \cos\phi + h_n^m \sin\phi) P_n^m(\cos\theta)\right\},$$
$$B_\theta = \sum_{n=1}^{N}\sum_{m=0}^{n}\left\{\left(\frac{1}{r}\right)^{n+2} (g_n^m \cos\phi + h_n^m \sin\phi) \frac{dP_n^m(\cos\theta)}{d\theta}\right\}, \tag{11}$$

$$B_\phi = -\sum_{n=1}^{N}\sum_{m=0}^{n}\left\{m\left(\frac{1}{r}\right)^{n+2} (g_n^m \sin\phi - h_n^m \cos\phi) \frac{P_n^m(\cos\theta)}{\sin\theta}\right\}. \tag{12}$$

As an example, one result is shown in Fig. 2b, which can be used to depict the field lines covering the sun. One advantage of the potential field extrapolation method is that the solution is relatively easily obtained; there are several techniques for doing this. On the other hand, the potential field is a minimum energy state that does not store the free magnetic energy released in the solar flares. This implies that the observed field lines in the area close to the PIL cannot be captured by the potential field. To convert the potential field into the dynamic phase of the solar flares, it is necessary to obtain the Poynting flux through the photosphere in order to obtain the free energy (Feynman and Martin 1995; van Ballegooijen and PMartens 1989).

Linear force-free field

The force-free Eq. (1) can be rewritten as

$$\nabla \times \boldsymbol{B} = \alpha \boldsymbol{B}, \tag{13}$$

where α is a coefficient. After taking the divergence of this equation, the left-hand side vanishes, and thus, we have

$$(\boldsymbol{B} \cdot \nabla)\alpha = 0, \tag{14}$$

which implies that the coefficient α is constant along all field lines. If the coefficient α is constant everywhere (not only along the field lines), Eq. (13) becomes a linear Equation that can be reduced to the Helmholtz equation,

$$\nabla^2 \boldsymbol{B} + \alpha \boldsymbol{B} = 0, \tag{15}$$

by taking the curl of Eq. (13). We call this solution the linear force-free field (LFFF), and it is also specified with an appropriate boundary condition.

For example, (Chiu and Hilton 1977) found the analytical general solution by using Green's functions:

$$B_i = \frac{1}{2\pi}\int_{-\infty}^{\infty} dx'dy' \left[\overline{G_i}(\boldsymbol{x},\boldsymbol{x}')B_z(x',y',0) + \tilde{G}_i(\boldsymbol{x},\boldsymbol{x}')C(x',y')\right], \tag{16}$$

where $C(x',y')$ is any finite integrable function (see Chiu and Hilton (1977)). $\tilde{G}_i(\boldsymbol{x},\boldsymbol{x}')$ is defined as

$$\overline{G_x} = \frac{x-x'}{R}\frac{\partial\overline{\Gamma}}{\partial z} + \alpha\overline{\Gamma}\frac{y-y'}{R},$$
$$\overline{G_y} = \frac{y-y'}{R}\frac{\partial\overline{\Gamma}}{\partial z} - \alpha\overline{\Gamma}\frac{x-x'}{R},$$
$$\overline{G_z} = -\frac{\overline{\Gamma}}{\partial R} - \frac{\overline{\Gamma}}{R},$$

where $R=(x-x')^2+(y-y')^2$, $\overline{\Gamma}$ is

$$\overline{\Gamma} = \frac{z}{Rr}\sin(\alpha r) - \frac{1}{R}\sin(\alpha z),$$

and $r=(x-x')^2+(y-y')^2+z^2$. Using these equations, if we are given B_z and the force-free α at the photosphere, then the LFFF is automatically determined.

Unlike the potential field, the LFFF can yield the free magnetic energy. In general, however, the observed force-free α measured in the photosphere varies in space. In particular, in solar active regions, the coefficient α attains high values close to the PIL and small values far from the

PIL. This implies that the LFFF is inappropriate for modeling solar active regions. Therefore, we need to obtain the NLFFF extrapolation by using the observed force-free α, i.e., we need to obtain not only the normal component of the magnetic field but also the horizontal components at the photosphere in order to reproduce the magnetic field of a solar active region.

Nonlinear force-free field

To demonstrate suitable magnetic fields in the solar active region, we consider solving the force-free Eq. (1) directly. However, because this equation contains nonlinearities that cannot be solved analytically, numerical techniques are necessary (i.e., Schrijver et al. (2006) or Metcalf et al. (2008)). Since important information can be obtained from observed photospheric magnetic fields, this becomes a boundary value problem. Below, we briefly describe several numerical methods that have been developed.

Vertical integration method. The algorithm of the vertical integration method is quite simple. The magnetic fields are integrated upward in the z direction, as originally proposed byNakagawa (1974) and further extended by Wu et al. (1990). Under the force-free assumption, the current densities of the horizontal components along the solar surface can be calculated as follows:

$$J_{x0} = \alpha_0 B_{x0},$$
$$J_{y0} = \alpha_0 B_{y0}, \tag{17}$$

where B_{x0} and B_{y0} are the horizontal components of the photospheric magnetic field, J_{x0} andJ_{y0} are the horizontal components of the current density, and α_0 is the force-free alpha obtained from J_{z0}/B_{z0}. Using Ampere's law, Eq. (3), and the solenoidal condition, Eq. (2), the following equations are obtained for the z-derivatives of the magnetic field:

$$\frac{\partial B_x}{\partial z} = J_{y0} + \frac{\partial B_{z0}}{\partial x},$$
$$\frac{\partial B_y}{\partial z} = \frac{\partial B_{z0}}{\partial y} - J_{x0},$$
$$\frac{\partial B_z}{\partial z} = -\frac{\partial B_{x0}}{\partial x} - \frac{\partial B_{y0}}{\partial y} \tag{18}$$

The integration, in which the information about the photospheric magnetic field is extended upward, is repeated, and the coronal magnetic field can be calculated in 3D. However the above algorithm is mathematically ill-posed, i.e., the calculation is not robust, as has been reported in several papers (e.g., Wiegelmann and Sakurai (2012)). For instance, once the nonphysical phenomena due to numerical errors appear during the integration, the magnetic field increases exponentially. One reason for this is that no restrictions are imposed on the top and side boundaries.

The Green's function method. A similar mathematical approach that uses the Green's function was developed by Yan (1995) and Yan and Sakurai (2000) but the magnetic field is assumed as follows: $B = O\left(\frac{1}{r^2}\right)$, i.e., B = 0 as r=>∞. They found the NLFFF solution based on Green's second identity, as follows:

$$c_i \boldsymbol{B}_i = \int_S \left(\boldsymbol{Y}\frac{d\boldsymbol{B}}{dn} - \frac{d\boldsymbol{Y}}{dn}\boldsymbol{B}_0 \right) dS, \tag{19}$$

where c_i=1 and c_i=1/2 correspond to points in the volume and at the boundary, respectively, $\boldsymbol{B}_0$is the measured photospheric magnetic field, and $\boldsymbol{Y}$ is a reference function,

$$\boldsymbol{Y}(r) = \frac{cos(\lambda_i \boldsymbol{r})}{4\pi|\boldsymbol{r} - \boldsymbol{r}'|}, \tag{20}$$

where $\boldsymbol{r}'$ is a fixed point and $\lambda(\boldsymbol{r}')$ is a parameter that depends on $\boldsymbol{r}'$. The reference function satisfies the Helmholtz equation,

$$\nabla^2 \boldsymbol{Y} + \lambda^2 \boldsymbol{Y} = \delta_i, \tag{21}$$

where δ_i is the Dirac delta function. The parameter λ_i can be obtained by solving

$$\int_V Y_i(\lambda_i^2 B_i - \alpha^2 B_i - (\nabla\alpha \times \boldsymbol{B}_i))dV = 0. \tag{22}$$

Although it has been pointed out that this technique is slow (Wiegelmann and Sakurai 2012), recently, the calculation speed has been dramatically accelerated by using a GPU (Wang et al. 2013).

Grad-Rubin method. Sakurai (1981) was the first to use the Grad-Rubin method for calculating the magnetic field in solar active regions, and this method was later extended, e.g., Amari et al. (2006). This technique follows directly from the force-free field property. First, the potential field is calculated based only on the normal components of the magnetic field. The force-free αcan be measured at the bottom surface as $\alpha=J_z/B_z$, and it can be distributed in 3D according to the following equation:

$$\boldsymbol{B}^k \cdot \nabla\alpha^k = 0, \tag{23}$$

where k is the iteration number and B^0 corresponds to the potential field. The magnetic field is updated according to

$$\nabla \times \boldsymbol{B}^{k+1} = \alpha^k \boldsymbol{B}^k, \tag{24}$$

and

$$\nabla \times \boldsymbol{B}^{k+1} = \alpha^k \boldsymbol{B}^k, \tag{25}$$

Since the vector potential $\boldsymbol{A}^k$ satisfying $\nabla\times\boldsymbol{A}^k=\boldsymbol{B}^k$ can be written as

$$\boldsymbol{A}^{k+1} = \int \frac{\boldsymbol{J}^k}{\sqrt{(x-x')^2 + (y-y')^2 + (z-z')^2}} dx'dy'dz', \tag{26}$$

the updated $\boldsymbol{B}$ automatically satisfies the solenoidal condition, and it is then substituted back into Eq. (23). This process is repeated until the magnetic field reaches a steady state. Although the force-free α can be determined at positive or negative polarity and will satisfy Eq. (23), the single-polarity information is neglected. Nevertheless, Régnier et al. (2002) and Canou and Amari (2010) were able to reconstruct magnetic fields that agree with the observations. Recently, the Grad-Rubin method has been improved by Amari et al. (2010); Wheatland and Régnier (2009), and (Wheatland and Leka 2011), who have obtained the unique solution by using two different solutions derived from different polarities, i.e., by changing the distribution of the force-free α at the bottom surface.

MHD relaxation method. In the MHD relaxation methods, the MHD equations are solved directly (in particular, this is the zero-beta MHD approximation (Mikić et al. 1988)); they solved

$$\frac{\partial \boldsymbol{v}}{\partial t} = -(\boldsymbol{v} \cdot \nabla)\boldsymbol{v} + \boldsymbol{J} \times \boldsymbol{B} + \nu \nabla^2 \boldsymbol{v}, \tag{27}$$

$$\frac{\partial \boldsymbol{B}}{\partial t} = \nabla \times (\boldsymbol{v} \times \boldsymbol{B} - \eta \boldsymbol{J}), \tag{28}$$

and

$$\boldsymbol{J} = \nabla \times \boldsymbol{B} \tag{29}$$

to find the force-free solution while keeping the photospheric magnetic field as the boundary condition. Here, $\boldsymbol{v}$ is the plasma velocity, and ν and η are the viscosity and resistivity, respectively. The zero-beta MHD is an extreme approximation of the low-beta solution. However, since a force-free state can be assumed in the zero-beta approximation, this method is valid. Several studies (Mikić and McClymont 1994; McClymont and Mikic 1994; Jiang and Feng 2012; Inoue et al. 2014b) have employed the potential field as the initial condition; consequently, the magnetic twist on the bottom surface is obtained by replacing the tangential components of the photospheric magnetic field above which the magnetic fields relaxes toward the force-free state through the MHD relaxation process. This process is called the stress-and-relaxation method (Roumeliotis 1996). In a simpler treatment, known as the magnetofrictional method, the equation of motion (27) is replaced with

$$\boldsymbol{v} = \mu \boldsymbol{J} \times \boldsymbol{B}, \tag{30}$$

where μ is a coefficient. This technique can also be used to find the force-free solution (Valori et al. 2005), and it has been applied to the photospheric magnetic field.

Note that if the three components of the photospheric magnetic field are fully satisfied at the solar surface and if the plasma velocity is zero there, these conditions are not consistent with the induction equation, which requires information about the differential value in the normal direction. Consequently, an error appears in $\nabla\cdot\boldsymbol{B}$. Therefore, the errors arising during the relaxation process should be eliminated, and several methods have been developed for eliminating them (Tóth 2000; Miyoshi and Kusano 2011). Often, the projection method is used, and this removes the errors derived from the potential component. We decompose the numerically obtained magnetic field $\boldsymbol{B}_N$ into $\boldsymbol{B}_p$ (the potential component) and $\boldsymbol{B}_{np}$ (the non-potential component), as follows:

$$\boldsymbol{B}_N = \boldsymbol{B}_p + \boldsymbol{B}_{np}. \tag{31}$$

In general, a vector field $\boldsymbol{B}$ can be described as

$$\boldsymbol{B} = \nabla \psi_p + \nabla \times \boldsymbol{A}_{np}, \tag{32}$$

where ψ_p and $\boldsymbol{A}_{np}$ are the scalar and vector potentials, respectively. Taking into account Eq. (5), $\nabla\psi_p$ and $\nabla\times\boldsymbol{A}_{np}$ correspond, respectively, to the potential and non-potential components of the magnetic field. Taking the divergence of Eq. (32), the equality $\nabla\cdot\nabla\times\boldsymbol{A}_{np}=\nabla\cdot\boldsymbol{B}_{np}=0$ is automatically satisfied. However, it is not guaranteed that $\nabla\cdot\nabla\psi_p=\nabla\cdot\boldsymbol{B}_p= 0$. If $\boldsymbol{B}_p$ contains a numerical error, we further decompose it into Bp′, which satisfies the solenoidal condition, and $\boldsymbol{B}_{error}$, the error, as follows:

$$\boldsymbol{B}_p = \boldsymbol{B}_{p'} + \boldsymbol{B}_{\text{error}}, \tag{33}$$

where, in general, $\boldsymbol{B}_{error}$ does not meet the solenoidal condition. However, taking the divergence of Eq. (33), the equation can be reduced to the Poisson equation,

$$\nabla \cdot \boldsymbol{B}_{\text{error}} = \nabla \cdot \boldsymbol{B}_N = \nabla^2 \psi_p. \tag{34}$$

Consequently, this equation can be solved, and the magnetic field satisfying the solenoidal condition $\boldsymbol{B}'$ can be updated as follows:

$$\boldsymbol{B}' = \boldsymbol{B}_N - \nabla \psi_p. \tag{35}$$

This technique has been widely used for eliminating errors (Tanaka 1995; Tóth 2000); however, solving the Poisson equation is computationally demanding. Therefore, numerical techniques for improving the calculation speed, e.g., a multigrid technique, are required (Inoue et al.2014b).

Another technique was proposed by Dedner et al. (2002), who introduced a modified induction equation,

$$\frac{\partial \boldsymbol{B}}{\partial t} = \nabla \times (\boldsymbol{v} \times \boldsymbol{B} - \eta \boldsymbol{J}) - \nabla \phi, \tag{36}$$

and a convenient equation for eliminating the errors derived from $\nabla\cdot\boldsymbol{B}$,

$$\frac{\partial \phi}{\partial t} + c_h^2 \nabla \cdot \boldsymbol{B} = -\frac{c_h^2}{c_p^2}\phi, \tag{37}$$

together with the equation of motion (27) and Ampere's law (29). Using Eq. (36), Eq. (37) can be changed to

$$\frac{\partial^2(\nabla \cdot \boldsymbol{B})}{\partial t^2} + \frac{c_h^2}{c_p^2}\frac{\partial(\nabla \cdot \boldsymbol{B})}{\partial t} = c_h^2 \nabla^2 (\nabla \cdot \boldsymbol{B}), \tag{38}$$

where c_h and c_p correspond to the advection and diffusion coefficients; this plays a role in propagating and diffusing the numerical errors of $\nabla\cdot\boldsymbol{B}$. The main advantage of this method is that it can be implemented very easily without significantly changing the numerical code. Another advantage is that this method is less computationally demanding than the projection method. These advantages were demonstrated by Inoue et al. (2014b).

The vector potential is specified to maintain the solenoidal condition. Using the vector potential, the induction equation can be written as

$$\frac{\partial \boldsymbol{A}}{\partial t} = -\boldsymbol{E} - \nabla \Phi, \tag{39}$$

where $E=\eta J-v\times B$ and Φ is the gage. Several papers have used the NLFFF extrapolation (e.g.,van Ballegooijen et al. (2000) and Cheung and DeRosa (2012)). In this case, the solution is sought under the proper boundary conditions and gage. Simply, B_z and J_z are fixed at the boundary (i.e., A_x and A_y are fixed), then A_z is obtained from $\nabla^2 A=J$ under the Coulomb gage $\nabla\cdot A=0$. A solution obtained by this method will completely satisfy the solenoidal condition. On the other hand, there is no guarantee that the horizontal components at the bottom surface, which are obtained by iteration, will match observed values.

The constrained transport (CT) method (Brackbill and Barnes 1980; Evans and Hawley 1988) uses a numerical differential approach to maintaining the solenoidal condition. When the magnetic field $\boldsymbol{B}$ and electric field $\boldsymbol{E}$ are defined at the face center and edge centers of each numerical cell, i.e.,

$$\frac{d}{dt}B_{xi+1/2,j} = -\frac{E_{zi+1/2,j+1/2} - E_{zi+1/2,j-1/2}}{\Delta y},$$
$$\frac{d}{dt}B_{yi,j+1/2} = \frac{E_{zi+1/2,j+1/2} - E_{zi-1/2,j+1/2}}{\Delta x}, \tag{40}$$

where symmetry is assumed in the z direction, then the solenoidal condition is automatically satisfied:

$$\frac{d}{dt}\left(\frac{B_{xi+1/2,j} - B_{xi-1/2,j}}{\Delta x} + \frac{B_{yi,j+1/2} - B_{yi,j-1/2}}{\Delta y}\right) = 0. \tag{41}$$

However, the solenoidal condition requires consistent interaction with the boundary condition, and thus, it might be difficult to use it with the NLFFF calculations, which require the three components of the photospheric magnetic field.

Optimization method. Wheatland et al. (2000) proposed an optimization method that was later improved by Wiegelmann (2004). This method iteratively minimizes a function L related to $\boldsymbol{J}\times\boldsymbol{B}$ and $\nabla\cdot\boldsymbol{B}$. First, we define a function L as

$$L = \int \left(\frac{1}{B^2}|(\nabla\times \boldsymbol{B})\times \boldsymbol{B}|^2 + |\nabla\cdot \boldsymbol{B}|^2\right) dV; \tag{42}$$

it is the sum of the Lorentz force and the solenoidal condition, and its value is prescribed to be zero in order to satisfy the force-free condition. The time derivative is expressed as

$$\frac{1}{2}\frac{dL}{dt} = -\int \frac{\partial \boldsymbol{B}}{\partial t}\cdot \boldsymbol{F} dV - \int \frac{\partial \boldsymbol{B}}{\partial t}\cdot \boldsymbol{G} dS, \tag{43}$$

where $\boldsymbol{F}$ and $\boldsymbol{G}$ are high-order differential equations in terms of $\boldsymbol{B}$. If the function $\boldsymbol{F}$ satisfies

$$\frac{\partial \boldsymbol{B}}{\partial t} = \mu \boldsymbol{F}, \tag{44}$$

and if the magnetic fields on the surface vanish at infinity, then the L monotonically decreases. The problem is then reduced to iteratively finding the steady state the time-dependent magnetic field $\boldsymbol{B}$ that satisfies Eq. (44).

NLFFF extrapolation using the observed images. van Ballegooijen (2004) modeled a filament by inserting a twisted magnetic flux tube, whose axis was along the observed filament, into a potential field, with the magnetofriction (van Ballegooijen et al. 2000) driving the system toward the force-free state. In this case, although the horizontal fields were not used, the filament and the sigmoid structure were satisfactorily reproduced (Bobra et al. 2008; Su et al. 2009; Savcheva et al. 2012). Rather than using the methods accounting for the photospheric horizontal fields, modeling the filaments in the quiet region would be very useful because the values are very weak and the directions are random, so this might depend on the observations. In an attempt to obtain consistent magnetic fields, several studies have considered the topology of the coronal loops obtained from images, in addition to accounting for the photospheric magnetic field (Aschwanden et al. 2014; Malanushenko et al. 2014).

Unfortunately, the NLFFF does not allow the full calculation of the coronal magnetic fields. First of all, because, in general, the photospheric magnetic field cannot satisfy the force-free state, there is a contradiction between the bottom and inner regions; consequently, the 3D-reconstructed field also deviates from the force-free state. Furthermore, although several methods have been developed for exploring the NLFFF, there are no guarantees that there is a unique solution that fits the photospheric magnetic field applied to a given boundary condition. In the NLFFF approach, there are several open problems related to the free magnetic energy or the topologies of the magnetic fields (Schrijver et al. 2008; De Rosa et al. 2009). Thus, there is a need for confirmation of the reliability of this approach.

NLFFF extrapolation applied to a reference field (Low and Lou 1990)

The above methods for the NLFFF reconstruction have been applied to the photospheric magnetic field observed in the solar active region. Most of these methods required knowledge of the reference magnetic field in order to determine to what extent the reconstructed field approaches the force-free state. One of the widely known solutions is a semi-analytical force-free field that was presented by (Low and Lou 1990). These authors found a force-free solution in spherical coordinates, where symmetry was assumed in the ϕ direction:

$$\boldsymbol{B} = \frac{1}{r\sin\theta}\left(\frac{1}{r}\frac{\partial A}{\partial \theta}r, -\frac{\partial A}{\partial r}\theta, Q\phi\right), \tag{45}$$

where A and Q are functions of r and θ. The force-free Eq. (1) can be rewritten as

$$\frac{\partial^2 A}{\partial r^2} + \frac{1-\mu^2}{r^2}\frac{\partial^2 A}{\partial \mu^2} + Q\frac{dQ}{dA} = 0, \tag{46}$$

where $\mu=\cos(\theta)$ and $\alpha=dQ/dA$. It can be further rewritten as

$$(1-\mu^2)\frac{\partial^2 P}{\partial \mu^2} + n(n+1)P + a^2\frac{1+n}{n}P^{1+2/n} = 0, \tag{47}$$

using a separable solution

$$A(r,\theta) = \frac{P(\mu)}{r^n}, \tag{48}$$

and as

$$Q(A) = aA^{1+2/n}. \tag{49}$$

These formulas are obtained under the assumption of a vanishing magnetic field as r→0, i.e., for positive n. Although we can write down the 1D differential equation with respect to P(μ), shown as Eq. (47), it cannot be solved analytically due to its nonlinearity. The solution of this equation, therefore, is obtained numerically. The boundary condition is that P = 0 at μ = −1 and 1, which was originally set by Low and Lou (1990), and the solution is called the Low and Lou solution. The boundary conditions are that B_θ and B_ϕ vanish along the axis, and the differential equation can be solved as a boundary value problem. One of the solutions is shown in Fig. 5a; here, the solution was transformed to Cartesian coordinates, and n = 1 and a^2 = 0.425 are assumed (see Low and Lou (1990) for details). The accuracy of the NLFFF was checked using this solution as the reference magnetic field.

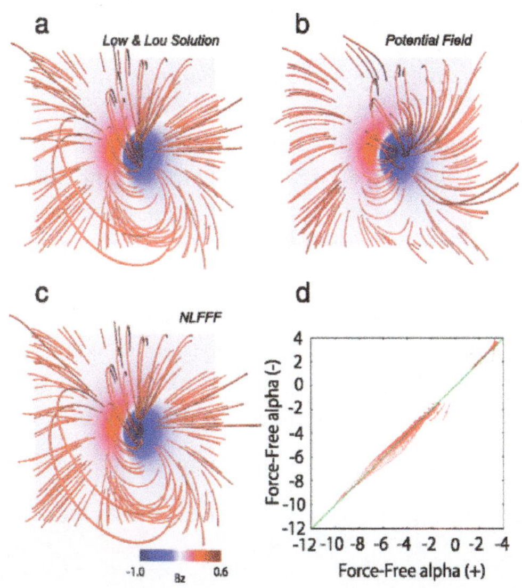

Figure 5: Semi-analytical Low and Lou solution and the NLFFF solution. **a** The magnetic field lines of the Low and Lou solution with the B_z distribution are shown in blue and red. **b** The potential field extrapolated from only the normal component of the magnetic field, using the Low and Lou solution on all boundaries. **c** The NLFFF solution based on the MHD relaxation method (Inoue et al. 2014b), extrapolated from all three components of the magnetic field of the Low and Lou solution on all boundaries. **d** Distribution of the force-free α from Inoue et al. 2014b, where the horizontal andvertical axes correspond to the force-free α measured at the field lines footpoints. The green line has a slope of unity (i.e., y=x). The image in (**d**) is copyright AAS and is reproduced by permission

Schrijver et al. (2006) estimated the accuracy of the NLFFF as reconstructed by various different methods; this included a semi-analytical force-free solution introduced by Low and Lou (1990). Their results suggest that the reconstruction accuracy is strongly method dependent, i.e., several methods satisfactorily captured the Low and Lou solution, although other methods failed. On the other hand, during the past decade, many efforts have been made to improve the numerical code for the NLFFF reconstruction (Amari et al. 2006; Valori et al. 2007; He and Wang 2008; Wheatland and Leka 2011; Jiang and Feng 2012; Inoue et al. 2014b).

Below, we review the results based on a recent extrapolation method that was proposed byInoue et al. (2014b) and is based on the MHD relaxation method. The potential field was reconstructed, based only on the normal component of the boundary magnetic field. This result is shown in Fig. 5b and differs significantly from the Low and Lou solution. Next, the reconstructed horizontal fields at the bottom surface were replaced by those of the Low and Lou solution, following which the magnetic fields in the domain were iteratively relaxed according to the equation of motion (27), the induction Eq. (36), Amperes law (29), and Eq. (37), which was used to correct the errors in $\nabla\cdot\boldsymbol{B}$.

During the iterations, at all boundaries, the vector $\boldsymbol{B}$ was fixed to be equal that in the Low and Lou solution, the velocity was set to zero, and the Neumann condition was imposed on ϕ, i.e.,$\partial\phi/\partial n$=0, where n is the direction perpendicular to the boundaries. In order to avoid a large discontinuity between the bottom and the inner domain, the velocity field was adjusted as follows. We defined $v^*=|\boldsymbol{v}|/|\boldsymbol{v}_A|$, and if v^* became larger than v_{max}, the velocity was modified as follows:

$$\boldsymbol{v} \Rightarrow \frac{v_{\max}}{v^*}\boldsymbol{v}, \tag{50}$$

where, v_{max}= 1.0. The resistivity was given as follows:

$$\eta = \eta_0 + \eta_1 \frac{|\boldsymbol{J} \times \boldsymbol{B}||\boldsymbol{v}|}{|\mathbf{B}|}, \tag{51}$$

where $\eta_0=3.75\times10^{-5}$ and $\eta_1=1.0\times10^{-3}$ (both are non-dimensional). The second term was introduced to accelerate the relaxation to the force free field, particularly in a weak-field region. In this study, c_h^2 and c_p^2 were set to 5.0 and 0.1, respectively; these values were selected by trial and error and depend on the boundary conditions, but it is best if the value of c_h is first set to account for the CFL condition. The viscosity was assumed as $\nu= 1.0\times10^3$; the viscosity also plays an important role in smoothly connecting the boundaries and nearby inner region, which indirectly helps our MHD calculation. A more detailed explanation of this was presented byInoue et al. (2014b).

Eventually, the final state obtained by using this method almost completely reproduced the Low and Lou solution, as shown in Fig. 5c. Quantitative results were also presented. The force-free αwas measured at both footpoints of all field lines, and this is shown in Fig. 5d. The force-free αmust be constant along the field lines, following Eq. (14), and from Fig. 5d, it can be concluded that this relation is satisfied. In addition, the authors quantitatively evaluated the accuracy by following Schrijver et al. (2006), evaluating

$$\begin{aligned}
C_{\text{vec}} &= \frac{\sum_i \boldsymbol{B}_i \cdot \boldsymbol{b}_i}{\sum_i |\boldsymbol{B}_i|^2 \sum_i |\boldsymbol{b}_i|^2},\\
C_{\text{cs}} &= \frac{1}{N}\sum_i \frac{\boldsymbol{B}_i \cdot \boldsymbol{b}_i}{|\boldsymbol{B}_i||\boldsymbol{b}_i|},\\
1 - E_{\text{M}} &= \frac{1}{N}\sum_i \frac{|\boldsymbol{b}_i - \boldsymbol{B}_i|}{|\boldsymbol{B}_i|},\\
1 - E_{\text{N}} &= \frac{\sum_i |\boldsymbol{b}_i - \boldsymbol{B}_i|}{\sum_i |\boldsymbol{B}_i|},\\
\epsilon &= \frac{\sum_i |\boldsymbol{b}_i|^2}{\sum_i |\boldsymbol{B}_i|^2},
\end{aligned} \tag{52}$$

where $\boldsymbol{B}$ and $\boldsymbol{b}$ are Low and Lou solution (reference solution) and the extrapolated solution, respectively, C_{vec} is the vector correlation, C_{cs} is the Cauchy-Schwarz inequality, E_M is the mean vector error, E_N is the normalized vector error, ε is the energy ratio, and N is the number of vectors in the field. Inoue et al. (2014b) obtained C_{vec} = 1.0, C_{cs} = 1.0, $1-E_N$ = 0.97, $1-E_M$ = 0.95,ε = 1.02, and these values were estimated over the entire region, which was divided into 64 × 64 × 64 grids (see Inoue et al. 2014b for details). They confirmed that the NLFFF can be reconstructed with high accuracy. Most of the recently developed methods allow for the recording of these values. Thus, it is possible to achieve force-free field extrapolation if the boundary condition completely satisfies the force-free condition.

NLFFF extrapolation applied to the solar active region

3D magnetic fields in the solar active region

In contrast to the NLFFF extrapolation using the Low and Lou solution, some problems arise when the bottom boundary is applied to the photospheric magnetic field. Schrijver et al. (2008)performed the NLFFF extrapolations by using the photospheric magnetic field observed by the Hinode satellite, corresponding to the period of 6 h before the X3.4-class flare that occurred in the solar active region 10930 on 13 December 2006. Different methods were applied for the NLFFF extrapolation. The authors pointed out a method-dependent accumulation of the free magnetic energy in the NLFFF. According to their calculations, a single NLFFF could yield sufficient free magnetic energy to produce an X-class flare. De Rosa et al. (2009) also performed the NLFFF extrapolation using different methods and for a different another active region (AR10953). They reported method-dependent configurations of the magnetic fields. From these results, it appeared that the NLFFF required further development.

Although the NLFFF remains problematic and does not enable the complete reproduction of the coronal magnetic field on the basis of photospheric data, several recent studies had roughly captured the field lines observed in EUV images, as well as processes involving stored-and-released magnetic energy, helicity, and flares (e.g., Canou and Amari (2010); Inoue et al. (2013);Vemareddy et al. (2013); Jiang and Feng (2013); Malanushenko et al. (2014); Aschwanden et al. (2014); Amari et al. (2014).

In what follows, we describe NLFFF results based on the MHD relaxation method developed byInoue et al. (2014b); note that the above equations are identical to those used by Low and Lou. The potential field is first reconstructed as the initial condition, and the boundary conditions are almost identical to those in the previous calculation, except that the potential fields are now fixed at the side and top boundaries. The following procedure is used to determine the bottom boundary. During the iterative process, the transverse components ($\boldsymbol{B}_{BC}$) at the bottom boundary are evaluated according to

$$\boldsymbol{B}_{BC} = \zeta \boldsymbol{B}_{obs} + (1-\zeta)\boldsymbol{B}_{pot}, \tag{53}$$

where $\boldsymbol{B}_{obs}$ and $\boldsymbol{B}_{pot}$ are the transverse components of the observational and the potential field, respectively, and ζ is a coefficient ranging from 0 to 1. R is introduced as an indication parameter for the force-free state, defined as $R=\int|J\times B|^2dV$; when it drops below a critical value, denoted by R_{min}, then ζ increases as $\zeta=\zeta+d\zeta$, where d ζ is given as a parameter. As ζ approaches unity, $\boldsymbol{B}_{BC}$ becomes consistent with the observational data. The vector fields include spurious forces that produce a sharp jump from the photosphere to the interior domain, and the above process can help to reduce their effects. In this study, $R_{min}=5.0\times10^{-3}$, $d\zeta=0.02$, and $v_{max}= 0.01$. In the MHD equations, c2hch2 and c2pcp2 are given as constant values, 0.04 and 0.1, respectively, and $\nu=1.0\times10^{-3}$. The resistivity is included in Eq. (51), with $\eta_0=5.0\times10^{-5}$ and$\eta_1=1.0\times10^{-3}$. For further details, see Inoue et al. (2014b).

Figure 6a shows the photospheric magnetic field 90 min before the M6.6-class flare that occurred on 13 February 2011. These data were obtained by a helioseismic and magnetic imager (HMI; Scherrer et al. (2012)) onboard the solar dynamics observatory (SDO) satellite (Pesnell et al. 2012). The upper and lower panels in Fig. 6b show enlarged views of the central area in Fig.6a; the arrows derived from the horizontal magnetic fields in the potential field are shown in the upper panel, and those derived from the observed one are shown in the lower panel. Figure 6c, d shows the magnetic field lines in the potential field and in the NLFFF approximation, respectively, superimposed on Fig. 6a. In particular, the central part of the NLFFF, in which strong sheared field lines build up and the current density is enhanced significantly, differs from that of the potential field. Figure 6e shows the 171 Å EUV images for the time period in Fig.6a; these were acquired by an atmospheric imaging assembly (AIA; Lemen et al. (2012)) on board SDO. The same field lines as in Fig. 6d were superimposed on Fig. 6e. Because it can be clearly seen that most of the field lines roughly correspond to these obtained from the EUV image, the NLFFF appears to satisfactory reproduce the field lines in the observed EUV image.

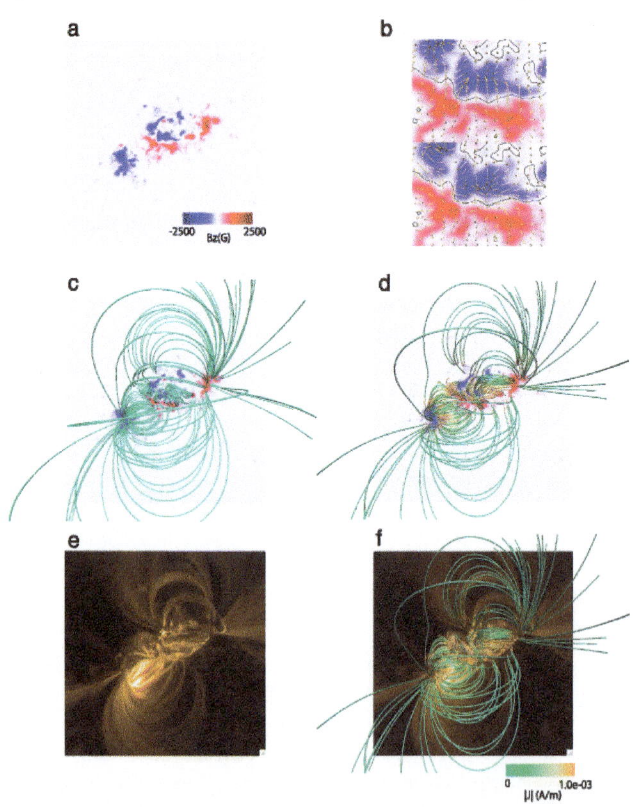

Figure 6: NLFFF for AR11158 at 16:00 UT on 13 February 2011 before a M6.6-class flare. **a** Photospheric magnetic field obtained by SDO/HMI, 90 min before the M6.6-class flare, with the B_z distribution plotted in red and blue. **b** The two panels show enlarged views of the central area in **a**; they show theB_z distribution and the horizontal fields with arrows, with the PIL in black. The upper and lower panels show the horizontal fields of the potential field and the observed fields, respectively. **c** The potential field (in green) is superimposed

on the data in **a**. **d** The NLFFF based on the MHD relaxation method (Inoue et al. 2014b) is plotted as in **c**, except that the strength of the current density is mapped onto the field line. **e** EUV images observed at 171 Å from the SDO/AIA at 16:00 UT on 13 February 2011. **f** The field lines, in the same format as in **d**, are superimposed on (**e**)

Stability analysis of the NLFFF

Magnetic field stability is an important issue in the studies of solar eruptions. Unfortunately, the photospheric magnetic field and the EUV or X-ray images do not allow for a quantitative stability analysis. On the other hand, the NLFFF might allow a quantitative analysis if we are given the 3D space information. One of the possible instabilities that can drive an eruption under the zero-beta assumption is a current driven, and one type current-driven instability is kink instability (Ji et al. 2003), which is determined by the magnetic twist of the poloidal field generated by the current in the flux tube (Kruskal and Kulsrud 1958). The magnetic twist (T_n) is related to the magnetic helicity; that is, the flux tube helicity is described by the following equation (Berger and Field 1984):

$$H = (T_n + W_r)\Phi^2, \tag{54}$$

where H is the magnetic helicity, Φ is the magnetic flux of the flux tube, and W_r is the magnetic writhe corresponding to the helical structure of the field line axis. The magnetic twist T_n indicates how much of the magnetic helicity is generated by the currents parallel to the flux tube (Berger and Prior 2006; Török et al. 2010); thus, T_n can be written as

$$T_n = \int \frac{dT_n}{ds} dl = \int \frac{J_{\|}}{4\pi |\boldsymbol{B}|} dl, \tag{55}$$

where || indicates the component parallel to the field line, and the line integral ∫dl is taken along the magnetic field line of the flux tube. Using $J_{\|}=\boldsymbol{J}\cdot\boldsymbol{B}/|\boldsymbol{B}|$, Eq. (55) can be further rewritten as

$$T_n = \frac{1}{4\pi} \int \frac{\boldsymbol{J}\cdot\boldsymbol{B}}{|\boldsymbol{B}|^2} dl. \tag{56}$$

If the magnetic fields meet the force-free condition, the magnetic twist can be written as

$$T_n = \frac{1}{4\pi} \int \alpha dl = \frac{1}{4\pi} \alpha L, \tag{57}$$

where α is the force-free alpha, and L is the length of the field line (Inoue et al. 2011; Inoue et al. 2012a). Inoue et al. (2012b) and Inoue et al. (2013) performed a stability analyses on the NLFFFs of AR10930 and AR11158, both of which produced X-class flares. Below, we describe the results of one of these twist analyses (for AR11158). AR 11158 produced an X2.2-class flare at 01:50 UT on 15 February 2011; it exhibited a quadruple field, as shown in Fig. 7a. The NLFFF based on the MHD relaxation method is shown in Fig. 7b; strong twisted lines were formed in the central region. The twist T_n was calculated for all field lines according to Eq. (56), and the result is shown in Fig. 7c. According to this result, most of the field lines were less than one turn, and none reached the critical twist of T_n=1.75, which is required for kink instability (Török et al. 2004). Therefore, it was concluded that the twisted lines prior to the X2.2-class flare produced by AR11158 would be stable with respect to kink instability. In another study, Jiang et al. (2014a) successfully reproduced a large twisted filament and checked its stability. It was reported that the twist did not reach the critical value required for kink instability. However, note that T_n in Eq. (56) is the local twist of an infinitesimal flux tube; this is not the same as the global twist of a macroscopic flux rope. In addition, there is no guarantee that the theoretical criteria are directly applicable to the NLFFF. In order to more strictly confirm the stability, a numerical stability analysis (Kusano and Nishikawa 1996; Inoue and Kusano 2006) and an MHD simulation would be useful.

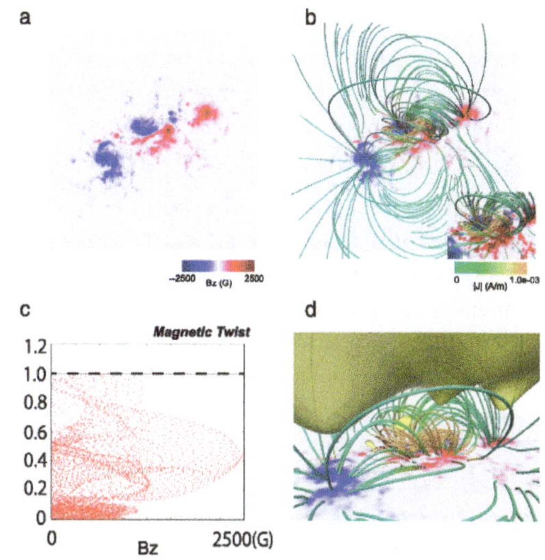

Figure 7: NLFFF for AR11158 at 00:00 UT on 15 February 2011 before a X2.2-class flare. **a** The B_z distribution of the photospheric magnetic field, approximately 2 h before the occurrence of the X2.2-class flare observed by SDO/HMI. **b** Magnetic field lines from the NLFFF are superimposed on **a**; the format of the field lines is the same as in Fig. 6d. The small inset corresponds to an enlarged view of the central area. **c** The magnetic twist distribution from Inoue et al. (2014a), where the vertical and horizontal axes are the twist and B_z, respectively. The dashed line corresponds to T_n = 1.0. The image is copyright AAS and reproduced by permission. **d** The magnetic field lines are plotted together with the surface corresponding to the critical height of the torus instability

Torus instability (Kliem and Török (2006); tested against observations by Liu (2008)) is also important for driving the flux tube into the upper corona, e.g., for triggering a CME (Isenberg and Forbes 2007; Aulanier et al. 2010; Démoulin and Aulanier 2010; Kliem et al. 2014). This instability is induced by a broken force balanced against the hoop force (Chen 1989), due to the flux tube current and the magnetic field suppressing the flux tube. The decay index,

$$n(z) = -\frac{z}{|\boldsymbol{B}|}\frac{\partial |\boldsymbol{B}|}{\partial z}, \tag{58}$$

is a convenient parameter (Kliem and Török 2006) because the location where this instability takes place is specified by n=1.5, which was already confirmed by several numerical studies (Török and Kliem 2007; Aulanier et al. 2010; Fan 2010). This stability analysis can be applied to the NLFFF analysis. For example, Guo et al. (2010) reconstructed the NLFFF using the optimization method (Wiegelmann 2004). In contrast to Inoue et al. (2011), they found strongly twisted lines over the critical twist of the kink instability and its writhe motion during the flare while a confined eruption was observed. They pointed out that even though the twisted lines in the NLFFF were not stable with respect to the kink instability, they were stable with respect to the torus instability, i.e., the flux tube remains within the magnetic field satisfying $n \leq 1.5$ during the eruption. Regarding the AR11158 studied by (Inoue et al. 2014a), the decay index at the twisted lines formed in the NLFFF cannot reach the critical value of the torus instability, as shown in Fig. 7d. Thus, the authors pointed out that the NLFFF was stable with respect to both torus instability and kink instability. On the other hand, for a different event, Jiang et al. (2014b) estimated the temporal evolution of the flux tube height obtained from the NLFFF in solar active region 11283, focusing on the X2.1-class flare that occurred at 22:20 UT on 6 September 2011. They found that the decay index at the flux rope axis reached the critical value for torus instability at the time at which the flare was generated, resulting in an instability-driven eruption.

As seen from these studies, the NLFFF enables us to quantitatively perform a stability analysis, which would be difficult to do based only on observations. Recently, highly accurate measurements of photospheric magnetic fields became available from two space satellites and ground observations; these have made the NLFFF a very useful tool for understanding the coronal magnetic field as well as for speculating on the onset and dynamics of solar flares.

MHD simulations of the solar eruptions based on the observational data

Necessity of MHD simulations combined with the NLFFF

Numerical modeling of the coronal magnetic field (potential field, LFFF, and NLFFF) successfully clarified many unknown issues with 3D magnetic fields that had not been revealed by observation. On the other hand, these models consider only the force-free equilibrium state, and they are thus not able to model dynamic states (in particular, energy-released processes) that occur during flare events, even though the buildup of energy occurs at a rate much slower than the Alfven time scale and thus can be handled by the NLFFF. MHD simulations can be used to reproduce such dynamic states.

The potential field does not strongly contribute to the magnetic field in the solar active region because there is no free energy available to induce dynamic behavior. For instance, Zuccarello et al. (2012) performed MHD simulations of solar eruptions, using the potential field as the initial condition. To obtain the solar eruption, the Poynting flux through the boundary was determined, and the authors provided the hypothetical shear and the convergence of the plasma on the solar surface. Consequently, the non-potential field was built up, and the sheared and converging motions helped to form the flux tube, resulting in an eruption (Figure 6 and Figure 8 in their paper). The hypothetical motions are important factors for building up the non-potential field, but these are much different from the observed ones. This means that there is a different process for the building up of energy, i.e., the magnetic field just prior to the onset of a flare deviates from the observed one. In contrast to this process, several studies inserted an analytical flux rope with a strong current and non-potentiality in a local area close to the PIL into the reconstructed potential field. Unfortunately, these flux tubes did not agree exactly with the observations, i.e., the boundary condition of the flux tube deviated greatly from the observations.

It might be possible to overcome the above problem by using MHD simulations with the NLFFF because the NLFFFs are constructed on the photospheric magnetic field, including the observed horizontal magnetic field on the solar surface. The motivations for using these simulations rather than the previous one are as follows: (i) It is likely that the artificial energy buildup process is not required by the existence of twisted motions because it already accounts for the observed twisting in the NLFFF. Although an additional process is required (discussed below) to create a new state that deviates from the NLFFF and produces eruptions, compared to the previous simulations, that process does not greatly deform the initial state. Therefore, MHD simulations can be performed under the photospheric magnetic field constraint. (ii) These simulations allow for the study of complex nonlinear dynamics, which could not be done previously. (iii) The results obtained from these simulations can be compared more exactly with observations, even indirect ones. Thus, these results contribute to confirming the reliability or to improving the MHD model. This field of study is emerging (Jiang et al. 2013), and only a few papers have yet been published. Below, we briefly discuss several of the pioneering studies.

MHD models of the solar eruptions, combined with the NLFFF

Overview of the recent studies. Jiang et al. (2013) were the first to perform the MHD simulation using the NLFFF to reproduce the X2.1-class flare in solar active region 11283. Their NLFFF, which was reconstructed by using the MHD relaxation method constructed in the modern MHD scheme (Feng et al. 2010), successfully captured the sigmoid structure of the magnetic field observed before the flare and demonstrated that the eruption was driven by the torus instability (Fig. 8a). An important advantage of this study seems to be that the same algorithm was used in both the NLFFF and MHD simulations. Kliem et al. (2013) also studied this eruption by setting the NLFFF as the initial condition of their MHD simulation (Fig. 8b). The NLFFF was reconstructed using the magnetic field observed on 8 April 2010, using the flux rope insertion and the magnetofrictional method. The NLFFF of this active region was thoroughly studied bySu et al. (2011). Kliem et al. (2013) found a critical value of the axial flux in the flux rope determined the stability. They reported that the criteria for the onset of a flare is that the axial flux be in the range of 5×10^{20} to 6×10^{20} Mx; in this case, the decay index is in the range of 1.3 to 1.8. For this eruption, the simulation results were in good agreement with some of the observations, such as those during the initial rising phase leading to the eruption. Amari et al. (2014) also successfully demonstrated a flux tube eruption in their MHD simulations, as shown in Fig. 8c. The flux tube was reconstructed by using the Grad-Rubin type method (Amari and Aly 2010) combined with the photospheric magnetic field observed by the Hinode solar optical telescope (SOT; Tsuneta et al. (2008)) 6 h before the X3.4-class flare in AR10930 at 02:40 UT on 13 December 2006. The authors found that 6 h before the flare, the NLFFF was destabilized with flux cancellation, the gas motion in characteristic of a sunspot moat flow or photospheric turbulent diffusion, and this resulted in the eruption. On the other hand,

2 days before the flare, the NLFFF predicted no eruption for the same situation. The authors pointed out the importance of the formation of a significantly large flux tube and the moving out from equilibrium.

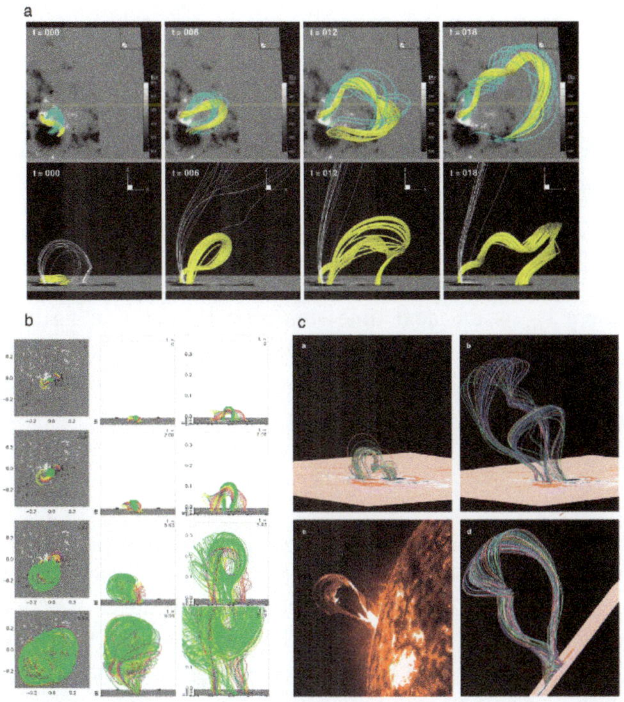

Figure 8: 3D MHD simulation based on the photospheric magnetic field. **a** The MHD modeling of the solar eruption from AR11283 associated with an X2.1-class flare observed on 11 September 2011 performed by Jiang et al. (2013). The image is copyright AAS and are reproduced by permission. **b** The MHD modeling of the solar eruption from AR11060 associated with a B3.7-class flare observed on 8 April 2010, performed by Kliem et al. (2013). The image is copyright AAS and reproduced by permission. **c**The MHD modeling of the solar eruption from AR10930 associated with an X3.4-class flare observed on 13 December 2006, performed by Amari et al. (2014). The images are from Nature reprinted by permission from Nature Publishing Group

MHD modeling of the solar eruption on 15 February 2011. Inoue et al. (2014a) and Inoue et al. (2015) studied the magnetic field dynamics during the X2.2-class flare produced by solar active region 11158 on 15 February 2011 (Schrijver et al. 2011; Janvier et al. 2014; Yang et al. 2014), by using MHD simulations combined with the NLFFF. Figure 7b shows the NLFFF structure approximately 2 h before the X2.2-class flare on 15 February 2011; note that strongly sheared magnetic fields lines are clearly visible at the PIL of the central sunspot. The stability analysis was discussed in a previous section. Based on these results, the NLFFF was quite stable, which implies that an additional process is required to drive the twisted lines. For instance, in a detailed data analysis, (Bamba et al. 2013) observed an increase in the small flux emerging at the PIL before the flare, and they suggested that this could destroy the stable magnetic field, as in the scenario described by Kusano et al. (2012).

The dynamics were investigated in the zero-beta MHD approximation, i.e., the density, pressure, and gravity were neglected. In such a situation, although the thermodynamics during the flare cannot be investigated, the magnetic field dynamics can be considered (Inoue et al. 2014a). This is the case because, during the flare, the magnetic energy converts into kinetic energy and thermal energy, which are the main factors for the energy store-and-release process in the solar corona. Therefore, in the early phase of a solar eruption that is not strongly compressible, zero-beta plasma is a good approximation, as demonstrated by Inoue and Kusano (2006). An advantage of this approximation is that it can neglect the sound waves, which often highly influence the rarefied CFL condition. As discussed in the previous section, the NLFFF reconstructed 2 h before the X2.2-class flare shows a stable equilibrium state, and the dramatic dynamics that appear in observations are not evident. Therefore, some additional process is required to break the stable equilibrium. Here,Inoue et al. (2014a) and Inoue et al. (2015) introduced an anomalous resistivity imposed on the strong current region, and the MHD relaxation was performed by using the NLFFF as the initial condition where the velocity adjustment defined in Eq. (50) was removed. We would expect the anomalous resistivity to induce reconnection in the region of strong current density (Yokoyama and Shibata 2001) and to produce long twisted lines in the NLFFF. After an additional iteration, since there is no guarantee that this new state can remain in equilibrium, a newly created flux tube might escape from the solar surface, as was shown in Amari et al. (2014). The anomalous resistivity was

$$\eta = \begin{cases} \eta_0 & J < j_c, \\ \eta_0 + \eta_2 \left(\frac{J - j_c}{j_c}\right)^2 & J > j_c, \end{cases} \tag{59}$$

where η_0 is the background resistivity and j_c is the threshold current necessary to excite the second term in Eq. (59) (Yokoyama and Shibata 1994). In this study, $\eta_0=1.0\times10^{-5}$, $\eta_2=1.0\times10^{-4}$, and $J_c = 30$. It can initiate and enhance the reconnection in the strong current region when the current is greater than the critical value, J_c. This value depends on the normalized value of the coronal magnetic field defined in each study.

Figure 9a shows two bundles of the twisted lines formed in the NLFFF; a strong current region was formed and sandwiched by these bundles. The side view is shown in Fig. 9b. We expect that reconnection takes place between the two bundles of the twisted lines, and long, strongly twisted lines are formed, which might break the equilibrium. After additional iterations with the anomalous resistivity, the single long bundle of strongly twisted lines shown in Fig. 9c, d was produced by the reconnection between the twisted lines formed in the NLFFF (shown in Fig.9a); which is reminiscent of the tether-cutting reconnection shown in Fig. 3e. The small inset shown in Fig. 9c shows the contour of one twist superimposed on the B_z map, where the footpoints for the part of the selected field lines plotted in Fig. 9c, d are anchored inside this contour. Note that there is no guarantee that this new state can remain in equilibrium because a flux tube composed of strongly twisted lines can escape from the solar surface (Amari et al. 2000;Kusano et al. 2012; Kliem et al. 2013).

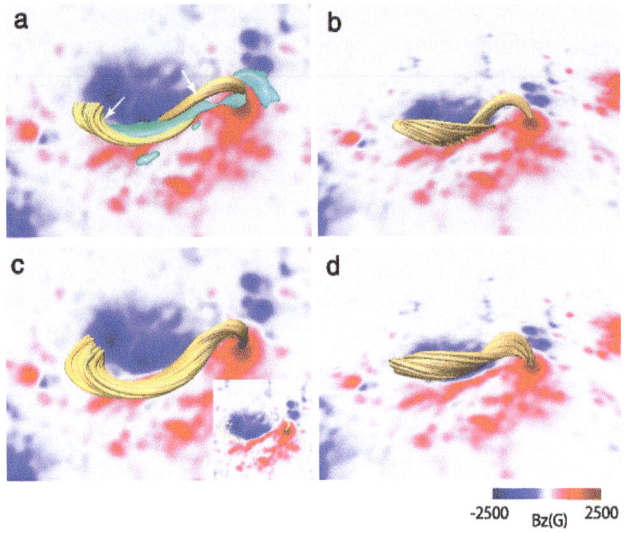

Figure 9: Twisted lines in the NLFFF and magnetic fields after further MHD relaxation process **a** The twisted lines in the NLFFF, reconstructed for 00:00 UT on 15 February 2011, together with the B_z distribution. The green surface corresponds to the isosurface of the current density $\boldsymbol{J}$ = 30, which is sandwiched by the twisted lines of the NLFFF. **b** The side view of **a**. **c** Strongly twisted lines are formed after the subsequent MHD relaxation process, which includes the anomalous resistivity. The small inset shows the contour (yellow) of one turn twist superimposed on the B_z map. **d** The side view of (**c**)

Next, an MHD simulation was executed using this new state, as shown in Fig. 9c; note that at the boundary, all components of the velocity are fixed to zero, and the normal component of $\boldsymbol{B}$ is fixed, while the horizontal one may vary, i.e., it is determined by the induction equation according to the dynamics. Consequently, as shown in Fig. 10, the equilibrium was broken, and a larger flux tube was formed and launched into the upper corona. Note that, in this process, the strongly twisted lines (in orange) that were formed during the initial state do not extend directly into the upper corona. Rather, they reconnect with the ambient field lines (in blue) that convert into the large flux tube. Interestingly, the strongly twisted lines in the initial state appear to convert into the post-flare loops often observed after a flare.

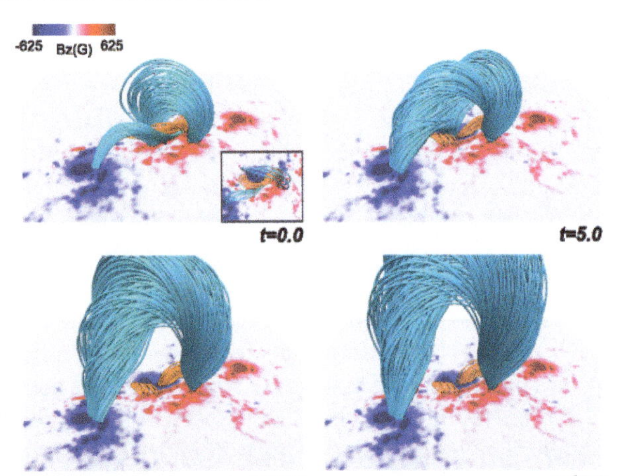

Figure 10: 3D dynamics of the flux tube during an X2.2-class flare. 3D dynamics of the flux tube during an X2.2-class flare obtained from our MHD simulation; the field lines with more (less) than one turn at t = 0 are depicted in orange (blue). The B_z distribution is shown in red and blue. The inset at t = 0 shows the top view of the field lines; the number of field lines with less than one turn has been reduced

These simulation results were compared with observations. The authors first confirmed that their simulation captures the shape of the two-ribbon flares. Following the CSHKP model, two-ribbon flares are generally considered to be due to the distribution of the footpoints of the reconnected field lines. Therefore, those could be reproduced to trace the reconnected field lines during a simulation. To achieve this, the authors traced the reconnected field lines by using the following equation:

$$\delta(\boldsymbol{x}_0, t_n) = |\boldsymbol{x}_1(\boldsymbol{x}_0, t_{n+1}) - \boldsymbol{x}_1(\boldsymbol{x}_0, t_n)|,$$

where t_{n+1} is the next time step after t_n, and $\boldsymbol{x}_1(\boldsymbol{x}_0,t_n)$ is the location of one footpoint of each field line at time t_n, which is traced from another footpoint at $\boldsymbol{x}_0$. Eventually, we calculate

$$\Delta(\boldsymbol{x}_0, t) = \int_0^t \delta(\boldsymbol{x}_0, t_n) dt_n, \qquad (60)$$

where $\Delta(\boldsymbol{x}_0,t)$ is a location where the length of a field lines is changed, meaning that the enhanced region corresponds to one in which there was a dramatic reconnection in the twisted lines. Figure 11a shows a 3D view of the field lines at t = 4.0, when the large flux tube has been formed during the initial launching phase. We first confirmed that the sheared two-ribbon profiles observed initially were reproduced in our simulation. Figure 11b shows the two-ribbon flares during the X2.2-class flare, observed by Hinode/SOT, at 01:50 UT, corresponding to the initial phase of the flare. Figure 11c shows the numerically calculated two-ribbon flares, following Eq. (60), at t = 4, reproduced in this simulation where Δ is chosen from the region in which T_n>0.3. The shape of the numerically calculated two-ribbon flares matches the observed one.

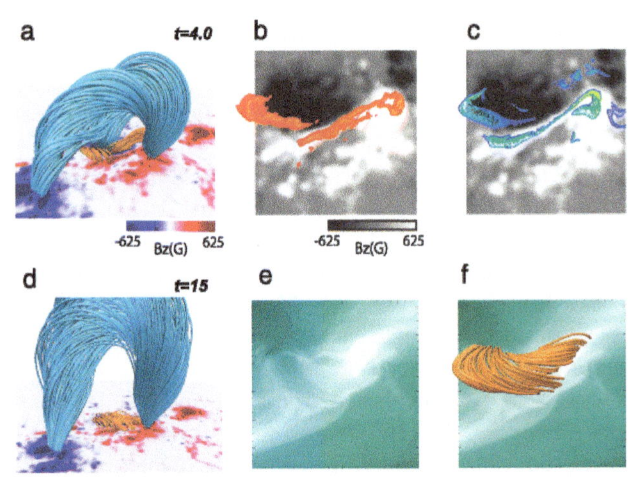

Figure 11: Comparison with observations, two-ribbon flares and the EUV image. **a** 3D magnetic field at t= 4.0 in the MHD simulation of Inoue et al. (2014a), showing the large flux tube with post-flare loops under it. These simulations tried to reproduce the observed sheared two-ribbon flares by using this initial launching phase. **b** Two-ribbon flares observed by Hinode/FG at 01:51 UT on 15 February 2011 during an X2.2-class flare from (Inoue et al. 2014a). The gray scale encodes the B_z distribution. **c** Two-ribbon flares reproduced by the MHD simulation of Inoue et al. (2014a) at t = 4.0 in **a**, in accordance with Eq. (60). **d** 3D magnetic field at t= 15 in the MHD simulation of Inoue et al. (2014a), showing the large ascending flux tube with post-flare loops under it. **e** The EUV image after an X2.2-class

flare at 02:29:50 UT on 15 February 2011, at 94 Å, obtained by SDO/AIA. **f** The field lines obtained from the MHD simulation by Inoue et al. (2014a) are superimposed on the data in **e**. Panels **b**, **c**, **e**, and **f** are copyright AAS and reproduced by permission

These simulation results were further compared with the EUV image data obtained fromSDO/AIA. Figure 11d shows the 3D magnetic structure at t = 15, clearly revealing the post-flare loops above which the large eruptive flux tube is ascending. We confirmed that the post-flare loops can capture the field lines in the EUV image, using simulation data at t= 15. Figures 11eshows the EUV image observed after the flare by 94 Å of SDO/AIA, and Fig. 11f shows the field lines at t = 10 superimposed on the EUV image. The field lines observed in the EUV image were successfully captured.

Finally, enhancement of the horizontal magnetic field B_t was discussed by Inoue et al. (2015). As shown in Fig. 12a, Wang et al. (2012) found a rapid enhancement of the horizontal field on the PIL in the photosphere, and they suggested that this was due to the reconnection. The simulation of Inoue et al. (2015) also indicated this enhancement, and the result is shown in Fig.12c calculated for the area shown in Fig. 12b. It was pointed out that the post-flare loops can be observed even during an early phase in which the horizontal fields are enhanced, and the new post-flare loops are subsequently produced through the above reconnection. Consequently, it was suggested that this enhancement is due to the accumulation of post-flare loops suppressing the pre-existing loops. Therefore, this enhancement would be strongly related to the reconnection. However, since this simulation was performed in the zero-beta MHD, to support this conclusion, it is necessary to have a more detailed analysis and discussion under more realistic assumptions, including high- β regimes corresponding to the chromosphere and the photosphere.

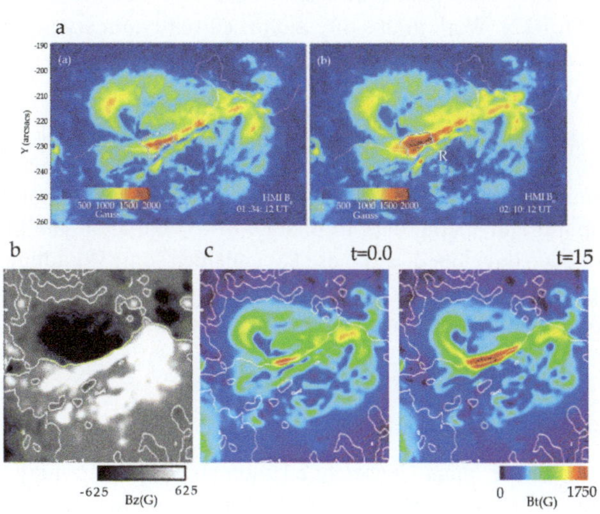

Figure 12: Enhancement of B_t during an X2.2-class flare. **a** The observations of B_t enhancement in the photosphere during an X2.2-class flare, reported by Wang et al. (2012). **b** The B_z distribution for which the B_t enhancement was measured in the MHD simulation by Inoue et al. (2015). The white contour line corresponds to the PIL. **c**B_t enhancements as observed in the MHD simulation by Inoue et al. (2015). All images are copyright AAS and are reproduced by permission

A summary of the dynamics is shown in Fig. 13. (i) The NLFFF is quite stable for the current-driven ideal MHD instability, so it is necessary to have a trigger process (here, the tether-cutting reconnection) to break the equilibrium. (ii) The tether-cutting reconnection creates strongly twisted lines in the NLFFF. Consequently, it breaks the equilibrium and reconnects with the ambient field lines. The result is that a large flux tube is formed as it ascends. (iii) Eventually, the flux tube will grow into a CME if the threshold of the torus instability is exceeded or equilibrium is lost.

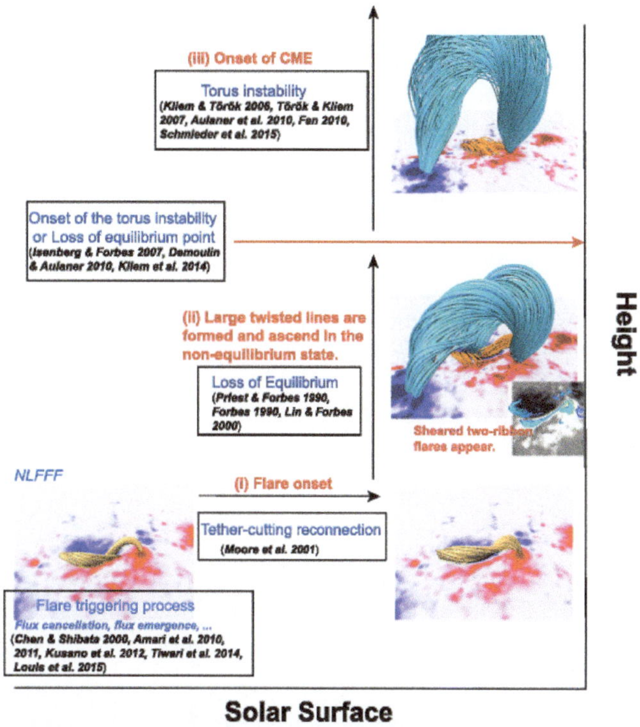

Figure 13: Summary of the dynamics of the magnetic field during an X2.2-class solar flare. Summary of the magnetic field dynamics during an X2.2-class solar flare, obtained from Inoue et al. (2014a) and Inoue et al. (2015)

CONCLUSIONS

The solar physics satellites Hinode and SDO, together with modern ground-based telescopes, provide photospheric magnetic field data with unprecedented accuracy. This enables us to reconstruct the 3D coronal magnetic field with high accuracy, such that it includes the potential field from the normal component not only of the photospheric magnetic field but also of the NLFFF, which contains both the normal and the horizontal magnetic fields. Because the NLFFF is reconstructed to include information about the horizontal magnetic fields at the photosphere, it can yield a 3D magnetic field close to that observed in the active regions instead of the one similar to that of the potential field, and it can show the accumulation of free magnetic energy and helicity that is required to produce a flare. In addition, the force-free α is given as a function of space, and so it is not an LFFF approximation. Therefore, the NLFFF can yield the magnetic configuration both before and after the flare, and several papers have reported various important physical quantities obtained from the NLFFF, including the free magnetic energy (Sun et al. 2012; Jiang et al. 2014b), the magnetic helicity (Thalmann et al. 2011; Valori et al. 2012;Pevtsov et al. 2014), and the magnetic twist and topology (Inoue et al. 2011; Guo et al. 2013;Inoue et al. 2013; Zhao et al. 2014). These quantities quantify the NLFFF stability, which cannot be obtained from observations. On the other hand, there is a problem in the NLFFF itself. Using the same format as in Fig. 5d, Fig. 14 shows the distribution of the force-free α measured at both footpoints of each field line for the NLFFF in Fig. 7b and the temporal evolution of $\int|\nabla\cdot B|2\int|\nabla\cdot B|2dV$ during the iteration of the NLFFF. Although the value of $\int|\nabla\cdot B|2\int|\nabla\cdot B|2dV$ is reduced to fourth order, the distribution of the force-free α is scattered. Therefore, an unexpected physical element, the residual force, remains; this is inevitably produced near the boundary in the NLFFF, due to the contradiction between the boundary and the inner domain. In addition, it should be noted that the coronal magnetic fields cannot be correctly reproduced only by the NLFFF. Peter et al. (2015) pointed out several limitations on the free energy and accumulated currents. Furthermore, reconstruction of the geometry of bright loops requires methods more advanced than the NLFFF (Aschwanden et al. 2014; Malanushenko et al. 2014). Therefore, a model more advanced than the NLFFF is required to construct the equilibrium state with high accuracy and overcome these limitations.

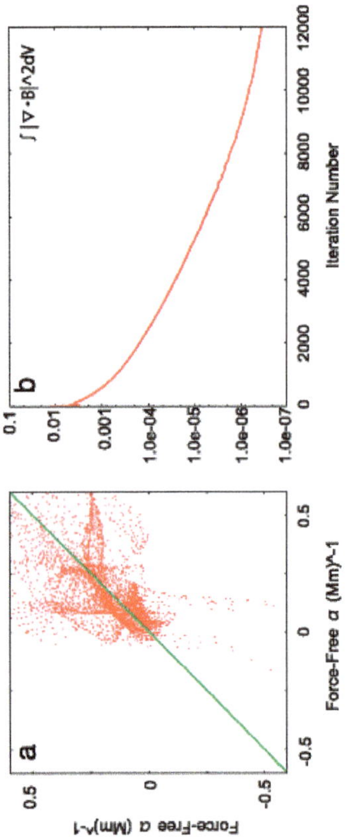

Figure 14: Force-freeness of the NLFFF. **a** Distribution of the force-free α measured at both footpoints of each magnetic field line for the NLFFF in Fig. 7b, that is from Inoue et al. (2014a). The image is copyright AAS and reproduced by permission. **b** The temporal evolution of $\int\nabla\cdot BdV\int\nabla\cdot BdV$ during the iteration in which the NLFFF is attained

Although the NLFFF yields the 3D properties of the magnetic field, this method does not reveal the dynamics of the solar flares. To determine the dynamics in a realistic situation, NLFFF results have been used as initial conditions for MHD simulations (Jiang et al. 2013; Kliem et al. 2013; Inoue et al. 2014a; Amari et al. 2014; Inoue et al. 2015). Because these simulations were constrained by the photospheric magnetic field, there are large artificial processes causing the buildup of energy; these likely yield twist and sheared motions, which were not assumed. Important and realistic physical processes are also being revealed, including the critical value for the flux of the flux tube for an eruption (Kliem et al. 2013) or the formation of a large flux tube producing a CME (Inoue et al. 2014a; 2015). Furthermore, the reliability of these simulations can be confirmed because they can be more precisely compared with the observations likely by Inoue et al. (2014a) and Inoue et al. (2015), in contrast to previous simulations that described hypothetical situations. Note that several can be indirectly compared, e.g., the two-ribbon flares discussed in this study. In order to provide a strict confirmation, however, a direct comparison is required (e.g., (Mikić et al. 2013)).

Some problems and questions related to these simulations still remain to be answered. For instance, as discussed above, the reconstructed field does not completely achieve a force-free state, and so the residual force must be treated carefully. If these residual forces are sufficiently strong, they may affect the magnetic field dynamics, and the interpretation of the dynamics becomes difficult. In addition to this, as Inoue et al. (2015) pointed out, the magnetic twist accumulated in the NLFFF might be gradually reduced throughout the numerical diffusion and also on the solar surface because the NLFFF returns to a lower energy level without retaining the observed horizontal magnetic fields. Furthermore, it is important to account for the observed process that triggers the solar flares in order to understand the conversion of the stable magnetic field into a dynamic one. Recently, the triggering was observed by using state-of-art data (e.g., Green et al. (2011); Bamba et al. (2013); Louis et al. (2015)). These data must be incorporated into simulations. Although most simulations start from an NLFFF that is already composed of twisted and sheared field lines, some studies attempted to recover the processes leading from the buildup to the release of energy; in this data-driven simulation, the coronal magnetic field was driven by the time-dependent photospheric magnetic field e.g., Cheung and De Rosa (2012). Work in this direction is currently underway, and this will be extended in the future.

With advanced computational resources now more readily available, more-refined 3D numerical MHD models of solar eruptions are being developed and improved. Recently, techniques combining simulations with highly resolved temporal and spatial data from state-of-the-art solar satellites have been developed, and these have yielded some preliminary results. In the future, it is likely to be necessary to further develop simulations of solar flares in order to more closely correspond to these observations.

ACKNOWLEDGEMENTS

We thank the Japan Geoscience Union (JPGU) for inviting us to the JPGU 2014 meeting. We are grateful to the referees for providing many constructive comments and for suggesting ways in which to improve this paper. We are grateful to one of the science editors, Dr. Tsutomu Nagatsuma, for encouraging us, and to the MPS members for useful discussions. We offer special thanks to Dr. Takahiro Miyoshi and Dr. Vinay Shankar Pandey for checking a part of this paper. S. I. thanks the Alexander Von Humboldt Foundation for supporting our work and for providing a precious opportunity to work in Germany. This work was also supported by JSPS KAKENHI Grant Number 15H05814 (PI: K. Kusano). The computational work was carried out within the computational joint research program at the Institute for Space-Earth Environmental Research, Nagoya University. The computer simulation was performed on the Fujitsu PRIMERGY CX400 system of the Information Technology Center, Nagoya University. Visualization was performed by VAPOR (Clyne and Rast 2005; Clyne et al. 2007). Finally, S. I. sincerely thanks Professor Kanya Kusano for providing an opportunity to engage in this work and for many constructive comments and discussions.

REFERENCES

1. Altschuler, MD, Newkirk G (1969) Magnetic fields and the structure of the solar corona. I: Methods of calculating coronal fields. Solar Phys 9: 131–149. doi:10.1007/BF00145734.
2. Amari, T, Aly J-J (2010) Observational constraints on well-posed reconstruction methods and the optimization-Grad-Rubin method. Astronomy Astrophys 522: A52. doi:10.1051/0004-6361/200913058.
3. Amari, T, Aly J-J, Luciani J-F, Mikic Z, Linker J (2011) Coronal mass ejection initiation by converging photospheric flows: toward a realistic model. Astrophys J Lett 742: L27–L32. doi:10.1088/2041-8205/742/2/L27.
4. Amari, T, Aly J-J, Mikic Z, Linker J (2010) Coronal mass ejection initiation: on the nature of the flux cancellation model. Astrophys J Lett 717: L26–L30. doi:10.1088/2041-8205/717/1/L26.
5. Amari, T, Boulmezaoud TZ, Aly JJ (2006) Well posed reconstruction of the solar coronal magnetic field. Astronomy Astrophys 446: 691–705. doi:10.1051/0004-6361:20054076.
6. Amari, T, Canou A, J-Aly J (2014) Characterizing and predicting the magnetic environment leading to solar eruptions. Nature 514: 465–469.
7. Amari, T, Luciani JF, Aly JJ, Mikic Z, Linker J (2003a) Coronal mass ejection: initiation, magnetic helicity, and flux ropes. I. Boundary motion-driven evolution. Astrophys J 585: 1073–1086. doi:10.1086/345501.
8. Amari, T, Luciani JF, Aly JJ, Mikic Z, Linker J (2003b) Coronal mass ejection: initiation, magnetic helicity, and flux ropes. II. Turbulent diffusion-driven evolution. Astrophys J 595: 1231–1250. doi:10.1086/377444.
9. Amari, T, Luciani JF, Mikic Z, Linker J (2000) A twisted flux rope model for coronal mass ejections and two-ribbon flares. Astrophys J Lett 529: L49–L52. doi:10.1086/312444.
10. An, JM, Magara T (2013) Stability and dynamics of a flux rope formed via flux emergence into the solar atmosphere. Astrophys J 773: 21. doi:10.1088/0004-637X/773/1/21.
11. Antiochos, SK, DeVore CR, Klimchuk JA (1999) A model for solar coronal mass ejections. Astrophys J 510: 485–493. doi:10.1086/306563.
12. Archontis, V, Hood AW, Tsinganos K (2014) Recurrent explosive eruptions and the "sigmoid-to-arcade" transformation in the Sun driven by dynamical magnetic flux emergence. Astrophys J Lett 786: L21. doi:10.1088/2041-8205/786/2/L21.
13. Asai, A, Ishii TT, Kurokawa H, Yokoyama T, Shimojo M (2003) Evolution of conjugate footpoints inside flare ribbons during a great two-ribbon flare on 2001 April 10. Astrophys J 586: 624–629. doi:10.1086/367694.
14. Aschwanden, MJ, Sun X, Liu Y (2014) The Magnetic field of active region 11158 during the 2011 February 12–17 flares: differences between photospheric extrapolation and coronal forward-fitting methods. Astrophys J 785: 34. doi:10.1088/0004-637X/785/1/34.
15. Aulanier, G, Janvier M, Schmieder B (2012) The standard flare model in three dimensions. I. Strong-to-weak shear transition in post-flare loops. Astron Astrophys 543: AA110. doi:10.1051/0004-6361/201219311.
16. Aulanier, G, Török T, Démoulin P, DeLuca EE (2010) Formation of torus-unstable flux ropes and electric currents in erupting sigmoids. Astrophys J 708: 314–333. doi:10.1088/0004-637X/708/1/314.
17. Bamba, Y, Kusano K, Yamamoto TT, Okamoto TJ (2013) Study on the triggering process of solar flares based on Hinode/SOT observations. Astrophys J 778: 48. doi:10.1088/0004-637X/778/1/48.
18. Benz, AO (2008) Flare observations. Liv Rev Sol Phys 5: 1.doi:10.1292/lrsp-2008-1.
19. Berger, MA, Field GB (1984) The topological properties of magnetic helicity. J Flu Mec 147: 133–148. doi:10.1017/S0022112084002019.
20. Berger, MA, Prior C (2006) The writhe of open and closed curves. J Phys A: Math General39: 8321–8348. doi:10.1088/0305-4470/39/26/005.
21. Bobra, MG, van Ballegooijen AA, DeLuca EE (2008) Modeling nonpotential magnetic fields in solar active regions. Astrophys J 672: 1209–1220. doi:10.1086/523927.
22. Brackbill, JU, Barnes DC (1980) The effect of nonzero product of magnetic gradient and B on the numerical solution of the magnetohydrodynamic equations. J Comp Phys 35: 426–430. doi:10.1016/0021-9991(80)90079-0.
23. Canou, A, Amari T (2010) A twisted flux rope as the magnetic structure of a filament in active region 10953 observed by Hinode. Astrophys J 715: 1566–1574. doi:10.1088/0004-637X/715/2/1566.
24. Carmichael, H (1964) A process for flares In: AAS/NASA Symposium on physics of solar flares, NASA SP-50, 451–456.. NASA, Washington, DC.
25. Carrington, RC (1859) Description of a singular appearance seen in the sun on September 1, 1859. MNRAS 20: 13–15.
26. Chen, J (1989) Effects of toroidal forces in current loops embedded in a background plasma. Astrophys J 338: 453–470. doi:1989ApJ...338..453C.
27. Chen, PF (2011) Coronal mass ejections: Models and their observational basis. Living Rev Solar Phys 8: 1. doi:10.12942/lrsp-2011-1.
28. Chen, PF, Shibata K (2000) An emerging flux trigger mechanism for coronal mass ejections. Astrophys J 545: 524–531. doi:10.1086/317803.
29. Cheng, X, Zhang J, Ding MD, Liu Y, Poomvises W (2013) The driver of coronal mass ejections in the low corona:

a flux rope. Astrophys J 763: 43. doi:10.1088/0004-637X/763/1/43.

30. Cheung, MCM, DeRosa ML (2012) A method for data-driven simulations of evolving solar active regions. Astrophys J 757: 147. doi:10.1088/0004-637X/757/2/147.
31. Chiu, YT, Hilton HH (1977) Exact Green's function method of solar force-free magnetic-field computations with constant alpha. I. Theory and basic test cases. Astrophys J 212: 873–885. doi:10.1086/155111.
32. Clyne, J, Mininni P, Norton A, Rast M (2007) Interactive desktop analysis of high resolution simulations: application to turbulent plume dynamics and current sheet formation. New J Phys 9: 301. doi:10.1088/1367-2630/9/8/301.
33. Clyne, J, Rast MM (2005) A prototype discovery environment for analyzing and visualizing terascale turbulent fluid flow simulations. Astrophysical and Planetary Sciences Faculty Contributions 5669: 284–294. Paper 22. Proceedings of SPIE-IS andT Electronic Imaging. doi:10.1117/12.586032.
34. Dedner, A, Kemm F, Kröner D, Munz CD, Schnitzer T, Wesenberg M (2002) Hyperbolic divergence cleaning for the MHD equations. J Comput Phys 175: 645–673. doi:10.1006/jcph.2001.6961.
35. Démoulin, P, Aulanier G (2010) Criteria for flux rope eruption: non-equilibrium versus torus instability. Astrophys J 718: 1388–1399. doi:10.1088/0004-637X/718/2/1388.
36. De Rosa, ML, Schrijver CJ, Barnes G (2009) A critical assessment of nonlinear force-free field modeling of the solar corona for active region 10953. Astrophys J 696: 1780–1791. doi:10.1088/0004-637X/696/2/1780.
37. Evans, CR, Hawley JF (1988) Simulation of magnetohydrodynamic flows: a constrained transport method. Astrophys J 332: 659–677. doi:10.1086/166684.
38. Fan, Y (2005) Coronal mass ejections as loss of confinement of kinked magnetic flux ropes. Astrophys J 630: 543–551. doi:10.1086/431733.
39. Fan, Y (2010) On the eruption of coronal flux ropes. Astrophys J 719: 728–736. doi:10.1088/0004-637X/719/1/728.
40. Feng, X, Yang L, Xiang C, ST Wu, Zhou Y, Zhong DK (2010) Three-dimensional solar WIND modeling from the sun to the earth by a SIP-CESE MHD model with a six-component grid. Astrophys J 723: 300–319. doi:10.1088/0004-637X/723/1/300.
41. Feynman, J, Martin SF (1995) The initiation of coronal mass ejections by newly emerging magnetic flux. J Geophys Res 100: 3355–3367. doi:10.1029/94JA02591.
42. Forbes, TG (1990) Numerical simulation of a catastrophe model for coronal mass ejections. J Goephys Res 95: 11,919–11,931. doi:10.1029/JA095iA08p11919.
43. Forbes, TG (2000) A review on the genesis of coronal mass ejections. J Goophys Res 105: 23,153–23,165. doi:10.1029/2000JA000005.
44. Forbes, TG, Priest ER (1995) Photospheric magnetic field evolution and eruptive flares. Astrophys J 446: 377–389. doi:10.1086/175797.
45. Gary, GA (2001) Plasma beta above a solar active region: rethinking the paradigm. Sol Phys 203: 71–86. doi:10.1023/A:1012722021820.
46. Golub, L, Deluca E, Austin G, et al. (2007) The X-ray telescope (XRT) for the Hinode mission. Sol Phys 243: 63–86. doi:10.1007/s11207-007-0182-1.
47. Green, LM, Kliem B, Wallace A (2011) Photospheric flux cancellation and associated flux rope formation and eruption. Astron Astrophys 526: A2-. doi:10.1051/0004-6361/201015146.
48. Guo, Y, Ding MD, Cheng X, Zhao J, Pariat E (2013) Twist accumulation and topology structure of a solar magnetic flux rope. Astrophys J 779: 157. doi:10.1088/0004-637X/779/2/157.
49. Guo, Y, Ding MD, Schmieder B (2010) Driving mechanism and onset condition of a confined eruption. Astrophys J Lett 725: L38–L42. doi:10.1088/2041-8205/725/1/L38.
50. He, H, Wang H (2008) Nonlinear force-free coronal magnetic field extrapolation scheme based on the direct boundary integral formulation. J Geophys Res Space Phys 113: A05S90. doi:10.1029/2007JA012441.
51. Hirayama, T (1974) Theoretical model of flares and prominences. I: Evaporating flare model. Sol Phys 34: 323–338. doi:10.1007/BF00153671.
52. Inoue, S, Hayashi K, Magara T, Choe GS, Park YD (2014a) Magnetohydrodynamic simulation of the X2.2 solar flare on 2011 February 15. I. Comparison with the observations. Astrophys J 788: 182. doi:10.1088/0004-637X/788/2/182.
53. Inoue, S, Hayashi K, Magara T, Choe GS, Park YD (2015) Magnetohydrodynamic simulation of the X2.2 solar flare on 2011 February 15. II. Dynamics connecting the solar flare and the coronal mass ejection. Astrophys J 803: 73. doi:10.1088/0004-637X/803/2/73.
54. Inoue, S, Hayashi K, Shiota D, Magara T, Choe GS (2013) Magnetic structure producing X- and M-class solar flares in solar active region 11158. Astrophys J 770: 79. doi:10.1088/0004-637X/770/1/79.
55. Inoue, S, Kusano K (2006) Three-dimensional simulation study of flux rope dynamics in the solar corona. Astrophys J 645: 742–756. doi:10.1086/503153.
56. Inoue, S, Kusano K, Magara T, Shiota D, Yamamoto TT (2011) Twist and connectivity of magnetic field lines in the solar active region NOAA 10930. Astrophys J 738: 161. doi:10.1088/0004-637X/738/2/161.
57. Inoue, S, Magara T, Pandey VS (2014b) Nonlinear force-free extrapolation of the coronal magnetic field based on the magnetohydrodynamic relaxation method. Astrophys J 780: 101–. doi:10.1088/0004-637X/780/1/101.
58. Inoue, S, Magara T, Watari S, Choe GS (2012a) Nonlinear force-free modeling of a three-dimensional sigmoid observed on the sun. Astrophys J 747: 65. doi:10.1088/0004-637X/747/1/65.
59. Inoue, S, Shiota D, Yamamoto TT (2012b) Buildup and release of magnetic twist during the X3.4 solar flare of 2006 December 13. Astrophys J 760: 17.

doi:10.1088/0004-637X/760/1/17.

60. Isenberg, PA, Forbes TG (2007) A three-dimensional line-tied magnetic field model for solar eruptions. Astrophys J 670: 1453–1466. doi:10.1086/522025.

61. Janvier, M, Aulanier G, Bommier V (2014) Electric currents in flare ribbons: observations and three-dimensional standard model. Astrophys J 788: 60. doi:10.1088/0004-637X/788/1/60.

62. Janvier, M, Aulanier G, Pariat E, Démoulin P (2013) The standard flare model in three dimensions. III. Slip-running reconnection properties. Astron Astrophys 555: A77. doi:10.1051/0004-6361/201321164.

63. Ji, H, Wang H, Schmahl EJ, Moon YJ, Jiang Y (2003) Observations of the failed eruption of a filament. Astrophys J Lett 595: L135–L138.

64. Jiang, C, Feng X (2012) A new implementation of the magnetohydrodynamics-relaxation method for nonlinear force-free field extrapolation in the solar corona. Astrophys J 749: 135. doi:10.1088/0004-637X/749/2/135.

65. Jiang, C, Feng X (2013) Extrapolation of the solar coronal magnetic field from SDO/HMI magnetogram by a CESE-MHD-NLFFF. Code Astrophys J 769: 144. doi:10.1088/0004-637X/769/2/144.

66. Jiang, C, Feng X, Wu ST, Hu Q (2013) Magnetohydrodynamic simulation of a sigmoid eruption of active region 11283. Astrophys J Lett 771: L30. doi:10.1088/2041-8205/771/2/L30.

67. Jiang, C, Wu ST, Feng X, Hu Q (2014a) Nonlinear force-free field extrapolation of a coronal magnetic flux rope supporting a large-scale solar filament from a photospheric vector magnetogram. Astrophys J Lett 786: L16. doi:10.1088/2041-8205/786/2/L16.

68. Jiang, C, Wu ST, Feng X, Hu Q (2014b) Formation and eruption of an active region sigmoid. I. A study by nonlinear force-free field modeling. Astrophys J 780: 55. doi:10.1088/0004-637X/780/1/55.

69. Karpen, JT, Antiochos SK, DeVore CR (2012) The mechanisms for the onset and explosive eruption of coronal mass ejections and eruptive flares. Astrophys J 760: 81. doi:10.1088/0004-637X/760/1/81.

70. Kataoka, R, Sato T, Kubo Y (2014) Radiation dose forecast of WASAVIES during ground-level enhancement. Space Weather 12: 380–386. doi:10.1002/2014SW001053.

71. Kliem, B, Lin J, Forbes TG, Priest ER, Török T (2014) Catastrophe versus instability for the eruption of a toroidal solar magnetic flux rope. Astrophys J Let 789: 46. doi:10.1088/0004-637X/789/1/46.

72. Kliem, B, Su YN, van Ballegooijen AA, DeLuca EE (2013) Magnetohydrodynamic modeling of the solar eruption on 2010 April 8. Astrophys J 779: 129. doi:10.1088/0004-637X/779/2/129.

73. Kliem, B, Török T (2006) Torus instability. Phys Rev Lett 96: 255002. doi:10.1103/PhysRevLett.96.255002.

74. Kopp, RA, Pneuman GW (1976) Magnetic reconnection in the corona and the loop prominence phenomenon. Sol Phys 50: 85–98. doi:10.1007/BF00206193.

75. Kosugi, T, Matsuzaki K, Sakao T (2007) The Hinode (solar-B) mission: an overview. Sol Phys 243: 3–17. doi:10.1007/s11207-007-9014-6.

76. Kruskal, MD, Kulsrud RM (1958) Equilibrium of a magnetically confined plasma in a toroid. Phis Fuluid 1: 265–274. doi:10.1063/1.1705884.

77. Kusano, K, Bamba Y, Yamamoto TT (2012) Magnetic field structures triggering solar flares and coronal mass ejections. Astrophys J 760: 31. 10.1088/0004-637X/760/1/31.

78. Kusano, K, Nishikawa K (1996) Bifurcation and stability of coronal magnetic arcades in a linear force-free field. Astrophys J 461: 415–423. doi:10.1086/177070.

79. Leake, JE, Linton MG, Antiochos SK (2014) Simulations of emerging magnetic flux. II: The formation of unstable coronal flux ropes and the initiation of coronal mass ejections. Astrophys J 787: 46. doi:10.1088/0004-637X/787/1/46.

80. Lemen, JR, Title AM, Akin DJ (2012) The atmospheric imaging assembly (AIA) on the solar dynamics observatory (SDO). Sol Phys 275: 17. doi:10.1007/s11207-011-9776-8.

81. Lin, J, Forbes TG (2000) Effects of reconnection on the coronal mass ejection process. J Gophys Res 105: 2375–2392. doi:10.1029/1999JA900477.

82. Liu, Y (2008) Magnetic field overlying solar eruption regions and kink and torus instabilities. Astrophys J 679: L151-. doi:10.1086/589282.

83. Liu, Y, Luhmann JG, Müller-Mellin R, et al. (2008) A comprehensive view of the 2006 December 13 CME: from the sun to interplanetary space. Astrophys J 689: 563–571. doi:10.1086/592031.

84. Louis, RE, Kliem B, Ravindra B, Chintzoglou G (2015) Triggering an eruptive flare by emerging flux in a solar active-region complex. Solar Phys 290: 3641–3662. doi:10.1007/s11207-015-0726-8.

85. Low, BC, Lou YQ (1990) Modeling solar force-free magnetic fields. Astrophys J 352: 343–352. doi:10.1086/168541.

86. Lynch, BJ, Antiochos SK, DeVore CR, Luhmann JG, Zurbuchen TH (2008) Topological evolution of a fast magnetic breakout CME in three dimensions. Astrophys J 683: 1192–1206. doi:10.1086/589738.

87. Malanushenko, A, Schrijver CJ, ML DeRosa, Wheatland MS (2014) Using coronal loops to reconstruct the magnetic field of an active region before and after a major flare. Astrophys J 783: 102. doi:10.1088/0004-637X/783/2/102.

88. Manchester, WB, Gombosi TI, Roussev I (2004) Modeling a space weather event from the sun to the earth: CME generation and interplanetary propagation. J Geophys Res Space Phys 109: A02107. doi:10.1029/2003JA010150.

89. Masuda, S, Kosugi T, Hara H, Tsuneta S, Ogawara Y (1994) A loop-top hard X-ray source in a compact solar flare as evidence for magnetic reconnection. Nature 371: 495–497. doi:10.1038/371495a0.

90. McClymont, AN, Mikic Z (1994) Thickness variations along coronal loops inferred from vector magnetograph

data. Astrophys J 422: 899–905. doi:10.1086/173781.

91. McKenzie, DE, Canfield RC (2008) Hinode XRT observations of a long-lasting coronal sigmoid. AandA 481: L65–L68. doi:10.1051/0004-6361:20079035.
92. Metcalf, TR, De Rosa ML, Schrijver CJ, et al (2008) Nonlinear force-free modeling of coronal magnetic fields. II. Modeling a filament arcade and simulated chromospheric and photospheric vector fields. Sol Phys 247: 269–299. doi:10.1007/s11207-007-9110-7.
93. Miyoshi, T, Kusano K (2011) A comparative study of divergence-cleaning techniques for multi-dimensional MHD schemes. Plasma Fusion Res 6: 1124. doi:10.1585/pfr.6.2401124.
94. Mikić, Z, Barnes DC, Schnack DD (1988) Dynamical evolution of a solar coronal magnetic field arcade. Astrophys J 328: 830–847. doi:10.1086/166341.
95. Mikić, Z, Lionello R, Mok Y, Linker JA, Winebarger AR (2013) The importance of geometric effects in coronal loop models. Astrophys J 773: 94. doi:10.1088/0004-637X/773/2/94.
96. Mikić, Z, McClymont AN (1994) Deducing coronal magnetic fields from vector magnetograms. In solar active region evolution: comparing models with observations, ed. KS Balasubramaniam and GW Simon, 225–232. Astronomical Society of thePacific Conference Series, volume 68. Astronomical Society of the Pacific, San Francisco.
97. Moore, RL, Sterling AC, Hudson HS, Lemen JR (2001) Onset of the magnetic explosion in solar flares and coronal mass ejections. Astrophys J 552: 833–848.
98. Nakagawa, Y (1974) Dynamics of the solar magnetic field. I. Method of examination of force-free magnetic fields. Astrophys J 190: 437–440. doi:10.1086/152895.
99. Nindos, A, Patsourakos S, Vourlidas A, Tagikas C (2015) How common are hot magnetic flux ropes in the low solar corona? A statistical study of EUV observations. Astrophys J 808: 117. doi:10.1088/0004-637X/808/2/117.
100. Nishida, K, Nishizuka N, Shibata K (2013) The role of a flux rope ejection in a three-dimensional magnetohydrodynamic simulation of a solar flare. Astrophys J Lett 775: L39. doi:10.1088/2041-8205/775/2/L39.
101. Parenti, S (2014) Solar prominences: observations. Liv Rev Sol Phys 11: 1. doi:10.12942/lrsp-2014-1.
102. Pesnell, WD, Thompson BJ, Chamberlin PC (2012) The solar dynamics observatory (SDO). Sol Phys 275: 3–15. doi:10.1007/s11207-011-9841-3.
103. Peter, H, Warnecke J, Chitta LP, Cameron RH (2015) Limitations of force-free magnetic field extrapolations: revisiting basic assumptions. Astron Astrophys 584: A68. doi:10.1051/0004-6361/201527057.
104. Pevtsov, AA, Berger MA, Nindos A, Norton AA, van Driel-Gesztelyi L (2014) Magnetic helicity, tilt, and twist. Space Sci Rev 186: 285–324. doi:10.1007/s11214-014-0082-2.
105. Priest, E (2014) Magnetohydrodynamics of the sun. Cambridge University Press, UK.
106. Priest, ER, Forbes TG (1990) Magnetic field evolution during prominence eruptions and two-ribbon flares. Sol Phys 126: 319–350. doi:10.1007/BF00153054.
107. Priest, ER, Forbes TG (2002) The magnetic nature of solar flares. Astron Astophys Rev 10: 313–377. doi:10.1007/s001590100013.
108. Régnier, S, Amari T, Kersalé E (2002) 3D Coronal magnetic field from vector magnetograms: non-constant-alpha force-free configuration of the active region NOAA 8151. Astron Astrophys 392: 1119–1127.
109. Roumeliotis, G (1996) The "stress-and-relax" method for reconstructing the coronal magnetic field from vector magnetograph data. Astrophys J 473: 1095–1103. doi:10.1086/178219.
110. Sakurai, T (1981). Calculation of force-free magnetic field with non-constant α Sol Phys 69: 343–359. doi:10.1007/BF00149999.
111. Sakurai, T (1982) Green's function methods for potential magnetic fields. Sol Phys 76: 301–321. doi:10.1007/BF00170988.
112. Sakurai, T (1989) Computational modeling of magnetic fields in solar active regions. Spa Sci Rev 51: 11–48. doi:10.1007/BF00226267.
113. Savcheva, AS, van Ballegooijen AA, DeLuca EE (2012) Field topology analysis of a long-lasting coronal sigmoid. Astrophys J 744: 78. doi:10.1088/0004-637X/744/1/78.
114. Scherrer, PH, Schou J, Bush RI, et al. (2012) The helioseismic and magnetic imager (HMI) investigation for the solar dynamics observatory (SDO). Sol Phys 275: 207–227. doi:10.1007/s11207-011-9834-2.
115. Schmieder, B, Aulanier G, Vršnak B (2015) Flare-C models: an observational perspective ME (invited review). Solar Phys 290: 3457–3486. doi:10.1007/s11207-015-0712-1.
116. Schrijver, CJ, Aulanier G, Title AM, Pariat E, Delannée C (2011) The 2011 February 15 X2 flare, ribbons, coronal front, and mass ejection: interpreting the three-dimensional views from the solar dynamics observatory and STEREO guided by magnetohydrodynamic flux-rope modeling. Astrophys J 738: 167. doi:10.1088/0004-637X/738/2/167.
117. Schrijver, CJ, Derosa ML, Metcalf TR, et al. (2006) Nonlinear force-free modeling of coronal magnetic fields part I: a quantitative comparison of methods. Sol Phys 235: 161–190. doi:10.1007/s11207-006-0068-7.
118. Schrijver, CJ, Derosa ML, Metcalf T, et al. (2008) Nonlinear force-free field modeling of a solar active region around the time of a major flare and coronal mass ejection. Astrophys J 675: 1637–1644. doi:10.1086/527413.
119. Shibata, K (1999) Evidence of magnetic reconnection in solar flares and a unified model of flares. Astrophys Space Sci 264: 129–144.
120. Shibata, K, Magara T (2011) Solar flares: magnetohydrodynamic processes. Liv Rev Sol Phys 8: 6. doi:10.12942/lrsp-2011--6.
121. Shiota, D, Isobe H, Chen PF, Yamamoto TT, Sakajiri T, Shibata K (2005) Self-consistent magnetohydrodynamic

modeling of a coronal mass ejection, coronal dimming, and a giant cusp-shaped arcade formation. Astrophys J 634: 663–678. doi:10.1086/496943.

122. Shiota, D, Kusano K, Miyoshi T, Shibata K (2010) Magnetohydrodynamic modeling for a formation process of coronal mass ejections: interaction between an ejecting flux rope and an ambient field. Astrophys J 718: 1305–1314. doi:10.1088/0004-637X/718/2/1305.

123. Shiota, D, Tsuneta S, Ito H, Kusao K, Nishikawa N, Suzuki TK (2012) Fine structure in three dimensional magnetic field in polar region. Hinode-3: The 3rd Hinode Science Meeting, Proceedings of the conference held 1–4 December 2009 at Hitotsubashi Memorial Hall, ASP Conference Series, Vol. 454, 375–378. Astronomical Society of the Pacific, San Francisco.

124. Sterling, AC, Hudson HS (1997) Yohkoh SXT observations of X-ray "dimming" associated with a halo coronal mass ejection. Astrophys J Lett 491: L55–L58. doi:10.1086/311043.

125. Sturrock, PA (1966) Model of the high-energy phase of solar flares. Nature 211: 695–700. doi:10.1038/211695a0.

126. Su, Y, Golub L, van Ballegooijen A, et al. (2007) Evolution of the sheared magnetic fields of two X-class flares observed by Hinode/XRT. Pub Astron Sci Japan 59: 785–791. doi:10.1093/pasj/59.sp3.S785.

127. Su, Y, Surges V, van Ballegooijen A, DeLuca E, Golub L (2011) Observations and magnetic field modeling of the flare/coronal mass ejection event on 2010 April 8. Astrophys J 734: 53. doi:10.1088/0004-637X/734/1/53.

128. Su, Y, van Ballegooijen A, Lites BW, Deluca EE, Golub L, Grigis PC, Huang G, Ji H (2009) Observations and nonlinear force-free field modeling of active region 10953. Astrophys J 691: 105. doi:10.1088/0004-637X/691/1/105.

129. Sun, X, Hoeksema JT, Liu Y, Wiegelmann T, Jayashi K, Chen Q, Thalmann J (2012) Evolution of magnetic field and energy in a major eruptive active region based on SDO/HMI observation. Astrophys J 748: 77. doi:10.1088/0004-637X/748/2/77.

130. Tanaka, T (1995) Generation mechanisms for magnetosphere-ionosphere current systems deduced from a three-dimensional MHD simulation of the solar wind-magnetosphere-ionosphere coupling processes. J Geophys 100: 12,057–12,074. doi:10.1029/95JA00419.

131. Thalmann, JK, Inhester B, Wiegelmann T (2011) Estimating the relative helicity of coronal magnetic fields. Sol Phys 272: 243–255. doi:10.1007/s11207-011-9826-2.

132. Titov, VS, Démoulin P (1999) Basic topology of twisted magnetic configurations in solar flares. Astrophys Astron 351: 707–720.

133. Tiwari, SK, Alexander CE, Winebarger AR, Moore RL (2014) Trigger mechanism of solar subflares in a braided coronal magnetic structure. Astrophys J Let 795: L24. doi:10.1088/2041-8205/795/1/L24.

134. Toriumi, S, Iida Y, Bamba Y, Kusano K, Imada S, Inoue S (2013) The magnetic systems triggering the M6.6 class solar flare in NOAA active region 11158. Astrophys J 773: 128. doi:10.1088/0004-637X/773/2/128.

135. Török, T, Kliem B, Titov VS (2004) Ideal kink instability of a magnetic loop equilibrium. Astron Astrophys 413: L27–L30. doi:10.1051/0004-6361:20031691.

136. Török, T, Kliem B (2005) Confined and ejective eruptions of kink-unstable flux ropes. Astrophys J Lett 630: L97–L100. doi:10.1086/462412.

137. Török, T, Kliem B (2007) Numerical simulations of fast and slow coronal mass ejections. Astron Nach 328: 743–746. doi:10.1002/asna.200710795.

138. Török, T, Aulanier G, Schmieder B, Reeves KK, Golub L (2009) Fan-spine topology formation through two-step reconnection driven by twisted flux emergence. Astrophys J 704: 485–495. doi:10.1088/0004-637X/704/1/485.

139. Török, T, Berger MA, Kliem B (2010) The writhe of helical structures in the solar corona. Astron Astrophys 516: A49. doi:10.1051/0004-6361/200913578.

140. Török, T, Panasenco O, Titov VS, Mikić Z, Reeves KK, Velli M, Linker JA, De Toma G (2011) A model for magnetically coupled sympathetic eruptions. Astrophys J Lett 739: L63. doi:10.1088/2041-8205/739/2/L63.

141. Tóth, G (2000) The $\nabla \cdot \boldsymbol{B}=0$ constraint in shock-capturing magnetohydrodynamics codes. J Comput Phys 161: 605–652. doi:10.1006/jcph.2000.6519.

142. Tóth, G, Sokolov IV, Gombosi TI, et al (2005) Space weather modeling framework: a new tool for the space science community. J Geophys Res 110: A12226. doi:0.1029/2005JA011126.

143. Tsuneta, S (1996) Structure and dynamics of magnetic reconnection in a solar flare. Astrophys J 456: 840–849. doi:10.1086/176701.

144. Tsuneta, S, Hara H, Shimizu T, et al. (1992) Observation of a solar flare at the limb with the YOHKOH soft X-ray telescope. Pub Astron Soc Japan 44: L63–L69.

145. Tsuneta, S, Ichimoto K, Katsukawa Y, et al. (2008) The solar optical telescope for the Hinode mission: an overview. Sol Phys 249: 167–169. doi:10.1007/s11207-008-9174-z.

146. Valori, G, Démoulin P, Pariat E (2012) Comparing values of the relative magnetic helicity in finite volumes. Sol Phys 278: 347–366.

147. Valori, G, Kliem B, Fuhrmann M (2007) Magnetofrictional extrapolations of Low and Lou's force-free equilibria. Sol Phys 245: 263–285.

148. Valori, G, Kliem B, Keppens R (2005). Extrapolation of a nonlinear force-free field containing a highly twisted magnetic loop Astrophsys Astron 433: 335–347. doi:10.1051/0004-6361:20042008.

149. van Ballegooijen, AA (2004) Observations and modeling of a filament on the sun. Astrophys J 612: 519–529. doi:10.1086/422512.

150. van Ballegooijen, AA, PMartens CH (1989) Formation and eruption of solar prominences. Astrophys J 343: 971–984. doi:10.1086/167766.

151. van Ballegooijen, AA, Priest ER, Mackay DH (2000) Mean field model for the formation of filament channels on the Sun. Astrophys J 539: 983–994.

doi:10.1086/309265.

152. van Driel-Gesztelyi, L, Baker D, Török T, et al. (2014) Coronal magnetic reconnection driven by CME expansion: the 2011 June 7 event. Astrophys J 788: 85. doi:10.1088/0004-637X/788/1/85.

153. Vemareddy, P, Ambastha A, Wiegelmann T (2013) Magnetic structure of solar active region NOAA 11158. Bull Astron Soc India 41: 183.

154. Vemareddy, P, Zhang J (2014) Initiation and eruption process of magnetic flux rope from solar active region NOAA 11719 to earth-directed CME. Astrophys J 797: 80. doi:10.1088/0004-637X/797/2/80.

155. Wang, H, Liu C (2015) Structure and evolution of magnetic fields associated with solar eruptions. Res Astron Astrophys 15: 145–174. doi:10.1088/1674-4527/15/2/001.

156. Wang, S, Liu C, Liu R, Deng N, Liu Y, Wang H (2012) Response of the photospheric magnetic field to the X2.2 flare on 2011 February 15. Astrophys J Lett 745: L17. doi:10.1088/2041-8205/745/2/L17.

157. Wang, R, Yan Y, Tan B (2013) Three-dimensional nonlinear force-free field reconstruction of solar active region 11158 by direct boundary integral equation. Sol Phys 288: 507–529. doi:10.1007/s11207-013-0422-5.

158. Wheatland, MS, Leka KD (2011) Achieving self-consistent nonlinear force-free modeling of solar active regions. Astrophsys J 728: 112. doi:10.1088/0004-637X/728/2/112.

159. Wheatland, MS, Régnier S (2009) A self-consistent nonlinear force-free solution for a solar active region magnetic field. Astrophys J Lett 700: L88–L91. doi:10.1088/0004-637X/700/2/L88.

160. Wheatland, MS, Sturrock PA, Roumeliotis G (2000) An optimization approach to reconstructing force-free fields. Astrophys J 540: 1150–1155. doi:10.1086/309355.

161. Wiegelmann, T (2004) Optimization code with weighting function for the reconstruction of coronal magnetic fields. Sol Phys 219: 87–108. doi:10.1023/B:SOLA.0000021799.39465.36.

162. Wiegelmann, T, Sakurai T (2012) Solar force-free magnetic fields. Liv Rev Sol Phys 9: 5. doi:10.12942/lrsp-2012-5.

163. Wu, ST, Sun MT, Chang HM, Hagyard MJ, Gary GA (1990) On the numerical computation of nonlinear force-free magnetic fields. Astrophys J 362: 698–708. doi:10.1086/169307.

164. Yan, Y (1995) The 3-D boundary element formulation of linear force-free magnetic fields with finite energy content in semi-infinite space. Sol Phys 159: 97–113. doi:10.1007/BF00733034.

165. Yan, Y, Sakurai T (2000) New boundary integral equation representation for finite energy force-free magnetic fields in open space above the Sun. Sol Phys 195: 89–109. doi:10.1023/A:1005248128673.

166. Yang, Y-H, Chen PF, Hsieh M-S, et al. (2014) Characteristics of the photospheric magnetic field associated with solar flare initiation. Astrophys J 786: 72. doi:10.1088/0004-637X/786/1/72.

167. Yokoyama, T, Akita K, Morimoto T, Inoue K, Newmark J (2001) Clear evidence of reconnection inflow of a solar flare. Astrophys J Lett 546: L69–L72. doi:10.1086/318053.

168. Yokoyama, T, Shibata K (1994) What is the condition for fast magnetic reconnection?Astrphys J Lett 436: L197–L200. doi:10.1086/187666.

169. Yokoyama, T, Shibata K (2001) Magnetohydrodynamic simulation of a solar flare with chromospheric evaporation effect based on the magnetic reconnection model. Astrophys J 549: 1160–1174. doi:10.1086/319440.

170. Zhao, J, Li H, Pariat E, Schmieder B, Guo Y, Weigelmann T (2014) Temporal evolution of the magnetic topology of the NOAA active region 11158. Astrophys J 787: 88. doi:10.1088/0004-637X/787/1/88.

171. Zuccarello, FP, Meliani Z, Poedts S (2012) Numerical modeling of the initiation of coronal mass ejections in active region NOAA 9415. Astrophys J 758: 117. doi:10.1088/0004-637X/758/2/117.

CHAPTER 6 FROM ELF TO COMPRESSIBILITY IN SOLIDS

Julia Contreras-García, Miriam Marqués, José Manuel Menéndez and José Manuel Recio

Keywords: ELF, compressibility, energy models

INTRODUCTION

A thorough understanding of many important electronic and mechanical properties of a crystalline material does not require the knowledge of the total energy of the solid, but just the energetics of its bonding network. Many phenomenological approaches have been developed to illustrate the advantage of this view [1]. These so-called bond energy models are useful for properties prediction and have become rather popular in the last years as trusty guides for material design. Specifically to the field of materials strength, Gilman's book [1] collects a number of valuable correlations between a variety of observable elastic moduli of different type of solids and electronic densities coming only from the active electrons of the valence space. Based on the ideas developed by Gilman [2], Gao et al. "initialized a link" between macroscopic observables and first principles calculations with the aim to evaluate the indentation hardness of multicomponent crystals [3]. Li et al. resorted to bond electronegativity in order to evaluate the bond energy and thus figure out the most stable bond combination for hard materials [4]. In conjunction with evolutionary algorithms [5], Li's model is theoretically generalized by Lyakhov and Oganov making use of the bond-valence model and graph theory [6].

Inherent to all these semi-empirical models, there are still elusive concepts to quantum-mechanical formalisms, as ionicity/covalent scales, electronegativities, covalent radii, etc. Efforts from the theoretical chemistry community have been devoted to reformulate these concepts from first principles. As an example, we can cite recent reformulations of electronegativity [7,8,9]. These chemical concepts do not derive directly from wavefunction analysis, but a posteriori models are needed to extract them. The bondon [10] and the reformulation of chemical bonding as a quantum condensate are some of the formulations that have been recently proposed [11]. Another approach that has gained many followers is the topological formalism. This approach divides three-dimensional space into chemically meaningful disjoint regions following the gradient of a scalar function, such as the electron density [12,13] or the electron localization function (ELF) [14,15]. Since we are interested in valence properties, our function of choice is the latter one. ELF is able to recover Lewis entities, such as bonds and lone pairs and quantify their properties (volume, charge, etc.). Due to its direct link with Lewis theory, we have chosen the model by Borkman and Parr [16] as our starting point to merge topology and bond energy models.

Briefly, our study is based on the success of the bond charge model [16,17,18] in predicting molecular properties, and the ability of the electron localization function [14,15] to provide the required parameters for its free from empirical data application. The bulk modulus (inverse of the compressibility) will be the focus of this first principles-like extension of BCM to the solid state.

The remainder of the paper is organized as follows. Firstly, the original definition of the bond charge model for homo and heteronuclear binary molecules and covalent solids will be presented. Secondly, the computational details for the calculations performed over a broad group of diamond-type and zinc-blende-type solids from the IV, III–V and II–VI groups are given. In Section 3, the basic parameters of this ab-initio BCM (hereafter, NEWBCM) are derived from first principles. The ability of this model to predict solid state properties (compressibility) will be tested against experimental data in Section 4. At the end of the section, this new insight into macroscopic properties will be used to understand macro-micro requirements for superhard materials. The article finishes with the main conclusions and guidelines for prospective work.

Citation: Julia Contreras-García, Miriam Marqués, José Manuel Menéndez, and José Manuel Recio, From ELF to Compressibility in Solids, Int. J. Mol. Sci. 2015, 16, 8151-8167; doi:10.3390/ijms16048151.

THEORETICAL BACKGROUD

In this section, the two approaches to be merged are introduced: the bond charge model and the ELF topological formalism.

The Bond Charge Model: Original Definition

Homo and Heteronuclear Diatomic Molecules

Within the BCM formalism [16], a homonuclear diatomic molecule (A_2) can be approximated by two cores holding a q/2 positive charge each, and a bond charge in between them, -q, that moves freely along a bond length $R_B = vR$ which is a fraction ($v \leq 1$) of the interatomic distance R (see Figure 1 (left)). The energy associated to this model at distances close to the equilibrium ($R \simeq R_e$) is given in atomic units by:

$$E(q, R) = E_0 + E_1 + E_2 = 2E_A - \frac{Cq^2}{R} + \frac{D'q}{R_B^2} \quad (1)$$

where E_0 represents the core energy (which equals to $2E_A$ for an A_2 molecule), E_1 refers to the coulombic attractive interactions and E_2 is related to the kinetic energy of the bond electrons approximated as particles in a box of length R_B. The constants, which are system independent, are then given by (see [16] for details):

$$C = \frac{7}{4}, \quad D' = \frac{\pi^2}{2m} \quad (2)$$

and the parameters q and v are inferred from empirical measures of the harmonic force constant, k_e, and the equilibrium distance, R_e.

In order to adapt this model to heteronuclear diatomic molecules, AX, or even to homonuclear molecules in an asymmetric surrounding, two considerations have to be taken into account. On the one hand, the polarity of the bond clearly influences the situation of the bond charge (Figure 1 (right)), being r_1 and r_2 the distances to its position from atoms A and X, respectively. On the other hand, the positive charge lying on each of the two nuclei will be different from $\frac{1}{2}q$ by $\frac{1}{2}\delta$ and $-\frac{1}{2}\delta$ [18]. The changes induced by the electronegativity difference, $\Delta\chi = \chi_A - \chi_X$, in E_2 are negligible, but the bond charge shift gives rise to noticeable changes in the coulombic interactions:

$$E_1 = \frac{q^2}{2}\left[\frac{(1-\delta^2)}{2R} - \frac{(1+\delta)}{r_1} - \frac{(1-\delta)}{r_2}\right] \quad (3)$$

so that the total energy reads as follows:

$$E = E_A(q, \delta) + E_X(q, \delta) + \frac{q^2}{2}\left[\frac{(1-\delta^2)}{2R} - \frac{(1+\delta)}{r_1} - \frac{(1-\delta)}{r_2}\right] + \frac{D'q}{R_B^2} \quad (4)$$

The determination of the model parameters, q, δ, r_1/r_2 and R_B has become a matter of choice. Probably, the most extended method has been to take into account the electronegativity of the atoms as shown by Parr et al. [19,20].

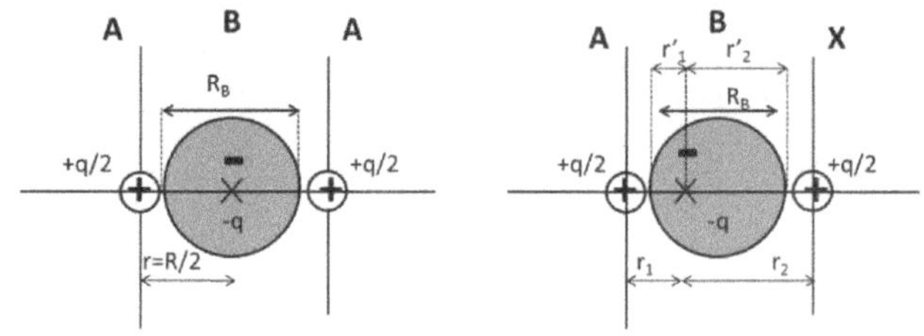

Figure 1: Bond charge model for homonuclear (**left**) and heteronuclear (**right**) molecules, A_2 and AX, respectively. The meaning of the main parameters that are needed in each case is shown in the figures: q and R_B stand for neutral homonuclear molecules as q, δ, r_1/r_2 and R_B do for heteronuclear ones; B label stands for the bond (see text for more details). r'1 and r'2 are also shown for clarity.

Binary Covalent Solids

The model by Parr and Borkman was extended by Martin [21] to describe the energy of covalent crystals. For simplicity, we consider a zinc-blende-type solid, AX, with a formula unit AXB_4 accounting for the four-fold coordination of the two elements in the unit cell, B represents the bond entity. According to BCM, the energy of the covalent solid per couple of atoms AX is given by

$$E_{AX} = E_A(q_B, \delta) + E_X(q_B, \delta) + E_B(q_B, \delta, R_B) \quad (5)$$

$$E_B = -\frac{Mq_B^2}{R_B} + \frac{4D'q_B}{R_B^2} \quad (6)$$

where M is the Madelung constant for these lattices with charges $2q_B(1+\delta)$ and $2q_B(1-\delta)$ at A and X atomic sites, and $-q_B$ at each bond position, as dictated by the multiplicity and the charge neutrality condition. We should recall that E_A and E_X account for the energy of the corresponding core electrons not involved in the bonding, and, in practice, are independent of geometrical parameters. The greatest modification introduced in the application of BCM to solids is the presence of an infinite sum, accounting for all A, X, and B charged species, which is subsumed into the Madelung constant. It also important to highlight that due to the inclusion of the bond charge in the summation, this Madelung constant is system dependant since the position of the bond charge will not correspond to a special Wyckoff position (invariable) except for homonuclear solids. The Madelung constants have been calculated thanks to the environ code [22] and are collected in Table 1.

Table 1. Parameters of the non-empirical new bond charge model (NEWBCM), [q_B, R_B,r_1/r_2], for a broad group of diamond-type and zinc-blende-type solids from the IV, III–V and II–VI groups. $r_1(r_2)$ stands for the distance between the charge, q, and the nucleus $A(X)$. $r'_1\,(r'_2)$ stands for the part of R_B belonging to $r_1(r_2)$, i.e., $r'_1=r_1-r_A$ and $r'_2=r_2-r_X$, where r_A is the core size of atom A and r_X is the core size of atom X. M and E_B stand for the Madelung constant and the bonding energy of the crystal, respectively. The cell parameter, a, and the electronegativity difference [23], $\Delta\chi = \chi_X - \chi_A$, are also shown for reference. Lengths are in Å, charges in electrons, energy in hartrees.

AX	a	$\Delta\chi$	r_1/r_2	r'_1/r'_2	R_B	q_B	M	E_B	δ
C	3.557	0.0	1.00	1.00	0.938	1.950	10.856	−11.644	0.000
Si	5.431	0.0	1.00	1.00	1.132	1.950	8.577	−7.623	0.000
Ge	5.658	0.0	1.00	1.00	1.003	2.100	7.414	−8.625	0.000
SiC	4.360	0.7	1.17	0.75	0.985	1.950	9.356	−9.556	0.000
BN	3.616	1.0	1.07	0.86	0.936	1.975	11.006	−12.136	0.241
BP	4.538	0.1	0.96	1.21	1.050	2.000	9.662	−9.739	0.250
BAs	4.777	0.0	0.70	0.82	1.023	2.050	8.711	−9.468	0.268
BSb	5.156	0.1	0.62	0.88	0.999	1.988	8.554	−8.954	0.245
AlP	5.464	0.6	1.14	0.97	1.122	1.975	8.925	−8.206	0.266
AlAs	5.661	0.5	0.96	0.83	1.080	2.025	8.028	−8.068	0.284
AlSb	6.136	0.4	0.81	0.77	1.093	1.963	7.304	−6.810	0.261
GaAs	5.653	0.4	1.19	1.18	0.992	2.100	7.662	−9.008	0.238
GaSb	6.096	0.3	1.00	1.12	0.986	2.038	6.764	−7.538	0.215
InP	5.869	0.4	1.64	1.38	1.012	1.970	8.494	−8.615	0.269
InAs	6.058	0.3	1.35	1.11	0.958	2.020	7.268	−8.187	0.287
CdS	5.818	0.8	1.89	1.48	0.976	1.975	10.016	−10.512	0.494
CdSe	6.077	0.7	1.56	1.27	0.938	2.025	8.591	−9.697	0.506
ZnS	5.410	0.9	1.67	1.50	1.001	2.025	10.130	−10.976	0.457
ZnSe	5.668	0.8	1.37	1.30	0.970	2.075	8.631	−10.134	0.470
ZnTe	6.104	0.5	1.15	1.25	0.971	2.038	7.415	−8.389	0.460

From Equation (6), the bond energy of a zinc-blende-type lattice can be finally expressed within the framework of BCM as:

$$E_B = -\frac{Mq_B^2}{R_B} + \frac{4D'q_B}{R_B^2} \tag{7}$$

It is interesting to highlight here that this expression is similar to the classical Madelung energy expression for ionic crystals. However, in our case the equation is done in terms of Lewis entities (cores, bonds) instead of ions. The expression will then hold as long as the orthogonality of these entities holds too. In solids such as diamond, where for example, good localized orbitals can be found, Equation (7) will give an accurate description of the bonding energy in the crystal. However, as the localized picture fails (e.g., down the periodic table), we should expect the fits to worsen. This equation will progressively fail as we go down the periodic table or move towards metallic solids, due to a less obvious separability.

If we reasonably assume that $(\partial E_B/\partial_R) \simeq 0$ at equilibrium then

$$\left(\frac{\partial E_B}{\partial R_B}\right)_{R=R_e} = \frac{Mq_B^2}{R_B^2} - \frac{8D'q_B}{R_B^3} = 0 \tag{8}$$

where we have assumed that the ratio r_1/r_2 remains constant upon compression, not introducing changes in M (see discussion in last section and Figure 5). It is very important to notice that all the parameters have to be evaluated at equilibrium. We then have for the constant M:

$$M = \frac{8D'}{q_B R_B} \tag{9}$$

Hence, the ground state bond energy at equilibrium is given by:

$$E_{B,eq} = -\frac{4D'q_B}{R_B^2} = -\frac{Mq_B^2}{2R_B} \tag{10}$$

where the negative value reflects the stabilization introduced by the chemical bond (see Table 1). This equation, just like in the original bond charge model [16,18], is a manifestation of the virial theorem. Upon substitution of Equation (9) in Equation (7), it can be seen that at equilibrium, the electrostatic interaction is twice as big as the kinetic energy term.

ELF Topology

The Electron Localization Function, ELF, was originally designed by Becke and Edgecombe to identify "localized electronic groups in atomic and molecular systems" [14]. Several interpretations have been given to this function. For example, ELF has been generalized and reinterpreted in the light of Markovian processes [24,25]. A very intuitive interpretation was given by Savin [15], according to which the ELF core, χ, can be understood as a local measure of the excess of local kinetic energy of electrons due to the Pauli principle, t_p:

$$\chi_\sigma(r) = \frac{t_p(\vec{r})}{t_{HEG}(\vec{r})} \tag{11}$$

relative to the homogeneous electron gas kinetic energy density, $t_{HEG}(\vec{r}) = \frac{3}{5}(6\pi^2)^{2/3}\rho_\sigma^{5/3}(\vec{r})$. t_p is calculated as the difference between the definite positive kinetic energy, $t(\vec{r})$, and the von Weizsäcker kinetic energy functional:

$$t_p(\vec{r}) = t(\vec{r}) - \frac{1}{4}\frac{|\rho_\sigma(\vec{r})|^2}{\rho_\sigma(\vec{r})} \tag{12}$$

ELF values are confined in the [0,1] range by a Lorentzian transformation which facilitates its interpretation:

$$ELF = \frac{1}{1+\chi^2} \tag{13}$$

ELF maxima appear associated with localized electron pairs, such as bonds, lone pairs and atomic shells, thus matching the Valence Shell Electron Pair Repulsion [26] model. These maxima can be used to partition the system into regions following the gradient flux (this would be like partitioning a set of mountains by the valleys) [12]. This partitioning gives rise to a set of non overlapping regions (or basins), labelled as Ωi. Their properties can be determined by integration over the region. Most commonly, the volume integration of the density, ρ, assigns a basin population, $\bar{N}(\Omega_i)$:

$$\bar{N}(\Omega_i) = \int_{\Omega_i} \rho(\vec{r})d\vec{r} \tag{14}$$

The properties of these basins have a chemical meaning inherited from the partition, so that the integration of the density over a basin associated with a bond maximum, can be understood as the bond charge.

COMPUTATIONAL DETAILS

Electronic structure calculations for a number of IV, III–V and II–VI solids with the diamond and zinc-blende-type structures were performed within the density-functional theory formalism. These compounds were selected to ensure the existence of a basin associated with the bond.

We have used the all-electron full potential linearised augmented-plane wave (FP-LAPW) approach as implemented in the ELK (for electrons in k-space) code [27]. The FP-LAPW method is among the most precise DFT-based methods for crystal structures [28]. It divides space into an interstitial region and non-overlapping muffin-tin (MT) spheres centered at the atomic sites. In the interstitial region, the basis set is described by plane-waves, whereas in the MT spheres, the basis set is described by radial solutions of the one particle Schrödinger equation and their energy derivatives multiplied by spherical harmonics. A high plane-wave cut-off $K_{max} R_{MT} = 9$, where R_{MT} is the smallest muffin-tin radius of the corresponding atomic species and K_{max} is the maximum size of the reciprocal lattice vectors, guaranteed convergence in all the studied systems. Moreover, spherical harmonics within each atomic sphere were expanded up to the maximum angular quantum number $l_{max} = 14$, to avoid discontinuities of the electron density and its derivatives at the muffin boundaries. Lattice parameters were set to the room temperature experimental values of the corresponding structures [29]. We have used both, the Perdew-Wang [30] local density approximation (LDA) and the Perdew-Burke-Ernzerhof [31] generalized gradient approximation (GGA) for the exchange-correlation functional. Brillouin-zone integrations were approximated using 8 × 8 × 8 Monkhorst-Pack grids, and the the self-consistent iteration process was repeated until the absolute change in total energy was lower than 1×10^{-5} hartrees.

In a second step, the topology of the ELF functions coming from the previous calculations is investigated using CRITIC, a code developed by some of the authors [32,33]. CRITIC is able not only to find ELF maxima and saddle points but it can also construct basins and integrate properties.

The training set has been chosen to lay in the shared-electron range. This characteristic is necessary so that there is a bond basin linkable to each pair of nearest neighbors in the crystal. In other words, it is possible to assign bonding parameters within the ELF framework.

Bond Length: R_B, r_1/r_2

The most common approach in the analysis of a bonding structure by means of the ELF approach is to use its 3D topology. However, the BCM uses a 1D definition of the chemical bond. Thus, we have devised an alternative route to obtain the BCM parameters from ELF. We have resorted to the 1D-ELF topology: ELF is analyzed along the straight line between pairs of atoms. In some cases, specially as going down the periodic table, this line might differ both from the AIM and ELF bonding lines, but it enables us to stick to the BCM formalism. On the one hand, the position of the bond charge within this length, r_1/r_2, (shifted from the center by the relative electronegativity of atoms) is determined by the local ELF maximum along the same direction. On the other hand, the distance between the two closest minima to the bond maximum along the internuclear direction provide the bond length, R_B (See Figure 2).

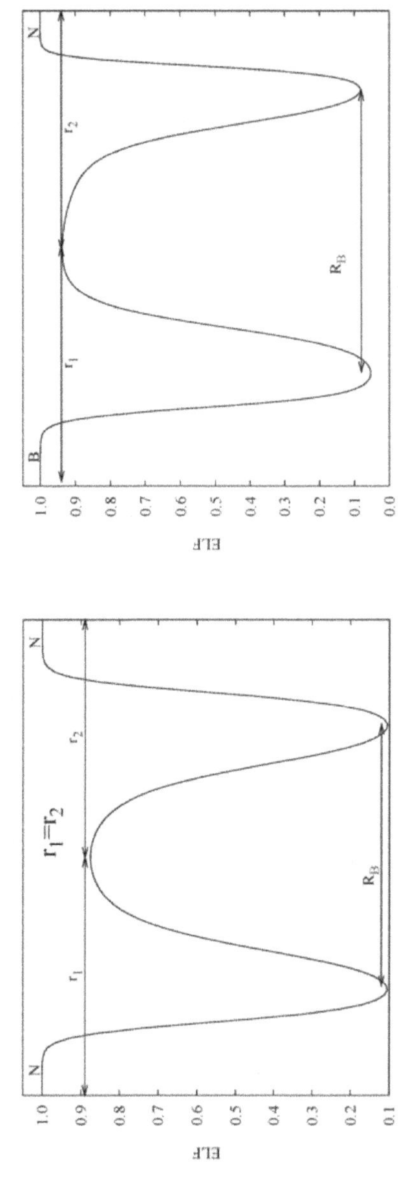

Figure 2: Definition of length-related bond charge model parameters, [R_B, r_1/r_2], for a homonuclear molecule, N_2 (left), and a heteronuclear molecule, BN (right).

DEFINING THE PARAMETERS FROM FIRST PRINCIPLES

ELF profiles along the internuclear line were constructed for each solid [14]. These profiles were then analyzed in order to find the position of the local maxima and minima, which enable the quantitative determination of ab initio R_B, r_1 and r_2 values according to the previous definition. Results are collected in Table 1.

In homonuclear cases the local maximum is located in the center of R_B and $r_1/r_2 = 1.0$ by definition (see N_2 in Figure 2(left)). Instead, in heteronuclear cases, the ELF profile is deformed and the maximum of the bond (or attractor) is no longer placed in the middle of R_B (see BN in Figure 2 (right)).

A similar value obtained for R_B is obtained in all solids, with numbers clustering around 1 Å. This means that the size of the bond (at least along the interbonding direction) depends neither on the lattice parameter of the crystal (which spreads a range of more than 2.5 Å), nor on the two atoms involved in the bond.

Secondly, we observe that $r_1/r_2 \simeq 1$ ratios are obtained for the purely covalent families and those whose $\Delta\chi \simeq 0$. There is one remarkable outlier: BN shows $r_1/r_2 \simeq 1$ with $\Delta\chi = 1$. In order to understand this fact, an extra column has been added toTable 1 with the displacement of the attractor within the bond length RB, that is, taking away core contributions: $r'_1 = r_1 - r_A, r'_2 = r_2 - r_X$.

This means that we now decompose the bond length into $R_B = r'_1 + r'_2$, so that $R=r_A+r'_1+r'_2+r_X$. The physical meaning of these quantities is depicted in Figure 1 (right). This quantity enables us to understand that although the charge in BN is nearly at the same distance from B and from N, it is displaced within R_B, as expected from their electronegativity difference (see Figure 2 (right)). This decomposition also helps to get rid of the core size in the comprehension of the system. Let's take the case of BP: since P belongs to the second row and B to the first one, we have $r_X >> r_A$, so that eventhough the center of the bond is displaced within the bond towards P ($r'_1 > r'_2$), once the core radii are taken into account, the distance relationship is inversed ($r_2 > r_1$).

Bond Charge: q_B

Obtaining the charge associated with ELF entities through the integration over the corresponding basins might be considered in principle a straightforward and accurate method. However, for heavy elements, some ill-defined cases appear for integration, where the low symmetry of the basins and the appearance of spurious critical points at the muffin surfaces may lead to serious difficulties if we seek a quantitative analysis of its topology.

Based on pseudopotential principles, we assume that the core electronic distributions are practically unaltered by the environment, and thus q_B values can be directly obtained by mere difference between the total and core number of electrons. Core values for the elements up to the 4th row involved in the solids under this study have been extracted from the ELF integrations gathered in [34]. Beyond the 4th row, calculations were perfomed for the atoms with the PBE functional [31] and the 6-31G* basis set, using the Gaussian 03 program [35]. Their core/valence charge, which is very stable to the level of calculation, was calculated with the TopMod program [36]. This assumption has the added value of providing extremely fast calculations. It has also been validated by comparison with the charges obtained by the integration over the corresponding 3D-basins with the CRITIC code [37]. For instance, integrated values for q_B of 1.952 and 1.948 for C and BN, respectively, are in good agreement with the corresponding data in Table 1. They are also independent of the exchange-correlation functional used.

It has been shown that ELF charges lie very close to chemical intuition values [38]. This is further seen in Table 1. Nearly 2 electrons are recovered in all cases for the bonds, in agreement with their bond order of 1. This is different from the typical BCM charges obtained from fitting, which in many cases deviate from the expected values [17,18]. Furthermore, this image of q_B correlating to the bond order does not agree with the definition of the q_B parameter in the traditional model. The step forward that ELF provides concerns also the evaluation of the charge transfer (δ). In all cases (see Table 1), δ also shows reasonable values following the expected trends for the IV, III–V, and II–VI crystal families dictated by their electronegativity [23]. Keep in mind that δ is calculated with respect to valence configurations, so that for example carbon and silicon, having the same configuration, [34] give rise to $\delta = 0$ in SiC.

This is an important characteristic of this Non Empirical New Bond Charge Model (NEWBCM), parameters are not only obtained ab initio, but they are closer to chemical expectations and recover/predict values in good agreement with experience (at difference with the initial BCM parameters).

TESTING THE MODEL: RESPONSE TO PRESSURE

Since pressure can introduce significant changes in the stability of phases, it is most useful in the synthesis of novel phases and metastable materials. Pressure allows precise tuning of a fundamental parameter, the interatomic distance, which controls the electronic structure and virtually all the interatomic interactions that determine materials properties [39,40,41]. With pressure tuning, properties can often be more rapidly and cleanly optimized than with chemical tuning, which necessitates the synthesis of a large number of different materials and can induce disorder, phase separation, and other undesirable effects. Pressure tuning is therefore a useful tool in the search for new solid state materials with tailored properties.

Bulk Modulus

Among the description of the responses to pressure, compressibility (κ) or its inverse, the bulk modulus (B_0), probably constitute the most widely used parameters. The inverse relationship between bulk modulus and volume has been widely explored [42,43,44,45], enabling to split the response of solids under pressure into local contributions (see for example [43,46,47,48]).

Following our BCM model, the valence charge is distributed along the chemical bonds, so that the compressibility of solids can be approximated to a good extent by the compressibility of the bonds, κ_B, whereas cores are assumed to remain untouched: $\kappa \simeq \kappa_B$ [43,49]. The bulk modulus of the solid is then given by (see S.I. for the development):

$$B_{0,B} = \left[V_B\left(\frac{\partial^2 E_B}{\partial V_B^2}\right)\left(\frac{\partial V_B}{\partial V}\right)\right]_0 = \frac{M q_B^2}{32\sqrt{3} R_B R^3} \tag{15}$$

where $V_B = \frac{1}{4}[V_{AXB_4} - (V_A + V_X)]$ is the bond's volume within the AXB_4 unit.

Equation (15) for our IV, III–V and II–VI compounds is analyzed in Figure 3. It is worth highlighting that a very good agreement is obtained for all data. The slope for the bulk modulus is around 1 (1.14), and the offset is near 9 GPa (8.99 GPa) which departures from zero but it is not too large if we compare it with the range (up to 500 GPa) displayed by B_0 in these compounds. The Pearson regression coefficient R^2 is

0.9774. A similar agreement was obtained by Gilman et al. [1] in an enlightening analysis from empirical measures where the relationship was analyzed in terms of valence density (i.e., q_B/V). The coherency between these two results can be perfectly understood from the analysis of Equation (15) if we take into account that $R = \frac{\sqrt{3}}{4}a$ and $V = a^3$ for these crystals:

$$B_{0,B} = \frac{2M}{9R_B}\frac{q_B^2}{V} \quad (16)$$

In his fit, Gilman made use of the cell volume and an average valence occupation q_B /V as would be expected from Lewis theory (i.e., a default bond occupation $q_B = 2$). Whereas the bond occupation has been found in our calculations to be very close to two, the dependence on the cell size can be improved by taking calculated bond lengths into account. Indeed, Equation (16) shows that the dependence on the lenght parameters should be $B_0 \propto (R_B V)^{-1}$ (i.e., $B_0 \propto (R_B R^3)^{-1}$, exponent nearly –4), whereas Gilman took $B_0 \propto V^{-1}$ alone (exponent on R = –3). Indeed, a logarithmic fit of the B_0 experimental data vs. R gives a slope of 3.8 (±0.1) (regression coefficient $R^2 = 0.9863$) which explains the need to increase the length dependence from R^{-3} to $R^{-3}R_B^{-1}$. As far as the bond charge is concerned, default occupations are very close to the values obtained with ELF, however, the quantitative values provided by electronic structure calculations enable us to obtain a better linear behavior. This is shown in Figure 4, where experimental and NEWBCM values have been included (in order to represent experimental values, the bond length dependence has been set to R^4). It is interesting to note that, although calculations provide a better linear behavior (R^2(NEWBCM) = 0.9927, R^2(Exptal.) = 0.9830), the use of default or nominal ($q_B = 2$) values already provides all the physics.

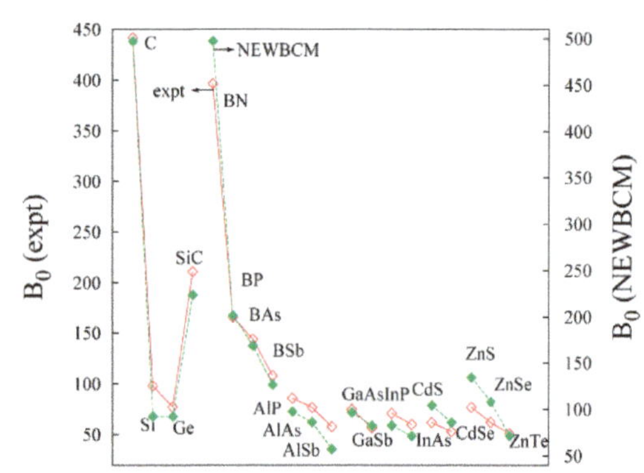

Figure 3: Experimental vs. non empirical new bond charge model (NEWBCM) bulk modulus (GPa).

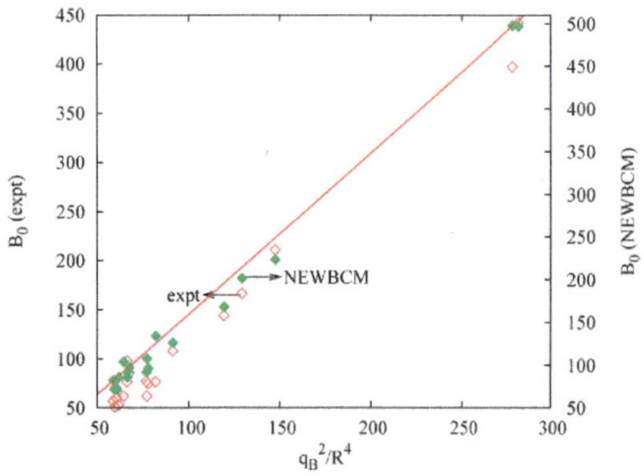

Figure 4: Experimental (empty symbols) and NEWBCM (full symbols) bulk moduli (in GPa) for IV, III–V and II–VI compounds. Bulk moduli in GPa, charge in electrons, R in atomic units.

From Micro- to Macroscopic: Understanding Compressibility

Understanding the inherent relationship between (quantifiable) microscopic properties and macroscopic properties is the corner stone of material design.

We have shown that the concepts derived from ELF can be used to understand the microscopic origin of solids compressibility. If we analyze the dependencies of Equation (15), we see that the least compressible materials are found within the first period, where smaller interatomic distances, $R \propto V^{1/3}$, and bond lengths, R_B, are found. While these results were already known from experience, the microscopic approach can also reveal new information. With this aim in mind, it is interesting to study the evolution of NEWBCM parameters (r_1, r_2, R_B and q_B) upon compression.

Let's start with R_B. Since $R_B = \nu R$ can be expressed as $R_B = R - r_A - r_X$, and the core radii remain practically untouched upon compression, a linear relationship necessarily holds for R_B (which virtually subsumes all the changes in the cell parameter) with R for a given material.

Furthermore (and less intuitively), the linear relationship also holds for r_1 and r_2 with R at constant δ. According to Parr's electronegativity approach [20], the electronegativity in the molecule AX, χ_{AX}, can be expressed in terms of the radii of the isolated atoms, r_A and r_X, as: [19]

$$\chi_{AX} = \frac{\chi_A r_A + \chi_X r_X}{R} \quad (17)$$

but also in terms of their properties in the molecule AX:

$$\chi_{AX} = K\frac{q+\delta}{r_1} \quad (18)$$

where K is a constant.

If we combine Equations (17) and (18) and reorganise, we see that:

$$r_1 = \frac{K(q+\delta)}{r_A\chi_A + r_X\chi_X}R = \alpha R \quad (19)$$

where K, q, r_A, r_X, χ_A and χ_X are constants which only depend on the system, so that α itsef is a constant as long as δremains constant. In other words, r_1 will scale linearly with the interatomic distance, as long as the charge transfer does not change upon pressurization. Of course, similar equations apply to r_2, with $r_2 = \beta R$.

In order to check this hypothesis, we have repeated our DFT-ELF procedure upon compression of the zero pressure reference cell (no optimization needed due to the symmetry of the cell). Two prototypical examples of covalent (BAs, $\Delta_\chi = 0.14$) and polar (BN, $\Delta_\chi = 1$) solids were chosen to compare their corresonding compression with respect to their bonding type. Figure 5 collects the results upon compression for r_1, r_2 and R_B. Negligible deviations from linearity are found in both cases, leading to three main conclusions:

- R_B is the quantity showing the greatest compression

of the model since it represents the valence size, i.e. the chemically sensitive part as pressure is applied [43,49].

- α, β and ν are constant upon compression for the various types of bonding considered. Further analysis shows that α, βand ν can certainly be related to hardness, since they determine which ion compresses the most. Indeed a look at Figure 5 shows that in the case of BN, where the hardness of atoms is more similar, r_1 and r_2 compress at similar rates, whereas in BAs, the compression of As overcomes that of B.

- As a consequence of the previous observation and from Equation (19), we can consider that the charge transfer is also constant upon compression for the ranges considered.

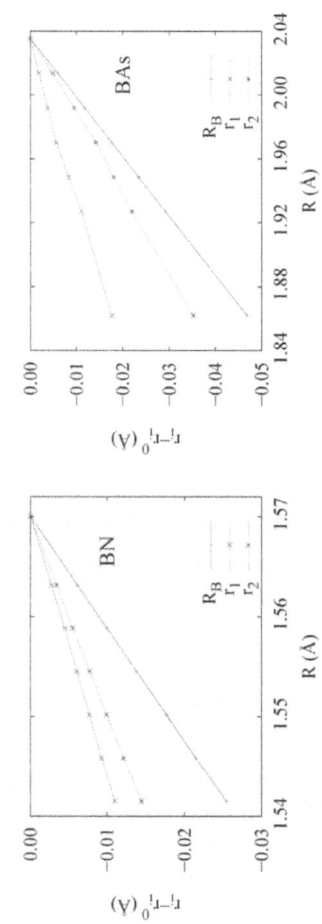

Figure 5: Evolution of the geometrical parameters, r_1, r_2 and R_B, upon compression of the interatomic distance R for BN (**left**) and BAs (**right**). Data relative to the corresponding equilibrium values denoted by a 0 superscript. All numbers in Å.

If we analyze these constants, it is observed that the hardest materials, diamond and BN, not only have the smallest R and the smallest R_B, but also the highest $\nu = R_B/R$ ratio (see Table 1).

The size of the parameters r_1 and r_2 (or α and β) also plays an important role. Since these parameters are related to the relative hardness of atoms, the smaller α and β are, the less compressible the whole system becomes. What is most important, both parameters should be similarly small. Otherwise, the softest ion will subsume the pressure. Ideally, for a given r_1, the compressibility is minimized when the relationship $\alpha = \beta$ is fulfilled. This condition is equivalent to locating the bond position in the middle of both atoms (see Figure 5 (left)).

These dependencies enable us to establish new relationships for materials with low compressibility, directly derived from microscopic conditions. From Table 1 we see that there are two cases which deserve our attention as far as their r_1/r_2 ratio is concerned: diamond and BN represent the two different ways of achieving $\alpha = \beta$ (i.e., $r_1 = r_2$):

- Perfectly covalent case (diamond): in this case A = X (or $\Delta_\chi = 0$), so that all lengths are the same.

- Polar case (BN): in this case A ≠ X ($\Delta_\chi \neq 0$). However, a relationship holds between the core and the bond location displacement. It is found that the cores differential size, $\epsilon = r_X - r_A$ for $\epsilon > 0$ is equivalent to the displacement of the bond location from the center toward the smaller atom A, $r'_1 = r'_2 + \epsilon$ (remember that $r_1 = r'_1 + r_A$ and $r_2 = r'_2 + r_X$). In BN, a good combination of atomic sizes and polarity compensate to provide $r_1/r_2 = 1.03$.

SUMMARY AND CONCLUSIONS

We have introduced the Electron Localization Function as the source of the necessary parameters for the analysis of the Bond Charge Model in solid state applications, named NEWBCM. The availability of charges and bond location has enabled us to develop a new BCM-based model in which, parameters are not only obtained ab initio, but they are closer to chemical intuition and recover good agreement with experience.

The ability of NEWBCM to describe compressibility trends among these crystal families has been exploited in order to settle the directions for novel superhard materials design. Some of these hints are not related to macroscopic properties but to microscopic ones. Thus, quantitative relationships between electronic structure and macroscopic properties, such as compressibility, can now be proposed.

The micro-macroscopic properties that guide the quest for superhard materials can be summarized as follows: (i) Small atoms; (ii) Small bond; (iii) High ν; (iv) Hard atoms; and (v) Atoms of similar hardness. Although most of these guidelines are related (e.g., (i) and (iv)) and extremely intuitive, some of them, such as (iii) would not be deduced at first glance, and provide useful quantitative insight into the search and prediction of novel superhard materials from microscopic grounds.

As an example, it has been shown that the hardest compounds within our study set, diamond and BN, both follow these principles, and more specifically, both show an r_1/r_2 ratio very close to one in spite of belonging to two very different bonding types (covalent and very polar, respectively). These conditions thus unite their different macroscopic properties into one relevant microscopic condition for superhardness.

Although this model is only applicable to zinc-blende types of structures, work is in progress to extend it to other structural types. This would, for example, enable us to understand the driving forces for chemical changes in phase transitions.

ACKNOWLEDGMENTS

Financial and computational support from the MALTA-Consolider Ingenio-2010 program under project CSD2007-00045 is acknowledged. JCG and MM thank the Spanish Ministry of Education and Principado de Asturias (FICYT), respectively, for postdoctoral grants. MM, JMM, and JMR thank the Spanish Ministry of Education for financial support (CTQ2009-14596-C02-02 and CTQ2012-38599-C02-02).

REFERENCES

1. Gilman, J.J. Electronic Basis of the Strength of Materials; Cambridge University Press: Cambridge, UK, 2003.
2. Gilman, J.J. Why silicon is hard. Science 1993, 261, 1436–1439.
3. Gao, F.; He, J.; Wu, E.; Liu, S.; Yu, D.; Li, D.; Zhang, S.; Tian, Y. Hardness of covalent crystals. Phys. Rev. Lett. 2003, 91, 015502.
4. Li, K.; Wang, X.; Zhang, F.; Xue, D. Electronegativity identification of novel superhard materials. Phys. Rev. Lett. 2008,100, 235504.
5. Oganov, A.R.; Glass, C.W. Crystal structure prediction using ab initio evolutionary techniques: Principles and applications. J. Chem. Phys. 2006, 124, 244704.
6. Lyakhov, A.O.; Oganov, A.R. Evolutionary search for superhard materials: Methodology and applications to forms of carbon and TiO2. Phys. Rev. B 2011, 84, 092103–1.
7. Parr, R.; Pearson, R.G. Absolute hardness: Companion parameter to absolute electronegativity. J. Am. Chem. Soc. 1983,105, 7512–7516.
8. Putz, M.V. Electronegativity: Quantum observable. Int. J. Quantum Chem 2009, 109, 733–738.
9. Putz, M.V. Chemical action and chemical bonding. J. Mol. Stru.: Theochem 2009, 900, 64–70.
10. Putz, M.V. The Bondons: The quantum particles of the chemical bond. Int. J. Mol. Sci. 2010, 11, 4227–4256.
11. Putz, M.V. Density functional theory of bose-einstein condensation: Road to chemical bonding quantum condensate.Struc. Bond. 2012, 149, 1–50.
12. Bader, R.F.W. Atoms in Moleculess, A Quantum Theory; Oxford University Press: Oxford, UK, 1990.
13. Martín Pendás, A.; Costales, A.; Luaña, V. The topology of the electron density in ionic materials. III. Geometry and ionic radii. J. Phys. Chem. B 1998, 102, 6937–6948.
14. Becke, A.D.; Edgecombe, K.E. A simple measure of electron localization in atomic and molecular systems. J. Chem. Phys. 1990, 92, 5397–5403.
15. Savin, A.; Jepsen, O.; Flad, J.; Andersen, L.K.; Preuss, H.; von Schnering, H.G. Electron localization in solid-state structures of the elements: The diamond structure. Angew. Chem. Int. Ed. Engl. 1992, 31, 187–188.
16. Borkman, R.F.; Parr, R.G. Toward an understanding of potential-energy functions for diatomic Molecules. J. Chem. Phys. 1968, 48, 1116–1126.
17. Parr, R.G.; Borkman, R.F. Simple bond-charge model for potential-energy curves of homonuclear diatomic molecules.J. Chem. Phys. 1968, 49, 1055–1058.
18. Borkman, R.F.; Simons, G.; Parr, R. Simple bond-charge model for potential-energy curves of heteronuclear diatomic molecules. J. Chem. Phys. 1969, 50, 58–65.
19. Parr, R.; Yang, W. Density Functional Theory of Atoms and Molecules; Oxford University Press: Oxford, UK, 1989.
20. Ray, N.K.; Samuels, L.; Parr, R.G. Studies of electronegativity equalization. J. Chem. Phys. 1979, 70, 3680–3684.
21. Martin, R.M. A simple bond charge model for vibrations in covalent crystals. Chem. Phys. Lett. 1968, 2, 268–270.
22. Oviedo's Quantum Chemistry Group. Available online: http://azufre.quimica.uniovi.es/software.html (accessed on 8 April 2015).
23. Pauling, L. The Nature of the Chemical Bond, 3rd ed.; Cornell Univ. Press: Ithaca, NY, USA, 1960.
24. Putz, M.V. Markovian approach of the electron localization functions. Int. J. Quantum Chem. 2005, 105, 1–11.
25. Putz, M.V. Density functionals of chemical bonding. Int. J. Mol. Sci. 2008, 9, 1050–1095.
26. Gillespie, R.J.; Nyholm, R.S. Inorganic stereo-chemistry. Quart. Rev. 1957, 11, 339–380.
27. ELK Code. Available online: http://elk.sourceforge.net/ (accessed on 8 April 2015).
28. Drief, F.; Tadjer, A.; Mesri, D.; Aourag, H. First principles study of structural, electronic, elastic and optical properties of MgS, MgSe and MgTe. Catal. Today 2004, 89, 343–355.
29. Semiconductors information web-site. Available online: www.semiconductors.co.uk (accessed on 8 April 2015).
30. Perdew, J.P.; Wang, Y. Accurate and simple analytic representation of the electron-gas correlation energy. Phys. Rev. B1992, 45, 13244–13249.
31. Perdew, J.P.; Burke K., *!!! REPLACE !!!*; Ernzerhof, M. Generalized gradient approximation made simple. Phys. Rev. Lett. 1996, 77, 3865–3868.
32. Contreras-García, J.; Martin Pendás, A.; Silvi, B.; Recio, J.M. Useful applications of the electron localization function in high-pressure crystal chemistry. J. Phys. Chem. Solids 2008, 69, 2204–2207.
33. Contreras-García, J.; Silvi, B.; Martín Pendás, A.; Recio, J.M. Computation of local and global properties of the electron localization function topology in crystals. J. Chem. Theory Comput. 2009, 5, 164–173.
34. Kohout, M.; Savin, A. Atomic shell structure and electron numbers. Int. J. Quant. Chem. 1996, 60, 875–882.
35. Frisch, M.J.; Trucks, G.W.; Schlegel, H.B.; Scuseria, G.E.; Robb, M.A.; Cheeseman, J.R.; Montgomery, J.A., Jr.; Vreven, T.; Kudin, K.N.; Burant, J.C.; et al. Gaussian 03, Revision C.02; Gaussian, Inc.: Wallingford, CT, USA, 2004.
36. Noury, S.; Krokidis, X.; Fuster, F.; Silvi, B. TopMoD Package; Universite Pierre et Marie Curie: Paris, France, 1997.
37. Otero-de-la-Roza, A.; Blanco, M.A.; Martín Pendás, A.; Luaña, V. Critic: A new program for the topological

analysis of solid-state electron densities. Comput. Phys. Commun. 2009, 180, 157–166.

38. Kohout, M.; Savin, A.; Preuss, H. Contribution to electron density analysis. I. Shell structure of atoms. J. Chem. Phys.1991, 95, 1928–1942.
39. Ackland, G.J.; Marqués, M.; Contreras-García, J.; McMahon, M.I. Origin of incommensurate modulations in the high-pressure phosphorus IV phase. Phys. Rev. B 2008, 78, 054120.
40. Marqués, M.; Ackland, G.J.; Lundegaard, L.F.; Contreras-García, J.; McMahon, M.I. Potassium under pressure: A pseudobinary ionic compound. Phys. Rev. Lett. 2009, 103, 115501.
41. Marqués, M.; Santoro, M.; Guillaume, C. L.; Gorelli, F.; Contreras-García, J.; Howie, R.; Goncharov, A.F.; Gregoryanz, E. Optical and electronic properties of dense sodium. Phys. Rev. B 2011, 83, 184106.
42. Anderson, D.L.; Anderson, O.L.J. The bulk modulus-volume relationship for oxides. J. Geophys. Res. 1970, 75, 3494–3500.
43. Contreras-García, J.; Mori-Sánchez, P.; Silvi, B.; Recio, J.M.A. Quantum chemical interpretation of compressibility in solids. J. Chem. Theory Comput. 2009, 5, 2108–2114.
44. Glasser, L. Volume-based thermoelasticity: Compressibility of inorganic solids. Inorg. Chem. 2010, 49, 3424–3427.
45. Glasser, L. Volume-based thermoelasticity: Compressibility of mineral-structured materials. J. Phys. Chem. C 2010, 114, 11248–11251.
46. Glasser, L.; Jenkins, H.D.B. Internally consistent ion volumes and their application in volume-based thermodynamics.Inorg. Chem. 2008, 47, 6195–6202.
47. Martín Pendás, A.; Costales, A.; Blanco, M.A.; Recio, J.M.; Luaña, V. Local compressibilities in crystals. Phys. Rev. B2000, 62, 13970–13978.
48. Ouahrani, T.; Menendez, J.M.; Marqués, M.; Contreras-García, J.; Baonza, V.G.; Recio, J.M. Local pressures in Zn chalcogenide polymorphs. Europhys. Lett. 2012, 98, 56002.
49. Contreras-García, J.; Recio, J.M. On bonding in ionic crystals. J. Phys. Chem. C 2011, 115, 257–263.

CHAPTER 7

THREE DIMENSIONAL THERMAL-SOLUTE PHASE FIELD SIMULATION OF BINARY ALLOY SOLIDIFICATION

P.C. Bollada, C.E. Goodyer, P.K. Jimack, A.M.Mullis, and F.W. Yang

Keywords: Phase field, Dendrite, Solidification, Adaptive mesh, Multigrid, Nonlinear PDEs

INTRODUCTION

We here present our computational approach to simulating, at the meso-scale, three dimensional, non-isothermal alloy solidification from an initial small, spherical seed into a mature, dendritic crystal. A feature of a mature dendrite is the geometric complexity of its evolving two-dimensional surface (see Fig. 1 for a typical snapshot in time). This makes tracking of the surface a difficult task in sharp interface models. A phase-field model avoids this by making use of a phase field, $\phi(\mathbf{x},t)\in[-1,1]$, to represent, at its two extremes, the liquid and solid state respectively and the evolution of the phase boundary, $\phi=0$, is the surface of interest. This solves one problem, but at a cost of introducing another. The computation requires an extra variable, the phase, which varies rapidly over a small region about the interface. Taking the thickness of the interface to be ≈ 1, we find the size of a mature dendrite grows to ~ 300, requiring the domain size to be significantly greater still (depending on the thermal field this may need to be $O(1000)$, or even more), we see that the interface region of interest is very much smaller than the overall domain. Of major concern in phase field models is the dependence of the computed results on the interface width. To address this, Karma [1], analysed the problem in the thin interface limit to produce a phase field formulation that is independent of the interface width up to several orders of magnitude larger than a physically real value, although this is less clear when a thermal field is coupled. That said, we adopt an interface width that is of physically realistic order.

Figure 1: Snapshot of the solid-liquid interface for a typical dendrite. This image was obtained from a simulation with Le = 40, $\Delta = 0.525$ and $\Delta x = 0.78$.

There are two coupled driving forces for growth: the alloy concentration, governed by a diffusion parameter D_c and, secondly, a temperature field governed by a diffusion parameter D_θ. The ratio of these two parameters gives the Lewis number, $Le=D_\theta/D_c$, which for many metallic alloys approaches 10000 . In two dimensions Lewis numbers of this magnitude have been realised by [2]. However, there are no prior results, even for the interface widths permitted by Karma's model, for even very moderate Lewis number in three dimensions. This paper seeks to demonstrate that such results are feasible provided the appropriate numerical techniques are employed.

Citation: P.C. Bollada, C.E. Goodyer, P.K. Jimack, A.M. Mullis, F.W. Yang, Three dimensional thermal-solute phase field simulation of binary alloy solidification, Journal of Computational Physics, Volume 287, 15 April 2015, Pages 130-150, ISSN 0021-9991, http://dx.doi.org/10.1016/j.jcp.2015.01.040.

The numerical solution to this phase field model (described in detail in the following section) requires methods to solve a time-dependent, highly nonlinear system of PDEs, of parabolic type, and capable of resolving varying length and time scales. A feature that adds another level of difficulty to this problem is that we are particularly interested in the tip radius and speed of growth of the dendrite only when it is fully mature, and these two numbers are steady. We select these parameters because experimental evidence [3]shows that, when scaled by the radius of curvature at the tip, all dendrites are self-similar. Consequently, fundamental studies of dendritic growth, whether analytical (see e.g.[4] and [5]) or computational, have focused on calculating the growth velocity and radius of curvature at the tip as a function of undercooling and the thermo-physical properties of the melt, these parameters being sufficient to uniquely describe the growth.

In summary, the computational problem is: non-linear, three-dimensional, stiff, involves multiple length scales to capture small phase and large temperature fields, multi-time scale associated with the Lewis number and, to establish a mature dendrite, requires a long simulation time.

The computational techniques we employ are: use of very fine meshing in the region around the moving boundary where phase field and solute field resolution is critical, and coarse meshing away from the boundary where only the slowly changing temperature field requires resolution; implicit time stepping to allow much larger time steps than would otherwise be possible; nonlinear smoothing in conjunction with a nonlinear multi-grid solver; and parallel processing with up to 1024 cores as the simulation progresses. The combination of all these techniques allows an almost optimal solution process to be developed, in which the number of degrees of freedom is evolved with the dendrite, to maintain the required resolution as the interface grows, and the solution time at each time step is approximately proportional to the number of degrees of freedom. Furthermore, the use of a parallel implementation ensures that sufficient primary memory is available to support a mesh resolution which is fully converged whilst maintaining a tractable solution time.

The particular phase field model we employ is an extension of [6], and is based on the three dimensional thermal phase field model of [7] and two dimensional thermal-solutal phase field model of [8]. One feature of the physical problem is that it is purely dissipative, or entropy increasing, as all natural relaxational phenomena are. The resulting PDEs are of Allen–Cahn [9] and Carn–Hilliard type [10]. That is to say, the model involves time derivatives of the three fields coupled to forms involving variational derivatives of some functional – typically the free energy functional. As the dendrite grows the free energy reduces monotonically with time but never achieves equilibrium if the domain boundary is far from the dendrite. Although we have listed some of the difficult aspects of this model, the relaxational aspect is typically an asset and results in stable numerical schemes: there is no convection, for example (at least in the absence of flow in the melt).

The variational form of the mathematical model is, of course, identical to the two-dimensional model in form. However, on realising the variational derivatives the resulting equations are more complex and nonlinear in the higher dimension. This is largely because of surface energy related anisotropy associated to alignment at the molecular scale. In two dimensions anisotropy is conveniently formulated using a single angle parameter, but in three dimensions we prefer to use a normal given in terms of Cartesian gradients.

In addition to the Lewis number, another key parameter in the simulations is the undercooling, Δ, which sets the temperature of the liquid›s initial and far boundary condition below its freezing point. As this parameter becomes larger the under-cooling becomes more severe, the solidification more rapid and fractal in appearance: and, also, correspondingly more difficult to simulate.

The other field, not hitherto discussed, is the solute field. For a binary alloy of initial concentration, c_∞, the concentration of an alloy component at any point is represented by a value of $c(\mathbf{x},t)\in[0,1]$. The requirement for equilibrium at the solid-liquid interface means that the concentration in the solid and the concentration in the liquid at the interface will be unequal. In a sharp interface model this results in a discontinuous jump inc at the interface, while in phase field models it results in a steep, but continuous, increase in c across the diffuse interface region, where there is some advantage in reformulating the problem to remove this. We show this in the next section.

GOVERNING EQUATIONS

The governing equations for dendritic growth of an under-cooled binary alloy are here presented in full, in both their variational form and in the (equivalent) form of PDEs for numerical implementation. See [8] for a full description of the model we employ, derived in turn from [7], which provides analysis of the anti-trapping term, which was extended to isothermal dilute binary alloys in [1]. The origin of the anisotropy on the left hand side of Eq. (2a), for example, is discussed in [8] and is vital to this formulation in eliminating interface kinetic effects. The non-dimensional equations for the phase field, ϕ, the solute concentration, c and the dimensionless temperature, θ, are given via a specification of the free energy.

$$F \equiv \int_V \tfrac{1}{2}A(\mathbf{n})^2\nabla\phi\cdot\nabla\phi + f(\theta,\phi)\,\mathrm{d}V \tag{1}$$

and the relations

$$\tau(c,\phi)A^2(\mathbf{n})\dot{\phi} = -\frac{\delta F}{\delta\phi}, \tag{2a}$$

$$\dot{c} = \nabla\cdot\left(K(\phi)\nabla\frac{\delta F}{\delta c} - \mathbf{j}\right). \tag{2b}$$

$$\dot{\theta} = D_\theta\nabla^2\theta + \tfrac{1}{2}\dot{\phi}. \tag{2c}$$

The solute diffusion parameter is given by

$$K = D_c\tfrac{1}{2}(1-\phi). \tag{3}$$

The parameter D_c is a diffusion constant, thus K=0 in the solid (ϕ=1) and $K=D_c$ in the liquid (ϕ=−1). Thus, in general, $\phi\in[-1,1]$. D_θ is the temperature diffusion coefficient (assumed constant). The normal to the interface is given by

$$\mathbf{n} = \frac{\nabla\phi}{|\nabla\phi|}, \tag{4}$$

which is well defined around $\phi=0$, and the anisotropy function for cubic symmetry (growth is preferred along the normals to the faces) is given for three dimensions by[7],

$$A(\mathbf{n}) \equiv A_0\left[1+\tilde{\epsilon}\left(n_x^4+n_y^4+n_z^4\right)\right] \tag{5}$$

where $\mathbf{n}=[n_x, n_y, n_z]$, $A_0=1-3\epsilon$, $\tilde{\epsilon}=4\epsilon/(1-3\epsilon)$ and $\epsilon \approx 0.02$ governs the amount of anisotropy. The reason for this arrangement of constants is to compare with the two dimensional form

$$A(\mathbf{n}) \equiv A_0\left[1+\tilde{\epsilon}\left(n_x^4+n_y^4\right)\right] \equiv 1+\epsilon\cos 4\psi \tag{6}$$

where the angle, ψ, is given by

$$\tan\psi = \frac{\partial\phi}{\partial y}\Big/\frac{\partial\phi}{\partial x}. \tag{7}$$

The dimensionless relaxation time function is defined by

$$\tau(c,\phi) \equiv \frac{1}{Le} + Mc_\infty\left[1+(1-k_E)U\right], \tag{8}$$

where the Lewis number $Le=D_\theta/D_c$ and

$$U \equiv \frac{1}{1-k_E}\left(\frac{2c/c_\infty}{1+k_E-(1-k_E)\phi}-1\right). \tag{9}$$

Here k_E is the equilibrium partition coefficient, c_∞ is the far boundary condition force$\in[0,1]$. The anti-trapping current $\mathbf{j}$, appearing in the solute equation, Eq. (2b), is prescribed by

$$\mathbf{j} = -\frac{1}{2\sqrt{2}}\left[1+(1-k_E)\right]U\dot{\phi}\mathbf{n}. \tag{10}$$

The profile of c exhibits a spike at the interface, which can present computational difficulties. Following [8], this is largely overcome by rewriting the solute equation using the variable U:

$$\left(\frac{1+k_E}{2}-\frac{1-k_E}{2}\phi\right)\frac{\partial U}{\partial t} = \nabla\cdot\left\{D_c\frac{1-\phi}{2}\nabla U+\mathbf{j}\right\}+\frac{1}{2}\left[1+(1-k_E)U\right]\frac{\partial\phi}{\partial t}. \tag{11}$$

The physical temperature field, T, can be recovered by the relationship

$$\theta = \frac{T-T_M-mc_\infty}{L/C_p}, \tag{12}$$

where L and C_p are the latent heat of the phase transition and heat capacity respectively. The slope of the liquidus line is given by $m=ML/[C_p(1-\kappa_E)]$ and T_M is the melting temperature of the alloy.

Finally the bulk free energy density is given by

$$f(\theta,\phi,U) \equiv \frac{\phi^2}{2}\left(\frac{\phi^2}{2}-1\right)+\lambda(\theta+c_\infty U)\left(\phi-\frac{2\phi^3}{3}+\frac{\phi^5}{5}\right). \tag{13}$$

We solve the system of Eqs. (2a), (2c) and (11) plus initial, typically small, solid seed (see Eq. (40) and Section 3.2) and far boundary conditions

$$\begin{aligned} \phi|_{\text{far}} &= -1 \\ U|_{\text{far}} &= 0 \quad (\equiv c|_{\text{far}} = c_\infty) \\ \theta|_{\text{far}} &= -\Delta \end{aligned} \tag{14}$$

where Δ is the given under-cooling. The equation for temperature is a standard diffusion equation[1] with a heating term, $\dot{\phi}$, proportional to the solidification rate (or cooling if melting). The driving force for the phase equations is given by $f(\theta,\phi,U)$, consisting of a double well potential having stable minima at $\phi=\pm1$ and a maximum at $\phi=0$ and a function of θ to create conditions for moving the phase boundary. For example a negative value of θ creates conditions favourable for solidification. The parameter, λ, is proportional to the interface width, which in turn is chosen as the characteristic length scale.

Parameter values

For the purposes of this paper we choose a selection of parameters to use as default values for the simulations below in Table 1. Any deviation from these parameter values is explicitly noted in the text.

Table 1: Table of parameter values used for the simulations in this paper.

Physical property	Symbol	Value
Anisotropy	ϵ	0.02
Boundary concentration	Mc_∞	0.05
Equilibrium partition coefficient	κ_E	0.3
Dimensionless interface width	λ	2
Ratio of solute diffusivity to characteristic diffusivity	D_c	1.2534
Lewis number - D_θ/D_c	Le	40 and 100
Dimensionless undercooling at the far boundary	Δ	0.25 to 0.80
Initial nuclear radius	R_0	5.0
Computational property	**Symbol**	**Value**
Finest grid size	Δx	0.195 to 0.78
Computation domain size	L^3	800×800×800

Anisotropic calculations

Note that the phase equation, Eq. (2a), is made considerably more complicated by the presence of the anisotropy term, $A(\mathbf{n})$, in the free energy functional. The variational derivative of a functional not involving gradients is simply the partial derivative of the density with respect to that variable. Thus

$$\begin{aligned} \frac{\delta}{\delta\phi}\int f(T,\phi)\,dV &= \frac{\partial f}{\partial\phi} \\ &= \phi^3-\phi+\lambda(\theta+c_\infty U)(1-2\phi^2+\phi^4). \end{aligned} \tag{15}$$

The variational derivative of the pure gradient part of the functional is given by

$$\frac{\delta G}{\delta\phi} \equiv \frac{\delta}{\delta\phi}\int g(\nabla\phi)\,dV = -\nabla\cdot\left(\frac{\partial}{\partial\mathbf{r}}g(\mathbf{r})\Big|_{\mathbf{r}=\nabla\phi}\right) \tag{16}$$

where, in our model,

$$g(\mathbf{r}) \equiv \tfrac{1}{2}A(\mathbf{n})^2|\mathbf{r}|^2, \text{ for } \mathbf{r}\in\mathbb{R}^3 \tag{17}$$

In order to expand Eq. (16) and thus, Eq. (2a), we first introduce the notation: $\phi_{,i}\equiv\partial_i\phi\equiv\frac{\partial\phi}{\partial x^i}$ etc., for Cartesian differentiation, and subscripts for differentiation on function space. Thus

$$g_i \equiv \frac{\partial g}{\partial r_i},$$
$$g_{ij} \equiv \frac{\partial}{\partial r_i}\frac{\partial g}{\partial r_j}. \tag{18}$$

This enables us to write

$$-\frac{\delta G}{\delta \phi} = \partial_i g_i$$
$$= \phi_{,ij} g_{ij} \quad \text{(using the chain rule)},$$

which written out in full reads

$$-\frac{\delta G}{\delta \phi} \equiv -\frac{\delta}{\delta \phi}\int g(\nabla\phi)\,dx^3 = \frac{\partial^2 \phi}{\partial x^i \partial x^j}\left(\frac{\partial}{\partial r_i}\frac{\partial g}{\partial r_j}\right)\Big|_{\mathbf{r}=\nabla\phi}.$$

Note that, g_{ij}, is a function of only first derivatives of ϕ. To avoid expanding the above in terms of the components, $\phi_{,i}$ we first introduce the substitutions $q \equiv |\nabla\phi|^2$ and $\mathbf{X} \equiv [X_1, X_2, X_3] \equiv [\phi_{,1}^2/q, \phi_{,2}^2/q, \phi_{,3}^2/q]$, to write the anisotropy

$$A = A_0\left(1 + \tilde{\epsilon}\sum_{i=1}^{3} X_i^2\right). \tag{19}$$

For an arbitrary function $h(\mathbf{r}) = \tilde{h}(\mathbf{r}, q, \mathbf{X}, A)$ we use the chain rule to write

$$\frac{\partial h}{\partial r_i} = \left(\frac{\partial}{\partial r_i} + \frac{\partial q}{\partial r_i}\frac{\partial}{\partial q} + \frac{\partial X_j}{\partial r_i}\frac{\partial}{\partial X_j} + \frac{\partial A}{\partial r_i}\frac{\partial}{\partial A}\right)\tilde{h}$$
$$= \left(\frac{\partial}{\partial r_i} + 2r_i\frac{\partial}{\partial q} + \frac{2r_i}{q}(\delta_{ij} - X_j)\frac{\partial}{\partial X_j} + \frac{2r_i}{q}[2A_0\tilde{\epsilon}X_i - 2(A - A_0)]\frac{\partial}{\partial A}\right)\tilde{h}, \tag{20}$$

where we have used

$$\frac{\partial q}{\partial r_i} = 2r_i,$$
$$\frac{\partial X_j}{\partial r_i} \equiv \frac{\partial}{\partial r_i}\left(\frac{r_j^2}{q}\right) = \frac{2r_i\delta_{ij}}{q} - \frac{2r_i r_j^2}{q^2} = \frac{2r_i}{q}(\delta_{ij} - X_j),$$
$$\frac{\partial A}{\partial r_i} = \frac{\partial X_j}{\partial r_i}\frac{\partial A}{\partial X_j} = \frac{2r_i}{q}(\delta_{ij} - X_j)\frac{\partial A}{\partial X_j}$$
$$= \frac{2r_i}{q}(\delta_{ij} - X_j)2A_0\tilde{\epsilon}X_j$$
$$= \frac{4r_i}{q}(A_0\tilde{\epsilon}X_i - A + A_0) \tag{21}$$

using, on the last simplification, the identity $A - A_0 \equiv A_0\tilde{\epsilon}\sum_j X_j^2$. This allows us to compute

$$g_i \equiv \frac{\partial}{\partial r_i}\left(\tfrac{1}{2}A^2 q\right)|_{\mathbf{r}=\nabla\phi} = \phi_{,i}A^2 + 4\phi_{,i}(A_0 X_i \tilde{\epsilon} - A + A_0)A \tag{22}$$

and further differentiation gives the concise forms

$$g_{ii} = (24X_i - 3)A^2 + (-48X_i^2\tilde{\epsilon} + 12X_i\tilde{\epsilon} - 40X_i + 4)A_0 A + 16X_i(X_i\tilde{\epsilon} + 1)^2 A_0^2$$
$$g_{ij} = \frac{\phi_{,i}\phi_{,j}}{g}\left[24A^2 + (-24X_i\tilde{\epsilon} - 24X_j\tilde{\epsilon} - 40)A_0 A + (16(X_j\tilde{\epsilon} + 1))(X_i\tilde{\epsilon} + 1)A_0^2\right], \quad i \neq j. \tag{23}$$

The above expressions are not only much more concise than the expanded equivalent as a function of $\phi_{,i}$, but are functions of X_i and A which are of order unity in size and thus minimise floating point errors (the expanded equivalent contains tenth order polynomials of $\phi_{,i}$). We note also that, in the absence of anisotropy, $\tilde{\epsilon} = 0$, reduces g_{ij} to δ_{ij} so that $\phi_{,ij}g_{ij} = \nabla^2\phi$. In the situation where g_{ij} is ill defined due to $|\nabla\phi| \to 0$we set.

$$\phi_{,ij}g_{ij}|_{|\nabla\phi|\to 0} = A_0^2(1 + \tilde{\epsilon})^2\nabla^2\phi \tag{24}$$

or, equivalently $\mathrm{tr}(\mathbf{g}) \equiv \delta_{ij}g_{ij}$. In practice we use this expression only when $|\nabla\phi|=0$, to machine precision without difficulty.

Using the notation $\mathrm{tr}(\mathbf{g}) \equiv \delta_{ij}g_{ij}$, the rearrangement

$$\phi_{,ij}g_{ij} \equiv \tfrac{1}{3}(\phi_{,11} + \phi_{,22} + \phi_{,33})(g_{11} + g_{22} + g_{33}) + (\phi_{,ij} - \tfrac{1}{3}\phi_{,kk}\delta_{ij})g_{ij}$$
$$\equiv \tfrac{1}{3}\nabla^2\phi\,\mathrm{tr}(\mathbf{g}) + (\phi_{,ij} - \tfrac{1}{3}\phi_{,kk}\delta_{ij})g_{ij}, \tag{25}$$

allows the dominant term to be isolated and has advantage, because the Laplacian can be discretised to minimise grid induced anisotropy. We will also see that the term $(\phi_{,ij} - \frac{1}{3}\phi_{,kk}\delta_{ij})$, like g_{ij}, on discretisation with a compact stencil at a discrete node, **p**, only has contributions from the nodes surrounding **p**. This affords simplification for the non-linear solver later discussed.

System summary

Writing, $M_{ij} \equiv \phi_{,ij} - \frac{1}{3}\phi_{,kk}\delta_{ij}$ we summarise the nonlinear PDE system that forms our mathematical model as

$$\tau(c,\phi)A(\mathbf{n})^2\dot{\phi} = \tfrac{1}{3}\nabla^2\phi\,\mathrm{tr}(\mathbf{g}) + M_{ij}g_{ij} - \frac{\partial f}{\partial \phi} \tag{26}$$

where $\tau(c,\phi)$ is given by Eq. (8), $\frac{\partial f}{\partial \phi}$ is given in Eq. (15), g_{ij} in Eq. (23), the solute is solved via Eq. (11),

$$\left(\frac{1+k_E}{2} - \frac{1-k_E}{2}\phi\right)\frac{\partial U}{\partial t} = \nabla\cdot\left\{D_c\frac{1-\phi}{2}\nabla U + \mathbf{j}\right\} + \frac{1}{2}[1 + (1-k_E)U]\frac{\partial \phi}{\partial t}, \tag{27}$$

and the temperature by Eq. (2c),

$$\dot{\theta} = D_\theta\nabla^2\theta + \tfrac{1}{2}\dot{\phi}. \tag{28}$$

DISCRETISATION

The approach taken to discretisation is based upon a cell centred finite difference scheme, in that the nodes of the domain are located at the centre of cubic cells, and thus, we use the term 'node' and 'cell centre' interchangeably. One consequence of this is that there are no nodes on the domain boundary, thus making the use of Dirichlet boundary conditions non-trivial. The scheme makes use of the PARAMESH library to support mesh adaptivity in parallel [11] and [12]. The meshes obtained by this approach take the form of an oct tree of regular blocks, within which the mesh is uniform, and it is the spatial discretisation on any one of these blocks that we discuss here. Subsequently we will discuss adaptive mesh refinement and the implicit temporal discretisation scheme that is employed.

Spatial discretisation

Compact finite difference stencils (3×3×3), are used to discretise the first and second derivatives. Denoting these 27 points by **Q** and defining a generic 27 point Laplacian stencil, W_{abc}, around a point $\mathbf{p}=[i,j,k]$ by

$$\nabla^2 u|_{\mathbf{Q}} = \frac{1}{(\Delta x)^2}\sum_{a=-1}^{1}\sum_{b=-1}^{1}\sum_{c=-1}^{1} W_{abc}u_{\mathbf{p}+[a,b,c]} \tag{29}$$

where Δx is the physical distance between nearest neighbours, we can recover the 7 point Laplacian stencil, built from only the centre node, **p** and the 6 nearest neighbours ($a^2+b^2+c^2=1$)

$$\nabla^2 u|_{\mathbf{Q}} = \frac{-6u_{i,j,k} + u_{i+1,j,k} + u_{i-1,j,k} + u_{i,j+1,k} + u_{i,j-1,k} + u_{i,j,k+1} + u_{i,j,k-1}}{(\Delta x)^2} \tag{30}$$

by setting the weights

$$W_{abc} = \begin{cases} -6 & a^2+b^2+c^2=0 \\ 1 & a^2+b^2+c^2=1 \\ 0 & \text{otherwise} \end{cases}. \tag{31}$$

However, this stencil is more prone to grid anisotropy than the following 27 point Laplacian stencil (see [13]), with weights

$$W_{abc} = \begin{cases} -128/30 & a^2+b^2+c^2=0 \\ 14/30 & a^2+b^2+c^2=1 \\ 3/30 & a^2+b^2+c^2=2 \\ 1/30 & a^2+b^2+c^2=3 \end{cases} \tag{32}$$

(26), in space it is necessary to approximate $\phi_{,ij}g_{ij}$ about the point **p**. Using the above notation we obtain:

$$(\phi_{,ij}g_{ij})|_{\mathbf{Q}} = \tfrac{1}{3}\nabla^2\phi|_{\mathbf{Q}}\,\mathrm{tr}(\mathbf{g})|_{(\mathbf{Q-p})} + M_{ij}|_{(\mathbf{Q-p})}g_{ij}|_{(\mathbf{Q-p})}, \tag{33}$$

where we use the notation, $|\mathbf{Q\_p}$, to denote that the central node is not used. The discretisation of M_{ij} and g_{ij} are detailed in Appendix A. Consequently, Eq. (33) has the property that only $\nabla^2\phi|\mathbf{Q}$ contains a contribution from the central node, $\phi\mathbf{p}$ and thus

$$\frac{\partial}{\partial\phi_{\mathbf{p}}}(\phi_{,ij}g_{ij})|_{\mathbf{Q}} = \tfrac{1}{3}\mathrm{tr}(\mathbf{g})|_{\mathbf{Q-p}}\frac{\partial}{\partial\phi_{\mathbf{p}}}\nabla^2\phi|_{\mathbf{Q}} = -\frac{128}{90}\,\mathrm{tr}(\mathbf{g})|_{\mathbf{Q-p}}. \tag{34}$$

This is important for the Jacobi linearisation described in the next section. The PDE for ϕ is thus approximated by ODEs at each point, **p**, by

$$\dot{\phi}_{\mathbf{p}} = F^{\phi}_{\mathbf{p}}(\phi_{\mathbf{Q}}, U_{\mathbf{p}}, \theta_{\mathbf{p}}) \tag{35}$$

where

$$F^{\phi}_{\mathbf{p}} \equiv \frac{\tfrac{1}{3}\nabla^2\phi|_{\mathbf{Q}}\mathrm{tr}(\mathbf{g})|_{\mathbf{Q-p}} + M_{ij}|_{\mathbf{Q-p}}g_{ij}|_{\mathbf{Q-p}} - \dfrac{\partial f}{\partial\phi}(\phi_{\mathbf{p}}, U_{\mathbf{p}}, \theta_{\mathbf{p}})}{\tau(c_{\mathbf{p}}, \phi_{\mathbf{p}})A^2|_{\mathbf{Q-p}}}. \tag{36}$$

In the above $\tau(c\mathbf{p},\phi\mathbf{p})$ is given by Eq. (8) and $\frac{\partial f}{\partial\phi}|_{\mathbf{p}}$ is given by Eq. (15) using the values for ϕ,U,θ at point **p**. The Laplacian $\nabla^2\phi|\mathbf{Q}$ is given by Eq. (29) with weights Eq. (32). The indexed functions M_{ij} are given by Eq. (A.1) and the functions g_{ij} are given by Eq. (23), where $A, X_i, |\nabla\phi|^2$ are all functions of $\phi_{,i}$ approximated by second order differences, Eq. (A.2). The PDE for U, Eq. (11), is approximated by the ODEs

$$\dot{U}_{\mathbf{p}} = F^{U}_{\mathbf{p}}(\dot{\phi}_{\mathbf{p}}, \phi_{\mathbf{p}}, U_{\mathbf{Q}}, \theta_{\mathbf{p}}) \tag{37}$$

where

$$F^{U}_{\mathbf{p}} = \frac{\nabla\cdot\left\{D_c\frac{1-\phi_{\mathbf{p}}}{2}\nabla U|_{\mathbf{Q}} + \mathbf{j}\right\} + \tfrac{1}{2}[1+(1-k_E)U_{\mathbf{p}}]\dot{\phi}_{\mathbf{p}}}{\left(\frac{1+k_E}{2} - \frac{1-k_E}{2}\phi_{\mathbf{p}}\right)} \tag{38}$$

which is expanded in full (shown in Section 3.3) with the same derivative discretisation scheme used for U as ϕ.

Finally the θ term is given by

$$\dot{\theta}_{\mathbf{p}} = F^{\theta}_{\mathbf{p}}(\theta_{\mathbf{Q}}, \dot{\phi}_{\mathbf{p}}) \equiv D_\theta\nabla^2\theta|_{\mathbf{Q}} + \tfrac{1}{2}\dot{\phi}_{\mathbf{p}}. \tag{39}$$

Boundary and initial conditions

We use zero Neumann boundary conditions for all variables. This is easily implemented by imposing values to the centre of ghost cells of all blocks adjacent to the Neumman boundary, that are equal to the centre of the cell values of those cells next to the boundary (see below for more discussion of ghost cells). In the discretisation, this sets all boundary derivatives equal to zero. In exploitation of the symmetry in the problem this is interpreted as reflective symmetry on the planes x=0, y=0, z=0 and, provided the domain is sufficiently large, as equivalent to Dirichlet conditions on the far boundary for all variables. If, during a simulation, the normal derivative of any of the dependent variables begins to deviate from zero by more than a prescribed tolerance then we may allow the domain to expand so as to ensure we retain a zero normal derivative on the revised boundary (see below for more details of the mesh adaptivity that facilitates this). For a fixed domain, it is important that the dimensions are sufficient not only to represent the growing dendrite but also the temperature field throughout the simulation, which typically extends well ahead of the phase interface (especially for large values of the Lewis number).

The physical initial conditions for this problem are not fully understood. One of the sources of theoretical difficulty in modelling these materials is modelling the process of nucleation by which crystallites form via fluctuations [14]. We adopt the approach of using as small an initial seed as possible thereby allowing the initial phase profile, temperature and concentration field to adjust in the very early stages of the simulation. The temperature field can take longer to adjust to a profile which is near the melting point of the alloy inside the solid if started at a constant field value and, so we anticipate this with the condition given below (see Eq. (41)).

The initial condition for the phase field with seed radius given by R is prescribed by

$$\phi(t=0, \mathbf{x}(\mathbf{p})) = -\tanh[\alpha(\sqrt{\mathbf{x}\cdot\mathbf{x}} - R)], \tag{40}$$

where we employ the factor $\alpha=0.6$, the precise value of which is not important as the solver smooths the phase profile if α is large and conversely sharpen the profile if α is too small within reason. It is not found necessary to normalise this profile so that $\phi(t=0,\mathbf{x}=0)=1$. The initial solute condition is U=0 and the temperature profile used is

$$\theta(t=0, \mathbf{x}(\mathbf{p})) = -\Delta + \tfrac{1}{2}\Delta(\phi+1). \tag{41}$$

The most significant parameter in the initial conditions, in terms of the sensitivity of the subsequent calculations, is the radius of the initial nucleus. It has been shown that the transient behaviour of the evolving dendrite can be affected by this value well into the simulation [15] (though the final geometry and velocity of the dendrite tip is much less sensitive). To this end we choose the smallest value of R such that the dendrite does not melt (melting can occur if there is insufficient solid $\phi=1$ in the nucleus due the encroachment of the diffuse interface near the nucleus centre). We find the smallest value to be $R\approx5$.

Temporal discretisation

Due to the stiffness of the nonlinear system of ODEs that arises following the spatial discretisation we employ the second order Backward Differentiation Formula (BDF2 [16]) time stepping, so that at a point, **p** in the grid domain at the centre the 3^3 points, **Q**, the phase field variable system is approximated by

$$\phi_{\mathbf{p}}^{n+1} - r_2\phi_{\mathbf{p}}^{n} + r_3\phi_{\mathbf{p}}^{n-1} = r_1 \Delta t^{n+1} F_{\mathbf{p}}^{\phi} \quad (42)$$

In practice, we introduce another variable

$$\phi_{\mathbf{p}}^{*} \equiv r_2\phi_{\mathbf{p}}^{n} - r_3\phi_{\mathbf{p}}^{n-1} \quad (43)$$

and write Eq. (42) as

$$\phi_{\mathbf{p}}^{n+1} - \phi_{\mathbf{p}}^{*} = r_1 \Delta t^{n+1} F_{\mathbf{p}}^{\phi}(\phi_{\mathbf{Q}}^{n+1}, U_{\mathbf{p}}^{n+1}, \theta_{\mathbf{p}}^{n+1}). \quad (44)$$

The right hand side, $F_{\mathbf{p}}^{\phi}$, is defined by Eq. (36).

For constant time step, r_1=2/3, r_2=4/3, r_3=1/3. For a growing dendrite it is essential to use a small time step at the initial (imposed) state. Thereafter the time step is increased (see Section 4.2 for detail) to fully exploit the implicit time stepping. The BDF2 for adaptive time stepping is given by

$$\begin{aligned} r_1 &\equiv (r+1)/(2r+1), \\ r_2 &\equiv (r+1)^2/(2r+1), \\ r_3 &\equiv r^2/(2r+1) \\ r &\equiv \Delta t^{n+1}/\Delta t^{n} \end{aligned} \quad (45)$$

where r is the ratio of the current to previous time step. The heat equation is discretised similarly. With $\theta_{\mathbf{p}}^{*} \equiv r_2\theta_{\mathbf{p}}^{n} - r_3\theta_{\mathbf{p}}^{n-1}$ we write

$$\theta_{\mathbf{p}}^{n+1} - \theta_{\mathbf{p}}^{*} = r_1 \Delta t^{n+1} F_{\mathbf{p}}^{\theta}(\phi_{\mathbf{p}}^{n+1}, \phi_{\mathbf{p}}^{*}, \theta_{\mathbf{Q}}^{n+1}, \theta_{\mathbf{p}}^{*}) \quad (46)$$

where

$$F_{\mathbf{p}}^{\theta}(\phi_{\mathbf{p}}, \phi_{\mathbf{p}}^{*}, \theta_{\mathbf{Q}}, \theta_{\mathbf{p}}^{*}) \equiv D_\theta \nabla^2\theta|_{\mathbf{Q}} + \frac{1}{2}\frac{\phi_{\mathbf{p}} - \phi_{\mathbf{p}}^{*}}{r_1\Delta t^{n+1}} \quad (47)$$

The solute equation is more complicated and is given in Appendix B.

Adaptive mesh and block tree structure

The domain is first divided into a number of mesh blocks each of which contains N×N×N hexahedral cells, where we typically choose N=8. We employ a domain of L^3=800^3, which is large enough for Lewis numbers of the order of 100. We divide this domain into 4^3 blocks, so that when N=8, each cell is of size 25^3 and refer to this as level 1. The adaptive mesh strategy then imposes a hierarchical sub-division of some or all these blocks, and their descendants, based upon an oct tree structure. This subdivision aims to concentrate cells where gradients of the oct tree variables are highest and to ensure neighbour blocks differ by at most one level. The finest grid we work with is at level 7, with a corresponding Δx=25/2^7=0.1953125. We find that the minimum finest level necessary to obtain qualitatively reasonable results is level 5, corresponding toΔx=0.78125. As noted previously, throughout this work we exploit cell-centred finite differences in our discretisation, in which a single unknown is associated with the centre of each hexahedral cell for each of the dependent variables.

In order to discuss further the tree structure of the blocks we denote any block by its label,i and its contents/properties, B_i:

$$B(i) = [l, s, p, \mathbf{c}, \mathbf{x}] \quad (48)$$

where l∈[1,n] is the level, s∈[1,8] is the sibling number (i.e. an index for which child of p the block is), p is the parent index, c_i,i∈[1,8] are the 8 child indices, and**x**=[x,y,z] is the Cartesian position coordinates of the block origin. Any one of these properties can be accessed by the block number, i . Some examples: p(i) is the block number of the parent of block i ; **x**(p(i)) is the position of the parent's origin; $c_{s(i)}$(p(i))=i is an identity. A natural position scheme for the child blocks is

$$\begin{aligned} \mathbf{x}(c_1(i)) &= \mathbf{x}(i) + \tfrac{1}{2}\Delta x[-1,-1,-1] \\ \mathbf{x}(c_2(i)) &= \mathbf{x}(i) + \tfrac{1}{2}\Delta x[1,-1,-1] \\ \mathbf{x}(c_3(i)) &= \mathbf{x}(i) + \tfrac{1}{2}\Delta x[-1,1,-1] \\ \mathbf{x}(c_4(i)) &= \mathbf{x}(i) + \tfrac{1}{2}\Delta x[1,1,-1] \\ \mathbf{x}(c_5(i)) &= \mathbf{x}(i) + \tfrac{1}{2}\Delta x[-1,-1,1] \\ \mathbf{x}(c_6(i)) &= \mathbf{x}(i) + \tfrac{1}{2}\Delta x[1,-1,1] \\ \mathbf{x}(c_7(i)) &= \mathbf{x}(i) + \tfrac{1}{2}\Delta x[-1,1,1] \\ \mathbf{x}(c_8(i)) &= \mathbf{x}(i) + \tfrac{1}{2}\Delta x[1,1,1], \end{aligned} \quad (49)$$

where Δx is the grid size associated with level i. A complete specification of all the blocks in the oct tree is then specified by the list:

$$\mathbf{B} = \{B(i), i \in [1, B_N]\} \quad (50)$$

where B_N is the total block number. Moreover, a childless block, i , can be indicated by specifying, **c**(i)=0 and so the oct tree also can be specified by a listing of just the leaf blocks

$$\mathbf{B} = \{B(i), i \in [1, B_N] : \mathbf{c}(i) = \mathbf{0}\}. \quad (51)$$

There is no adaptive meshing within each block, which always contains N×N×N cells and the adaptive strategy is further restricted by only allowing blocks at level n adjacent to blocks of n−1,n and n+1. Thus, even though a block may be flagged for coarsening, this (latter) restriction often prevents this happening. Conversely, blocks flagged for refinement must, if necessary, be accompanied by refinement on neighbouring blocks. Blocks are flagged for refinement if, for any point, **p**, in the block, the following criterion is satisfied:

$$e \equiv \max\{e_\phi|\phi_{\mathbf{p}} - \phi_{\mathbf{p-q}}|, e_U|U_{\mathbf{p}} - U_{\mathbf{p-q}}|, e_T|T_{\mathbf{p}} - T_{\mathbf{p-q}}|\} > \eta, \quad (52)$$

where we use, for tolerance, η~1 and

$$|\phi_{\mathbf{p}} - \phi_{\mathbf{p-q}}| \equiv \sqrt{(\phi_{\mathbf{p}} - \phi_{\mathbf{p}-[1,0,0]})^2 + (\phi_{\mathbf{p}} - \phi_{\mathbf{p}-[0,1,0]})^2 + (\phi_{\mathbf{p}} - \phi_{\mathbf{p}-[0,0,1]})^2}, \quad (53)$$

etc. Typically the weights, e_ϕ,e_U and e_T are chosen to sum to unity. Sometimes it is convenient to set e_U to zero to suppress unnecessary refinement within the solid. Ife<0.1η, the block is flagged for derefinement.

Although each block is logically defined to be of dimension N×N×N the PARAMESH implementation actually allocates a block of dimension (N+2G)×(N+2G)×(N+2G), where G is the number of guard cells (sometimes referred to as ghost cells) around each block. When G=1, as used in this paper, the first and last cells in each dimension are guard cells – an update function may be called at any time in order to populate these guard cells with the corresponding values from the interior of each of the neighbouring blocks (with a separate treatment required to impose boundary conditions for blocks at the edge of the domain, as discussed previously). The application of a discrete stencil on any block requires access to neighbouring blocks via the guard cell nodes. When the neighbouring block is coarser the guard cell of the coarse block is found by a weighted average of the 8 surrounding coarse nodes (some of which are in the parent block). Using a tri-linear function of x,y and z, gives the weightings, in order of nearest neighbours first

$$\mathbf{w} = \left[\frac{27}{64}, \frac{9}{64}, \frac{9}{64}, \frac{9}{64}, \frac{3}{64}, \frac{3}{64}, \frac{3}{64}, \frac{1}{64}\right]. \quad (54)$$

The process is known as prolongation. For example, the value of ϕ at a fine node is prolonged by

$$\phi_{\text{fine}} = \sum_{i=\text{coarse cube centres}} w_i \phi_i. \quad (55)$$

The reverse process, of finding a guard cell value for a coarse block when one or more neighbours is refined is known as restriction and is the simple average of the eight nearest, one level finer, cell centres. Both operations, restriction and prolongation using Eq. (54), are also required for multigrid as detailed in the next section.

SOLVER METHOD

The discretisation above produces a system of nonlinear algebraic equations for ϕ_p^{n+1}, U_p^{n+1} and θ_p^{n+1} at each time step, t^{n+1}. In this section we describe the solution algorithm that is used to solve these systems, based upon a nonlinear multigrid (Full Approximation Scheme (FAS) [17]) approach. Initially we describe the nonlinear smoother upon which the multigrid is built, before exploring how the multigrid solver combines this with the hierarchical mesh adaptivity introduced in the previous section. Finally, in Section 4.3, we explain the key features of the parallel implementation, including the issues associated with parallel dynamic load-balancing.

Nonlinear smoother

The non-linear system of algebraic equations at the end of Section 3.3 can be written using the generic vector notation $\mathbf{v}_p^{n+1} \equiv [\phi_p^{n+1}, \theta_p^{n+1}, U_p^{n+1}]$ by

$$\mathbf{A}_p(\mathbf{v}_Q^{n+1}) = 0 \quad (56)$$

where

$$\mathbf{A}_p(\mathbf{v}_Q^{n+1}) \equiv \mathbf{v}_p^* - \mathbf{v}_p^{n+1} + r_1 \Delta t^{n+1} \mathbf{F}_p(\mathbf{v}_Q^{n+1}) \quad (57)$$

for each node, **p**, in the grid. Recall, that the appearance of **Q** indicates coupling between points **p** and neighbouring nodes, and the BDF2 notation $\mathbf{v}_p^*$ in combination with r_1 are defined in Section 3.3. Using an iteration method, with $\mathbf{v}_p^{n+1}$ approximated by $\mathbf{v}_p^{n+1,m}$, we define the defect

$$\mathbf{d}_p^{n+1,m} = -\mathbf{A}_p(\mathbf{v}_Q^{n+1,m}). \quad (58)$$

The pointwise Newton update for this iteration is given by

$$\mathbf{v}_p^{n+1,m+1} = \mathbf{v}_p^{n+1,m} - \omega \tilde{\mathbf{d}}_p^{n+1,m}, \quad (59)$$

where $\tilde{\mathbf{d}}_p^{n+1,m}$ is found by solving the 3×3 system

$$\mathbf{J}_p^{n+1,m} \tilde{\mathbf{d}}_p^{n+1,m} \equiv \mathbf{d}_p^{n+1,m} \quad (60)$$

with the 3×3 Jacobian matrix defined by

$$\mathbf{J}_p^{n+1,m} \equiv \frac{\partial \mathbf{d}_p^{n+1,m}}{\partial \mathbf{v}_p^{n+1,m}}. \quad (61)$$

In practice we typically select, $\omega \approx 0.9$ and find that off diagonal terms of $\mathbf{J}_p^{n+1,m}$ are not essential to obtain a convergent iteration.

The precise form of the entries of $\mathbf{J}_p^{n+1,m}$ may be deduced from the above. However, we illustrate this in detail for one of the diagonal terms for the sake of clarity. Denoting the diagonal entries of $\mathbf{J}_p^{n+1,m}$ by $[J_p^{\phi,n+1,m}, J_p^{\theta,n+1,m}, J_p^{U,n+1,m}]$, using Eqs. (44) and (36), and treating terms not including ϕ**p** as constant we find

$$J_p^{\phi,n+1,m} \equiv \frac{\partial d_p^{\phi}(\mathbf{v}_Q^{n+1,m})}{\partial \phi_p} = \left[1 + r_1 \Delta t^{n+1} \frac{1}{3} \frac{g_{11} + g_{22} + g_{33}}{\tau(c,\phi) A^2} \frac{128}{30(\Delta x)^2}\right]_p, \quad (62)$$

where we note that the contribution from the central node to $\nabla^2\phi$ is

$$\frac{\partial \nabla^2 \phi|_Q}{\partial \phi_p} = -\frac{128}{30(\Delta x)^2}. \quad (63)$$

The new updated solution for ϕ at time t^{n+1} at iteration m and point **p** is given by

$$\phi_p^{n+1,m+1} = \phi_i^{n+1,m} - \omega \frac{d_p^{\phi,n+1,m}}{J_p^{\phi,n+1,m}}. \quad (64)$$

The term $J_p^{\theta,n+1,m} = 1 + r_1 \Delta t^{n+1} D_\theta \frac{128}{30(\Delta x)^2}$ and, though the term $J_p^{U,n+1,m}$ has many components, the linearity of F_p^U in **Up** leads to straightforward updates for the U components, noting that gradients involving U|**Q_p** are treated as constant.

The above describes a pointwise non-linear Jacobi smoother. The Jacobi approach lends itself to parallel implementation since the computation at each point may be completed using neighbouring values from the previous iteration only. Consequently, only one ghost cell update is required prior to each sweep through the mesh. This keeps inter-processor communication to a minimum. Using the stencils described above it is possible to complete the updates using just one layer of ghost cells. Hence if a block size of 8×8×8(say) is used in PARAMESH then a 10×10×10 block is actually allocated to accommodate the ghost layer of each block edge.

Having derived the point-wise Jacobi smoother, in the following subsection we show how, this may be used as part of a non-linear geometric multigrid solver combined with local mesh adaptivity. Discussion of the parallel implementation is postponed until Section 4.3.

Nonlinear multigrid

Although the Jacobi smoother described above gives a convergent iteration for the system Eq. (56) (for sufficiently small Δt and good initial guess), the convergence is far too slow to be of any practical use. Fortunately, however, the iteration also satisfies a smoothing property which means that it damps out the highest frequency components of the error (defect) far more quickly than the rest, see the early chapters of [18] for discussion and analysis of this. This makes it ideally suited for use as part of a nonlinear multigrid scheme. In this work we make use of the Full Approximation Scheme (FAS) of Brandt [17]. We denote the value of variables at point, **p**, time t^{n+1}, iteration, m, and level/grid size, h by $\mathbf{v}_{\mathbf{p}(h)}^{n+1,m}$ and coarser level $\mathbf{v}_{\mathbf{p}(2h)}^{n+1,m}$. The restriction operation, $\mathbf{I}_{2h}^{h}(\mathbf{v}_{\mathbf{p}(h)})$, is the assignment to $\mathbf{vp}_{(2h)}$ of the simple average of the 8 surrounding nodes, $\mathbf{vp}_{(h)}$. Prolongation, $\mathbf{I}_{h}^{2h}\mathbf{v}_{\mathbf{p}(2h)}$, is an assignment to $\mathbf{vp}_{(h)}$ given by applying Eq. (54) to give a weighted average of the values at the 8 nearest coarse nodes $\mathbf{vp}_{(2h)}$. In solving Eq.(56), FAS computes a defect from the restricted defect and variables to give a modified **A(v)=f** on these lower levels. See Fig. 2, where we detail FAS for our notation.

FAS algorithm to solve $\mathbf{A}(\mathbf{v}) = \mathbf{0}$

1. h = finest grid (top level)
2. $\mathbf{v}_{\mathbf{p}(h)}^{n+1,0} \leftarrow \mathbf{v}_{\mathbf{p}(h)}^{n}$: set initial guess equal to value at last time step
3. $\mathbf{v}_{\mathbf{p}(h)}^{n+1,\nu+1} \leftarrow$ **V-cycle**$(\mathbf{v}_{\mathbf{p}(h)}^{n+1,\nu}, \mathbf{0}, h)$: application of V-cycle, $\nu = 0, 1, \ldots$ until convergence

Recursive Function V-cycle$(\mathbf{v}_{\mathbf{p}(h)}^{0}, \mathbf{f}_{\mathbf{p}(h)}, h) \rightarrow \mathbf{v}_{\mathbf{p}(h)}^{m+1}$ (solves $\mathbf{A}(\mathbf{v}) = \mathbf{f}$)

1. $\mathbf{v}_{\mathbf{p}(h)}^{m+1} \leftarrow \mathbf{S}(\mathbf{v}_{\mathbf{p}(h)}^{m}, \mathbf{f}_{\mathbf{p}(h)}, h) = \begin{cases} \mathbf{d}_{\mathbf{p}(h)}^{m} \leftarrow \mathbf{f}_{\mathbf{p}(h)} - \mathbf{A}(\mathbf{v}_{\mathbf{Q}(h)}^{m}) \\ \mathbf{v}_{\mathbf{p}(h)}^{m+1} \leftarrow \mathbf{v}_{\mathbf{p}(h)}^{m} - \omega \left[\frac{d_{\mathbf{p}(h)}^{\phi,m}}{J_{\mathbf{p}(h)}^{\phi,m}}, \frac{d_{\mathbf{p}(h)}^{T,m}}{J_{\mathbf{p}(h)}^{T,m}}, \frac{d_{\mathbf{p}(h)}^{U,m}}{J_{\mathbf{p}(h)}^{U,m}} \right] \end{cases}, \quad m = 0, \ldots, M-1$
   (pre) smooth M times (typically we use $M \sim 4$)
2. $\mathbf{d}_{\mathbf{p}(h)}^{m+1} \leftarrow \mathbf{f}_{\mathbf{p}(h)} - \mathbf{A}_{\mathbf{p}}(\mathbf{v}_{\mathbf{Q}(h)}^{m+1})$
   $\mathbf{v}_{\mathbf{p}(2h)}^{0} \leftarrow \mathbf{I}_{2h}^{h}(\mathbf{v}_{\mathbf{p}(h)}^{m+1})$
   $\mathbf{d}_{\mathbf{p}(2h)}^{0} \leftarrow \mathbf{I}_{2h}^{h}(\mathbf{d}_{\mathbf{p}(h)}^{m+1}) + \mathbf{A}(\mathbf{v}_{\mathbf{p}(2h)}^{0})$: restriction
3. **if** not coarsest level
   $\mathbf{v}_{\mathbf{p}(2h)}^{m+1} \leftarrow$ **V-cycle**$(\mathbf{v}_{\mathbf{p}(2h)}^{0}, \mathbf{d}_{\mathbf{p}(2h)}^{0}, 2h)$.
   else (bottom level)
   $\mathbf{v}_{\mathbf{p}(2h)}^{m+1} \leftarrow \mathbf{S}(\mathbf{v}_{\mathbf{p}(2h)}^{m}, \mathbf{d}_{\mathbf{p}(2h)}^{0}, 2h), \quad m = 0, \ldots, N$ (typically we use $N \sim 4$)
   end if
4. $\mathbf{d}_{\mathbf{p}(2h)}^{0} \leftarrow \mathbf{v}_{\mathbf{p}(2h)}^{m+1} - \mathbf{v}_{\mathbf{p}(2h)}^{0}$: compute the correction (now on upward part of v-cycle)
5. $\mathbf{v}_{\mathbf{p}(h)}^{0} \leftarrow \mathbf{v}_{\mathbf{p}(h)}^{m+1} + \mathbf{I}_{h}^{2h}(\mathbf{d}_{\mathbf{p}(2h)}^{0})$: prolong and correct fine grid solution
6. $\mathbf{v}_{\mathbf{p}(h)}^{m+1} \leftarrow \mathbf{S}(\mathbf{v}_{\mathbf{p}(h)}^{m}, \mathbf{0}, h), \quad m = 0, \ldots, M_{post} - 1$: (post) smooth with this new value $M_{post} \sim 4$ times

Figure 2: The FAS algorithm.

Note that in our work local mesh adaptivity is an essential feature. This has been described in Section 3.4, where PARAMESH block data structures are used as nodes of an oct-tree. In this work we also use the oct-tree as part of the geometric multigrid solver by developing an implementation of the Multi Level Adaptive Technique (MLAT) (see[17]). MLAT allows us to use the Jacobi smoother on each block without modification, provided the prolongation and restriction operators are adapted to deal with guard cells of the interface between too differing levels of refinement. Briefly, the smoother is only applied in the regions of the domain that contain fine level blocks, but the coarse grid correction takes place on all parts of the domain that contain the coarse level blocks (though the modified right hand side associated with the FAS scheme only contributes to those coarse grid regions that have fine grids on them).

For the results presented in the following sections we are primarily interested in obtaining solutions at large times. Hence we choose a time-stepping strategy with this in mind, based upon the number of nonlinear V-cycles that are required to achieve convergence (for an alternative strategy, which uses a local error estimate to control Δt, see [19]). The principle is simple: if the nonlinear multigrid converges easily at a give time step then increase Δt, whereas if it converges slowly (or fails to converge) then decrease Δt(repeating the time step in the case of failure). Convergence is deemed to have occurred when the infinity norm of the defect (possibly weighted differently for each dependent variable), d , satisfies $d<d_{max}$, for a user-defined stopping parameter d_{max}. If this is not satisfied in V_{fail} V-cycles then Δt is halved and the step is retaken. If convergence occurs in V_{min} V-cycles or less then Δt is increased by 10% however if convergence occurs in more than V_{max} V-cycles then Δt is halved for the next step. Fig. 3 shows a typical evolution of the time step size for three different choices of V_{min}, based upon initial $\Delta t=10^{-6}$, $d_{max}=10^{-10}$ and $V_{max}=10$. Note that although the oscillation in Δt is aesthetically undesirable it has no adverse effect on the solution quality nor (so long as $V_{max}<V_{fail}$) the overall efficiency.

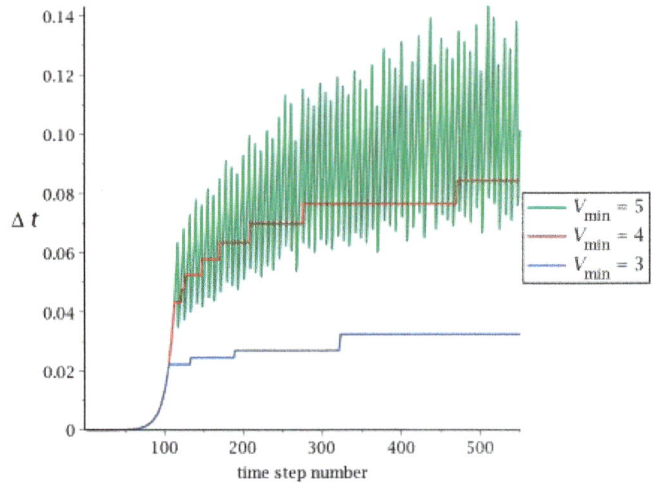

Figure 3: An illustration, for Le = 40, $\Delta = 0.525$ of the evolution of the time step, Δt , from 10^{-6} by V-cycle control. (For interpretation of the references to colour in this figure legend, the reader is referred to the web version of this article.)

Parallel implementation

Our implementation requires communication between blocks both on the same level (to apply the smoothing steps) and between levels (for prolongation and restriction). For parallel processing, clearly, communication between cores needs to be kept to a minimum, but an important secondary consideration is that each core has as near as possible equal load. For a uniform mesh an allocation of each core to an equal volume of the computational domain results in an obvious fair division of labour. On the other hand, for an adaptive mesh, such an approach fails since the loading between cores will differ enormously.

Given the label, i, of each block in Eq. (48), we present the Morton ordering, $M(i) \in [1, B_N]$, of an adaptive mesh

$$M(i) = \begin{cases} 1 & l = 1,\ s = 1 \text{ (bottom level)} \\ M(p(i)) + 1, & s = 1,\ l > 1 \text{ (one plus parent's label)} \\ M(s(i-1)) + 1, & s > 2, \text{ and } l = \text{ leaf level} \\ M(c_8(i)) + 1, & \text{otherwise, i.e. } s = 2 \end{cases}, \quad (65)$$

where 'leaf level' refers an unrefined level (parent without child) and includes the finest level. The blocks may then be put into a Morton ordered list:

$$\mathbf{M(B)} \equiv \{B_i : j = M(i),\ j \in [1, B_N]\}$$
$$\equiv \{\ldots, B_{M(i)}, B_{M(j)}, \ldots : M(i) < M(j)\} \quad (66)$$

We write the form given in the second line of Eq. (66) to highlight the difference between Morton ordering and the alternative ordering we adopt below, Eqs. (67) and(68).

Load balancing for B_N blocks on processors p_i, i=1,2,…,c_N follows the same order with approximately B_N/c_N blocks per core. The resulting ordering is well known to exhibit parallel inefficiency for non-uniform meshes. This is because Morton ordering on adaptive meshes leaves the majority of the top level blocks on a small fraction of cores and since multigrid advances from the top to bottom level and back sequentially, the majority of the cores will be idle at the top level. In three dimensions this problem is acute because there is a factor of 8 between levels.

We adopt the following ordering, which may be termed Morton-Level ordering.

$$\mathbf{L}(\mathbf{B}) \equiv \{B_i : j = L(i),\ j \in [1, B_N]\},$$
$$\equiv \{\mathbf{L}_k(\mathbf{B}), k \in [1, n]\}, \tag{67}$$

where the subsets, $\mathbf{L}_k(\mathbf{B})$ on each level, k are defined

$$\mathbf{L}_k(\mathbf{B}) \equiv \{\ldots, B_{M(i)}, B_{M(j)}, \ldots : M(i) < M(j), l(i) = l(j) = k\} \tag{68}$$

In summary, we implement a depth first traversal of the blocks and then, using this numbering, divide the work load at each level in turn between all processors. For a uniform mesh this strategy results in a near optimum allocation to cores. For adaptive meshes the communication on any level is also optimum, but communication between level is compromised.

COMPUTATIONAL RESULTS

This section provides a selection of computational results that are designed to validate and assess our proposed solution algorithm. Since this is the first attempt to produce three dimensional results for the solidification of a non-isothermal alloy using a realistic interface width we have no external simulations against which to validate our code. Hence the approach taken here has been to firstly validate a two-dimensional restriction of our implementation against our own two-dimensional solver, implemented completely independently and described in [19]. These tests show that we are indeed able to reproduce results from [2] (e.g. the tip radius and velocities are in excellent agreement) even though the two code bases are completely independent, e.g. in [2] fourth order accurate stencils were used.

Furthermore, we have also successfully validated two 3-d simplifications of our implementation. In [20] we consider a thermal-only restriction (i.e. a pure metal, so no concentration equation present), where we show quantitative agreement with results obtained using the 3-d, explicit, thermal-only approach of [21] and [22]. Similarly, in [23]we show quantitative agreement between an isothermal version of our solver (i.e. no temperature equation present) and another 3-d isothermal solidification code described in [21]. To validate the 3-d non-isothermal simulations we now rely on having mesh convergence and multigrid performance.

The remainder of this section is divided into two subsections. The first of these considers the mesh convergence of our implementation, showing that large time solutions obtained on a sequence of finer levels of maximum refinement do indeed appear to converge to particular dendrite geometries, as tested for a selection of parameter values. The second subsection considers the numerical performance of the solver. In particular it is shown that optimal performance is achieved, whereby the time required to complete a time step grows almost linearly with the total number of degrees of freedom. The capability improvements associated with the distributed memory parallel implementation are also discussed.

Mesh convergence

In order to gain further confidence in our computational approach we have undertaken a number of mesh convergence tests, in which we considered computational simulations in which the maximum level of mesh refinement is systematically increased. In order to appreciate the need for very fine grids at the phase interface Fig. 4 illustrates a cross section along the x-axis of a typical solution. This corresponds to the same parameter values as used to compute the dendrite illustrated in Fig. 1. It is clear that the phase variable, ϕ, changes +1 (bulk solid) to −1 (bulk melt) over a very small distance, similarly the solute concentration varies rapidly both at, and immediately ahead of, the interface. The temperature variable, θ, decays much more smoothly however – though a large domain is required to ensure that the boundary effect does not contaminate the solution. In addition to the interface width a further feature of significant interest is the geometry of the dendrite tip. Fig. 5 shows the computed evolution of the tip radius for three different undercoolings (0.325,0.525 and 0.725) at Le=40. The parameter, $d_0 = 5\sqrt{2}/(8\lambda)$ is the (non-dimensional) capillary length as a function of the interface width. Our value for λ=2 gives, d_0=0.44 indicating that the interface width in our simulation is of the order of the physical width. We use as our characteristic time scale, $t_0 = 0.80(\tilde{d}_0^2\lambda^3)/\tilde{D}_c$ where $\tilde{d}_0$ and $\tilde{D}_c$ are the physically dimensioned capillary length and solute diffusivity coefficient respectively. Note that the results for Fig. 5 were computed using a fine mesh spacing of Δx=0.39. It is an important question to ask if this is sufficiently fine for the solution to be insensitive to further mesh refinement.

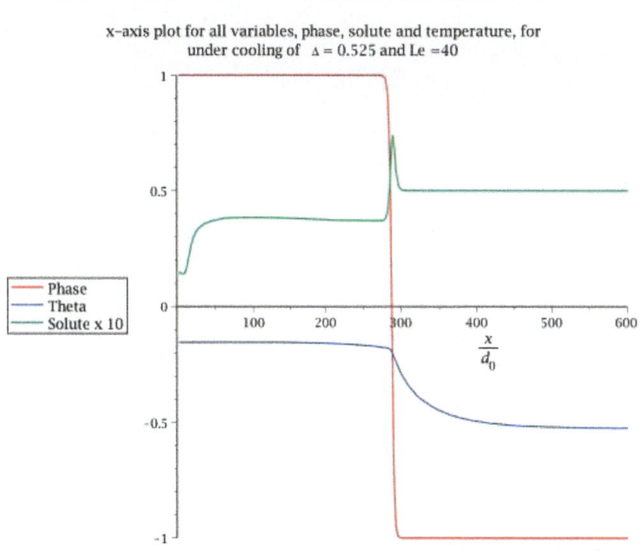

Figure 4: This plot shows the variables: phase ϕ, solute concentration c, and dimensionless temperature θ. At Lewis number of 40 and under cooling of 0.525 the temperature diffusion zone extends to about 300 in a domain size of 800^3. (For interpretation of the references to colour in this figure legend, the reader is referred to the web version of this article.)

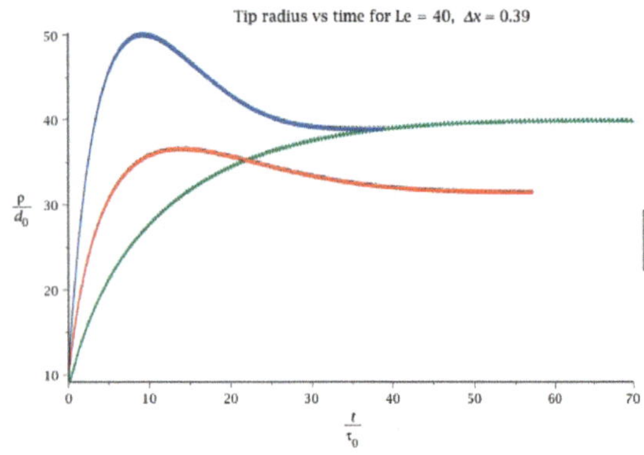

Figure 5: Tip radius at dx = 0.39 and Le = 40 for a range of under cooling Δ. Even though the radius at Δ = 0.325 takes longer to reach steady state, this simulation is faster than the others due to greater stability resulting in fewer, larger time steps. (For interpretation of the references to colour in this figure legend, the reader is referred to the web version of this article.)

Fig. 6 shows the computed tip radius as a function of time for three different maximum refinement levels (Δx=0.78,0.39 and 0.195) for case Le=40 and Δ=0.325. The tip radius involves estimating a second derivative of ϕ in the region where ϕ=0. This is undertaken by estimating the radius on the x-axis at ϕ=0 using the phase field:

$$r = \left.\frac{\phi_{,x}}{\phi_{,uu}}\right|_{\phi(x,y,z)=0,\ y=0,\ z=0} \tag{69}$$

where $\frac{\partial}{\partial u} = \frac{1}{\sqrt{2}}\left(\frac{\partial}{\partial y} + \frac{\partial}{\partial z}\right)$, $\phi_{,x} = \frac{\partial \phi}{\partial x}$ and $\phi_{,uu} = \frac{\partial^2 \phi}{\partial u^2}$ (as will become clear shortly, the u direction is more convenient than the y or z directions). Expression Eq. (69) comes from the definition

$$r=1/\kappa \tag{70}$$

where the curvature, κ, is defined in the normalised direction **u** to be

$$\kappa=\mathbf{u}\cdot\nabla\mathbf{n}\cdot\mathbf{u}. \tag{71}$$

On the x-axis, ignoring the z direction, and with a directional derivative in the ydirection, we find, using ϕ to compute the normal $\mathbf{n}\equiv\nabla\phi/|\nabla\phi|$, that

$$\kappa = \frac{\partial n_2}{\partial y} = \frac{\partial}{\partial y}\frac{\phi_{,y}}{\sqrt{\phi_{,x}^2+\phi_{,y}^2}} = \frac{\phi_{,yy}}{\phi_{,x}} \tag{72}$$

where, by symmetry $\phi_{,xy}|_{y=0}=\phi_{,y}|_{y=0}=0$ on the axis. Again, by symmetry, this relation holds for any normalised parameter and so

$$\kappa = \frac{\phi_{,yy}}{\phi_{,x}} = \frac{\phi_{,zz}}{\phi_{,x}} = \frac{\phi_{,uu}}{\phi_{,x}} \tag{73}$$

In practice, of course, the value ϕ=0 lies between two successive nodes on the x-axis,i and i+1. Furthermore, the x-axis lies, by definition, on y=z=0 where there are no grid points. We compute the derivatives only using the points

$$[i,2,2],[i+1,2,2],[i,3,3],[i+1,3,3] \tag{74}$$

which relate to the physical points

$$[i,j,k]\rightarrow[O_x,0,0]+\Delta x[i-3/2,j-3/2,k-3/2] \tag{75}$$

where O_x is the x coordinate of the block origin. Thus, we know by symmetry, that the values at these points can be equated to the image nodes (which we do not use explicitly)

$$\phi_{[i,1,1]} = \phi_{[i,2,2]},$$
$$\phi_{[i,0,0]} = \phi_{[i,3,3]}. \tag{76}$$

We compute the radius of curvature using the direction u by

$$r = \left.\left(\frac{\phi_x}{\phi_{uu}} + \frac{\phi}{\phi_x}\right)\right|_{[i,2,2]} \tag{77}$$

where the last term is a correction to compensate for, in general, ϕ≠0 at the point[i,2,2]. This is discretised by

$$\phi_x = \frac{\phi_{i+1,2,2} - \phi_{i-1,2,2}}{2\Delta x},$$
$$\phi_{uu} = \frac{-\phi_{i,2,2} + \phi_{i,3,3}}{2(\Delta x)^2} \tag{78}$$

It is the nature of this approximation that leads to the oscillatory results that are observed on the least fine simulation (Δx=0.78). Nevertheless it is clear that the results for Δx=0.39 and Δx=0.195 are almost indistinguishable at the scale used here and so we have a good degree of confidence that a converged solution is obtained by Δx=0.39. Similar computations of the large-time tip radius, on different levels of grid refinement, have been undertaken for two other cases, as shown in Table 2. Whilst the convergence is not so mature in every case, the evidence that results atΔx=0.39 are of sufficient accuracy to be of quantitative validity is very strong.

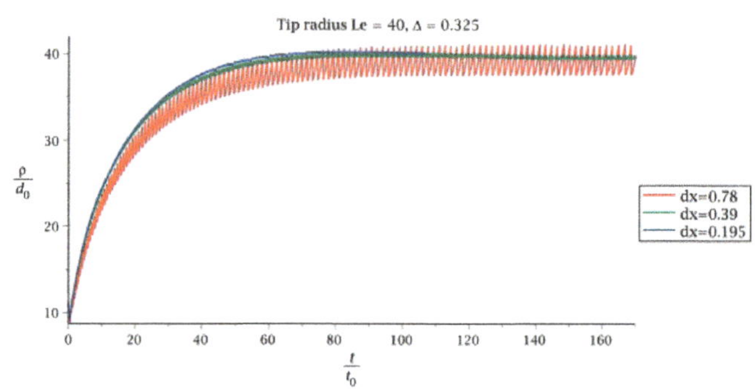

Figure 6: Convergence test on tip radius with grid sizes dx = {0.78,0.39,0.195}. The plot shows, for Δ = 0.325, the full transient behaviour of the tip radius for all grid resolutions. In this case the results for all three grids are in good agreement. Higher under cooling makes the coarser grid less reliable and even dx = 0.39 becomes less reliable. (For interpretation of the references to colour in this figure legend, the reader is referred to the web version of this article.)

Table 2: Table of tip radius results for two under-coolings and two Lewis numbers. For this particular selection of parameters there is reasonable agreement even on the coarser mesh Δx = 0.78. But the slight discrepancy shown for the higher under-cooling is symptomatic of the necessity for a finer mesh, Δx > 0.39, in general.

Parameters		Tip radius		
Le	Δ	Δx = 0.78	Δx = 0.39	Δx = 0.195
40	0.325	39.0 ± 2	39.7 ± 0.3	39.7 ± 0.1
40	0.525	29.7 ± 2	31.5 ± 0.3	31.8 ± 0.1
100	0.325	40.3 ± 2	39.8 ± 0.3	39.8 ± 0.1

Note that the Δx=0.195 results for the other cases shown in Table 2 were not obtained by undertaking complete runs at this maximum refinement level. Instead, the large-time simulation computed using a maximum level of Δx=0.39 was restarted with a maximum level of Δx=0.195 and executed until a steady state tip radius was reached. This was tested for the (Le=40, Δ=0.325) case, and shown to give identical results. The approach is illustrated for Le=100, Δ=0.325 in Fig. 7 and Le=40, Δ=0.525 in Fig. 8. Note that there is a change of scaling between these plots and the amplitude of oscillation is in fact identical for a given mesh size, dx , for both Le=40 and Le=100and is approximately proportional to Δx. The oscillation itself is an artifact of computing a second derivative using a low order approximation scheme at a physical point not coincident with a grid point.

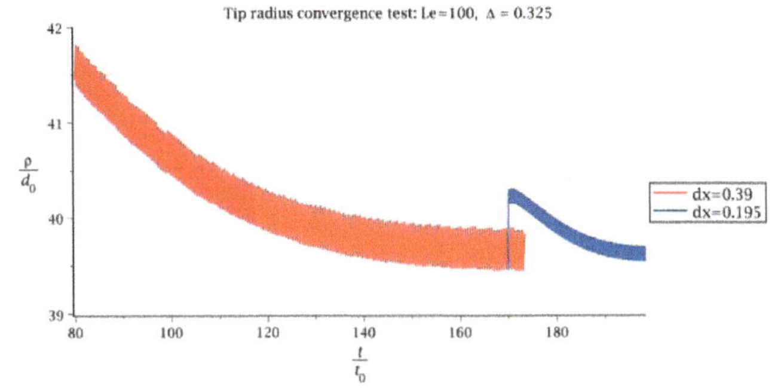

Figure 7: Convergence test on tip radius with grid sizes dx ∈ [0.39,0.195]. A checkpoint file at dx = 0.39 is used as an initial condition for a dx = 0.195 to test convergence. The restart recovers from an initial transient before settling to a value very similar to the steady state at the coarser, dx = 0.39 run. (For interpretation of the references to colour in this figure legend, the reader is referred to the web version of this article.)

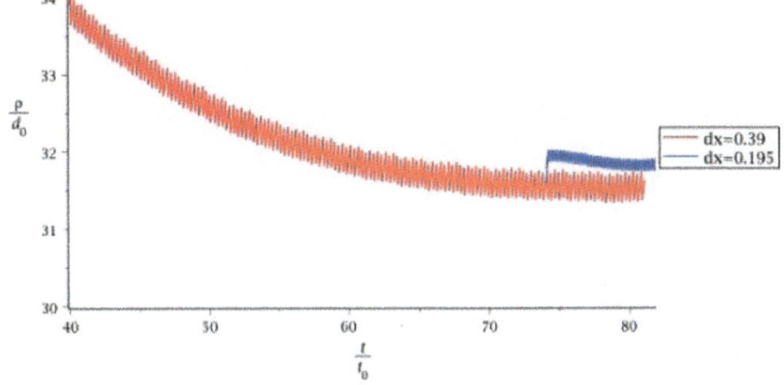

Figure 8: Convergence test on tip radius with grid sizes dx ∈ [0.39,0.195]. A checkpoint file at dx = 0.39 is used as an initial condition for a dx = 0.195 to test convergence. (For interpretation of the references to colour in this figure legend, the reader is referred to the web version of this article.)

Numerical performance

Having demonstrated the mesh convergence of our proposed technique in the previous subsection, we now consider the computational performance of the implementation used.

The most important feature of any successful multigrid implementation is that it should enable solutions of systems of algebraic equations to be obtained in a run time that is close to O(N) as N→∞ where N is the number of degrees of freedom. In this simulation we begin with a small solid seed at the origin and this grows (under that right parameter conditions) in time. As it grows the region of maximum mesh refinement gets larger and larger, as the isosurface, ϕ=0 has ever increasing area. This causes N to increase in time. Hence, an excellent test of our solver is whether the computed time to take each time step only grows in proportion to N. Unfortunately this test is harder to undertake than initially might be imagined since, as N increases the amount of memory required to compute a time step also increases (linearly). As described in Section 4 we deal with this through a distributed memory parallel implementation. We start the execution, with a small seed, using a modest number of cores (16 or 32 say) and increase these as the execution progresses. Furthermore, as discussed in the previous subsection, we also wish to consider different choices for the maximum level of mesh refinement, which also impacts on the number of computational nodes to be used and therefore the total memory requirement.

Fig. 9 shows a selection of timings for a set of computations for a single, representative, test case (Le=40, Δ=0.525) using different cores. The vertical axis shows the execution time for a single implicit time step and the horizontal axis shows the number of degrees of freedom (both on log scales). Timings are taken just as the dendrite is starting to form (early dendrite see Fig. 10), part way through its formation (mid dendrite Fig. 11), and once the dendrite is clearly formed (late dendrite Fig. 1). Timings are also taken using different maximum refinement levels (either 2 or 3 per case). It is very clear from Fig. 9that multigrid performance is achieved throughout these different stages of the evolution of the dendrite and at different maximum refinement levels. This can be seen from the excellent proximity of the points to the time line of slope one ($t = DoF/100\,000$) on the log–log scale.

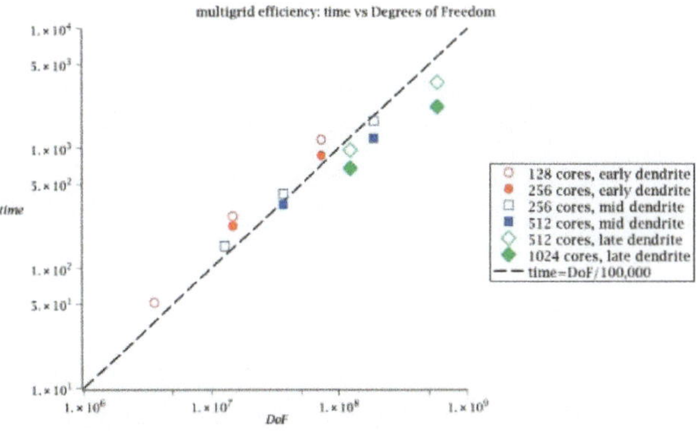

Figure 9: Multigrid efficiency. A demonstration of the linearity of solve time with the number of degrees of freedom. In all 6 cases the corresponding points fit well (for a single time step) to the line of slope 1.

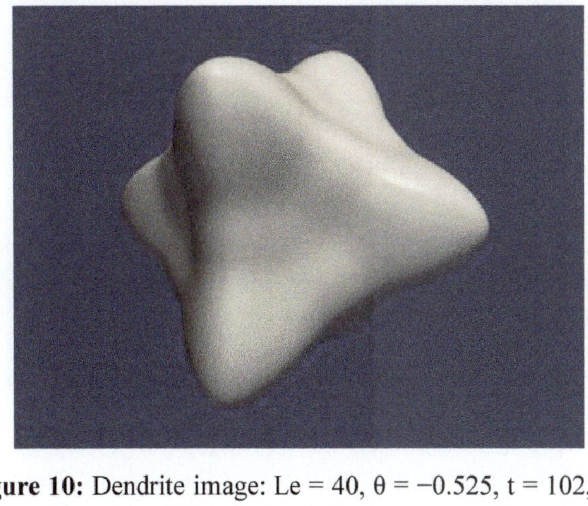

Figure 10: Dendrite image: Le = 40, θ = −0.525, t = 102, dx = 0.78. This simulation took 12 hours on a 12 core machine.

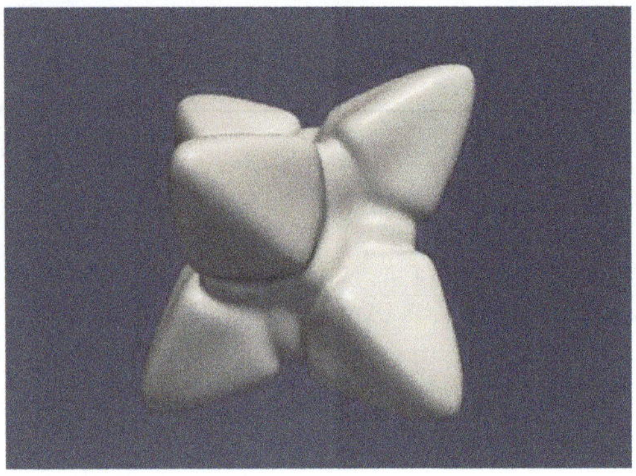

Figure 11: Dendrite image: Le = 40, θ = −0.525, t = 186, Δx = 0.78. This simulation is of a dendrite with the coarsest maximum refinement and took about 40 hours to simulate on a 12 core machine. The last 10% of the run (a time interval of t = 18) took about 10 hours.

Note that our use of distributed memory parallel computing throughout this work has been aimed primarily at providing the capacity to solve large systems (up to and beyond a billion degrees of freedom per implicit time step) in a computationally efficient manner. It is clear from Fig. 12 that the strong parallel scalability of our implementation is not optimal. Nevertheless it is apparent that, as well as providing additional memory capacity to allow larger problems to be tackled, out parallel implementation also continues to improve the speed of the execution each time the core count is increased so long as the number of degrees of freedom is sufficiently large.

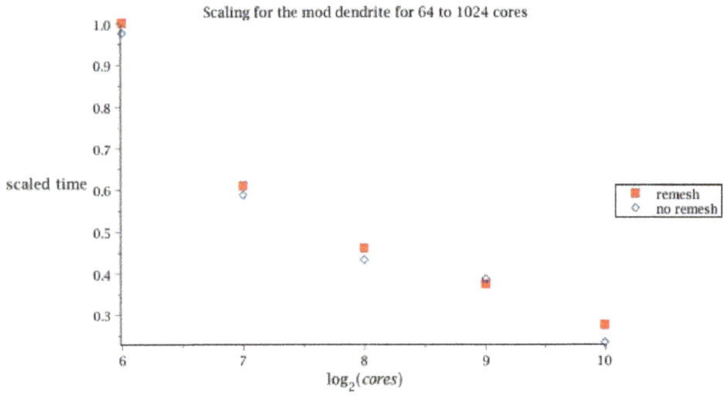

Figure 12: Plot of the relative wall-clock time to undertake 10 time steps at the mid dendrite stage (see Fig. 11), for mesh dx = 0.39, using different numbers of cores (64 to 1024).

CONCLUSIONS

We have presented, in detail, the mathematical model and methods employed to simulate, for the first time, a three dimensional, fully coupled thermal-solute-phase field model for dendritic growth. This was achieved through the coupling of multiple numerical techniques from compact discrete finite difference stencils, AMR, MLAT and parallel execution.

For moderate Lewis numbers we have been able to obtain results at sufficient grid resolution, which we have confidence can provide quantitative accuracy in 3-d for the first time. This breakthrough into three dimensional fully coupled thermal simulation is of importance, since heat generation at the growing two dimensional surface is very much an integral part of the natural physical process and two dimensional results therefore have little quantitative value. Our next goal is to undertake a systematic simulation of the effect of increasing Lewis number. We end with the observation that, though tip radius is fully converged, the tip velocity is only near steady state. We give the companion plot to Fig. 5 for the tip velocities for the same parameters in Fig. 13.

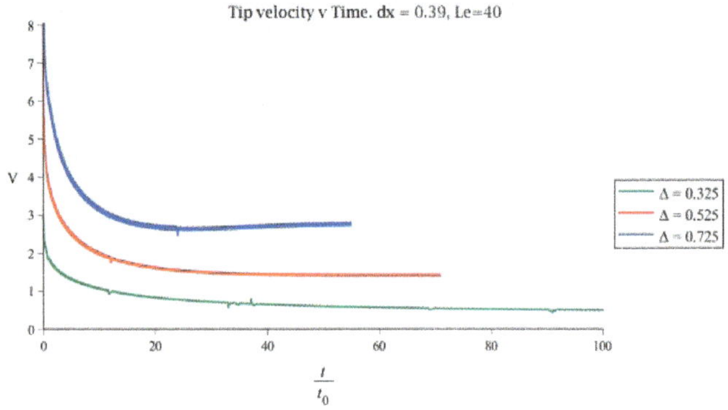

Figure 13: This is a companion plot to the tip radius plot 5: The tip velocity at dx = 0.39 and Le = 40 for a range of under cooling Δ = {0.325,0.525,0.725}. Though the tip radius has reached steady state in each of these cases the tip velocity for the highest undercooling cases is still increasing. (For interpretation of the references to colour in this figure legend, the reader is referred to the web version of this article.)

ACKNOWLEDGEMENTS

This research was funded by EPSRC grant number EP/H048685. We are also grateful for the use of the HECToR UK National Supercomputing Service.

APPENDIX

Appendix A. Discretisation of M_{ij}, $\nabla\phi$ and g_{ij}

We discretise, M_{ij} as follows

$$\Delta|_{\mathbf{Q-p}} \equiv \phi_{\mathbf{p}+[1,0,0]} + \phi_{\mathbf{p}+[-1,0,0]} + \phi_{\mathbf{p}+[0,1,0]} + \phi_{\mathbf{p}+[0,-1,0]} + \phi_{\mathbf{p}+[0,0,1]} + \phi_{\mathbf{p}+[0,0,-1]}$$

$$M_{11}|_{\mathbf{Q-p}} = \frac{1}{\Delta x^2}\left(\phi_{\mathbf{p}+[1,0,0]} + \phi_{\mathbf{p}+[-1,0,0]} - \tfrac{1}{3}\Delta|_{\mathbf{Q-p}}\right),$$

$$\begin{aligned}
M_{22}|_{\mathbf{Q-p}} &= \frac{1}{\Delta x^2}\left(\phi_{\mathbf{p}+[0,1,0]} + \phi_{\mathbf{p}+[0,-1,0]} - \tfrac{1}{3}\Delta|_{\mathbf{Q-p}}\right),\\
M_{33}|_{\mathbf{Q-p}} &= \frac{1}{\Delta x^2}\left(\phi_{\mathbf{p}+[0,0,1]} + \phi_{\mathbf{p}+[0,0,-1]} - \tfrac{1}{3}\Delta|_{\mathbf{Q-p}}\right),\\
M_{12}|_{\mathbf{Q-p}} = M_{21}|_{\mathbf{Q-p}} &= \frac{1}{4\Delta x^2}\left(\phi_{\mathbf{p}+[1,1,0]} + \phi_{\mathbf{p}+[-1,-1,0]} - \phi_{\mathbf{p}+[1,-1,0]} - \phi_{\mathbf{p}+[-1,1,0]}\right)\\
M_{23}|_{\mathbf{Q-p}} = M_{32}|_{\mathbf{Q-p}} &= \frac{1}{4\Delta x^2}\left(\phi_{\mathbf{p}+[0,1,1]} + \phi_{\mathbf{p}+[0,-1,-1]} - \phi_{\mathbf{p}+[0,-1,1]} - \phi_{\mathbf{p}+[0,1,-1]}\right)\\
M_{31}|_{\mathbf{Q-p}} = M_{13}|_{\mathbf{Q-p}} &= \frac{1}{4\Delta x^2}\left(\phi_{\mathbf{p}+[1,0,1]} + \phi_{\mathbf{p}+[-1,0,-1]} - \phi_{\mathbf{p}+[1,0,-1]} - \phi_{\mathbf{p}+[-1,0,1]}\right),
\end{aligned} \tag{A.1}$$

where |Q−p denotes use of some or all the 33 − 1 surrounding nodes, p + [1, 0, 0],p + [0, 1, 0],...,p + [1, 1, 1]. The matrix elements, g_{ij}, are functions of the components of ∇φ only:

$$\begin{aligned}
\phi_{,1}|_{\mathbf{Q-p}} &= \frac{1}{2\Delta x}\left(\phi_{\mathbf{p}+[1,0,0]} - \phi_{\mathbf{p}+[-1,0,0]}\right)\\
\phi_{,2}|_{\mathbf{Q-p}} &= \frac{1}{2\Delta x}\left(\phi_{\mathbf{p}+[0,1,0]} - \phi_{\mathbf{p}+[0,-1,0]}\right)\\
\phi_{,3}|_{\mathbf{Q-p}} &= \frac{1}{2\Delta x}\left(\phi_{\mathbf{p}+[0,0,1]} - \phi_{\mathbf{p}+[0,0,-1]}\right)
\end{aligned} \tag{A.2}$$

Appendix B. Discretisation of U

Similarly to the treatment of θ and φ we have $U^*_{\mathbf{p}} \equiv r_2 U^n_{\mathbf{p}} - r_3 U^{n-1}_{\mathbf{p}}$, and write

$$U^{n+1}_{\mathbf{p}} - U^*_{\mathbf{p}} = r_1 \Delta t^{n+1} F^U_{\mathbf{p}} t(\phi^{n+1}_{\mathbf{Q}}, \phi^*_{\mathbf{p}}, U^{n+1}_{\mathbf{Q}}, U^*_{\mathbf{p}}, \theta^{n+1}_{\mathbf{p}}) \tag{B.1}$$

where

$$F_{\mathbf{p}}^{U}(\phi_{\mathbf{Q}},\phi_{\mathbf{p}}^{*},U_{\mathbf{Q}},U_{\mathbf{p}}^{*},\theta_{\mathbf{p}}) \equiv \left\{ \frac{\nabla\cdot\left(D_c\frac{1-\phi}{2}\nabla U+\mathbf{j}\right)+\frac{1}{2}[1+(1-k_E)U]\frac{\partial\phi}{\partial t}}{\left(\frac{1+k_E}{2}-\frac{1-k_E}{2}\phi\right)} \right\}\Bigg|_{\mathbf{Q}}$$
$$\equiv \frac{D_c\frac{1-\phi_{\mathbf{p}}}{2}\nabla^2U|_{\mathbf{Q}}+\nabla(D_c\frac{1-\phi_{\mathbf{p}}}{2})|_{\mathbf{Q-p}}\cdot\nabla U|_{\mathbf{Q-p}}+\nabla\cdot\mathbf{j}|_{\mathbf{Q}}+\frac{1}{2}[1+(1-k_E)U_{\mathbf{p}}]\frac{\phi_{\mathbf{p}}-\phi_{\mathbf{p}}^{*}}{r_1\Delta t^{n+1}}}{\left(\frac{1+k_E}{2}-\frac{1-k_E}{2}\phi_{\mathbf{p}}\right)}. \quad \text{(B.2)}$$

The divergence of the anti-trapping current is given by

$$\nabla\cdot\mathbf{j}|_{\mathbf{Q}} = -\frac{1}{2\sqrt{2}}[1+(1-k_E)]\nabla\cdot(U\dot{\phi}\mathbf{n})|_{\mathbf{Q}} \quad \text{(B.3)}$$

where

$$\nabla\cdot(U\dot{\phi}\mathbf{n})|_{\mathbf{Q}} = \dot{\phi}\nabla U\cdot\mathbf{n}+U\nabla\dot{\phi}\cdot\mathbf{n}+U\dot{\phi}\nabla\cdot\mathbf{n}$$
$$= \frac{\phi_{\mathbf{p}}^{n+1}-\phi_{\mathbf{p}}^{*}}{r_1\Delta t^{n+1}}\nabla U|_{\mathbf{Q-p}}\cdot\mathbf{n}|_{\mathbf{Q-p}}+U_{\mathbf{p}}\frac{\nabla\phi^{n+1}|_{\mathbf{Q-p}}-\nabla\phi^{*}|_{\mathbf{Q-p}}}{r_1\Delta t^{n+1}}\cdot\mathbf{n}|_{\mathbf{Q-p}}+U_{\mathbf{p}}\frac{\phi_{\mathbf{p}}^{n+1}-\phi_{\mathbf{p}}^{*}}{r_1\Delta t^{n+1}}\nabla\cdot\mathbf{n}|_{\mathbf{Q}} \quad \text{(B.4)}$$

and

$$\nabla\cdot\mathbf{n}|_{\mathbf{Q}} \equiv \frac{\partial}{\partial x^a}\left(\frac{\phi_{,a}}{\sqrt{\phi_{,b}\phi_{,b}}}\right)|_{\mathbf{Q}} = \frac{1}{\sqrt{\nabla\phi\cdot\nabla\phi}|_{\mathbf{Q-p}}}\left(\nabla^2\phi|_{\mathbf{Q}}-\frac{\phi_{,a}\phi_{,ab}\phi_{,b}}{\nabla\phi\cdot\nabla\phi}\Big|_{\mathbf{Q}}\right)$$
$$\equiv \frac{1}{\sqrt{\nabla\phi\cdot\nabla\phi}|_{\mathbf{Q-p}}}\left(\nabla^2\phi|_{\mathbf{Q}}-n_a|_{\mathbf{Q-p}}\phi_{,ab}|_{\mathbf{Q}}n_b|_{\mathbf{Q-p}}\right). \quad \text{(B.5)}$$

REFERENCES

1. Alain Karma, Phase-field formulation for quantitative modeling of alloy solidification, Phys. Rev. Lett. 87 (August 2001) 115701.
2. J. Rosam, P.K. Jimack, A.M. Mullis, Quantitative phase-field modeling of solidification at high lewis number, Phys. Rev. E 79 (March 2009) 030601.
3. U. Bisang, J.H. Bilgram, Phys. Rev. E 54 (1996) 5309.
4. J. Lipton, W. Kurz, R. Trivedi, Acta Metall. 35 (1987) 957.
5. E. Brener, D. Temkin, Phys. Rev. E 51 (1995) 351.
6. A.M. Mullis, C.E. Goodyer, P.K. Jimack, Towards a three-dimensional phase-field model of dendritic solidification with physically realistic interface width, Trans. Indian Inst. Metals 65 (6) (2012) 617–621.
7. Alain Karma, Wouter-Jan Rappel, Phase-field simulation of three-dimensional dendrites: is microscopic solvability theory correct?, J. Cryst. Growth 174 (1–4) (1997) 54–64. American Crystal Growth 1996 and Vapor Growth and Epitaxy 1996.
8. J.C. Ramirez, C. Beckermann, A. Karma, H.-J.J. Diepers, Phase-field modeling of binary alloy solidification with coupled heat and solute diffusion, Phys. Rev. E, Stat. Nonlinear Soft Matter Phys. 69 (May 2004) 051607.
9. Samuel M. Allen, John W. Cahn, A microscopic theory for antiphase boundary motion and its application to antiphase domain coarsening, Acta Metall. 27 (6) (1979) 1085–1095.
10. John W. Cahn, On spinodal decomposition, Acta Metall. 9 (9) (1961) 795–801.
11. Peter MacNeice, Kevin M. Olson, Clark Mobarry, Rosalinda de Fainchtein, Charles Packer, Paramesh: a parallel adaptive mesh refinement community toolkit, Comput. Phys. Commun. 126 (3) (2000) 330–354.
12. K. Olson, Paramesh: a parallel adaptive grid tool, in: A. Deane, et al. (Eds.), Parallel Computational Fluid Dynamics 2005: Theory and Applications, Elsevier, 2006.
13. W.F. Spotz, G.F. Carey, A high-order compact formulation for the 3d poisson equation, Numer. Methods Partial Differ. Equ. 12 (1996) 235–243.
14. L. Gránásy, T. Pusztai, T. Börzsönyi, G. Tóth, G. Tegze, J.A. Warren, J.F. Douglas, Phase field theory of crystal nucleation and polycrystalline growth: a review, J. Mater. Res. 21 (02) (2006) 309–319.
15. J. Rosam, P.K. Jimack, A.M. Mullis, An adaptive, fully implicit multigrid phase-field model for the quantitative simulation of non-isothermal binary alloy solidification, Acta Mater. 56 (17) (2008) 4559–4569.
16. A. Iserles, A First Course in the Numerical Analysis of Differential Equations, Cambridge University Press, 1996.
17. A. Brandt, Multi-level adaptive solutions to boundary-value problems, Math. Comput. 31 (1977) 333–390.
18. U. Trottenberg, C. Oosterlee, A. Schüller, Multigrid, Elsevier Academic Press, 2001.
19. J. Rosam, P.K. Jimack, A.M. Mullis, A fully implicit fully adaptive time and space discretization method for phase-field simulation of binary alloy solidification, J. Comput. Phys. 225 (2007) 1271–1287.
20. J. Green, P.K. Jimack, A.M. Mullis, J. Rosam, An adaptive, multilevel scheme for the implicit solution of three-dimensional phase-field equations, Numer. Methods Partial Differ. Equ. 27 (2011) 106–120.
21. Jun-Ho Jeong, Nigel Goldenfeld, Jonathan A. Dantzig, Phase field model for three-dimensional dendritic growth with fluid flow, Phys. Rev. E 64 (September 2001) 041602.
22. N. Goldenfeld N. Provatas, J.A. Dantzig, Adaptive mesh refinement computation of solidification microstructures using dynamic data structures, J. Comput. Phys. 148 (1999) 265–290.
23. C.E. Goodyer, P.K. Jimack, A.M. Mullis, Y. Xie, H. Dong, On the fully implicit solution of a phase-field model for binary alloy solidification in three dimensions, Adv. Appl. Math. Mech. 4 (2012).

CHAPTER 8

MATHEMATICS IN PAPER - FROM FIBER SUSPENSION FLUID DYNAMICS TO SOLID STATE PAPER MECHANICS

Jari Hämäläinen, Taija Hämäläinen , Teemu Leppänen, Heidi Niskanen and Joonas Sorvari

Keywords: Papermaking, multi-phase flow, fiber orientation, fiber flocculation, solid mechanics, fiber network rheology of paper, Computational Fluid Dynamics (CFD), Finite Element Analyses (FEA)

INTRODUCTION

Paper is made from wood fibers with some chemicals and fillers. Papermaking starts from wood handling and stock preparation. Fibers are mixed in water in a low concentration (order of 1%) when entering the first unit process of a paper machine, that is, the headbox (see Figure 1). Fibers tend to align with streamlines of a carrying phase water flow and they form fiber flocs in favorable flow conditions. By means of turbulence, both fiber orientation and flocculation can be controlled - stronger turbulence leads to more random orientation and less flocculated fiber suspension flow, which, in turn, influences the fibrous structure of the finished paper. The strength properties of the paper depend strongly on fiber orientation, which also determines the dimensional stability of the paper. Fiber flocs make denser 'spots' in a paper sheet, which may cause some variation in ink absorption, for example. Therefore, the control of fluid dynamics in the beginning of the papermaking process is a key issue in papermaking. Furthermore, the finished paper with its different end-use purposes can be studied by using mathematical modeling based on either fiber-level network models or on the continuum approach. These aspects are discussed more detailed in this paper than in essay [1].

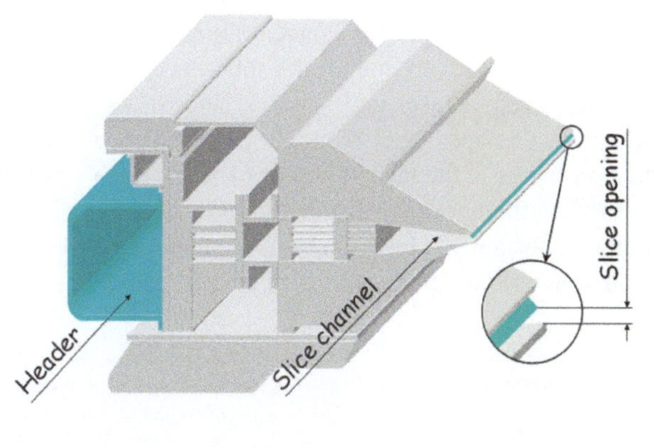

Figure 1: Paper machine headbox.

The paper is organized as follows: The second section deals with specific features of fiber suspension flows, namely fiber orientation and flocculation and how they can be modeled. The third section is devoted to the microscopic level solid mechanics of paper as a fiber network. Finally, the last section discusses the rheology of paper and its macroscopical modeling.

FLUID DYNAMICS OF FIBER SUSPENSION FLOWS

Fiber orientation and flocculation

Both fiber orientation and fiber flocculation are characteristic fluid dynamical phenomena of fiber suspension flows at low concentrations in the headbox and the forming section. The main purpose of a headbox is to create a thin and even jet which then enters the forming section. In the headbox, the suspension is mixed in the turbulence generator and then accelerated in the

Citation: Jari Hämäläinen, Taija Hämäläinen, Teemu Leppänen, Heidi Niskanen and Joonas Sorvari, Mathematics in paper - from fiber suspension fluid dynamics to solid state paper mechanics. Journal of Mathematics in Industry 2014, 4:14, 10.1186/2190-5983-4-14.

contracting slice channel. After the slice opening, the jet enters into free air and hits the moving fabrics called "wires", where the water removal and the formation of the fiber network are initiated. Subsequent water removal processes by pressing and drying make paper more solid, but the basic fibrous structure of paper does not change noticeably.

The structure of paper can be characterized by numerous mechanical properties. Fiber orientation describes how the fibers are aligned in the flow and finally in the finished paper as a fibrous network. Fibers are mainly aligned with streamlines of the headbox fluid flow; more precisely, the alignment of fibers forms a varying distribution which depends on the dominating flow conditions. A measured flow field with fibers is presented in Figure 2. Many mechanical properties of paper are affected by fiber orientation, for example, the strength properties of paper as it is stronger in the fiber direction than in the cross direction. Further, moisture changes cause fibers to swell, leading to cockling and curling of the finished paper. Therefore, headbox flows need to be controlled precisely so as not to cause harmful cross-directional flows in the headbox, such as those illustrated in Figure 3 due to a non-even flow rate distribution from the tapered header. The second mechanical measure of paper is its basis weight defining how many grams it weights per square meter. Large-scale weight variations are controlled by headbox actuators, but small-scale variations are due to fiber flocculation (see Figure 4 for a photo of a paper sample in centimeter scale). Flocculation describes the tendency of fibers to form aggregates. Fibers used in papermaking tend to tangle together even in low concentration suspensions. This may affect the local variation of the solid mass distribution of paper and thereby causes undesired variance in the properties of the produced paper sheet. Non-uniformity of properties may induce problems in the later stages of papermaking or printing processes, as the unevenly distributed mass contains different quantities of water, and thus, expands unequally, or absorbs ink unevenly, deteriorating the quality of printed photos.

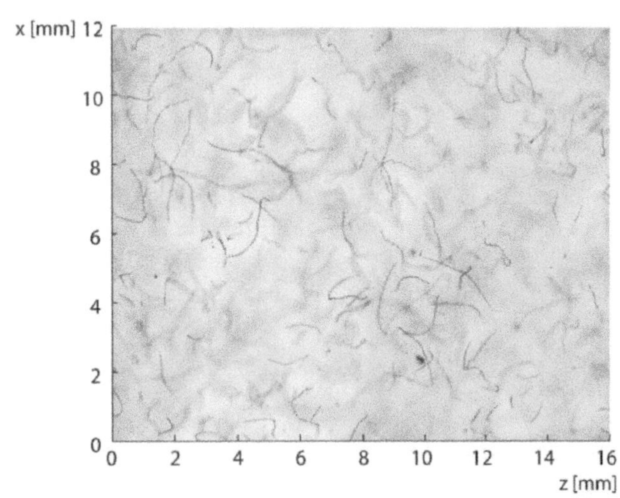

Figure 2: Fibers in a flow.

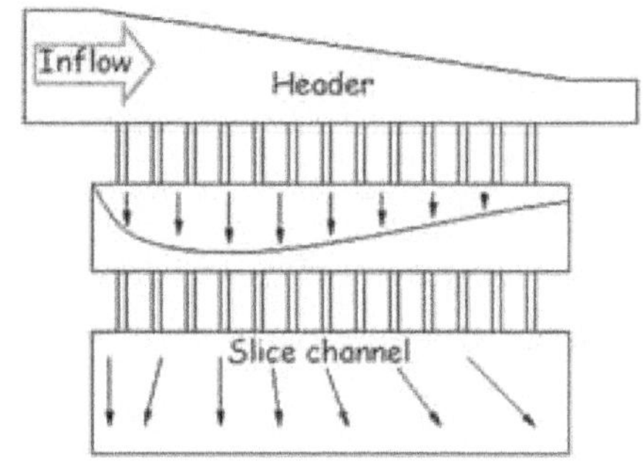

Figure 3: Illustration of fluid flows in a headbox.

Figure 4: Basis weight variation called formation.

Formulation of the fiber orientation probability distribution model

When multiphase flows are concerned, there are namely two different methods to model the particles in such a flow. The first option is the Lagrangian approach, which models the motion of individual fibers. However, as accurate as it may be, this approach is computationally very expensive and thus it is not practical to be used in such a large scale flow as takes place in the headbox. The other option is the Eulerian approach, which takes into account an undefined number of fibers and predicts their statistical behavior. The latter one is the basic idea behind the fiber orientation probability distribution (FOPD) model.

The basis of the probability distribution modeling approach is the conservation of the probability flux and the use of diffusion-convection or the Fokker-Planck equation. This approach has been widely used (e.g. [2–5]) in studying the development of fiber orientation in contracting channel flows. As a result, the fiber orientation distribution may be obtained.

It is common to simplify the model and to consider only the planar orientation in vertical or horizontal planes. However, by doing so, one degree of freedom is lost and the results may not be as realistic as they would be when the motion of fiber is described in three-dimensional space as it is in reality. Nevertheless, the model is able to illustrate the most probable orientation distribution in the headbox flow under turbulent conditions (e.g. [6, 7]) and near walls [8]. The evolution of the fiber orientation distribution Ψ can be written as

$$\frac{\partial \Psi}{\partial t} - D_t \Delta \Psi - D_r \Delta_{S^2} \Psi + \nabla \cdot (\mathbf{v}\Psi) + \nabla_{S^2} \cdot (\mathbf{w}\Psi) = 0, \quad (1)$$

where the notations Δ_{S^2} and ∇_{S^2} refer to the Laplace and divergence operators on a unit sphere, respectively. Furthermore, **v** is the velocity of the fluid and **w** is the rotational velocity of the fiber. Coefficients Dt and Dr are the translational and rotational diffusion coefficients respectively, describing the effects of velocity fluctuations on how the fibers are distracted from their state of orientation at a certain moment. It is usual to assume that there is no time-dependence, and thus, the first term of the equation can be ignored. The rotational velocity of the fiber is determined as (see e.g. [9–11])

$$\mathbf{w} = \omega\mathbf{p} + \lambda\epsilon\mathbf{p} - \lambda\langle\mathbf{p}, \epsilon\mathbf{p}\rangle\mathbf{p}, \tag{2}$$

where $\lambda = (r^2 - 1)/(r^2 + 1)$) is a parameter related to the fiber aspect ratio r=L/d, L and d being the length and the diameter of the fiber, respectively, and **p** gives the orientation of the fiber at the surface of a unit sphere. The origin of the vector **p**, defined as

$$\mathbf{p} = \begin{pmatrix} \cos\theta\sin\phi \\ \sin\theta\sin\phi \\ \cos\phi \end{pmatrix}, \tag{3}$$

is placed at the center of the fiber. The notation $\langle\cdot,\cdot\rangle$ refers to the inner product of the vectors. Furthermore,

$$\epsilon = \frac{1}{2}(\nabla\mathbf{u} + (\nabla\mathbf{u})^T) \quad \text{and} \quad \omega = \frac{1}{2}(\nabla\mathbf{u} - (\nabla\mathbf{u})^T) \tag{4}$$

are the strain rate and vorticity tensors, respectively. Resulting orientation distributions are illustrated in Figure 5. The figure shows the development of fiber orientation along a contracting channel achieved using a planar approximation of the model. Figure 6 shows the FOPD in three-dimensional surface, that is, a 3D solution for a 2D contracting channel and for one projected angle.

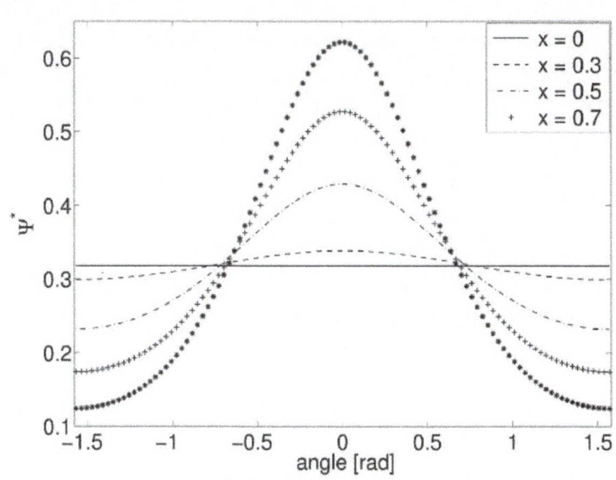

Figure 5: Fiber orientation distributions along a contracting channel. Modeled fiber orientation distributions. The channel outlet is located at x=0.7 m and assumption of random orientation distribution is used at the channel inlet at x=0 m.

The earliest studies of non-spherical particles were conducted by Jeffery already in 1922 [12] when he derived the equations of motion for a single ellipsoid, non-Brownian particle in Newtonian fluid with simple shearing and showed that the particles execute rotational motion in periodic, closed orbits. Jeffery's work has been the basis for current modeling and has been widely applied.

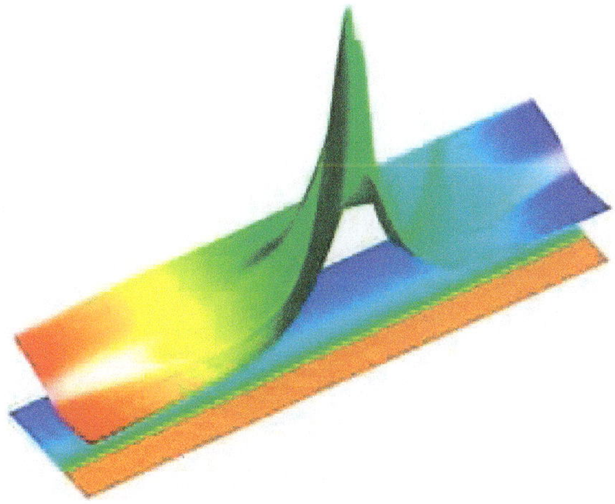

Figure 6: Visualization of the fiber orientation distribution.

The model given above is based on the assumption that the fibers are rigid. However, it can also be used in describing the orientation of flexible fibers at least from a statistical point of view as discussed in [5] even though it does not take into account the flexibility of the fibers. Some studies have also taken into account the fiber-fluid coupling through an orientation tensor including momentum exchange between the fiber and the fluid phases [13]. This type of approach to model the most probable orientation state of the fibers on different flow conditions has shown to be rather convenient when studying the flows of industrial scales. Naturally, there are some assumptions e.g. about the low enough concentration and rigidity of the fibers which does not necessarily meet the conditions in reality. The latter assumption is violated in the case of papermaking, where the wood fibers are long and flexible, thus causing additional complexity such as possible flocculation for the flow. However, the commonly taken approach gives practical information about the real-life phenomena present under the flows of papermaking and is of importance in trying to improve the papermaking process.

Modeling of fiber floc dynamics by using a population balance approach

Fiber flocculation has not been widely studied due to the challenges it creates both for measuring and computational methods. The simplest models treat fiber suspension as a one phase flow ([14–16]) although the interaction of the two fibers should be taken into account, as shown in [17] when comparing these results to measurements [18]. More accurate models [19,20] use the Eulerian two-phase flow approach and are based on the population balance model also taking into account the occurrence of flocs of different sizes.

Modeling of fiber flocculation inside the headbox requires taking into account the two-phase nature of the flow and the interactions between the carrying phase, water, and the dispersed phase, flocs. Moreover, to describe the real situation in the paper machine headbox and in the forming section, one has to model not only flocs of one size, but several sizes. This can be done by using the population balance approach

$$\frac{\partial}{\partial t}n(v,t) + \frac{\partial}{\partial x^i}\left(U^i(v,t)n(v,t)\right) = B_B - D_B + B_C - D_C, \tag{5}$$

where n(v,t) represents the number density of particles of size v at time t and B_B, D_B, B_C and D_C represent the birth rate due to the break-up of larger particles, the death rate due to the break-up into smaller particles, the birth rate due to the coalescence of smaller particles, and the death rate due to the coalescence with other particles, respectively. The carrying phase velocity field is denoted by U. B_B, D_B, B_C and D_C are defined as

$$B_B = \int_v^{\infty} g(v;s)n(s)\,ds, \tag{6}$$

$$D_B = n(v)\int_0^{v} g(v;s)\,ds, \tag{7}$$

$$B_C = \frac{1}{2}\int_0^{v} Q(v-s;s)n(v-s)n(v)\,ds, \tag{8}$$

$$D_C = n(v)\int_0^{\infty} Q(v;s)n(s)\,ds, \tag{9}$$

where g(v;s) and Q(v;s) represent the specific break-up rate and the specific coalescence rate, respectively. For the break-up and coalescence models, the equations developed for bubble flows [21, 22] can be used, if the model parameters are adequately tuned and verified with measurements.

As the physical characteristics of wood fibers used in papermaking may vary significantly depending on the paper grade produced, it is intuitively clear that flocculation phenomena must depend on fiber characteristics. When thinking about a fiber network formed of long, flexible fibers or a network of short, stiff fibers, it is easy to see that breaking the network of long fibers requires more force than the short fiber case: If the concentrations of these two suspensions are the same, the suspension of long fibers has more interfiber connections to be broken. A practical approach to study this characteristic is an old concept - the crowding factor [23] defined as

$$N_{cf} = \frac{2}{3}C_v\left(\frac{L}{d}\right)^2 \approx \frac{5C_m L^2}{\omega}, \tag{10}$$

where C_v is the volumetric concentration, L is the average fiber length, d is the average fiber diameter, C_m is the pulp concentration, and ω is the fiber coarseness. The crowding factor represents the number of fibers within the rotational sphere of influence of a single fiber. It was shown in [19] that the internal strength of flocs is related to this factor.

The population balance approach allows the detailed study of the development of the floc size distribution, as shown in Figure 7. This example demonstrates the dynamics of a floc population of ten flow size groups between 1 and 10 mm in a headbox contracting channel. It shows that the smaller flocs are coalescing and their number density decreases whereas the number density of the largest flocs increases. For more detailed analyses, we refer to [19].

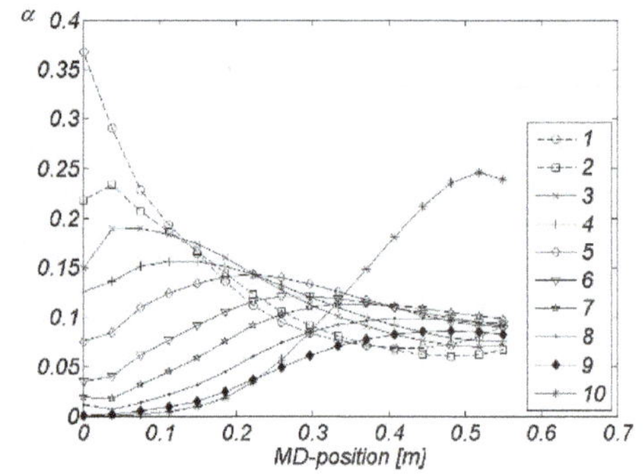

Figure 7: Number densities of ten different floc size groups in the slice channel.

When fiber flocculation is modeled further in the paper machine in the forming section, more physical couplings are needed. In the forming section, most of the water included in the suspension is removed with different types of suction elements and the fiber concentration is significantly increased. This means more fiber-fiber interactions and less freedom for fibers to move. In addition, when the water is removed through the moving wires, the fiber network starts building up, further increases the resistance of the wires and thus hinders the water removal. This effect can be taken into account by adding the solids pressure force [24]

$$F_{d,i}^{SP} = -G_0 e^{C_{SP}(\alpha_d - \alpha_{d,\max})}\frac{\partial \alpha_d}{\partial x_i} \tag{11}$$

to the momentum equation of the dispersed phase. In Eq. (11) G_0 and C_{SP} are the reference elasticity modulus and the compaction modulus, respectively, and αd and $\alpha_{d,max}$ represent the local volume fraction of the fibrous phase and the maximum packing parameter, respectively. Now by selecting suitable value for the coefficient C_{SP}, the solids pressure gradient is only activated in the regions close to the maximum packing.

As discussed earlier in this paper, the flocculated state of the suspension is inherited to the variation of the solid mass distribution of the paper produced. Taking into account the phenomena in the forming section, the evolution of the fiber floc size can also be connected to the solid mechanical modeling of the finished paper sheet [20]. The shear forces and the water removal have significant effect on the floc size of the suspension, as illustrated in Figure 8. The figure illustrates the computational domain including the contracting channel with some turbulence lamellae causing the wakes in the flow, and the free jet from the headbox to the moving wire.

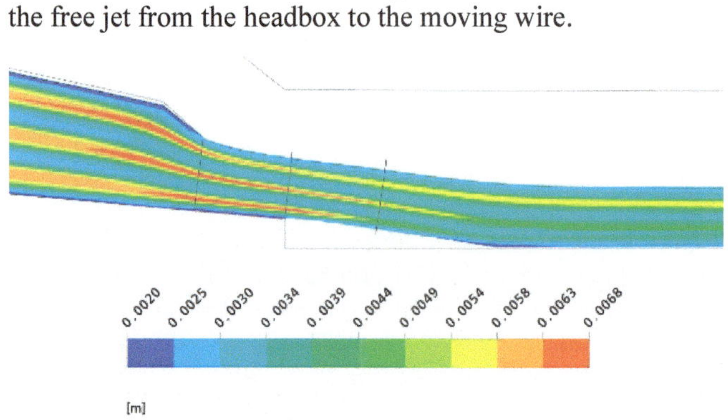

Figure 8: Mean floc size in the free jet.

The following sections continue from the paper web obtained in the forming section and further dewatering in the papermaking process.

PAPER AS FIBROUS MATERIAL

Paper physics

Stress-strain curves yield important information on materials behavior. For a typical paper grade, such as newsprint, the stress-strain curve is characterized by an initial linear region which is followed by a non-linear region. The transition from the linear to the non-linear region takes place smoothly. In engineering literature, the transition point is known as the proportional limit. The proportional limit is commonly associated with the yield limit which is the point in the stress-strain curve after which plastic or permanent strains are developed. In the non-linear region gradient of the curve decreases but is still positive. The degree of non-linearity depends on humidity conditions. In dry conditions, the curve is rather linear even after the proportional limit. However, in humid conditions clear strain hardening is observed. In this sense, paper is brittle in dry conditions and ductile in humid conditions. The stress-strain curve depends also on the rate of loading. Both the elastic modulus and tensile strength increase with the strain rate. The viscous nature of paper can also be observed from a relaxation or creep test.

Since paper is a fibrous network in which fibers are connected by hydrogen bonds, we may ask what is the influence of fiber and bond properties on the mechanical properties of a paper sheet. It is intuitively clear that the elastic modulus of paper must depend on the elastic modulus of fibers. Due to the porous structure of paper and random orientation of fibers, the modulus is less than the corresponding modulus of fibers. In a seminal paper by Cox [25] a network theory for the tensile behavior of paper was presented. In the network model, long, straight and non-interacting fibers were considered. In Cox's theory, the elastic modulus of a paper sheet is given by $E = \frac{1}{3}\rho/\rho_f E_f$, where E_f is the elastic modulus of fibers and ρ and ρf are densities of the sheet and the fibers, respectively. For highly bonded networks, Cox's theory seems to be in reasonable agreement with the experimental observations. However, for less dense networks the modulus can be lower than predicted by Cox's theory [26].

The non-linear irreversible response of a paper sheet under elongation has been attributed to the progressive disruption of inter-fiber bonds and to intra-fiber mechanisms. Rance [27] argued that the breaking of bonds results in increased deformation due to more localized stress concentrations. However, according to Alava [28], the inelasticity of paper arises from plastic deformations that take place in the fibers when microfibrils slide relative each other.

It is also still unclear whether the time-dependence of paper arises from fiber properties or from the behavior of inter-fiber bonds. In an in-depth review by Haslach [29], it was concluded that most likely the time-dependency is controlled by a mechanism which occurs within the inter-fiber bonds, namely the release of microcompression in the inter-fiber bonds.

Although there is currently no comprehensive theory of the inelasticity deformation of paper, there is consensus between paper physicists that the fracture strength of paper is governed by the failure of bonds. Only in highly bonded networks is the sheet strength determined by the tensile failure of fibers.

Fiber network modeling

It is obviously in the interest of papermakers to gain information on the relationship between the mechanical properties of paper and the constituent parts of paper. Extensive efforts have been made to construct realistic fiber networks and to simulate their behavior. Heyden [30] classifies network modeling approaches into uniform strain models, semi-analytical models and to computer simulation models. In this paper, we are interested in the latter and especially in simulation models in which the finite element method (FEM) is utilized.

The first step in the fiber network modeling is to construct a realistic network, which is a challenging task especially if 3D structures are considered. The generation begins by placing fibers randomly inside a cell [30] or by depositing fibers on a rectangular surface one-by-one [31] until the desired network density or grammage is attained. Fiber parameters such as length, orientation and curvature may vary according to a probability distribution. In the network generation, geometric intersection or closest point problems are typically addressed in order to find possible inter-fiber bonding points. Free fibers or fiber clusters which are not connected to the rest of the network are typically removed from the network in order to obtain positive definite system matrix in finite element calculations. Network connectedness can be examined with aid of graph search algorithms [30]. The graph nodes are the intersection and boundary points and free fiber segments represent the graph edges. It should be noted that network systems are large. In an area of one square millimeter there are roughly 100-200 fibers, and every fiber has approximately 10-40 inter-fiber bonds. Furthermore, the number of the fibers increases quadratically with the area of the network. The size of the network is still the limiting factor when considering industrially important problems. We are still in the millimeter scale when dense three-dimensional networks are considered. To increase the size of the networks, efficient algorithms and coding methods are needed.

Network generation is only one part of the problem. In order to perform simulations, a finite element model of the generated network must be created. Although it is possible to model fibers with solid continuum elements [32], beam elements are typically used [30, 31] since they reduce the size of the discrete finite element system. Fibers are commonly assumed to follow the linearly elastic constitutive law [30], although inelastic models [32] can and should be used if extensive deformations are considered. Considering the physical behavior of paper networks, it seems that the proper modeling of inter-fiber bonding is critical in order to model paper networks realistically. For small deformations, ideal rigid bonding, as used in Cox's theory, may be adequate. However, to model the non-linear response or failure of a paper sheet, deformable bonding models which allow bond failure should be utilized. In Heyden's [30] network model, fibers are bonded together with an elastic spring, the stiffness and strength of which reduces in a stepwise manner until complete failure occurs. There has been growing interest to take into account the subsequent sliding of fibers after bond failure has occurred

[31, 33]. The inter-fiber frictional sliding may be particularly important for wet networks since water breaks hydrogen bonds so that fibers can more easily slide against one another.

The resulting discrete finite element problem of fiber networks is typically very large and non-linear due to possible material and geometric non-linearity and frictional sliding. Thus, tools of high performance computing must be utilized.

A fiber network model

For illustrative purposes we consider here the problem of simulating the behavior of a fiber network in an extension test. Recently, Lavrykov et al. [32] have presented an alternative way to generate fiber networks. In their model, the sheet structure is obtained by compressing randomly placed fibers between two rigid plates. The compressing was simulated with the finite element method. A similar approach is used here.

In our deposition model, fibers are initially modeled as straight rectangular cuboids with a uniform length. The fibers are randomly placed on a plane and are not to allowed to intersect. At this stage, the fibers are parallel to the plane. The in-plane orientation of the fibers varies according to a uniform distribution. After fibers can no longer be placed on a given plane, we proceed to the next plane, and this procedure is continued until the desired grammage is attained. In other words, fibers are placed in a cell plane by plane. After the initial structure has been obtained we proceed to FEM simulation where we discretize the fibers with hexahedron elements and place the structure between two rigid plates. Then, the fibers are allowed to drop under gravity to the bottom stationary rigid plate. The upper rigid plate is moving and compresses the fibers so that the prescribed thickness of the network is attained. The forming of the sheet structure, i.e. dropping of the fibers and compression of the network, is simulated with aid of the FEM. In the simulation, fiber-to-fiber contacts are taken into account. After we have obtained the initial sheet structure, fiber-to-fiber contact points are searched and in the contact points bonds are formed. Bonds are modeled as multiaxial springs that can fail if a given threshold value is reached. An example of a generated network is shown in Figure 9.

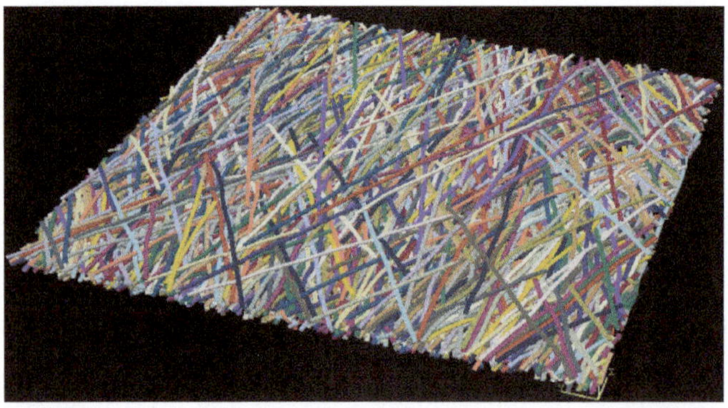

Figure 9: Example network geometry. Example of a generated fiber network.

The network generation procedure was implemented in ABAQUS finite element software using Python scripting. In this way, we can combine finite element model making and simulations with useful numerical tools. In the initial intersection testing and in the search for fiber-to-fiber contact Gilbert-Johnson-Keerthi (GJK) collision detection algorithm [34] of convex objects is utilized. For a non-convex object, the same GJK algorithm can be used to detect the collision of the convex hulls of the objects. In the contact searching, bounding volume hierarchies and hierarchy traversal collision testing are used. In a bounding volume hierarchy, a geometric object is partitioned into a tree structure. The bounding volume hierarchy of a fiber is constructed using convex hulls. The root node is the convex hull of the fiber. At each node, the fiber part is subdivided into two nearly equal size parts and the convex hulls of the parts are stored in the child nodes. This procedure is continued until we end up in a structure in which the leaf nodes contain the convex hull of a single element. The fiber-to-fiber intersection tests are conducted using their bounding volume hierarchies. In the intersection tests, both hierarchies are descended simultaneously using a depth-first approach [35]. After all contacts have been found, the network connectedness is studied using the breadth-first search (BFS) strategy.

Figure 10 illustrates the effect of the bonding properties of fibers on the stress-strain behavior of a paper sheet. We simulated, using the FEM, an extension test using the same network but varied the bonding properties of fibers. The properties of the network are given in Table 1. If the bonds are rigid, the stress-strain curve is rather linear. The behavior is different from the case where the bonds are modeled with stiff breakable bonds in which case we see strain hardening as bonds deform and break. Figure 11 shows that for several fibers which are located near the loading edge, many of the bonds are broken so that the fibers are more free to move with the displaced edge.

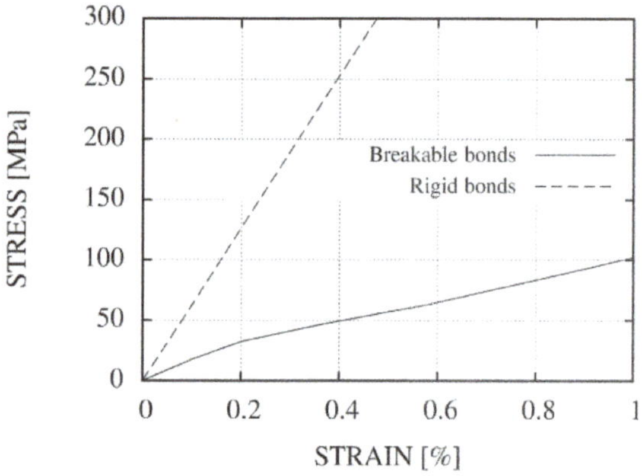

Figure 10: Stress-strain curve of the network. Effect of bond properties on the stress-strain behavior of a paper sheet.

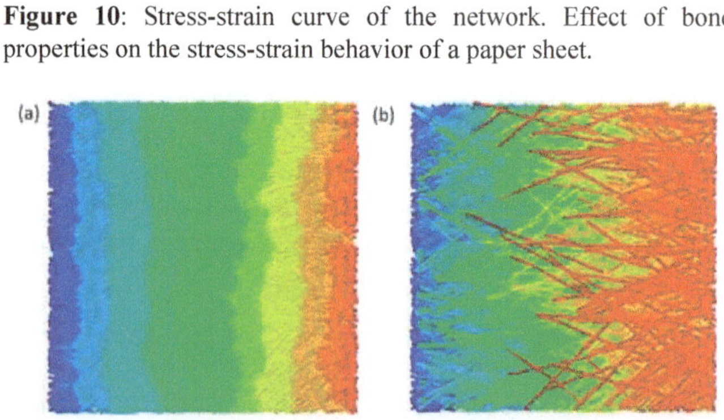

Figure 11: MD displacement fields. Effect of bond properties on the MD displacement fields. (a) Rigid bonds. (b) Stiff breakable bonds. Red color indicates maximum displacement, blue color minimum displacement.

Table 1: Fiber network properties

Fiber length	[mm]	2
Fiber width	[μm]	30
Fiber thickness	[μm]	10
Fiber basis weight	[g/m^2]	8
Elastic modulus of fibers	[GPa]	35
Network size	[mm^2]	4×4
Network thickness	[μm]	100
Number of fibers		1,400

PAPER AS CONTINUOUS MATERIAL

Continuum description of paper

At the macroscopic level, the paper structure can be considered as continuous solid material obeying the laws of continuum mechanics. The equilibrium equations together with constitutive laws form the basis of solid continuum mechanics. The continuum description of paper is a convenient approach to solve industrially relevant problems. However, paper is a heterogeneous and anisotropic viscoplastic material which makes it challenging to model even when the continuum description is used. In modeling applications, some simplifications are commonly needed. On a sufficiently large scale, paper can be considered to exhibit homogeneous material behavior. In addition, the anisotropic behavior of paper is often handled by assuming paper to behave orthotropically. An orthotropic material has three orthogonal planes of symmetry which can be chosen as coordinate planes. In the case of machine made paper, the principal material directions coincide with MD, CD and the thickness direction (ZD), as illustrated in Figure 12. Paper is very orthotropic, being much stiffer in MD than in CD or ZD. The value for the ratio between the elastic modulus in MD and CD is between 2 and 4 and for MD and ZD around 100.

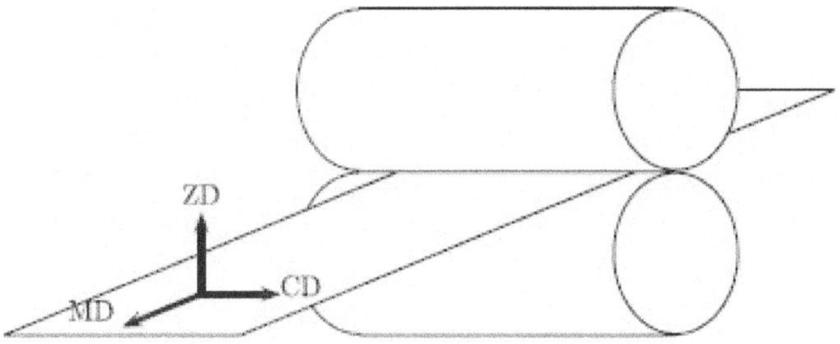

Figure 12: Principal material directions of paper. Principal material directions of machine made paper: MD, CD and ZD.

Although the uniaxial inelasticity of paper has been long recognized [36] and is currently relatively well understood [29, 37], only very recently have well calibrated multiaxial inelastic constitutive models been proposed. Classical plasticity models have been proposed by Xia et al. [38], Mäkelä and Östlund [39] and more recently by Harryson and Ristinmaa [40]. Integral based viscoelastic constitutive models have been proposed by Uesaka et al. [41] and Lif [42]. Apparently, we are still missing a model which combines the viscous and plastic properties of paper.

The continuum description of paper is widely used in modeling applications. A typical application is FEM modeling related to papermaking or to printing processes. Examples include web runnability [43], calendering [44], winding [45] and printing nip mechanics [46].

Modeling the fracture of paper

As an example of an application of a continuum description, we consider modeling the fracture of paper. As the continuum description is used, paper is treated as a solid continuum. However, material properties are allowed to vary locally. To be precise, paper is divided into small volumes each having their own principal material directions. In addition, every volume has its own directional stress-strain behavior depending on, for example, the fiber orientation anisotropy, dry solids content and basis weight. More information about this method, the locally varying fiber orientation and fiber orientation anisotropy can be found from reference [47]. As an example, Figure 13 shows how the tensile strength depends on fiber orientation anisotropy ξ and dry solids content β in the example presented in this paper. A detailed and sophisticated presentation of material parameter dependencies is available in reference [48].

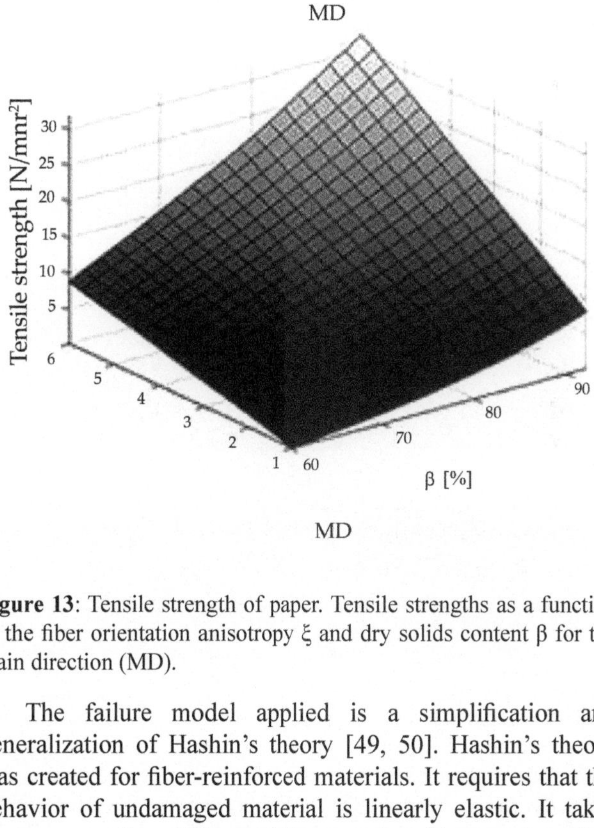

Figure 13: Tensile strength of paper. Tensile strengths as a function of the fiber orientation anisotropy ξ and dry solids content β for the main direction (MD).

The failure model applied is a simplification and generalization of Hashin's theory [49, 50]. Hashin's theory was created for fiber-reinforced materials. It requires that the behavior of undamaged material is linearly elastic. It takes into account four different failure modes: fiber tension, fiber compression, matrix tension, and matrix compression. When this theory is used, two different stages of the fracture process should be defined: damage initiation and damage evolution.

In the case of paper, the linear elasticity requirement in Hashin's theory is a substantial weakness; the stress-strain behavior of paper is commonly nonlinear and inelastic, as stated earlier. From that point of view, the failure model is a generalization of Hashin's theory since no restrictions are imposed for material behavior. Actually, if fracture simulations are considered, the essential issue is the determination of the tensile strength (and stress-strain behavior) dependencies (see Figure 13). Whereas in Hashin's theory four different failure modes are considered, in the failure model applied only two failure modes are considered; the failure of fibers and matrix are not separated. From that point of view, the present model is a simplification compared to Hashin's theory. Also the damage evolution is much simpler in the present model than in Hashin's theory: when the stress level reaches the tensile strength, the stiffness of the material is decreased to a thousandth. The failure model was implemented in the ABAQUS user subroutine USDFLD. Numerical problems were addressed by a two step stabilization process. The first step is directly controlled by the user and the second step is adaptively process controlled by ABAQUS.

Figure 14 presents a simulated fracture of a paper sample in an extension test. The sample was loaded in MD. The basis weight of the sample was 45 g/m^2 and the constant dry solids content and basis weight was assumed. The only variation taken into account was the variation of fiber orientation in all three directions: MD, CD and ZD. As can be seen from the MD strains in Figure 14, the location of the initial fracture is not a simple matter. It is not determined only by one location - also the surrounding areas of one location influences the stress level of the location. In addition, the whole stress-strain behavior varies in all directions. That is, a higher MD strain level in some location does not mean that the fracture starts from that location.

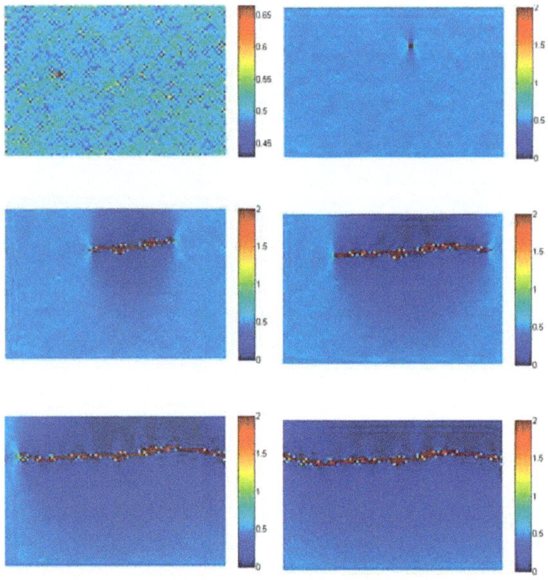

Figure 14: Simulated fracture process. Simulated MD strain during the fracture process when the variation of fiber orientation is taken into account. In figures, MD corresponds to the vertical direction and the fracture proceeds from left to right and from top to bottom. A 90% dry solids content was used in the simulation.

CONCLUSIONS

This paper aimed at giving a general view on mathematical modeling in papermaking. Many modeling challenges are excluded from this paper: Fluid-Structure Interactions (FSI) for the running paper web, coating of the paper web, the rheology of fluids and solids, heat transfer during the drying process, to mention a few. However, modeling approaches presented in this paper may be used separately or together in several paper production, converting or end-use situations. For example, by fluid dynamics the variation of fiber orientation and basis weight may be predicted and by solid mechanics the importance of these factors on paper behavior can be studied. The future work concentrates to the collaboration of the presented modeling approaches, for example, stress-strain behavior used in continuum mechanical approach could be obtained by fiber network modeling.

REFERENCES

1. Hämäläinen J: Mathematics in paper. J Math Ind 2012., 2: Article ID 2 Article ID 2
2. Olson JA: The motion of fibres in turbulent flow, stochastic simulation of isotropic homogeneous turbulence. Int J Multiph Flow 2001, 27: 2083–2103. 10.1016/S0301-9322(01)00050-7MATHCrossRef
3. Hyensjö M, Dahlkild A: Study of the rotational diffusivity coefficient of fibres in planar contracting flows with varying turbulence levels. Int J Multiph Flow 2008,34: 894–903. 10.1016/j.ijmultiphaseflow.2008.02.005CrossRef
4. Krochak PJ, Olson JA, Martinez DM: The orientation of semidilute rigid fiber suspensions in a linearly contracting channel. Phys Fluids 2008., 20: Article ID 073303 Article ID 073303
5. Niskanen H, Eloranta H, Tuomela J, Hämäläinen J: On the orientation probability distribution of flexible fibres in a contracting channel flow. Int J Multiph Flow2011, 37: 336–345. 10.1016/j.ijmultiphaseflow.2010.11.006CrossRef
6. Olson JA, Kerekes RJ: The motion of fibres in turbulent flow. J Fluid Mech 1998,377: 47–64.MATHCrossRef
7. Olson JA, Frigaard I, Chan C, Hämäläinen J: Modelling turbulent fibre suspension flowing in a planar contraction: the one-dimensional headbox. Int J Multiph Flow 2004, 30: 51–66. 10.1016/j.ijmultiphaseflow.2003.10.006MATHCrossRef
8. Krochak PJ, Olson JA, Martinez DM: Near-wall estimates of the concentration and orientation distribution of a semi-dilute rigid fibre suspension in Poiseuille flow. J Fluid Mech 2010, 653: 431–462.MATHCrossRef
9. Shin M, Koch DL: Rotational and translational dispersion of fibres in isotropic turbulent flows. J Fluid Mech 2005, 540: 143–173. 10.1017/S0022112005005690MATHCrossRef
10. Parsheh M, Brown M, Aidun C: On the orientation of stiff fibres suspended in turbulent flow in planar contraction. J Fluid Mech 2005, 545: 245–269. 10.1017/S0022112005006968MATHCrossRef
11. Phelps JH, Tucker CL III: An anisotropic rotary diffusion model for fiber orientation in short- and long-fiber thermoplastics. J Non-Newton Fluid Mech2009, 156: 165–176. 10.1016/j.jnnfm.2008.08.002MATHCrossRef
12. Jeffery GB: The motion of ellipsoidal particles immersed in a viscous fluid. Proc R Soc Lond A 1922, 102: 161–179. 10.1098/rspa.1922.0078CrossRef
13. Krochak PJ, Olson JA, Martinez M: Fiber suspension flow in a tapered channel: the effect of flow/fiber coupling. Int J Multiph Flow 2009, 35: 676–688. 10.1016/j.ijmultiphaseflow.2009.03.005CrossRef

14. Steen M: Turbulence and flocculation in fibre suspensions. PhD thesis. University of Trondheim, Norway; 1990. Steen M: Turbulence and flocculation in fibre suspensions. PhD thesis. University of Trondheim, Norway; 1990.
15. Kuhn DCS, Sullivan PE: Analysis and measurement of the flocculation intensity of flowing pulp suspensions. TAPPI Papermakers Conference 2001.
16. Plikas A, Kuhn DCS, Sullivan PE: A numerical model of fibre suspensions in a grid generated turbulent flow. Proceedings of ASME FEDSM'00 2000, 991–996.
17. Hyensjö M: On fibre suspension flow modelling: mechanical fibre flocculation and fibre orientation. PhLic thesis. Royal Institute of Technology, Stockholm, Sweden; 2005. Hyensjö M: On fibre suspension flow modelling: mechanical fibre flocculation and fibre orientation. PhLic thesis. Royal Institute of Technology, Stockholm, Sweden; 2005.
18. Karema H, Salmela J, Tukiainen M, Lepomäki H: Prediction of paper formation by fluidisation and reflocculation experiments. 12th Fundamental Research Symposium 2001, 559–589.
19. Hämäläinen T, Hämäläinen J: Fibre floc evolution model, part I: flocculation in a headbox. Nord Pulp Pap Res J 2010,25(1):39–47. 10.3183/NPPRJ-2010-25-01-p039-047CrossRef
20. Hämäläinen T, Hämäläinen J: Fibre floc evolution model, part II: influence of the jet-to-wire ratio to resulting fibre floc evolution in the initial drainage zone.Nord Pulp Pap Res J 2010,25(1):48–55. 10.3183/NPPRJ-2010-25-01-p048-055CrossRef
21. Luo H, Svendsen HF: Theoretical model for drop and bubble breakup in turbulent dispersions. AIChE J 1996,42(5):1225–1233. 10.1002/aic.690420505CrossRef
22. PrinceMJ,BlanchHW:Bubblecoalescenceandbreak-upin airsparged bubble columns. AIChE J 1990,36(10):1485–1499. 10.1002/aic.690361004CrossRef
23. Kerekes RJ, Schell CJ: Characterization of fibre flocculation regimes by a crowding factor. J Pulp Pap Sci 1992,18(1):32–38.
24. Gidaspow D: Multiphase Flow and Fluidization. Academic Press, London; 1994.MATH
25. Cox HL: The elasticity and strength of paper and other fibrous materials. Br J Appl Phys 1952, 3: 72–79. 10.1088/0508-3443/3/3/302CrossRef
26. Sampson WW: Materials properties of paper as influenced by its fibrous architecture. Int Mater Rev 2009, 54: 134–156. 10.1179/174328009X411154CrossRef
27. Rance HF: The formulation of methods and objectives appropriate to the rheological study of paper. Tappi J 1956, 2: 104–115.
28. Alava M, Niskanen K: The physics of paper. Rep Prog Phys 2006, 69: 669–723. 10.1088/0034-4885/69/3/R03CrossRef
29. Haslach HW Jr.: The moisture and rate-dependent mechanical properties of paper: a review. Mech Time-Depend Mater 2000, 4: 169–210. 10.1023/A:1009833415827CrossRef
30. Heyden S: Network modelling for the evaluation of mechanical properties of cellulose fluff. PhD thesis. Lund University, Department of Mechanics and Materials; 2000. Heyden S: Network modelling for the evaluation of mechanical properties of cellulose fluff. PhD thesis. Lund University, Department of Mechanics and Materials; 2000.
31. Kulachenko A, Uesaka T: Direct simulations of fiber network deformation and failure. Mech Mater 2012, 51: 1–14.CrossRef
32. Lavrykov S, Lindström SB, Singh KM, Ramarao BV: 3D network simulations of paper structure. Nord Pulp Pap Res J 2012, 27: 256–263. 10.3183/NPPRJ-2012-27-02-p256-263CrossRef
33. Wilbrink DV, Beex LAA, Peerlings RHJ: A discrete network model for bond failure and frictional sliding in fibrous materials. Int J Solids Struct 2013, 50: 1354–1363. 10.1016/j.ijsolstr.2013.01.012CrossRef
34. Gilbert EG, Johnson DW, Keerthi SS: A fast procedure for computing the distance between complex objects in three-dimensional space. IEEE J Robot Autom 1988,4: 193–203. 10.1109/56.2083CrossRef
35. Mukundan R: Advanced Methods in Computer Graphics: With Examples in OpenGL. Springer, London; 2012. CrossRef
36. Steenberg B: Paper as a visco-elastic body. Sven Papperstidn 1947, 50: 127–140.
37. Coffin DW: The creep response of paper. 13th Fundamental Research Symposium2005, 651–747.
38. Xia QS, Boyce MC, Parks DM: A constitutive model for the anisotropic elastic-plastic deformation of paper and paperboard. Int J Solids Struct 2002, 39: 4053–4071. 10.1016/S0020-7683(02)00238-XMATHCrossRef
39. Mäkelä P, Östlund S: Orthotropic elastic-plastic material model for paper materials. Int J Solids Struct 2003, 40: 5599–5620. 10.1016/S0020-7683(03)00318-4MATHCrossRef
40. Harryson A, Ristinmaa M: Large strain elasto-plastic model of paper and corrugated board. Int J Solids Struct 2008, 45: 3334–3352. 10.1016/j.ijsolstr.2008.01.031CrossRef
41. Uesaka T, Kodaja I, Okushima S, Fukuchi R: History-dependent dimensional stability of paper. Rheol Acta 1989, 28: 238–245. 10.1007/BF01332856CrossRef
42. Lif J: In-plane hygro-viscoelasticity of paper at small deformations. Nord Pulp Pap Res J 2005, 20: 139–149. 10.3183/NPPRJ-2005-20-02-p139-149CrossRef
43. Lif J: Hygro-viscoelastic stress analysis in paper web offset printing. Finite Elem Anal Des 2006, 42: 341–366. 10.1016/j.finel.2005.08.002CrossRef
44. Wikström M: Finite element modelling of calendering - some aspects of the effects of temperature gradients and structure inhomogeneities. J Mater Sci1996, 31: 3159–3166. 10.1007/BF00354662CrossRef
45. Ärölä K, von Hertzen R: An elastoplastic continuum model for a wound roll with interlayer slippage. Finite Elem Anal Des 2006, 42: 503–517. 10.1016/j.

finel.2005.10.005CrossRef

46. Kariniemi M, Parola M, Kulachenko A, Sorvari J, von Hertzen L: Effect of blanket properties on web tension in offset printing. Advances in Printing and Media Technology 2010.
47. Lipponen P, Erkkilä A-L, Leppänen T, Hämäläinen J: On the importance of in-plane shrinkage and through-thickness moisture gradient during drying on cockling and curling phenomena. 14th Pulp and Paper Fundamental Research Symposium2009, 389–436.
48. Erkkilä A-L, Leppänen T, Hämäläinen J: Empirical plasticity models applied for paper sheets having different anisotropy and dry solids content levels. Int J Solids Struct 2013, 50: 2151–2179. 10.1016/j.ijsolstr.2013.03.004CrossRef
49. Hashin Z, Rotem A: A fatigue criterion for fiber-reinforced materials. J Compos Mater 1973, 7: 448–464. 10.1177/002199837300700404CrossRef
50. Hashin Z: Failure criteria for unidirectional fiber composites. J Appl Mech 1980,47: 329–334. 10.1115/1.3153664

SECTION II: COMPUTATIONAL MANY-BODY PROBLEM, CONDENSED MATTER AND MOLECULAR DYNAMICS

CHAPTER 9

MACROSCOPIC TRANSPORT EQUATIONS IN MANY-BODY SYSTEMS

Marta Galanti, Duccio Fanelli and Francesco Piazza

Keywords: Transport equations, nonlinear diffusion, macromolecular crowding, non-ideal fluids, stochastic processes

INTRODUCTION

Diffusive transport is central in many areas of physics, chemistry, biology, and soft matter [1–4]. However, while the mathematics of diffusive processes in dilute and simple media is fairly well developed and understood [1], many interesting and relevant diffusive processes take place in strongly non-ideal conditions. These include a wealth of different highly dense media, from non-ideal plasmas [5] to biological membranes [6], media with complex topological structures, including porous media [7–9] and living cells [10, 11] and strongly confining environments [3,12–16].

Crowding and confinement effects on diffusion-influenced phenomena still pose fundamental yet unanswered questions. Several computational [17] and experimental [18, 19] evidences exist of anomalous diffusion under dense crowded conditions [20]. Anomalous diffusion refers to phenomena that lead to a nonlinear growth of particles' mean square displacement. Deviation from the standard linear behavior arise, for example, when obstacles impede the motion of particles [17,21, 22] or when distinguishable species compete for the available spatial resources [10, 23–27]. These conditions are certainly met when studying the molecular mobility inside a cell [28, 29]. Cells are occupied for over 30% of their volume by membrane delimited organelles and different sorts of cytoskeletal structures. In this respect, living cells behave much like fractal or otherwise disordered systems [30, 31]. However, strong evidences also exist in favor of normal (Brownian) diffusion, crowding, and confinement resulting in this scenario in (often nontrivial) modifications of the diffusion coefficient [10, 27, 32, 33]. Moreover, as shown in Galanti et al. [34], the effect of crowding can result in crossovers between normal and anomalous diffusion, leading to different descriptive scenarios which appear to depend on the selected initial conditions and on the specific time scale of observation.

Another related issue is that of diffusion-limited reactions [35], which are ubiquitous in many domains in biology and chemistry, touching upon problems such as association, folding and stability of proteins [13, 36] and bimolecular reactions in solution [37–41], including enzyme kinetics [42], but also the dynamics of *active* agents [43, 44]. Many theoretical studies have tackled these and related problems under different angles [13, 39, 42, 44–47]. Nevertheless, a full theoretical comprehension of transport in non-ideal media remains an elusive task, Fick›s law itself and the very notion of effective diffusion coefficient being questionable in a disordered medium [31].

In this paper we review the subtle link between macroscopic transport equations, such as the diffusion equation, and microscopic processes, modeling the stochastic dynamics of some agents. For a classical and comprehensive account on diffusion in disordered media the reader can refer to Haus and Kehr [48] and references therein [49–56].

The purpose of our study is to contribute to the debate with a two-fold approach. One the one hand, we wish to understand in greater depth the delicate procedure of obtaining mean-field transport equations from microscopic, agent-based stochastic models. The idea is that sometimes it may prove simpler or more effective to describe a complex transport process (or a simple one occurring in a complex medium) at the microscopic level. On the contrary, it is sometimes better to deal with macroscopic equations. It is thus important to investigate how the two levels of description interface with each other. Furthermore, we will elaborate on the reasons why considerable information can be eventually lost when passing from the microscopic stochastic process to the macroscopic mean-field description, and draw attention on the distinct

Citation: Galanti M, Fanelli D and Piazza F (2016) Macroscopic Transport Equations in Many-Body Systems, doi: 10.3389/fphy.2016.00033.

notions of *point-like* and*extended* crowding, this latter bearing potentially interesting applications, only partially explored in the relevant literature.

The paper is organized as follows. In Section 2 we discuss the general framework of simple exclusion processes (SEPs), which constitute the basic tool of the microscopic description, as well as the process of obtaining mean-field equations from SEPs in the continuum limit. We show that SEPs in inhomogeneous media can be constructed in apparently equivalent manners as for the site-dependence of the hopping rates. However, different advection-diffusion equations can ensue in the continuum limit, rendering this operation a rather subtle one in disordered systems [57]. Moreover, we show that the effect of crowding, enforced in the microscopic description through an excluded volume effect, is recognizable in the continuum limit only if some degree of inhomogeneity is introduced. In Section 3, we move a step forward and consider microscopic exclusion processes involving agents characterized by a *finite* size, as opposed to standard SEPs. In the last section we draw the conclusions and we summarize the different extent to which the crowding fine-tunes deviation from the classical picture.

FROM MICROSCOPIC PROCESSES TO MACROSCOPIC EQUATIONS

SEPs are space-discrete, agent-based stochastic processes modeling some kind of transport according to specific rules and bound to the constraint that no two agents can ever occupy the same site. SEPs occupy a central role in non-equilibrium statistical physics [58, 59]. While the first theoretical ideas underlying such processes can be traced back to Boltzmann's works [60], SEPs were introduced and widely studied in the 70s as simplified models of one-dimensional transport for phenomena like hopping conductivity [61] and kinetics of biopolymerization [5]. Along the same lines, the asymmetric exclusion process (ASEP), originally introduced by Spitzer [62], has become a paradigm in non-equilibrium statistical physics [63–66] and has now found many applications, such as the study of molecular motors [67], transport through nano-channels [68] and depolymerization of microtubules [69].

The most general SEP in one dimension is described by a stochastic jump process on a $1D$ lattice with inequivalent sites in the presence of a field

$$\begin{aligned} n_i(k+1) - n_i(k) &= z_{i-1}^{+} n_{i-1}(k)[1 - n_i(k)] \\ &+ z_{i+1}^{-} n_{i+1}(k)[1 - n_i(k)] \\ &- z_i^{+} n_i(k)[1 - n_{i+1}(k)] \\ &- z_i^{-} n_i(k)[1 - n_{i-1}(k)] \end{aligned} \tag{1}$$

Equation (1) is to be regarded as the update rule for a Monte Carlo process, where $n_i(k)$ is the occupancy of site i at time $t = k\Delta t$, which can be either zero or one. The quantities $z_i^{\pm}$ are variables which have the value 0 or 1 according to a random number ξ_i which has a uniform distribution between 0 and 1. By defining the jump probabilities $q_j^{\pm}$ $(j = i, i \pm 1)$ one can formally write:

$$\begin{aligned} z_{i-1}^{+} &= \theta(\xi_i) - \theta(\xi_i - q_{i-1}^{+}) \\ z_{i+1}^{-} &= \theta(\xi_i - q_{i-1}^{+}) - \theta(\xi_i - q_{i-1}^{+} - q_{i+1}^{-}) \\ z_i^{+} &= \theta(\xi_i - q_{i-1}^{+} - q_{i+1}^{-}) - \theta(\xi_i - q_{i-1}^{+} - q_{i+1}^{-} - q_i^{+}) \\ z_i^{-} &= \theta(\xi_i - q_{i-1}^{+} - q_{i+1}^{-} - q_i^{+}) - \theta(\xi_i - 1) \end{aligned} \tag{2}$$

where $\theta(\cdot)$ stands for the Heaviside step function and where we are assuming that $q_{i-1}^{+} + q_{i+1}^{-} + q_i^{+} + q_i^{-} = 1$. Note that the ordering of appearance of the $q_i^{\pm}$ in the above expressions is arbitrary. Equations (2) entail that $\langle z_j^{\pm} \rangle = q_j^{\pm}$, where $\langle\cdot\rangle$ indicates an average over many values of ξ_i, for a given configuration $\{n_i\}$. The above process is fully determined by the fields $q_i^{\pm}$, specifying the probability of jumping from site i to site $i+1\,(q_i^{+})$ or to site $i-1\,(q_i^{-})$ in a time interval Δt.

A (discrete-time) master equation for the above SEP can be obtained by averaging over many Monte Carlo cycles performed according to rule (1)

$$\begin{aligned} P_i(k+1) - P_i(k) &= q_{i-1}^{+}[P_{i-1}(k) - P_{i,i-1}(k)] \\ &+ q_{i+1}^{-}[P_{i+1}(k) - P_{i,i+1}(k)] \\ &- q_i^{+}[P_i(k) - P_{i,i+1}(k)] \\ &- q_i^{-}[P_i(k) - P_{i,i-1}(k)] \end{aligned} \tag{3}$$

where we have defined the one-body and two-body site occupancy probabilities

$$P_i(k) = \langle\langle n_i(k) \rangle\rangle \tag{4}$$

$$P_{i,i\pm1}(k) = \langle\langle n_i(k) n_{i\pm1}(k) \rangle\rangle \tag{5}$$

Here $\langle\langle\cdot\rangle\rangle$ denotes averages over many independent Monte Carlo cycles performed until time $k\Delta t$, starting from the same initial condition. We emphasize that the same equation has been derived through a slightly different procedure by Richards in 1977 [61].

Mean-Field Equations

With the aim of deriving macroscopic transport equations from the microscopic stochastic process described by Equation (1), it is customary to assume a mean-field (MF) factorization,

$$\begin{aligned} P_{i,i\pm1}(k) &\equiv \langle\langle n_i(k) n_{i\pm1}(k) \rangle\rangle = \langle\langle n_i(k) \rangle\rangle \langle\langle n_{i\pm1}(k) \rangle\rangle \\ &= P_i(k) P_{i\pm1}(k) \end{aligned} \tag{6}$$

With the help of Equation(6), the master Equation (3) becomes

$$\begin{aligned} P_i(k+1) - P_i(k) &= q_{i-1}^{+} P_{i-1}(k)[1 - P_i(k)] \\ &+ q_{i+1}^{-} P_{i+1}(k)[1 - P_i(k)] \\ &- q_i^{+} P_i(k)[1 - P_{i+1}(k)] \\ &- q_i^{-} P_i(k)[1 - P_{i-1}(k)] \end{aligned} \tag{7}$$

Nonlinear mean-field equations for exclusion process of this type have been used since the 70s to investigate one-dimensional transport in solids [70]. In fact, despite mean-field descriptions for the inhomogeneous ASEP are known to provide imperfect descriptions of certain non-equilibrium observables in one dimension, e.g., the current-density relation and critical exponents [71], continuum descriptions can be employed reliably to track the time-evolution of large-wavelength density fluctuations [34, 72–78]

Site-Dependent Rates: Three Jump Processes Yield Three Transport Equations

Let a be the lattice spacing and let us define a *reversal* probability ϵ_i, such that

$$q_i^+ = Q_i \qquad q_i^- = Q_i - \epsilon_i \tag{8}$$

The condition (Equation 8) (with $\epsilon_i > 0$) amounts to considering a field introducing a bias in the positive x direction. In order to take the continuum limit $\lim_{a,\Delta t \to 0} P_i(k) = P(x,t)$, we must require

$$\lim_{a,\Delta t \to 0} \frac{Q_i a^2}{\Delta t} = D(x) \tag{9}$$

$$\lim_{a,\Delta t \to 0} \frac{\epsilon_i a}{\Delta t} = v(x) \tag{10}$$

Equation (9) defines the position-dependent diffusion coefficient, while Equation (10) defines the field-induced drift velocity. Note that we are assuming that the reversal probability vanishes linearly with a.

A pondering pause is required at this point before carrying out the continuum limit. In fact, a moment's thought is enough to realize that there are different ways one can enforce quenched disorder, corresponding to spatially varying hopping rates, in a jump process with the aim of modeling propagation in a inhomogenous medium. The master Equation (7) reflects only one of the possible choices. This observation has been made and thoroughly discussed in Painter and Sherratt [57]. In this paper the authors investigate the movements of cells capable of sensing strategies which depend on environmental factors.

To illustrate this interesting mathematical property, let us consider the jump from site i to site $i+1$. The probability of an agent taking an $i \rightarrow i+1$ leap can be equally well taken as (1) proportional to q_i^+, (2) proportional to q_{i+1}^+ or (3) proportional to $(q_i^+ + q_{i+1}^+)/2$. All three cases correspond to the same space-dependent function $D(x)$ in the continuum limit, as prescribed by Equation (9). However, as we shall see in the following, depending on whether rule (1), (2), or (3) is chosen, one is led to totally different advection-diffusion equations in the continuum limit.

No-Exclusion Processes

In order to illustrate this subtle point, let us start with jump processes in the presence of quenched disorder but with no exclusion constraints on the allowed moves. Following the same reasoning that led us to Equation (7), it is not difficult to realize that the three possible choices (1), (2), and (3) referred above lead to the following master equations

$$\begin{aligned} P_i(k+1) - P_i(k) &= q_{i+1}^- P_{i+1}(k) + q_{i-1}^+ P_{i-1}(k) \\ &\quad - (q_i^+ + q_i^-) P_i(k) \end{aligned} \tag{11}$$

$$\begin{aligned} P_i(k+1) - P_i(k) &= q_i^- P_{i+1}(k) + q_i^+ P_{i-1}(k) \\ &\quad - (q_{i+1}^+ + q_{i-1}^-) P_i(k) \end{aligned} \tag{12}$$

$$\begin{aligned} P_i(k+1) - P_i(k) &= \left(\frac{q_{i+1}^- + q_i^-}{2}\right) P_{i+1}(k) + \left(\frac{q_{i-1}^+ + q_i^+}{2}\right) P_{i-1}(k) \\ &\quad - P_i(k)\left[\left(\frac{q_{i+1}^+ + q_i^+}{2}\right) + \left(\frac{q_{i-1}^- + q_i^-}{2}\right)\right] \end{aligned} \tag{13}$$

In the first case, Equation (11) (already obtained above), the rate for a given jump depends on the starting site, while the second, Equation (12), is the opposite, the rate depends on the index of the target site. The third case, Equation (13) is an intermediate, symmetric situation where jump rates are associated with *links* rather than with *nodes*. In the continuum limit $a, \Delta t \rightarrow 0$ one has

$$Q_{i\pm 1} \approx Q(x) \pm \frac{\partial Q}{\partial x} a + \frac{1}{2}\frac{\partial^2 Q}{\partial x^2} a^2 + \ldots \tag{14}$$

$$\epsilon_{i\pm 1} \approx \epsilon(x) \pm \frac{\partial \epsilon}{\partial x} a + \ldots \tag{15}$$

$$P_{i\pm 1}(k) \approx P(x,t) \pm \frac{\partial P}{\partial x} a + \frac{1}{2}\frac{\partial^2 P}{\partial x^2} a^2 + \ldots \tag{16}$$

Therefore, recalling Equations (9) and (10), we have from Equations (11), (12) and (13), respectively,

$$\frac{\partial P(x,t)}{\partial t} = -\frac{\partial J_1(x,t)}{\partial x} \tag{17}$$

$$\frac{\partial P(x,t)}{\partial t} = -\frac{\partial J_2(x,t)}{\partial x} \tag{18}$$

$$\frac{\partial P(x,t)}{\partial t} = -\frac{\partial J(x,t)}{\partial x} \tag{19}$$

where

$$J_1(x,t) = -D(x)\frac{\partial P(x,t)}{\partial x} + \left[v(x) - \frac{\partial D(x)}{\partial x}\right] P(x,t) \tag{20}$$

$$J_2(x,t) = -D(x)\frac{\partial P(x,t)}{\partial x} + \left[v(x) + \frac{\partial D(x)}{\partial x}\right] P(x,t) \tag{21}$$

$$J(x,t) = -D(x)\frac{\partial P(x,t)}{\partial x} + v(x)P(x,t) \tag{22}$$

We see that the stochastic processes (11) and (12) correspond to diffusion with drift (or, equivalently, in an external potential). The drift velocity comprises two contributions: the difference between the right-bound and left-bound jump rate fields and a contribution arising from the spatial variation of the diffusion coefficient. Interestingly, the latter term has the opposite sign depending on whether jumps at the microscopic level are controlled by the rates evaluated at the start or target sites. This means that in these cases, even a symmetric jump process results in diffusion with drift. Conversely, considering symmetrized jump rates does not result in the appearance of such additional term in the drift velocity. Working in this setting one hence recover the standard diffusion equation.

Enforcing the Excluded-Volume Constraint

We can now come back to our original aim, i.e., taking the continuum limit of the master Equation (1). It is now clear that, if we want to consider an inhomogeneous medium, we must not restrict to the prescription leading to Equation (1), but we must also consider the other two kinds of processes described above in the absence of exclusion. Extending the reasoning that led us to Equation (1) and recalling Equations (12) and (13), the three master equations with excluded-volume constraints read

$$P_i(k+1) - P_i(k) = [q_{i+1}^- P_{i+1}(k) + q_{i-1}^+ P_{i-1}(k)][1 - P_i(k)] - P_i(k)\{q_i^+[1 - P_{i+1}(k)] + q_i^-[1 - P_{i-1}(k)]\} \quad (23)$$

$$P_i(k+1) - P_i(k) = [q_i^- P_{i+1}(k) + q_i^+ P_{i-1}(k)][1 - P_i(k)] - P_i(k)\{q_{i+1}^+[1 - P_{i+1}(k)] + q_{i-1}^-[1 - P_{i-1}(k)]\} \quad (24)$$

$$\begin{aligned} P_i(k+1) - P_i(k) = {} & \left(\frac{q_{i+1}^- + q_i^-}{2}\right) P_{i+1}(k)[1 - P_i(k)] \\ & + \left(\frac{q_{i-1}^+ + q_i^+}{2}\right) P_{i-1}(k)[1 - P_i(k)] \\ & - \left(\frac{q_{i+1}^+ + q_i^+}{2}\right) P_i(k)[1 - P_{i+1}(k)] \\ & - \left(\frac{q_{i-1}^- + q_i^-}{2}\right) P_i(k)[1 - P_{i-1}(k)] \end{aligned} \quad (25)$$

The mean field limit $a, \Delta t \to 0$ of the above master equations is readily obtained by introducing as above the Taylor expansions of $q(x)$ and $P(x, t)$. By doing this, we find

$$\frac{\partial P(x,t)}{\partial t} = -\frac{\partial \mathcal{J}_1(x,t)}{\partial x} \quad (26)$$

$$\frac{\partial P(x,t)}{\partial t} = -\frac{\partial \mathcal{J}_2(x,t)}{\partial x} \quad (27)$$

$$\frac{\partial P(x,t)}{\partial t} = -\frac{\partial \mathcal{J}_3(x,t)}{\partial x} \quad (28)$$

where

$$\mathcal{J}_1(x,t) = -[1 - P(x,t)]\frac{\partial}{\partial x}[D(x)P(x,t)] - D(x)P(x,t)\frac{\partial P(x,t)}{\partial x} + v(x)P(x,t)[1 - P(x,t)] \quad (29)$$

$$\begin{aligned} \mathcal{J}_2(x,t) = {} & P(x,t)\frac{\partial}{\partial x}\{D(x)[1 - P(x,t)]\} \\ & - D(x)[1 - P(x,t)]\frac{\partial P(x,t)}{\partial x} \\ & + v(x)P(x,t)[1 - P(x,t)] \end{aligned} \quad (30)$$

$$\mathcal{J}_3(x,t) = -D(x)\frac{\partial P(x,t)}{\partial x} + v(x)P(x,t)[1 - P(x,t)] \quad (31)$$

We see that even in the case of excluded-volume interactions, the mean-field equations can be cast in the form of continuity equations with suitably defined currents given by Equations (29), (30) and (31). We note that, as it happens with the symmetric exclusion process without quenched disorder, the microscopic exclusion constraint disappears in taking the continuum limit of the master Equation (25) in the absence of external fields, which yields a transport equation, Equation (28), identical to its counterpart with no exclusion, Equation (19) in the case $v(x) \equiv 0$.

Equations (26), (27) and (28) are nonlinear advection-diffusion equation, appropriate for describing the continuum limit of a microscopic exclusion process occurring on a lattice of inequivalent sites in the presence of a field. It is interesting to note that in the case of equivalent sites, which translates to a constant diffusion coefficient, the diffusive parts become linear, i.e., the microscopic exclusion rule is *lost* in the diffusive part. In the case of zero field, one then simply recovers the ordinary diffusion equation for the three jump processes, which, as it is widely known, can be derived from a microscopic jump process with no exclusion rules. This curious observation has been first reported by Huber [70]. If both the diffusion coefficient and the drift velocity are constant, Equations (26), (27) and (28) all reduce to

$$\frac{\partial P}{\partial t} = D\nabla^2 P - v\frac{\partial}{\partial x}[P(1-P)] \quad (32)$$

an equation already obtained recently in Simpson et al. [78].

Equations (26), (27) and (28) contain the single-particle probability field $P(x, t)$, which is a number between zero and one. The value $P = 1$ should correspond to the maximum density ρ_M allowed in the system. Thus, more *physical* equations can be obtained by introducing the agent density

$$\rho(x,t) \equiv \rho_M P(x,t) = \frac{\phi_M}{v_1(\sigma/2)} P(x,t) \quad (33)$$

where

$$v_1(r) = \frac{(\pi^{1/2} r)^d}{\Gamma(1 + d/2)} \quad (34)$$

is the volume of a d-dimensional sphere[1] of radius r and ϕ_M is the maximum packing fraction for systems of d-dimensional hard spheres, $\phi_M = 1$ $(d = 1)$, $\phi_M = \pi/\sqrt{12} \approx 0.907$ $(d = 2)$ and $\phi_M = \pi/\sqrt{18} \approx 0.740$ $(d = 3)$ [79]. With these definitions, and using a more general vector notation, Equations (26), (27) and (28) become

$$\frac{\partial \rho(x,t)}{\partial t} = -\nabla \cdot \mathcal{J}_1(x,t) \quad (35)$$

$$\frac{\partial \rho(x,t)}{\partial t} = -\nabla \cdot \mathcal{J}_2(x,t) \quad (36)$$

$$\frac{\partial \rho(x,t)}{\partial t} = -\nabla \cdot \mathcal{J}_3(x,t) \quad (37)$$

with

$$\mathcal{J}_1(x,t) = -\left(1 - \frac{\rho}{\rho_M}\right)\nabla[D(x)\rho(x,t)] - D(x)\left(\frac{\rho}{\rho_M}\right)\nabla\rho(x,t)$$

$$+ v(x)\rho(x,t)\left(1-\frac{\rho}{\rho_M}\right) \tag{38}$$

$$\mathcal{J}_2(x,t) = \rho(x,t)\nabla\left[D(x)\left(1-\frac{\rho}{\rho_M}\right)\right] - D(x)\left(1-\frac{\rho}{\rho_M}\right)\nabla\rho(x,t) + v(x)\rho(x,t)\left(1-\frac{\rho}{\rho_M}\right) \tag{39}$$

$$\mathcal{J}_3(x,t) = -D(x)\nabla\rho(x,t) + v(x)\rho(x,t)\left(1-\frac{\rho}{\rho_M}\right). \tag{40}$$

EXTENDED CROWDING

In the previous section we described how to derive macroscopic mean-field equations starting from microscopic master equations which account for exclusion effects. In our description we did not take into account size and shape of the agents by prescribing that each molecule occupies a single lattice site and can not move if the target site is already occupied by another particle. To recover a macroscopic continuous description, we performed the limit for vanishing lattice spacing. This strategy amounts to considering agents of vanishing size in the continuum limit. We term this peculiar situation in the macroscopic world *point-like crowding*. As we have shown in the previous section, considerable microscopic information is lost in the continuum limit with point-like agents. The point-like characterization has to be the reason why the mean-field approximation loses the memory of the microscopic exclusion constraint and the diffusion equation is recovered for equivalent sites in the absence of a field. Starting from these premises, we set to work in the*extended crowding* framework, where the finite size of the particles is explicitly accounted for. Operating within this scenario, the exclusion constraint shall be detectable in the mean-field limit even for unbiased motion and homogeneous domain. A reason for employing the *extended crowding* philosophy is that in many biological contexts one has to model the interaction between agents displaying complicated shapes, which are not well represented by spherical particle occupying a single lattice site. Biological macromolecules diffusing in the cytoplasm, or even proliferating cells, for example, are often elongated and rod-shaped, a property that inspired the derivation of microscopic models using *hard rods* as the individual units [80, 81]. This is for example the case of the human peritoneal mesothelial cells modeled in Simpson et al. [80] as hard rods of aspect ratio equal to four. Moreover, when modeling the interaction between multiple species diffusing in the same environment, the differences in shape and dimension of the agents do have an influence on the extent of the collective motion.

The first model for the diffusion of extended particles with exclusion interactions on a one-dimensional lattice was described in Schönherr and Schütz [82] for a general process involving symmetric, as well as asymmetric, hopping dynamics of the rods (the theory is named L-ASEP). Referring to rods of length L, where L has to be interpreted as the aspect ratio between the dimensions of the elongated agent, the authors derive a mean-field equation for the one-dimensional exclusion process. In the absence of a field and for equivalent sites the equation for the density of particles reads

$$\frac{\partial\rho}{\partial t} = D_0\frac{\partial^2}{\partial x^2}\left[\frac{\rho}{1-(L-1)\rho}\right]. \tag{41}$$

The previous equation shows that, at least in one dimension, the extended-crowding procedure yields a modified diffusion term even in the absence of external fields or spatial inhomogeneities. By defining a density-dependent diffusion coefficient, which has to be regarded as a collective diffusivity $D(\rho)=D_0/[(1-(L-1)\rho]^2$, eq. (41) can be reformulated as a nonlinear transport equation:

$$\frac{\partial\rho}{\partial t} = \frac{\partial}{\partial x}\left[D(\rho)\frac{\partial\rho}{\partial x}\right]. \tag{42}$$

The nonlinear diffusion Equation (41) has been derived through an ingenious but complicate change of variables based on a quantitative mapping between the L-ASEP and the zero-range process [82]. However, it is interesting to note that it can be regarded as the local-density approximation (LDA) of a simple general property of one-dimensional exclusion processes. As pointed out in 1967 by Lebowitz and Percus [83] concerning bulk properties:

"*For many purposes, however, adding a finite diameter does not introduce any new complications; it merely requires the replacement in certain expressions of the actual volume per particle* ρ^{-1} *by the reduced volume* $\rho^{-1}-\sigma$, *i.e.*, $\rho\rightarrow\rho/(1-\sigma\rho)$."

In that sense, the quantity $\rho/[1-(L-1)\rho]$ is recognized as an effective density in Ferreira and Alcaraz [84] within the analysis of the velocity of finite sized particles which occupy L units of lattice spacing in one dimension. By performing the substitution

$$\rho(x,t) \rightarrow \frac{\rho(x,t)}{[1-(L-1)\rho(x,t)]}$$

in Fick's law, one recovers (41). Point-like crowding in the mean field approximation corresponds to systems of fully penetrable spheres, while extended-size crowding yields a transport equation suitable for systems of totally impenetrable (hard) objects. We stress that the case of point-like crowding is recovered for agents of aspect ratio $L = 1$. The discussion above which leads to Equation (41) applies to one dimensional systems. Starting from this setting, one can raise the question whether similar arguments might be employed to obtain a modified nonlinear equation accounting for excluded volume effects in the diffusion of hard spheres in two and three dimensions. Unfortunately, the strategy used to recover Equation (41) in Schönherr and Schütz [82] can not be employed to provide a description of the extended crowding in higher dimensions. Several other models have been proposed for two and three dimensional domains [76, 80, 85], starting from stochastic processes enforced with different microscopic rules. Depending on the shape of the agents (hard rods or hard spheres), on the prescribed hopping rules, on the allowed mechanisms for changing the orientation of the agents (rotation or reptation), and on the mean field assumptions made to recover the macroscopic picture, different equations are derived for the density of extended particles in higher dimensions. All models yield however the same qualitative behavior at low densities, suggesting that the diffusion coefficient should increase linearly with the concentration amount.

SUMMARY AND DISCUSSION

Crowding and confinement certainly affect diffusion-driven phenomena, and potentially impact a large plethora of distinct applications. Particularly relevant is the study of mobility inside the cell: the complex cellular environment is in fact populated by obstacles and compartments, which impede the particles flowing. Moreover, the cytoplasm is quite densely packed, different molecular species competing for the available spatial resources. Given these premises, it is important to elaborate on plausible theoretical pictures that enable one to adequately capture the constraints imposed by many-body interactions or reflecting the specificity of the embedding, confining environment.

In this paper, we have reviewed a general framework allowing to obtain macroscopic transport equation accounting for excluded volume effects starting from microscopic stochastic exclusion processes. The aim of this procedure is to derive mean-field equations suitable for describing transport processes in many-body systems in highly non-ideal conditions. Two strategies can be identified to carry out this task. The first route, termed *point-like* crowding, leads from standard SEPs to the mean-field Equations (35), (36) and (37). For a homogenous medium in the absence of a field these reduce to a simple diffusion equation, which is why this scenario can be dubbed *point-like* crowding. Only for inequivalent sites and/or in the presence of a field the microscopic exclusion constraint does survive in the mean-field limit. Different is the case of multiple species [86] or recognizable agents [34], where cross-terms appear in the mean field limit due to the fact that particles are made distinguishable. In general, we can state that crowding reflects in the macroscopic limit of the point-like description only if some degree of inhomogeneity is enforced at the microscopic level. Moreover, we have highlighted that in the presence of quenched disorder, i.e., inequivalent sites, there are (at least) three different ways to assign the site dependence of the hopping rates. These correspond to different microscopic stochastic processes and lead to as many different nonlinear advection-diffusion equations in the continuum limit.

The second strategy, named *extended-crowding*, takes inspiration from a modified microscopic exclusion process in one dimension involving extended agents, the so-called L-ASEP [82]. Qualitatively similar equations for the continuous density are found in higher dimensions at moderate concentrations for different shapes of the individual units. Remarkably, for extended objects characterized by a finite size, the blueprint of the crowding persists at the macroscopic level, even for systems defined on homogeneous domains and without external potentials. This observation marks a clear distinction between the two approaches to which we alluded above and points to the potential interest of the extended crowding setting, which, to the best of our knowledge, remains to be largely explored.

CONFLICT OF INTEREST STATEMENT

The authors declare that the research was conducted in the absence of any commercial or financial relationships that could be construed as a potential conflict of interest.

REFERENCES

1. Crank J. *The Mathematics of Diffusion*. New York, NY: Oxford University Press (1975).
2. Dhont J. *An Introduction to Dynamics of Colloids*. Amsterdam: Elsevier (1996).
3. McGuffee SR, Elcock AH. Diffusion, crowding & protein stability in a dynamic molecular model of the bacterial cytoplasm.*PLoS Comput Biol.* (2010) 6:e1000694. doi: 10.1371/journal.pcbi.1000694
4. Caspi A, Granek R, Elbaum M. Diffusion and directed motion in cellular transport. *Phys Rev E* (2002) 66:011916. doi: 10.1103/PhysRevE.66.011916
5. Kremp D, Schlanges M, Bonitz M, Bornath T. Reaction and diffusion in dense nonideal plasmas. *Phys Fluids B Plasma Phys.*(1993) 5:216–29.
6. Saffman PG, Delbrück M. Brownian motion in biological membranes. *Proc Natl Acad Sci USA.* (1975) 72:3111–3.
7. Talmon Y, Shtirberg L, Harneit W, Rogozhnikova OY, Tormyshev V, Blank A. Molecular diffusion in porous media by PGSE ESR. *Phys Chem Chem Phys.* (2010) 12:5998–6007. doi: 10.1039/b922060g
8. Mitra PP, Sen PN, Schwartz LM, Le Doussal P. Diffusion propagator as a probe of the structure of porous media. *Phys Rev Lett.* (1992) 68:3555–8.
9. Whitaker S. Diffusion and dispersion in porous media. *AIChE J.* (1967) 13:420–7.
10. Dix JA, Verkman AS. Crowding effects on diffusion in solutions and cells. *Annu Rev Biophys.* (2008) 37:247–63. doi: 10.1146/annurev.biophys.37.032807.125824
11. Phair RD, Misteli T. High mobility of proteins in the mammalian cell nucleus. *Nature* (2000) 404:604–9. doi: 10.1038/35007077
12. Condamin S, Benichou O, Tejedor V, Voituriez R, Klafter J. First-passage times in complex scale-invariant media. *Nature*(2007) 450:77–80. doi: 10.1038/nature06201
13. Zhou H-X. Protein folding and binding in confined spaces and in crowded solutions. *J Mol Recognit.* (2004) 17:368–75. doi: 10.1002/jmr.711
14. Minton AP. Implications of macromolecular crowding for protein assembly. *Curr Opin Struct Biol.* (2000) 10:34–9. doi: 10.1016/S0959-440X(99)00045-7
15. Zimmerman SB, Minton AP. Macromolecular crowding: biochemical, biophysical, and physiological consequences. *Annu Rev Biophys Biomol Struct.* (1993) 22:27–65.
16. Ando T, Skolnick J. Crowding and hydrodynamic interactions likely dominate *in vivo* macromolecular motion. *Proc Natl Acad Sci USA.* (2010) 107:18457–62. doi: 10.1073/pnas.1011354107
17. Ghosh SK, Cherstvy AG, Metzler R. Non-universal tracer diffusion in crowded media of non-inert obstacles. *Phys Chem Chem Phys.* (2015) 17.3:1847–58. doi: 10.1039/c4cp03599b
18. Pastor I, Vilaseca E, Madurga S, Garcés JL, Cascante M, Mas F. Diffusion of α-chymotrypsin in solution-crowded media. A fluorescence recovery after photobleaching

study *J Phys Chem B* (2010) 114.11:4028–34. doi: 10.1021/jp910811j

19. Axelrod D, Koppel DE, Schlessinger J, Elson E, Webb WW. Mobility measurement by analysis of fluorescence photobleaching recovery kinetics. *Biophys J.* (1976) 16.9:1055.
20. Metzler R, Jeon JH, Cherstvy A, Barkai E. Anomalous diffusion models and their properties: non-stationarity, non-ergodicity, and ageing at the centenary of single particle tracking newblock. *Phys Chem Chem Phys.* (2014) 16:24128–64. doi: 10.1039/C4CP03465A
21. Sekhar Burada P, Hänggi P, Marchesoni F, Schmid G, Talkner P. Diffusion in confined geometries. *ChemPhysChem.* (2009)10.1:45–54. doi: 10.1002/cphc.200800526
22. Stauffer D, Schulze C, Heermann DW. Superdiffusion in a model for diffusion in a molecularly crowded environment. *J Biol Phys.* (2007) 33.4:305–312. doi: 10.1007/s10867-008-9075-2
23. Foffi G, Pastore A, Piazza F, Temussi PA. Macromolecular crowding: chemistry and physics meet biology. *Phys Biol.* (2013)10.4:040301. doi: 10.1088/1478-3975/10/4/040301
24. Rivas G, Ferrone F, Herzfeld J. Life in a crowded world. *Phys Biol.* (2004) 5.1:23–7. doi: 10.1038/sj.embor.7400056
25. Ellis RJ. Macromolecular crowding: obvious but underappreciated. *Trends Biochem Sci.* (2001) 26.10:597–604. doi: 10.1016/S0968-0004(01)01938-7
26. Zhou HX, Rivas G, Minton AP. Macromolecular crowding and confinement: biochemical, biophysical, and potential physiological consequences. *Annu Rev Biophys.* (2008) 37:375. doi: 10.1146/annurev.biophys.37.032807.125817
27. Konopka MC, Shkel IA, Cayley S, Record MT, Weisshaar JC. Crowding and confinement effects on protein diffusion *in vivo.J Bacteriol.* (2006) 188:6115–23. doi: 10.1128/JB.01982-05
28. Balbo J, Mereghetti P, Herten DP, Wade RC. The shape of protein crowders is a major determinant of protein diffusion.*Biophys J.* (2013) 104:1576–84. doi: 10.1016/j.bpj.2013.02.041
29. Weiss M, Elsner M, Kartberg F, Nilsson T. Anomalous subdiffusion is a measure for cytoplasmic crowding in living cells.*Biophys J.* (2004) 87:3518–24. doi: 10.1529/biophysj.104.044263
30. Ben-Avraham D, Havlin S. *Diffusion and Reactions in Fractals and Disordered Systems*. Cambridge, UK: Cambridge University Press (2000).
31. Bouchaud J-P, Georges A. Anomalous diffusion in disordered media: statistical mechanisms, models and physical applications. *Phys Rep.* (1990) 195:127–293.
32. Novak IL, Kraikivski P, Slepchenko BM. Diffusion in cytoplasm: effects of excluded volume due to internal membranes and cytoskeletal structures. *Biophys J.* (2009) 97:758–67. doi: 10.1016/j.bpj.2009.05.036
33. Nicholson C, Phillips JM. Ion diffusion modified by tortuosity and volume fraction in the extracellular microenvironment of the rat cerebellum. *J Physiol.* (1981) 321:225–57.
34. Galanti M, Fanelli D, Maritan A, Piazza F. Diffusion of tagged particles in a crowded medium. *Europhys Lett.* (2014)107:20006. doi: 10.1209/0295-5075/107/20006
35. von Smoluchowski M. Drei vortrage ubër diffusion brownsche molekular bewegung und koagulation von kolloidteichen.*Phys Z* (1916) 17:557–71.
36. Cheung MS, Klimov D, Thirumalai D. Molecular crowding enhances native state stability and refolding rates of globular proteins. *Proc Natl Acad Sci USA.* (2005) 102:4753–8. doi: 10.1073/pnas.0409630102
37. Dzubiella J, McCammon JA. Substrate concentration dependence of the diffusion-controlled steady-state rate constant. *J Chem Phys.* (2005) 122:184902. doi: 10.1063/1.1887165
38. Piazza F, Rios PdeL, Fanelli D, Bongini L, Skoglund U. Anticooperativity in diffusion-controlled reactions with pairs of anisotropic domains: a model for the antigen-antibody encounter. *Eur Biophys J.* (2005) 34:899–911. doi: 10.1007/s00249-005-0460-5
39. Fanelli D, McKane AJ. Diffusion in a crowded environment. *Phys Rev E* (2010) 82:021113. doi: 10.1103/PhysRevE.82.021113
40. Schmit JD, Kamber E, Kondev J. Lattice model of diffusion-limited bimolecular chemical reactions in confined environments. *Phys Rev Lett.* (2009) 102:218302. doi: 10.1103/PhysRevLett.102.218302
41. Tachiya M, Traytak SD. Concentration dependence of fluorescence quenching by ionic reactants. *J Phys Condens Matter*(2007) 19:065111. doi: 10.1088/0953-8984/19/6/065111
42. Agrawal M, Santra SB, Anand R, Swaminathan R. Effect of macromolecular crowding on the rate of diffusion-limited enzymatic reaction. *PRAMANA-J Phys.* (2008) 71:359–68. doi: 10.1007/s12043-008-0169-y
43. Ricci N, Barbanera F, Erra F. A quantitative approach to movement, displacement, and mobility of protozoa. *J Eukaryot Microbiol.* (1998) 45:606–11.
44. Galanti M, Fanelli D, Piazza F. Conformation-controlled binding kinetics of antibodies. *Sci Rep.* (2016) 6:18976. doi: 10.1038/srep18976
45. Cecconi F, De Los Rios P, Piazza F. Diffusion-limited unbinding of small peptides from PDZ domains. *J Phys Chem B* (2007)111:11057–063. doi: 10.1021/jp0730390
46. Dorsaz N, De Michele C, Piazza F, Foffi G. Inertial effects in diffusion-limited reactions. *J Phys* (2010) 22:104116. doi: 10.1088/0953-8984/22/10/104116
47. Felderhof BU, Deutch JM. Concentration dependence of the rate of diffusion-controlled reactions. *J Chem Phys* (1976)64:4551. doi: 10.1063/1.432087
48. Haus JW, Kehr KW. Diffusion in regular and disordered lattices. *Phys Rep.* (1987) 150:263–406.
49. Shlesinger MF, West BJ. *Random Walks and Their Applications in the Physical and Biological Sciences.*

New York, NY: American Institute of Physics (1984).

50. Spitzer F. *Principles of Random Walk*. The University Series in Higher Mathematics. Princeton, NJ; Toronto, ON; London: D. Van Nostrand Co., Inc. (1984).

51. Goel NS, Richter-Dyn N. *Stochastic Models in Biology*. Elsevier (2016).

52. Montroll EW, Weiss GH. Random walks on lattices. *J Math Phys* (1965) 6.2:167–81. doi: 10.1063/1.1704269. Available online at: http://scitation.aip.org/content/aip/journal/jmp/6/2/10.1063/1.1704269

53. Lennard-Jones JE. Processes of adsorption and diffusion on solid surfaces. *Trans Faraday Soc* (1932) 28:333–59.

54. Bernasconi J, Alexander S, Orbach R. Classical diffusion in one-dimensional disordered lattice. *Phys Rev Lett.* (1978)41.3:185.

55. Van Beijeren H. Transport properties of stochastic Lorentz models. *Rev Mod Phys.* (1982) 54.1:195.

56. Termonia Y. Random walk on a strange attractor. *Phys Rev Lett.* (1984) 53.14:1356.

57. Painter KJ, Sherratt JA. Modelling the movement of interacting cell populations. *J Theor Biol.* (2003) 225:327–39. doi: 10.1016/S0022-5193(03)00258-3

58. Privman V, editor. *Nonequilibrium Statistical Mechanics in One Dimension*. Cambridge University Press (1997).

59. Liggett TM. *Stochastic Interacting Systems: Contact, Voter and Exclusion Processes*. Berlin; Heidelberg: Springer-Verlag (1999).

60. Boltzmann L. *Vorlesungen über Gastheorie (Lectures on Gas Theory)*. Vol. I and II. Transl. by S. G. Brush in 1966. Berkeley: University of California Press (1896).

61. Richards PM. Theory of one-dimensional hopping conductivity and diffusion. *Phys Rev B* (1977) 16:1393–409.

62. Spitzer F. Interaction of Markov processes. *Adv Math.* (1970) 5:246–90.

63. Derrida B, Evans MR, Hakim V, Pasquier V. Exact solution of a 1D asymmetric exclusion model using a matrix formulation. *J Phys A Math Gen.* (1993) 26:1493.

64. Schütz G, Domany E. Phase transitions in an exactly soluble one-dimensional exclusion process. *J Stat Phys.* (1993) 72:277–96.

65. Derrida B. An exactly soluble non-equilibrium system: the asymmetric simple exclusion process. *Phys Rep.* (1998) 301:65–83.

66. Harris RJ, Schütz GM. Fluctuation theorems for stochastic dynamics. *J Stat Mech Theory Exp.* (2007) 2007.07:P07020. doi: 10.1088/1742-5468/2007/07/P07020

67. Golubeva N, Imparato A. Efficiency at maximum power of interacting molecular machines. *Phys Rev Lett.* (2012)109:190602. doi: 10.1103/PhysRevLett.109.190602

68. Zilman A, Bel G. Crowding effects in non-equilibrium transport through nano-channels. *J Phys Condens Matter* (2010)22:454130. doi: 10.1088/0953-8984/22/45/454130

69. Reese L, Melbinger A, Frey E. Crowding of molecular motors determines microtubule depolymerization. *Biophys J.* (2011)101:2190–200. doi: 10.1016/j.bpj.2011.09.009

70. Huber DL. Particle kinetics on one-dimensional lattices with inequivalent sites. *Phys Rev B* (1977) 15:533–8.

71. Nossan JS. Disordered exclusion process revisited: some exact results in the low-current regime. *J Phys A Math Theor.*(2013) 46:315001. doi: 10.1088/1751-8113/46/31/315001

72. Lakatos G, O'Brien J, Chou T. Hydrodynamic mean-field solutions of 1D exclusion processes with spatially varying hopping rates. *J Phys A Math Gen.* (2006) 39:2253–64. doi: 10.1088/0305-4470/39/10/002

73. Fernando AE, Landman KA, Simpson MJ. Nonlinear diffusion and exclusion processes with contact interactions. (2010)*Phys Rev E* 81:011903. doi: 10.1103/PhysRevE.81.011903

74. Landman KA, Fernando AE. Myopic random walkers and exclusion processes: single and multispecies. *Phys A Stat Mech Appl.* (2011) 390:3742–53. doi: 10.1016/j.physa.2011.06.034

75. Penington CJ, Hughes BD, Landman KA. Building macroscale models from microscale probabilistic models: a general probabilistic approach for nonlinear diffusion and multispecies phenomena. *Phys Rev E* (2011) 84:041120. doi: 10.1103/PhysRevE.84.041120

76. Penington CJ, Korvasová K, Hughes BD, Landman KA. Collective motion of dimers. *Phys Rev E* (2012) 86:051909. doi: 10.1103/PhysRevE.86.051909

77. Penington CJ, Hughes BD, Landman KA. Interacting motile agents: taking a mean-field approach beyond monomers and nearest-neighbor steps. *Phys Rev E* (2014) 89:032714. doi: 10.1103/PhysRevE.89.032714

78. Simpson MJ, Landman KA, Hughes BD. Pathlines in exclusion processes. *Phys Rev E* (2009) 79:031920. doi: 10.1103/PhysRevE.79.031920

79. Torquato S. *Random Heterogenous Materials. Interdisciplinary Applied Mathematics*. New York, NY: Springer-Verlag (2002).

80. Simpson MJ, Baker RE, McCue SW. Models of collective cell spreading with variable cell aspect ratio: a motivation for degnerate diffusion models. *Phys Rev E* (2011) 83:021901. doi: 10.1103/PhysRevE.83.021901

81. Baker RE, Simpson MJ. Models of collective cell motion for cell populations with different aspect ratio: diffusion, proliferation and travelling waves. *Phys A Stat Mech Appl.* (2012) 391:3729–50. doi: 10.1016/j.physa.2012.01.009

82. Schönherr G, Schütz GM. Exclusion process for particles of arbitrary extension: hydrodynamic limit and algebraic properties. *J Phys A Math Gen.* (2004) 37:8215. doi: 10.1088/0305-4470/37/34/002

83. Lebowitz JL, Percus JK. Kinetic equations and density expansions: exactly solvable one-dimensional system. *Phys Rev.*(1967) 155:122–38.

84. Ferreira AA, Alcaraz FC. Anomalous tag diffusion in the asymmetric exclusion model with particles of arbitrary sizes. *Phys Rev E* (2002) 65:052102. doi: 10.1103/

PhysRevE.65.052102

85. Bruna M, Chapman SJ. Excluded-volume effects in the diffusion of hard spheres. *Phys Rev E* (2012) 85:011103. doi: 10.1103/PhysRevE.85.011103

86. Fanelli D, McKane AJ, Pompili G, Tiribilli B, Vassalli M, Biancalani T. Diffusion of two molecular species in a crowded environment: theory and experiments. *Phys Biol.* (2013) 10:045008. doi: 10.1088/1478-3975/10/4/045008

CHAPTER 10 MODELING POTENTIAL ENERGY SURFACES

Pietro Ballone

Keywords: Atomistic modeling, bond-order potentials, ab initio methods

INTRODUCTION

Most, if not all, of computer simulations using particles require the specification of the system potential energy as a function of particles' coordinates [1]. The most ab initio methods, such as those discussed in [2], represent systems as made of electrons and atomic nuclei, and Coulomb's law is sufficient to account for every interaction. In all other cases, particlesrepresent composite objects, such as atoms or atomic nuclei, dressed by core electrons, possibly embedded into a sea of valence electrons described at some approximate level of a many-body theory. Then, all the relevant interactions need to be worked out on a case by case basis, and the effort required to determine inter-particle forces may represent a sizeable fraction of the work to be done to investigate condensed matter systems [3].

The sections that follow contain an overview of modeling approaches and a discussion of their relative merits and limitations. Needless to say, the variety of systems and methods, together with the shear size of the knowledge accumulated over decades, impose strict limits to the scope of this presentation. First of all, the focus is on atomistic models, i.e., models in which the number and geometry of interaction centers follows the distribution of atoms closely. A second major branch of modeling, concerning coarse graining approaches, is the subject of a separate contribution (see [4]).

Moreover, again, for limitations of space, the discussion that follows mainly concerns the most restrictive picture of interatomic interactions, based on the assumption that the potential energy of a system of N atoms can be expressed as a single-valued function of their 3N coordinates $\{\mathbf{R}_i, i = 1, ..., N\}$, which represents the so-called potential energy surface (PES) of the system. This assumption relies, first of all, on the so-called Born-Oppenheimer approximation [5], whose validity is loosely attributed to the ~3–4 orders of magnitude difference in the mass of electrons and atomic nuclei, giving rise to a clear separation of the characteristic energy and time scales for the motion of electrons and atomic nuclei. Then, for any given instantaneous configuration of the atomic cores, electrons will be able to reach their electronic ground state, justifying the single-value assumption for the system potential energy. Experience shows that this "adiabatic assumption" is fairly well justified for a wide variety of systems and thermodynamic conditions. To be precise, it turns out that some cases are left out of this picture and often represent systems and phenomena of great interest. Methods suitable to deal with these cases are discussed in [6].

Computational science and simulation, in particular, always have a practical and an algorithmic aspect to them, and a central theme of research is the development of efficient ways to approximate and represent PESs. The availability of simple and computationally-convenient models of inter-particle interactions, for instance, has been instrumental in the dawning of computer simulation. Since then, the two complementary stages of determining the relevant interactions and of working out their structural, thermodynamic and dynamical consequences have cross fertilized each other, so much that the terms,modeling and simulation, often appear together in the title of books, papers, conferences, workshops and funding proposals.

Nowadays, the general perception of atomistic modeling is that of an overwhelmingly important and successful field, steadily expanding its reach towards more complex systems, which in this context means systems combining a wider variety of chemical bonds. In this respect, it is clear that much remains to be done, for instance, to bring under the cover of simulation heterogeneous systems and interfaces at which organic, semiconducting and metal phases meet each other or to

Citation: Pietro Ballone, Modeling Potential Energy Surfaces, 16(1), 322-349; doi:10.3390/e16010322.

model systems in which chemical transformations take place. During the last few decades, ab initio simulation methods have progressively come to play the role of the elephant in the (modeling) room. Methods, such as density functional theory [7,8] and ab initio molecular dynamics [9], could, in principle, replace all other approaches, reducing the variety of modeling problems to just one, concerning the effective and accurate representation of the energy of valence electrons in the field of atomic nuclei or ionic cores.

Up to now, this replacement has not been pervasive, mainly because of the size and time limitations of ab initio methods running on present day computers and partly because the approximations that make ab initio computations feasible still somewhat limit their accuracy on the energy scale of thermal motion, especially for molecular systems whose properties are determined by weak interactions among closed shell molecules. Ab initio modeling, however, is progressing and extending its reach. For what concerns atomistic simulation, therefore, empirical and semi-empirical models might eventually be squeezed out by the combination of ab initio methods and coarse-grained approaches. Simple models of atom-atom interactions, however, are likely to retain their appeal, because of their unique ability to represent and rationalize the microscopic forces underlying the properties and behaviors of condensed matter systems.

THE POTENTIAL ENERGY SURFACE (PES) OF A MANY-ATOM SYSTEM

From a physicist point of view, ordinary matter consists of an assembly of electrons and atomic nuclei, evolving according to the laws of quantum mechanics. The non-relativistic limit is adequate for many of the systems and properties of interest for the present discussion, and unless differently specified, we shall restrict ourselves to this case.

Let us therefore consider a system made of N electrons and K nuclei, and let $\{\mathbf{r}_i, i = 1, ..., N\}$ and $\{\mathbf{R}_\alpha, \alpha = 1, ..., K\}$ be the coordinates of electrons and nuclei, respectively. The corresponding linear momenta are denoted by $\{\mathbf{p}_i\}$ and $\{\mathbf{P}_\alpha\}$. In the absence of external fields, the system Hamiltonian is:

$$\hat{H}_0 = \sum_{\alpha=1}^{K} \frac{\mathbf{P}_\alpha^2}{2M\alpha} + \sum_{i=1}^{N} \frac{\mathbf{p}_i^2}{2m} + \frac{1}{2}\sum_{\alpha\neq\beta} \frac{Z_\alpha Z_\beta e^2}{|\mathbf{R}_\alpha - \mathbf{R}_\beta|} - \sum_{i,\alpha} \frac{Z_\alpha e^2}{|\mathbf{r}_i - \mathbf{R}_\alpha|} + \frac{1}{2}\sum_{i\neq j} \frac{e^2}{|\mathbf{r}_i - \mathbf{r}_j|} \quad (1)$$

that, for the sake of simplicity, we re-write as:

$$\hat{H}_0 = T_{ion} + T_{ele} + V_{ion-ion} + V_{ion-ele} + V_{ele-ele} \quad (2)$$

with an obvious correspondence between Equations (1) and (2). The Hamiltonian does not depend on the spin of electrons and nuclei, since we restrict ourselves to the non-relativistic limit, and we do not include any spin-orbit interaction into our Hamiltonian. Unless differently specified, Hartree atomic units ($\hbar = e^2 = m = 1$) are used in this section.

Let us assume that the system is described by a many-body wave function, $\Psi(\mathbf{r}_1, ..., \mathbf{r}_N; \mathbf{R}_1, ..., \mathbf{R}_k; t)$, whose time evolution is determined by the time-dependent Schrodinger equation:

$$i\hbar\frac{\partial\Psi(\{\mathbf{r}_i\};\{\mathbf{R}_\alpha\};t)}{\partial t} = \hat{H}_0\Psi(\{\mathbf{r}_i\};\{\mathbf{R}_\alpha\};t) \quad (3)$$

with appropriate boundary conditions in space and in time. Since the Hamiltonian is time independent, let us turn to the equivalent version of this same problem, concerned with the stationary states, $\Psi_k(\{\mathbf{r}_i\};\{\mathbf{R}_\alpha\})$ of $\hat{H}_0$.

The first important step towards the definition of a potential energy surface for the atomic nuclei is provided by the Born-Oppenheimer approximation (BO), which, under suitable and often verified conditions, opens the way to a separate description of the time evolution of electrons and nuclei [5]. The intuitive justification of BO is the observation that the motion of electrons and nuclei takes place over different time scales, since M_α/m is at least $M_n/m \sim 1, 800$, and usually approaches $2Z_\alpha M_n/m$, where M_n is the mass of a nucleon (proton or neutron). Moreover, the ratio of vibrational and rotational excitations is again $\sim \sqrt{M_\alpha/m}$. Experimental data confirm that, indeed, typical electronic excitations are of the order of a few eV; vibrational energies reach up to a few hundred meV, and even for small molecules, the separation of rotational levels is of the order of 1 meV. The conclusion is that the excitation of electrons, because of vibrational or rotational motion, is very unlikely. We can therefore represent the motion of electrons as taking place in the slowly varying field of the nuclei. Consistently with these qualitative arguments, the BO approximation breaks down whenever the energy of relevant electronic excitations becomes comparable to typical vibrational energies (or, much less likely, comparable to rotational energies). In those cases, vibrational and electronic excitations need to be considered on the same footing.

The core of the so-called adiabatic approximation can be given a semi-rigorous mathematical formulation in the following way [5]. Let us re-write $\hat{H}_0$ as:

$$\hat{H}_0 = \hat{T}_{ion} + \hat{H}_{ele} \quad (4)$$

where $\hat{H}_{ele} = T_{ele} + V_{ion-ion} + V_{ion-ele} + V_{ele-ele}$. The energy term, $V_{ion-ion}$, commutes with all other terms in $\hat{H}_{ele}$, and its inclusion in the electronic part is just a matter of convenience.

For every choice of the nuclear coordinates, $\{\mathbf{R}_\alpha, \alpha = 1, ..., K\}$, the eigenvalue problem:

$$\hat{H}_{ele}\psi_j(\{\mathbf{r}_i\} \mid \{\mathbf{R}_\alpha\}) = E_j(\{\mathbf{R}_\alpha\})\psi_j(\{\mathbf{r}_i\} \mid \{\mathbf{R}_\alpha\}) \quad (5)$$

is well defined and provides a sequence of eigenvalues, $E_j(\{\mathbf{R}_\alpha\})$, and eigenfunctions $\psi_j(\{\mathbf{r}_i\}|\{\mathbf{R}_\alpha\})$. At this stage, nuclei are "clamped", i.e., they are no longer treated as particles embodied with a mass and a momentum, but only as sources of the potential acting on the electrons. The notation, $(\mathbf{r}_i \mid \mathbf{R}_\alpha)$, means that ψ_j is an explicit function of $\mathbf{r}_i$ and depends parametrically on the nuclear coordinates, $\{\mathbf{R}_\alpha\}$.

The functions, ψ_j, are a basis for the Hilbert space spanned by the electron coordinates, and we can represent Ψ_k as follows:

$$\Psi_k(\{\mathbf{r}_i\}, \{\mathbf{R}_\alpha\}) = \sum_i \psi_j(\{\mathbf{r}_i\} \mid \{\mathbf{R}_\alpha\})\chi_j^{(k)}(\mathbf{R}_\alpha) \quad (6)$$

where, at this stage, $\chi_j^{(k)}(\mathbf{R}_\alpha)$ is simply the coefficient expressing the projection of Ψ_k on ψ_j:

$$\chi_j^{(k)}(\{\mathbf{R}_\alpha\}) = \int \psi_j^*(\{\mathbf{r}_\mathrm{i}\} \mid \{\mathbf{R}_\alpha\})\Psi_k(\{\mathbf{r}_\mathrm{i}\}, \{\mathbf{R}_\alpha\})\Pi_{i=1}^N d\mathbf{r}_\mathrm{i} \tag{7}$$

The equation for Ψ_k becomes:

$$\begin{aligned}&\hat{H}_0\Psi_k(\{\mathbf{r}_\mathrm{i}\}, \{\mathbf{R}_\alpha\}) = (\hat{T}_{ion} + \hat{H}_{ele})\Psi_k(\{\mathbf{r}_\mathrm{i}\}, \{\mathbf{R}_\alpha\})\\ &= \sum_j \chi_j^{(k)}(\{\mathbf{R}_\alpha\})E_j(\mathbf{R}_\alpha)\psi_j(\{\mathbf{r}_\mathrm{i}\} \mid \{\mathbf{R}_\alpha\}) + \psi_j(\{\mathbf{r}_\mathrm{i}\} \mid \{\mathbf{R}_\alpha\})\hat{T}_{ion}\chi_j^{(k)}(\{\mathbf{R}_\alpha\})\\ &+\chi_j^{(k)}(\{\mathbf{R}_\alpha\})\hat{T}_{ion}\psi_j(\{\mathbf{r}_\mathrm{i}\} \mid \{\mathbf{R}_\alpha\}) = \mathcal{E}_k\Psi_k(\{\mathbf{r}_\mathrm{i}\}, \{\mathbf{R}_\alpha\})\end{aligned} \tag{8}$$

Let us now multiply on the left by $\psi_m^*(\{\mathbf{r}_\mathrm{i}\} \mid \{\mathbf{R}_\alpha\})$ and integrate over the electron coordinates. One obtains in this way a set of coupled partial differential equations for the $\chi_m^{(k)}(\{\mathbf{R}_\alpha\})$ functions:

$$E_m(\{\mathbf{R}_\alpha\})\chi_m^{(k)}(\{\mathbf{R}_\alpha\}) + \hat{T}_{ion}\chi_m^{(k)}(\{\mathbf{R}_\alpha\}) + \sum_j \chi_j^{(k)}(\{\mathbf{R}_\alpha\})\langle\psi_m \mid \hat{T}_{ion} \mid \psi_j\rangle = \mathcal{E}_k\chi_m^{(k)}(\{\mathbf{R}_\alpha\}) \tag{9}$$

where $\mathcal{E}_k$ is the eigenvalue of the full, i.e., electrons and ions Hamiltonian $\hat{H}_0$, and the relation, $\langle\psi_m \mid \psi_j\rangle = \delta_{mj}$, has been used. The coupling among the equations is due to the non-diagonal part of $\langle\psi_m \mid T_{ion} \mid \psi_j\rangle$:

$$\langle\psi_m \mid \hat{T}_{ion} \mid \psi_j\rangle = \sum_\alpha \frac{1}{M_\alpha}\int \left[-i\frac{\partial\psi_m(\{\mathbf{r}_\mathrm{i}\} \mid \{\mathbf{R}_\alpha\})}{\partial\mathbf{R}_\alpha}\right]^* \left[-i\frac{\partial\psi_j(\{\mathbf{r}_\mathrm{i}\} \mid \{\mathbf{R}_\alpha\})}{\partial\mathbf{R}_\alpha}\right]\Pi_{i=1}^N d\mathbf{r}_\mathrm{i} \tag{10}$$

whose computation requires the parametric dependence of $\chi_m(\mathbf{R}_\alpha)$ on the $\{\mathbf{R}_\alpha\}$ coordinates to be continuous and differentiable.

Neglecting these non-diagonal terms, the equations for the electronic and ionic coordinates are decoupled, and the picture emerging from this manipulation of Equation (6) is that of nuclei evolving on the potential energy surfaces $U_j[\{\mathbf{R}_\alpha\}] = E_j(\{\mathbf{R}_\alpha\}) + \langle\psi_j \mid T_{ion} \mid \psi_j\rangle$. This last expression, corresponding to the so-called Born-Huang approximation [10], represents, in fact, an upper bound for the system's potential energy. A lower bound, instead, is given by the original BO approximation, i.e., $U_j[\{\mathbf{R}_\alpha\}] = E_j(\{\mathbf{R}_\alpha\})$.

The nuclear motion in general is quantum mechanical, and, depending on initial conditions, it might occur on any of theU_j potential energy surfaces (PESs). More precisely, since the equations for different j's are separated, it will take place on a single surface of index j, provided the starting point is consistent with this choice. This condition, that we identify with adiabatic motion, underlies most of the simulations that are routinely carried out in computational-condensed matter physics. Moreover, again, in most cases, but with noticeable exceptions, the relevant PES corresponds to the electronic ground state, and the scale of times and energies of interest allows the usage of classical dynamics instead of quantum mechanics [6].

The following sections are devoted to the discussion of the general properties of PESs, and of computationally tractable approaches to approximate them. Before doing that, it might be interesting to consider briefly when the BO approximation and the conditions for adiabatic motion are no longer valid.

An estimate of the $\langle\psi_m \mid T_{ion} \mid \psi_j\rangle$ terms can be obtained by perturbation theory, showing that the strength of the non-diagonal coupling is proportional to:

$$\langle\psi_m \mid \hat{T}_{ion} \mid \psi_j\rangle \propto \frac{1}{E_m - E_j}\langle\psi_m \mid [\mathbf{P}_\alpha, \hat{H}_{ele}] \mid \psi_j\rangle \tag{11}$$

Moreover, the matrix element of the commutator can be shown to depend primarily on the properties of individual atoms and to be only moderately dependent on the $\{\mathbf{R}_\alpha\}$ coordinates. Then, the major factor determining the coupling strength among different adiabatic surfaces is the energy gap separating different PESs. Whenever $(E_m - E_j)$ becomes comparable to the typical energies of the atomic motion, the BO decoupling is no longer valid, the electronic and ionic motion are intimately intertwined and both need to be treated quantum mechanically. The range of quantum mechanical features that become relevant in the non-BO case go beyond delocalization and diffraction, but includes the appearance of geometric (Berry-Pancharatnam) phases [11].

Far from being the exception, violations of the BO approximation are pervasive. They occur often, but not exclusively, at the so-called conical intersections [11], playing a major role in chemical reactions and, for instance, challenging our ability to model catalysis [12]. Apparent non-BO effects are routinely highlighted by clever experiments [13,14].

Metals, whose occupied states are immediately contiguous in energy to the empty states, may appear as the most obvious candidates for large deviations from the BO picture. In the vicinity of the Fermi surface, however, single particle excitations are the only relevant excitations, but the coupling of each of these excitations to the nuclear motion (through Equation (11)) is vanishingly small. Collective electron excitations, such as plasmons, couple to the atomic motion, but their energies are of the order of several eV and, thus, are comparable to, if not higher than, those of closed shell atoms and molecules. As a result, vibrational properties of metals are generally well described by adiabatic dynamics. Exceptions are represented by Kohn anomalies, resulting from the nesting of reciprocal lattice vectors with the Fermi surface. Metals also provide the setting for a type of BO violation qualitatively different from those considered until now, represented by superconductors, in which the coupling of the electron and nuclear motion changes the symmetry of the ground state.

The isolated system picture underlying the BO decoupling has been generalized in [15–17] to the case of electrons and nuclei evolving in an external time-dependent potential. It was shown, in particular, that the full wave function can be factorized exactly into an electronic and a nuclear wave function, again opening the way to the definition of a time-dependent PES. The picture is less simple than in the static case, since it involves the introduction of a Berry vector potential and of Berry-Pancharatnam geometric phases [18,19] into the problem. This approach has already provided the basis for the real-time simulation of molecular systems in strong (laser) external fields. For completeness, I mention that some details of the formal framework might still need to be worked out for a fully rigorous treatment [20].

PROPERTIES OF POTENTIAL ENERGY SURFACES

Basic features of the PES can be anticipated even without an explicit solution of the standard electronic problem inEquation (5). A surprisingly realistic intuition of what a PES looks like was outlined in elegant Latin prose long before quantum mechanics [21], based on an atomistic hypothesis and on the assumption that the still undiscovered atoms felt each other mainly at short distances.

The modern interpretation confirms this picture and adds a wealth of microscopic detail. The direct Coulomb repulsion among nuclei, unscreened by electrons at short distances, prevents the close contact of atoms and their eventual collapse. The kinetic energy of the electrons tightly bound to the nuclei will provide an additional repulsive contribution, resulting from the need to preserve the Pauli principle. On the other hand, the formation of chemical bonds gives rise to attractive potentials, binding atoms together. Even in the case of inert species, subtle quantum mechanical effects give rise to dispersion forces, which provide a weak, but pervasive, attraction.

Arguably, the simplest and most intuitive picture of atomic interactions is provided by pair potential models, in which the system energy is written as:

$$U[\{\mathbf{R}_\alpha\}] = \frac{1}{2}\sum_{\alpha,\beta} \phi_{\alpha\beta}(|\ \mathbf{R}_\alpha - \mathbf{R}_\beta\ |) \tag{12}$$

where the α, β label on $\phi_{\alpha,\beta}$ indicates that the interaction depends on the chemical identity of particles α and β. A spherically symmetric potential has been assumed for the sake of simplicity.

Computations and comparison with experiments have shown that an expression of this kind is suitable for rare gases [22] and for simple ionic compounds [23]. Systems and models of this kind have been instrumental in establishing computer simulation as a quantitative research tool in condensed matter and in chemical physics.

Needless to say, the scope of pair potentials is very narrow, and limitations of this model were already apparent well before the dawn of computer simulation, based on the results of lattice dynamics models in metals and semiconductors.

One could think of the pair potential expression as being only the lowest order approximation of the PES into an n-body expansion of the form:

$$U[\{\mathbf{R}_\alpha\}] = \frac{1}{2!}\sum_{\alpha,\beta} V_2(\mathbf{R}_\alpha, \mathbf{R}_\beta) + \frac{1}{3!}\sum_{\alpha,\beta\gamma} V_3(\mathbf{R}_\alpha, \mathbf{R}_\beta, \mathbf{R}_\gamma) + \ldots \tag{13}$$

For a system made of a finite and constant number of particles, such an expression can always be written down. For instance, one could define V_2 as the interaction energy of two isolated atoms, V_3 as the corresponding energy of trimers, minus the symmetrized combination of V_2 contributions, etc. Such an expansion, however, is useful only if it converges within a few terms, at least because the cost of evaluating successive n body terms grows rapidly with increasing n. Moreover, it contributes to the physical understanding of the system behavior only when its convergence is absolute, i.e., it does not require the cancellation of contributions of alternating sign, whose amplitude is constant or even increasing with increasing order. Model computations based on a tight binding Hamiltonian [24], however, show that even for simple systems, the expansion in Equation (13) is not well behaved and, thus, is seldom useful for practical computations.

More fruitful than the systematic expansion of Equation (13) has been the introduction of the cluster potential idea [25,26], loosely and sometimes more closely based on the bond-order concept introduced by Pauling [27]. In this approach, a fixed and low number of terms is retained; the expression looses its character of a systematic series to become an asymptotic expansion. Each of the few terms that are retained describe low-order potentials whose strength depends on the local environment. Approaches of this kind have given origin to the most popular family of potentials used to simulate metals and metallic alloys and also to some important approaches to approximate the PES of semi-conductors, which are discussed in the following sections.

MANY-BODY INTERACTIONS: METALS AND METAL ALLOYS

Metals and their alloys posed an early challenge to the pair or few-body potential picture, since their basic properties manifest essential many-body interactions [28].

The successful and physically-motivated incorporation of these effects into tractable models in the early eighties of the last century has spawned a vast simulation activity, aiming, at first, at reproducing phase diagrams, then at analyzing in detail surfaces and interfaces and further progressing towards the prediction of mechanical properties through multi-scale approaches. Physical metallurgy is currently one of the most active and productive subfields of atomistic simulation [29,30].

Many-body interactions in metals were first identified by the analysis of their elastic properties. For instance, the elastic constants of cubic materials consisting of atoms interacting via spherically symmetric pair potentials have to satisfy the so-called Cauchy relations, stating, for instance, that $C_{12} = C_{44}$. The violation of this relation, known in the solid state literature as a Cauchy anomaly, is the rule more than the exception in metals, unambiguously pointing to a deviation from the pair potential picture.

These features were first rationalized by considering the basic representation of a metal, as made of ions embedded into a sea of valence electrons. Since the major ingredient, i.e., the homogeneous electron gas could be solved analytically, and, at least for sp metals, the electron-ion interaction is weak, the full problem could be attacked by perturbation theory [28,31]. Carried up to the second order, this approach provides an expression for the system total energy that consists of a large volume (or, equivalently, density) term and a pair potential contribution. The volume term is able to account for the Cauchy anomaly. In simple metals, such as the alkalis, the pair potential is relatively soft at short distances and oscillates at large distances, reflecting Friedel oscillations. These features explain the bccstructure of these systems at normal conditions and provide a clue to understand more complex structures adopted by the lighter alkali metals at very low temperature or found in slightly more complex systems, such as alloys, or heavier sp metals, such as gallium, indium or tin. Approaches of this kind are now mainly of historical interest, since most of the cases relevant for applications involve transition metals, and in those systems, the valence electron-ion interaction is by no means weak; the perturbation expansion cannot be limited to the second order and becomes rapidly untreatable beyond that point [32]. Besides these fundamental problems, other practical difficulties concern the definition and the zero-order solution of an electron gas problem suitable for inhomogeneous systems and for alloys. Electron gas perturbative approaches, therefore, could not solve problems, such as the inward relaxation of crystal surfaces, the quantitative description of stacking faults or the overestimation by pair potentials of the vacancy formation energy in metals.

To overcome these problems, new models have been proposed in [33–35], conforming to the cluster-potential idea [26], and representing low-order approximations to a bond-order potential. The embedded atom model (EAM) of [33,34], loosely based on density functional theory, has the broadest appeal, and for this reason, it is used here as a representative of a wider class of models.

According to EAM, each metal ion, i, at position $\mathbf{R}_i$ gains an energy, $E[\rho_e(\mathbf{R}_i)]$, upon being immersed into the valence electron distribution at density $\rho_e(\mathbf{R}_i)$ and interacts with neighboring ions by a short range repulsive pair potential, $V_2(R)$. The energy of N metal atoms, therefore, is:

$$U[\{\mathbf{R_i}\}] = \frac{1}{2}\sum_{i \neq j}^{N} V_2(|\mathbf{R_i} - \mathbf{R_j}|) + \sum_{i=1}^{N} E[\rho_e(\mathbf{R_i})] \tag{14}$$

The picture is completed by a prescription to compute the electron density, ρ_e, at the position, $\mathbf{R}_i$, of each atomic core. EAM represents such a density as the sum of contributions from every other atom:

$$\rho_e(\mathbf{R_i}) = \sum_{j \neq i} t_j(|\mathbf{R_i} - \mathbf{R_j}|) \tag{15}$$

where the $t_j(R)$ are again relatively short-range functions, mimicking the tail of the electron distribution around an isolated atom. Since it introduces a local embedding density, this prescription overcomes most of the limitations of the free electron models, which instead rely on a global definition of the valence electron density.

Parameters and auxiliary functions, such as t(R), $E[\rho_e]$ and $V_2(R)$, could be computed from first principles [36], but this approach has been only moderately successful. Far more effective has been the strategy of adopting the EAM potential energy expression as a general framework, relying on fitting experimental quantities to tune a few parameters distributed into the functional form.

The success of EAM has been due to its ability to overcome the limitations of simpler models, easily accounting for the Cauchy anomaly, the reduced value of the vacancy formation energy, the inward relaxation of compact metal surfaces and the reconstruction of more open ones. Its broad acceptance relies also on the many and physically appealing properties of the model, discussed in a number of publications, such as the ease of extending EAM to alloys or the close relation with pair potentials in the case of homogeneous systems at constant volume.

From the computational point of view, the efficiency of EAM is due to the pair potential form of both the repulsive contribution, V_2, and the embedding density expression in Equation (15). The time required to carry out a simulation based on EAM is expected to be twice that of a pair potential model, since a pass on all atom pairs is required to compute the repulsive potentials and the embedding density, while a second pass is needed to compute forces on atoms arising from the embedding energy. With suitable lists of neighbors, and depending on the range of $V_2(R)$ and of t(R), EAM can be used to carry out MDsimulations for systems of 10^4 atoms over several nanoseconds using laptops or inexpensive PCs. Supercomputers extend these ranges to several million atoms, and μs time scales.

Needless to say, an empirical and approximate approach, such as EAM, cannot provide the final answer to the problem of modeling metals, and transition metals, in particular. A comprehensive discussion of inaccuracies and limitations identified during thirty years of applications is beyond the scope of this short review, and only two examples are briefly mentioned here. Phonons in transition metal crystals, a property routinely measured by inelastic neutron scattering, are not well reproduced by EAM. The elastic constants usually enter the fitting of the potential, and thus, the low-frequency acoustic phonons close to the Γ-point of the first Brillouin zone are usually well reproduced. Higher frequency modes at the zone boundary, however, turn out to be too soft with respect to the experimental data (see Figure 1). Transition metal clusters from a few to several thousand atoms are important for catalysis and represent a basic ingredient of nanotechnology. EAM neglects the details of the electronic structure of the atoms, leaving out quantum mechanical effects, such as Jahn-Teller. Thus, EAM is unable to quantitatively reproduce the structure and cohesive properties of the very small aggregates as provided by density functional computations. Beyond ~100 atoms, cluster properties are expected to evolve more continuously with size, approaching those of bulk phases beyond 10^4 atoms. EAM has been used extensively to investigate clusters across this range, but a quantitative validation of the model is still lacking and difficult to achieve, since more ab initio computations become too expensive to carry out, and experiments find it difficult to probe this range of cluster sizes.

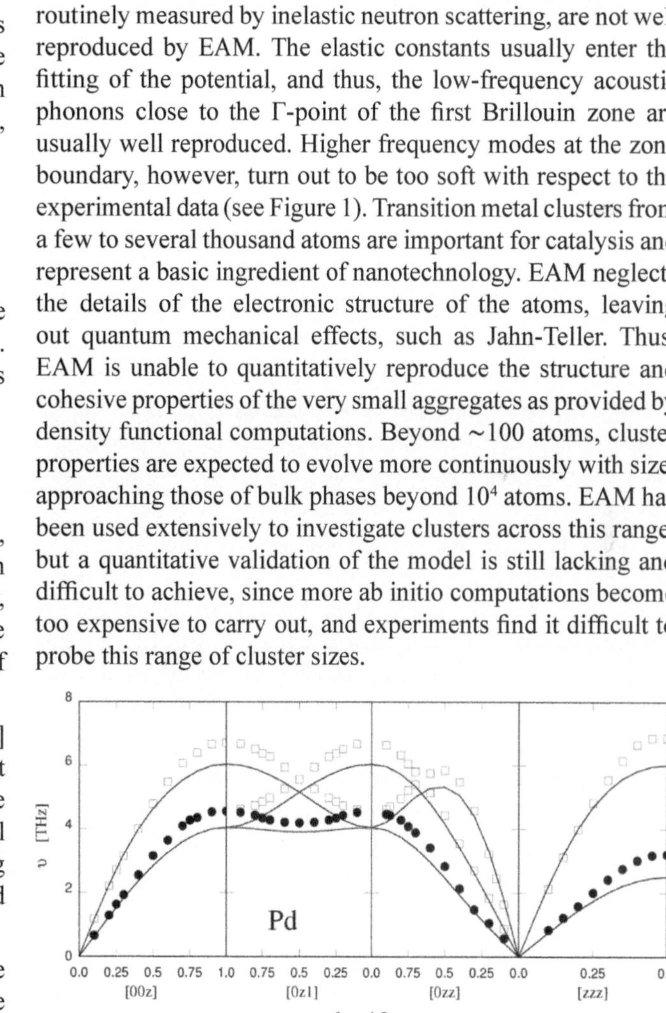

Figure 1: Phonon frequencies of fccpalladium from experiments (symbols, see [38]) and from the embedded atom model (EAM) model of [33].

A step beyond EAM, needed to quantitatively model the fine details of the structure, thermodynamics and dynamics of transition metal systems, requires the introduction of explicit angular terms into the potential energy expression. This can be achieved through a conceptually simple extension of EAM, known as modified EAM (MEAM) [34], or resorting to a chemically accurate bond-order potential model, including the directionality of d and f electron orbitals, as well as the distinction of σ, π, δ, ..., bonding, anti-bonding and non-bonding orbitals [37].

The MEAM is somewhat more complex to use than EAM, and probably for this reason, it has been less extensively applied. Moreover, its ability to quantitatively overcome the limitations of the simpler model is not always so apparent. The other approaches, more closely based on the bond order approach, appear to be cumbersome to use in simulations, and the number of applications based on these models has been limited.

Because of the inclusion of angularly dependent forces, the scope of MEAM could, in principle, cover semiconductors. Successful applications have been published [34], but more specific models, described in the following section, have received broader attention in this subfield.

SEMICONDUCTORS AND INSULATORS

Semiconductor materials, exemplified by silicon, germanium, gallium arsenite, etc., are characterized by fairly open and complex structures of relatively low coordination, stabilized by sizeable angular forces, arising from the directionality of covalent bonds. Apart from elemental systems, most inorganic semiconductors are characterized, in fact, by a combination of covalent and ionic bonding. Several of these systems, most notably silicon and germanium, turn into metals upon melting.

Despite the difficulty of reproducing these properties by few-body potentials, the urgency of investigating the elements and compounds that fueled the electronic revolution stimulated the first bold attempts. The two- and three-body potential for silicon proposed by Stillinger and Weber [39] arguably has been the most representative example of this first generation of models.

Despite their interest, approaches of this kind have been only moderately successful, and once again, the bond-order concept [27] proved more fruitful. Its application to semiconductors was first discussed by Abell [25] before being used in a more empirical setting by Tersoff [40,41] and extended by Brenner [42] to a wider class of systems and problems.

According to these models, the potential energy of an assembly of N atoms of coordinates $\{\mathbf{R}_i\}$ is written as:

$$E_N = \sum_{i \neq j} \left[A \exp\left(-\lambda_1 R_{ij}\right) - B_{ij} \exp\left(-\lambda_2 R_{ij}\right) \right] \tag{16}$$

where $R_{ij} = |\mathbf{R}_i - \mathbf{R}_j|$. The first term, representing the short-range repulsion, is a genuine pair potential. The second term contains many-body contributions via the dependence of B_{ij} on the local environment around the interacting pair, ij.

This form has obvious analogies with the EAM case. The difference is that B_{ij} not only counts neighbors, as the embedding density does, but takes into account also the angular correlation among their mutual positions. This addition is required to enforce the dominance of tetrahedral sp^3 coordination, but also to carve a secondary role for other structures, from the sp^2bonding of graphite, to the octahedral coordination of liquid silicon and germanium [40,41]. Parallel to the EAM case for metals, potentials of this type replaced previous models and established a new standard in modeling semiconducting systems. Success, however, has been somewhat less pervasive than in the case of EAM, for reasons that are relatively easy to identify. First of all, interactions in semiconductors are more complex and propagate at a longer range, since screening is not as effective as in metals. Moreover, semiconducting alloys and compounds give rise to partially Coulombic interactions, whose combination with covalent bonding has seldom been modeled, even by bond-order potentials.

Furthermore, in this case, the systematic improvement beyond the semi-empirical Tersoff and Brenner potentials has to rely on the analytical development of chemically accurate bond-order models [43]. Work along these lines is underway and has shown promising developments, but current models still appear fairly difficult to implement in molecular dynamics or Monte Carlo packages.

An important development of Brenner's scheme has been the introduction of reactive force fields, able to describe chemical transformations in the system under consideration. The majority of the parameterizations and applications published until now concern organic systems, but potentials of this kind are mentioned here for their similarity with models first introduced for semiconductor systems. Prototypical examples of a reactive force field are the so-called ReaxFF [44] and the REBOpotential [45]. Both models require a massive parametrization effort, and for this reason, they appear to be fairly ad hocand system specific.

A different line of attack to modeling semiconducting systems is suggested by the observation that in many cases, force fields of the form currently used to model organic systems and consisting on stretching, bending and torsion might indeed provide a good representation of structural and dynamical properties of semiconductors and of network insulators, such as silica. Models of this kind, in fact, were developed well before the age of computer simulation, and extensively used in lattice dynamics studies of semiconductors and insulators [46]. The problem of these models is that, mainly because of the established tradition, the topology of bonds is kept fixed, bonds are harmonic and can neither form nor break. These models, therefore, describe only low amplitude oscillations around a pre-assigned minimum of the potential energy surface. Removing these inessential constraints by introducing rules to break, form and interchange bonds results in a far more realistic picture. It was shown, for instance, that such a reactive force field model of silica undergoes melting at approximately the right conditions [47] (see Figure 2), and the same model has been used to provide an intriguing view of the amorphous silica surface at length and time scales unachievable by other methods [48].

Progressively increasing the electronegativity difference in compound semiconductors enhances the charge transfer among atoms, widening the band gap and turning the system into an ionic insulator. In the limit of strongly ionic materials, of course, pair potentials are adequate, but only a few compounds belong to this class, such as, for instance, alkali-halides or the oxides and chlorides of Group IIA and Group IIB metals. In between ionic insulators and polar semiconductors, there is a vast number of systems, including technologically relevant compounds, such as ceramics, transition metal oxides, ferroelectric and ferroid materials, minerals and bio-minerals, in particular, for which no current model is fully satisfactory. One of the major issues for these systems is the inclusion of polarizability into ionic and polar models [49]. Unfortunately, simulation approaches using polarizable models require either the minimization at every step of a polarization energy functional or the inclusion into the model of charged shells [50]. These last represent electronic degrees of freedom and react to electric fields on a time scale much faster than that of ionic vibrations [51]. Both methods are significantly at a disadvantage with respect to cases in which the potential energy is an explicit function of the atomic coordinates, and the simulation of systems bound by a combination of covalent and ionic forces appears to be split between oversimplified pair potential models and ab initio approaches.

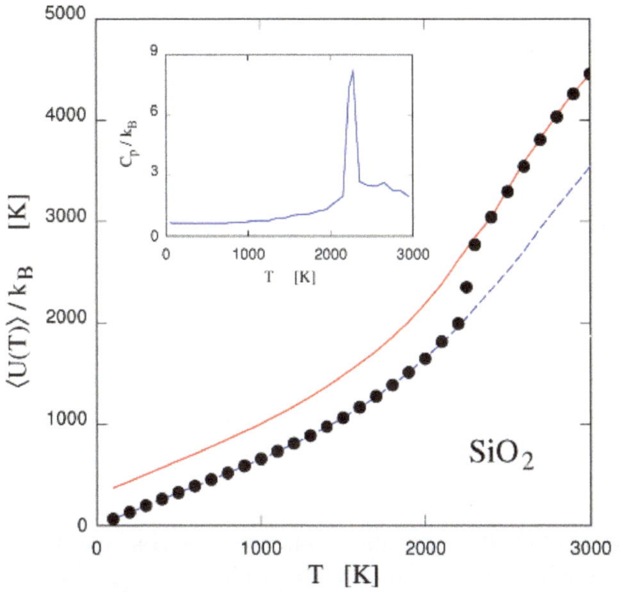

Figure 2: Average potential energy per atom $\langle U(T)\rangle/K_B$ of SiO_2 computed by the force field of [47]. k_B is the Boltzmann constant, introduced to express energies in temperature units (K). Solid dots: heating a β-cristobalite sample. Solid line: cooling the same sample from high temperature. The potential energy contribution, C_p, to the constant pressure specific heat computed on heating the full model is shown in the inset. The peak in C_p and the anomaly in $\langle U(T)\rangle$ are around the same temperature point to a melting transition at $T_M \sim 2150$ K.

FORCE FIELDS FOR MOLECULAR SYSTEMS

Although every material ultimately consists of atoms, many systems are more easily understood as being made of molecules.

Modeling the PES of small and relatively unreactive species, such as N_2,O_2, CO, CO_2, but, also, PF_6, BF_4, BH_4, etc., requires only a slight extension of the pair-potential picture. Each molecule is represented by a small number of interaction centers, which may or may not coincide with atoms in number and position. The intra-molecular configuration is enforced by constraints representing rigid bonds or, less often, by harmonic springs, while centers on different molecules interact pair-wise. Because of their simplicity, models for small inorganic molecules have been used since the early days of computer simulation. Perhaps the most remarkable observation concerning these systems is that the quantitative details of their PES are still under investigations and require surprisingly sophisticated models to be reproduced [52,53]. Conspicuously absent in the list of small unreactive and supposedly simple molecules is water, whose peculiar properties and special role have motivated an extraordinary modeling effort, which is discussed separately in Section 7.

A specialized subfield of modeling simple species concerns systems in which a weakly bound molecular fluid is physisorbed on an inert solid surface, such as MgO, mica, graphite and flat or stepped transition metal surfaces. In this case, the effect of the solid substrate on the molecular fluid often is represented as an external field. In the case of crystal surfaces, the in-plane dependence of the field strength can be expanded in plane waves, whose wave vectors reflect the periodicity and symmetry of the surface lattice [54].

Organic Molecular Systems

In many respects, organic molecular systems are not so different from any other molecular systems, but the range and impact of their applications together with the explosive expansion of simulation in bio-physics and bio-chemistry amply justify a separate discussion. Systems of interest in this context include polymers, hydrocarbons, sugars, cellulose, etc., but also the endless variety of biological molecules, from phospholipids to proteins and nucleic acids. Other molecular organic systems of biological interest include drugs, simple nutrients, signal molecules, such as hormones, metabolic species, such as ATP, GTP, NADP, coenzymes, including vitamins, and prosthetic groups.

The modeling and simulation of systems of this kind arguably is the computational condensed matter activity with the largest economic relevance, both directly via the commercialization of packages and force fields and indirectly through the impact it has on applied research.

Despite the complexity of the structures they form, the PESs of organic systems turns out to be approximated fairly well by simple analytic expressions. First of all, the organic and biological species of interest are made primarily of light elements, forming strong covalent bonds through their s and p orbitals, giving origin to closed shell molecules. Systems of this kind, therefore, can be thought of as consisting of atoms connected by a fixed topology of bonds, with inter-molecular, i.e., non-bonded, interactions consisting of pair-wise Coulomb and dispersion forces. Because of their sp character, intra-molecular angular forces are relatively simple. Whenever d electron metals are involved, as in metal centers and in prosthetic groups, modeling becomes far more challenging.

In the standard cases, the PES of organic and biological systems is written as the sum of contributions from bonded (U_b) and non-bonded (U_{nb}) interactions:

$$U = U_b + U_{nb} \tag{17}$$

The bonded energy, in turn, is given by the sum of two-, three- and four-body terms from atoms joined by one ({ij}), two ({ijk}) and three ({ijkl}) consecutive covalent bonds:

$$U_b = \frac{1}{2}\sum_{\{ij\}} K^s_{ij}[R_{ij} - \bar{R}_{ij}]^2 + \frac{1}{2}\sum_{\{ijk\}} K^b_{ijb}[\theta_{ijk} - \bar{\theta}_{ijk}]^2 + \frac{1}{2}\sum_{\{ijkl\}} K^{\tau}_{ijkl}\left[1 + \cos\left(n\phi_{ijkl} - \bar{\phi}_{ijkl}\right)\right] \tag{18}$$

K^s_{ij}, K^b_{ijk} and K^{τ}_{ijkl} are suitable force constants; R_{ij}, $\bar{\theta}_{ijk}$, $\bar{\phi}_{ijkl}$ and n reflect the length, bending and dihedral angles of unstrained bonds. The sub-indices, ij, etc., indicate that each of these parameters depends on the chemical identity of the atoms involved. The form for the dihedral contribution in Equation (18) is just one of a few different expressions used in popular force fields, while the choice for stretching and bending terms is more uniform.

Non-bonded interactions are written as.

$$U_{nb} = \frac{1}{4\pi\epsilon_0}\sum_{i\neq j}{}' \frac{q_i q_j}{R_{ij}} + \sum_{i\neq j}{}' 4\epsilon_{ij}\left[\left(\frac{\sigma_{ij}}{R_{ij}}\right)^{12} - \left(\frac{\sigma_{ij}}{R_{ij}}\right)^{6}\right] \tag{19}$$

where the $\{q_i\}$ are atomic charges, Coulomb forces are assumed to be acting in vacuum and σ_{ij} and ϵ_{ij} are suitable coefficients for the dispersion interaction. The prime on each sum indicates that pairs of atoms separated by one and two consecutive bonds are excluded, and the contribution from pairs separated by three consecutive bonds might be reduced.

The remarkable and, to same extent, unique property of the PES of organic and biological systems is that the bonds, whose properties are described in Equation (18), are fairly transferable, meaning that the equilibrium length, stiffness, etc., of a given organic bond is nearly the same in a large number of homologous compounds. Highlighting these similarities and exploiting them to endow the model with broad transferability is the most challenging and most rewarding part of modeling organic molecular systems.

The parametrization and, especially, validation of these potentials may require sizeable computations and are the playground of large collaborations, since it requires the convergence of several types of complementary expertise. Any single system might be analyzed by ab initio computations to derive intra-molecular force constants and atomic charges. These need to be complemented by suitable coefficients for the dispersion part, which are usually obtained by fitting measured properties, such as the equilibrium density and enthalpy per molecule or the molecular diffusion constant.

Generic potentials covering large classes of compounds and widely used by the community include Amber [55], CHARMM [56], OPLS [57] and Gromos [58]. More specialized parameterizations, tuned on the properties of specific families of compounds, are too many to be listed.

In many respects, the most uncertain part of the parametrization is the choice of coefficients for the non-bonded interactions. The definition of atomic charges is not unique, and different methods provide fairly different results. The most popular approach [59] attributes charges by fitting the electrostatic potential outside gas-phase molecules, as provided by ab initio computations. The method is physically sound, but the fit becomes ill conditioned whenever the molecular size exceeds ~ 15–20 atoms or when the geometry is compact, thus reducing the number of multipolar momenta whose modulus is significantly different from zero. Constraints and minimum conditions on the size of individual charges do improve the fit [60], but the choice of these parameters remains fairly uncertain. For each individual system, the error introduced by the choice of the charge may be compensated for by the selection of the dispersion coefficients. In fact, it has been observed many times that it was possible to accurately reproduce the target properties of condensed phases such as the density or the molecular diffusion even starting from the fairly different charges provided by different methods. Unfortunately, this cancellation of errors limits the transferability of the potential, since an equivalent compensation might not occur when a given organic molecule is transferred into a different environment.

Especially for large biological systems, computational cost considerations have motivated approximations and shortcuts that might reduce the size of the simulated system. One obvious saving is obtained by representing CH_2 and CH_3 groups in aliphatic chains by a single particle. This united-atoms approximation is fairly well justified, since these groups are small and and the non-bonded potential arising from them is fairly spherical. Moreover, the motion of hydrogen in each of these groups is frozen by quantum effects up to fairly high temperature.

A second more drastic approximation concerns systems in solution. Since, especially in biochemistry, one is interested in the properties of the solute, implicit solvent models [61] have been developed to replace the effect of the solvent by suitable modifications of the solute force field. In many respects, implicit solvent models are a special case of coarse graining and, as such, are left out of our discussion.

In summary, the force field modeling of organic and biological systems is a largely successful enterprise, validated by a vast number of applications and supporting the research of a large portion of the simulation community. Furthermore, in this case, and almost needless to say, the vast simulation activity has highlighted many cases of inaccuracies or outright failures. The general feeling, however, is that the scale of most of these simulations is too large to allow, at present, the usage of significantly more sophisticated and more expensive approaches. Polarizability is likely to be the single most relevant missing ingredient, but the available methods to include it into simulations are still fairly expensive, and for this reason, explicitly polarizable models have been used only for a limited number of large-scale studies.

At present, a very active research field is the development of force fields for organo-metallic complexes, which represent prosthetic groups in proteins or active groups in a variety of organic opto-electronic devices and are important also for homogeneous catalysis. Peculiar difficulties are represented by the variety of coordination numbers, sometimes corresponding to different spin states, thus pointing to multiple PESs fairly close in energy. Moreover, the structure of organo-metallic complexes is characterized by the importance of quantum mechanical effects, such as Jahn-Teller, or by the so-called trans influence, defined as the "tendency of a ligand to selectively weaken the bond trans to itself" [62]. Models to include these effects in empirical PES models might turn out to be too complex to be used in practice. A more promising alternative is provided by QM/MMapproaches, using classical force fields for most of the system and resorting to ab initiomethods for the challenging portion around the metal center.

An intriguing subset of mainly, but not exclusively, organic compounds is represented by the so-called room temperature ionic liquids [63], defined as molecular ionic systems whose melting temperature is below 100°. Prototypical systems are made by an alkane substituted imidazolium cation, joined to an organic or inorganic anion. Systems of this kind are relevant here, not only because of the intense simulation activity that concerns them, but mainly because they provide a bridge between different classes of bonding and, thus, pose special modeling problems. The bulk of the extensive simulation work carried out at present relies on Amber-like force fields, with specialized parameterizations (see, for instance, [64,65]). Models of this kind are fairly successful, but issues concerning polarizability and the attribution of partial charges to atoms become particularly important for these systems. Despite these difficulties, a number of simulations have successfully addressed the properties of very complex systems, consisting of room temperature ionic liquids in combination with a variety of solvents and neutral organic compounds, including bio-molecular species (seeFigure 3).

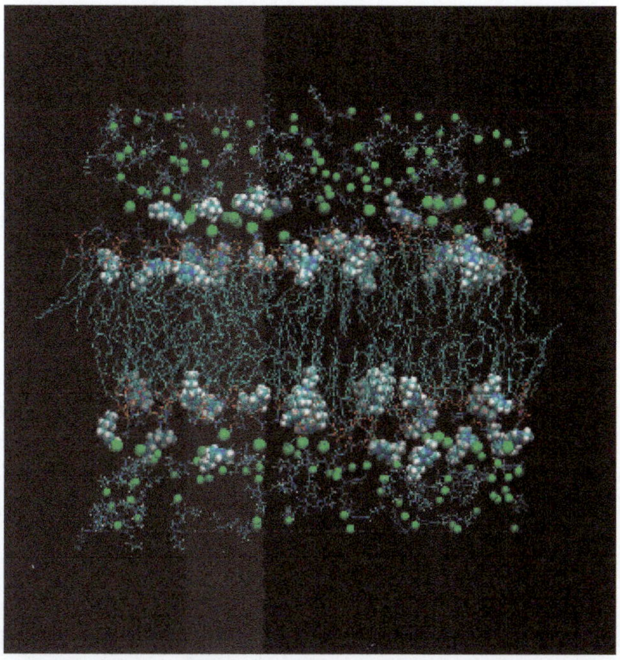

Figure 3: Snapshot from a molecular dynamics simulation of a room temperature ionic liquid/water solution at 0.5 M concentration in contact with a POPCphospholipid bilayer [66]. Green balls: $[Cl]^{-}$; gray-silver molecules: $[bmim]^{+}$. wireframe molecules: POPC. Water has been removed to highlight the incorporation of $[bmim]^{+}$ cations into the phospholipid bilayer.

A few carbon systems, such as fullerenes, carbon nanotubes and graphene, lie at the boundary between inorganic and organic species and even blur the distinction between covalent and metal character. Not surprisingly, systems of this kind have been represented by a variety of models, from Tersoff-Brenner to a molecular force field, such as those described in this section.

WATER

Because of its fundamental role in life and of its widespread and generally benign presence in nature, water has always been the object of interest and fascination. In this respect, computational physicists and chemists are no exception, although the reasons for their interest are somewhat different from those of the rest of humankind. A number of measurements have highlighted a wide variety of peculiarities, if not anomalies, in the properties of water [67]. These include the surprising expansion of water upon freezing, the density anomaly observed at 4 °C at ambient pressure, and, more in general, the non-monotonic variation of several physics-chemical properties in the vicinity of this remarkable density maximum. Other peculiar features consist in the wide temperature range of super-cooling, the high liquid-vapor critical temperature and the large value of the latent heat of the liquid water-ice transition.

To a large extent, these anomalous behaviors are embodied into the PES of water systems and arise from the strength and directionality of the hydrogen bond network that provides the bulk of water cohesion. In part, however, they are due to the light mass of the water molecule, causing non-negligible quantum effects that influence the properties of hydrogen bonds. Heavy water, for instance, is already somewhat different from ordinary water, so much that D_2O is known to have peculiar and generally adverse biological effects. This duality of potential energy versus quantum mechanical effects poses apparent and significant problems to modeling [68]. Potentials tuned on the exact PES of water do not reproduce its properties when used in a classical simulation. On the other hand, potentials tuned on experimental properties of water do not necessarily reflect the details of the exact PES.

Work to provide a quantitative and comprehensive description of water properties is still in progress [69,70]. In the meantime, a vast number of simulations in which water is the unique or an essential component are being carried out with a variety of simple potentials, reflecting the basic atomistic and electronic structure of the water molecule. Two major families are in use: TIPnP[71–73], with n = 3, 4 and 5, and SPC [74–77], both based on fixed charges (rigid ions) and centers of short range interactions, joined by rigid or harmonic bonds.

Models of this kind allow the routine simulation by MD of systems of 50×10^3 water molecules solvating whole proteins, covering times well in excess of 100 ns. Results are generally good, and a large number of successful applications clearly validate these models, at least up to the accuracy needed for these large-scale applications. However, it is fair to say that no single model of the rigid ion type is able to provide a uniformly satisfactory account of water properties over a wide range of regimes and thermodynamic conditions. Several of these models, in particular, do not display the experimental density maximum of water or place it at (P, T)conditions far from the experimental ones [69,70]. The liquid-vapor coexistence curve is also poorly predicted by rigid ion models, unless the potential parameters are explicitly adjusted for this purpose. In such a case, however, the accurate description of some other quantity might need to be sacrificed. The description of critical properties, that are accurately known from measurements, are only moderately well reproduced [78].

Water clusters and droplets are another, distinct subfield of water research. Thermodynamic and spectroscopic data are available from experiments, but are not sufficiently detailed to provide a full description of structural and dynamical properties. In this case, state-of-the-art quantum chemistry computations supplement the experimental information [79]. Once again, it turns out that rigid ion models are only moderately successful in predicting their properties and usually fail to reproduce the reduced binding of very small clusters. The oxygen-oxygen equilibrium distance in the water dimer, for instance, is greatly underestimated by popular models, and its cohesive energy is correspondingly overestimated. These discrepancies decrease in importance with increasing cluster size, but the convergence to the bulk cohesive properties, reliably described by current DFTmodels of water, is fairly slow (See Table 1). In these small systems, the rigid-ion assumption, or, in other terms, the lack of polarizability, again seems to be the major problem. The molecular dipole moment of water, for instance, changes from $\mu = 1.855$ D in the gas phase molecule, to nearly $\mu = 3$ D in ice and in liquid water, but rigid ion models cannot reproduce this change. Moreover, within rigid-ion models, hydrogen bonds have only a Coulombic origin, contradicting the results of experiments and quantum chemistry computations showing that both Coulomb and covalent contributions are important [80] and change in slightly different ways upon changing the aggregation state of water.

Table 1: Cohesive energy (kJ/mol per water molecule) of $(H_2O)_2$, of cyclic water clusters $(H_2O)_n$, n = 3, 4, 5, 6, and of the cubic D_{2d} form of $(H_2O)_8$ computed by an SPC, rigid ion model (SPC/Fw, [77]). Deviations from dispersion-corrected [81] DFT [82] results are given in parentheses. Data are from [83].

n	2	3	4	5	6	8
PBE+vdW	12.06	25.95	34.48	36.09	36.82	45.28
SPC/Fw	14.35	26.66	33.44	35.10	35.67	40.69
	[18.99%]	[2.7%]	[−3.0%]	[−2.7%]	[−3.1%]	[−10.1%]

Somewhat surprisingly, the inclusion of polarizability into simple models has not resulted yet into the systematic improvement of the description of the properties for extended water systems [84], while it has been more successful for clusters.

All these difficulties have stimulated a large number of new attempts. It might be worth mentioning the representation of electron polarizability via classical [85] and quantum [86] Drude oscillators, the application to water [87] of the empirical valence band (EVB) theory [88] and the usage of polarizable Thole models [89].

Ab initio modeling, discussed in more detailed below, will eventually provide the method of choice to study water [90]. Until now, however, approaches of this kind using standard approximations for the exchange-correlation energy (see next section) have given rather mixed results [91].

THE *AB INITIO* ROUTE

Over the last twenty years, the art of representing PES as a function of atomic coordinates has seen its role increasingly challenged by the explosive growth of ab initio simulation methods.

As discussed in Section 2, the exact PES of a system made by N electrons evolving in the field of K nuclei can be determined point by point by computing the energy eigenvalues of the $\hat{H}_{ele}$ Hamiltonian:

$$E_k(\{\mathbf{R}_\alpha\}) = \frac{\hat{H}_{ele}\psi_k(\{\mathbf{r_i}\} \mid \{\mathbf{R}_\alpha\})}{\psi_k(\{\mathbf{r_i}\} \mid \{\mathbf{R}_\alpha\})} \tag{20}$$

For any single choice of the $\{\mathbf{R}_\alpha\}$ coordinates, a fairly extended array of quantum chemistry ab initio methods, such as configuration interaction, Møller-Plesset perturbation theory or coupled clusters, are available to find all or a few of the lowest energy eigenvalues and eigenvectors of this so-called standard problem in electronic structure computations.

For what concerns the direct application of ab initio methods to simulation, however, progress came primarily through the advent of density functional theory, whose recognized theoretical and practical foundation is provided by the Hohenberg-Kohn (HK) theorem [92] and by the seminal paper by Kohn and Sham (KS) [93]. In a very schematic way, density functional theory in the popular Kohn-Sham formulation represents the ground state electron density, $\rho(\mathbf{r})$, in terms of an auxiliary set of non-interacting electron orbitals $\{\phi_i(\mathbf{r}),\ i = 1, ..., K\}$, generally known as the Kohn-Sham orbitals:

$$\rho(\mathbf{r}) = \sum_{i=1}^{K} \mid \phi_i(\mathbf{r}) \mid^2 \tag{21}$$

To reproduce the exact density, the (unspecified) potential acting on the non-interacting electrons has to be different from the one acting on their interacting counterpart. The properties of such a potential and, in particular, its local, i.e., multiplicative nature are a corollary of the HKtheorem.

Then, according to KS, the system ground state energy is the minimum of the unique and universal functional:

$$E_{KS}[\rho \mid \{\mathbf{R}_\alpha\}] = -\frac{1}{2}\sum_{i=1}^{K}\langle\phi_i \mid \nabla^2 \mid \phi_i\rangle + \frac{1}{2}\int\frac{\rho(\mathbf{r})\rho(\mathbf{r}')}{\mid \mathbf{r}-\mathbf{r}' \mid}d\mathbf{r}d\mathbf{r}' - \sum_{\alpha=1}^{K}Z_\alpha\int\frac{\rho(\mathbf{r})d\mathbf{r}}{\mid \mathbf{r}-\mathbf{R}_\alpha \mid} + U_{XC}[\rho] \tag{22}$$

where $U_{XC}[\rho]$ is the so-called exchange correlation energy, a functional of the electron density, $\rho(\mathbf{r})$, which also contains a small fraction of the kinetic energy of the interacting electrons. Minimization of Equation (22) under the constraint of orthonormality for the Kohn-Sham orbitals results in a set of coupled partial differential equations for $\{\phi_i\}$.

Methods to solve this problem have been developed and discussed in a vast numbers of papers and textbooks [7,8]. The accuracy of the solution depends on the functional used to approximate $U_{XC}[\rho]$, and on the choice of the basis used to represent the orbitals. Popular choices for the exchange-correlation energy are generalized gradient corrections, such as PBE [82], or hybrid functionals, such as B3LYP [94]. Basis sets range from atomic orbitals to wavelets, but plane waves [95,96] and Gaussian functions [97] are probably the most widely used choice for implementations tuned on molecular dynamics applications.

The solution of the standard problem in Equation (5) obtained through Equation (22) is restricted to the ground state PES. Even within this limited scope, the PES itself can only be determined point by point. Nevertheless, the KS energy expression can be used to evolve the atomic positions in time, thus opening the way to MD, provided one can: (i) minimize Equation (22) fast enough; and (ii) evaluate forces on the atoms through:

$$\mathbf{F}_\beta = -\nabla_{\mathbf{R}_\beta} E_{KS}[\rho \mid \{\mathbf{R}_\alpha\}] \tag{23}$$

Towards this goal, the work of Car and Parrinello [9] has truly represented the single most important breakthrough, whose major innovation consisted of the introduction of direct minimization approaches for Equation (22), exploiting the close similarity of the electronic configuration at two successive steps of MD. Evaluation of forces, moreover, was greatly eased by the choice of plane waves as the basis set to represent KS orbitals, whose unbiased coverage of the entire space allows the application of the Hellmann-Feynman theorem in its simplest form to compute gradients of the ground state energy

[95,98]. Atoms evolve on the adiabatic PES implicitly defined by Equation (22) classically or quantum mechanically. The validity of a classical time evolution for the atoms according to Newton's equations relies on conditions discussed in detail in Chapter [6]. Outside these conditions, one could resort to a path integral approach, as done, for instance, in [99].

The method can be extended to simulate the atomic dynamics on the single PES of an electronically excited state [100], provided the different symmetry of the ground and excited state allows a meaningful definition of both PESs by density functional methods. As apparent from the discussion of the Born-Oppenheimer approximation, multiple PESs close in energy make it impossible to disentangle the ionic and electron dynamics, and in these cases, resorting to semiclassical or to more accurate quantum mechanical approaches [6] is mandatory.

Somewhat simplified versions of the density-functional-based MD, resorting to localized bases and relying on a self-consistent tight-binding approach have been developed [101,102] and provide a cheaper and popular alternative to unrestricted DFT methods. The price to be paid is a slight limitation in the quality of the solution, as well as occasional failures of the method.

The amazing success of density-functional-based simulation methods is due to the fact that they represent the only method endowed with truly predictive power, which can be used for systems of several hundred atoms, with up to a few thousand valence electrons. ab initio simulation, therefore, is the method of choice whenever we cannot guess a suitable representation of the PES or when we need an accuracy that cannot be provided by the empirical models that are available. Ab initio simulation is also strictly required for systems whose structure is affected by electronic effects, such as Jahn-Teller, and also enjoys a clear advantage in describing spin-polarization effects or systems undergoing chemical transformations and non-stoichiometric compounds exhibiting different valence states.

Well known drawbacks are represented by the computational cost that limits the size and especially the time scale of ab initio simulations, even though the reach of the method is constantly expanding. At present, large computations running on state-of-the-art facilities may involve ~ 1, 000 atoms and ~4, 000–5, 000 valence electrons. Early problems with metals have been progressively eased by approaches relying on the accurate step-by-step minimization of the KS energy functional. Problems, however, remain with transition and, especially, rare-earth metals, for which standard exchange-correlation approximations give unsatisfactory results, and quantum chemistry hybrid methods fail fairly spectacularly [103]. Progress is being achieved with methods incorporating strong correlation at some approximate level, such as LSD+U [104].

Difficulties remain also in the limit of weakly interacting molecular systems. Furthermore, in this case, early methods lacked essential components, such as the dispersion interaction, which in molecular systems provide a good portion of cohesion. Dispersion interactions are now increasingly included in ab initio simulations [81], especially for molecular systems and for water, in particular. Results are encouraging, although not yet in full quantitative agreement with experiments. However, the accuracy, reliability and computational efficiency of these methods are improving rapidly.

The major problem in current MD applications of ab initio methods arguably is that achieving accurate results for difficultsystems, such as transition metals and oxides or molecular systems, still require an extensive preliminary calibration stage and system-specific exchange correlation approximations [105], effectively spoiling the ab initio character of these methods. Perhaps more importantly, these adjustments of the model decrease their reliability for systems exhibiting different bonding types, since the improvement on one type might worsen the description of the other type.

Most of the cost of KS-DFT computations is due to the representation of the density in terms of KS orbitals. Approaches relying on genuine density functional formalism, such as a refined Thomas-Fermi method, could enjoy a huge computational advantage, but no successful scheme has emerged during the years, and only very idealized Gordon-Kim approaches [106] have been used with some success.

CONCLUSIONS

Explicit or implicit expressions of the PES of condensed matter systems represent the basis of our ability to simulate them, possibly understanding and sometimes predicting their properties by purely computational methods. For this reason, the development of approximations and efficient representations of PES is the focus of an intense research effort, involving a sizable portion of the computational community.

Such a modeling activity is an art as much as a science. It is a science in the systematic derivation of interatomic forces from more fundamental interactions. It is an art in the invention of effective ways to incorporate new ideas in physically transparent and computationally efficient mathematical expressions. Like many other forms of art, it relies on a big deal of craftsmanship, required in the stage of parameterizing force fields, validating them and incorporating them into widely used computer packages, using sophisticated programming techniques, tuned on state-of-the-art computational hardware.

It should be apparent from the discussion of the previous sections that the last thirty years have seen an amazing enhancement of our ability to model a wide variety of systems at the atomistic level, fueling the explosive growth of simulation studies, while, at the same time, being driven by it. Equally amazing, however, is the extent of what we are still unable to model satisfactorily. Interfaces between different materials, for instance, are intrinsically difficult to describe by simple approaches. Excluding ab initio, no reliable, general and widely accepted model is available to simulate water and electrolyte solutions in contact with neutral or charged electrodes, organic and biological molecules on solid surfaces or the junction of metal and semiconducting phases. Even homogeneous phases, such as non-stoichiometric oxides, still represent a formidable challenge for models suitable for simulating 10^4 atoms over 100 ns or more. Systems undergoing chemical transformations are another sore point, even though methods, such as ReaxFF and REBO, are achieving progress in this direction.

At this stage, strategic decisions on the directions and aims of the modeling effort have to take into account the rapid growth of ab initio methods, which easily account for the intermixing of different bonding categories, cover electrostatic polarizability, provide information on excited state PES and may include magnetic interactions and spin effects through their approximate description of exchange.

The rapid progress of methods and computational equipment implies that the foreseeable future spans at most ten to fifteen years from now. Over this time, empirical models of PES will continue to play an important and useful role in the atomistic simulation of large systems ($N \simeq 10^4$ atoms) over times in excess of 100 ns. Most biochemistry and biophysics simulations fall into this class.

On the longer run, however, the general picture of modeling might indeed change. First of all, the domain proper to atomistic modeling concerns the investigation of the microscopic details underlying larger-scale phenomena. In this context, the scales of interest rarely exceed $\sim 10^4$ atoms and correspondingly short times of less than ~ 10 ns. Beyond this range, simulation may become the exclusive domain of coarse graining and multi-scale approaches, provided refined versions of these methods are developed over the next few years.

Ab initio methods already represent the method of choice for systems for which we do not have reliable approximations of their PES, for phenomena that can be represented by 100 to 1, 000 atoms and that take place within a 50–100 ps time span. Mixed QM/MM approaches extend this reach and represent the most appealing method to treat systems, such as protein reaction centers, organometallic catalysts, etc., in which a small portion of a large system needs to be represented in full chemical detail.

The parallel development of ab initio and of refined coarse graining and multi-scale methods, therefore, could greatly shrink the role of empirical PES approximations in atomistic simulation. Even these likely developments, however, might not mark the end of atomistic potential models, since simple and transparent representations of PES will continue to provide the conceptual basis to rationalize the properties of condensed matter systems in terms of atoms, of molecules and of their microscopic interactions.

ACKNOWLEDGMENTS

I thank Carlo Pierleoni for useful discussions and for a careful reading of the manuscript.

REFERENCES

1. Frenkel, D.; Smit, B. Understanding Molecular Simulation, 2nd ed; Academic Press: San Diego, CA, USA, 2002.
2. Morales, M.A.; Clay, R.; Pierleoni, C.; Ceperley, D.M. First-principle methods: A perspective from quantum Monte Carlo. Entropy 2014, 16, 287–321.
3. Finnis, M. Interatomic Forces in Condensed Matter; Oxford Series on Materials Modeling: Oxford, UK, 2010.
4. Delle Site, L. What is a multiscale problem in molecular dynamics? Entropy 2014, 16, 23–40.
5. Ziman, J.M. Electrons and Phonons; Oxford University Press: Oxford, UK, 1960; Chapter 5.
6. De Carvalho, F.F.; Bouduban, M.E.F.; Curchod, B.F.E.; Tavernelli, I. Nonadiabatic molecular dynamics based on trajectories. Entropy 2014, 16, 62–85.
7. Martin, R.M. Electronic Structure, Basic Theory and Practical Methods; Cambridge University Press: Cambridge, UK, 2004.
8. Kaxiras, E. Atomic and Electronic Structure of Solids; Cambridge University Press: Cambridge, UK, 2003.
9. Car, R.; Parrinello, M. Unified approach for molecular dynamics and density-functional theory. Phys. Rev. Lett 1985,55, 2471–2474.
10. Born, M.; Huang, K. Dynamical Theory of Crystal Lattices; Oxford University Press: Oxford, UK, 1954.
11. Yarkony, D.R. Diabolical conical intersections. Rev. Mod. Phys 1996, 68, 985–1013.
12. Kroes, G.J.; Gross, A.; Baerends, E.J.; Scheffler, M.; McCormack, D.A. Quantum theory of dissociative chemisorption on metal surfaces. Acc. Chem. Res 2002, 35, 193–200.
13. Bowman, J.M. Beyond born-oppenheimer. Science 2008, 319, 40.
14. White, J.D.; Chen, J.; Matsiev, D.; Auerbach, D.J.; Wodtke, A.M. Conversion of large-amplitude vibration to electron excitation at a metal surface. Nature 2005, 433, 503–505.
15. Abedi, A.; Maitra, N.T.; Gross, E.K.U. Exact factorization of the time-dependent electron-nuclear wave function. Phys. Rev. Lett 2010, 105, 123002.
16. Abedi, A.; Maitra, N.T.; Gross, E.K.U. Correlated electron-nuclear dynamics: Exact factorization of the molecular wavefunction. J. Chem. Phys 2012, 137, 22A530.
17. Hunter, G. Conditional probability amplitudes in wave mechanics. Int. J. Quantum Chem 1975, 9, 237–242.
18. Berry, M.V. Quantal phase factors accompanying adiabatic changes. Proc. R. Soc. Lond. Ser. A 1984, 392, 45–57.
19. Pancharatnam, S. Generalized theory of interference, and its applications. Proc. Indian Acad. Sci. A 1956, 44, 247–262.
20. Sutcliffe, B. Is there an exact potential energy surface? Theor. Chem. Acc 2012, 131.
21. Boscovich, R.J. Theoria Philosophiae Naturalis; Venice, Italy, 1758.
22. Lennard-Jones, J.E. On the determination of molecular fields. Proc. R. Soc. Lond. A 1924, 106, 463–477.
23. Fumi, F.G.; Tosi, M.P. Ionic sizes and Born repulsive parameters in the NaCl-type alkali halides. J. Phys. Chem. Solids1964, 25, 31–43.
24. Carlsson, A.E.; Ashcroft, N.W. Pair potentials from band theory: Application to vacancy-formation energies. Phys. Rev. B 1983, 27, 2101–2110.
25. Abell, G.C. Empirical chemical pseudopotential theory of molecular and metallic bonding. Phys. Rev. B 1985, 31, 6184–6196.
26. Carlsson, A.E. Beyond Pair Potentials in Elemental Transition Metals and Semiconductors. In Solid State Physics: Advances in Research and Applications;

Ehrenreich, H., Turnbull, D., Eds.; Academic Press: Boston, MA, USA, 1990; Volume 43, pp. 1–91.

27. Pauling, L. The Nature of the Chemical Bond, 3rd ed.; Cornell University Press: Ithaca, NY, USA, 1960.
28. Hafner, J. From Hamiltonians to Phase Diagrams; Springer: Berlin, Germany, 1987.
29. Handbook of Materials Modeling; Yip, S., Ed.; Springer: Berlin, Germany, 2005.
30. In Comprehensive Nuclear Materials; Basic Aspects of Radiation Effects in Solids/Basic Aspects of Multi-Scale Modeling. Konings, R.J.M., Ed.; Elsevier: Amsterdam, The Netherlands, 2012; Volume 1.
31. Dagens, L.; Rasolt, M.; Taylor, R. Charge densities and interionic potentials in simple metals. Phys. Rev. B 1975, 11, 2726–2734.
32. Moriarty, J.A. First-principles interatomic potentials in transition metals. Phys. Rev. Lett 1985, 55, 1502–1505.
33. Daw, M.S.; Baskes, M.I. Semiempirical, quantum mechanical calculation of hydrogen embrittlement in metals. Phys. Rev. Lett 1983, 50, 1285–1288.
34. Baskes, M.I. Application of the embedded-atom method to covalent materials: A semiempirical potential for silicon.Phys. Rev. Lett 1987, 59, 2666–2669.
35. Rosato, V.; Guillope, M.; Legrand, B. Thermodynamical and structural properties of f.c.c. transition metals using a simple tight-binding model. Philos. Mag. A 1989, 59, 321–336.
36. Jacobsen, K.W.; Nørskov, J.K.; Puska, M.J. Interatomic interactions in the effective-medium theory. Phys. Rev. B 1987,35, 7423–7442.
37. Drautz, R.; Pettifor, D.G. Valence-dependent analytic bond-order potential for transition metals. Phys. Rev. B 2006, 74, 174117.
38. Miiller, A.P.; Brockhouse, B.N. Crystal dynamics and electronic specific heats of palladium and copper. Can. J. Phys1971, 49, 704–723.
39. Stillinger, F.; Weber, T. Computer simulation of local order in condensed phases of silicon. Phys. Rev. B 1985, 31, 5262–5271.
40. Tersoff, J. New empirical model for the structural properties of silicon. Phys. Rev. Lett 1986, 56, 632–635.
41. Tersoff, J. New empirical approach for the structure and energy of covalent systems. Phys. Rev. B 1988, 37, 6991–7000.
42. Brenner, D.W. Empirical potential for hydrocarbons for use in simulating the chemical vapor deposition of diamond films. Phys. Rev. B 1990, 42, 9458–9471.
43. Pettifor, D.G.; Oleinik, I.I. Analytic bond-order potentials beyond Tersof-Brenner. I. Theory. Phys. Rev. B 1999, 59, 8487–8499.
44. Van Duin, A.C.T.; Dasgupta, S.; Lorant, F.; Goddard, W.A., III. ReaxFF: A reactive force field for hydrocarbons. J. Phys. Chem. A 2001, 105, 9396–9409.
45. Brenner, D.W.; Shenderova, O.A.; Harrison, J.A.; Stuart, S.J.; Ni, B.; Sinnott, S.B. A second-generation reactive empirical bond order (REBO) potential energy expression for hydrocarbons. J. Phys.: Condens. Matter 2002, 14, 783–802.
46. Keating, P.N. Effect of invariance requirements on the elastic strain energy of crystals with application to the diamond structure. Phys. Rev 1966, 145, 637–645.
47. Cabriolu, R.; Del Popolo, M.G.; Ballone, P. Melting of a tetrahedral network model of silica. Chem. Phys. Phys. Chem2009, 11, 10820–10823.
48. Cabriolu, R.; Ballone, P. Thermodynamic properties and atomistic structure of the dry amorphous silica surface from a reactive force field model. Phys. Rev. B 2010, 81, 155432.
49. Ponder, J.W.; Wu, C.; Ren, P.; Pande, V.S.; Chodera, J.D.; Schnieders, M.J.; Haque, I.; Mobley, D.L.; Lambrecht, D.S.; DiStasio, R.A.; et al. Current status of the AMOEBA polarizable force field. J. Phys. Chem. B 2010, 114, 2549–2564.
50. Dick, B.G.; Overhauser, A.W. Theory of the dielectric constants of alkali halide crystals. Phys. Rev 1958, 112, 90–103.
51. Mitchell, P.J.; Fincham, D. Shell model simulations by adiabatic dynamics. J. Phys.: Condens. Matter 1993, 5, 1031–1038.
52. Rezaei, M.; Sheybani-Deloui, S.; Moazzen-Ahmadi, N.; Michaelian, K.H.; McKellar, A.R.W. Spectroscopic evidence for a planar cyclic CO trimer. J. Chem. Phys 2013, 138, 071102.
53. Visser, G.W.M.; Hesselmann, A.; Jansen, G.; Wormer, P.E.S.; van der Avoird, A. New CO-CO interaction potential tested by rovibrational calculations. J. Chem. Phys 2005, 122, 054306.
54. Righi, M.C.; Ferrario, M. Potential energy surface for rare gases adsorbed on Cu(111): Parameterization of the gas/metal interaction potential. J. Phys.: Condens. Matter 2007, 19, 305008.
55. Cornell, W.D.; Cieplak, P.; Bayly, C.I.; Gould, I.R.; Merz, K.M., Jr; Ferguson, D.M.; Spellmeyer, D.C.; Fox, T.; Caldwell, J.W.; Kollman, P.A. A second generation force field for the simulation of proteins, nucleic acids, and organic molecules. J. Am. Chem. Soc 1995, 117, 5179–5197.
56. Patel, S.; Brooks, C.L., III. CHARMM fluctuating charge force field for proteins: I parameterization and application to bulk organic liquid simulations. J. Comput. Chem 2004, 25, 1–16.
57. Jorgensen, W.L.; Maxwell, D.S.; Tirado-Rives, J. Development and testing of the OPLS all-atom force field on conformational energetics and properties of organic liquids. J. Am. Chem. Soc 1996, 118, 11225–11236.
58. Schuler, L.D.; Daura, X.; van Gunsteren, W.F. An improved GROMOS96 force field for aliphatic hydrocarbons in the condensed phase. J. Comp. Chem 2001, 22, 1205–1218.
59. Chandra Singh, U.; Kollman, P.A. An approach to computing electrostatic charges for molecules. J. Comput. Chem1984, 5, 129–145.
60. Cornell, W.D.; Cieplak, P.; Bayly, C.I.; Kollman, P.A. A well-behaved electrostatic potential based method using

charge restraints for deriving atomic charge. J. Phys. Chem 1993, 97, 9620–9631.

61. Roux, B.; Simonson, T. Implicit solvent models. Biophys Chem 1999, 78, 1–20.
62. Pidcock, A.; Richards, R.E.; Venanzi, L.M. $^{195}Pt^{31}P$ nuclear spin coupling constants and the nature of the trans-effect in platinum complexes. J. Chem. Soc. A 1966, 1707–1710.
63. Welton, T. Room-temperature ionic liquids. Solvents for synthesis and catalysis. Chem. Rev 1999, 99, 2071–2084.
64. Canongia Lopes, J.N.; Deschamps, J.; Padua, A.A.H. Modeling ionic liquids using a systematic all-atom force field. J. Phys. Chem. B 2004, 108, 2038–2047.
65. Canongia Lopes, J.N.; Deschamps, J.; Padua, A.A.H. Modeling ionic liquids using a systematic all-atom force field. J. Phys. Chem. B 2004, 108, 11250.
66. Bingham, R.; Ballone, P. Computational study of room-temperature ionic liquids interacting with a POPC phospholipid bilayer. J. Phys. Chem. B 2012, 116, 11205–11216.
67. Brovchenko, I.; Oleinikova, A. Multiple phases of liquid water. Chem Phys Chem 2008, 9, 2660–2675.
68. Billeter, S.R.; King, P.M.; van Gunsteren, W.F. Can the density maximum of water be found by computer-simulation. J. Chem. Phys 1994, 100, 6692–6699.
69. Guillot, B.; Guissani, Y. How to build a better pair potential for water. J. Chem. Phys 2001, 114, 6720–6733.
70. Guillot, B. A reappraisal of what we have learnt during three decades of computer simulations on water. J. Mol. Liquids 2002, 101, 219–260.
71. Jorgensen, W.L.; Chandrasekhar, J.; Madura, J.D.; Impey, R.W.; Klein, M.L. Comparison of simple potential functions for simulating liquid water. J. Chem. Phys 1983, 79, 926–935.
72. Mahoney, M.W.; Jorgensen, W.L. A five-site model for liquid water and the reproduction of the density anomaly by rigid, nonpolarizable potential functions. J. Chem. Phys 2000, 112, 8910–8922.
73. Abascal, J.L.F.; Sanz, E.; Garcia Fernandez, R.; Vega, C. A potential model for the study of ices and amorphous water: TIP4P/Ice. J. Chem. Phys 2005, 122, 234511.
74. Berendsen, H.J.C.; Postma, J.P.M.; von Gunsteren, W.F.; Hermans, J. Intermolecular Forces; Pullman, P., Ed.; Reidel: Dordrecht, Holland, 1981.
75. Berendsen, H.J.C.; Grigera, J.R.; Straatsma, T.P. The missing term in effective pair potentials. J. Phys. Chem 1987, 91, 6269–6271.
76. Dang, L.X.; Pettitt, B.M. Simple intermolecular model potentials for water. J. Phys. Chem 1987, 91, 3349–3354.
77. Wu, Y.; Tepper, H.L.; Voth, G.A. Flexible simple point-charge water model with improved liquid-state properties. J. Chem. Phys 2006, 124, 024503.
78. Chialvo, A.A.; Cummings, P.T. Molecular-based modeling of water and aqueous solutions at supercritical conditions.Adv. Chem. Phys 1999, 109, 115–205.
79. Temelso, B.; Archer, K.A.; Shields, G.C. Benchmark structures and binding energies of small water clusters with anharmonicity corrections. J. Phys. Chem. A 2011, 115, 12034–12046.
80. Khaliullin, R.Z.; Bell, A.T.; Head-Gordon, M. Electron donation in the water-water hydrogen bond. Chem.-Eur. J 2009,15, 851–855.
81. Grimme, S. Semiempirical GGA-type density functional constructed with a long-range dispersion correction. J. Comput. Chem 2006, 27, 1787–1799.
82. Perdew, J.P.; Burke, K.; Ernzerhof, M. Generalized gradient approximation made simple. Phys. Rev. Lett 1996, 77, 3865–3868.
83. Bingham, R.J.; Ballone, P. Energy, structure and vibrational modes of small water clusters by a simple many-body potential mimicking polarization effects. Mol. Phys 2013.
84. Chen, B.; Potoff, J.J.; Siepmann, I.J.I. Adiabatic nuclear and electronic sampling Monte Carlo simulations in the Gibbs ensemble: Application to polarizable force fields for water. J. Phys. Chem. B 2000, 104, 2391–2401.
85. Lamoureux, G.; Harder, E.; Vorobyov, I.V.; Roux, B.; MacKerell, A.D., Jr. A polarizable model of water for molecular dynamics simulations of biomolecules. Chem. Phys. Lett 2006, 418, 245–249.
86. Jones, A.; Cipcigan, F.; Sokhan, V.P.; Crain, J.; Martyna, G.J. Electronically coarse-grained model for water. Phys. Rev. Lett 2013, 110, 227801.
87. Day, T.J.F.; Soudackov, A.V.; Cuma, M.; Schmitt, U.W.; Voth, G.A. A second generation multistate empirical valence bond model for proton transport in aqueous systems. J. Chem. Phys 2002, 117, 5839–5849.
88. Aqvist, J.; Warshel, A. Simulation of enzyme-reactions using valence-bond force-fields and other hybrid quantum-classical approaches. Chem. Rev 1993, 93, 2523–2544.
89. Thole, B.T. Molecular polarizabilities calculated with a modified dipole interaction. Chem. Phys 1981, 59, 341–350.
90. Zhang, C.; Wu, J.; Galli, G.; Gygi, F. Structural and vibrational properties of liquid water from van der Waals density functionals. J. Chem. Theory Comput 2011, 7, 3054–3061.
91. Yoo, S.; Zeng, X.C.; Xantheas, S.S. On the phase diagram of water with density functional theory potentials: The melting temperature of ice Ih with the PBE and BLYP functionals. J. Chem. Phys 2009, 130, 221102.
92. Hohenberg, P.; Kohn, W. Inhomogeneous electron gas. Phys. Rev 1964, 136, B864–B871.
93. Kohn, W.; Sham, L.J. Self-consistent equations including exchange and correlation effects. Phys. Rev 1965, 140, A1133–A1138.
94. Becke, A.D. Density-functional thermochemistry. III. The role of exact exchange. J. Chem. Phys 1993, 98, 5648–5652.
95. Marx, D.; Hutter, J. Modern Methods and Algorithms of Quantum Chemistry; Grotendorst, J., Ed.; John von Neumann Institute for Computing: Jülich, Germany, 2000; NIC Series; Volume 1, pp. 301–449.

96. Quantum Espresso is an Open Source Distribution. Available online: http://www.quantum-espresso.org (accessed on 15 October 2013).

97. Cp2k is an Open Source Distribution. Available online: http://www.cp2k.org (accessed on 15 October 2013).

98. Srivastava, G.P.; Weaire, D. The theory of the cohesive energies of solids. Adv. Phys 1997, 36, 463–517.

99. Morrone, J.A.; Car, R. Nuclear quantum effects in water. Phys. Rev. Lett 2008, 101, 017801.

100. Frank, I.; Hutter, J.; Marx, D.; Parrinello, M. Molecular dynamics in low-spin excited states. J. Chem. Phys 1998, 108, 4060–4069.

101. Soler, J.M.; Artacho, E.; Gale, J. D; García, A.; Junquera, J.; Ordejón, P.; Sánchez-Portal, D. The SIESTA method for ab initio order-N materials simulation. J. Phys.: Condens. Matter 2002, 14, 2745–2779.

102. Porezag, D.; Frauenheim, Th.; Köhler, Th.; Seifert, G.; Kaschner, R. Construction of tight-binding-like potentials on the basis of density-functional theory: Application to carbon. Phys. Rev. B 1995, 51, 12947–12957.

103. Paier, J.; Marsman, M.; Kresse, G. Why does the B3LYP hybrid functional fail for metals? J. Chem. Phys 2007, 127, 024103.

104. Anisimov, V.I.; Zaanen, J.; Andersen, O.K. Band theory and Mott insulators: Hubbard U instead of Stoner I. Phys. Rev. B 1991, 44, 943–954.

105. Tonigold, K.; Gross, A. Dispersive interactions in water bilayers at metallic surfaces: A comparison of the PBE and RPBE functional including semiempirical dispersion corrections. J. Comp. Chem 2012, 33, 695–701.

106. Tabacchi, G.; Hutter, J.; Mundy, C.J. A density-functional approach to polarizable models: A Kim-Gordon response density interaction potential for molecular simulations. J. Chem. Phys **2005**, 123, 074108.

CHAPTER 11 COMPUTATIONAL METHODS IN MATERIALS SCIENCE

Martin O. Steinhauser and Stefan Hiermaier

Keywords: Computational physics, modeling and simulation, multiscale methods, polymers, biopolymers, dendrimers, shock waves, lithotripsy, molecular dynamics

INTRODUCTION

Some of the most fascinating problems in all fields of science involve multiple temporal or spatial scales. Many processes occurring at a certain scale govern the behavior of the system across several (usually larger) scales. This notion and practice of multiscale modeling can be traced back to the beginning of modern science, see e.g., the discussions in [1–3]. In many problems of materials science this notion arises quite naturally as a structure-property paradigm: The basic microscopic constituents of materials are atoms, and their interactions at the microscopic level (on the order of nanometers and femtoseconds) determine the behavior of the material at the macroscopic scale (on the order of centimeters and milliseconds and beyond), with the latter being the scale of interest for technological applications. The idea of performing material simulations across several characteristic length and time scales has therefore obvious appeal as a tool of potentially great effect on technological innovation.

With the increasing availability of very fast computers and concurrent progress in the development and understanding of efficient algorithms, numerical simulations have become prevalent in virtually any field of research [4–10]. Fast parallelized computer systems today allow for solving complex, non-linear many body problems directly, not involving any preceding mathematical approximations which is the normal case in analytical theory, where all but the very simplest problems of practical interest are too complex to be solved by pencil and paper. Computer simulations are not only a connecting link between analytic theory and experiment, allowing to scrutinize theories, but they can also be used as an exploratory research tool under physical conditions not feasible in real experiments in a laboratory. Computational methods have thus established a new, interdisciplinary research approach which is often referred to as "Computational Materials Science" or "Computational Physics". This approach brings together elements from diverse fields of study such as physics, mathematics, chemistry, biology, engineering and even medicine and has the potential to handle multiscale and multi-disciplinary simulations in realistic situations.

For example, simulations in material physics are focused on the investigation of lattice and defect dynamics at the atomic scale using MD and Monte Carlo (MC) methods, often using force-fields (physical potentials) that are derived from solving the non-relativistic Schrödinger equation for a limited number of atoms [11–13]. In contrast to this, materials-related simulations in the field of mechanical engineering typically focus on large-scale problems, often resorting to finite element methods (FEM) where the micro structure is homogenized by using averaging constitutive laws [14–16].

With powerful computational tools at hand, even simulations of practical interest in engineering sciences for product design and testing have become feasible. Material systems of industrial interest are highly heterogeneous and are characterized by a variety of defects, interfaces, and other microstructural features. As an example, Figure 1 displays a Scanning Electron Microscope (SEM) micrograph of the granular surface structure and the fracture surface of Aluminum Oxide (Al_2O_3) after planar impact load. Inorganic crystalline materials have structural features such as grain boundaries between crystals which are mm to μm in size (cf. Figure 1a), dislocations, and point defects such as vacancies on the atomic scale. Hence, these structures have to be studied from a hierarchical perspective [17].

Citation: Martin O. Steinhauser and Stefan Hiermaier, Computational Methods in Materials Science, doi:10.3390/e16010322

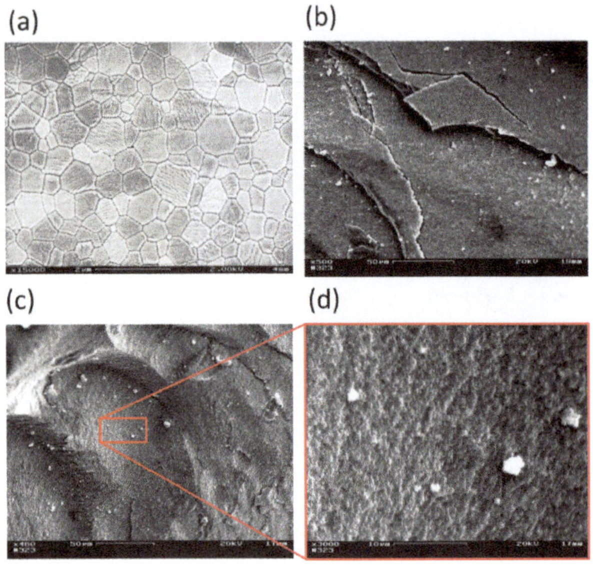

Figure 1: (**a**) Micrograph section of an etched Al_2O_3 ceramic surface exhibiting the granular structure on the microscale. (**b**) SEM photograph of the fracture surface of Al_2O_3 after edge-on impact experiment [18] with striking speed of $v \approx 400$m/s. (**c,d**) Microstructural details of the Al_2O_3 surface exhibiting structural hierarchies. Figure by M.O. Steinhauser, Fraunhofer EMI.

Recently, polycrystalline materials have been synthesized with a distribution of grain sizes less than 1 micron [19–23] (nanocrystalline materials). The small grain size, hence large interface area, gives rise to desirable properties such as superplasticity (in which large irreversible deformation can occur without fracture) [24], and improved strength and toughness. Small grain size also implies short diffusion distances, so that processes which depend on diffusion, such as sintering, are facilitated and can occur at lower temperatures than would otherwise be possible. Predicting the properties and performance of such materials under load is central for modern materials research and for product design in industry. However, due to the complexity of structural hierarchies in condensed matter on different scales, there is no single computational model or physical theory which can predict and explain all material behavior in one unified and all-embracing approach. Hence, the explicit micro structure of different important classes of materials such as metals, ceramics, or materials pertaining to soft matter (glasses or polymers) has to be incorporated in different models with delimited validity, cf. Figure 2.

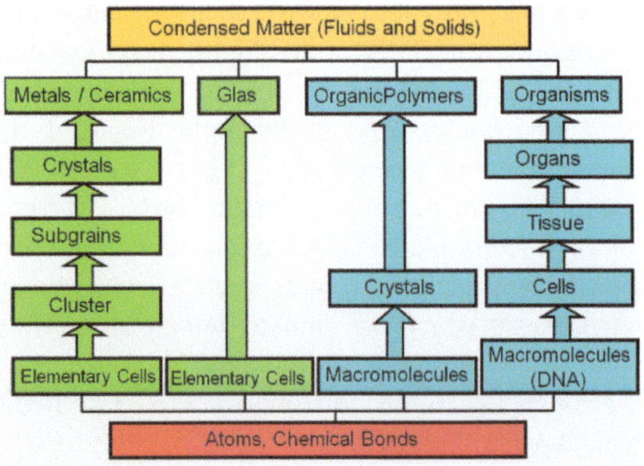

Figure 2: Schematic hierarchical view of structural properties of important classes of materials contrasting typical structural features of inorganic crystalline materials (engineering materials, green) and the structural features of self-organizing organic biological materials (blue). At the nanoscale, the basic constituents of all condensed matter are the atoms bound together in chemical bonds.

For example, in the analysis of fibrous composites where fibers are embedded in a matrix to form an anisotropic sheet or lamina (which in turn are bonded together to form a laminate), cf. Figure 3, the fibers and matrix are regarded as continuous media in the analysis of the single lamina [25,26]; the laminae are then regarded as continuous in the analysis of the laminate. The stacking sequence of laminae and the orientation of fibers within them governs the anisotropy of the composite. A similar continuum assumption is often used in the analysis of particulate composites [27] and of foams [28]. Many biopolymers such as collagen show a similar hierarchical structure where the amino acids are organized in triple-helical fibrous tropocollagen molecules which are about 300 nm long and 1.5nm in diameter and which in turn are organized in fibrils and fibers on the micrometer scale [29,30]. Most of the structural materials used by Nature are polymers or composites of polymers. Such materials would probably not be the first choice of an engineer intending to build very stiff and long-lived mechanical structures (see Figure 2). The engineer selects materials to fabricate a part according to an exact design. In contrast, Nature goes the opposite direction and grows both the material and the whole organism, e.g., a plant or an animal, using the principles of biologically controlled self-assembly. Additionally, biological structures are able to grow, remodel and adapt to changing environmental conditions during their whole lifetime, which even allows for self-repair [31].

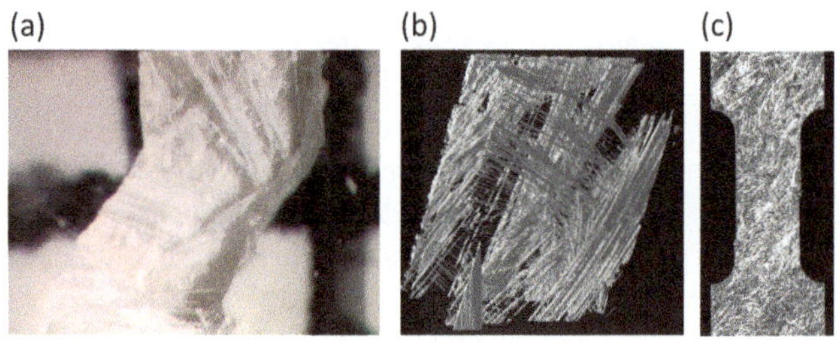

Figure 3: (a) Light microscopic view of a glass fiber reinforced sheet moulding compound (SMC) which is used in car industry as light-weight material. (b) CT-microscopic reconstruction of a section of the laminar glass fiber structure in the material. (c) Scanning Acoustic microscopic view of a tensile-test SMC specimen exhibiting the glass fiber bundles within the compound. Figure by M.O. Steinhauser, Fraunhofer EMI.

The typical hierarchical structural features of materials have to be taken into account when developing mathematical and numerical models which describe their behavior. With this respect, usually one of two possible strategies is pursued: In a "sequential modeling approach" one attempts to piece together a hierarchy of computational approaches in which large-scale models use the coarse-grained representations with information obtained from more detailed, smaller-scale models ("bottom-up" vs. "top-down" approach). This sequential modeling technique has proven effective in systems in which the different scales are weakly coupled. The vast majority of multiscale simulations that are actually in use is sequential. Examples of such approaches are abundant in literature, including practically all MD simulations whose underlying potentials are derived from ab initio calculations

[32]. The second strategy pursued in multiscale simulations is the "concurrent" or "parallel approach". Here, one attempts to link methods appropriate at each scale together in a combined model, where the different scales of the system are considered concurrently and often communicate with some type of handshaking procedure [33–35]. This approach is necessary for systems, whose behavior at each scale inherently depends strongly on what happens at the other scales, for example dislocations, grain boundary structure, or dynamic crack propagation in polycrystalline materials.

This review is organized as follows: In the next section we first discuss the relevance of physical model building for computer simulation. Then, in Section 3., a survey of typical simulation techniques for numerical simulation of condensed matter systems on different length and time scales is provided before we focus in a tutorial-like fashion on the MD method in Section 4. Here, typical numerical optimization techniques for the search for particle interactions are first reviewed, followed by general considerations on the efficiency and the run time of algorithms used for MD applications. We then review recent MD applications in shock wave physics in Section 5. and of polymer physics using coarse-grained particle models in Section 6. Here, we focus on some aspects of computer simulations of macromolecules which are of relevance for biopolymers such as DNA, polypeptides, or cell membranes. Finally, in Section 7, we briefly explore as a promising emerging interdisciplinary research field the application of concepts of polymer and shock wave physics to biological systems which may contribute to an improved understanding of medical applications such as non-invasive extracorporeal shock wave lithotripsy or tumor treatment. Our review ends with concluding remarks in Section 8.

PHYSICAL AND NUMERICAL MODELING

The span of length scales commonly pertaining to materials science comprises roughly 10 to 12 orders of magnitude and classical physics is sufficient to describe most of the occurring phenomena, cf. Figure 4. Yet, classical MD or MC methods are only valid down to length scales comparable to the typical size of atoms ($\approx 10^{-10}$m) and typically treat atoms as point particles or spheres with eigenvolume. In principle, the relativistic time-dependent Schrödinger equation describes the properties of molecular systems with high accuracy, but anything more complex than the equilibrium state of a few atoms cannot be handled at this ab initio level. Quantum theory as a model for describing materials behavior is valid also on the macroscopic scale, but the application of the (non-relativistic) Schrödinger equation to many particle systems of macroscopic size is completely in vain due to the non-tractable complexity of the involved calculations. Hence, approximations are necessary; the larger the complexity of a system and the longer the involved time span of the investigated processes are, the more severe the required approximations are. For example, at some point, the ab initio approach has to be abandoned completely and replaced by empirical parameterizations of the used model. Therefore, depending on the kind of question that one asks and depending on the desired accuracy with which specific structural features of the considered system are resolved, one has the choice between many different models which often can be usefully employed on a whole span of length and time scales.

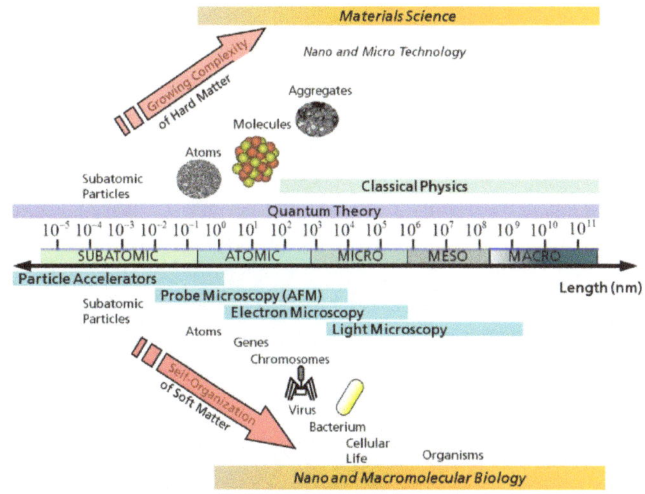

Figure 4: Schematic comparing the relevant length scales in materials science according to [3]. In the field of nano- and microtechnology one usually tries to approach the molecular level from larger scales, miniaturizing technical devices, whereas nature itself always seems to follow a bottom-up approach, assembling and self-organizing its complex (soft) structures from the atomic scale to complex cellular organisms. The typical scopes of important experimental methods using microscopes is displayed as well. The validity of classical physics is limited to length scales down to approximately the size of atoms which, in classical numerical schemes, are often treated as point particles or spheres with a certain eigenvolume.

Unfortunately, there is no simple "hierarchy" that is connected with a length scale according to which the great diversity of simulation methods could be sorted out. For example, Monte Carlo lattice methods can be applied at the femtoscale of Quantumchromodynamics (10^{-15} m) [36], at the Ångstrøm scale (10^{-10} m) of solid state crystal lattices [37], or at the micrometer scale (10^{-6} m), simulating grain growth processes of polycrystal solid states [38]. Thus, before getting started with computer simulations, as always in research, it is important to establish first, which phenomena and properties one is primarily interested in and which questions one is going to ask.

In many practical cases, the basic question which model shall be used for answering a specific question is the main problem. The next step for rendering the model accessible to an algorithmic description, is the discretization of the time variable (for dynamic problems) and of the spatial domain in which the constitutive equations of the problem are to be solved. Then appropriate algorithms for solving the equations of the mathematical model have to be chosen and implemented. Before trusting the output of a newly written computer program and before applying it to new problems, the code should always be tested to the effect whether it reproduces known analytic or experimental results. This is a necessity as the correctness and plausibility of the outcome of an algorithm (usually dimensionless numbers) cannot be predicted by simply looking at the source code. The success of a computer experiment in essence depends on the creation of a model which is sufficiently detailed such that the crucial physical effects are reproduced and yet is sufficiently simple (of small complexity) for the simulation still to be feasible.

Once a decision is made, a physical model is expressed as mathematical equations which are solved in a systematic fashion, i.e., in a way that can be formulated as a finite stochastic or deterministic algorithm and be implemented as a computer program. The numerical solutions of the governing equations associated with the physical and mathematical model are then interpreted and provide answers to the specific real system which was transformed into the model system. A comparison of the answers for a specific problem obtained by mathematically exploiting a specific model, finally provides some ideas about its general validity and quality as well as of the derivations and theoretical concepts associated with it. This principal procedure in physical and numerical modeling is illustrated schematically in the flowchart of Figure 5.

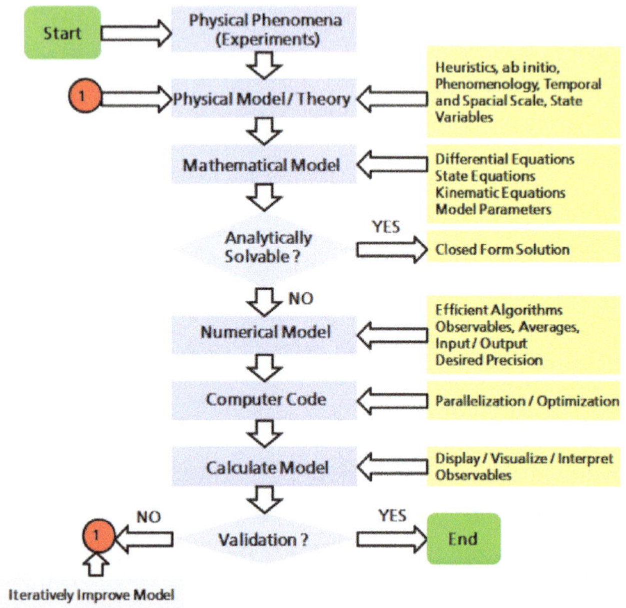

Figure 5: Physical, mathematical and numerical modeling scheme illustrated as flow chart. Starting from the experimental evidence one constructs physical theories for which a mathematical formulation usually leads to differential equations, integral equations, or master (rate) equations for the dynamic (i.e., time dependent) development of certain state variables within the system's abstract state space. Analytic solutions of these equations are very rarely possible, except when introducing simplifications usually involving symmetries. Thus, efficient algorithms for the treated problem have to be found and implemented as a computer program. Execution of the code yields approximate numerical solutions to the mathematical model which describes the dynamics of the physical "real" system. Comparison of the obtained numerical results with experimental data allows for a validation of the used model and subsequent iterative improvement of the model and of theory.

Computer Simulations as a Research Tool

Computer simulation is adding a new dimension to scientific investigation and has been established as an investigative research tool which is as important as the traditional approaches of experiment and theory. The experimentalist is concerned with obtaining factual information concerning physical states and dynamic processes. The theorist, challenged by the need to explain measured physical phenomena, invents idealized models which are subsequently translated in a mathematical formulation. As is common in theory, most mathematical analysis of the basic laws of nature as we know them, is too complex to be done in full generality and thus one is compelled to make certain model simplifications in order to make predictions. Hence, a comparison between a theoretical prediction and an experimental interpretation is frequently questioned because of the simplifying approximations with which the theoretical solution was obtained or because of the uncertainty of the experimental interpretation. For example, even the relatively "simple" laws of Newtonian mechanics become analytically unsolvable, as soon as there are more than two interacting bodies involved [39,40]. Most of materials science however deals with many ($N \sim 10^{23}$) particles, atoms, molecules or abstract constituents of a system.

Computer simulations, or computer experiments, are much less impaired by many degrees of freedom, lack of symmetries, or non-linearity of equations than analytical approaches. As a result, computer simulations establish their greatest value for those systems where the gap between theoretical prediction and laboratory measurements is large.

The principal design of practically all computer simulation programs for scientific purposes is displayed in Figure 6: Usually, during a pre-processing phase some administrative tasks are done (system setup, defining initial system structure, reading in simulation parameters, initializing internal variables, etc.) before the actual simulation run is started.

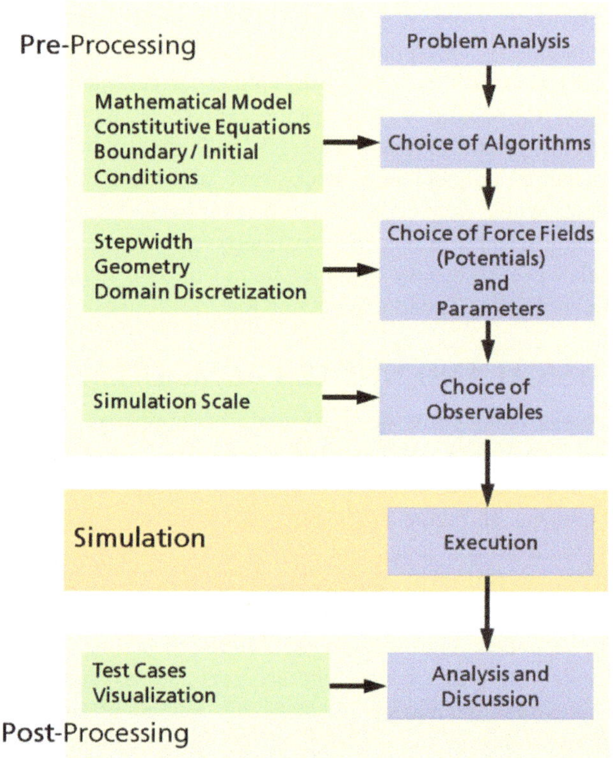

Figure 6: Principal design of a computer simulation. Usually, some pre-processing as preparation for the main simulation is done with a pre-processor. This piece of computer code might be integrated in the main source code or – in particular in commercial codes –is written as an extra piece of code, compiled and run separately from the main simulation code. During pre-processing, many administration tasks can be done which are not related to the actual simulation run. In large-scale simulations, data are stored during execution for later analysis. This analysis and visualization is done during post-processing. The advantage of separating pre- and post-processing from the actual simulation code is that the code design remains clearly arranged and the important issue of optimization for speed only has to be done for the relevant pieces of the main simulation code. However, in large-scale simulations, involving billions of particles, even the task of data analysis can become a major problem that needs optimization and fast parallelized algorithms [41].

Analyzing data "on-the-fly" during the simulation phase is usually too expensive; therefore, data snapshots of the system are stored during certain preset time intervals which can later be analyzed and visualized during the post-processing phase. Very often, the pre- and post-processing code is separated from the main simulation code and since the mid 1990s, Graphical User Interfaces (GUIs) are commonly used for these tasks. In UNIX environments, TCL/TK is a classical script language used to program GUIs. Since the mid 1990s, a C++ based graphical library—Qt—is available for Open Source developments under the GNU General Public Licence.

The starting point for a computer simulation is the invention of an idealized adequate model of the considered physical process. This model is written in the language of mathematics and determined by physical laws, state variables, initial and boundary conditions. The question as to when a model is "adequate" to a physical problem is not easy to answer. There are sometimes many concurrent modeling strategies and it is a difficult question which aspects are essential and which ones are actually unimportant or peripheral.

SIMULATION METHODS FOR DIFFERENT LENGTH AND TIME SCALES

The first scientific simulation methods ever developed and implemented on working electronic computers were MC and MD methods, fully rooted in classical physics [42–49]. Many problems of classical MD techniques lie in the restriction to small (atomistic and microscopic) length and time scales. In atomistic MD simulations of hard matter, i.e., crystalline systems which are mostly governed by their available energy states, the upper limit on today's hardware is typically a cube with an edge length of a few hundred nanometers simulated for a few nanoseconds. With coarse-grained models, where the individual MD particles represent complete clusters of atoms, molecules or other constituents of the system, this limit can be extended to microseconds or even seconds. In this respect, soft matter systems such as polymers, which are very long macromolecules, constitute a very interesting class of materials due to their intrinsic universal scaling features [50,51] which are a consequence of their fractal properties [52]. Macroscopic physical properties of materials can be distinguished in:

- static equilibrium properties, e.g., the radial distribution function of a liquid, the potential energy of a system averaged over many timesteps, the static structure function of a complex molecule, or the binding energy of an enzyme attached to a biological lipid membrane.

- dynamic or non-equilibrium properties, such as diffusion processes in biomembranes, the viscosity of a liquid, or the dynamics of the propagation of cracks and defects in crystalline materials.

Many different properties of materials are determined by structural hierarchies and processes on multiscales. An efficient modeling of the system under investigation therefore requires special simulation techniques which are adopted to the respective problems. Table 1 provides a general overview of different simulation techniques used on various length scales in materials science along with some typical applications. This division is primarily based on a spatial, rather than a physical classification.

Table 1: Customary classification of length scales. Displayed are also typical scopes of different simulation methods and some typical applications pertaining to the respective scale.

Scale (m)	Typical Simulation Methods	Typical Applications
Electronic/Atomistic		
$\sim 10^{-12} - 10^{-9}$	Self-Consistent Hartree-Fock (SC-HF) [53,54]	crystal ground states,
$\sim 10^{-12} - 10^{-9}$	Self-Consistent DFT [12,55,56]	NMR, IR, UV spectra,
$\sim 10^{-12} - 10^{-9}$	Car-Parinello (ab initio) Molecular Dynamics [13]	molecular geometry,
$\sim 10^{-12} - 10^{-9}$	Tight-Binding [57]	electronic properties,
$\sim 10^{-12} - 10^{-9}$	Quantum Monte Carlo (QMC) [58–60]	chemical reactions
Atomistic/Microscopic		
$\sim 10^{-9} - 10^{-6}$	Molecular Dynamics [45,46]	equations of state,
$\sim 10^{-9} - 10^{-6}$	Monte Carlo using *classical* force fields [42,44]	Ising model, DNA
$\sim 10^{-9} - 10^{-6}$	Hybrid MD/MC [61–63]	polymers, rheology,
$\sim 10^{-9} - 10^{-6}$	Embedded Atom Method [64–66]	transport properties,
$\sim 10^{-9} - 10^{-6}$	Particle in Cell [67,68]	phase equilibrium,
Microscopic/Mesoscopic		
$\sim 10^{-8} - 10^{-1}$	MD and MC using	complex fluids,
	effective force fields [51]	soft matter,
$\sim 10^{-9} - 10^{-3}$	Dissipative Particle Dynamics [69]	granular matter,
$\sim 10^{-9} - 10^{-3}$	Phase Field Models [70]	fracture mechanics,
$\sim 10^{-9} - 10^{-3}$	Cellular Automata [71]	grain growth,
$\sim 10^{-9} - 10^{-4}$	Mean Field Theory	phase transformations,
$\sim 10^{-6} - 10^{2}$	Finite Element Methods including	polycrystal elasticity,
	microstructural features [72–75]	polycrystal plasticity,
$\sim 10^{-6} - 10^{2}$	Smooth Particle Hydrodynamics [76,77]	diffusion,
$\sim 10^{-9} - 10^{-4}$	Lattice-Boltzmann [78]	interface motion,
$\sim 10^{-9} - 10^{-4}$	Dislocation Dynamics [79–82]	dislocations,
$\sim 10^{-6} - 10^{0}$	Discrete Element Method [83]	grain boundaries
Mesoscopic/Macroscopic		
$\sim 10^{-3} - 10^{2}$	Hydrodynamics [84]	macroscopic flow,
$\sim 10^{-3} - 10^{2}$	Computational Fluid Dynamics [85–87]	macroscopic elasticity,
$\sim 10^{-6} - 10^{2}$	Finite Element Methods [88–90]	macroscopic plasticity,
$\sim 10^{-6} - 10^{2}$	Smooth Particle Hydrodynamics [8,91,92]	fracture mechanics,
$\sim 10^{-6} - 10^{2}$	Finite Difference Methods [93,94]	aging of materials,
$\sim 10^{-6} - 10^{0}$	Cluster & Percolation Models	fatigue and wear

Electronic/Atomistic Scale

The sub-atomic electronic structure of a material yields information on molecular geometry, magnetic properties, (NMR, IR, or UV) spectroscopic data, quantum mechanical ground and excited states and on the chemistry of materials. Modeling materials on this scale needs to take into account the degrees of freedom of the electrons explicitly. Some basic simulation methods, so called ab initio methods were developed which solve the Schrödinger equation approximately, usually based on the Born-Oppenheimer approximation. With ab initio methods, the only information that has to be provided are the number of atoms and the positions of the atoms within the system. In contrast to this, semi-empirical or empirical approaches require a model of the interactions between the atoms to be supplied. The idea of "inventing" and designing new materials on demand just by entering the into a computer the elements one wants to use and the specific properties one wants to optimize is an ideal which is currently still far from reach, even with ab initio methods [95–97].

The Born-Oppenheimer Approximation

Quantum mechanical computer simulation methods based on Density Functional Theory (DFT) [12,55,56] were developed which calculate the ground state energy of many particle systems. Other ab initio methods combine DFT with classical MD in a way that the degrees of freedom of the electrons can be treated explicitly in contrast to using classical "effective potentials" between atoms which neglect the electronic movements. The basic idea of ab initio MD methods is to approximately solve the electronic Schrödinger equation in each timestep, thereby determining the potential hypersurface for the actual nuclear coordinates, i.e., the effective potential energy of the nuclei. The approximate solution is obtained either by solving the Hartree-Fock [53] or the Kohn-Sham [12] equations. Then one computes the forces on the nuclei and moves them according to Newton's equation of motion which yields these forces. This simulation strategy forms the basis of the Car-Parinello method [13,98]. Due to the large difference in mass between electrons and the atom cores ($m_c/m_e \gg 10^3$), the electrons are able to follow almost instantaneously the only slowly occurring change in the core positions. Thus, the electrons are assumed to always be in the ground state associated to the actual position of the nuclei. This is the reason why the degrees of freedom of the atom cores and the electrons can be separated (Born-Oppenheimer Approximation) [99].

In quantum mechanics, the Schrödinger equation replaces Newton's equations of motion. However, the Schrödinger equation is so complex that it only can be solved analytically for very few simple cases; the direct numerical solution using computers is also limited to very simple systems and very few numbers of atoms because of the high-dimensional phase space in which the Schrödinger equation is formulated. The time dependent quantum mechanical state function Ψ of a system consisting of N nuclei and K electrons can be written as

$$\Psi = \Psi(\vec{R}_1, ..., \vec{R}_N, \vec{r}_1, ..., \vec{r}_K, t) \tag{1}$$

where $\vec{R}_i$ and $\vec{r}_i$ denote positions of the ith nucleus and the ith electron, respectively. The variable t denotes the time. Using the abbreviations $\vec{R}$ and $\vec{r}$ for $(\vec{R}_1, ..., \vec{R}_N)$ and $(\vec{r}_1, ..., \vec{r}_N)$, respectively, one can write the probability density to find the system under consideration at time t in the volume element dV_1, ..., dV_{N+K} of configuration space centered at the point $(\vec{R}, \vec{r})$ as:

$$\Psi^*(\vec{R}, \vec{r}, t)\Psi(\vec{R}, \vec{r}, t)dV_1 \cdots dV_{N+K} \tag{2}$$

The movement of the nuclei during the adaptation of the electron movement is negligibly small in the sense of classical dynamics, thus one sets

$$\Psi(\vec{R}, \vec{r}, t) \approx \Psi^{BO}(\vec{R}, \vec{r}, t) = \sum_{n=0}^{\infty} \chi_n(\vec{R}, t)\phi_n(\vec{R}, \vec{r}) \tag{3}$$

i.e., one separates the full wave function $\Psi(\vec{R}, \vec{r}, t)$ into a simple product form, where χ_n is the nuclear wave function, and where the electronic wave function $\phi_n(\vec{R}, \vec{r})$ does not depend on time anymore but only on the nuclear coordinates $\vec{R}$. Using a Taylor expansion of the stationary Schrödinger equation and several approximations that rely on the difference in masses between electrons and nuclei, see for example Chapter 8 in [100], the stationary Schrödinger equation can be separated into two equations, the electronic Schrödinger equation and an equation for the nuclei. The first equation describes how the electrons behave when the position of the nuclei is fixed. Its solution leads to an effective potential that appears in the equation for the nuclei and describes the effect of the electrons on the interaction between the nuclei. Thus, the cores move within an energy landscape of the surrounding, fast moving electrons. After restriction to the ground stateφ_0 and further approximations (neglecting all coupling terms) [101], one obtains a classical-mechanical model for the core movements determined by the force [101]

$$\vec{F}_i(t) = M_i\ddot{\vec{R}}_i(t) = -\nabla_{\vec{R}_i} \min_{\phi_0} \left\{ \int \phi_0{}^*(\vec{R}(t), \vec{r})\mathcal{H}_e(\vec{R}(t), \vec{r})\phi_0(\vec{R}(t), \vec{r})d^3r \right\} \tag{4}$$

Only considering the ground state ϕ_0 associated with the ground state energy E_0 the electronic Hamilton operator HeHefulfills the eigenvalue equation

$$\mathcal{H}_e(\vec{R}(t), \vec{r})\phi_0(\vec{R}(t), \vec{r}) = E_0(\vec{R}(t))\phi_0(\vec{R}(t), \vec{r}) \tag{5}$$

The forces $\vec{F}_i(t) = M_i\ddot{\vec{R}}_i(t)$ act on the nuclei and their positions can be calculated according to the laws of classical mechanics.

Car-Parinello MD

In the mid 1980s a revolution occurred in the field of atomistic computer simulation with the introduction of "Car-Parinello" (CP) techniques [13]. The basic idea of this technique is based on calculating the interactions on the particles during the simulation run directly from the electronic structure instead of using previously parameterized potentials ("force fields", a term that is more common in the context of biological systems). Furthermore, multi-body contributions and polarization effects are included automatically. Successful realizations of this idea combine MD with density functional theory for electrons in the Kohn-Sham formulation [12,56]. The first ab initio simulation using this method was published in 1993 [102] considering water, more than twenty years after the groundbreaking work by Rahman and Stillinger [103]. In contrast to the latter work however, the essential empirical input parameter of the simulation is only the volume of the periodic simulation cell, with which the then simulated 32 oxygen and 64 hydrogen atoms yield the experimentally known density of 1kg/l. Everything else follows from theory. The CP method introduces a fictitious dynamic movement of the electronic degrees of freedom in terms of pseudo-Newtonian equations of motion in the form

$$\mu_i\ddot{\Psi}_i(t) = \mathcal{H}_e\Psi_i + \sum_j \Lambda_{ij}\Psi_i \tag{6}$$

with fictitious masses μ_i of the orbitals $\{\Psi_i\}$. When He is diagonalized in each timestep (He $\Psi_i = \varepsilon_i \Psi_i$ with $\varepsilon_i = \Lambda_{ij}$) the classical forces acting on the nuclei can be calculated and integrated according to the Newtonian equations of motion for the degrees of freedom of the nuclei:

$$M_i\ddot{\vec{R}}_i(t) = -\nabla\langle\Psi|\mathcal{H}_e|\Psi\rangle \tag{7}$$

where the total wave function Ψ is given as Slater-determinant of the occupied orbitals $\{\Psi_i\}$. Thus, CP generates a classical dynamics of the nuclei in phase space while the dynamics of the electrons is purely fictitious (only the solution of the time dependent Schrödinger equation generates the correct electronic motion). The self-consistent solution of the electronic problem is avoided and substituted in each MD timestep by a dynamic propagation of the orbitals, which are considered as classical fields with constraints. We note that the CP method in the limit of zero orbital masses μ_i yields the Born Oppenheimer result, so it is a controlled approximation for Born Oppenheimer dynamics.

There are many well-known software packages used in materials science and quantum chemistry that are available to academic and industrial users directly and free of cost (e.g., ACESII, AMPAC, CPMD, GAMESS, QUANTUM ESPRESSO, SIESTA) or through commercial vendors (e.g., VASP, CASTEP, GAUSSIAN, Molpro). Many of these codes are based on Density Functional Theory (DFT) but some also implement Hartree-Fock based models and were developed by different scientific teams during the past 20 years. The results of quantum mechanical calculations are often used in the design of classical molecular force fields, providing a connection to the next scale.

Atomistic/Microscopic Scale

Simulations performed at the atomistic or microscopic scale of molecules are much more diverse than those typical of quantum chemistry and a wide range of properties from thermodynamics to bulk transport properties of solids and fluids can be calculated. As a result of this diversity, researchers in a broad array of disciplines (e.g., physics, chemistry, chemical engineering, molecular biology, biochemistry or even geochemistry) contribute to the development and enhancement of methods on this length scale with typical associated time scales ranging roughly from 10^{-12} s to 10^{-6} s for the longest runs on the largest supercomputers.

For computer simulations using semi-empirical or classical force fields, there are several academic software packages freely available (e.g., CHARMM, DL POLY, GROMACS, NAMD, IMD, XMD) or through commercial licenses (e.g., GROMOS). Systems considered on the microscopic scale are still mainly determined in their behavior by their energy, albeit the motion of the electrons can be neglected. Thus, individual atoms or clusters of atoms can be described with methods based on classical interaction potentials. The two oldest used methods are classical MD and MC. Additional interaction potentials for modeling covalent bonds, Coulomb interactions, torsion and bending in molecules are only effectiveinteractions on this scale, as the quantum mechanical electronic contributions are neglected. Due to its simplicity and numerical efficiency, the Lennard-Jones potential is an often used generic model potential. For example, at the Ernst-Mach-Institute (EMI) in Freiburg, an efficient implementation of a code ("MD-Cube") providing modules for multiscale simulations is realized, cf. Figure 7, where new multi-physics modules can be added to the software tool using well-defined interfaces, and a kernel takes care of many administration tasks of the simulation.

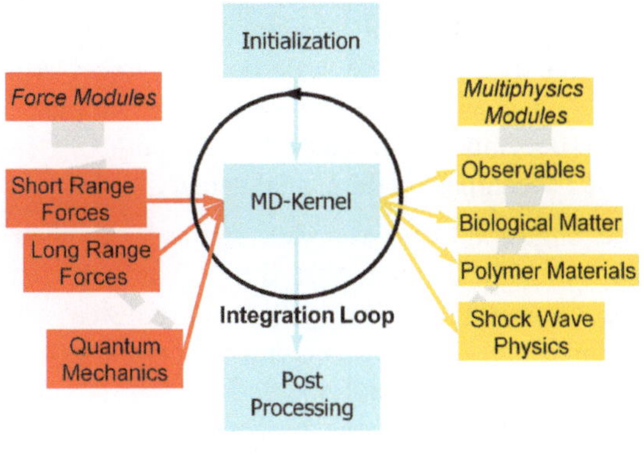

Figure 7: Design schematic of the particle-based multiscale simulation package "MD-Cube" at EMI. A kernel takes care of administrative tasks. Force modules can be added via defined interfaces as well as modules for the demands of different physical applications.

Microscopic/Mesoscopic Scale

Many real material systems have structures much larger than can be studied based on the atomistic/microscopic scale. For example, the properties of block-copolymer materials are strongly influenced by the molecular segregation into mesoscale domains with typical time scales raging from 10^{-8} s to 10^{-4} s. This is the typical domain of soft matter and biological systems, e.g., polymers, amphiphiles or colloidal systems. It is the scale on which self-organization of matter in biological systems, e.g., cells or membranes, occurs. These systems are driven by an interplay between their energy and entropy, as there are many configurations and packing densities available to these systems.

In solid states, dislocation dynamics, crystallizations and phase transformations typically occur on this scale and in polycrystalline systems nucleation and grain growth play a fundamental role. Particle-based methods on this length and time scale include many variations of MD and MC methods using effective interaction potentials or coarse-grained methods such as Dissipative Particle Dynamics (DPD) [69] or the Discrete Element Method (DEM) [83]. With these methods the individual particles do not represent "elementary" particles, i.e., atoms, but complete clusters of atoms or molecules that are treated as classical particles. Such coarse-grained models are used when one needs to study the behavior of a system containing very many molecules for a long time. For example, colloidal suspensions are dispersions of mesoscopic solid particles. These particles themselves consist of millions or billions of atoms. Furthermore, the number of solvent molecules per colloid is comparable or even larger. Clearly, a MD simulation that follows the behavior of several thousand colloids over an experimentally relevant time interval (milliseconds to seconds) would be prohibitively expensive. The DPD method lumps together the forces due to individual solvent molecules to yield an effective friction and a fluctuating force between moving fluid elements. While this approach does not provide a correct atomistic description of the molecular motion, it has the advantage that it does reproduce the correct hydrodynamic behavior on long length and time scales. However, at present, there exists no rigorous demonstration that this is true for an arbitrary DPD fluid, albeit all existing numerical studies suggest that, in the limit where the integration time step $\hat{t} \rightarrow 0$ 0, the large-scale behavior of the DPD fluid is described by the Navier-Stokes equation. The kinetic theory for the transport properties of DPD fluids [104] supports this conclusion. One interesting limit of the DPD model is the "dissipative ideal gas", i.e., a DPD fluid without the conservative forces. The static properties of this fluid are those of an ideal gas. However, its transport behavior is that of a viscous fluid. The advantage of DPD over conventional (atomistic) MD is that it involves a coarse-grained model. This makes the technique useful when studying the mesoscopic properties of complex fluids. However, if one is only interested in static properties, one can use the standard MD or MC techniques on a model with the same conservative forces, but without dissipation. The coarse-grained "bead-spring model" of macromolecular chains connects the particles by flexible, entropic springs and is widely used in polymer physics on this scale. The Lattice-Boltzmann Method [78] is a simulation technique which solves the Navier-Stokes equation of fluid flow on a lattice, and which considers a typical volume element of a fluid to be composed of a collection of particles that are represented by a particle velocity distribution function for each fluid component at each grid point, i.e., it is a hybrid particle/mesh method. Cellular Automata are discrete – that is lattice-based – dynamical systems that are typically used as sampling schemes for nucleation and crystallization simulations in engineering applications. Phase Field models, e.g., of Ginzburg-Landau type, are used to calculate diffusion and phase transitions on this scale. The mesoscopic scale is also the regime where methods, based on continuum theory are used. For engineering applications, mesh-based methods, e.g., FEM are used almost exclusively for the calculation of fluid flows, solid mechanics and coupled fluid/solid systems. These methods are also used for classical fracture mechanics, albeit particle based methods in recent years have been proved to be as viable as mesh-based methods in this respect, see e.g., [33,105–107]. A modern continuum method, which is based on the conservation equations of continuum theory but avoids many mesh distortion problems of FEM approaches, is the method of Smooth Particle Hydrodynamics (SPH) [8,76]. The idea of SPH is somewhat contrary to the concepts of conventional discretization methods, which discretize a continuum system into a discrete algebraic system. SPH is a particle-based mesh-free approach which is attractive in many applications especially in hydrodynamic simulations in which the density is field variable in the system equations. The computational frames in SPH are neither grid cells as in finite difference methods, nor mesh elements as in the FEM methods, but the moving particles in space. The basic idea is to replace the equations of fluid dynamics by equations for particles. In effect, one replaces the continuum equations by a set of particle equations that approximate the continuum and, at the same time, provide a rigorous model of the underlying, and more fundamental, molecular system. There is no need for any predefined connectivity between these particles. All one needs is an initial particle distribution. SPH approximates the particles in the current domain by introducing kernel functions which can serve as an interpolation field [108]. If one wishes to interpret the physical meaning of the kernel function as the probability of a particle's position, one is dealing with a probabilistic method, otherwise, it is just a smoothing technique carried out in the continuum. Thus, the essence of the method is to choose a smoothing function $W(\vec{r}, h)$ with particle position $\vec{r}$ and smoothing length h, the influence domain of a particle, and to use it to localize the strong form of a partial differential equation through a convoluted integration.

Mesoscopic/Macroscopic Scale

When averaging over many degrees of freedom one finally arrives at the phenomenological macroscopic scale. Here, continuum models are used which describe the (visco-)elastic behavior of solids and the properties of fluids based on the Navier-Stokes Equation. Mostly, mesh-based methods such as FEM, and other procedures of computational fluid dynamics for the solution of partial differential equations are used, which are typically closely connected to applications in engineering sciences. The FEM method is generally attributed to A. Hrennikov [109] and R. Courant [14]. The development of the FEM method has been restricted to engineering application for a long time until in the 1970s this method was standardized as a theory by mathematicians apt for the treatment of of partial differential equations in the formulation as variation problems. The salient feature of FEM is the discretization of the continuum into discrete elements. The individual elements are connected together by a topological map which is called a mesh. The finite element interpolation functions are then build upon the mesh, which ensures the compatibility of the interpolation. However, this procedure is not always advantageous, because the numerical compatibility condition is not the same as the physical compatibility condition of a continuum. For instance, in a Lagrangian type of computations, one may experience mesh distortion, which can either end the computation altogether or result in drastic deterioration of accuracy. In addition, FEM often requires a very fine mesh in problems with high gradients or a distinct local character, which can be computationally expensive. For this reason, adaptive remeshing procedures have become a necessity in FEM. Other numerical applications that are not linked to a specific kind of discretization such as hydrocodes or wave propagation codes which decouple the stress tensor in a deviatoric and hydrostatic component, are typically used for the simulation of crash and impact situations of materials [76,77,110–112]. All codes on this level of resolution are usually based on a solution of the continuum conservation equations of energy, mass and momentum and use explicit formulations of equations of state as well as of material response to external loading, so-called constitutive equations. In technical applications, one usually aims at a direct connection to macroscopically measured parameters without introducing any microscopic or molecular quantities.

THE KEY INGREDIENTS OF MOLECULAR DYNAMICS SIMULATIONS

In this section we focus in a tutorial-like fashion on some key issues of MD simulations including common optimization techniques, often found scattered in specialized conference proceedings or other publications and we include a short discussion of the efficiency of the algorithms typically used in MD applications.

One timestep in a MD simulation is typically of the order of femtoseconds ($\sim 10^{-15}$ s). With several million timesteps that are usually simulated in a MD run, the largest available length- and timescales for atomic systems are typically limited to the order of a few hundred nanometers simulated for a few hundred nanoseconds [113,114]. With larger computer systems, one will be able to simulate even larger systems; however, the available time scale does not necessarily grow with the number of available processors, as the time domain cannot be decomposed distributed over many CPUs as it is done when decomposing the spatial domain.

The largest hitherto reported atomistic simulation run was performed with more than 1.9×10^{10} particles [113], opening the route to investigations of physical structure-property phenomena on the micrometer scale, which overlaps with typical computational methods based on continuum theory. Figure 8 exhibits the necessary computer hardware necessary in modern computational science along with a photograph of the first electronic computer, the ENIAC, developed at the Los Alamos Laboratories and which started to operate in 1946.

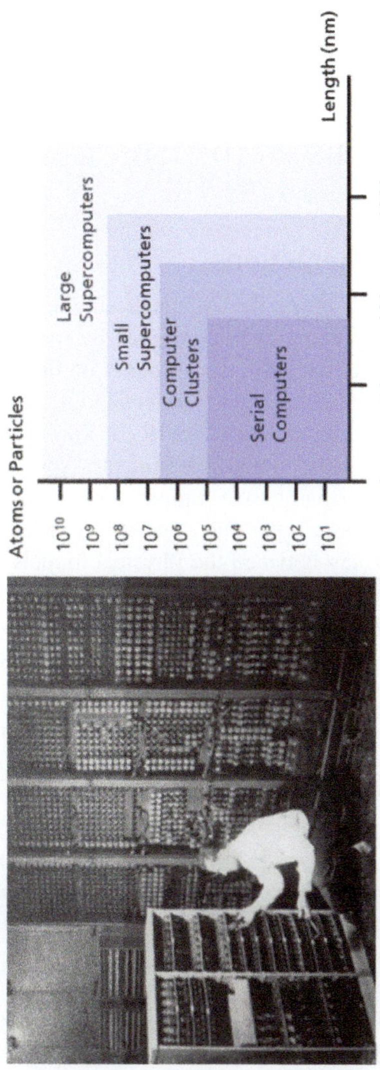

Figure 8: (a) The ENIAC (Electronic Numerical Integrator And Computer), one of the first electronic computers that started to operate in 1946. (The very first working electronic computer was the Z3 developed by Konrad Zuse in the 1930s in Germany). The ENIAC weight 30 tons used more than 18.000 vacuum tubes that can be seen in the picture and had a basic clock speed of 10^5 cycles per second. It was programmed by plugging cables and wires and setting switches using a huge plugboard that was distributed over the entire machine. US Army Photo. (b) Illustration of the available system size (edge length of a simulated cube of classical particles or atoms) and the necessary computer hardware for modern large-scale MD.

Molecular Dynamics (MD) simulations are carried out in an attempt to analyze the properties of a N-particle system, i.e., an assembly of atoms, molecules, or particles, in terms of their molecular structural properties. Macroscopic properties always arise as ensemble averages over a representative statistical ensemble (either equilibrium or non-equilibrium) of molecular systems. For molecular modeling, this has two important consequences:

- The knowledge of one single structure, even if it is the structure of the global energy minimum, is not sufficient. It is always necessary to generate a representative ensemble at a given temperature, in order to compute macroscopic properties.
- The atomic details of structure and motion obtained in molecular simulations, is often not relevant for macroscopic properties. This opens the route for simplifications in the description of interactions and averaging over irrelevant details. Statistical mechanics provides the theoretical framework for such simplifications.

For the generation of a representative equilibrium ensemble two methods are available: MC and MD simulations. While MC methods are much simpler than MD as they do not require the calculation of molecular forces, they do not yield significantly better statistics than MD in a given amount of computing time. For the generation of non-equilibrium ensembles and for the analysis of dynamics events, only MD is the appropriate and more universal technique. The model assumptions on the physical behavior of a many particle system investigated in MD simulations is put into the interaction potentials F, respectively the force fields $-\nabla\Phi$. MD simulations in their simplest form consist in the step-by-step numerical solution of the classical Newtonian equations of motion, which for N particles of mass m_i and position vectors $\vec{r}_i$ may be written as

$$\vec{F}_i = m_i \frac{d^2}{dt^2}\vec{r}_i = -\nabla_i \Phi(\vec{r}_i - \vec{r}_N). \tag{8}$$

Limitations of MD

After an initial equilibration phase, the system will usually reach an equilibrium state. By averaging over an equilibrium trajectory (coordinates over a function of time) many macroscopic properties can be extracted from the output. Common approximations (and therefore limitations) of MD simulations are:

- Artificial boundary conditions

  The system size that can be simulated with MD is very small compared to real molecular systems. Hence, a system of particles will have many unwanted artificial boundaries (surfaces). In order to avoid real boundaries one introduces periodic boundary conditions (see Section 4.3.) which can introduce artificial spatial correlations in too small systems. Therefore, one should always check the influence of system size on results.

- Cut off of long-range interactions

  Usually, all non-bonded interactions are cut-off at a certain distance in order to keep the cost of force computation (and the search effort for interacting particles) as small as possible. Due to the minimum image convention (see Section 4.4.) the cutoff range may not exceed half the box size. While this is large enough for most systems in practice, problems are only to be expected with systems containing charged particles. Here, simulations can go wrong badly and, e.g., lead to an accumulation of the charged particles in one corner of the box. Here, one has to use special algorithms such as the particle-mesh Ewald method [115,116].

- The simulations are classical

Using Newton's equations of motion implies the use of classical mechanics for the description of the atomic motion. All those material properties connected with the fast electronic degrees of freedom are not correctly described. For example, atomic oscillations (e.g., covalent C-C-bond oscillations in polyethylene molecules, or hydrogen-bonded motion in biopolymers such as DNA, proteins or biomembranes) are typically of the order 10^{14} Hz. The specific heat is another example which is not correctly described in a classical model as here, at room temperature, all degrees of freedom are excited, whereas quantum mechanically, the high-frequency bonding oscillations are not excited, thus leading to a smaller (correct) value of the specific heat than in the classical picture. A general solution to this problem is to treat the bond distances and bond angles as constraints in the equations of motion. Thus, the highest frequencies in the molecular motion are removed and one can use a much higher timestep in the integration [117].

- The electrons are in the ground state

Using conservative force fields in MD implies that the potential is a function of the atomic positions only. No electronic motions are considered, thus the electrons remain in their ground state and are considered to follow the core movements instantaneously. This means that electronically excited states, electronic transfer processes and chemical reactions cannot be treated.

- Approximative force fields

Force fields are not really an integral part of the simulation method but are determined from experiments or from a parameterization using ab initio methods. Also, most often, force fields are pair-additive (except for the long-range Coulomb force) and hence cannot incorporate polarizabilities of molecules. However, such force fields exist and there is continuous effort to generate such kind of force fields [118,119]. In most practical applications however, e.g., for biomacromolecules in aqueous solution, pair potentials are quite accurate mostly because of error cancellation. This does not always work, for example ab initio predictions of small proteins still yields mixed results and when the proteins fail to fold, it is often unclear whether the failure is due to a deficiency in the underlying force fields or simply a lack of sufficient simulation time [120,121].

- Force fields are pair additive

All non-bonded forces result from the sum of non-bonded pair interactions. Non pair-additive interactions such as the polarizability of molecules and atoms, are represented by averaged effective pair potentials. Hence, the pair interactions are not valid for situations that differ considerably from the test systems on which the models were parameterized. The omission of polarizability in the potential implies that the electrons do not provide a dielectric constant with the consequence that the long-range electrostatic interaction between charges is not reduced (as it should be) and thus overestimated in simulations.

Classics in MD that helped to develop the method are for example Alder and Wainwright [45,47] (1958, 1961), Rahman [103] (1964), Verlet [122,123] (1967, 1968), Weeks, Chandler and Andersen [124] (1979), Rahman and Stillinger [125] (1971), Parinello and Rahman [98,126] (1981, 1982), van Gunsteren and Berendsen [117] (1982), Hoover [127] (1986) and Allen and Tildesley [7] (1991).

Molecular Interactions

All macroscopic properties of materials, be it in the solid, fluid or gaseous state, are determined by the intermolecular forces acting between the constituents of matter. Important sources of measuring and understanding intermolecular forces are scattering experiments, IR- or Raman spectroscopy, thermophysical data (virial coefficients, specific heats) or NMR data. In general, one cannot measure intermolecular forces directly just with one type of experiment but it is rather an interplay between the physical models used for the interpretation of data and the derivation of a functional form of the sought-after intermolecular potential. The analytical form of the potential which is derived from theory is then consistently adjusted to the experimental findings [128,129].

Classical MD simulations are used in Polymer Physics to investigate the structure (e.g., form factors, pair correlation functions), the dynamics (e.g., transport coefficients, correlations) and the thermodynamics (e.g., phase diagrams and ensemble averages of observables of interest) of polymer molecules and complexes which are described as N-particle systems. In polymers the atoms are covalently bound in a fixed topological arrangement; thus one distinguishes non-bonded interactionsacting between all atoms of a system and bonded interactions which are only effective between atoms in each particular polymer chain or complex macromolecule.

Non-bonded Interactions

Various physical properties are determined by different regions of the potential hypersurface of interacting particles. Thus, for a complete determination of potential curves, widespread experiments are necessary. For a N–body system the total energy Φ_{nb}, i.e., the potential hypersurface of the non-bonded interactions can be written as [7]

$$\Phi_{nb}(\vec{r}) = \sum_{i}^{N} \phi_1(\vec{r}_i) + \sum_{i}^{N}\sum_{j>i}^{N} \phi_2(\vec{r}_i,\vec{r}_j) + \sum_{i}^{N}\sum_{j>i}^{N}\sum_{k>j>i}^{N} \phi_3(\vec{r}_i,\vec{r}_j,\vec{r}_k) + \cdots, \tag{9}$$

where $\phi_1, \phi_2, \phi_3, \cdots$ are the interaction contributions due to external fields (e.g., the effect of container walls) and due to pair, triple and higher order interactions of particles. In classical MD one often simplifies the potential by the hypotheses that all interactions can be described by pairwise additive potentials. Despite this reduction of complexity, the efficiency of a MD algorithm taking into account only pair interactions of particles is rather low (of order $\mathcal{O}(N^2)$) and several optimization techniques are needed in order to improve the runtime behavior to $\mathcal{O}(N)$. The simplest general Ansatz for a non-bonded potential for spherically symmetric systems, i.e.,

$\Phi(\vec{r}) = \Phi(r)$ with $r = |r_i - r_j|$ is a potential of the following form:

$$\Phi_{nb}(r) = \Phi_{Coulomb}(r) + \left(\frac{C_1}{r}\right)^{12} + \left(\frac{C_2}{r}\right)^{6}. \tag{10}$$

Parameters C_1 and C_2 are parameters of the attractive and repulsive interaction and the electrostatic energy $\Phi_{Coulomb}(r)$ between the particles with position vectors $\vec{r}_i$ and $\vec{r}_j$ is given by:

$$\Phi_{Coulomb}(r) = \frac{1}{\epsilon}k \cdot \sum_i \sum_{j>i} \frac{z_i z_j e^2}{|\vec{r}_i - \vec{r}_j|} \tag{11}$$

The constant k = 1 is in the cgs-system of units and ε is the dielectric constant of the medium, for example $\varepsilon_{air} = 1$ for air, $\varepsilon_{prot} = 4$ for proteins or $\varepsilon_{H2}0 = 82$ for water. The z_i denote the charge of individual monomers in the macromolecule and e is the electric charge of an electron.

The electrostatic interaction originates the dipolar character of water which is the basic requirement for the existence of life. Water is a dipole because of the higher electronegativity of oxygen which gives rise to a partial negative charge at the oxygen atom and partial positive charges at the H-atoms in the H_2O-molecule. If the electronegativity of one atom is large enough, it can attract the whole electron from the bonding partner. This is the case for example with NaCl where the initially electric neutral Cl atom becomes a Cl^--ion and Na turns into Na^+ accordingly. Chemical bonds which emerge from Coulomb attraction of ions are called ionic bonds. This type of chemical bond plays an important role for the formation of structures of biomolecules. For example, charged sidegroups may bind to receptors within the cell membrane or protein structures are be stabilized when a positively charged, protonated ammonium group ($^+NH_4$) and a negatively charged carboxyl group ($COOH^-$) form an ionic bonding.

The electric dipole moment is defined as $\vec{p}_{el} = |q|\vec{d}$ where q is the total charge of the dipole and $\vec{d}$ is the distance vector between the two charges. The electric field $\vec{E}$ of a dipole $\vec{p}_{el}$ is given by [130]

$$\vec{E}(\vec{r}) = \frac{1}{4\pi\epsilon\epsilon_0} \frac{3\vec{u}_r(\vec{p}_{el} \cdot \vec{u}_r) - \vec{p}_{el}}{|\vec{r}|^3} \tag{12}$$

where $\vec{u}_r$ is a unit vector pointing from the origin r = 0 into the direction of $\vec{r}$. Thus the potential energy of a dipole in an electric field $\vec{E}$ is given by

$$E_p = \vec{p}_{el} \cdot \vec{E} \tag{13}$$

Inserting Equation (12) in (13) one yields a characteristic leading $1/r^3$ term in the potential energy between two dipoles with moments $\vec{p}_1$ and $\vec{p}_2$.

Electric neutral molecules may also interact with each other due to thermal (Brownian) motion which may induce a mutual dipole moment in the molecules. These induced dipoles create an attractive interaction which is called Van-der-Waals interaction. The quantity that measures the ability of a neutral molecule to induce dipoles is the polarizability α which is given by

$$\vec{p}_{el} = \alpha\epsilon_0\vec{E} \tag{14}$$

For the potential energy E_p of a dipole which is induced by an electric field $\vec{E}$ one obtains by elementary integration:

$$E_p = -\int_0^E \vec{p}_{el} \cdot d\vec{E}' = -\frac{1}{2}\alpha\epsilon_0|\vec{E}|^2 \tag{15}$$

Thus, the characteristic interaction energy of two mutual induced electric dipoles is proportional to $1/r^6$ and consequently rather weak. The polarizability of water molecules gives rise to another, directed interaction which is called hydrogen bond. Hydrogen bonds are not only found in fluid and solid water but also in complex biopolymers and macromolecules, for example in proteins, where hydrogen bonds are responsible for the genesis of tertiary structures such as α-helices or β-sheets. Despite the directed nature of the hydrogen bond one often assumes a spherically symmetric analytic form of the type $(A \cdot r^{-12} - B \cdot r^{-6})$, but also a more precise form taking into account the non-linearity of the hydrogen bond by angle θ between N-H-O have been proposed [131]:

$$\Phi_{hb} = \sum_{ij} \cos\theta(-A \cdot r_{ij}^{-6} + B \cdot r_{ij}^{-12}) + (1 - \cos\theta)(-C \cdot r_{ij}^{-6} + D \cdot r_{ij}^{-12}) \tag{16}$$

Here, parameters A, B, C, D are constants depending on the considered atom pairs.

With decreasing distance of two dipoles the electronic repulsion of the atomic shells starts to dominate. The minimum of the non-bonded interaction is reached when the attraction just cancels repulsion. This distance is called Van-der-Waals radius σ_0.

The probably most commonly used form of the potential of two neutral atoms which are only bound by Van-der-Waals interactions, is the Lennard-Jones (LJ), or (a-b) potential which has the form [132]

$$\Phi_{a,b}(r) = \alpha\epsilon\left[\left(\frac{\sigma_0}{r}\right)^a + \left(\frac{\sigma_0}{r}\right)^b\right], \tag{17}$$

where

$$\alpha = \frac{1}{a-b}\left(\frac{a^a}{b^b}\right)^{\frac{1}{a-b}}, \quad \Phi_{min} = \epsilon \text{ and } \Phi(\sigma) = 0. \tag{18}$$

The most often used LJ-(6-12) potential for the interaction between two particles with a distance $r = |\vec{r}_i - \vec{r}_j|$ then reads (cf. Equation (10)):

$$\Phi_{LJ}(r) = 4\epsilon\left[\left(\frac{\sigma_0}{r}\right)^{12} + \left(\frac{\sigma_0}{r}\right)^{6}\right]. \tag{19}$$

Parameter ε determines the energy scale and σ_0 the length scale. In simulations one uses dimensionless reduced units which tend to avoid numerical errors when processing very small numbers, arising e.g., from physical constants such as the Boltzmann constant $k_B = 1.38 \cdot 10^{-23}$J/K. In these reduced (simulation) units, one MD time step is measured in units of τ□ = $(m\sigma^2/\varepsilon)^{1/2}$, where m is the mass of a particle and ε and σ_0 are often simply set to $\sigma_0 = \varepsilon = k_B T = 1$. Applied to real molecules, for example to Argon with $m = 6.63 \times 10^{-23}$kg, $\sigma_0 \approx 3.4 \times 10^{-10}$m and $\varepsilon/k_B \approx 120$K one obtains a typical MD time step $\hat{\tau} \approx 3.1 \times 10^{-13} s$.

Using an exponential function instead of the repulsive r^{-12} term, one obtains the Buckingham potential [133]:

$$\Phi(r) = b\exp(-ar) - \frac{c}{r^6} - \frac{d}{r^8}. \tag{20}$$

This potential however has the disadvantage of using a numerically very expensive exponential function and it is known to be unrealistic for many substances at small distances r and has to be modified accordingly.

For reasons of efficiency, a classical MD potential should be short-ranged in order to keep the number of force calculations between interacting particles at a minimum. Therefore, instead of using the original form of the potential in Equation (19), which approaches 0 at infinity, it is common to use a modified form, where the potential is simply cut off at its minimum value $r = r_{\text{min}} = \sqrt[6]{2}$ and shifted to positive values by ε such that it is purely repulsive and smooth at $r = r_{\text{cut}} = \sqrt[6]{2}$:

$$\Phi^{\text{cut}}_{LJ}(r) = \begin{cases} 4\epsilon\left\{\left(\frac{\sigma_0}{r}\right)^{12} - \left(\frac{\sigma_0}{r}\right)^{6}\right\} + \epsilon & r \le 2^{1/6}\sigma_0, \\ 0 & \text{otherwise.} \end{cases} \tag{21}$$

Another extension of the potential in Equation (21) is proposed in [51] where a smooth attractive part is introduced again, in order to allow for including different solvent qualities of the solvent surrounding the polymer:

$$\Phi_{cos}(r) = \left[\frac{1}{2}\cdot\cos(\alpha r^2 + \beta) + \gamma\right]\epsilon. \tag{22}$$

This additional term adds an attractive part to the potential of Equation (21) and at the same time – by appropriately choosing parameters α, β and γ – keeps the potential cutoff at r_{cut} smooth. The parameters α, β and γ are determined analytically such that the potential tail of Φ_{cos} has zero derivative at $r = 2^{1/6}$ and at $r = r_{cut}$, while it is zero at $r = r_{cut}$ and has value γ at $r = 2^{1/6}$, where γ is the depth of the attractive part. Further details can be found in [51]. When setting $r_{cut} = 1.5$ one sets $\gamma = -1$ and obtains α and β as solutions of the linear set of equations

$$2^{1/3}\alpha + \beta = \pi, \tag{23}$$

$$2.25\alpha + \beta = 2\pi \tag{24}$$

The total unbounded potential can then be written as:

$$\Phi_{Total}(r,\lambda) = \begin{cases} \Phi^{\text{cut}}_{LJ}(r) - \lambda\epsilon & 0 < r < 2^{1/6}\sigma_0, \\ \lambda\,\Phi_{cos}(r) & 2^{1/6}\sigma_0 \le r < r_{cut}, \\ \infty & \text{otherwise,} \end{cases} \tag{25}$$

where λ is a new parameter of the potential which determines the depth of the attractive part. Instead of varying the solvent quality in the simulation by changing temperature T directly (and having to equilibrate the particle velocities accordingly), one can achieve a phase transition in polymer behavior by changing λ accordingly, cf. Figure 9. Using coarse-grained models in the context of lipids and proteins, where each amino acid of the protein is represented by two coarse-grained beads, it has become possible to simulate lipoprotein assemblies and protein-lipid complexes for several microseconds [134].

The assumption of a short ranged interaction is usually fulfilled very well for all (uncharged) polymeric fluids. However, as soon as charged systems are involved this assumption breaks down and the calculation of the Coulomb force requires special numerical treatment due to its infinite range.

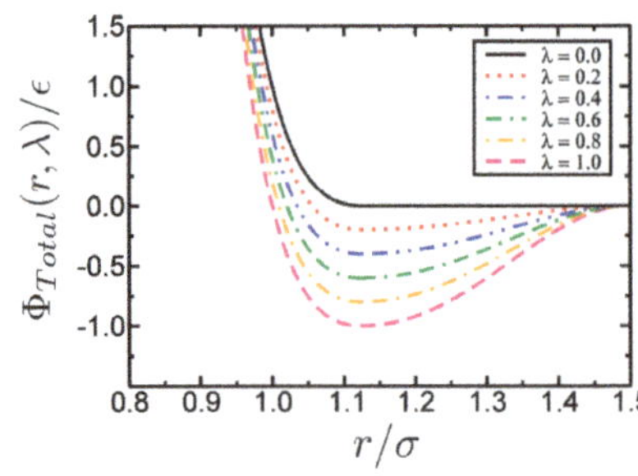

Figure 9: Graph of the total unbounded potential of Equation (25) which allows for modeling the effects of different solvent qualities.

Bonded Interactions

Using the notion of intermolecular potentials acting between the particles of a system one cannot only model fluids made of simple spherically symmetric particles but also more complex molecules with internal degrees of freedom (due to their specific monomer connectivity). If one intends to incorporate all aspects of the chemical bond in complex molecules one has to treat the system with quantum chemical methods, cf. Section 3.1. Usually, one considers the inner degrees of freedom of polymers and biomacromolecules by using generic potentials that describe bond lengths l_i, bond angles θ and torsion angles φ. When neglecting the fast electronic degrees of freedom, often bond angles and bond lengths can be assumed to be constants. In this case, the potential includes lengths l_0 and angles θ_0, φ_0 at equilibrium about which the molecules are allowed to oscillate, and restoring forces which ensure that the system attains these equilibrium values on average. Hence the bonded interactions Φ_{bonded} for polymeric macromolecular systems with internal degrees of freedom can be treated by using some or all parts of the following potential term:

$$\Phi_{bonded}(r,\theta,\phi) = \frac{\kappa}{2}\sum_i(|\vec{r}_i - \vec{r}_{i-1} - l_0|)^2 + \frac{k_\theta}{2}\sum_k(\theta_k - \theta_0)^2 + \frac{\beta}{2}\sum_m(\phi_m - \phi_0)^2. \tag{26}$$

Here, the summation indices sum up the number of bonds i at positions $\vec{r}_i$, the number of bond angles k between consecutive monomers along a macromolecular chain and the number of torsion angles m along the polymer chain. A typical value of $\kappa = 5000$ ensures that the fluctuations of bond angles are very small (below 1%). The terms l_0, θ_0 and ϕ_0 are the equilibrium distance, bond angle and torsion angle, respectively. In particular in polymer physics, very often a Finitely Extensible Non-linear Elastic (FENE) potential is used which, in contrast to a harmonic potential, restricts the maximum bond length of a polymer bond to a prefixed value R_0 [51]:

$$\Phi_{FENE}(r) = \begin{cases} -\frac{1}{2}\kappa R_0^2 \ln(1 - \frac{r^2}{R_0^2}) & r < R_0, \\ \infty & \text{otherwise.} \end{cases} \quad (27)$$

The FENE potential can be used instead of the first term on the right hand side of the bonded potential in Equation (26). Figure 10 illustrates the different parameters which are used in the description of bonded interactions in Equation (26). Further details on the use of potentials in macromolecular biology and polymer physics may be found in [3,135–137].

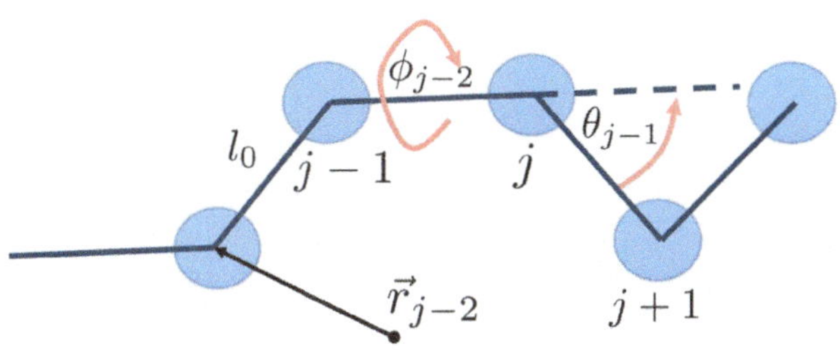

Figure 10: Illustration of the potential parameters used for modeling bonded interactions.

Periodic Boundary Conditions

In a MD simulation only a very small number of particles can be considered. To avoid the (usually) undesired artificial effects of surface particles which are not surrounded by neighboring particles in all directions and thus are exerted to non-isotropic forces, one introduces periodic boundary conditions. Using this technique, one measures the "bulk" properties of the system, due to particles which are located far away from surfaces. As a rule, one uses a cubic simulation box were the particles are located. This cubic box is periodically repeated in all directions. If, during a simulation run, a particle leaves the central simulation box, then one of its image particles enters the central box from the opposite direction. Each of the image particles in the neighboring boxes moves in exactly the same way, cf. Figure 11 for a two dimensional visualization.

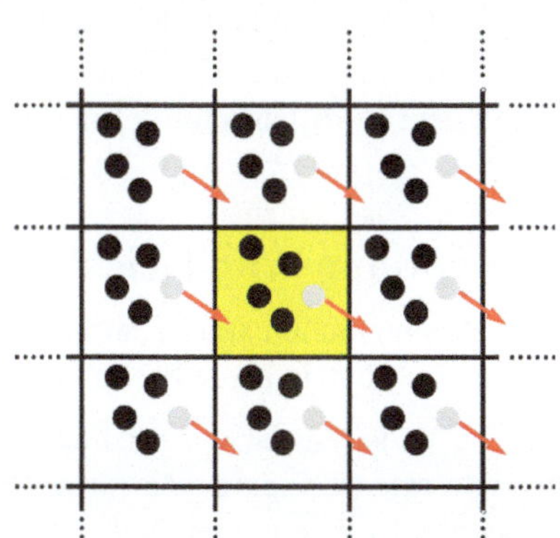

Figure 11: Two-dimensional schematic of periodic boundary conditions. The particle trajectories in the central simulation box are copied in every direction.

The cubic box is used almost exclusively in simulations with periodic boundaries, mainly due to its simplicity, however also spherical boundary conditions have been investigated were the three-dimensional surface of the sphere induces a non-Euclidean metric. The use of periodic boundary conditions allows for a simulation of bulk properties of systems with a relatively small number of particles.

Minimum Image Convention

The question whether the measured properties with a small, periodically extended system are to be regarded as representative for the modeled system depends on the specific observable that is investigated and on the range of the intermolecular potential. For a LJ potential with cut-off as in Equation (21) no particle can interact with one of its images and thus be exposed to the artificial periodic box structure which is imposed upon the system. For long range forces, also interaction of far away particles have to be included, thus for such systems the periodic box structure is superimposed although they are actually isotropic. Therefore, one only takes into account those contributions to the energy of each one of the particles which is contributed by a particles that lies within a cut-off radius that is at the most $1/2L_B$ with box length L_B. This procedure is called minimum image convention. Using the minimum image convention, each particle interacts with at the most (N – 1) particles. Particularly for ionic systems a cut-off has to be chosen such that the electro-neutrality is not violated. Otherwise, particles would start interacting with their periodic images which would render all calculations of forces and energies erroneous.

Force Calculation

The most crucial part of a MD simulation is the force calculation. At least 95% of a MD code is spent with the force calculation routine which uses a search algorithm that determines interacting particle pairs. Therefore this is the task of a MD program which has to be optimized first and foremost. We will review a few techniques that have become standard in MD simulations which enhance the speed of force calculations considerably and speed up the algorithm from $\mathcal{O}(N^2)$ run time to $\mathcal{O}(N)$ run time. Starting from the original LJ potential between two particles i and j with distance $r = |\vec{r}_i - \vec{r}_j|$ ofEquation (17), one obtains the potential function for N interacting particles as the following double sum over all particles:

$$\Phi(\vec{r}_1, ..., \vec{r}_N) = \sum_{i=1}^{N}\sum_{j=i+1}^{N} \Phi_{LJ}(r) = 4\epsilon \sum_{i=1}^{N}\sum_{j=i+1}^{N} \left(\frac{\sigma_0}{r}\right)^6 \times \left(\left(\frac{\sigma_0}{r}\right)^6 - 1\right). \quad (28)$$

The corresponding force $\vec{F}_i$ exerted on particle i by particle j is given by the gradient with respect to $\vec{r}_i$ as:

$$\vec{F}_i = -\nabla_{\vec{r}_i}\Phi_{LJ}(\vec{r}_1, ..., \vec{r}_N) = -24 \times \epsilon \sum_{j=1, j\neq i}^{N} \frac{1}{r^2} \times \left(\frac{\sigma_0}{r}\right)^6 \times \left(1 - 2 \times \left(\frac{\sigma_0}{r}\right)^6\right) \vec{r}_{ij}, \quad (29)$$

where $\vec{r}_{ij} = (\vec{r}_i - \vec{r}_j)$ is the direction vector between particles i and j at positions $\vec{r}_i$ and $\vec{r}_j$, and r= $|\vec{r}_i - \vec{r}_j|$. Hence, in general, the force $\vec{F}_i$ on particle i is the sum over all forces $\vec{F}_{ij} := -\nabla_{\vec{r}_i}\Phi$ between particle i and all other particles j:

$$\vec{F}_i = \sum_{i=1, j\neq i}^{N} \vec{F}_{ij} \qquad (30)$$

The least favorable method of looking for interacting pairs of particles and for calculating the double sum in Equations (28) and (29) is the "brute force" method that simply involves taking a double loop over all particles in the (usually) cubic simulation box, thus calculating $\frac{1}{2}N(N-1)$ interactions with a N^2 efficiency. This algorithm becomes extremely inefficient for systems of more than a few thousand particles, cf. Figure 12a.

(a) (b) (c)

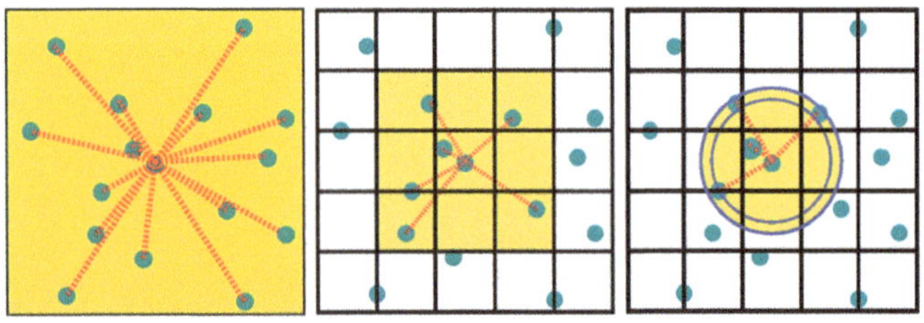

Figure 12: MD Optimization schemes for the search of potentially interacting particles. (a) The least efficient all particle "brute force" approach with run time $\mathcal{O}(N^2)$ (b) The linked-cell algorithm which reduces the search effort to $\mathcal{O}(N)$. (c) The linked-cell algorithm combined with neighbor lists which further reduces the search effort by using a list of potentially interacting neighbor particles which can be used for several timesteps before it has to be updated. In this 2D representation, the radius of the larger circle is $r_{cut} + r_{skin}$ and the inner circle, which contains actually interacting particles, has radius r_{cut}.

Linked-Cell Algorithm

In general, in molecular systems, the potential as well as the corresponding force decays very fast with the distance rbetween the particles. Thus, for reasons of efficiency, in molecular simulations one often uses the modified LJ potential ofEquation (21) which introduces a cutoff r_{cut} for the potential. The idea here is to neglect all contributions in the sums inEquations (28) and (29) that are smaller than the threshold r_{cut} which characterizes the range of the interaction. Thus, in this case the force $\vec{F}_i$ on particle i is approximated by

$$\vec{F}_i \approx -24\times\epsilon \sum_{\substack{j=1,j\neq i\\ 0<r\leq r_{cut}}}^{N} \frac{1}{r^2}\times\left(\frac{\sigma_0}{r}\right)^6\times\left(1-2\times\left(\frac{\sigma_0}{r}\right)^6\right)\vec{r}_{ij} \qquad (31)$$

Contributions to the force on particle i that stem from particles j with $r \leq r_{cut}$ are neglected. This introduces a small error in the computation of the forces and the total energy of the system, but it reduces the overall computational effort from $\mathcal{O}(N^2)$ to $\mathcal{O}(N)$. For systems with short-ranged or rapidly decaying potentials, a very efficient algorithm for the search of potentially interacting particles, i.e., those particles that are within the cutoff distance r_{cut} of a particle i, has been developed [4]. In MD this algorithm can be implemented most efficiently by geometrically dividing the volume of the (usually cubic) simulation box into small cubic cells whose sizes are slightly larger than the interaction range r_{cut} of particles, cf. Figure 12b. The particles are then sorted into these cells using the linked-cell algorithm (LCA). The LCA owes its name to the way in which the particle data are arranged in computer memory, namely as linked list for each cell. For the calculation of the interactions it is then sufficient to calculate the distances between particles in neighboring cells only, since cells which are further than one cell apart are by construction beyond the interaction range. Thus, the number of distance calculations is restricted to those particle pairs of neighboring cells only which means that the sums in Equation (30) are now split into partial sums corresponding to the decomposition of the simulation domain into cells. For the force $\vec{F}_i$ on particle i in cell number n one obtains a sum of the form

$$\vec{F}_i = \sum_{\substack{cell\ m\\ m\in\Omega(n)}} \sum_{\substack{j\in\{\text{all particles in cell } m\}\\ j\neq i}} \vec{F}_{ij} \qquad (32)$$

where $\Omega(n)$ denotes cell n itself together with all cells that are direct neighbors of cell n. The linked-cell algorithm is a simple loop over all cells of the simulation box. For each cell there is a linked list which contains a root pointer that points to the first particle in the respective cell which then points to the next particle within this particular cell, until the last particle is reached which points to zero, indicating that all particles in this cell have been considered. Then the algorithm switches to the root pointer of the next cell and the procedure is repeated until all interacting cells have been considered, cf. Figure 12.

Assuming the average particle density in the simulation box as $\langle\rho\rangle$ then the number of particles in each one of the subcells is $\langle\rho\rangle r_{cut}^3$. The total number of subcells is $N/\langle\rho\rangle r_{cut}^3$ and the total number of neighbor cells of each subcell is 26 in a cubic lattice in three dimensions (3D). Due to Newton's third law only half of the neighbors actually need to be considered. Hence, the order to which the linked-cell algorithm reduces the search effort is given by:

$$\frac{26}{2}\left(\langle\rho\rangle r_{cut}^2\right)\frac{N}{\langle\rho\rangle r_{cut}^3} = 13(\langle\rho\rangle r_{cut}^3)^2 N \qquad (33)$$

For this method to function, the size of the simulation box has to be at least $3r_{cut}$, cf. Figure 12. For simulations of dense melts with many particles, this requirement is usually met. Consequently, by this method, the search-loop effort is reduced to $\mathcal{O}(N)$, but with a pre-factor that still can be very large, depending on the density of particles $\langle\rho\rangle$ and the interaction range r_{cut}.

Linked-Cell Algorithm With Neighbor-Lists

If one compares the interaction sphere around one particular particle to the volume that is actually scanned around it by this algorithm, one finds that only a fraction $\frac{4\pi}{81} \approx \frac{1}{6}$ of the particles really interact, cf. Figure 12c for a 2D picture. To avoid these wasted calculations one should approximate the interaction sphere more closely. One way to do it would be to make the cells smaller and simultaneously increase the number of neighbor cells (next nearest, etc.) which one has to scan. This method has the severe drawback of increasing the loop iterations dramatically, while most of the cells will be empty. Thus, in a naive implementation the cost of the loop overhead will eventually overwhelm the benefit of the saved distance calculations. However, there is a different method available: Along with the cell subdivision a neighbor list of potentially interacting particle pairs is constructed. Here, only particles within the inner sphere of radius r_{cut} surrounding a specific particle i actually interact with each this particle, cf. Figure 12c. In order to speed up the search algorithm for identifying interacting particles, the volume between the outer sphere of radius ($r_{cut} + r_{skin}$) and the smaller one is additionally covered by the neighbor list for particle i. Thus, this list contains not only actually interacting particles at some specific point in time, but it also contains all particles that might enter the interaction range of the inner sphere within the next few timesteps. This greatly speeds up the simulation, because the list of potentially interacting particles will be valid for several timesteps, in the order of 5–15, before it has to be rebuilt. The interval, at which list-reconstruction has to be done, depends upon r_{cut}, the particle density ρ and the skin radius r_{skin}. Once a particle has moved a distance larger than $d^2 = \left(\frac{r_{skin}}{2}\right)^2$, the update is due. The accumulated distance that each particle moved can be readily monitored during the distance calculation. Tests of this method with the bonded potential of Equation (25) and the FENE potential of Equation (27) for flexible macromolecules with $r_{cut} = 1.5\sigma$ and $\rho = 0.85\sigma$ show that a radius of $r_{skin} \approx 0.35\sigma$ to 0.40σ is the optimal choice [3,51]. The d_i length of the cells in each direction is given by $L_i/\mathrm{modulo}(L_i/r_{cut}^{max})$ with L_i being the box size in each direction and r_{cut}^{max} being the largest cutoff of all potentials that are used. The cells are numbered, beginning with the one in the lower left corner of the simulation box where the origin of the coordinate system is located. Each time when an update of the neighbor list is due, the particles are periodically back-folded into the simulation box and then sorted into the different cells according to their coordinates. Subsequently, only the distances of particles of neighbor cells are calculated, with each cell having 26 neighbors in three dimensions. Again, due to Newton's third axiom only half of them have to be considered.

Ghostparticles

In a MD simulation one can only investigate the properties of a relatively small number of particles compared to a real system. Therefore one introduces the periodic boundary conditions as described in Section 4.3. The periodic boundary conditions are realized by performing distance calculations taking into account the minimum image convention according to which the real distance between any two particles is given by the shortest distance of any of their images. Once a particle has crossed the boundaries it is periodically back-folded into the simulation box of volume $V = L_x \times L_y \times L_z$.. This periodic wrap-around is done in the innermost loop of the force calculation and therefore is extremely expensive in terms of simulation time.

Consequently, another method of gaining speed in a MD simulation is to remove any mentioning of periodic boundaries in the force calculations. This can be done by using the concept of ghost particles, see e.g., [3].

The idea with ghost particles is the following: All particles that are in cells which are on the surface of the simulation box are being duplicated right away into the extra ghost cells surrounding the whole box. These ghost particles are now used for distance calculations instead of the original coordinates. As a result, the periodic back-folding only has to be done for a relatively small number of surface particles in the outermost cells, but not anymore for the many particles in the innermost force loop.

As a result of Newton's third axiom, one actually needs to consider only half of the ghost cells surrounding the original box. The schematic in Figure 13 displays how the individual ghost layers in each direction are set up: The number of adjacent neighbor cells now depends upon the location of the considered cell. In a cube there are 18 different cases and again due to Newton' s third axiom only half of them need to be considered. The sub-routine that contains the search algorithm which examines adjacent cells has to make provision for all 9 different cases. The number of adjacent cells for these different cases is fixed and can be written in a static array.

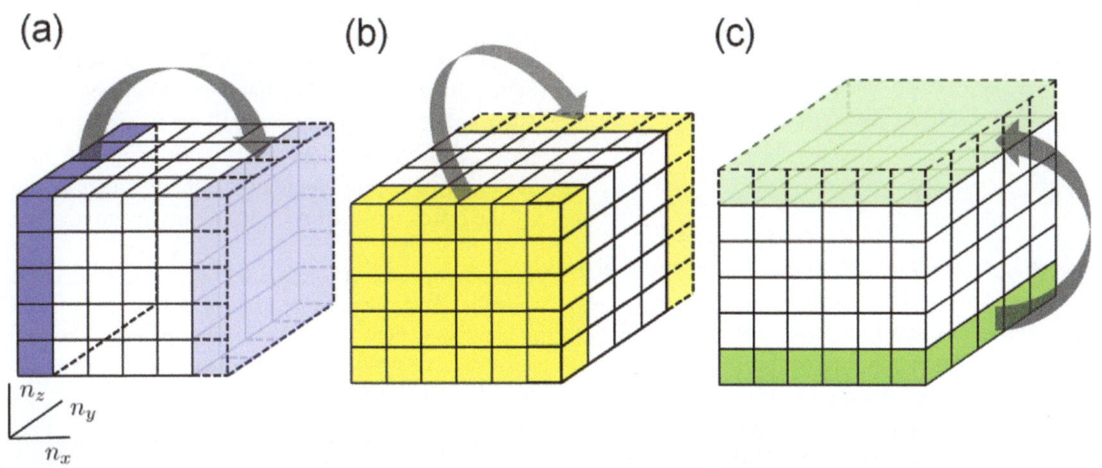

Figure 13: Schematic of the sequential construction of different ghost cell layers. An individual cell of the simulation box can be identified by the three integers (i_x, i_y, i_z). The original box in this example has $n_x = n_z = 5$ and $n_y = 4$ cells in each direction, i.e., there is no need for the simulation box to have cubic shape. (a) In a first step, all particles of cells with indices ($i_x = 1$; $i_y = 1, ..., n_y$; $i_z = 1, ..., n_z$) are copied into the first layer of ghostparticles. The second layer of ghostparticles then contains all particles pertaining to cells ($i_x = 1, ..., n_x$; $i_y = 1$; $i_z = 1, ..., n_z$) including the ghostparticles of the ghostcells from the first ghostlayer. Finally, the third layer of ghostcells is constructed by copying the particles of the cells with ($i_x = 1, ..., n_x$; $i_y = 1, ..., n_y$; $i_z = 1$) as indicated in the figure.

The additional advantage of using ghost cells lies in the fact, that the effort of setting up the cells and the effort of book-keeping decreases with system size, as the share of cells in the outermost layers of the simulation box decrease. For example, in a system with $N = 10^3$ particles, on average $\approx 73\%$ of all particles are ghosts, whereas this number has decreased to an average value of $\approx 13\%$ for a system with the same density, but $N = 2 \times 10^5$. Using this technique for larger system can result in a overall speed-up of up to a factor of 2.

Efficiency of the MD Method

The computing time of an algorithm is measured by the number of elementary steps (ES) that this algorithm needs until it stops, i.e., until the problem is solved. Examples of elementary steps are:

- Testing an if-condition,
- Assigning a value, i.e., changing the contents of a memory,
- Executing one of the elementary operations (+, −, ×, DIV, MOD),
- Initializing a loop variable.

As an example. we consider the following piece of pseudocode which gets as input an array a[1, . . . , n] and sorts this array according to the size of its elements.

```
START
for i := 1 TO N - 1 DO
for j := 1 TO N DO
if a[i] > a[j] then h = a[i]; a[i] = a[j]; a[j] = h
END
```

The kernel of the two loops in this algorithm consists of one if-condition (1 ES) and – if the condition is true – three assignments (3 ES) which switch a[i] with a[j]. Each i- and j-loop counts 1 ES. Thus, one can directly calculate the number of ES:

$$\sum_{k=1}^{N-1}\left(1+\sum_{l=k+1}^{N}(1+4)\right) = (N-1)+\sum_{k=1}^{N-1}\sum_{l=k+1}^{N}5 \tag{34}$$

$$= (N-1)+5\times\frac{N\times(N-1)}{2} \tag{35}$$

$$= 2.5\times N^2-1.5\times(N-1)\,. \tag{36}$$

Assuming that one ES on an average computer takes 10^{-9} seconds, one can sort arrays containing 2×10^5 elements within one second. Often however, one is only interested in how the run time of an algorithm depends on the number of input elements N, only considering the leading term in the computation time. In the example above one would speak of a "quadratic", or "order N^2" runtime and write symbolically $\mathcal{O}(N^2)$. The meaning of this symbolic $\mathcal{O}$-notation is the following:

A function g(N) is of order $f(N), i.e., g(N) = \mathcal{O}[f(N)]$ if there are constants c and N_0 such that for all $N \geq N_0$: $g(N) \leq c \times f(N)$. For example, the function $3N^2 + 4N$ is of order N^2, or in symbolic notation: $3N^2+4N=\mathcal{O}(N^2)$, as one can choose $c = 3$. Then $3N^2 + 4N \leq 3N^2$ for all $N > 5$. Thus, the previous relation is true for e.g., $N_0 = 6$.

In Table 2 we review five different algorithms A_1–A_5 with corresponding run times N, N^2, N^3, 2^N, N!, where N is the considered system size, e.g., the number of atoms, particles, nodes or finite elements in some simulation program. We again assume that one elementary step takes 10^{-9} seconds on a real computer.

Table 2: Overview of typical run times of algorithms occurring in materials science applications. Depicted are the number of ES and the corresponding real times for the different algorithms under the assumption that one ES takes 10^{-9} seconds.

Algorithm	run time	$N=10$	$N=20$	$N=50$	$N=100$
		10 ES	10 ES	10 ES	10 ES
A_1	N	$10^{-8}\,s$	$2\times10^{-8}\,s$	$5\times10^{-8}\,s$	$10^{-7}\,s$
		100 ES	400 ES	2.5×10^3 ES	10^5 ES
A_2	N^2	$10^{-7}\,s$	$4\times10^{-7}\,s$	$2.5\times10^{-6}\,s$	$10^{-5}\,s$
		1000 ES	8×10^3 ES	10^5 ES	10^6 ES
A_3	N^3	$10^{-6}\,s$	$8\times10^{-6}\,s$	$10^{-4}\,s$	$0.001\,s$
		1024 ES	10^5 ES	10^{15} ES	10^{30} ES
A_4	2^N	$10^{-6}\,s$	$10^{-3}\,s$	13 days	$\sim10^{13}$ years
		$\sim10^6$ ES	$\sim10^{18}$ ES	$\sim10^{64}$ES	10^{158} ES
A_5	$N!$	$3\times10^{-3}\,s$	77 years	10^{48} years	$\sim10^{141}$ years

The division of algorithms according to their run time in Table 2 allows for classifying algorithms into efficient and inefficient ones. It is obvious that exponential run times (algorithms A_4 and A_5) are not acceptable for all practical purposes. For these algorithms, even with very small system sizes N one reaches run times which are larger than the estimated age of the universe (10^{10} years). Algorithm A_5 could, for example, be a solution of the traveling salesman problem. If the first point out of N has been visited, there are (N – 1) choices for the second one. This finally results in an exponential run time of at the least N! steps. A runtime 2^N as in A_4 is typical for problems where the solution space of the problem consists of a subset of a given set of N objects; There are 2^N possible subsets of this basis set. The "efficient" algorithms A_1, A_2, A_3 with run times of at the most N^3 are the most commonly used ones in computational materials science.

Table 3 shows why algorithms A_1, A_2 and A_3 are considered to be efficient: Assuming that the available computer systems—due to a technology jump—will be 10 or 100 times faster than today, then the efficiency of algorithms A_1, A_2 and A_3 will be shifted by a factor, whereas for the exponential algorithms A_4, A_5 the efficiency will be shifted only by an additive constant.

Table 3: Speedup of the runtime of different algorithms assuming a hardware speedup factor of 10 and 100. The runtime of efficient polynomially bounded (class **P**) algorithms will be shifted by a factor while exponential (class **NP**) algorithms are only improved by an additive constant.

Algorithm	run time	efficiency	CPU speedup factor 10	CPU speedup factor 100
A_1	N	E_1	$10\times N_1$	$100\times N_1$
A_2	N^2	E_2	$\sqrt{10}\times N_2=3.16\times N_2$	$\sqrt{100}\times N_2=10\times N_2$
A_3	N^3	E_3	$\sqrt[3]{10}\times N_3=2.15\times N_3$	$\sqrt[3]{100}\times N_3=4.64\times N_3$
A_4	2^N	E_4	$\log_2(10\times N_4)=N_4+3.3$	$\log_2(100\times N_4)=N_4+6.6$
A_5	$N!$	E_5	$\approx N_5+1$	$\approx N_5+2$

Algorithms A_1, A_2 and A_3 have polynomial run times. An algorithm is said to be efficient if its runtime—which depends on some input N—has a polynomial upper bound. For example, the runtime function $2N^4(\log_2 N)^4 + 3\sqrt{N}$ has a polynomial upper bound (for large N), e.g., N^5. In $\mathcal{O}$-notation this is expressed as $\mathcal{O}(N^k)$ with k being the degree of the polynomial. Algorithms A_4 and A_5 on the other hand have no polynomial upper limit. Thus, they are called inefficient. In computer science, the class of problems that can be solved with efficient algorithms (i.e., algorithms that are polynomially bounded) are denoted with the letter **P**, cf. Figure 14a. As the set of polynomials is closed under addition, multiplication and composition, **P** is a very robust class of problems: Combining several polynomial algorithms results into an algorithm which again exhibits a polynomial runtime.

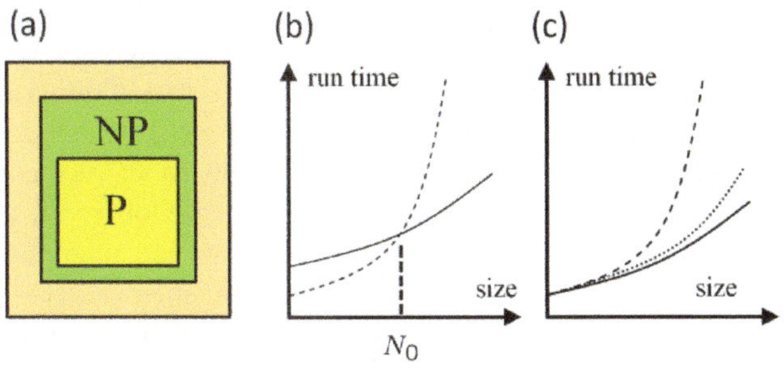

Figure 14: (a) Venn diagram of the class **P** (efficiently solvable problems), class **NP**(non-deterministic polynomial, i.e., inefficiently solvable problems), and undecidable problems (orange box) for which no algorithms are known. Today, it is generally assumed that all problems in **P** are contained in the class **NP**, cf. Figure 14. So far, no proof that decides whether **P** = **NP** or **P** ≠ **NP** is known. (b) An inefficient algorithm (dashed line) can – for some small values N up to an input number N_0 – be more efficient than a polynomially bounded algorithm (solid line). (c) A real algorithm (dotted line) will always have a run time behavior somewhat in between the worst-case (dashed line) and best-case (solid line) run time.

Due to the robustness of the definition of the class **P** of efficient algorithms, an inefficient algorithm can have a shorter runtime than its efficient counterpart, up to a certain system size N_0. For example, an algorithm with a runtime $1000 \times N^{1000}$falls into the class **P** whereas an algorithm with a runtime 1.1^N is exponential and thus inefficient. However, the exponential algorithm only exhibits longer runtime than the efficient one for system sizes up to N ~ 123, 000, cf. Figure 14b.

In the example of the sort algorithm above, the "worst-case" run time is considered assuming that the if-condition within the loop of the algorithm is true and thus, three elementary steps are always executed. In the "best-case"—e.g., if the array has been sorted – this if-condition is not true and there are only (N – 1) + N(N – 1) = N^2 – 1 ES. For a randomly shuffled array one can show that the expectation value for the number of elementary steps is $\sum_{k=2}^{N}(1/k) \approx \ln N$ [138]. Thus, with a randomly sorted array the total number of ES in this example is roughly $N^2 + 3N \ln N$. Hence, the actual runtime of an algorithm lies somewhere between the worst-case and the average-case runtime behavior, cf. Figure 14c.

In particle based simulations with assumed pairwise additive interactions the total force on the particles (or atoms) in a system depends on the current position of two particles only. If this assumption breaks down and has to abandoned in a simulation model, contributions of more complex non-additive interactions to the total potential have to be considered. For example, a suitable form of three-body interactions was introduced for the first time by Axilrod and Teller [139]. Such a potential depends on the position of at least three different particles. Solving the Schrödinger equation in ab initiosimulations also leads to a N^3 or even higher polynomial dependency of the run time. This is the main reason why ab initiomethods are restricted to very small system sizes.

Solving the classical Newtonian equations of motion with a "brute-force" strategy (cf. Figure 12a) leads to a $\mathcal{O}(N^2)$ run time as 1/2 × N × (N – 1) particle distances have to be calculated. This is also generally true in finite element codes where special care has to be taken when elements start to penetrate each other. Usually one uses so-called contact-algorithms which use a simple spring model between penetrating elements. The spring forces try to separate the penetrating elements again and the core of the contact algorithm is a lookup-table of element knots which is used to decide whether two elements penetrate each other or not. This algorithm in its plain form has an efficiency of $\mathcal{O}(N^2)$. As an $\mathcal{O}(N^2)$ efficiency of an algorithm still restricts the system size to very small systems of a few thousand particles one uses several methods to speed-up the efficiency of algorithms in computer simulations. Usually, this is done by using sorted search tables which can then be processed linearly (and thus reaching an efficiency of $\sim \mathcal{O}(N \log N)$. Hence, when it comes to the efficiency of algorithms in materials science, one will always try to minimize the effort to $\mathcal{O}(N)$.

Amdahl's Law

In principle, one can achieve a further speedup in the execution of a MD program by parallelizing it. Here, the upper limit of a possible optimization is given by Amdahl's law [140]: Let T_1 the execution time for a sequential program. If a fraction f of this program can be parallelized using M processors, then the theoretical execution time is determined by the sum of the time $T_s = (1 - f)T_1$ which is needed for the serial part and the time $T_p = (f \cdot T_1)/M$ needed for the parallelized program part. The maximum speedup S (f, M) of a parallelized code is thus given by:

$$S(f, M) = \frac{T_1}{T_s + T_p} = \frac{T_1}{(f \cdot T_1/M + (f/M) \cdot T_1} = \frac{1}{1 - f + f/M} \tag{37}$$

which is called Amdahl's law. Analyzing this equation for different pairs of values (f, M) shows that the actual speedup of a parallelized program is always smaller than the theoretical value as the parallelization itself is expensive. Also, for fixed f, the speedup does not grow linearly with M but approaches a limiting value. This is particularly important for massive-parallel program implementations with thousands of processors.

APPLICATION: SIMULATING THE EFFECT OF SHOCK WAVES IN POLYCRYSTALLINE SOLID STATES

In this section we review recent numerical applications in the field of shock wave physics based on the numerical methods that have been introduced and discussed in the previous sections. We start our discussion with a succinct introduction into shock wave physics and then focus on modeling polycrystalline solids such as high-performance ceramics. Results of simulating shock wave propagation in such materials using both, FEM, and a concurrent multiscale particle-based model are presented.

Nowadays, shock wave processes and their numerical simulation cover a spectrum that ranges from gas-dynamics of super-sonic objects, over air-blast-waves originating from detonations including their interaction with deformable structures, to the effects of shocks in structures, e.g., induced by projectiles or meteorites. Shock waves in soft matter have increasingly attracted interest in the field of medical treatment of inflammations or of nephroliths.

The specific characteristics distinguishing shock waves from ordinary acoustic waves are the extremely short rise times (in the range of nanoseconds, in contrast to microseconds with acoustic waves) and their dissipative nature. Reason for the formation of a shock wave is either the super-sonic motion of an object and the related wave-superposition or the pressure-dependency of sound speed which again leads to wave superposition and steepening of the wave front. An example of the tremendous effect shock waves may have in solids is exhibited in Figure 15. Here, a solid aluminum block was impacted by a 10mm diameter aluminum projectile at an impact velocity of 7km/s. In the vicinity of the impact location the high pressure amplitudes lead to the formation of crater lips under hydrodynamic pressure conditions. As a result, no phase transition of the material occurs, but rather only the high-pressure shock waves are responsible for the lip formation. As the initiated shock travels further into the material, it is reflected at the free surfaces shaping release waves.

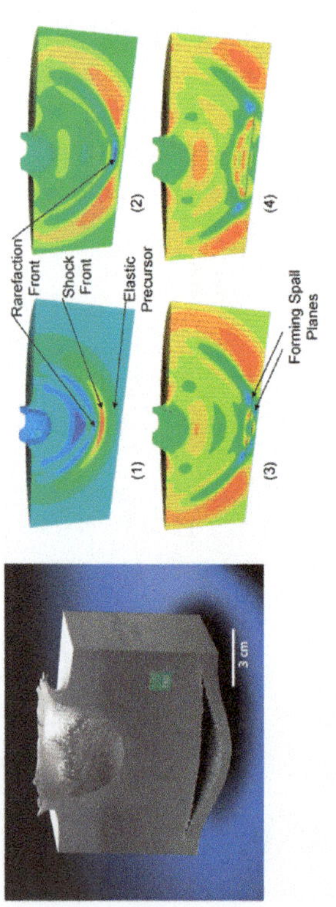

Figure 15: Spallation failure in a semi-infinite aluminium target after impact by a 10mmaluminum sphere at 7km/s. (a) Target after impact experiment cut into half. (b) Hydrocode simulation of the impact and related shock propagation leading to the formation of spallation planes.

At locations where several release waves are superimposed, a tensile pressure state is established which can lead to instantaneous failure, called spallation.

Until the early 1940s, investigations of shock wave formation and propagation were restricted almost completely to gaseous media. Nevertheless, the achievements in gas dynamics set the basis for fundamental work on shock waves in solids. During the course of his studies on finite-amplitude waves in solids, Riemann [141] invented the method of characteristics which became the tool of choice for the investigation of wave propagation, until almost a century later von Neumann and Richtmyer [142] introduced the idea of artificial viscosity (AV) which refers to the transformation of kinetic energy into heat through the narrow shock transition zone. Although AV was introduced for numerical reasons, it is an elective addition in hydrocodes used to modify a physical process so that it can be more easily computed. If the AV is too small, velocity oscillations about the correct mean value are observed to develop behind a shock. The proper formulation and magnitude needed for an AV has undergone many refinements over the years and culminated in the method pioneered by Godunov in 1959 [143], in which a local elementary wave solution is used to capture the existence and propagation characteristics of shock and rarefaction waves.

Rankine [144] and Hugoniot [145,146] set the basics for the thermodynamics of shock waves. Treating a one-dimensional shock wave as a distortion moving at a shock velocity υ_S one can relate the conditions ahead and behind the shock to each other via the conservation equations for mass, momentum and energy. With the thermodynamic conditions described by the mass density ρ, the pressure p and the specific internal energy e, and using index 0 for the initial and 1 for the shocked state, respectively, the Rankine-Hugoniot equations describe the jump conditions to be:

$$\rho_0 v_S = \rho_1 \left(v_S - v_1 \right), \tag{38}$$

$$\rho_0 v_S v_1 = p_1 - p_0, \tag{39}$$

$$p_1 v_1 = \left(e_1 - e_0 \right) \rho_0 v_S + \frac{1}{2} \rho_0 v_S v_1^2, \tag{40}$$

where the material ahead of the shock is assumed to be at rest. A fourth equation is needed to find a solution for the Riemann problem described by Equations (38–40). If known, the material specific Equation of State (EOS) p(ρ, e) can be utilized for that purpose. On the other hand, a relation between any other pair of the involved variables can be employed to identify the EOS. The latter method has become the classic approach for deriving high pressure EOS's for solid materials. It involves an experiment, e.g., the flyer-plate test, where a planar shock in an arbitrary material is investigated to measure its shock velocity υ_S along with the particle velocity υ_1. Thus, the measured relation between shock velocity and particle velocity

$$v_S = v_S \left(v_1 \right) \tag{41}$$

allows for a derivation of the governing EOS and represents an application of the Riemann problem in solid state physics, originally formulated in the field of gas dynamics. For most crystalline materials, specifically for metals, relation (41) is linear. Porosity of materials however, leads to significant non-linearities in the shock-particle velocity relation. Experimental investigations of highly porous and inhomogeneous materials face specific complexity concerning a precise representative velocity measurement. Therefore, meso-mechanical simulation of the shock propagation in composite materials on the basis of known component EOS data have become a useful characterization tool (see for example [92]).

So far, the outlined approach describes one specific curve in (p, ρ, e)-space. Performing a shock experiment leads to the so called principal Hugoniot curve representing all possible thermodynamic states available to a material when loaded by shock waves of various amplitudes. In order to find a mathematical description of high pressure states in its vicinity, the Hugoniot curve is utilized as reference curve. A typical EOS formulation of that kind is for example modeled by an equation of Mie-Grüneisen type.

$$p(V,e) = p_{ref}(V)\frac{\Gamma(V)}{V}\left[e - e_{ref}(V)\right], \tag{42}$$

where p_{ref} is given by the Hugoniot curve and the Grüneisen coefficient is $\Gamma = V \left.\frac{\partial p}{\partial e}\right|_V$.

The predictive capability of numerical simulation of shock processes such as high- and hypervelocity impact scenarios strongly depends on the quality of the employed EOS. A wide spectrum of materials has been characterized experimentally in terms of their high-pressure EOS over the last decades. Shock compression experiments [147,148] as well as, more recently, isentropic compression tests [149] are well established for the identification of reference curves.

A fundamental requirement for the shock wave initiation and its stable propagation is the convexity of the related EOS. Bethe [150] formulated the following two conditions for the existence of shock waves:

$$\left.\frac{\partial^2 p}{\partial V^2}\right|_S > 0, \tag{43}$$

$$\Gamma = V\left.\frac{\partial p}{\partial e}\right|_V > -2, \tag{44}$$

along with the criterion for stable propagation:

$$\left.\frac{\partial p}{\partial V}\right|_e < 0. \tag{45}$$

Modeling Polycrystalline Solids Using Power Diagrams

Understanding the micro-structural features of polycrystalline materials such as high-performance ceramics (HPCs) or metals is a prerequisite for the design of new materials with desired superior properties, such as high toughness or strength. Thus, for enhancing simulation models used for the prediction of material properties on multiscales, there exists a simultaneous need for characterization and ever more realistic representations of micro structures. On the length scale of a few microns to a few hundred microns, many materials exhibit a polyhedral granular structure which is known to crucially influence their macroscopic mechanical properties. Today, one is compelled to search for the optimal micro structure for a specific application by intricate and expensive experimental "trial-and-error" studies. In order to overcome this situation by numerical simulations, a detailed and realistic modeling of the available experimental structures is a basic requirement. With numerical investigations taking explicitly into account the micro structural details, one can expect to achieve a considerably enhanced understanding of the structure-property relationships of such materials [151,152]. With ceramics, the specific shape and size of these polycrystalline grain structures is formed during a sintering process where atomic diffusion on the nanometer scale plays a dominant role. Usually, the sintering process results in a dense micro structure with grain sizes of up to several hundred micrometers. Using a nano-sized fine-grained granulate as a green body along with an adequate process control it is possible to minimize both, the porosity (< 0.05% in volume), as well as the generated average grain size (< 1 μm). It is known that both leads to a dramatic increase in hardness which outperforms most metal alloys at considerably lower weight. Producing very small grain sizes in the making of HPCs below 100nm results again in decreasing hardness [153]. Hence, there is no simple connection between grain size and (macroscopic) hardness of a polycrystalline material.

The micro structure of densely sintered ceramics can be considered in very good approximation as a tessellation of $\mathbb{R}^2$ with convex polyhedra, i.e., as a polyhedral cell complex, cf. Figure 16a. A direct, primitive discretization of the micro-photograph into equal-spaced squares in a 2D mesh can be used for a direct simulation of material properties, cf. Figure 16b. However, with this modeling approach, the grain boundaries on the micrometer scale have to be modeled explicitly with very small elements of finite thickness. Thus, the influence of the area of the interface is unrealistically overestimated in light of the known fact that grain boundaries, which constitute an area of local disorder, often exhibit only a thickness of a few layers of atoms [154]. Moreover, a photomicrograph is just one 2D sample of the real micro structure in 3D, hence the value of its explicit rendering is very questionable. Finally, with this approach there is no 3D information available at all. While experimentally measured micro structures in 3D are generally not available for ceramic materials, only recently first reports about measured micro structures of steel have been published [155,156]. Nevertheless, these experiments are expensive and their resolution as well as the number of measured grains still seem to be poor [156].

(a) (b)

Figure 16: (a) Micrograph of a HPC (Al_2O_3) exhibiting the mi-

cro structure with an average grain size of 0.7 μm. (b) 2D FEM simulation of a primitive model of this micro structure with a shock impulse traveling through the material from left to right. The plane of the micrograph has been sectioned into 601 × 442 equal-spaced squares which are used as finite elements. The nodes of the upper and lower edge have been assigned $\vec{v} = 0$ as boundary condition, whereas the leftmost element nodes of the sample are given an initial speed of $v_x = 500$ m/s. The color code exhibits the pressure profile.

A different way of generating micro structures, is based on classical Voronoi diagrams in d-dimensional Euclidean space $\mathbb{E}^d$ and their duals—the Delaunay triangulations—which both constitute important models in stochastic geometry and have been used in various scientific fields for describing space-filling, mosaic-like structures resulting from growth processes. Voronoi diagrams are geometric structures that deal with proximity of a set of points (or more general objects). Often one wants to know details about proximity: Who is closest to whom? who is furthest and so on. The origin of this concept dates back to the 17th century. In his book on the principles of philosophy [157], R. Descartes claims that the solar system consists of vortices. His illustrations show a decomposition of space into convex regions, each consisting of matter revolving round one of the fixed stars. Even though Descartes has not explicitly defined the extension of these regions, the underlying idea seems to be the following: Let a space $\mathbb{T}$ and a set S of sites p in $\mathbb{T}$ be given, together with the notion of the influence a site pexerts on a point x of $\mathbb{T}$. Then the region of p consists of all points x for which the influence of p is the strongest, over all $t \in \mathbb{T}$. This concept has independently emerged, and proven useful, in various fields of science. Different names particular to the respective field have been used, such as medial axis transform in biology or physiology, Wiegner-Seitz zones in chemistry and physics, domains of action in crystallography, and Thiessen polygons in meteorology. The mathematicians Dirichlet (1850) [158], and Voronoi (1908) [159] were the first to formally introduce this concept. They used it for the study of quadratic forms; here the sites are integer lattice points, and influence is measured by the Euclidean distance. The resulting structure has been called Dirichlet tesselation or Voronoi diagram, cf. Figure 17a, which has become its standard name today. Voronoi was the first to consider the dual of this structure, where any two point sites are connected whose regions have a boundary in common, cf. Figure 17b. Later, Delauney [160] obtained the same by defining that two point sites are connected if (and only if) they lie on a circle whose interior contains no point of $\mathbb{T}$. After him, the dual of the Voronoi diagram has been denoted Delaunay tesselation or Delaunay triangulation.

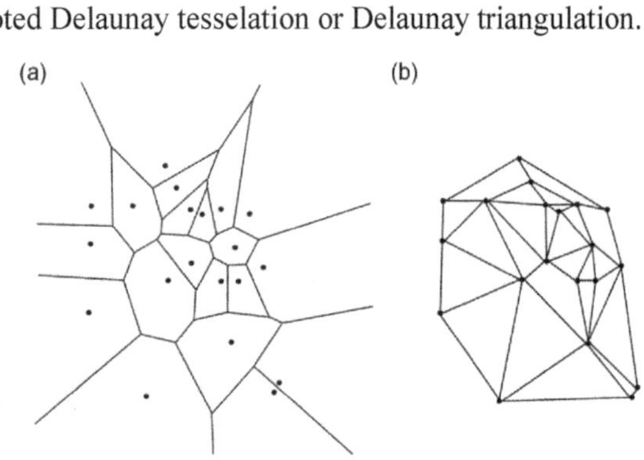

Figure 17: (a) Voronoi diagram of N=20 sites. For a finite set of generator points $\mathbb{T} \subset \mathbb{T}^d$ the Voronoi diagram maps each $p \in \mathbb{T}$ onto its Voronoi region R(p) consisting of all $x \in \mathbb{T}^d$ that are closer to p than to any other point in $\mathbb{T}$. (b) Delaunay triangulation for the sites in (a).

Voronoi tessellations in $\mathbb{R}^2$ have been used in many fields of materials science, e. g. for the description of biological tissues or polymer foams [161]. Ghosh et al. [162] utilized Voronoi cells to obtain stereologic information for the different morphologies of grains in ceramics and Espinoza et al. [163] used random Voronoi tessellations for the study of wave propagation models that describe various mechanisms of dynamic material failure at the micro scale. However, these models have major drawbacks such as limitations to two dimensions and a generic nature of the structures as they are usually not validated with actual experimental data. Besides its applications in other fields of science, the Voronoi diagram and its dual can be used for solving numerous, and surprisingly different, geometric problems. Moreover, these structures are very appealing, and a lot of research has been devoted to their study (about one in every 16 papers in computational geometry), ever since Shamos and Hoey [164] introduced them to this field. The reader interested in a complete overview over the existing literature should consult the book by Okabe et al. [165] who list more than 600 papers, and the survey by Aurenhammer [166].

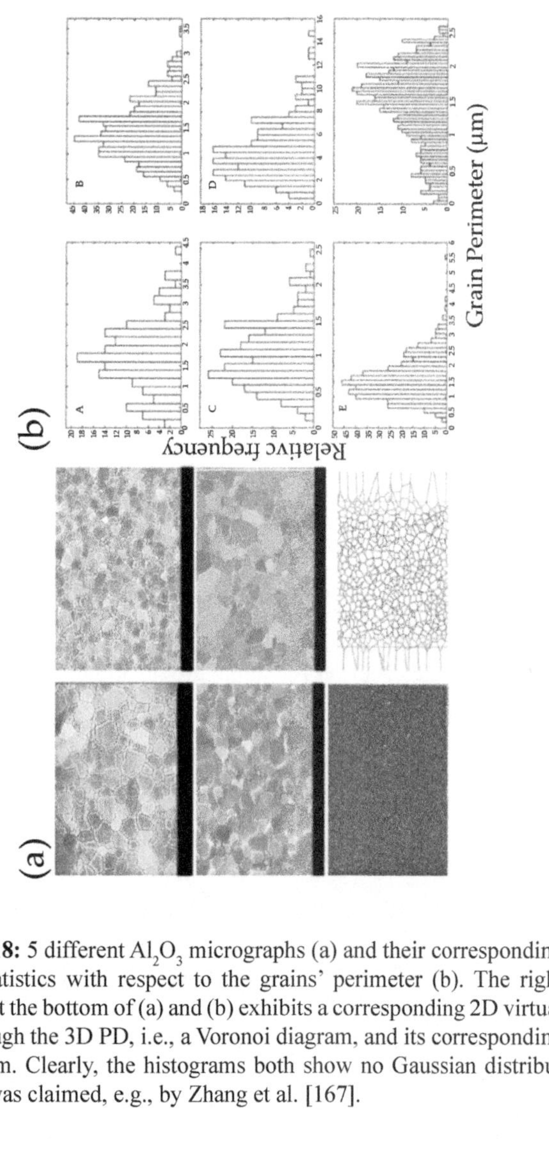

Figure 18: 5 different Al_2O_3 micrographs (a) and their corresponding grain statistics with respect to the grains' perimeter (b). The right picture at the bottom of (a) and (b) exhibits a corresponding 2D virtual cut through the 3D PD, i.e., a Voronoi diagram, and its corresponding histogram. Clearly, the histograms both show no Gaussian distribution as was claimed, e.g., by Zhang et al. [167].

In a recent approach to micro structural modeling of polycrystalline solids it was suggested to use power diagrams (PDs) along with a new optimization scheme for the generation of realistic 3D structures [169]. PDs are a well studied generalization of Voronoi diagrams for arbitrary dimensions [165] and have some major advantages over Voronoi diagrams as outlined in [168]. The suggested optimization is based on the statistical characterization of the grains in terms of the distribution of the grain areas A and the grain perimeters P obtained from cross-section micro-photographs, cf. Figure 18. An important result obtained using this method is that neither the experimental area nor the perimeter distribution obey a Gaussian statistics which is contrary to what was claimed e.g., by Zhang et al. [167,168].

The optimization scheme for the generation of realistic polycrystalline 3D structures is based on comparing all polyhedral cells (typically at least 10.000) inside a cube of a given PD in 3D with the 2D experimental data. This comparison is performed for each coordinate axis by generating a set of parallel, equidistant 2D slices (typically 500 slices for each of the three coordinate directions) through the cube and perpendicular to the respective axis, see Figure 19. For each 2D slice the grain sizes A are calculated and combined into one histogram. The same is done for the perimeter P. Then, the calculated histograms are compared with the experimental histograms A_i^{exp} and P_i^{exp} by calculating the first k central moments of the area and perimeter distributions A_i and P_i, respectively. A figure of merit m of conformity is defined according to which the PDs are optimized [168]:

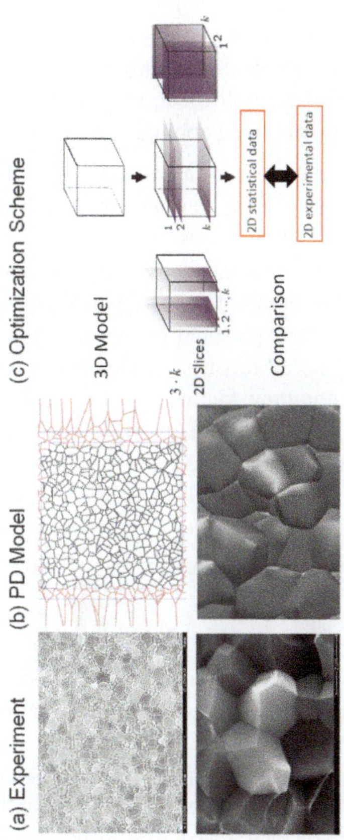

Figure 19: Optimization scheme as suggested in [168]. (a) 2D experimental photomicrograph (top) and an SEM picture of the 3D crystalline surface structure of Al_2O_3 (bottom). (b) 2D virtual slice of a power diagram (top) and the corresponding 3D surface structure obtained from this model. (c) Comparison between 2D experimental data of (a) and the 3D model of (b).

$$m = \sum_{i=1}^{k} \left(\frac{P_i - P_i^{exp}}{P_i^{exp}} \right)^2 + \left(\frac{A_i - A_i^{exp}}{A_i^{exp}} \right)^2 . \qquad (46)$$

The figure of merit m in Equation (46) is first calculated from the initial PD generated by a Poisson distribution of generator points. Using a reverse Monte-Carlo scheme, one generator point is chosen at random, its position modified and m is checked again. If m has decreased, the MC move is accepted, otherwise it is rejected. The modification of generator points is continued until m has reached a given threshold, typically 10^{-1}. If m = 0 is reached, the first k central moments of the experimental distributions agree completely with the model. In Figure 20 we present the resulting histogram of an optimized PD for Al_2O_3 and show the time development of the figure of merit m for this sample, following the proposed optimization scheme described above.

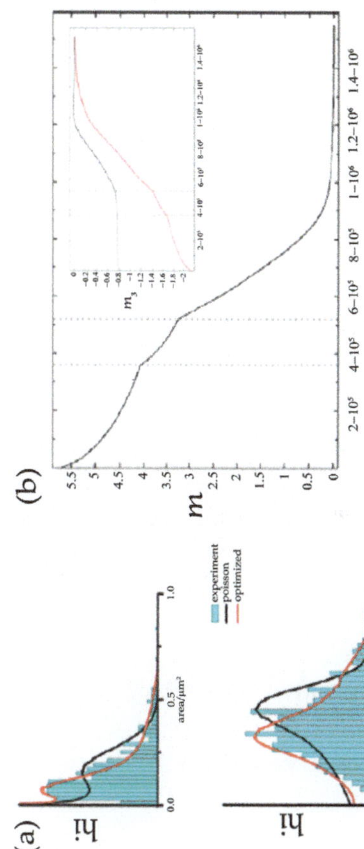

Figure 20: (a) Area (top) and perimeter (bottom) distribution of one of the Al_2O_3micrographs of Figure 18 before (black) and after (red) optimization. The bar graphs show the respective histograms of experimental data. (b) Time development of the figure of merit m during the optimization for the structure of (a). After 358000 and 512000 optimization steps, the maximum step size of the reverse MC algorithm (changing the position of a generator point) was increased, which shows a direct influence on the speed of optimization. After 1.5 million steps the deviation between the model and experiment has dropped below 1.3×10^{-4}. The inset shows the corresponding time development of m of the perimeter (red) and (area) distribution of the third central moment.

Having an efficient means to generate realistic polycrystalline structures, they can be meshed and be used for a numerical FEM analysis, cf. Figure 22. For simulations of macroscopic material behavior, techniques based on a continuum approximation, such as FEM or SPH are almost exclusively used. Figure 21 shows a 3D tile of a meshed PD. In a continuum approach the considered grain structure of the material is typically subdivided into smaller (finite) elements, e.g., triangles (in 2D) or tetrahedra in 3D. Tetrahedral elements at the surface can either be cut, thus obtaining a smooth surface, or they can represent (a more realistic) surface coarseness. Also displayed is an enlarged section of the 3D tetrahedral mesh at the surface of the virtual specimen. Upon failure, the elements are separated according to some predefined failure modes, often including a heuristic Weibull distribution [170,171] which is artificially imposed upon the system.

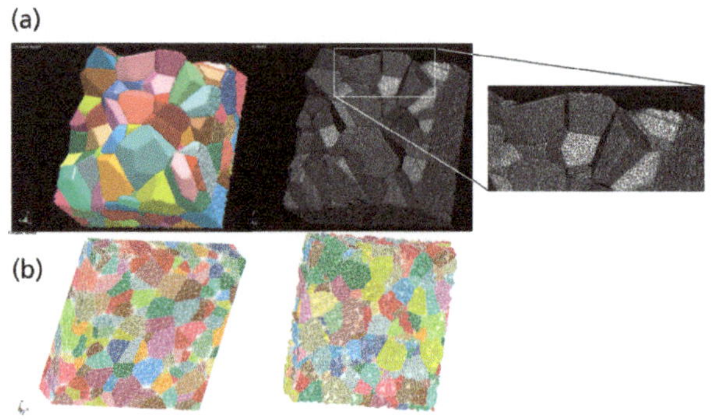

Figure 21: 3D structures of a meshed PD. In (a) the granular surface structure, its mesh and a detailed augmented section of the mesh at the surface are displayed. (b) A different realization of a 3D structure displaying the possibility of either leaving a (more realistic) rough surface micro structure, or smoothing the surface and thus obtaining a model body with even surface [168].

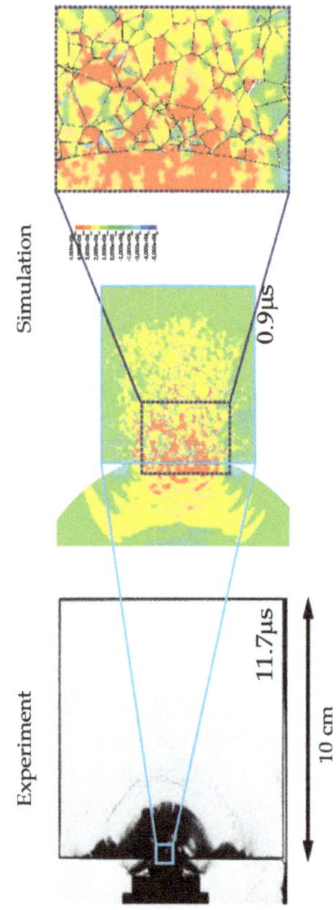

Figure 22: Illustration of the multiscale problem. With concurrent FEM methods which include micro structural details, only a very small part of a real system can actually be simulated due to the necessary large number of elements. Figure taken from [172].

Figure 22 illustrates the disadvantages and the multiscale problem associated with FEM simulations in which micro structural details are included. On the left, a high-speed camera snapshot of an edge-on impact experiment 11.7 μs after impact is shown, where a macroscopic steel sphere impacts the edge of an Aluminum Oxinitride (AlON) ceramic tile of dimension (10 × 10 × 0.2) cm. The enlargements in the middle and on the right show the small size of the region that is actually accessible to FEM analysis in a concurrent multiscale simulation approach. With FEM only a very small part of a macroscopic system can actually be simulated due to the necessary large number of elements. This is why in FEM simulations of polycrystalline materials, in order to be able to simulate a sufficient number of grains, often only two dimensions are considered in the first place. For most codes, an element number exceeding a few dozen millions is the upper limit which is still feasible in FEM simulations on the largest super computer systems. More severe, the constitutive equations for the material description which are needed in a phenomenological description, are derived from experiments with idealized load conditions. This often leads to many fit parameters in models, which diminishes their physical value. In addition, FEM generally has many computational problems (numerical instabilities) when it comes to very large element distortions in the vicinity of the impact region where the stresses, strain rates, and deformations are very large. The time scale of a multiscale FEM simulation does not a priori fit to the timescale of the experiment; thus, parameter adaptations of the included damage model are necessary (but are often unphysical). Also, the contact algorithms implemented in common engineering codes such as "pamcrash" or "lsdyna3D" which ensure that elements cannot penetrate each other in impact situations, where high strain rates occur, are often unphysical, very inefficient, and thus not well suited for parallelized applications.

The multiscale problem associated with FEM simulations described in Figure 22 is further worsened by the fact that the results of FEM analyses of highly dynamic processes are often strongly influenced by mesh resolution and mesh quality [173,174], which, from a physical point of view, is not acceptable, since the physical properties of a system should be invariant to the arbitrarily chosen spatial resolution of the problem. This common feature of FEM and related methods (such as SPH) is illustrated in Figure 23, where the same geometry is simulated as in Figure 22, illustrating the strong dependence of FEM and SPH on mesh resolution.

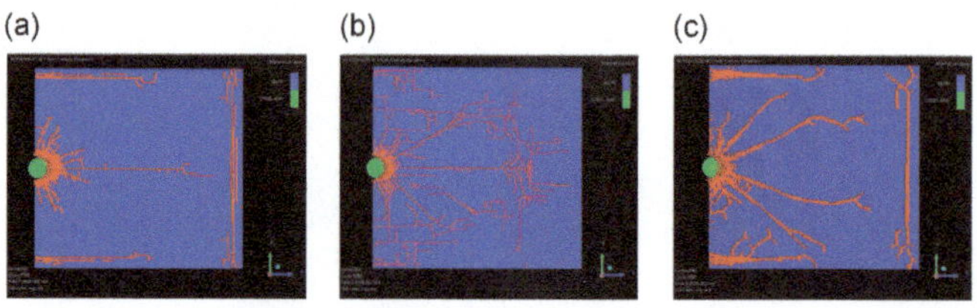

Figure 23: Snapshots of simulations of the edge-on impact system of Figure 22 using a primitive discretization of the geometry of the system in terms of hexahedral elements. (a) FEM with mesh resolution of 0.5mm. (b) FEM with mesh resolution of 1.0mm. (c) SPH with mesh resolution of 0.5mm. All computational results are obtained using a commercial code (Autodyn-3D) and are different in terms of the damage pattern of the cracks propagating through the material 3 μs after impact.

A Particle Model for Simulating Shock Wave Failure in Solids

Investigations of materials which involve multiple structure levels, such as nano- and polycrystalline solids, require large ensembles of atoms to accurately reflect the structures on the atomic and microscopic levels. For systems of reasonable size, atomistic simulations are still limited to following the dynamics of the considered systems only on time scales of nanoseconds. Such scales are much shorter than what is needed to follow many dynamic phenomena that are of experimental interest [80,175].

Whether a material under load displays a ductile, metal-like behavior or ultimately breaks irreversibly, depends on the atomic crystal structure and on the propagation of defects in the material. Broken atomic bonds (cracks) and dislocations are the two major defects determining mechanical properties on the atomic scale. Molecular dynamics investigations of this type using generic models of the solid state have lead to a basic understanding of the processes that govern failure and crack behavior, such as the instability of crack propagation [24,176–181], the dynamics of dislocations [33,80,182,183], the limiting speed of crack propagation [35,184,185], the brittle-to-ductile transition [35,154,186,187], or the universal features of energy dissipation in fracture [188].

Most metals are crystalline in nature, i.e., they are solids composed of atoms arranged in a regularly ordered repeating pattern. When crystals form, they may solidify into either a polycrystalline solid or a single crystal. In a single crystal, all atoms are arranged into one lattice or a crystal structure. The structure of single crystals makes them ideal for studies of material response to shock loading. When a highly ordered material, such as a metal crystal, is put under a planar shock, the crystal is compressed along the direction of the shock propagation, see Figure 24a. This uniaxial response can remain elastic so that, once the disturbance is removed, the lattice will relax back to its original configuration. However, under high-stress conditions, the configuration of atoms in the lattice may be changed irreversibly. Irreversible changes in phase and the development of defects at the atomic level lead to macroscopic changes, such as plasticity, melting, or solid-to-solid phase transformations. When the dynamic compression is removed, the shock-modified micro structure may influence the formation and growth of voids, cracks, and other processes that may cause the material to fail, see Figure 24b. These atomistic changes can dramatically affect a materials behavior, such as its thermodynamic state, strength, and fracture toughness. Few data are available on the phase transformations that occur under highly dynamic stress conditions or on the defects and voids that may form and grow as a result. MD methods for typical engineering applications on dislocation dynamics, ductility and plasticity, failure, cracks and fracture under shock loading in solids were extended to large-scale simulations of more than 10^8 particles during the late 1990s by Abraham and Coworkers [33,34], Holian and Lomdahl [189], Zhou [175] and others [190,191]. Today, many-particle MD simulations taking into account the degrees of freedom of several billion particles have been simulated in atomistic shock wave and brittle to ductile failure simulations [192–194].

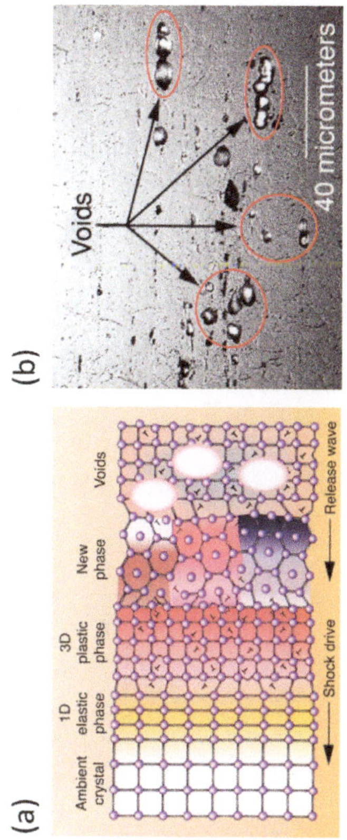

Figure 24: (a) Schematic of a crystal placed under shock loading. Initially it will compress uniaxially and then relax plastically through defects on the nanoscale, a process known as the one-dimensional to three-dimensional transition. The material may also undergo a structural transformation, represented here as a cubic to hexagonal change. The transformation occurs over a characteristic time scale. The new phase may be polycrystalline solid or melt. Once pressure is released, the microvoids that formed may grow, leading to macroscopic damage that causes the solid to fail. (b) This micrograph shows the voids that occur when a polycrystalline aluminium alloy is shocked and recovered. As the shock wave releases, the voids grow and may coalesce, resulting in material failure.

In the following we discuss a recently proposed concurrent multiscale approach for the simulation of failure and cracks in brittle materials which is based on mesoscopic particle dynamics, the Discrete Element Method (DEM), but which allows for simulating macroscopic properties of solids by fitting only a few model parameters [195].

Instead of trying to reproduce the geometrical shape of grains on the microscale as seen in two-dimensional (2D) micrographs, in the proposed approach one models the macroscopic solid state with soft particles, which, in the initial configuration, are allowed to overlap, cf. Figure 25a. The overall system configuration, see Figure 25b, can be visualized as a network of links that connect the centers of overlapping particles, cf. Figure 25c. The degree of particle overlap in the model is a measure of the force that is needed to detach particles from each other. The force is imposed on the particles by elastic springs. This simple model can easily be extended to incorporate irreversible changes of state such as plastic flow in metals on the macro scale. However, for brittle materials, where catastrophic failure occurs after a short elastic strain, in general, plastic flow behavior can be completely neglected. Additionally, a failure threshold is introduced for both, extension and compression of the springs that connect the initial particle network. By adjusting only two model parameters for the strain part of the potential, the correct stress-strain relationship of a specific brittle material as observed in (macroscopic) experiments can be obtained. The model is then applied to other types of external loading, e.g., shear and high-speed impact, with no further model adjustments, and the results are compared with experiments performed at EMI.

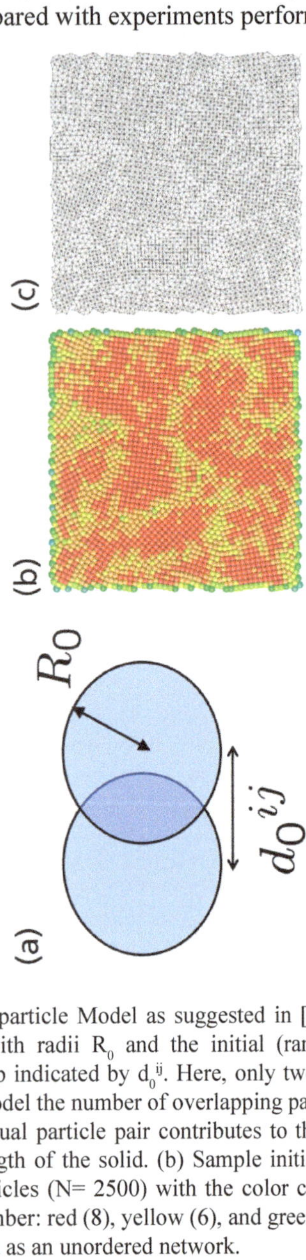

Figure 25: The particle Model as suggested in [195]. (a) Overlapping particles with radii R_0 and the initial (randomly generated) degree of overlap indicated by d_0^{ij}. Here, only two particles are displayed. In the model the number of overlapping particles is unlimited and each individual particle pair contributes to the overall pressure and tensile strength of the solid. (b) Sample initial configuration of overlapping particles (N= 2500) with the color code displaying the coordination number: red (8), yellow (6), and green (4). (c) The same system displayed as an unordered network.

Model Potentials

The main features of a coarse-grained model in the spirit of Occam's razor with only few parameters, are the repulsive forces which determine the materials resistance against pressure and the cohesive forces that keep the material together. A material resistance against pressure load is introduced by a simple Lennard Jones type repulsive potential Φ^{ij}_{rep} which acts on every pair of particles {ij} once the degree of overlap d^{ij} decreases compared to the initial overlap d_0^{ij}:

$$\phi^{ij}_{rep}(\gamma, d^{ij}) = \begin{cases} \gamma R_0^3 \left(\left(\frac{d_0^{ij}}{d^{ij}}\right)^{12} - 2\left(\frac{d_0^{ij}}{d^{ij}}\right)^{6} + 1 \right) & : \; 0 < d^{ij} < d_0^{ij} \\ 0 & : \; d^{ij} \geq d_0^{ij} \end{cases} . \tag{47}$$

Parameter γ scales the energy density of the potential and prefactor R_0^3 ensures the correct scaling behavior of the calculated total stress $\sum_{ij}\sigma^{ij} = \sum_{ij}F^{ij}/A$ which, as a result, is independent of N. Figure 26 shows that systems with all parameters kept constant, but only N varied, lead to the same slope (Young's modulus) in a stress-strain diagram. In Equation (47)R_0 is the constant radius of the particles, $d^{ij} = d^{ij}(t)$ is the instantaneous mutual distance of each interacting pairs {ij} of particles, and $d_0^{ij} = d^{ij}(t = 0)$ is the initial separation which the pair {ij} had in the starting configuration. Every single pair {ij} of overlapping particles is associated with a different initial separation d_0^{ij} and hence with a different force. The minimum of each individual particle pair {ij} is chosen such that the body is force-free at the start of the simulation.

When the material is put under a low tension the small deviations of particle positions from equilibrium will vanish as soon as the external force is released. Each individual pair of overlapping particles can thus be visualized as being connected by a spring, the equilibrium length of which equals the initial distance d_0^{ij}. This property is expressed in the cohesive potential by the following equation:

$$\Phi^{ij}_{coh}\left(\lambda, d^{ij}\right) = \lambda R_0 \left(d^{ij} - {d_0}^{ij}\right)^2, \; d^{ij} > 0. \tag{48}$$

In this equation, λ (which has dimension [energy/length]) determines the strength of the potential and prefactor R_0 again ensures a proper intrinsic scaling behavior of the material response. The total potential is the following sum:

$$\Phi_{tot} = \Sigma_{ij}\left(\phi^{ij}_{rep} + \phi^{ij}_{coh}\right). \tag{49}$$

The repulsive part of Φ_{tot} acts only on particle pairs that are closer together than their mutual initial distance d_0^{ij}, whereas the harmonic potential Φ_{coh} either acts repulsively or cohesively, depending on the actual distance d^{ij}. Failure is included in the model by introducing two breaking thresholds for the springs with respect to compressive and to tensile failure, respectively. If either of these thresholds is exceeded, the respective spring is considered to be broken and is removed from the system. A tensile failure criterium is reached when the overlap between two particles vanishes, i.e., when:

$$d^{ij} > (2R_0). \tag{50}$$

Failure under pressure load occurs when the actual mutual particle distance is less by a factor α (with $\alpha \in (0, 1)$) than the initial mutual particle distance, i.e., when

$$d^{ij} < \alpha \cdot {d_0}^{ij}. \tag{51}$$

Particle pairs without a spring after pressure or tensile failure still interact via the repulsive potential and cannot move through each other.

An appealing feature of this model, as opposed to many other material models used for the description of brittle materials, see e.g., [83,196–201], is its simplicity. The proposed model has a total of only three free parameters: γ and λ for the interaction potentials and α for failure. These model parameters can be adjusted to mimic the behavior of specific materials. The initial particle structure of the system is generated in a warmup process before the actual simulation, during which the simulation box is filled with particles which are allowed to grow for an expansion time τ, until the desired particle radii R_0 and overall particle density Ω are reached. From the initial particle pair distance distribution $\langle d_0^{ij} \rangle$ one can derive a maximal expectation value for σ_{max} [195]:

$$\sigma_{max} = 1 - \frac{\langle d_0{}^{ij} \rangle}{2R_0}. \tag{52}$$

The maximum tensile strength does not appear as a parameter in the model; it follows from the initial particle configuration and the resulting initial density Ω, see Figure 27. The closer $\langle d_0^{ij} \rangle$ is to $2R_0$, the less is the maximum tensile strength. The random distribution of initial particle distances ultimately determines the system's stability upon load, as well as its failure behavior and the initiation of cracks.

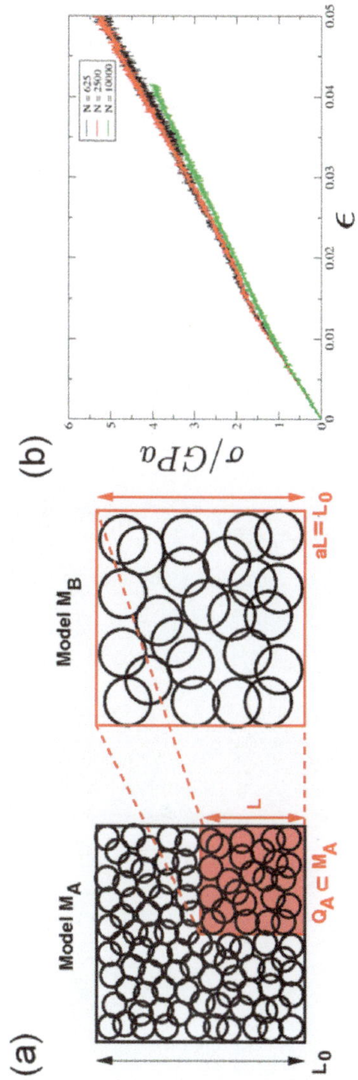

Figure 26: (a) Schematic of the intrinsic scaling property of the proposed material model. Here, only the 2D case is shown for simplicity. The original system (Model M_a) with edge length L_0 and particle radii R_0 is downscaled by a factor of 1/a into the subsystem Q_A of M_A (shaded area) with edge length L, while the particle radii are upscaled by factor a. As a result, model M_B of size $aL = L_0$ is obtained containing much fewer particles, but representing the same macroscopic solid, since the stress-strain relation (and hence, Young's modulus E) upon uni-axial tensile load is the same in both models. (b) Young's modulus E of systems with different number of particles N in a stress-strain (σ–ε) diagram. In essence, E is indeed independent of N.

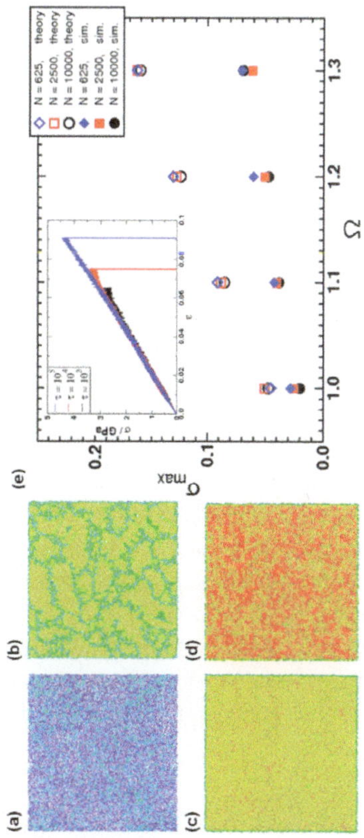

Figure 27: Sample configurations of systems with N = 10000 particles and different particle densities Ω. The color code displays the coordination numbers: blue (0), green (4), yellow (6), red (8). (a) Ω = 0.7. (b) Ω = 0.9 (c) Ω = 1.3 (d) Ω = 1.7 (e) Breaking strength σ_b for different system sizes N (filled symbols) as a function of particle density Ω, compared with the theoretical breaking strength σ_{max} (open symbols). The inset shows the stress-strain (σ – ε) relation for systems with three different initial expansion times τ. In essence, the larger the expansion time for the generation of the random initial overlap of particles, the larger is the material strength σ_{max}.

Strain, Shear and Impact Load

In the following, some simulation snapshots of applications of the proposed multiscale model adapted from [195] are displayed. Here, the model is applied to simulate a virtual material specimen of macroscopic dimensions under tensile, shear and impact load. First, the model parameters have been fitted so as to reproduce Young's modulus and the strength of a typical ceramic (Al_2O_3) in a quasistatic uniaxial tensile load simulation. Then, without any further fitting of parameters, the system is sheared and an impact experiment, as described in Figures 22 and 23 is performed.

In quasi-static load experiments, the involved physical processes occur on time scales small enough so that the system under investigation is always very close to equilibrium. The material is strained instantaneously on two opposite sides and then allowed to relax by means of a MD integration scheme with timestep $\Delta t = 0.001\hat{\tau}$ in reduced simulation units. During relaxation the particles may move and bonds may break. Equilibrium is reached per definition as soon as no bonds break during 1000 consecutive timesteps. After reaching equilibrium, the external strain is increased again and the whole procedure is iterated until ultimate failure occurs. For the actual simulation adapted to Al_2O_3, Ω = 1.1, ρ = 3.96 and κ = 350 are chosen; these values corresponding to a typical experimental situation of 99% volume density, ρ = 3.96 g/cm^3 and E = 370GPa. Results of the simulations are displayed in the picture series of Figure 28, which shows 4 snapshots of the fracture process, in which the main features of crack instability, as pointed out by Sharon and Fineberg [184] (onset of branching at crack tips, followed by crack branching and fragmentation) are well captured. At first, many micro-cracks are initiated all over the material by failing particle pair bonds. These micro-cracks lead to local accumulations of stresses in the material until crack tips occur where local tensions accumulate to form a macroscopic crack. This crack ultimately leads to a macroscopic, catastrophic failure of the model solid, which corresponds very well to the fracture pattern of a brittle material. Similar FEM simulations, see Figure 29, using about 50 million elements, still exhibit a strong dependence of the number of elements and of element size [202]. One advantage of the proposed particle model in contrast to FEM models is, that many systems with the same set of parameters (and hence the same macroscopic properties) but statistically varying micro-structure (i.e., initial arrangement of particles) can be simulated, which is very awkward to attain using FEM. By way of performing many simulations a good statistics for corresponding observables can be achieved.

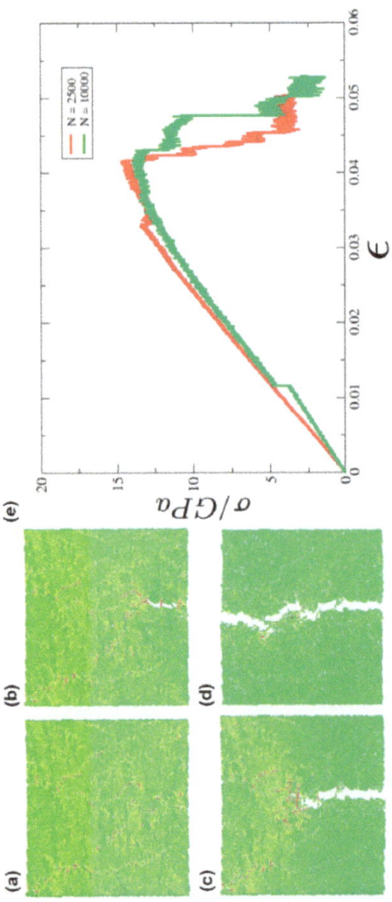

Figure 28: Crack initiation and propagation in the virtual macroscopic material sample upon uni-axial tensile load using N = 10^4 particles. The color code is: Force-free bonds (green); Bonds under tensile load (red). (a) Initiation of local tensions. (b) Initiation of a crack tip with local tensions concentrated around this crack tip. (c) Crack propagation including crack instability. (d) Failure. (e) Averaged stress-strain (σ-ε) relation. For N = 2500 (green curve) 10 different systems were averaged, and for N = 10000 (red curve) the stress-strain relations of 5 different initial particle configurations obtained in uni-axial load simulations were averaged.

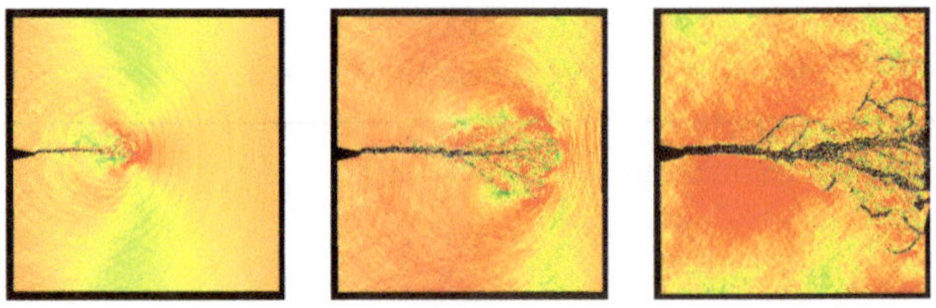

Figure 29: Snapshots of a 2D FEM simulation of the fracture process of a PMMA (poly-methyl methacrylate) plate which is subject to an initial uniform strain rate in vertical direction. Here, there is no statistical variability of the microstructure modeled with more than 50 × 10^6 elements. Thus, the model solid has to be artificially pre-notched and it still exhibits a strong dependence on the mesh size and the number of elements. Adapted from [202].

The results of uni-axial shear simulations are presented in Figure 30. Only the color-coded network of particles is shown: stressfree (green), tension (red) and pressure (blue). Starting from the initially unloaded state, the top and bottom layer of particles is shifted in opposite directions. At first, in Figure 30a, the tension increases over the whole system. Then, as can be seen from Figure 30b and c, shear bands form and stresses accumulate, until failure due to a series of broken bonds occurs.

Figure 30: Quasi-static shear loading of a virtual material specimen with N = 2500. The color code is the same as in Figure 28 except for particle bonds under pressure which are coded in blue. (a) Onset of shear tensile bands and (orthogonal) shear pressure bands in the corners of the specimen. (b) Shear bands traversing the whole specimen. (c) Ultimate failure.

Finally, in Figure 31 non-equilibrium MD simulation (NEMD) results for systems with varying shock impact velocities are presented and compared with high-speed impact experiments performed at EMI with different ceramic materials

($Al_2O_3$and SiC). These oxide and non-oxide ceramics represent two major classes of ceramics that have many important applications. The impactor hits the target at the left edge. This leads to a local compression of the particles in the impact area. The top series of snapshots in Figure 31a shows the propagation of a shock wave through the material. The shape of the shock front and also the distance traveled by it correspond very well to the high-speed photographs in the middle ofFigure 31a. These snapshots were taken at comparable times after the impact had occurred in the experiment and in the simulation, respectively. In the experiments which are used for comparison, specimens of dimensions ($100 \times 100 \times 10$)mmwere impacted by a cylindrical blunt steel projectile of length 23mm, mass m = 126 g and a diameter of 29.96mm [18]. After a reflection of the pressure wave at the free end of the material sample, and its propagation back into the material, the energy stored in the shock wave front finally disperses in the material. One can study in great detail the physics of shock waves traversing the material and easily identify strained or compressed regions by plotting the potential energies of the individual pair bonds. Also failure in the material can conveniently be visualized by plotting only the failed bonds as a function of time, cf. the bottom series of snapshots in Figure 31a. A simple measure of the degree of damage is the number of broken bonds with respect to the their total initial number. This quantity is calculated from impact simulations of$Al_2O_3$and SiC, after previously adjusting the simulation parameters γ, λ and α accordingly. Figure 31b exhibits the results of this analysis. For all impact speeds the damage in the SiC-model is consistently larger than in the one for Al_2O_3 which is also seen in the experiments.

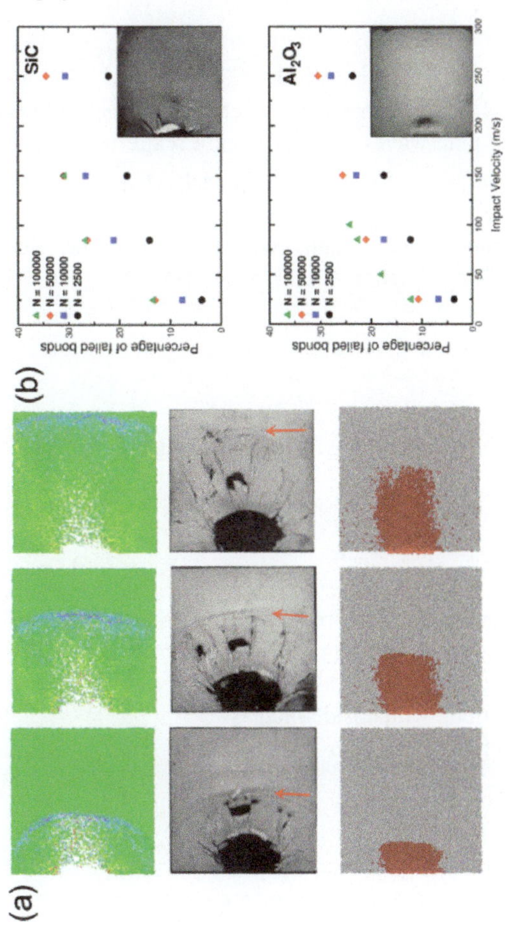

Figure 31: Results of a simulation of the edge-on-impact (EOI) geometry, cf. Figure 23, except this time, the whole macroscopic geometry of the experiment can be simulated while still including a microscopic resolution of the system. The impactor is not modeled explicitly, but rather its energy is transformed into kinetic energy of the particle bonds at the impact site. (a) Top row: A pressure wave propagates through the material and is reflected at the free end as a tensile wave (not shown). Middle row: The actual EOI experiment with a SiC specimen. The time interval between the high-speed photographs is comparable with the simulation snapshots above. The red arrows indicate the propagating shock wave front. Bottom row: The same simulation run but this time only the occurring damage in the material with respect to the number of broken bonds is shown. (b) Number of broken bonds displayed for different system sizes N, showing the convergence of the numerical scheme. Simulation parameters (α, γ, λ) are chosen such that the correct stress-strain relations of two different materials (Al_2O_3 and SiC) are recovered in the simulation of uniaxial tensile load. The insets show high-speed photographs of SiC and Al_2O_3, respectively, 4 μs after impact.

The impactor is not modeled explicitly, but rather its total kinetic energy is transformed into kinetic energy of the particles in the impact region. Irreversible deformations of the particles such as plasticity or heat are not considered in the model, i.e., energy is only removed from the system by broken bonds. Therefore, the development of damage in the material is generally overestimated.

In summary, the discussed concurrent multiscale particle model reproduces the macroscopic physics of shock wave propagation in brittle materials very well while at the same time allowing for a resolution of the material on the micro scale and avoiding typical problems (large element distortions, element-size dependent results) of FEM. The observed failure and crack pattern in dynamic impact simulations can be attributed to the random initial distribution of particle overlaps which generates differences in the local strength of the material. By generating many realizations of systems with different random initial overlap distribution of particles, the average values obtained from these many simulations lead to the fairly accurate results when compared with experimental studies.

COARSE-GRAINED MD SIMULATIONS OF SOFT MATTER: POLYMERS AND BIOMACRO-MOLECULES

The field of simulation of polymers and of biological macromolecular structures has seen an exciting development over the past 30 years, mostly due to the emergence of physical approaches in the biological sciences, which lead to the investigation of soft biomaterials, structures and diseases as well as to the development of new medical treatments and diagnostic methods [203,204]. The challenge faced by material scientists, namely that structure and dynamics of materials are characterized by extremely broad spectra of length and time scales, is particularly apparent with polymeric materials. Due to their length, polymers can attain a large number of conformations which contribute to the total free energy of a macromolecular system [205]. The structure ofmacromolecules is thus determined by an interplay between energetic and entropic contributions. The hydrodynamic interactions of polymers with solvent molecules, covalent interactions, intra- and intermolecular interactions and – particularly in biological macromolecules – the Coulomb interactions between charged monomers along with hydrogen bonds add up with the entropic forces to build a very complex system of interacting constituents. The enormous range of characteristic time and length scales accounts for their widespread use in technological applications and furthermore the most important biomolecular structures are polymers. It is clear that the treatment of such systems calls for hierarchical multiscale modeling approaches which can efficiently sample the complex potential energy hypersurface of polymeric materials, ensuring equilibration over the relevant length and time scales. Here, we provide a few examples of a very powerful MD method that is widely used in polymer physics for extending the available time and length scales in simulations, namely coarse-grained MD. With this method, the explicit chemical structure of the macromolecules is neglected by summarizing many atoms within one monomer in the simulation. With this approach, a focus is put on the structure due to the specific connectivity of the monomers and on general universal scaling laws.

Coarse-Grained Polymers

With polymer systems, many properties can be successfully simulated by only taking into account the basic and essential features of the chains, thus neglecting the detailed chemical structure of the molecules. Such coarse grained models, cf. Figure 32, are used because atomistic models of long polymer chains are usually intractable for time scales beyond nanoseconds, but which are important for many physical phenomena and for comparison with real experiments. Also, the fractal properties of polymers render this type of model very useful for the investigation of scaling properties. Fractality means, that some property X of a system (with polymers, e.g., the radius of gyration R_g) obeys a relation $X \propto N^k$, where $N \in \mathbb{N}$ is the size of the system, and $k \in \mathbb{Q}$ is the fractal dimension which is of the form $k = p/q$ with $p \neq = q$, $q \in \mathbb{N}$ and $p \varepsilon \mathbb{N}$. The basic properties that are sufficient to extract many structural static and dynamic features of polymers are:

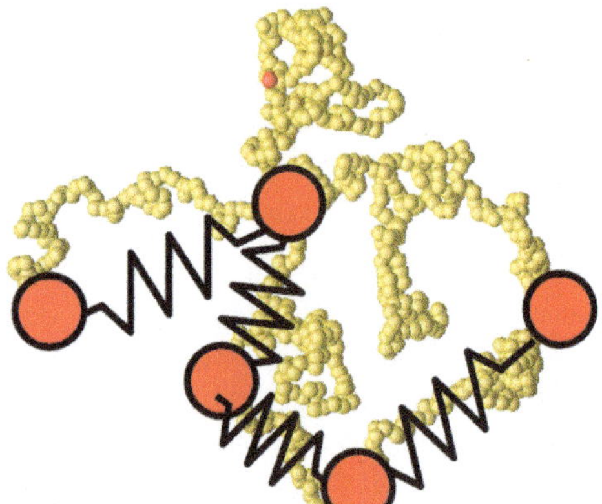

Figure 32: A coarse-grained model of a polymer chain where some groups of the detailed atomic structure (yellow beads) is lumped into one coarse-grained particle (red). The individual particles are connected by springs (bead-spring model).

- The connectivity of monomers in a chain.
- The topological constraints, e.g., the impenetrability of chain bonds.
- The Flexibility or stiffness of monomer segments.

Using coarse-grained models in the context of lipids and proteins, where each amino acid of the protein is represented by two coarse-grained beads, it has become possible to simulate lipoprotein assemblies and protein-lipid complexes for several microseconds [134].

Scaling of Linear, Branched and Semiflexible Macromolecules

As for bulk condensed matter in general, analysis of the microscopic structure of polymer systems is mostly carried out by scattering experiments. Depending on the system under study and the desired resolution, photons in the X-ray and light scattering range, or neutrons are used. The general set up of a scattering experiment is very simple and indicated schematically in Figure 33 [206]: One uses an incident beam of monochromatic radiation with wavelength λ and initial intensity I_0. This beam becomes scattered by a sample and the intensity I of the scattered waves is registered by a detector (D) at a distance d, under variation of the direction of observation. The scattering vector q⃗ q→ is defined as

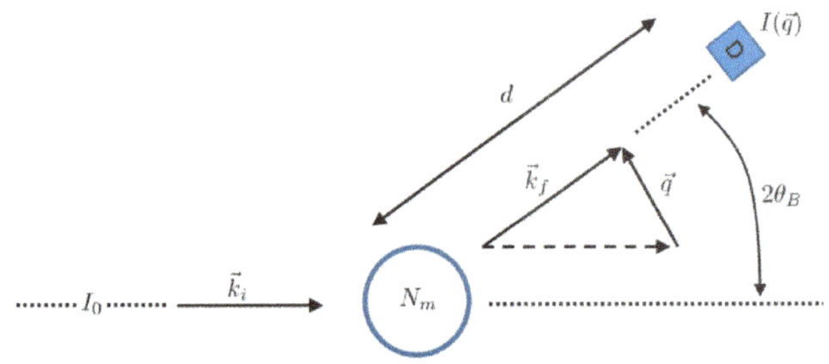

Figure 33: General set up of a scattering experiment according to [206]. Details are described in the main text.

$$\vec{q} = \vec{k}_f - \vec{k}_i \tag{53}$$

where $\vec{k}_f$ and $\vec{k}_i$ denote wave vectors of the incident and the scattered plan waves. The result of a scattering experiment is usually expressed by giving the intensity distribution $I(\vec{q})$ in $\vec{q}$-space, cf. Figure 33. In the majority of scattering experiments on polymers the radiation frequency remains practically unchanged, thus one has

$$|\vec{k}_f| \approx |\vec{k}_i| = \frac{2\pi}{\lambda} \tag{54}$$

where $\vec{q}$ is related to the Bragg scattering angle θ_B by

$$|\vec{q}| = \frac{4\pi}{\lambda} \sin\theta_B \tag{55}$$

If the scattering can be treated as being due to just one class of materials, the scattering properties can be described by the interference function $S(\vec{q})$, also called scattering function or structure function., which is defined as:

$$S(\vec{q}) = \frac{I(\vec{q})}{I_m N_m} \tag{56}$$

Here N_m represents the total number of particles (or monomers in the case of macromolecules) in the sample and I_m is the scattering intensity produced by one particle, if placed in the same incident beam. The scattering function expresses the ration between the actual intensity which would be measured and the intensity which would be measured, if all particles in the sample were to scatter incoherently. Scattering diagrams generally emerge from the superposition and interference of the scattered waves emanating from all particles in the sample. The total scattering amplitude measured at the detector is then given by

$$C = \sum_{i=1}^{N_m} \exp i\phi_i \tag{57}$$

Simple geometric considerations [206] show that the phases ϕ_i are determined by the particle position $\vec{R}_i$ and the scattering vector $\vec{q}$ only, being given by

$$\phi_i = \vec{q}\vec{r}_i \tag{58}$$

Thus, the scattering amplitude produced by a set of particles at locations $\vec{r}_i$ may be formulated as a $\vec{q}$-dependent function in the form

$$C(\vec{q}) = \sum_{i=1}^{N_m} \exp i\vec{q}\vec{r}_i \tag{59}$$

It is a well-known fact from electrodynamics that the scattering intensity is proportional to the squared total scattering amplitude, i.e.,

$$I(\vec{q}) \propto \langle |C(\vec{q})|^2 \rangle \tag{60}$$

The brackets indicate an ensemble average which involves, as always in statistical treatments of physical systems, all microscopic states of the sample. For ergodic systems the time average carried out by the detector equals the theoretical ensemble average. As the normalization of the amplitudes of the single scattered waves is already implied in the definition of the structure function, Equation (56), Equation (60) can be written as

$$S(\vec{q}) = \frac{1}{N_m} \langle |C(\vec{q})|^2 \rangle \tag{61}$$

which is a basic equation of general validity that may serve as starting point for the derivation of other forms of scattering, e.g., for isotropic systems, where $S(\vec{q}) = S(q)$ with $q = |\vec{q}|$, cf. Equation (68). Inserting Equation (59) into (61) one obtains

$$S(\vec{q}) = \frac{1}{N_m} \sum_{i,j=1}^{N_m} \langle \exp i\vec{q}(\vec{r}_i - \vec{r}_j) \rangle \tag{62}$$

Equation (62) can be calculated directly in molecular computer simulations as the positions $\vec{r}$ of all scattering particles N_m are exactly known at all times (within the boundaries of numerical errors when using floats in double precision as a representation of real numbers).

Polymers usually do not exist in vacuum but in solution. A solvent is referred to as good when the prevailing form of the effective interaction between polymer segments in this solvent is the repulsive part of the potential energy at shorter distances. In this case the chains tend to swell and the size R of the polymer (e.g., the end-to-end distance R_e in the case of linear chains or the radius of gyration R_g) scale with an exponent $\nu = 3/5$. In the opposite case of a poor solvent, polymers tend to shrink and R scales with $\nu = 1/3$. The point were the repulsive and attractive interactions just cancel each other defines the θ–point and θ–temperature, respectively. Here, the chain configuration is that of a Gaussian random coil with an exponent $\nu = 1/2$. There are still three-body and higher order interactions present in a θ–solvent, but their contribution to the free energy is negligibly small [52]. For the description of the distance of temperature T from the θ–temperature, a dimensionless parameter is used, the reduced temperature ζ which is defined as:

$$\zeta = \frac{|T - T_\theta|}{T_\theta}. \tag{63}$$

A crossover scaling function f serves for the description of the scaling behavior of polymer chains in different solvent conditions [52]. The argument of f is given by $\zeta\sqrt{N}$. At θ–temperature,

$$f(\zeta\sqrt{N}) \simeq 1, \quad \zeta\sqrt{N} \ll 1, \quad R = R_0 \propto N^{1/2}. \tag{64}$$

At $T < T_\theta$,

$$f(\zeta\sqrt{N}) \simeq (\zeta\sqrt{N})^{1/3}, \quad \zeta\sqrt{N} \gg 1, \quad R \propto N^{1/3}\zeta^{1/3}. \tag{65}$$

At $T > T_\theta$,

$$f(\zeta\sqrt{N}) \simeq (\zeta\sqrt{N})^{3/5}, \quad \zeta\sqrt{N} > 1, \quad R \propto N^{3/5}. \tag{66}$$

In experiments, it is rather difficult to obtain complete and conclusive results in the study of the collapse transition of chains, because one is often restricted to the three distinct limiting cases of polymer solutions, the extreme dilute regime, theθ-point, and the regime of very high polymer concentrations [207].

At the θ –temperature the chains behave as $\langle R_g^2 \rangle \propto \langle R_e^2 \rangle \propto (N-1)^{2\nu_\theta}$ with $\nu_\theta = 0.5$ besides logarithmic corrections in 3D. Therefore, one expects that a plot of $\langle R^2 \rangle / (N-1)$ vs. T for different values of N shows a common intersection point at $T = T_\theta$ where the curvature changes: for $T > T_\theta$ the larger N, the larger the ratio $\langle R^2 \rangle / (N-1)$, while for $T < T_\theta$ the larger N, thesmaller the ratio $\langle R^2 \rangle / (N-1)$. Using the model potential of Equation (25) – instead of varying temperature (which involves rescaling of the particle velocities), different solvent qualities are obtained by tuning the interaction parameter λ. The corresponding transition curves are displayed in Figure 34 which show a clear intersection point at roughly $\lambda = \lambda_\theta \approx 0.65$. Moreover it can be seen that the transition becomes sharper with increasing chain length N. The different curves do not intersect exactly at one single point, but there is an extended region in which the chains behave in a Gaussian manner. The size of this region is $\propto N^{-1/2}$ [52]. There is a very slight drift of the intersection point toward a smaller value of λ with increasing chain length N.

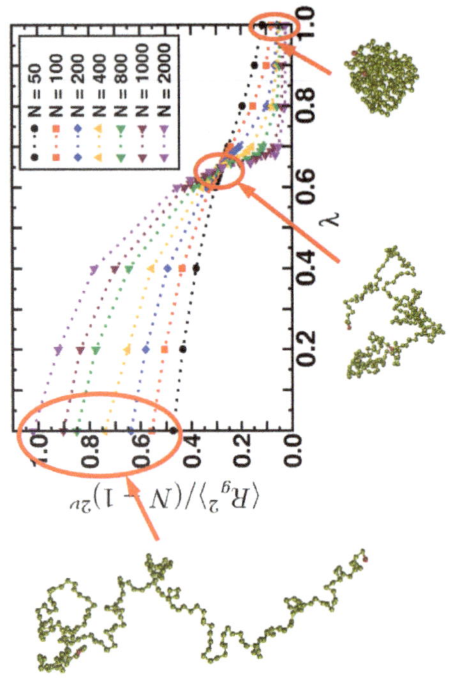

Figure 34: Coil-to-globule transition from good to bad solvent behavior of a polymer chain. Plot of $\langle R_g^2\rangle / (N-1)^{2\nu}$ vs. interaction parameter λ for linear chains. The points represent the simulated data and the dotted lines are guides to the eye. $\nu = \nu_\theta = 0.5$. Also displayed are simulation snapshots of linear chains for the three cases of a good,θ-, and a bad solvent.

Therefore, to obtain a more precise estimate of the θ–temperature in the limit of ($N \to \infty$), one has to chose a different graph that allows an appropriate extrapolation. If one draws straight horizontal lines in Figure 34 the intersection points of these lines with the curves are points at which the scaling function $f(\sqrt{N}\varsigma)$ of Equation (65) is constant. Plotting different intersection points over $N^{-1/2}$ therefore yields different straight lines that intersect each other exactly at $T = T_\theta$ and $\lambda = \lambda_\theta$ respectively. This extrapolation ($N \to \infty$) is displayed in Figure 35. The different lines do not intersect exactly at $N^{-1/2} = 0$ which is due to the finite length of the chains. As a result of these plots one yields the value of λ for which the repulsive and attractive interactions in the used model just cancel each other:

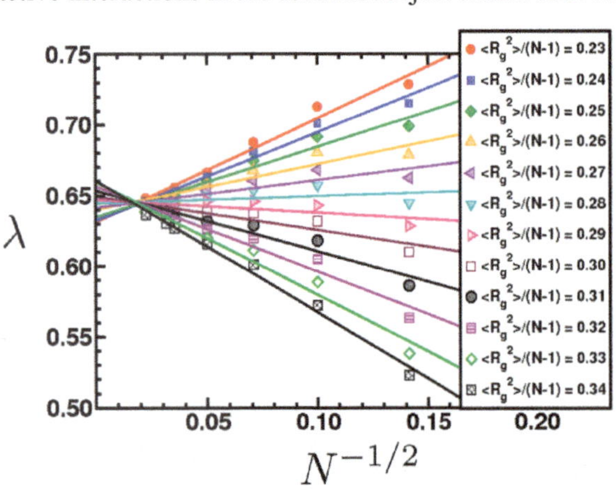

Figure 35: Interaction parameter λ of Equation (25) vs. $N^{-1/2}$ for different values of the scaling function. Data points are based on the radius of gyration of linear chains [51].

$$\lambda_\theta = 0.65 \pm 0.02. \tag{67}$$

An important property of individual chains is the structure factor S(q) the spherical average of which is defined as [3]:

$$S(q) = \left\langle \frac{1}{N^2} \left| \sum_{i=1}^{N} e^{-i\vec{q}\vec{r}_i} \right|^2 \right\rangle_{|\vec{q}|}, \tag{68}$$

with subscript $|\vec{q}|$ denoting the average over all scatter vectors $\vec{q}$ of the same magnitude $|k| = q$, $\vec{r}_i$ being the position vector to the ith monomer and N denoting the total number of monomers (scatter centers). For different ranges of the scatter vectors the following scaling relations hold [52]:

$$S(q) = \begin{cases} (1 - 1/3\, q^2 \langle R_g{}^2 \rangle)\, N & (2\pi)^2/\langle R_g{}^2\rangle \gg q^2, \\ q^{-1/\nu} & (2\pi)^2/\langle R_g{}^2\rangle \ll q^2 \ll (2\pi)^2/l_b{}^2, \\ 1/N & (2\pi)^2/l_b{}^2 \ll q^2, \end{cases} \tag{69}$$

where l_b is the (constant) bondlength of the monomers (often also called segment length). The importance of S(q) lies in the fact that it is directly measurable in scattering experiments. For ideal linear chains the function S(q) can be explicitly calculated and is given by the monotonously decreasing Debye function.

$$S(x) = \frac{2}{x^2}\left(x - 1 + e^{-x}\right), \tag{70}$$

where the quantity x is given by $x = q^2 \langle R_g{}^2\rangle_0$ with index 0 denoting θ –conditions. For small values of x, corresponding to large distances between scattering units, the Debye function S(x) also provides a good description of a linear chain in a goodsolvent with the scaling variable x describing the expansion of the chain. For very small scattering vectors q one obtains theGuinier approximation [51] by an expansion of S(q), which is used in experiments to calculate the radius of gyration $\langle R_g{}^2\rangle$. In the intermediate regime of scattering vectors, S(q) obeys a scaling law which, in a double-logarithmic plot, should yield a slope of $-1/\nu$. For large q-values finally, S(q) is expected to behave as 1/N. The overall expected behavior of S(q) is summarized in Equation (69).

In the vicinity of the θ–region, the scaling exponent equals $\nu = \nu_\theta = 0.5$. Therefore $q^2S(q)$, plotted against wave vector q, which is called a Kratky plot should approach a constant value. Figure 36 displays this behavior for different chain lengths with high resolution in terms of λ. The respective dotted horizontal line is a guide to the eye. The larger the chains are, the smaller is the λ-range at which the chains display ideal (Gaussian) random–walk behavior. For large values of λ the chains are collapsed and form compact globules the local structure of which is also reflected in the structure function by several distinct peaks for larger q-values. These peaks become the more pronounced the longer the chains are, reflecting the fact that the transition curves become ever sharper with increasing chain length. Hence, longer chains are already in the collapsed regime for values of λ at which the smaller chains still exhibit Gaussian behavior. The structure function of the largest system in Figure 36 for λ = 1.0 already resembles very much the scattering pattern of a sphere.

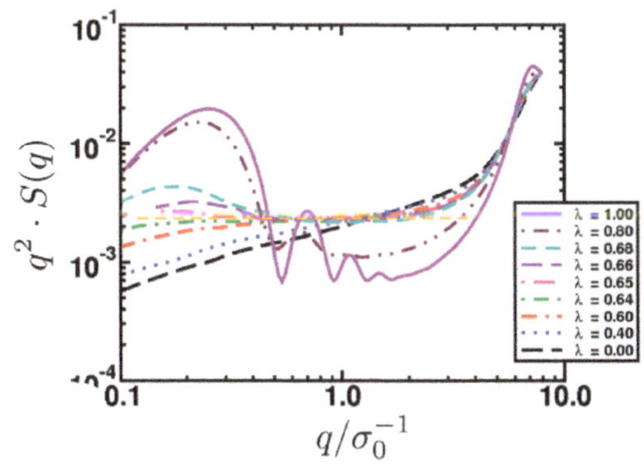

Figure 36. Kratky plot of S(q) of linear chains (N = 2000) for different values of the interaction parameter λ.

In Figure 37 the scaling of $\langle R_g^2\rangle$ for different star polymers as a function of N and for different functionalities f is displayed. Functionality f = 2 corresponds to linear chains, f = 3 corresponds three-arm star polymers and so on. The star polymers were generated with the MD simulation package "MD-Cube" developed by Steinhauser [51,63] which is capable of handling a very large array of branched polymer topologies, from star polymers to dendrimers, H-polymers, comb-polymers or randomly hyperbranched polymers. Details of the set-up of chains which works the same way for linear and branched polymer topologies can be found in [208]. Figure 37a shows a double-logarithmic plot from which one obtains the scaling exponents of R_g for stars with different numbers of arms. The results for linear chains are displayed as well, for which chain lengths of up to N = 5000 were simulated. Within the errors of the simulation, the exponents do not depend on the number of arms, as expected from theory. The obtained scaling exponents are summarized in Table 4 and exhibit a reasonable agreement with theory.

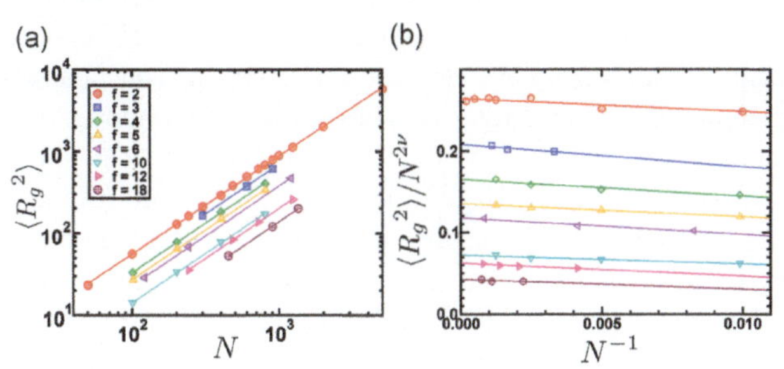

Figure 37: (a) Log-Log plot of $\langle R_g^2\rangle$ vs. N of star polymers with different arm numbers f. For comparison, data for linear chains (f = 2) are displayed as well. (b) Scaling plot of the corrections to scaling of $\langle R_g^2\rangle$ (f) plotted vs. N^{-1} in good solvent. For clarity, the smallest data point of the linear chains (f = 2, N = 50) is not displayed.

Table 4: Obtained scaling exponents v for star polymers in simulations with different arm numbers f.

f	2	3	4	5	6	10	12	18
ν	0.5989	0.601	0.603	0.614	0.617	0.603	0.599	0.601

In Figure 37b it is exhibited, that the corrections to scaling due to the finite size of the chains are ∝ N^{-1}. A plot with exponents −2 or −1/2 leads to worse correlation coefficients. This result is consistent with lattice-MC simulations on a fcc-lattice [209]. More details on finite-size scaling can be found in [50,51].

The fundamentals of the dynamics of fully flexible polymers in solution or in the melt were worked out in the pioneering works of Rouse [210] and Zimm [211], as well as of Doi and Edwards [212] and de Gennes [52]. In contrast to fully flexible polymers, the modeling of semiflexible and stiff macromolecules has received recent attention, because such models can be successfully applied to biopolymers such as proteins, DNA, actin filaments or rodlike viruses [213,214]. Biopolymers are wormlike chains with persistence lengths l_p (or Kuhn segment lengths l_K) comparable to or larger than their contour lengthL and their rigidity and relaxation behavior are essential for their biological functions.

Using the second term of the bonded potential of Equation (26), a bending rigidity $\Phi_{bend}(\theta)$ can be introduced. Rewriting this term by introducing the unit vector $\vec{u}_j = (\vec{r}_{j+1} - \vec{r}_j)/|\vec{r}_{j+1} - \vec{r}_j|$ along the macromolecule, cf. Figure 10, one obtains:

$$\Phi_{bend}(\theta) = \frac{k_\theta}{2}\sum_{j=1}^{N-1}(\vec{u}_{j+1} - \vec{u}_j)^2 = k_\theta \sum_{j-1}^{N-1}(1 - \cos\theta_j), \quad (71)$$

where θ_i is the angle between $\vec{u}_j$ and $\vec{u}_{j+1}$. The crossover scaling from coil-like, flexible structures on large length scales to stretched conformations at smaller scales can be seen in the scaling of S(q) when performing simulations with different values of k_θ [208]. Results for linear chains of length N = 700 are displayed in Figure 38a. The chains show a scaling according to q^ν. The stiffest chains exhibit a q^{-1}–scaling which is characteristic for stiff rods. Thus, by varying parameter k_θ, the whole range of bending stiffness of chains from fully flexible chains to stiff rods can be covered. The range of q–vectors for which the crossover from flexible to semiflexible and stiff occurs shifts to smaller scatter vectors with increasing stiffnessk_θ of the chains. The scaling plot in Figure 38b shows that the transition occurs for $q \approx 1/l_K$, i.e., on a length scale of the order of the statistical Kuhn length. In essence, only the fully flexible chains (red data points) exhibit a deviation from the master curve on large length scales (i.e., small q–values), which corresponds to their different global structure compared with semi-flexible macromolecules. Examples for snapshots of stiff and semiflexible chains are finally displayed in Figure 39.

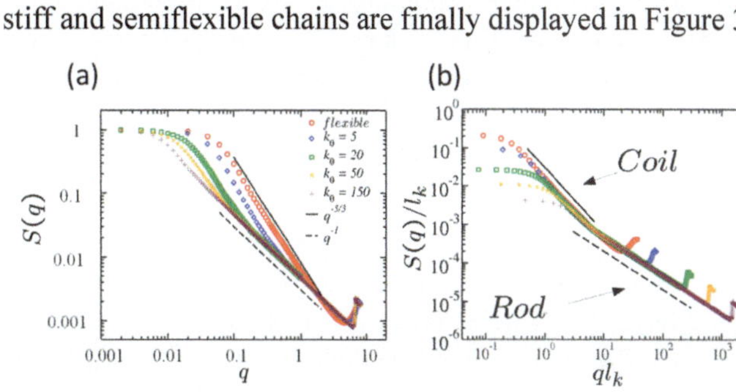

Figure 38: (a) S(q) of single linear chains with N = 700 and varying stiffness k_θ. The scaling regimes (fully flexible and stiff rod) are indicated by a straight and dashed line, respectively. (b) Scaling plot of $S(q)/l_K$ versus $q \cdot l_K$ using the statistical segment lengthl_K adapted from [208].

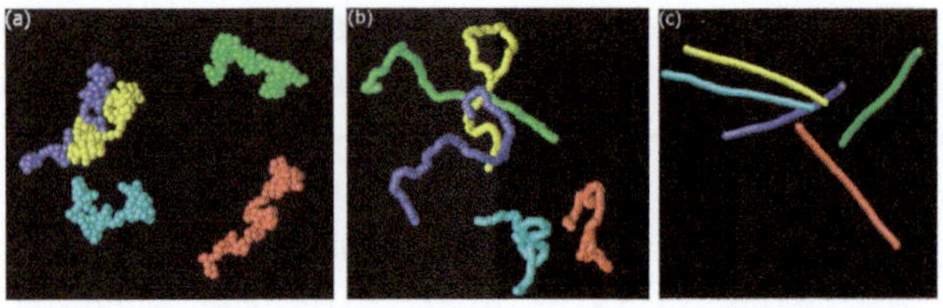

Figure 39: Simulation snapshots of (a) flexible chains ($k_\theta = 0$), (b) semiflexible chains ($k_\theta = 20$), (c) stiff, rod-like chains ($k_\theta = 50$).

Polyelectrolytes

A large variety of synthetic and biological macromolecules are polyelectrolytes [215]. The most well-known polyelectrolyte biopolymers, proteins, DNA and RNA, are responsible in living systems for functions which are incomparably more complex and diverse than the functions usually discussed for synthetic polymers present in the chemical industry. For example, polyacrylic acid is the main ingredient for diapers and dispersions of copolymers of acrylamide or methacrylamide and methacrylic acid are fundamental for cleaning water [216]. In retrospect, during the past 30 years, despite the tremendous interest in polyelectrolytes, unlike neutral polymers [52,217], the general understanding of the behavior of electrically charged macromolecules is still rather poor. The contrast between our understanding of neutral and charged polymers results form the long range nature of the electrostatic interactions which introduce new length and time scales that render an analytical treatment beyond the Debye-H¨ckel approximation very complicated [218,219]. Here, the traditional separation of scales, which allows one to understand properties in terms of simple scaling arguments, does not work in many cases. Experimentally, often a direct test of theoretical concepts is not possible due to idealizing assumptions in the theory, but also because of a lack of detailed control over the experimental system, e.g., in terms of the molecular weight. Quite recently, there has been increased interest in hydrophobic polyelectrolytes which are water soluble, covalently bonded sequences of polar (ionizable) groups and hydrophobic groups which are not [220]. Many solution properties are known to be due to a complex interplay between short-ranged hydrophobic attraction, long-range Coulomb effects, and the entropic degrees of freedom. Hence, such polymers can be considered as highly simplified models of biologically important molecules, e.g., proteins or lipid bilayers in cell membranes. In this context, computer simulations are a very important tool for the detailed investigation of charged macromolecular systems. A comprehensive review of recent advances which have been achieved in the theoretical description and understanding of polyelectrolyte solutions can be found in [221].

The investigation of aggregates between oppositely charged macromolecules plays an important role in technical applications and in in particular biological systems. For example, DNA is associated with histone proteins to form the chromatin. Aggregated of DNA with cationic polymers or dendrimers are discussed in the context of their possible application as DNA vectors in gene therapies [224,225]. Here, we present MD simulations of two flexible, oppositely charged polymer chains and illustrate the universal scaling properties of the formed polyelectrolyte complexes that are formed when the chains collapse and build compact, cluster-like structures which are constrained to a small region in space [223]. The properties are investigated as a function of chain length N and interaction strength ξ. Starting with Equation (10) and using k = 1 (cgs-system of units) the dimensionless interaction parameter

$$\xi = \xi_B k_B T / \epsilon\sigma \tag{72}$$

can be introduced, where the Bjerrum length ξ_B is given by:

$$\xi_B = e^2 / \epsilon k_B T, \tag{73}$$

where k_B is the Boltzmann constant, T is temperature, epsilon is the energy scale from the Lennard-Jones potential ofEquation 21, σ defines the length scale (size of one monomer) and e is the electronic charge. The interaction parameter for the presented study is chosen in the range of $\xi = 0, \ldots, 100$ which covers electrically neutral chains ($\xi = 0$) in good solvent as well as highly charged chain systems ($\xi = 100$). The monomers in the chains are connected by harmonic bonds using the first term of the bonded potential of Equation (26). The interaction with the solvent is taken into account by a stochastic force $(\vec{\Gamma}_i)$ and a friction force with a damping constant χ, acting on each mass point. The equations of motion of the system are thus given by the Langevin equations

$$m\ddot{\vec{r}}_i = \vec{F} - \chi m \dot{\vec{r}}_i + \vec{\Gamma}_i. \tag{74}$$

The force $\vec{F}_i$ comprises the force due to the sum of potentials of Equation 21 with cutoff $r_{cut} = 1.5$, Equation 11 with k = 1 and $z_{i/j} = \pm 1$, and the first term on the right-hand side of the bonded potential in Equation 26 with $\kappa = 5000\varepsilon/\sigma$ and bond length $l_0 = \sigma_0 = 1.0$. The stochastic force $\vec{\Gamma}_i$ is assumed to be stationary, random, and Gaussian (white noise). The electrically neutral system is placed in a cubic simulation box and periodic boundary conditions are applied for the intermolecular Lennard-Jones interaction according to Equation (21), thereby keeping the density $\rho = N/V = 2.1 \times 10^{-7}/\sigma^3$ constant when changing the chain length N. The number of monomers N per chain was chosen as N = 10, 20, 40, 80 and 160 so as to cover at least one order of magnitude. For the Coulomb interaction a cutoff that is half the box length $r_{cut} = 1/2L_B$ was chosen. This can be done as the eventually collapsed polyelectotlyte complexes which are analyzed are confined to a small region in space which is much smaller than r_{cut}. In the following we discuss exemplarily some scaling properties of charged linear macromolecules in the collapsed state. The simulations are started with two well separated and equilibrated chains in the simulation box. After turning on the Coulomb interactions with opposite charges $z_{i/j} = \pm 1$ in the monomers of both chains, the chains start to attract each other. In a first step during the aggregation process the chains start to twist around each other and form helical like structures as presented in Figure 40. In a second step, the chains start to form a compact globular structure because of the attractive interactions between dipoles formed by oppositely charged monomers, see the snapshots in Figure 41a.

Figure 41a exhibits the universal scaling regime of R_g obtained for intermediate interaction strengths ξ and scaled by $(N - 1)^{2/3}$. Here, the data of various chain lengths fall nicely on top of each other. This scaling corresponds to the scaling behavior of flexible chains in a bad solvent and is also in accordance with what was reported by Shrivastava and Muthukumar [226]. The change of R_g is connected with a change of the density ρ of the polyelectrolyte aggregate. However, in Figure 41b, which presents an example of ρ for $\xi = 4$, only a slight dependence of the density on the chain length N can be observed. ρmeasures the radial monomer density with respect to the center of mass of the total system. For longer chains, there is a plateau while for short chains there is a pronounced maximum of the density for small distances from the center of mass. While this maximum vanished with decreasing ξ it appears also at higher interaction strengths for longer chains. Monomers on the outer part of the polyelectrolyte complex experience a stronger attraction by the inner part of the cluster than the monomers inside of it, and for smaller ξ, chains of different lengths are deformed to different degrees which leads to a chain length dependence of the density profile.

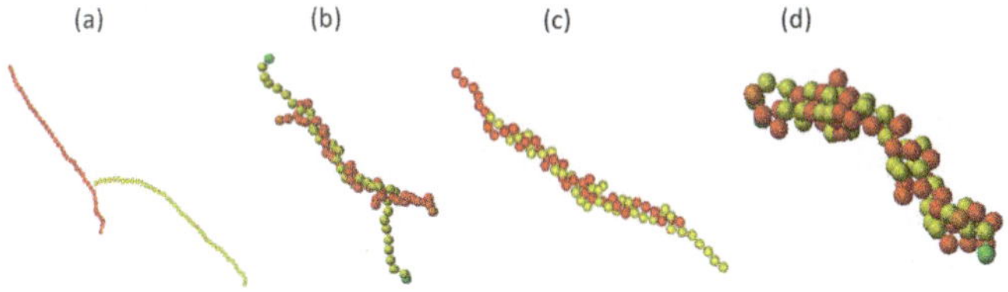

Figure 40: Twisted, DNA-like polyelectrolyte complexes formed by electrostatic attraction of two oppositely charged linear macromolecules with N = 40 at $\hat{\tau} = 0$ (a), $\hat{\tau} = 10500$ (b), $\hat{\tau} = 60000$ (c) and $\hat{\tau} = 120000$ (d), where $\hat{\tau}$ is given in Lennard-Jones units. The interaction strength is $\xi = 8$ [222,223].

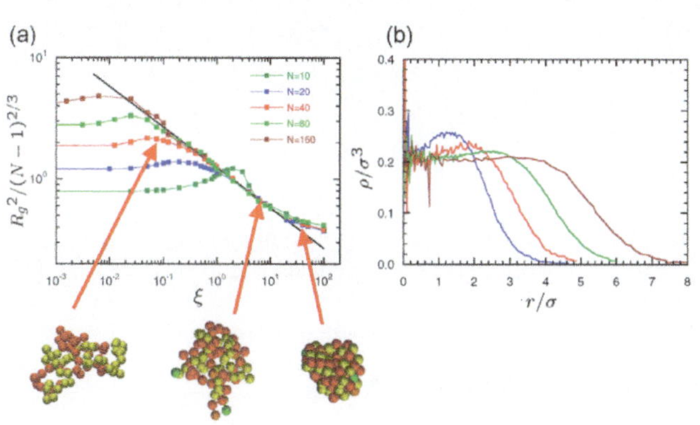

Figure 41: (a) Radii of gyration as a function of the interaction strength ξ for various chain lengths according to [223]. The radius of gyration R_g^2 is scaled by $(N-1)^{2/3}$, where $(N-1)$ is the number of bonds of a single chain. Also displayed are sample snapshots of the collapsed globules with N = 40 and interaction strengths ξ = 0.4, 4, 40. (b) Radial density of monomers with respect to the center of mass of a globule and interaction strength ξ = 4 for different chain lengths, N = 20 (blue), N = 40 (red), N = 80 (green) and N = 160 (brown).

EMERGING COMPUTATIONAL APPLICATIONS IN BIOPHYSICS AND MEDICAL TUMOR TREATMENT

Biology is becoming one the most fascinating interdisciplinary field where physics, chemistry, mathematics and materials science meet. The insight that was gained at the cellular and subcellular scales during the last few decades offer a possibility to develop a formal description of biology using the quantitative methods of physics. Simulation studies of biological materials now range from electronic structure calculations of DNA, molecular simulations of proteins, biomembranes and biomacromolecules like actin to continuum theories of collagenous tissues. Several collagen-related diseases occur in humans, with osteoporosis and tendinitis being complex examples. Both are linked to the mechanical properties of collagen and its higher order, structurally related form: collagen fibrils and fibers, as well as macroscopic tissues, tendon, and bone, cf. Figure 42.

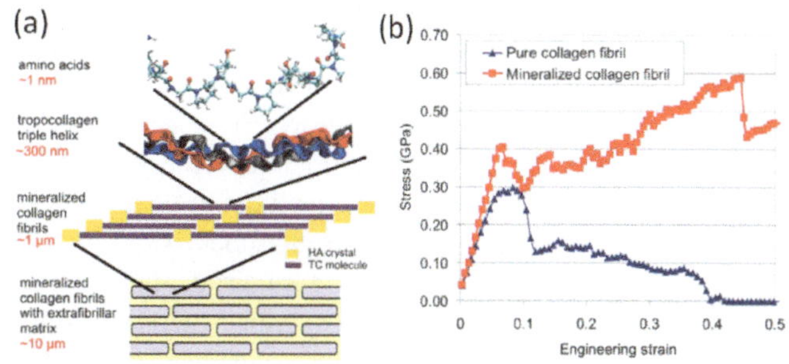

Figure 42: (a) Hierarchical features of collagen which determines the mechanical properties of cells, tissues, bones and many other biological systems, from atomistic to microscale. Three polypeptide strands arrange to form a triple helical tropocollagen molecule. Tropocollagen (TP) molecules assemble into collagen fibrils which mineralize by formation of hydroxipatite (HA) crystals in the gap regions which exist due to a staggered array of molecules. (b) Stress-strain response of a mineralized collagen fibril exhibiting larger strength than a non-mineralized, pure collagen fibril. Both figures adapted from [227] with permission.

Soft biological matter is the constituent element of life, and therefore, the integration of concepts from different disciplines to develop a quantitative understanding of matter in living systems has become an exciting new area of research. Characterization of soft biological materials within a physical approach is focused towards the elucidation of the fundamental principles of self-assembly, deformation and fracture. Deformation and fracture in turn are closely linked to the atomic micro structure of a material. While in hard matter like polycrystalline solids mechanisms such as dislocation propagation, random distribution of cracks or crack extension occur, biomacromolecules exhibit molecular folding or unfolding, rupture of hydrogen or other chemical bonds, intermolecular entanglements or cross-links. The role of mechanics is even more decisive on the cellular level, because on this level, several basic questions can be addressed: How is cell stress and energy transmitted to the environment; what are the forces and interactions that give rise to cell deformations or cell rupture; What exactly is the role of hydrogen bonds when it comes to the stability of cell membranes? While standard biology answers these questions in an empirical, merely descriptive manner, a physical approach tries to establish a framework for a quantitative description [228,229]. At larger length scales the interaction of molecules with cells or different types of tissue become dominant and play a major role in the mechanical material behavior.

Here, the application of shock waves is of particular interest for medical applications. Since the introduction of shock wave therapy for the treatment of gall- and kidneystones as a new non-invasive technique to disintegrate concrements, i.e., extracorporeal shock wave lithotripsy (ESWL), in the 1980s [232], a multitude of new indications for ESWL have arisen. Apart from physical disintegration of calculi as an approved standard therapy in humans, enthesopathies like tennis elbow or bone pathologies (pseudoarthroses and delayed unions) represent classical indications for ESWL, see e.g., [233] and the references therein. The antibacterial effects of extracorporeal shock waves could offer new applications in some difficult-to-treat infections [234]. At the beginning of ESWL a major problem was that kidney or gall stones were destroyed by waves only, when the stone was exactly on target. Thus, it became necessary to locate the stones in 3D. Initially this was considered a major technological problem and was solved with several independent X-rays; today, on uses magnetic resonance (MR) imaging which allows live monitoring of ESWL treatments. Despite the therapeutic success of ESWL as therapeutic modality without the need of surgical risks and surgical pain, the exact mechanisms of shock wave therapy, e.g., effects on tissue damage, rupture or damage of cell membranes, remain widely unknown. Albeit the term "shock waves" is commonly used in the context of medical treatments, usually the waves that are generated by an underwater high-voltage spark discharge and subsequently focused using an elliptical reflector, cf. Figure 43, are merely high-pressure ultrasound waves (highly focused ultrasound, HIFU) which do not attain the physical characteristics and short rise times of real shock waves. Thus, up to now, the full potential of shock waves for medical treatments remains to be fully explored.

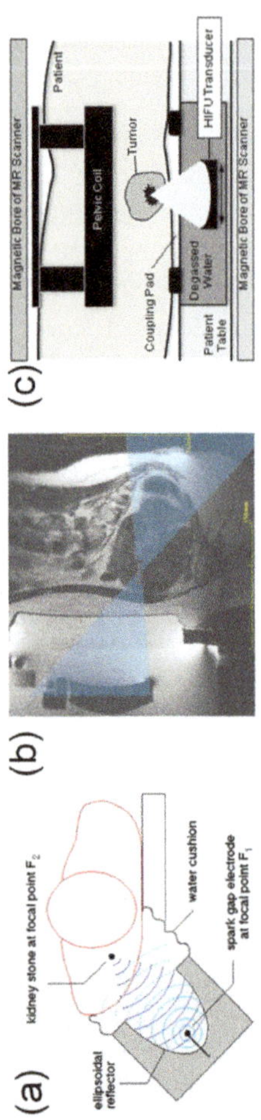

Figure 43: (a) Schematic of the principle of ESWL. (b) Image of human body with symptomatic fibroids. Sonication pathway is superimposed on the image and the spot where irreversible thermal damage is expected is also superimposed onto this image. Adapted from [230] with permission. (c) Schematic of a magnetic resonance guided HIFU equipment for the possible treatment of tumor tissue. Figure adapted from [231] with permission.

MR-guided HIFU is able to disintegrate kidney stones and to cure non-unions and certain soft tissue disorders. The effect of the shock wave in urology and orthopaedic applications seems to be different. Currently, at least two different mechanisms of shock waves are noted. On the one hand, the positive pressure is responsible for a direct shock wave effect, and on the other hand, tensile waves cause cavitation, which is called indirect shock wave effect [236], cf. Figure 44. Reflections at interfaces and damping in the material leads to energy loss of the wave. The disintegrating effect of the shock wave however very strongly depends on the plasticity of the material; for example, a wave with sufficient energy to disintegrate a kidney stone has minimal effect on an intact bone [235]. Many questions in the medical application of shock waves, e.g., the possibility for the successful treatment of tumor cells using ESWL and MR-guided HIFU are still unanswered, opening a wide research field which requires the application of sophisticated experimental and numerical techniques not only on multi-scales but also on multi-fields of research.

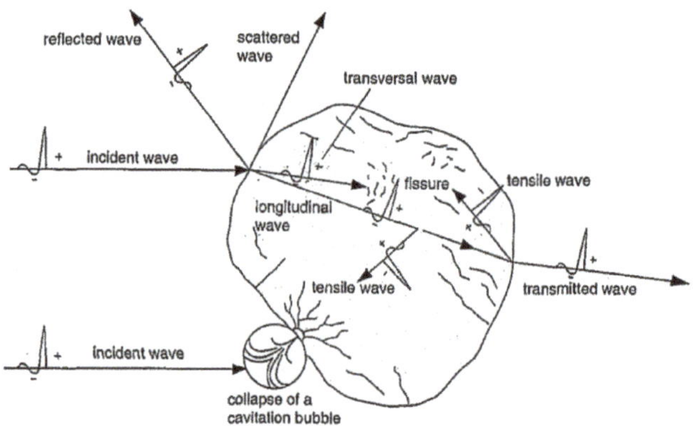

Figure 44: Scheme of the disintegration of a kidney stone as a result of direct and indirect shock wave effects. Figure adapted from [235] with permission.

CONCLUDING REMARKS

In summary we have attempted to provide a survey of numerical methods used on different time and length scales. We discussed first the importance of physical modeling in computational physics and then focused on the MD method which is the most used simulation technique on the molecular scale. As typical applications we presented methods for the generation of realistic microstructures of polycrystalline solids in 3D which can be used for FEM analysis on the macroscale including micro structural details. Results of MD and NEMD shock impact simulations employing a multiscale model based onmicroscopic soft particles were presented and the potential of the model for the description of macroscopic properties of brittle materials was demonstrated. Also, MD applications in polymer physics were presented which build the framework for modeling biological macromolecules and for modeling more complex systems such as the application of shock waves in cellular structures and biological tissue. Eventually we discussed new exciting emerging applications of shock wave and polymer physics in medical research where many physical principles which make the application of ESWL and HIFU successful, are still widely unexplored.

The improvement of simulation techniques which allow for a coupling of different length and time scales and for simulating larger systems on mesoscopic and macroscopic scales has been a very active field of research in computer simulation during the last ten years and many research programs devoted to intensive methodological development have been initiated in many research areas, see e.g., [237]. Developing efficient algorithms, and coupling strategies, as well as extending the available physical length and time scales in simulations up to at least micrometers and -seconds and beyond, remains one of the top motivations in all different fields of basic and applied research using supercomputers, cf. Figure 8. The theoretical progress in understanding hierarchical biological materials will facilitate to develop new materials technologies through the use of multiple hierarchies in an efficient and controlled manner, that is, lead to a structural design of materials on many scales concurrently, instead of trial-and-error approaches.

Since there is tremendous continuing progress in multiscale computational techniques, this review is by no means exhaustive. We hope that by presenting several recent research examples from shock wave and polymer physics, which is triggered by our own research interests, and by highlighting emerging applications of these established computational approaches in medical engineering, we have conveyed the message that multiscale modeling is an enterprise of truly multidisciplinary nature which combines the skills of physicists, materials scientists, chemists, mechanical, chemical and medical engineers, applied mathematicians and computer scientists. The breaking down of traditional barriers between different scientific disciplines represents the actual power and the great promise of computational multiscale approaches for enhancing our understanding, and our ability to controlling complex physical phenomena, even in the field of life sciences, biology and medicine.

ACKNOWLEDGMENTS

The authors acknowledge financial support of the Fraunhofer-Gesellschaft e.V. Germany under grant No. 100128 "Hochfeste Kunststoffe" and under grant No. 600016 of the Vintage class program. Martin Steinhauser acknowledges many stimulating discussions with Alexander Blumen and Martin Kühn.

REFERENCES

1. Phillips, R. Crystals, Defects and Microstructures—Modeling Across Scales; Cambridge University Press: Cambridge, UK, 2001.
2. Yip, S, Ed.; Handbook of Materials Modeling; Springer: Berlin, Germany, 2005.
3. Steinhauser, MO. Computational Multiscale Modeling of Solids and Fluids—Theory and Applications; Springer: Berlin/Heidelberg/New York, Germany/USA, 2008.
4. Hockney, R; Eastwood, J. Computer Simulation Using Particles; McGraw-Hill: New York, NY, USA, 1981.
5. Ciccotti, G; Frenkel, G; McDonald, I. Simulation of Liquids and Solids; North-Holland: Amsterdam, The Netherlands, 1987.
6. Hockney, R. The Potential Calculation and Some Applications. Methods Comp. Phys **1970**, 9, 136–211.
7. Allen, M; Tildesly, D. Computer Simulation of Liquids; Oxford University Press: Oxford, UK, 1991.
8. Liu, GR; Liu, MB. Smoothed Particle Hydrodynamics A Meshfree Particle Method; World Scientific Co. Pte. Ltd: Singapore, 2003.
9. Gates, T; Odegard, G; Frankland, S; Clancy, T. Computational Materials: Multi-Scale Modeling and Simulation of Nanostructured Materials. Compos. Sci. Technol **2005**, 65, 2416–2434.
10. Steinhauser, M; Kühn, M. Numerical Simulation of Fracture and Failure Dynamics in Brittle Solids. In Anisotropy, Texture, Dislocations, Multiscale Modeling in Finite Plasticity and Viscoplasticity and Metal Forming; Khan, A, Kohei, S, Amir, R, Eds.; Neat Press: Fulton, Maryland, USA, 2006; pp. 634–636.
11. Finnis, M; Sinclair, J. A Simple Empirical N-Body Potential for Transition Metals. Phil. Mag. A **1984**, 50, 45.
12. Kohn, W; Sham, L. Self-Consistent Equations Including Exchange and Correlation Effects. Phys. Rev **1965**, 140, 1133–1138.
13. Car, R; Parinello, M. Unified Approach for Molecular Dynamics and Density-Functional Theory. Phys. Rev. Lett **1985**,55, 2471.
14. Courant, R. Variational Methods for the Solution of Problems of Equilibrium and Vibrations. Bull. Amer. Math. Soc**1943**, 49, 1–23.
15. Bishop, J; Hill, R. A Theoretical Derivation of the Plastic Properties of a Polycrystalline Face-Centered Material. Philos. Mag **1951**, 42, 414–427.
16. Markenscoff, X, Gupta, A, Eds.; Collected Works of J.D. Eshelby; Springer: New York, NY, USA, 2006.
17. Smith, C. Structural Hierarchy in Inorganic Systems; American Elsevier: New York, NY, USA, 1969.
18. Steinhauser, MO; Grass, K; Thoma, K; Blumen, A. A Nonequilibrium Molecular Dynamics Study on Shock Waves.Europhys. Lett **2006**, 73, 62.
19. Eastman, J; Siegel, R. Nanophase Synthesis Assembles Materials from Atomic Clusters. Res. Develop **1989**, 31, 56–60.
20. Siegel, R; Hahn, H. Nanophase Materials; World Scientific: Singapore, 1987.
21. Hahn, H; Averback, R. The Production of Nanocrystalline Powders by Magnetron Sputtering. J. Appl. Phys **1990**, 67, 1113–1115.
22. Sawaguchi, A; Toda, K; Nihara, K. Mechanical and Electrical Properties of Silicon Nitride Silicon Carbide Nanocomposite Material. J. Am. Ceram. Soc **1991**, 74, 1142–1144.
23. Krell, A; Blank, P; Ma, H; Hutzler, T. Processing for High-Density Submicrometer Al_2O_3 for New Applications. J. Am. Ceram. Soc **2003**, 86, 546–553.
24. Buehler, M; Abraham, F; Gao, H. Hyperelasticity Governs Dynamic Fracture at a Critical Length. Nature **2003**, 426, 141–146.

25. Hashin, Z; Shtrickman, S. A Variational Approach to the Theory of the Elastic Behavior of Multiphase Materials. J. Mech. Phys. Solids **1963**, 11, 127–140.
26. Christensen, R. Mechanics of Composite Materials; John Wiley: New York, NY, USA, 1979.
27. Ahmed, S; Jones, F. A Review of Particulate Reinforcement Theories for Polymer Composites. J. Mater. Sci **1990**, 25, 4933–4942.
28. Gibson, L; Ashby, M. Cellular Solids; Pergamon Press: Oxford, UK, 1988.
29. Kadler, K; Holmes, D; Chapman, J. Collagen Fibril Formation. Biochem. J **1996**, 316, 1–11.
30. Gautieri, A; Buehler, MJ; Redaelli, A. Deformation Rate Controls Elasticity and Unfolding Pathway of Single Tropocollagen Molecules. J. Mech. Behav. Biomed. Mater **2009**, 2, 130–137.
31. Wu, DY; Meure, S; Solomon, D. Self-Healing Polymeric Materials: A Review of Recent Developments. Porog. Polym. Sci **2008**, 33, 479–522.
32. Bazant, M; Kaxiras, E. Modeling of Covalent Bonding in Solids by Inversion of Cohesive Energy Curves. Phys. Rev. Lett **1996**, 77, 4370–4373.
33. Abraham, F; Brodbeck, D; Rafey, RA; Rudge, W. Instability Dynamics of Fracture. A Computer Simulation Investigation. Phys. Rev. Lett **1994**, 72, 272–275.
34. Abraham, FF; Brodbeck, D; Rudge, WE; Broughton, JQ; Schneider, D; Land, B; Lifka, D; Gerner, J; Rosenkranz, M; Skovira, J; Gao, H. Ab Initio Dynamics of Rapid Fracture. Model. Simul. Mater. Sci. Eng **1998**, 6, 639.
35. Abraham, F; Broughton, J; Bernstein, N; Kaxiras, E. Spanning the Length Scales in Dynamic Simulation. Comput. Phys**1998**, 12, 538.
36. Cabibbo, N; Iwasaki, Y; Schilling, K. High Performance Computing in Lattice QCD. Parallel Comput **1999**, 25, 1197–1198.
37. Evertz, H. The Loop Algorithm. Adv. Phys **2003**, 52, 1–66.
38. Holm, E; Bateille, C. The Computer Simulation of Microstructural Evolution. JOM-J. Min. Met. Mat. Soc **2001**, 53, 20–23.
39. Sundman, K. Memoire Sur Le Probleme Des Trois Corps. Acta Math **1913**, 36, 105–179.
40. Saari, D. A Visit to the Newtonian N-Body Problem via Elementary Complex Variables. Am. Math. Monthly **1990**, 89, 105–119.
41. Ahrens, J; Brislawn, K; Martin, K; Geveci, B; Law, C; Papka, M. Computer Grafics and Applications. IEEE Comp. Sci. Press **2001**, 21, 34–41.
42. Metropolis, N; Ulam, S. The Monte Carlo Method. J. Am. Stat. Assoc **1949**, 44, 335–341.
43. Potts, R. Some Generalized Order-Disorder Transformations. Proc. Cambidge Phil. Soc **1952**, 48, 106–109.
44. Metropolis, N; Rosenbluth, A; Rosenblut, M; Teller, A; Teller, E. Equation of State Calculations by Fast Computing Machines. J. Chem. Phys **1953**, 21, 1087–1092.
45. Alder, B; Wainwright, T. Phase Transition for a Hard Sphere System. J. Chem. Phys **1957**, 27, 1208–1029.
46. Alder, B; Wainwright, T. Molecular Dynamics by Electronic Computers. Proceedings of the International Symposium on Transport Processes in Statistical Mechanics, Brussels, Belgium, August 27–31, 1958; Prigogine, I, Ed.; Interscience Publishers, Inc: New York, NY, USA, 1960; pp. 97–131.
47. Alder, B; Wainwright, T. Phase Transition in Elastic Disks. Phys. Rev **1962**, 127, 359–361.
48. Metropolis, N. The Beginning of the Monte Carlo Method. Los Alamos Science Special Issue **1987**, 12, 125–130.
49. Binder, K. Monte Carlo and Molecular Dynamics Simulations in Polymer Science; Oxford University Press: Oxford, UK, 1995.
50. Dünweg, B; Steinhauser, MO; Reith, D; Kremer, K. Corrections to Scaling in the Hydrodynamics of Dilute Polymer Solutions. J. Chem. Phys **2002**, 117, 914–924.
51. Steinhauser, MO. A Molecular Dynamics Study on Universal Properties of Polymer Chains in Different Solvent Qualities. Part I: A Review of Linear Chain Properties. J. Chem. Phys **2005**, 122, 094901.
52. De Gennes, PG. Scaling Concepts in Polymer Physics; Cornell University Press: Ithaca, London, UK, 1979.
53. Hartree, D. The Wave Mechanics of an Atom with a Non-Coulomb Central Field. Proc. Cambridge Phil. Soc **1928**, 24, 89–132.
54. Fock, V. Näherungsmethoden zur Lösung des quantenmechanischen Mehrkörperproblems. Z. Physik **1932**, 61, 126–148.
55. Hohenberg, P; Kohn, W. Inhomogeneous Electron Gas. Phys. Rev **1964**, 36, 864–871.
56. Kohn, W. Density Functional and Density Matrix Method Scaling Linearly with the Number of Atoms. Phys. Rev. Lett**1996**, 76, 3168–3171.
57. Slater, J; Koster, G. Simplified LCAO Method for the Periodic Potential Problem. Phys. Rev **1954**, 94, 1498–1524.
58. Hammond, M, Lester, W, Reynolds, P, Eds.; Monte Carlo Methods in Ab Initio Quantum Chemistry; World Scientific: Singapore, 1994.
59. Nightingale, M, Umrigar, C, Eds.; Quantum Monte Carlo Methods in Physics and Chemistry; Springer: New York, NY, USA, 1999.
60. Ballone, P; Andreoni, W. Equilibrium Structures and Finite Temperature Properties of Silicon Microclusters from ab initio Molecular-Dynamics Calculations. Phys. Rev. Lett **1988**, 60, 271–274.
61. Binder, K; Heermann, D. Monte Carlo Simulations in Statistical Physics; Springer Verlag Berlin: Heidelberg/New York/Tokyo, Germany/USA/Japan, 1988.
62. Binder, K. Applications of Monte Carlo Methods to Statistical Physics. Rep. Progr. Phys **1997**, 60, 487.
63. Steinhauser, M. Computational Methods in Polymer Physics. Recent Res. Devel. Physics **2006**, 7, 59–97.
64. Daw, M; baskes, M. Embedded-Atom Method:

Derivation and Application to Impurities, Surfaces, and Other Defects in Metals. Phys. Rev. B **1983**, 29, 6443.

65. Foiles, S; baskes, M; Daw, M. Embedded-Atom-Method Functions for the Fcc Metals Cu, Ag, Au, Ni, Pd, Pt, and Their Alloys. Phys. Rev. B **1986**, 33, 7983–7991.
66. Daw, MS. Model of Metallic Cohesion: The Embedded-Atom Method. Phys. Rev. B **1988**, 39, 7441–7452.
67. Harlow, F. A Machine Calculation Method for Hydrodynamic Problems. . Technical report, Los Alamos Scientific Laboratory: New Mexico, USA, 1955.
68. Dawson, J. Particle Simulation of Plasmas. Rev. Mod. Phys **1983**, 55, 403.
69. Hoogerbrugge, P; Koelman, J. Simulating Microscopic Hydrodynamic Phenomena with Dissipative Particle Dynamics. Europhys. Lett **1992**, 19, 155–160.
70. Chan, J; Hilliard, J. Free Energy of a Non-Uniform System I: Interfacial Energy. J. Chem. Phys **1958**, 28, 258–266.
71. Wolfram, S. Undecidability and Intractability in Theoretical Physics. Phys. Rev. Lett **1985**, 54, 735–738.
72. Hugh, PM; Asaro, R; Shih, C. Computational Modeling of Metal Matrix Composite Materials-I. Isothermal Deformation Patterns in Ideal Microstructures. Acta Metall. Mater **1993**, 41, 1461–1476.
73. Hugh, PM; Asaro, R; Shih, C. Computational Modeling of Metal Matrix Composite Materials-II. Isothermal Stress-Strain Behavior. Acta Metall. Mater **1993**, 41, 1489–1499.
74. Hugh, PM; Asaro, R; Shih, C. Computational Modeling of Metal Matrix Composite Materials-III. Comparisons with Phenomenological Models. Acta Metall. Mater **1993**, 41, 1489–1499.
75. Hugh, PM; Asaro, R; Shih, C. Computational Modeling of Metal Matrix Composite Materials-IV. Thermal Deformations. Acta Metall. Mater **1993**, 41, 1501–1510.
76. Lucy, L. A Numerical Approach to the Testing of the Fission Hypothesis. Astron. J **1977**, 82, 1013.
77. Gingold, R; Monaghan, J. Kernel Estimates as a Basis for General Particle Method in Hydrodynamics. J. Comp. Phys**1982**, 46, 429–453.
78. Ladd, A. Short-Time Motion of Colloidal Particles—Numerical Simulation via Fluctuating Lattice-Boltzmann Equation. Phys. Rev. Lett **1993**, 70, 1339.
79. Devincre, B. Three Dimensional Stress field Expressions for Straight Dislocation Segments. Solid State Commun **1993**,93, 875–878.
80. Zhou, S; Preston, D; Lomdahl, P; Beazley, D. Large-Scale Molecular Dynamics Simulations of Dilocation Intersection in Copper. Science **1998**, 279, 1525–1527.
81. Yang, W; Cai, J; Ing, YS; Mach, CC. Transient Dislocation Emission from a Crack Tip. J. Mech. Phys. Solids **2001**, 49, 2431–2453.
82. Cai, W; Arsenlis, A; Weinberger, CR; Bulatov, VV. A Non-Singular Continuum Theory of Dislocations. J. Mech. Phys. Solids **2006**, 54, 561–587.
83. Cundall, P; Strack, O. A Discrete Numerical Model for Granular Assemblies. Geotechnique **1979**, 29, 47–65.
84. Guyon, E; Hulin, JP; Petit, L; Mitescu, CD. Physical Hydrodynamics; Oxford University Press: Oxford, UK, 2001.
85. Griebel, M. Numerical Simulation in Fluid Dynamics; ; SIAM Monographs on mathematical modeling and computation. Society for Industrial and Applied Mathematics: Philadelphia, USA, 1997.
86. Chorin, A; Marsden, J. A Mathematical Introduction to Fluid Dynamics; Springer: New York, NY, USA, 1979.
87. Crank, J; Nicholson, P. A Practical Method for the Numerical Evaluation of Solutions of Partial Differential Equations of the Heat-Conduction Type. Proc. Cambridge Philos. Soc **1947**, 43, 50–67.
88. Bathe, K. Finite Element Procedures in Engineering Analysis; Prentice Hall: Cambridge, Country, 1982.
89. Benson, D. Computational Methods in Lagrangian and Eulerian Hydrocodes. Comput. Meth. Appl. Meth. Engng **1992**,99, 235.
90. Belytschko, T; Hughes, T. Computational Methods for Transient Analysis; Elsevier Science Limited: New York, NY, USA, 1983.
91. Liu, W; Hao, S; Belytschko, T; Li, S; Chang, C. Multiple Scale Meshfree Methods for Damage Fracture and Localization. Comp. Mat. Sci **1999**, 16, 1997–2005.
92. Hiermaier, S. Structures Under Crash and Impact; Springer: New York, NY, USA, 2008.
93. Cohen, A. Numerical Analysis; McGraw-Hill: London, UK, 1962.
94. Abramovitz, M; Segun, I. Handbook of Mathematical Functions; Dover Publications: New York, NY, USA, 1964.
95. Carter, EA. Challenges in Modelling Materials Properties Without Experimental Input. Science **2008**, 321, 800–803.
96. Rambasubramaniam, A; Carter, EA. Coupled Quantum-Atomistic and Quantum-Continuum Mechanics Methods in Materials Research. MRS Bull **2007**, 32, 913–918.
97. Peng, Q; Zhang, X; Hung, L; Carter, EA; Lu, G. Quantum Simulation of Materials at Micron Scales and Beyond. Phys. Rev. B **2008**, 78, 054118.
98. Parinello, M; Rahman, A. Polymorphic Transitions in Single Crystals: A New Molecular Dynamics Method. J. Appl. Phys **1981**, 52, 7182.
99. Born, M. Quantenmechanik der Stossvorgänge. Z. Physik **1926**, 38, 803–827.
100. Scherz, U. Quantenmechanik: Eine Einführung mit Anwendungen auf Atome, Moleküle und Festkörper; Teubner: Stuttgart, Leipzig, Germany, 1999.
101. Kosloff, R. Time-Dependent Quantum-Mechanical Methods For Molecular Dynamics. J. Chem. Phys **1988**, 92, 2087–2100.
102. Laasonen, K; Sprik, M; Parinello, M. "Ab Initio" Liquid Water. J.Chem. Phys **1993**, 99, 9080.
103. Rahman, A. Correlations in the Motion of Atoms in Liquid Argon. Phys. Rev **1964**, 136, 405–411.
104. Marsh, C; Warren, P. Kinetic Thoery For Dissipative Particle Dynamics: The Importance of

Collisions. Europhys. Lett**1999**, 48, 1–7.

105. Kschischo, M; Kern, R; Gieger, C; Steinhauser, M; Tolle, R. Automatic Scoring and SNP Quality Assessment Using Accuracy Bounds for FP-TDI SNP Genotyping Data. Applied Bioinformatics **2005**, 2, 75–84.
106. Cleri, F; Yip, S; Wolf, D; Phillpot, S. Atomic-Scale Mechanism of Crack-Tip Plasticity: Dislocation Nucleation and Crack-Tip Shielding. Phys. Rev. Lett **1997**, 79, 1309–1312.
107. Komanduri, R; Chandrasekaran, N; Raff, LM. Molecular Dynamic Simulations of Uniaxial Tension at Nanoscale of Semiconductor Materials for Micro-Electro-Mechanical Systems (MEMS) Applications. Mater. Sci. Eng., A **2003**, 340, 58–67.
108. Li, S; Liu, WK. Meshfree and Particle Methods and Their Applications. Appl. Mech. Rev **2002**, 55, 1–34.
109. Hrennikoff, H. Solutions of Problems in Elasticity by the Framework Method. J. Appl. Mech. A **1941**, 8, 169–175.
110. Monaghan, J. Smoothed Particle Hydrodynamics. Annu. Rev. Astron. Astrophys **1992**, 30, 543–574.
111. Liberski, L; Petschek, A; Carney, T; Hipp, J. High Strain Lagrangian Hydrodynamics. J. Comput. Phys **1993**, 109, 67–75.
112. Liberski, LD; Petschek, A. Smoothed Particle Hydrodynamics with Strength of Materials. In The Next Language Conference; Springer: New York, NY, USA, 1991; pp. 248–257.
113. Kadau, K; Germann, T; Lomdahl, P. Large-Scale Molecular Dynamics Simulation of 19 Billion Particles. J. Modern. Phys. C **2004**, 15, 193–201.
114. Voter, A; Montelenti, F; Germann, T. Extending the Time Scale in Atomistic Simulations of Materials. Annu. Rev. Mater. Res **2002**, 32, 321–346.
115. Darden, T; York, D; Pedersen, L. Particle Mesh Ewald: An N-log N Method for Sums in Large Systems. J. Chem. Phys**1993**, 103, 8577–8592.
116. Essmann, U; Perera, L; Berkowitz, M; Darden, T; Lee, H; Pedersen, L. A Smooth Paritcle Mesh Ewald Potential. J. Chem. Phys **1995**, 103, 8577–8592.
117. van Gunsteren, W; Berendsen, H. Algorithms for Brownian Dynamics. Molec. Phys **1982**, 45, 637.
118. Schutz, C; Warshel, A. What are the Dielectric Constants of Proteins and How to Validate Electrostatic Models?Proteins **2001**, 44, 400–417.
119. Warshel, A; Sharma, P; Kato, M; Parson, W. Modeling Electrostatic Effects in Proteins. Biochim. Biophys. Acta Mater**2006**, 1764, 1647–1676.
120. Freddolino, PL; Liu, F; Gruebele, M; Schulten, K. Ten-Microsecond Molecular Dynamics Simulation of a Fast-Folding WW Domain. Biophys. J **2008**, 94, L75–L77.
121. Freddolino, PL; Park, S; Roux, B; Schulten, K. Force Filed Bias in Protein Folding Simulations. Biophys. J **2009**, 96, 3772–3780.
122. Verlet, L. Computer "Experiments" on Classical Fluids. I. Thermodynamical Properties of Lennard-Jones Molecules.Phys. Rev **1967**, 159, 1098–1003.
123. Verlet, L. Computer Experiments on Classical Fluids. II. Equilibrium Correlation Functions. Phys. Rev **1968**, 159, 201.
124. Weeks, J; Chandler, D; Andersen, H. Role of Repulsive Forces in Forming the Equilibrium Structure of Simple Liquids.J. Chem. Phys **1971**, 54, 5237–5247.
125. Rahman, A; Stillinger, F. Molecular Dynamics Study of Liquid Water. Phys. Rev **1971**, 55, 3336.
126. Parinello, M; Rahman, A. Crystal Structure and Pair Potentials: A Molecular-Dynamics Study. Phys. Rev. Lett **1980**, 45, 1196–1199.
127. Hoover, W. Molecular Dynamics; Springer Verlag: Berlin/New York, Germany/USA, 1986.
128. Maitland, G; Rigby, M; Smith, E; Wakeham, W. Intermolecular Forces - Their Origin and Determination; Clarendon Press: Oxford, UK, 1981.
129. Hobza, P; Zahradnik, R. Weak Intermolecular Interactions in Chemistry and Biology; Academia: Prague, Czechoslovakia, 1980.
130. Jackson, JD. Classical Electrodynamics; John Wiley & Sons, Inc: New York, NY, USA, 1975.
131. Daune, M. Moleckulare Biophysik; Vieweg Verlag: Braunschweig, Wiesbaden, Germany, 1997.
132. Haberland, R; Fritsche, S; Peinel, G; Heinzinger, K. Molekulardynamik—Grundlagen und Anwendungen; Friedrich VIeweg & Sohn Verlagsgesellschaft mbH: Braunschweig, Wiesbaden, Germany, 1995.
133. Buckingham, R. The Classical Equation of State of Gaseous Helium, Neon and Argon. Proc. Roy. Soc **1938**, A106, 264.
134. Shih, AY; Arkhipov, A; Freddolino, PL; Schulten, K. Coarse Grained Protein-Lipid Model with Application to Lipoprotein Particles. J. Phys. Chem. B **2006**, 110, 3674–3684.
135. Schlenkrich, M; Brinckmann, J; MacKerell, A; Karplus, M. Empirical Potential Energy Function for Phospholipids: Criteria for Parameter Optimization and Applications; Birkhäuser: Boston, MA, USA, 1996.
136. Feller, S. An Improved Empirical Potential Energy Function for Molecular Simulations of Phospholipids. J. Phys. Chem. B **2000**, 104, 7510–7515.
137. Siu, SWI; Vácha, R; Jungwirth, P; Böckmann, RA. Biomolecular Simulations of Membranes: Physical Properties from Different Force Fields. J. Chem. Phys **2008**, 125, 125103.
138. Horowitz, E; Sahni, S; Rajasekaran, S. Computer Algorithms; Computer Science Press: New York, NY, USA, 1998.
139. Axilrod, B; Teller, E. Interaction of the van der Waals Type between Three Atoms. J. Chem. Phys **1943**, 11, 299.
140. Amdahl, G. Validity of the Single Processor Approach to Achieve Large Scale Computing Capabilities. AFIPS Spring Computer Conference, Atlantic City, NJ, USA, April 18–20, 1967; 30, pp. 483–485.
141. Riemann, B. Über die Fortpflanzung ebener Luftwellen von endlicher Schwingungsweite. Abhandlungen der Königlichen Gesellschaft der Wissenschaften zu

Göttingen **1860**, 8, 43–65.

142. von Neumann, J. A Method for the Numerical Calculation of Hydrodynamic Shocks. J. Appl. Phys **1950**, 21, 232–237.

143. Godunov, S. A Difference Method for Numerical Calculation of Discontinuous Solutions of the Equations of Hydrodynamics. Math. Sb **1959**, 47, 271–306.

144. Rankine, W. On the Thermodynamic Theory of Waves of Finite Longitudinal Disturbances. Phil Trans **1870**, 160 II, 277–288.

145. Hugoniot, P. Sur la propagation du mouvement dans les corps et plus specialement dans les gaz parfaits - i. Journal de l'Ecole Polytechnique **1887**, 57, 3–97.

146. Hugoniot, P. Sur la propagation du mouvement dans les corps et plus specialement dans les gaz parfaits - ii. Journal de l'Ecole Polytechnique **1889**, 58, 1–125.

147. Asay, JR; Kerley, G. The Response of Materials to Dynamic Loading. Int. J. Imp. Eng **1987**, 5, 69.

148. Davison, L; Graham, R. Shock Compression of Solids. Phys. Rep **1979**, 55, 257.

149. Knudson, M. Shock Wave Science and Technology Reference Library; Solids I; Springer: New York, NY, USA, 2007.

150. Bethe, H. Classic Papers in Shock Compression Science; Springer: New York, NY, USA, 1942.

151. Chen, M; McCauley, J; Hemker, K. Shock-Induced Localized Amorphization in Boron Carbide. Science **2003**, 299, 1563–1566.

152. Steinhauser, MO; Grass, K. Failure and Plasticity Models of Ceramics—A Numerical Study. The 11th Int. Symposium on Plasticity and Current Applications, (PLASTICITY 2005), Kauai, Hawaii, January 03–08, 2005, In Dislocations, Plasticity, Damage and Metal Forming: Materials Response and Multiscale Modeling; Khan, A, Kohei, S, Amir, R, Eds.; Neat Press: Fulton, MD, USA, 2005; pp. 370–373.

153. Krell, A; Blank, P; Ma, H; Hutzler, T; van Bruggen, MP; Apetz, R. Transparent Sintered Corundum with High Hardness and Strength. J. Am. Ceram. Soc **2003**, 1, 12–18.

154. Bringa, EM; Caro, A; Wang, Y; Victoria, M; McNaney, JM; Remington, BA; Smith, RF; Torrala, BR; Swygenhoven, HV. Ultrahigh Strength in Nanocrastalline Materials under Shock Loading. Science **2005**, 309, 1838–1841.

155. Lauridsen, E. Approaches for 3D Materials Characterization. JOM **2006**, 58, 12.

156. Lewis, A; Suh, S; Stukowski, M; Geltmacher, A; Spanos, G; Rajan, K. Quantitative Analysis and Feature Recognition in 3D Microstructural Data Sets. JOM **2006**, 12, 51.

157. Descartes, R. Principia Philosophiae; Ludovicus Elzevirus: Amsterdam, The Netherlands, 1644.

158. Dirichlet, G. Über die Reduktion der positiven quadratischen Formen mit drei unbestimmten ganzen Zahlen. J. Reine und Angew. Math **1850**, 40, 209–227.

159. Voronoi, G. Nouvelles applications des paramétres continus à la théorie des formes quadratiques. deuxième Mèmoire: Recherches sur les parallélloèdres primitifs. J. Reine und Angew. Math **1908**, 134, 198–287.

160. Delaunay, B. Sur la sphére vide. A la memoire de Georges Voronoi. Izv. Akad. Nauk SSSP, Otdelenie Mathematicheskih i Estestvennyh Nauk **1934**, 7, 793–800.

161. Stoyan, D; Kendall, WS; Mecke, J. Stochastic Geometry and its Applications; Wiley: Chichester, UK, 1987.

162. Gosh, S; Yunshan, L. Voronoi Cell Finite Element Model Based on Micropolar Theory of Thermoelasticity for Heterogeneous Materials. Int. J. Num. Meth. Eng **1995**, 38, 1361–1368.

163. Espinosa, H; Lee, S. Modeling of Ceramic Microstructures: Dynamic Damage Initiation and Evolution. In CP505, Shock Compression of Condensed Matter; Furnish, MD, Chhabildas, LC, Hixson, RS, Eds.; American Institute of Physics: Melville, NY, USA, 1999.

164. Shamos, M; Hoey, D. Closest-Point Problems. In Proc 16th Annu IEEE Sympos Found Comput Sci; IEEE Computer Society Press: Los Angelos, 1975; pp. 151–162.

165. Okabe, A; Boots, B; Sugihara, K. Spatial Tesselations - Concepts and Applications of Voronoi Diagrams; John Wiley & Sons: Chichester, West Sussex, England, 1992.

166. Aurenhammer, F. Power Diagrams: Properties, Algorithms and Applications. SIAM J. Comput **1987**, 16, 78–96.

167. Zhang, KS; Wu, MS; Feng, R. Simulation of Microplasticity-Induced Deformation in Uniaxially Strained Ceramics by 3-D Voronoi Polycrystal Modeling. Int. J. Plasticity **2005**, 21, 801–834.

168. Kühn, M; Steinhauser, MO. Modelling and Simulation of Microstrucutres Using Power Diagrams: Proof of the Concept. Appl. Phys. Lett **2008**, 93, 1–3.

169. Steinhauser, MO; Kühn, M. Anisotropy, Texture, Dislocations, Multiscale Modeling in Finite Plasticity and Viscoplasticity and Metal Forming; Neat Press: Maryland, USA, 2006; ; chapter Numerical Simulation of Fracture and Failure Dynamics in Brittle Solids, pp. 634–636.

170. Weibull, W. A Statistical Distribution Function of Wide Applicability. J. Appl. Mech **1951**, 18, 293–297.

171. Zavattieri, PD; Espinoza, HD. An Examination if the Competition between Bulk Behavior and Interfacial Behavior of Ceramics Subjected to Dynamics Pressure-Shear Loading. J. Mech. Phys. Solids **2003**, 51, 607.

172. Steinhauser, M; Kühn, M. Modeling of Shock-Wave Failure in Brittle Materials. 3rd Intl Conference on Multiscale Materials Modeling (MMM), Freiburg, Germany, September 18–22, 2006, In MMM Multiscale Materials Modeling; Gumbsch, P, Ed.; Fraunhofer IRB Verlag: Stuttgart, Germany, 2006; pp. 380–382.

173. Xu, XP; Needleman, A. Numerical Simulations of Dynamics Interfacial Crack Growth Allowing for Crack Growth away from the Bond Line. Int. J. Fracture **1995**, 74, 253.

174. D'Addetta, GA; Kun, F; Ramm, E. On the Application

of a Discrete Model to the Fracture Process of Cohesive Granular Materials. Granul. Matter **2002**, 4, 77.

175. Zhou, SJ; Beazley, DM; Lomdahl, PS; Holian, BL. Large-Scale Molecular Dynamics Simulations of Three-Dimensional Ductile Fracture. Phys. Rev. Lett **1997**, 78, 479.
176. Fineberg, J; Gross, SP; Marder, M; Swinney, HL. Instability in the Propagation of Fast Cracks. Phys. Rev. B **1992**, 45, 5146–5154.
177. Marder, M. New Dynamical Equation for Cracks. Phys. Rev. Lett **1991**, 66, 2484–2487.
178. Abraham, FF. Unstable Crack Mtion is Predictable. J. Mech. Phys. Solids **2004**, 53, 1071–1078.
179. Abraham, FF; Gao, H. How Fast Can Cracks Propagate? Phys. Rev. Lett **2000**, 84, 3113.
180. Buehler, MJ; Gao, H. Dynamical Fracture instabilities Due to Local Hyperelasticity at Crack Tips. Nature **2006**, 439, 307–310.
181. Holland, D; Marder, M. Cracks and Atoms. Adv. Mater **1999**, 11, 793–806.
182. Bulatov, V; Abraham, F; Kubin, L; Devrince, B; Yip, S. Connecting Atomistic and Mesoscale Simulations of Crystal Plasticity. Nature **1998**, 391, 669–672.
183. Bulatov, VV; Hsiung, LL; Tang, M; Arsenlis, A; Bartelt, MC; Cai, W; Florando, JN; Hirtani, M; Rhee, M; Hommes, G; Pierce, TG; de la Rubia, T. Dilocation Multi-Junctions and Strain Hardening. Nature **2006**, 440, 1174–1178.
184. Sharon, E; Fineberg, J. Conifrming the Continuum Theory of Dynamic Brittle Fracture for Fast Cracks. Natrue **1998**,397, 333–335.
185. Abraham, FF. The Atomic Dynamics of Fracture. J. Mech. Phys. Solids **2001**, 49, 2095–2111.
186. Abraham, FF. Dynamics of Brittle Fracture With Variable Elasticity. Phys. Rev. Lett **1996**, 77, 869–872.
187. Abraham, FF. Spanning the Continuum to Quantum Length Scales in a Dynamic Simulation of Brittle Fracture.Europhys. Lett **1998**, 44, 783–787.
188. Gross, SP; Fineberg, J; Marder, M; McCormick, WD; Swinney, HL. Acoustic Emissions from Rapidly Moving Cracks.Phys. Rev. Lett **1993**, 71, 3162.
189. Holian, B; Lomdahl, P. Plasticity Induced by Shock-Waves in Nonequilibrium Molecular-Dynamics Simulations.Science **1998**, 280, 2085.
190. Holian, B. Molecular Dynamics Comes of Age for Shock Wave Research. Shock Waves **2004**, 13, 489–495.
191. van Swygenhoven, H; Derlet, P; Feichtinger, D; Hasnaoui, A; Samaras, M. Atomistic Simulations in Nanocrystalline Materials. . Technical report, EPFL Supercomputing Review, Ecole Polytechnique Fédéral Lausanne: Lausanne, Switzerland, 2002.
192. Abraham, FF; Walkup, R; Gao, H; Duchaineau, M; Rubia, TDDL; Seager, M. Simulating Materials Failure by Using up to One Billion Atoms and the World's Fastest Computer: Brittle Fracture. Proc. Natl. Acad. Sci. USA **2002**, 99, 5777–5782.
193. Gao, H; Huang, Y; Abraham, F. Continuum and Atomistic Studies of Intersonic Crack Propagation. J. Mech. Phys. Solids **2001**, 49, 2113–2132.
194. Buehler, M; Hartmaier, A; Gao, H; Duchaineau, M; Abraham, F. Atomic Plasticity: Description and Analysis of a One-Billion Atom Simulation of Ductile Materials Failure. Comput. Methods Appl. Engrg **2004**, 193, 5257–5282.
195. Steinhauser, MO; Grass, K; Strassburger, E; Blumen, A. Impact Failure of Granular Materials—Non-Equilibrium Multiscale Simulations and High-Speed Experiments. Int. J. Plasticity **2009**, 25, 161–182.
196. Walton, OR; Braun, RL. Viscosity, Granular-Temperature, and Stress Calculation for Shearing Assemblies of Inelastic, Frictional Disks. J. Rheol **1986**, 39, 949.
197. Gröger, T; Tüzün, U; Heyes, DM. Modeling and Measuring of Cohesion in Wet Granular Materials. Powder Technol**2003**, 133, 203.
198. Leszczynski, JS. A Discrete Model of a Two-Particle Contact Applied to Cohesive Granular Materials. Granul. Matter**2003**, 5, 91.
199. Kadau, K. Molekulardynamik-Simulationen von strukturellen Phasenumwandlungen in Festkörpern, Nanopartikeln und ultradünnen Filmen, . PhD thesis, Gerhard-Mercator-University: Duisburg, Germany, 2001.
200. Kadau, K; Bartels, G; Brendel, L; Wolf, DE. Contact Dynamics Simulations of Compacting Cohesive Granular Systems.Comp. Phys. Comm **2002**, 147, 190.
201. Sadd, MH. Contact Law Effects on Wave-Propagation in Particulate Materials Using Discrete Element Modelling. J. Non-Linear Mech **1993**, 28, 251.
202. Arias, I; Knap, J; Pandolfi, A; Ortiz, M. Massively Parallel Simulations of Dynamic Fracture and Fragmentation of Brittle Solids. XXI ICTAM, Warsaw, Poland, 15–21 August 2004; 2004.
203. Brendel, W. Shock Waves: A New Physical Principle in Medicine. Eur. surg. Res **1986**, 18, 177–180.
204. Wang, CJ. An Overview of Shock-Wave Therapy in Musculoskeletal Disorders. Med. J **2003**, 26, 220–230.
205. Rubinstein, M; Colby, RH. Polymer Physics; Oxford University Press: Oxford, NY, USA, 2003.
206. Phillips, R; Lax, P. Scattering Theory; Academic Press: New York, NY, USA, 1967.
207. Chu, B; Xu, R; Zuo, J. Transition of Polystyrene in Cyclohexane from Theta to the Collapsed State. Macromolecules**1988**, 21, 273–274.
208. Steinhauser, MO; Schneider, J; Blumen, A. Simulating Dynamic Crossover Behavior of Semiflexible Linear Polymers in Solution and in the Melt. J. Chem. Phys **2009**, 130, 164902.
209. Binder, K. Monte Carlo Methods in Condensed Matter Physics; Springer: Berlin, Germany, 1992; ; Topics Applied Physics.
210. Prince, E; Rouse, J. A Theory of the Linear Viscoelastic Properties of Dilute Solutions of Coiling Polymers. J. Chem. Phys **1953**, 21, 1281–1286.
211. Zimm, BH. Dynamics of Polymer Molecules in Dilute

Solution: Viscoelasticity, Flow Birefringence and Dielectric Loss.J. Chem. Phys **1956**, 24, 269–278.

212. Doi, M; Edwards, S. The Theory of Polymer Dynamics; Clarendon Press: Oxford, UK, 1986.
213. Bustamante, C; Marko, J; Siggia, E; Smith, S. Entropic Elasticity of Lambda-Phage DNA. Science **1994**, 265, 1599–1600.
214. Ober, CK. Shape Persistence of Synthetic Polymers. Science **2000**, 288, 448–449.
215. Manning, G. Limiting Laws and Counterion Condensation in Polyelectrolyte Solutions I. Colligative Properties. J. Chem. Phys **1969**, 51, 924–933.
216. Ramirez, ER. Purification of Industrial Waste Waters by Flotation. US Patent No. 3975269 **1976**.
217. Flory, P. Statistical Mechanics of Chain Molecules; Wiley: New York, NY, USA, 1969.
218. Debye, P; Hückel, E. The Theory of Electrolytes. I. Lowering of Freezing Point and Related Phenomena. Physikalische Zeitschrift **1923**, 24, 185–206.
219. Barrat, JL; Joanny, F. Theory of Polyelectrolyte Solutions. Adv. Chem. Phys **2007**, 94, 1–66.
220. Khoklov, AR; Khalatur, PG. Solution Properties of Charged Hydrophobic/Hydrophilic Copolymers. Curr. Opin. Colloid Interface Sci **2005**, 10, 22–29.
221. Holm, C; Joanny, JP; Kremer, K; Netz, R; Reineker, P; Seidel, C; Vilgis, TA; Winkler, RG. Polyelectrolyte Theory. Adv. Polym. Sci **2004**, 166, 67–111.
222. Steinauser, MO. Molekulardynamik-Simulationen von Polyelectrolyten unterschiedlicher Ladungsdichte: Polyelektrolyt-Komplexbildung. . Master's thesis. University of Ulm: Ulm, Germany, 1998.
223. Winkler, RG; Steinhauser, MO; Reineker, P. Complex Formation in Systems of Oppositely Charged Polyelectrolytes: A Molecular Dyamics Simulation Study. Phys. Rev. E **2002**, 66, 021802.
224. Yamasaki, Y; Teramoto, Y; Yoshikawa, K. Disappearance of the Negative Charge in Giant DNA with a Folding Transition. Biophys. J **2001**, 80, 2823–2832.
225. Gössl, I; Shu, L; Schlüter, A; Rabe, J. Complexes of DNA and Positively Charged Dendronized Polymers. J. Am. Chem. Soc **2002**, 124, 6860.
226. Srivastava, D; Muthukumar, M. Interpenetration of Interacting Polyelectrolytes. Macromolecules **1994**, 27, 1461–1465.
227. Buehler, MJ. Molecular Nanomechanics of Nascent Bone: Fibrillar Toughening by Mineralization. Nanotechnology**2007**, 18, 295102.
228. Stolarska, MA; Kim, Y; Othmer, HG. Multi-Scale Models of Cell and Tissue Dynamics. Phil. Trans. R. Soc. A **2009**, 367, 3525–3553.
229. Cowin, SC; Gailani, G; Benalla, M. Hierarchical Proelasticity: Movement of Interstitial Fluid between Porosity Levels in Bones. Phil. Trans. R. Soc. A **2009**, 367, 3401–3444.
230. Hindley, J; Gedyc, WM; Regan, L; Steward, E; Hynnen, K; Macdanold, N; Inbar, Y; Itzchak, Y; Rabinovici, J; Kim, K; Geschwind, JF; Hesley, G; Gostout, B; Ehrenstein, T; Hengst, S; Sklair-Levy, M; Shushan, A; Jolesz, F. MRI Guidance of Focused Ultrasound Therapy of Uterine Fibrosis: Early Results. AJR **2004**, 183, 1713–1719.
231. Tempany, CMC; Steward, EA; McDannold, N; QUade, BJ; Jolesz, FA; Hynynen, K. MR Imaging—Guided Focused Ultrasound Surgery of Uterine Leiomyomas: A Feasibility Study. Radiology **2002**, 226, 897–905.
232. Brendel, W; Chaussy, C; Forssmann, B; Schmiedt, E. A New Method of Non-Invasive Destruction of Renal Calculi by Shock Waves. Br. J. Surg **1979**, 66, 12.
233. Gerdesmeyer, L; Wagenpfeil, S; Haake, M. Extracoporeal Shock Wave Therapy For The Treatment of Chronic Calcifying Lendonitis of the Rotator Cuff: A Randomized Controlled Trial. JAMA **2003**, 290, 2573–2580.
234. Gerdesmeyer, L; von Eiff, C; Horn, C; Henne, M. Antibacterial Effects of Extracorporeal Shock Waves. Ultrasound Med. Biol **2005**, 31, 115–119.
235. Thiel, M. Application of Shock Waves in Medicine. Clin. Orthop. Related Res **2001**, 387, 18–21.
236. Krause, H. Extrakorporale Stosswellentherapie; Chapman & Hall GmbH: Weinheim, Germany, 1997; pp. 15–34.

CHAPTER 12 INTRAMOLECULAR INTERACTIONS

Robert Penfold, and Peter J. Wilde

Keywords: Computer simulation, Molecular modelling, Graph theory

INTRODUCTION

The implementation of computer code for realistically simulating the configurations and motion of molecular objects requires modelling of many-body through-bond interactions. In turn, it is necessary to identify the participating atoms in each interaction. This paper presents a novel and exhaustive enumeration procedure exploiting the line graph transformation of the graph that encodes the molecular structure. In principle, by virtue of the recursive nature of the algorithm, straightforward extension to arbitrarily high order interactions is possible.

Application of graph theoretic methods to the general study of molecular structure, and to equilibrium statistical mechanics in particular, are not new. Significant examples include the development of molecular branching rules [1], the enumeration of isomers and the definition of topological indexes [2], as well as the analysis of discrete lattice models [3] and Mayer's cluster decomposition of the 2-body configuration integral [4]. Modelling and theory distinguish between simple materials, comprised of weakly interacting elementary units, and complex materials including objects with internal structure characterized by relatively strong coupling and typically mimicking the covalent architecture of molecular species [5]. Furthermore, intramolecular interactions can be treated either quantum mechanically (Car–Parrinello method [6] with density functional theory of electronic structure [7]) or by a classical effective potential obtained through some more or less ad hoc coarse-graining procedure [8, 9]. In molecular dynamics or equilibrium Monte Carlo simulations, for example, the total intramolecular potential energy is typically decomposed as [10]

$$\mathcal{U}_{\text{intra}}(\{\mathbf{r}_s\}) = \underbrace{\sum_{\text{bonds}} \mathcal{U}_2(\mathbf{r}_i, \mathbf{r}_j)}_{\text{topology}} + \underbrace{\sum_{\text{bends}} \mathcal{U}_3(\mathbf{r}_i, \mathbf{r}_j, \mathbf{r}_k) + \sum_{\text{torsions}} \mathcal{U}_4(\mathbf{r}_i, \mathbf{r}_j, \mathbf{r}_k, \mathbf{r}_l) + \cdots}_{\text{flexibility}}, \quad (1)$$

where the sums range over suitable n-body potentials U_n that depend on the spatial configuration $\{r_s\}$ of the constituent atoms. Similar expansions are developed for non-bonded forces too, and may also include 1-body coupling to an external field, but the indiscriminate character of these through-space interactions usually means that the identification of participating atoms is determined by a simple range parameter. For intramolecular interactions, however, the enumeration and indexing of atoms involving in each sum of (1) is subject to the constraints of molecular topology. Indeed, the 2-body bond interactions U_2 *define*the molecular framework, while the higher order terms in (1) serve to model the more or less restricted molecular flexibility associated with bond hybridization or electronic delocalisation (e.g., aromaticity and resonance structures). The selection of these higher order potentials is often based on chemical intuition and are typically supplied to simulation software as user defined input. Mature and widely used simulation packages (e.g., GROMACS, LAMMPS, NAMD, etc.) invariably support highly optimized force fields (e.g., CHARMM, AMBER, OPLS, MMFF, etc.) that faithfully represent detailed atomistic structures. An alternative approach is to input only the molecular topology, then systematically generate all possible many-body index lists from this information and invite the user to select non-zero force constants and appropriate functional forms for the required potentials. This paradigm is more natural for the implementation of coarse-grained models derived by thermodynamic considerations (e.g., MARTINI [11]). To facilitate this semi-automatic procedure, a graph theoretic construction is developed here to exhaustively enumerate and index arbitrary n-body intramolecular interactions starting from the description of 2-body adjacency.

The central idea in this work observes the correspondence between the hierarchy of n-body intramolecular interactions

Citation: Robert Penfold and Peter J. Wilde, Intramolecular Interactions, DOI 10.1007/s10910-015-0510-x.

and iteration of the line graph [12] transformation L(G) on a connected "molecular" graph G.

GRAPH THEORY

A simple, connected and undirected graph $G = (V(G), E(G))$ is composed of a finite nonempty vertex set $V(G) = \{v_1, \ldots, v_p\}$ of $p \in \mathbb{N}$ vertices v_i and a finite edge set $E(G) = \{e_1, \ldots, e_q\}$ comprising $q \in \mathbb{N}$ distinct unordered pairs $e_\alpha = \{v_i, v_j\}$ of distinct vertices such that each element from V(G) appears at least once [13]. The number of vertices p=|V(G)| and edges q=|E(G)| are called the order and size of G, respectively. By restricting edges to ensure bond uniqueness (any vertex pair is joined by at most one edge) and forbid loops (each edge must join distinct vertices), molecular structures are naturally represented by such graphs G. Each vertex $v_i \in V(G)$ has a nonempty neighbor set $S_i(G) = \{v_j : \{v_i, v_j\} \in E(G)\}$ that lists its adjacent vertices $v_j \in V(G)$, and $\deg(v_i) = |S_i(G)|$ is called the degree of v_i. The p×p square, symmetric and binary vertex adjacency matrix A(G) of a graph G, defined by

$$A_{ij}(G) = \begin{cases} 1, & \text{if } \{v_i, v_j\} \in E(G) \\ 0, & \text{otherwise} \end{cases},$$

is in one-to-one correspondence with the molecular structure and is arguably the most natural algebraic representation of topological connectivity. An elementary inductive argument [14] establishes that $(A^m)_{ij}(G)$ is the total number of m length walks (a sequence of vertices joined consecutively by edges) between vertices v_i and v_j. In particular, the degree of vertex v_i corresponding to the valency of atom i in the molecular structure is

$$\deg(v_i) = (A^2)_{ii}(G) = \sum_{j=1}^{p} (A_{ij}(G))^2 = \sum_{j=1}^{p} A_{ij}(G) = \sum_{\alpha=1}^{q} Z_{i\alpha}(G),$$

so that the p × p diagonal degree matrix becomes

$$\mathbf{D}(G) = \sum_{i=1}^{p} \deg(v_i)\mathbf{e}_i\,\mathbf{e}_i^{\mathrm{T}} = \mathrm{diag}\big(\deg(v_1), \ldots, \deg(v_p)\big),$$

where the e_i are natural basis vectors in the coordinate vector space $\mathbb{R}^p$. Another useful representation of graph connectivity is the p × q binary vertex-edge incidence matrix Z(G) defined by

$$Z_{i\alpha}(G) = \begin{cases} 1, & \text{if } v_i \in e_\alpha \\ 0, & \text{otherwise} \end{cases}.$$

For each graph G there is an associated line graph L(G)(also called the "derivative" graph [15]) such that V L(G) is in bijective correspondence with E(G) and

$$\mathbf{A}(L(G)) = \mathbf{Z}^{\mathrm{T}}(G)\,\mathbf{Z}(G) - 2\mathbf{I}_q, \qquad (2)$$

where I_q is the q × q identity matrix. In other words, each edge of G is mapped to a vertex of L(G), while two vertices of the line graph are adjacent if and only if their corresponding edges are incident in G (that is, they share a common endpoint). Furthermore, the associated Kirchhoff matrix K(G) = D(G) − A(G) (also called the Laplacian of G) satisfies a similar relationship

$$-\mathbf{K}(G) = \mathbf{Z}(G)\,\mathbf{Z}^{\mathrm{T}}(G) - 2\mathbf{D}(G),$$

and the positive semidefinite signless Laplace matrix of G is [16]

$$\mathbf{Q}(G) = \mathbf{Z}(G)\,\mathbf{Z}^{\mathrm{T}}(G) = \mathbf{A}(G) + \mathbf{D}(G).$$

The spectrum of K(G) provides a useful consistency check since the algebraic multiplicity of the zero eigenvalue is equal to the number of connected components in G [16]. Hence, for a physically sensible molecular graph G, the rank of K(G) must be p − 1.

We recall the following definitions and terminology. A cycle graph C_r comprises p = r vertices, all of degree 2, connected in a closed chain by q = r edges. Removing a single edge produces a path graph P_r (of order p = r and size q = r − 1) with two terminal vertices of degree 1. The complete graph K_r on p = r vertices is maximally connected with $q = \frac{1}{2}r(r-1)$ edges such that $\{v_i, v_j\} \in E(K_r)$ for all distinct $v_i, v_j \in V(K_r)$. A graph G = V(G), E(G) is k-partite if the vertices can be partitioned into k disjoint sets, so that $V(G) = \cup_{r=1}^{k} V_r(G)$ where $V_r(G) \cap V_s(G) = \emptyset$ for $r \neq s$. The complete bipartite graph (k = 2) is denoted $K_{m,n}$ with $V(K_{m,n}) = V_1(K_{m,n}) \cup V_2(K_{m,n})$ and size p = m + n such that $m = |V_1(K_{m,n})|$ and $n = |V_2(K_{m,n})|$

We will also have occasion to consider directed graphs G = V(G), A(G) where edges are replaced by arrows specified by ordered pairs $(v_i, v_j) \in A(G)$ and oriented with the tail at vertex v_i pointing towards the head at vertex v_j. Associated with each vertex $v_i \in V(G)$ are two disjoint neighbor sets $S_i^-(G) = \{v_j : (v_i, v_j) \in A(G)\}$ and $S_i^+(G) = \{v_j(v_j, v_i) \in A(G)\}$ where $S_i = S_i^- \cup S_i^+$. At most one of S_i^- or S_i^+ may be empty. The indegree of vertex $v_i \in V(G)$ is $\deg^-(v_i) = |S_i^-(G)|$ and the outdegree is $\deg^+(v_i) = |S_i^+(G)|$. If $S_i^- = \emptyset$ so that $\deg^-(v_i) = 0$ then vertex v_i is called a source and, similarly, if $S_i^+ = \emptyset$ so that $\deg^+(v_i) = 0$ then vertex v_i is called a sink.

INTRAMOLECULAR INTERACTIONS

In a molecular structure, covalently bonded atom pairs are considered to be adjacent. Similarly, a pair of adjacent bonds sharing a common "hinge" atom form a more or less flexible bend. The corresponding 3-body interaction is often described by an effective potential in terms of the external bond angle θ supplementary to the angle subtended at the hinge atom [17]. A 4-body dihedral interaction is associated with a pair of adjacent bends that share a common bond. Two situations are possible [18] (see Fig. 1): "proper" torsions arise when both hinge atoms are distinct, so that one bend is rotated about the other through a dihedral angle □; while "improper" dihedral interactions link two bends through a common hinge atom, and are defined by a wag angle ω. Proper torsions typically account for geometric restrictions conferred by implicit substituents (usually protons) or lone electron pairs and may be alternatively characterized by a bond lying along the dihedral axis. Conversely, the dihedral axis of an improper torsion does not contain a bond and these interactions are used to constrain planar groups (like rings) or to hinder interconversion of stereocenters.

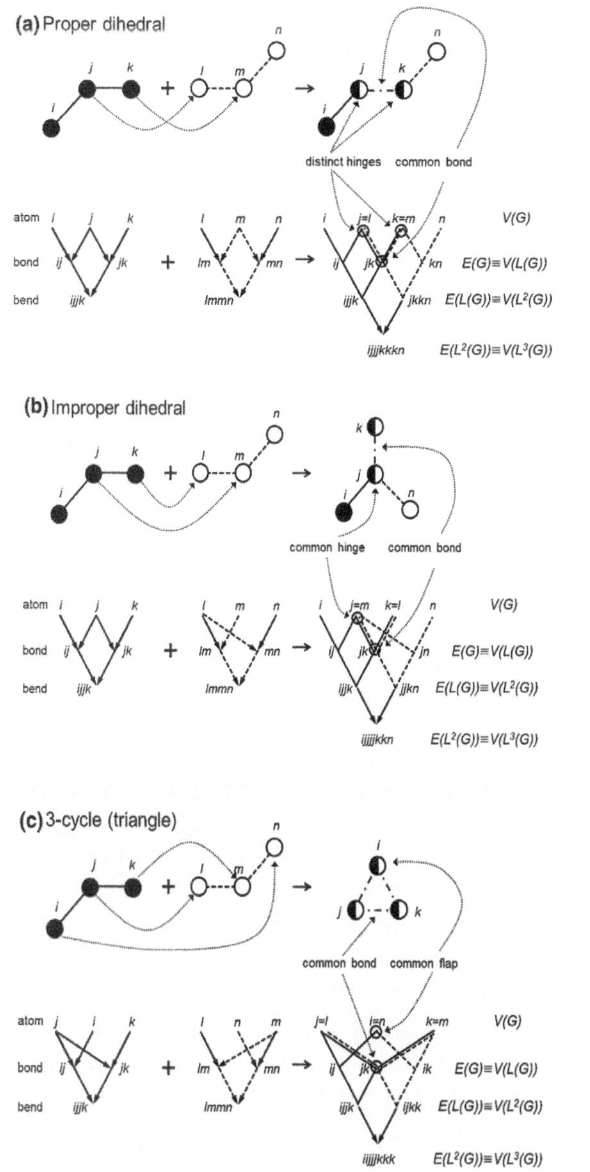

Figure 1: Generic structures of 4-body interactions represented on the directed acyclic graph Hv(G) are compared with the physical formation from the combination of two bends with atomic indexes ijk and lmn respectively. **a** Construction of the typical proper dihedral index sequence ijjjkkkn comprising two triplet repeats arising from the distinct hinge atoms j and k that form the shared bond. A similar sequence iijjjkkk arises for the degenerate 33-cycle structure as shown in (**c**), but where the "flap" atoms are also identified to close the ring. **b** The typical improper dihedral index sequence ijjjjkkn with only a single hinge atom j

In graph theoretical terms, this hierarchical organization of interactions is precisely captured by iterated application of the line graph construction inductively defined by

$$L^n(G) = \begin{cases} G\,, & \text{if} \quad n = 0 \\ L(L^{n-1}(G))\,, & \text{if} \quad n > 0 \end{cases}. \tag{3}$$

A graph G establishes adjacency of vertices (i.e., bonds); the graph L(G) encodes the adjacency of bonds (i.e., bends) in G; the graph $L(L(G))=L^2(G)$ encodes the adjacency of bends (i.e., dihedrals) in G; and so on. Generally, the graph $L^{\nu-2}(G)$ encodes the adjacency of ν-body interactions in G. It has been shown [19] that the sequence of graphs $L^n(G)$ with n=0, 1, 2, … has only four possible outcomes as $n\to\infty$:

- if $G \cong C_r$ (a cycle graph on r vertices), then $L^n(G) \cong G$ for all $n \in \mathbb{N}$ (cycle graphs are the only connected graphs for which L(G) is isomorphic to G);
- if $G \cong K_{1,3}$ (the complete bipartite "claw" graph), then $L^n(G) \cong C_3$ (a triangle) for all $n \in \mathbb{N}$;
- if $G \cong P_r$ (a path graph on r vertices), then $L^n(G) \cong P_{\max\{0,r-n\}}$ so each subsequent graph is a shorter path until eventually the sequence terminates at the trivial null graph;
- otherwise, G is a "prolific" graph [20] so that the sizes of the graphs in the sequence eventually increase without bound,

$$\left|V\left(L^n(G)\right)\right| \to \infty \quad \text{as} \quad n \to \infty.$$

Ghebleh and Khatirinejad [21] have proved an interesting and chemically relevant result concerning the smallest non-negative integer m such that $L^m(G)$ is nonplanar: the so-called line index $m=\xi(G)$ of a graph G. In particular, if G is not prolific it is easy to see that Ln(G)is planar for all $n\geq 0$, but for G prolific then $0\leq\xi(G)\leq 4$ and a complete characterization of these graphs is possible [21].

Enumeration

Given a molecular graph G, the total number of intramolecular interactions $N_n(G)$ involving $n\in N$ connected atoms (vertices on G) is generally given by

$$N_n(G) = \left|V\left(L^{n-1}(G)\right)\right| = \begin{cases} |V(G)|\,, & (n=1) \\ \frac{1}{2}\operatorname{Tr}\left(\mathbf{A}^2(L^{n-2}(G))\right), & (n=2,\,3,\,4,\,\ldots)\,. \end{cases}$$

Here, $N_1(G)$ simply counts the number of 1-body interactions in the presence of an external field. Elementary combinatorial arguments also establish the handshaking lemma [22]: the number of bonds is just half the total number of incident edges over all vertices so that

$$N_{\text{bond}}(G) = \frac{1}{2}\sum_{v\in V(G)} \deg(v) = \tfrac{1}{2}\operatorname{Tr}\left(\mathbf{D}(G)\right) = \tfrac{1}{2}\mathbf{u}^{\mathrm{T}}\mathbf{D}(G)\mathbf{u} = N_2(G),$$

where $\mathbf{u} = \sum_{i=1}^{p} \mathbf{e}_i$ is a pp-vector of ones. By summing the number of possible bond pairs over each vertex, the total bend count is obtained

$$\begin{aligned} N_{\text{bend}}(G) &= \sum_{v\in V(G)} \binom{\deg(v)}{2} = \frac{1}{2}\sum_{v\in V(G)} \deg(v)\left(\deg(v)-1\right) \\ &= \tfrac{1}{2}\operatorname{Tr}\left(\mathbf{D}(G)\left(\mathbf{D}(G)-\mathbf{I}\right)\right) = \tfrac{1}{2}\operatorname{Tr}\left(\mathbf{D}^2(G)\right) - N_2(G) \\ &= \tfrac{1}{2}\mathbf{u}^{\mathrm{T}}\mathbf{D}^2(G)\mathbf{u} - N_2(G) = N_3(G). \end{aligned}$$

Similarly reckoning the combinations of bond triplets over all vertices yields the number of improper dihedral interactions

$$N_{\text{impr}}(G) = \sum_{v \in V(G)} \binom{\deg(v)}{3} = \frac{1}{3!} \sum_{v \in V(G)} \deg(v)(\deg(v)-1)(\deg(v)-2)$$
$$= \tfrac{1}{6}\mathbf{u}^{\mathrm{T}}\mathbf{D}^3(G)\mathbf{u} - N_3(G) - \tfrac{1}{3}N_2(G).$$

Other possible arrangements of three contiguous bonds are just the proper dihedrals and triangular 3-cycles. In total, these arrangements can be enumerated as follows: for each bond, calculate the product of the number of free edges otherwise incident on each of the two vertices; then sum over all edges to get

$$N_{\text{prop}}(G) + 3N_{\text{3cyc}}(G) = \sum_{\{v_i, v_j\} \in E(G)} \left(\deg(v_i) - 1\right)\left(\deg(v_j) - 1\right)$$
$$= \frac{1}{2}\sum_{i=1}^{p}\sum_{j=1}^{p}(D_{ii}(G) - 1)A_{ij}(G)(D_{jj}(G) - 1)$$
$$= \tfrac{1}{2}\mathbf{u}^{\mathrm{T}}\mathbf{D}(G)\mathbf{A}(G)\mathbf{D}(G)\mathbf{u} - 2N_3(G) - N_2(G),$$

where we have observed the identity $\sum_{i=1}^{p} A_{ij}(G) = D_{ii}(G)$. The number of 3-cycles, however, can be obtained directly from the adjacency matrix as

$$N_{\text{3cyc}}(G) = \tfrac{1}{3!}\,\mathrm{Tr}\left(\mathbf{A}^3(G)\right).$$

The total 4-body interaction number N_4 includes torsions of both types as well as any 3-cycles present

$$N_4(G) = N_{\text{prop}}(G) + 3N_{\text{impr}}(G) + 3N_{\text{3cyc}}(G),$$

but over-counts the improper dihedrals and triangles by distinguishing the three rotational permutations of labels.

Combining these results gives the useful consistency checksums

$$N_1(G) = \tfrac{1}{2}\mathbf{u}^{\mathrm{T}}\mathbf{A}(G)\mathbf{u},$$
$$N_2(G) = \tfrac{1}{2}\mathbf{u}^{\mathrm{T}}\mathbf{D}(G)\mathbf{u},$$
$$N_3(G) = \tfrac{1}{2}\mathbf{u}^{\mathrm{T}}\mathbf{D}^2(G)\mathbf{u} - N_2(G),$$
$$N_4(G) = \tfrac{1}{2}\mathbf{u}^{\mathrm{T}}\mathbf{D}(G)\mathbf{Q}(G)\mathbf{D}(G)\mathbf{u} - 5N_3(G) - 2N_2(G),$$

that involve only the molecular graph G.

INDEXING

By definition of the line graph transformation, for $A_{ij}(L^n(G))=1$, the indexes $i=\alpha$ and $j=\beta$ point to the edges $e_\alpha, e_\beta \in E(L^{n-1}(G))$ that share a common vertex $v_l \in V(L^{n-1}(G))$. In turn, these edges $e_\alpha = \{v_k, v_l\}$ and $e_\beta = \{v_l, v_m\}$ are associated with the adjacency matrix of the previous graph in the recursive hierarchy so that $A_{kl}(L^{n-1}(G)) = A_{lm}(L^{n-1}(G)) = 1$. Successively backtracking to the source graph G yields, for each order n, a sequence of atomic indexes that participate in an n-body interaction. A formal pseudo-code implementation of this procedure is set out in Algorithm 1. Details of each sequence structure completely characterizes the interaction form.

Algorithm 1: Generate ν-body atom index list

```
Function List(ν, p, A)
  Data: ν ∈ N the order of multibody interactions.
        p = |V(G)| ∈ N, the order of simply connected graph G = (V, E).
        A(G) a p × p symmetric binary matrix.
  Result: r = |V(L^{ν−1}(G))| ∈ N, the order of the ν − 1 iterated line graph L^{ν−1}(G) of simply
          connected graph G = (V, E).
        edges, an r-length vector listing (for ν > 1) the edge set E(L^{ν−2}(G)) with elements
          expressed as a nested hierarchy of order pairs identifying the vertices of simply connected
          graph G = (V, E) associated with a ν-body intramolecular interaction. If ν = 1, then
          edges = V(G).
        A_L(L^{ν−1}(G)) a r × r symmetric binary matrix.
  if ν > 1 then
      (q, e_old, M) ← List(ν − 1, p, A)                      /* recursive call */
      α ← 0                                                 /* initialize edge count */
      for j = 1 to q do                                     /* loop over columns */
          for i = 1 to j − 1 do                             /* loop over rows */
              if M[i, j] = 1 then                           /* edge e_α = {v_i, v_j} here */
                  α ← α + 1                                 /* increment edge count */
                  edges[α] ← (e_old[i], e_old[j])           /* construct ordered pair */
      r ← α                                                 /* store edge count */
      Z[q, r] ← 0                                           /* initialize incidence matrix */
      α ← 0                                                 /* initialize edge count */
      for j = 1 to q do                                     /* loop over columns */
          for i = 1 to j − 1 do                             /* loop over rows */
              if M[i, j] = 1 then                           /* edge e_α = {v_i, v_j} here */
                  α ← α + 1                                 /* increment edge counter */
                  Z[i, α] ← 1                               /* set e_α */
                  Z[j, α] ← 1
      A_L ← Z^T Z − 2I_r                                    /* equation 2 */
  else                                                      /* base case */
      for i = 1 to p do                                     /* loop over rows */
          if A[i, i] ≠ 0 then break                         /* EXCEPTION: G not simple (loop) */
          edges[i] ← i                                      /* vertex of G */
      r ← p
      A_L ← A
  return (r, edges, A_L)
```

For concreteness, specify a maximum order ν for the multibody interactions of interest so that n-body terms with n=1, 2, …, ν are considered. Most commonly in molecular applications ν=4. A convenient representation of the recursive structure defined by (3) is itself a weakly connected ν-partite directed acyclic graph (DAG) denoted Hν(G) on the union of disjoint vertex sets $\cup_{n=1}^{\nu} V_n(G)$ where $V_n(G)=V(L^{n-1}(G))$. For clarity, define the vertex set $V(H_\nu(G))$ of $H_\nu(G)$ with elements $v_{jn} \in V_n(G)$. The edge directions induce a partial order relation on the vertices so that $v_{im} \leq v_{jn}$ only if $m<n$ where $v_{im} \in V_m(G)$ and $v_{jn} \in V_n(G)$. Sources of $H_\nu(G)$ with indegree 0 are just the atomic indexes of the molecular graph G. All other vertices on $H_\nu(G)$ have indegree 2 as a consequence of the defining property of an edge in $L^n(G)$. Similarly, the sinks of $H_\nu(G)$ with outdegree 0 are just the vertices of $L^{\nu-1}(G)$. At each level $n = 1, 2, \ldots, \nu - 1$ all other vertices $v_{jn} \in V_n(G)$ on $H_\nu(G)$ have outdegrees given by $\mathbf{D}(L^{n-1}(G))$. Moreover, the hierarchy of line graphs provides a natural topological order on $H_\nu(G)$. Further, the adjacency matrix takes the block tridiagonal form

$$\mathbf{A}(H_\nu(G)) = \begin{pmatrix} \mathbf{0} & \mathbf{Z}(G) & \mathbf{0} & \cdots & \mathbf{0} & \mathbf{0} \\ \mathbf{Z}^{\mathrm{T}}(G) & \mathbf{0} & \mathbf{Z}(L(G)) & \cdots & \mathbf{0} & \mathbf{0} \\ \mathbf{0} & \mathbf{Z}^{\mathrm{T}}(L(G)) & \mathbf{0} & \cdots & \mathbf{0} & \mathbf{0} \\ \vdots & \vdots & \vdots & \ddots & \vdots & \vdots \\ \mathbf{0} & \mathbf{0} & \mathbf{0} & \cdots & \mathbf{0} & \mathbf{Z}(L^{\nu-2}(G)) \\ \mathbf{0} & \mathbf{0} & \mathbf{0} & \cdots & \mathbf{Z}^{\mathrm{T}}(L^{\nu-2}(G)) & \mathbf{0} \end{pmatrix}.$$

By virtue of this DAG structure, each vertex $v_{jn} \in V(H_\nu(G))$ is associated with a unique sequence of 2^{n-1} atomic indexes v_{i1} that define each n-body interaction. For the trivial cases n = 1 and n = 2, these sequences correspond respectively with the individual atom list V(G) and the atom pairs E(G) defining the molecular bonds. Bends (n = 3) are signified by a pattern of $2^2 = 4$ atoms with a single repeated index identifying the common hinge atom. An 8-atom sequence classifies a 4-body interaction and distinguishes between three types as illustrated in Fig. 1. A pair of triply repeated atom indexes identifies the distinct hinge atoms of a proper torsion, provided the remaining two indexes are different (Fig. 1a). Otherwise, a common pair of "flap" atoms signals a degenerate 3-cycle where only three atom indexes appear (Fig. 1c). Each triangle generates ${}^3C_2 = 3$ such sequences by the arbitrary choice of a single common bond. The single common hinge of an improper dihedral corresponds to a fourfold repeated index among four distinct labels (Fig. 1b). Again, three sequences are generated for each improper dihedral by arbitrary assignment of the shared bond.

EXAMPLES

A toy model: methylcyclopropane

The molecular graph G indicated in Fig. 2 presents amongst other possibilities a plausible lumped model for methylcyclopropane where hydrogen atoms are absorbed onto the carbon backbone in the usual way. Direct inspection of G immediately establishes four 1-body interactions in the presence of an external field (that is, the number of "atoms" $N_1 = 4$) and also four 2-body bonds $N_2 = 4$. Atom 2 is the hinge for three distinct 3-body bends with a further two hinged at atoms 3 and 4 respectively, to give $N_3 = 5$. Among the 4-body interactions there are two proper torsions $N_{prop} = 2$ with distinct hinge atoms 2-3 and 2-4, respectively, as well as a single improper dihedral $N_{impr} = 1$ with common hinge atom 2. A single 3-cycle is present $N_{3cyc} = 1$ so that $N_4 = 2 + 3 \times (1 + 1) = 8$. Adjacency matrices for the iterated line graphs necessary for describing interactions up to the 4-body level are given by

$$\mathbf{A}(G) = \begin{pmatrix} 0 & 1^1 & 0 & 0 \\ 1^1 & 0 & 1^2 & 1^3 \\ 0 & 1^2 & 0 & 1^4 \\ 0 & 1^3 & 1^4 & 0 \end{pmatrix}, \quad \mathbf{A}(L(G)) = \begin{pmatrix} 0 & 1^1 & 1^2 & 0 \\ 1^1 & 0 & 1^3 & 1^4 \\ 1^2 & 1^3 & 0 & 1^5 \\ 0 & 1^4 & 1^5 & 0 \end{pmatrix},$$

$$\mathbf{A}(L^2(G)) = \begin{pmatrix} 0 & 1^1 & 1^2 & 1^4 & 0 \\ 1^1 & 0 & 1^3 & 0 & 1^6 \\ 1^2 & 1^3 & 0 & 1^5 & 1^7 \\ 1^4 & 0 & 1^5 & 0 & 1^8 \\ 0 & 1^6 & 1^7 & 1^8 & 0 \end{pmatrix}.$$

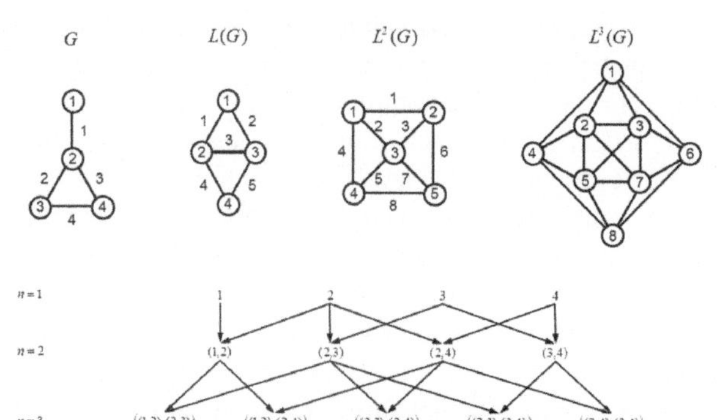

Figure 2: A plausible effective model of methylcyclopropane is represented by the graph G. The iterated line graphs $L^{n-1}(G)$ are also shown for n = 2, 3, 4. Vertices $v_{im} \in V_m(G) = V\,L^{m-1}(G)$ and edges $e_{km} = \{v_{im}, v_{jm}\}$ are labelled. The corresponding directed acyclic graph (DAG) generated by the line graph hierarchy is given where the vertices at each level n are denoted by the inherited sequence of atomic labels from G as indicated by the directed edges. Inspection of G confirms the 4-body interactions identified by the DAG sinks and comprise of: two proper torsions (denoted "p") with distinct hinge atoms 2-3 and 2-4, respectively; a single improper dihedral (denoted "i") with common hinge 2; and a single 3-cycle (denoted "c") of atoms 2, 3 and 4

For added clarity, the edge index associated with each adjacent vertex pair is indicated by the superscript. From these matrices, the DAG obtained that represents the line graph hierarchy is shown in Fig. 2. The atomic index sequences automatically generated on the DAG, particularly at the 4-body level (n = 4), confirm the informal analysis of the molecular graph. It is easy to show that the complete graph on five vertices K5 is a minor of $L^3(G)$, whence it follows from the theorem of Wagner [23] that $L^3(G)$ is nonplanar. All of G, L(G) and $L^2(G)$ are manifestly planar (see Fig. 2) so the line index $\xi(G) = 3$ in accord with the result of Ghebleh and Khatirinejad [21].

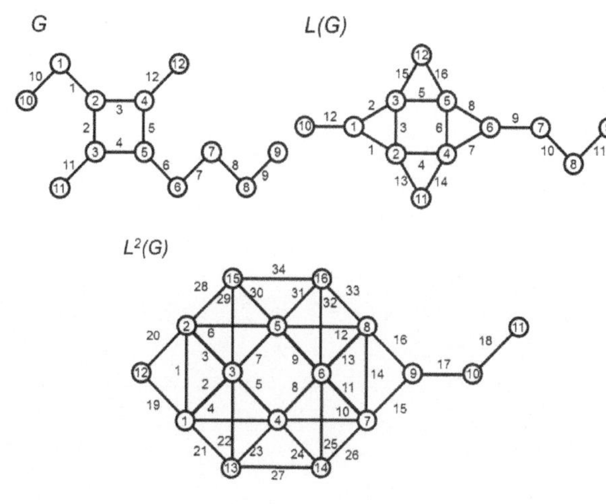

Figure 3: Effective coarse grained bile salt model of Vila Verde and Frenkel [24]. The molecular graph G is shown along with the iterated line graphs L(G) and $L^2(G)$. Vertices and edges are labelled

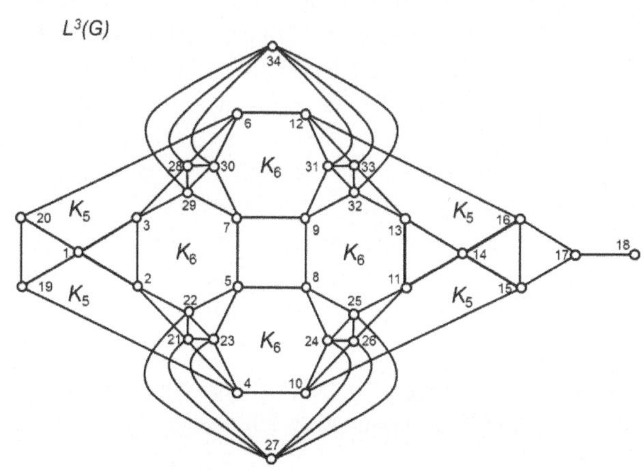

Figure 4: Effective coarse grained bile salt model of Vila Verde and Frenkel [24]. The molecular line graph $L^3(G)$. To simplify the diagram, subgraphs corresponding to complete graphs K_q on q vertices (q-cliques of $L^3(G)$) have been rendered as pentagons (q = 5) and hexagons (q = 6)

Table 1: Vertex sequences associated with the directed acyclic graph (DAG) description of the line graph hierarchy for the effective bile salt model of Vila Verde and Frenkel [24]

N_n	$n = 1$	$n = 2$	$n = 3$	$n = 4$	
1	1	(1, 2)	((1, 2), (2, 3))	(((1, 2), (2, 3)), ((1, 2), (2, 4)))	i
2	2	(2, 3)	((1, 2), (2, 4))	(((1, 2), (2, 3)), ((2, 3), (2, 4)))	i
3	3	(2, 4)	((2, 3), (2, 4))	(((1, 2), (2, 4)), ((2, 3), (2, 4)))	i
4	4	(3, 5)	((2, 3), (3, 5))	(((1, 2), (2, 3)), ((2, 3), (3, 5)))	p
5	5	(4, 5)	((2, 4), (4, 5))	(((2, 3), (2, 4)), ((2, 3), (3, 5)))	p
6	6	(5, 6)	((3, 5), (4, 5))	(((1, 2), (2, 4)), ((2, 4), (4, 5)))	p
7	7	(6, 7)	((3, 5), (5, 6))	(((2, 3), (2, 4)), ((2, 4), (4, 5)))	p
8	8	(7, 8)	((4, 5), (5, 6))	(((2, 3), (3, 5)), ((3, 5), (4, 5)))	p
9	9	(8, 9)	((5, 6), (6, 7))	(((2, 4), (4, 5)), ((3, 5), (4, 5)))	p
10	10	(1, 10)	((6, 7), (7, 8))	(((2, 3), (3, 5)), ((3, 5), (5, 6)))	p
11	11	(3, 11)	((7, 8), (8, 9))	(((3, 5), (4, 5)), ((3, 5), (5, 6)))	i
12	12	(4, 12)	((1, 2), (1, 10))	(((2, 4), (4, 5)), ((4, 5), (5, 6)))	p
13			((2, 3), (3, 11))	(((3, 5), (4, 5)), ((4, 5), (5, 6)))	i
14			((3, 5), (3, 11))	(((3, 5), (5, 6)), ((4, 5), (5, 6)))	i
15			((2, 4), (4, 12))	(((3, 5), (5, 6)), ((5, 6), (6, 7)))	p
16			((4, 5), (4, 12))	(((4, 5), (5, 6)), ((5, 6), (6, 7)))	p
17				(((5, 6), (6, 7)), ((6, 7), (7, 8)))	p
18				(((6, 7), (7, 8)), ((7, 8), (8, 9)))	p
19				(((1, 2), (2, 3)), ((1, 2), (1, 10)))	p
20				(((1, 2), (2, 4)), ((1, 2), (1, 10)))	p
21				(((1, 2), (2, 3)), ((2, 3), (3, 11)))	p
22				(((2, 3), (2, 4)), ((2, 3), (3, 11)))	p
23				(((2, 3), (3, 5)), ((2, 3), (3, 11)))	i
24				(((2, 3), (3, 5)), ((3, 5), (3, 11)))	i
25				(((3, 5), (4, 5)), ((3, 5), (3, 11)))	p
26				(((3, 5), (5, 6)), ((3, 5), (3, 11)))	p
27				(((2, 3), (3, 11)), ((3, 5), (3, 11)))	i
28				(((1, 2), (2, 4)), ((2, 4), (4, 12)))	p
29				(((2, 3), (2, 4)), ((2, 4), (4, 12)))	p
30				(((2, 4), (4, 5)), ((2, 4), (4, 12)))	i
31				(((2, 4), (4, 5)), ((4, 5), (4, 12)))	i
32				(((3, 5), (4, 5)), ((4, 5), (4, 12)))	p
33				(((4, 5), (5, 6)), ((4, 5), (4, 12)))	p
34				(((2, 4), (4, 12)), ((4, 5), (4, 12)))	i

At each level n, the inherited sequence of atomic labels from the molecular graph G is listed. Inspection of G confirms the 4-body interactions identified by the DAG sinks and comprise of: 22 proper torsions (denoted "p") and 4 improper dihedrals (denoted "i") with no 3-cycles

Bile salt: taurocholate

A chemically relevant example, central to the hydrolysis and solubilisation of lipid associated with food digestion in the human lower gastrointestinal tract, are the bile salt and bile acid surfactants. Vila Verde and Frenkel [24] have proposed an effective coarse-grained model of trihydroxy bile salts (taurocholate) for a molecular dynamics study of micelle formation that is related to the rate and extent of nutrient absorption by intestinal cells. The authors proposed a "three-to-one" mapping scheme, that groups three carbon or nitrogen atoms into a single bead, to arrive at the molecular graph G shown in Fig. 3. Iterated line graphs up to $L^3(G)$ are collected in Figs. 3 and 4. Table 1 lists the vertices for the corresponding DAG description of the hierarchy as sequences of bead indexes.

Overall, the counts of bonds, bends, proper torsions and improper dihedrals are obtained as follows,

$$N_{\text{bond}} = 12, \quad N_{\text{bend}} = 16, \quad N_{\text{prop}} = 22, \quad N_{\text{impr}} = 4,$$

where N4 = 22 + 3 × 4 = 34. Clearly, there are no 3-cycles in this example so $N_{cyc} = 0$. It is easy to verify that $L^2(G)$ admits the minor K_5 and hence, by Wagner's theorem [23], it follows that $L^2(G)$ is nonplanar. Both G and L(G) are manifestly planar so the line index $\xi(G) = 2$ in accord with the result of Ghebleh and Khatirinejad [21].

CONCLUSION

The line graph transformation provides a practical and elegant theoretical tool for exhaustively enumerating and indexing many-body intramolecular interactions. Given a suitable graphical representation of a molecular structure, an explicit pseudo-code implementation of the recursive line graph algorithm is given for automatically generating complete canonical lists of atomic indexes associated with each interaction order. No attempt has been made to computationally optimize this algorithm or the associated data structures. Instead, clarity of exposition is the main objective here. We anticipate the main application will involve embedding the algorithm within a Monte Carlo or Molecular Dynamics simulation code where other implementation details will determine the most efficient realization. In accord with common practice, intramolecular interactions up to order 4 have been considered (bonds, bends and dihedrals), but the method can be extended to arbitrarily many atomic centers. Higher order interactions will involve increasingly many sub-type variations and polycyclic structures. Two specific examples are discussed: a toy model of methylcyclopropane and a published effective potential model of taurocholate bile salt [24] that is relevant for the study of digestive processes in the human lower gastrointestinal tract.

ACKNOWLEDGMENTS

This work was financially supported by the Biotechnology and Biological Sciences Research Council through its core strategic grant to the Institute of Food Research. We also thank Andrew Watson for critical reading of the manuscript.

REFERENCES

1. D. Bonchev, J. Mol. Struct. (Theochem) **336**, 137 (1995)
2. D.H. Rouvray, A.T. Balaban, in *Applications of Graph Theory*, ed. by R.J. Wilson, L.W. Beinecke (Academic Press, London, 1979), p. 177
3. H.N.V. Temperley, in *Applications of Graph Theory*, ed. by R.J. Wilson, L.W. Beinecke (Academic Press, London, 1979), p. 121
4. J.-P. Hanson, I.R. McDonald, *Theory of Simple Liquids*, 2nd edn. (Academic Press, London, 1986), pp. 79–92
5. B. Kirchner, Phys. Rep. **440**, 1 (2007)
6. J. Hutter, WIRE: Comp. Mol. Sci. **2**, 604 (2012)
7. R.D. Kohn, A.D. Becke, R.G. Parr, J. Phys. Chem. **100**, 12974 (1996)
8. R. Penfold, S. Abbas, S. Nordholm, Fluid Phase Equilib. **120**, 39 (1996)
9. J.T. Padding, A.A. Louis, Phys. Rev. E **74**, 031402 (2006)
10. M.P. Allen, in *Computational Soft Matter: From Synthetic Polymers to Proteins*, ed. by N. Attig, K. Binder, H. Grubmüller, K. Kremer (John von Neumann Institute for Computing, Jülich, 2004), p. 1
11. S.J. Marrink, H.J. Risselada, S. Yefimov, D.P. Tieleman, A.H. de Vries, J. Phys. Chem. B.**111**, 7812 (2007)
12. R.L. Hemminger, L.W. Beinecke, in *Selected Topics in Graph Theory*, ed. by L.W. Beinecke, R.J. Wilson (Academic Press, London, 1978), p. 271
13. F. Harary, *Graph Theory* (Addison-Wesley, Reading, 1969)
14. J. Clark, D.A. Holton, *First Look at Graph Theory* (World Scientific, Singapore, 1991)
15. G. Sabidussi, Math. Zeitschr **76**, 385 (1961)
16. A.E. Brouwer, W.H. Haemers, *Spectra of Graphs* (Springer, New York, 2012)
17. P.J. Flory, *Statistical Mechanics of Chain Molecules* (Carl Hanser Verlag, Munich, 1989)
18. R.E. Tuzun, D.W. Noid, B.G. Sumpter, J. Comp. Chem. **18**, 1513 (1997)
19. A.C.M. van Rooij, H.S. Wilf, Acta Math. Hungar. **16**, 263 (1965)
20. M. Knor, P. Potočnik, R. Škrekovski, Discrete Appl. Math. **160**, 2234 (2012)
21. M. Ghebleh, M. Khatirinejad, Discrete Math. **308**, 144 (2012)
22. R.J. Wilson, *Introduction to Graph Theory*, 3rd edn. (Longman, Harlow, 1985)
23. R. Diestel, *Graph Theory*, 3rd edn. (Springer, Heidelberg, 2005)
24. A. Vila Verde, D. Frenkel, Soft Matter **6**, 3815 (2010)

SECTION III: COMPUTATIONAL MOLECULAR BIOPHYSICS, CELLULAR STRUCTURE PREDICTION AND REACTION–DIFFUSION SYSTEMS

CHAPTER 13

MATHEMATICAL AND COMPUTATIONAL MODELING IN BIOLOGY AT MULTIPLE SCALES

Jack A Tuszynski, Philip Winter, Diana White, Chih-Yuan Tseng, Kamlesh K Sahu, Francesco Gentile, Ivana Spasevska, Sara Ibrahim Omar, Niloofar Nayebi, Cassandra DM Churchill, Mariusz Klobukowski, and Rabab M Abou El-Magd

Keywords: Mathematical biology, Epidemiological models, Cellular physiological models, Cancer models, Maximum entropy, Molecular dynamics, Force fields, Solvation free energies, Quantum mechanics, Computational enzymology

INTRODUCTION

Mathematical, computational and physical methods have been applied in biology and medicine to study phenomena at a wide range of size scales, from the global human population all the way down to the level of individual atoms within a biomolecule. Concomitant with this range of sizes between global to atomistic, the relevant modeling methods span time scales varying between years and picoseconds, depending on the area of interest (from evolutionary to atomistic effects) and relevance. This review will cover some of the most common and useful mathematical and computational methods. Firstly, we outline the maximum entropy principle as an inference tool for the study of phenomena at different scales, from gene evolution and gene networks to protein-drug molecular interactions, followed with a survey of the methods used for large scale systems—populations, organisms, and cells—and then zooming down to the methods used to study individual biomolecules—proteins and drugs. To study the large systems, the most common and reliable mathematical technique is to develop systems of differential equations. At the molecular scale, molecular dynamics is often used to model biomolecules as a system of moving Newtonian particles with interactions defined by a force field, with various methods employed to handle the challenge of solvent effects. In some cases, pure quantum mechanics methods can and should be used, which describe molecules using either wave functions or electron densities, although computational costs in time and resources may be prohibitive, so hybrid classical-quantum methods are often more appropriate. Quantum methods can be particularly valuable in the study of enzymes and enzymatic reactions.

MAXIMUM ENTROPY IN BIOLOGY AND DRUG DISCOVERY

Two reasoning methods, deduction and inductive inference, have been utilized in the development of theories to interpret phenomena we observe in nature, and to make predictions about complex systems. Deduction allows us to draw conclusions when sufficient information is available, and is contrasted with inductive inference (also known as inductive logic or probable inference). Inductive inference provides a least biased way to reason when the available information is insufficient for deduction. It is called "inference" when we make estimates of quantities for which we do not have enough information to use deductive reasoning, and "induction" when we are generalizing from special cases [1].

When we deal with complex systems, for example either many-body interactions at the microscopic level, complicated regulatory protein-protein networks at the mesoscopic level, or population genetics at the macroscopic level, we never have enough knowledge to completely understand the system. Therefore, we normally rely on inductive inference based on the available information to infer the most preferred solution to problems related to these systems. Particularly, we are interested in a mathematical tool for inductive inference based on the Bayesian interpretation of probability, the rules of probability

Citation: Tuszynski et al., Mathematical and computational modeling in biology at multiple scales. Theoretical Biology and Medical Modelling 2014 11:52, doi:10.1186/1742-4682-11-52.

theory, and the concept of entropy. Bayesian interpretation treats probability as a degree of our knowledge about a system of interest, rather than the frequency of appearance of an event. Cox demonstrated that this type of probability can be manipulated by the rules of standard probability theory [2]. This forms the building blocks of inductive inference, termed Bayesian inference. Moreover, Caticha and Giffin have shown that Bayesian inference is a special case of entropy-based inference [3]. Therefore, our discussion in this section will be founded on entropy-based inference.

First, we briefly address the basics of entropy-based inference, which includes using entropy as an information measure and a tool for inductive inference, then we provide examples in the fields of biology and drug discovery to demonstrate that these fields benefit from the application of inductive inference. Regarding using entropy as an information measure, we consider two examples. The first example provides a clue to investigate genomic evolution through appropriate genomic sequence analysis [4]. The second one discusses robustness of biological networks from information point of view [5]. Regarding using entropy as a tool for inductive inference, we consider another two examples. The first one demonstrates the benefit of introducing this inference in virtual screening for drug discovery [6]. The second one then shows an application of this scheme in fragment-based drug design [7]. These examples also illustrate in a straightforward way how to extract information and unveil the global characteristics of the biological systems. They show that there exists a universal reasoning platform to solve any problem of interest that is independent of the specifics of a given type of problem. The key in this platform lies in the answer to the question, "What are the constraints in the system?" Once the constraints are determined, the maximum entropy principle provides a robust, universal and least biased prescription for information processing. Furthermore, it helps us to analyze problems and gain insights into the functioning of the underlying complex systems.

Entropy as an information measure

Shannon's pioneering work on the quantification of information loss during communication established a new viewpoint on entropy, which was until then only known as a measure of randomness in thermodynamics [8]. Since then, using entropy as an information measure has attracted much attention not only in signal processing, but also in the field of biology.

Information in genomic evolution

With the advance of genomic sequencing technology there is more and more genomic sequence data available for species across the three domains: Bacteria, Archaea, and Eukaryota. The question is, how do we compare complete genomes and extract useful information from the sequencing data?

To address the question of genome comparison, Chang et al. [4] proposed an entropy-based scheme for complete genome comparison. The foundation of their approach is to define the probability distribution that represents our current state of knowledge regarding the occurrence of different combinations of the four bases in DNA sequences. Chang et al. [4] specified that k-mer nucleotides in the sequence encode genetic information, where k is an arbitrary number. The occurrence of k-mers in a DNA sequence characterizes that sequence. Based on this definition, information in sequences can be quantified with Shannon information. Furthermore, Chang et al. [4] introduced the concept of reduced Shannon information, which is defined as the ratio of the Shannon information of the genome to the Shannon information of random sequences, so as to quantify to what extent the information contained in the genome is different from the information in a random DNA sequence. Note that this concept is similar to the concept of relative entropy, which is discussed in the next section. Based on reduced Shannon information (or relative entropy), a universal feature across three taxonomic domains was observed; namely, the effective root-sequence length of a genome, which is defined as the ratio of genome length and reduced Shannon information, linearly depended on k, and was a genome-independent constant. Furthermore, this study revealed a possible genome growth mechanism: at an early stage of evolution, a genome is likely to utilize random segmental duplication, which would maximize reduced Shannon information. These insights not only provide a clue to the origin of evolution but also may shed light on further questions, such as which genes are responsible for drug resistance.

Robustness of biological networks

The development of high throughput screening techniques, such as microarray technology, has generated numerous protein-protein interaction data to map out biological networks, and has revealed regulatory mechanisms of the biological entities involved in the networks. Many studies have suggested that the robustness of biological networks may be the key for identifying systems that can tolerate external perturbations and uncertainty triggered by external forces [9,10]. It has further been shown that the robustness of biological networks shares a global feature; these networks are scale-free networks, which means that one can observe a power-law degree distribution in these networks.

Therefore, there have been many endeavors to provide more insights into the origin of these power-law distributions. Bak et al. [8] proposed the mechanism of self-organized criticality, which leads to a scale-free structure in complicated systems. An entropy-based interpretation described by Dover [5] suggested a promising and intuitive way to understand the emergence of power-law distributions in complicated networks. According to Dover's studies on a toy model, the emergence of the power-law distributions is merely a consequence of the maximum entropy principle when the internal order of sub-networks of a complicated large network remained fixed. Note that the internal order was defined as the mean of the Boltzmann entropy over all sub-networks. In the framework of entropy-based inference, the power-law distributions of biological networks simply represent the most preferred choice that maintains the fixed internal order of the sub-networks.

Entropic scheme for inductive inference

In addition to the use of entropy as an information measure, the concept of entropy also plays a role in inductive inference. The inductive inference addressed here refers to two processes. The first process is the determination of the most likely state of knowledge about a system of interest based on the information in hand. The second process is the determination of the most likely updated state of knowledge when we acquire new information regarding the system. The foundation of inductive inference is the maximum entropy principle and relative entropy. The least biased inference one can make based on the information in hand is the one that maximizes the relative entropy of all possible new and old beliefs. For more details, the reader is referred to Caticha [1].

Entropy in molecular docking

Our first example of applying entropy for inductive inference is in silico drug discovery. Virtual screening has attracted much attention in the pharmaceutical industry [12, 13]. It provides a more economical way to screen diverse chemicals as drug candidates compared with a wet-lab approach. Basically, it consists of the creation of a chemical library, followed by searching optimal ligand-receptor binding modes through docking algorithms, and finally the evaluation of binding affinities. There are three criteria that are required to successfully identify drug candidates. First, the chemical library needs to be large and contain diverse chemical structures. Second, conformational search algorithms need to be able to search possible binding modes within a reasonable time. Third, an appropriate scoring function needs to be utilized to correctly evaluate the binding affinity of the chemical structures. In the framework of information theory, the first and third criteria are the fundamental information required in virtual screening process. The second criterion then can be treated as an information processing guideline. The efficiency and accuracy of this step will depend on the methods of information processing.

Genetic algorithms, which borrow from the concept of genomic evolution processes to search conformations of complex targets and chemical structures, are commonly used in docking protocols, such as AutoDock [14]. Chang et al. have offered a better alternative, MEDock [6]. Although MEDock did not completely exploit entropic-based inductive inference for searching, it does utilize the maximum entropy principle as a guideline to make decisions during this process. The fundamental question asked in MEDock is "What is the probability of finding the deepest energy valley in a ligand-target interaction energy landscape?" Maximum entropy provides a direction to update the initial guess of binding modes (described by an almost uniform distribution) to the optimal mode (a localized distribution around the global energy minimum).

Entropy in aptamer design

The second example of entropy for inductive inference is aptamer design. Aptamers are short nucleic acid sequences that are traditionally identified through an experimental technique, the Systematic Evolution of Ligands by Exponential Enrichment (SELEX) [15, 16]. Aptamers can bind to specific molecular targets including small molecules, proteins, nucleic acids, and phospholipids, and can also be targeted to complex structures such as cells, tissues, bacteria, and other organisms. Because of their strong and specific binding through molecular recognition, aptamers are promising tools in molecular biology and have both therapeutic and diagnostic clinical applications [15–18]. Unfortunately, some limitations of SELEX have slowed the progress of discovering specific aptamers for various applications [18]. With the help of entropy-based inductive inference, a fragment-based approach has been developed to design aptamers given the structure of the target of interest [18].

The concept of the fragment-based approach to aptamer design is to ask the question "Given the structural information about the target, what is the preferred probability distribution of having an aptamer that is most likely to interact with the target?" The solution was found using entropy-based inductive inference [7]. This approach initially determines the preferred probability distribution of first single nucleotide that likely interacts with the target. Subsequently, the approach iteratively updates the probability distribution as more nucleotides are added to the growing aptamer. The maximum entropy principle allows us to determine to what extent this update is sufficient, and what is the sequence of nucleotides that is most likely to bind to the target. This method has been applied to design aptamers to bind specifically to targets such as thrombin, phosphatidylserine [19] and galectin-3 (under experimental confirmation).

The maximum entropy principle and inductive inference just provide one reasoning platform to make the most preferable inference based on all kinds of information for understanding biological systems at different scales. In the next section, a variety of mathematical and computational models addressing other aspects that have been developed for biological and medical problems are surveyed.

MATHEMATICAL AND COMPUTATIONAL MODELS FOR BIOLOGICAL SYSTEMS

In recent years, mathematical biology has emerged as a prominent area of interdisciplinary scientific research. It is not a new area of research, but with recent advances in medical and computational methods, it has grown extensively, being applied to solve many health related problems across a spectrum of life sciences. Areas of mathematical biology where modeling has made contributions to biology and medicine include epidemiology, cell physiology, cancer modeling, genetics, cellular biology, and biochemistry. Because there is such a broad range of topics and methods that can be discussed, we limit ourselves to a discussion of how differential equations have been used to solve important biological problems in epidemiology, cell physiology, and cancer modeling, and briefly discuss some of the clinical advances that have arisen from such efforts. For a more extensive review on mathematical modeling for each of these branches of science, we refer the reader to recent books on these topics [20–23]. For the reader who is interested in learning more about mathematical biology from a beginner's perspective, books by Edelstein-Keshet [24], Murray [25, 26], and Britton [27] are also recommended.

Here, we highlight only a few of the models that have been developed to study epidemiological, physiological, and cancer problems. The reader is encouraged to look more extensively into the literature regarding other models that have been developed and successfully applied to improve present medical treatments.

Mathematical models in epidemiology

Epidemiology describes the study of patterns, cause and effect, and treatment of disease within a given population [28]. Here, we provide a brief introduction to epidemiological models used for studying the spread of various types of disease, many of which are outlined in [25, 27]. One of the first models to describe the dynamics of a disease caused by a viral infection is the so-called SIR model, an ordinary differential equation (ODE) model developed by Kermack and Mckendrick in 1927 [29].

$$\frac{dS}{dt} = -\beta IS \quad (1)$$

$$\frac{dI}{dt} = \beta IS - \mu I \quad (2)$$

$$\frac{dR}{dt} = \mu I \quad (3)$$

This model, given by equations (1), (2), and (3) (for all differential equation models we omit initial and boundary conditions for ease in reading), describes the rate of change of the number of susceptible (S), infected (I), and recovered (R) individuals in a population over time, where β describes the rate of transmission of disease, and μ describes the rate of removal of infected individuals (those that have recovered). An important feature of this model is that it incorporates recovered patients, meaning that an individual can acquire immunity, as is often the case for viral-type infections like influenza and measles. This model, although quite basic, provides important information to health care professionals interested in understanding how severe an outbreak is. For example, from these equations, the basic reproduction number given by

$$R_0 = \frac{\beta S(0)}{\mu} \quad (4)$$

describes the average number of secondary infections produced by one infected individual introduced into a completely susceptible environment. High values of R_0, corresponding to high numbers of initially susceptible individuals S(0), and/or high disease transmission rates β, result in the high probability of an outbreak. In particular, if R_0 is less than 1, the infection will not persist (and will eventually die out), whereas if R_0 is greater than 1, the infection will grow (and there will be an epidemic).

One key assumption of this model is that the total population N (N = S + I + R) is constant and that there is no death or birth. Many models have since been developed to include such population demographics [30–32], the first being completed by Soper [31] in an attempt to understand the dynamics of measles.

A number of extensions have been made to describe a wider class of infections. For example, the SIRS and SIS models allow for the movement of individuals back into a susceptible class S, meaning there may be no immunity to re-infection [30]. Such models are useful in studying bacterial-type infections like tuberculosis and gonorrhea. Other models, referred to as SEIR and SEIRS models, where "E" stands for a latent class of individuals (exposed but not showing symptoms), can be used to describe a disease where a delayed time of infection may exist [33]. For example, this is often the case with individuals suffering from malaria.

A disadvantage of ODE-based modeling is that it assumes the well mixing of large populations of individuals. Also, such models are deterministic, meaning that the outcome is determined solely on the initial conditions and the parameters that govern the dynamics. For some populations, where contacts and transmission rates between individuals may vary, agent based [34, 35] stochastic [36] or network type [36] models may be more useful. Also, age-structured models [37] may be more appropriate for diseases that depend on age, such as AIDS.

Another disadvantage of ODE-based modeling is that it does not describe the movement of individuals through space. This information is extremely important because a disease may not just spread within a single population, but may spread from one location to another. Examples of models that incorporate spatial dynamics include partial differential equations (PDEs). These models have been used to study the outbreak of rabies in continental Europe in the early 1940s [38], as well as to study the more recent outbreak of the West Nile Virus in 1999 in New York State [39]. Other models used to study the spatial spread of disease include patch models [40]. In the patch model of Lloyd and May [40] the authors consider an SEIR modeling approach. Here, the total population is broken up into subpopulations (patches), where S_i, E_i, I_i, and R_i denote the number of susceptible, exposed (latent), infected, and recovered individuals, in each patch i, respectively. The dynamics of each patch are governed by their own similar set of differential equations,

$$\frac{dS_i}{dt} = dN_i - dS_i - \sigma_i S_i, \quad (5)$$

$$\frac{dE_i}{dt} = \sigma_i S_i - (d + \varepsilon)E_i, \quad (6)$$

$$\frac{dI_i}{dt} = \varepsilon E_i - (d + \gamma)I_i, \quad (7)$$

$$\frac{dR_i}{dt} = \gamma I_i - dR_i. \quad (8)$$

In each patch, all model parameters are the same, except the infection rate σ_i, which depends on each connection between patches. Here, σ_i is called the force of infection, and is given by the mass action expression

$$\sigma_i = \sum_{j=1}^{n} \beta_{ij} I_j, \quad (9)$$

where n is the total patch number and β_{ij} is the rate of infection between patches i and j.

Mathematical models have influenced protocol in disease control and management. Now, such modeling is part of epidemiology policy decision making in many countries [41]. Some important modeling contributions include the design and analysis of epidemiology surveys, determining data that should be collected, identifying trends and forecasting outbreaks, as well as estimating the uncertainty in these outbreaks.

Physiological models at the cellular level: enzyme kinetics, ion channels, and cell excitability

The field of physiology is arguably the number one biological field where mathematics has had the greatest impact. Two broad areas of physiology where mathematics has made a profound impact are cell physiology and systems physiology.

Here, we focus on cell physiology, and restrict ourselves to the topics of enzyme kinetics, ion channels, and cell excitability. For an excellent review on systems physiology, the reader is referred to Keener and Sneyd [22].

The rate of change of a simple chemical reaction can be described by the law of mass action, a law that describes the behavior of solutions in dynamic equilibrium [42]. That is, for a simple reaction given by

$$A + B \xrightarrow{k_+} C, \quad (10)$$

where k_+ is the reaction rate constant, the rate of the production of product molecules is given by

$$\frac{d[C]}{dt} = k_+[A][B], \quad (11)$$

where [X] is the concentration of each species X = A, B, C. Equation (11) is commonly referred to as the law of mass action. The above formulation can only be used for the simplest of reactions involving a single step and only two reactants, although extensions are fairly straight-forward and have been developed over the past century to describe more complicated reactions [43]. One example is the model of Michaelis and Menten [44], used to describe reactions catalyzed by enzymes. Given the reaction scheme

$$S + E \underset{\leftarrow k_{-1}}{\overset{\rightarrow k_1}{}} C \xrightarrow{k_2} P + E, \quad (12)$$

where S is the substrate, E the enzyme, P the product concentration, and k_1, k_{-1}, and k_2 are the reaction rate constants, Michaelis and Menten describe this reaction by the following four ODEs,

$$\frac{ds}{dt} = k_{-1}c - k_1 se, \quad (13)$$

$$\frac{de}{dt} = (k_{-1} + k_2)c - k_1 se, \quad (14)$$

$$\frac{dc}{dt} = k_1 se - (k_2 + k_{-1})c, \quad (15)$$

$$\frac{dp}{dt} = k_2 c. \quad (16)$$

Assuming that the substrate is at equilibrium (ds/dt = 0), one can simplify this system and find explicit solutions. Also, without solving these equations, one can gain useful information about the process. For example, the velocity of the reaction (the rate at which the products are formed) is

$$V = \frac{dp}{dt} = \frac{V_{max} s}{K_s + s}, \quad (17)$$

where $V_{max} = k_2 e_o$ and $K_s = k_{-1}/k_1$ (K_s is called the equilibrium constant). Equation (17) is often referred to as the Michaelis–Menten equation. Also, the steady-state approximation simplifies the above system of equations so that we can find explicit solutions [45]. This approximation requires that the rates of formation and breakdown of the complex c are essentially always equal (dc/dt = 0). Further extensions of this model have been developed to describe other types of enzyme activity, such as cooperativity [46] and enzyme inhibition [47], and have had similar success.

Computational systems biology has been creating a series of tools that are useful for application to enzyme kinetics. This is particularly true in the area of parameter estimation where several algorithms have been shown to be useful to enzyme kinetics. The availability of increasingly sophisticated and standardized modeling and simulation software will undoubtedly benefit enzyme kinetics [48].

Biochemical networks are sets of reactions that are linked by common substrates and products. The dynamics of biochemical networks are frequently described as sets of coupled ODEs, similar to those given by equations (13) through (16), that represent the rate of change of concentrations of the chemical species involved in the network [48]. The right-hand side in these ODEs is typically the algebraic sum of the rate laws of the reactions that produce or consume the chemical species (positive when it is produced, negative when consumed). There is formally no difference between a biochemical network and an enzyme reaction mechanism, as both conform to this description. For systems biology studies, it is sufficient to represent each enzyme-catalyzed reaction as a single step and associate it with an appropriate integrated rate law [49]. The systems biologist should be cautioned, though, that mechanistic details may indeed affect the dynamics, as is the case with competitive versus uncompetitive inhibitor drugs [50–52].

The Systems Biology Markup Language (SBML) [53] is a standard format to encode the information required to express a biochemical network model including its kinetics. SBML is based on the Extended Markup Language (XML), which is itself a standard widely adopted on the Internet. After a series of progressive developments, there are now several compatible software packages available to model biochemical networks. Some are generic and provide many algorithms, while others are more specialized. This includes not only simulators [54], but also packages for graphical depiction and analysis of networks [43, 55, 56], and databases of reactions and kinetic parameters [57], to name but a few examples. In some cases these packages can even work in an integrated way, such as in the Systems Biology Workbench (SBW) suite [58].

Another area of cell physiology where mathematical modeling has been used to describe complex molecular-scale dynamics is the study of ion channels. Molecules (both large and small) move back and forth across a cell membrane, to ensure that conditions for homeostasis are met [42]. Some molecules are small enough (and soluble to lipids) to diffuse across the membrane, while others require energy, working against electrochemical gradients between the outside and the inside of the cell [42]. For example, differences in ionic potential across a cell membrane can drive ionic current. The Nernst equation,

$$V = \frac{RT}{zF}\ln\left(\frac{c_e}{c_i}\right), \quad (18)$$

describes the potential difference V across a cell membrane, where c_e and c_i are the external and internal ionic concentration (of a particular ion) respectively, R is the universal gas constant, Tis the absolute temperature, F is Faraday's constant, and z is the ion's charge. To determine the ionic current across a membrane, one can write the total potential drop across the cell membrane, V_T, as the addition of the Nernst potential V to the potential drop due to an electrical current rI_c (where r is the resistance), so that $V_T = V + rI_c$. Solving for the ionic current we arrive at

$$I_c = g(V_T - V), \qquad (19)$$

where g = 1/r is the membrane conductance. Ions only travel through small pores, referred to as ion channels, and the control of such ionic current is vital for proper cellular function.

A second model for determining the ionic flow across a cell membrane is given by the Goldman–Hodgkin–Katz equation,

$$I = P\frac{z^2F^2}{RT}V\frac{c_i - c_e\exp\left(\frac{-zFV}{RT}\right)}{1-\exp\left(\frac{-zFV}{RT}\right)}, \qquad (20)$$

where P is the permeability of the membrane to the ion [59]. Such an equation is derived under the assumption of a constant electric field. The use of either the linear equation (19) or nonlinear equation (20) in defining ionic current is often debated and depends on the underlying properties of the particular cell studied.

Using the fact that we can model the cell membrane as a capacitor (since it separates charge), and knowing that there is no net build-up of charge on either side of the membrane, the sum of the capacitive and ionic currents across a membrane should add up to zero,

$$C_m\frac{dV}{dt} + I_{\text{ion}}(V,t) = 0. \qquad (21)$$

Here, C_m is the capacitance, I_{ion} is the ionic current, and $V = V_i - V_e$. In 1952, through a series of pioneering papers, Hodgkin and Huxley presented experimental data and a comprehensive theoretical model that fit experimental findings, to describe the action potential across the giant axon of a squid [60–64]. The model given by equation (22) (and based on equation (21)), was awarded the Nobel Prize in Physiology and Medicine in 1963, and is possibly one of the greatest mathematical results in physiology to date:

$$c_m\frac{dV}{dt} = -g_{\text{Na}}(V-V_{\text{Na}}) - g_{\text{K}}(V-V_{\text{K}}) - g_{\text{L}}(V-V_{\text{L}}) + I_{\text{app}}. \qquad (22)$$

In equation (22), g_{Na}, g_K, and g_L describe the sodium, potassium, and leakage conductance (other ionic contributions, including the chloride current, that are small), respectively, V_{Na}, V_K, and V_L are their respective resting potentials, and I_{app} is a small applied current. Hodgkin and Huxley were able to measure the individual ionic currents, and to use this information to determine the functional forms for each of the conductances.

Much of the work completed on ion channels and cell excitability has been used to study diseases that are associated with malfunction of ion channels. As a result, such channels have become new targets for drug discovery [47]. One example of a disease caused by the disruption of the action potential of cardiac myocytes is cardiac arrhythmia. Certain drugs used in the treatment of arrhythmias, such as lidocaine and flecainide, are sodium channel blockers, and so interfere with open sodium channels. Although these drugs have been used in treatment for cardiac arrhythmias, their exact mode of action is not well understood. Current computational models are being developed to understand the function of these, as well as other anti-arrhythmia drugs [65]. Another example of a disease caused by the disruption of ion channels is cystic fibrosis (CF), which has been found to be associated with malfunctions in chloride channel operation [66]. Although there is still no cure for CF, new directions for treatment protocols are being developed [67].

Models of cancer growth and spread: avascular tumor growth, tumor-induced angiogenesis, and tumor invasion

The primary mathematical modeling techniques used in the area of cancer modeling, include the study of avascular tumor growth, angiogenesis, and vascular tumor growth (tumor invasion), and so in what follows we limit ourselves to these topics.

Growth and development of a solid tumor occurs in two stages: avascular growth and vascular growth (leading to tumor invasion), and has been studied extensively in a biological framework [68]. The modeling of avascular tumor growth is one of the earliest approaches taken by mathematicians to study cancers [69]. The simplest type of model that can be used to describe how cancer cells of a solid tumor change over time is the exponential growth law, given by equation (23).

$$\frac{dN}{dt} = rN \qquad (23)$$

Such an equation is limited in its application, since it suggests that tumors grow (with growth rate r) to a tumor of unbounded size. Other models, such as the logistic growth equation (24)

$$\frac{dN}{dt} = rN\left(1-\frac{N}{k}\right) \qquad (24)$$

have been used to describe tumor size saturation (a type of volume constraint). In particular, equation (24) describes the rate of change of cancer cells N in a tumor, where r is the growth rate and k is the carrying capacity (the maximum number of cancer cells within the tumor).

A limiting case of the logistic equation is the Gompertz model, given by equation (25).

$$\frac{dN}{dt} = r\log\left(\frac{k}{N}\right)N \qquad (25)$$

This model is one of the most commonly used tumor growth models to date, and was first used by Casey to fit real tumor growth in 1934 [70]. Such ODE-based models are useful because they can be easily analyzed. Also, these models have the advantage that they can be expanded to incorporate other cell types (such as proliferating, quiescent, and dead cells) by the inclusion of more ODEs, one for each cell type. Since these models are similar to the compartmental SIR model described in the section concerning epidemiology, they have similar limitations, one being a lack of spatial information (such as the location of each cell type within a tumor). Some models for solid tumor growth have included such information by incorporation of reaction-diffusion type equations, like equation (26). This latter model, studied by Murray [26], was developed to describe the spatio-temporal invasion of gliomas. Equation (26) is read as follows: cancer cells C grow exponentially with rate ρ, and diffuse at a rate that depends on whether cells are moving in white brain matter or gray brain matter (diffusion given by D(x)).

$$\frac{dC}{dt} = \nabla\cdot(D(x)\nabla C) + \rho C \qquad (26)$$

Other models by Cruywagen et al.[71] and Tracqui et al.

[72] have used a more realistic logistic growth term for cancer cell growth. However, the survival times calculated for these models are only slightly different from those calculated using model (26), and so using either logistic growth or exponential growth is appropriate in this modelling framework. One limitation of the Cruywagen et al. [71] and Tracqui et al. [72] is that they are constructed under the assumption of constant diffusion, thus neglecting to distinguish between grey and white matter in the brain. Simulation of models similar to (26), such as those completed by Swanson et al.[73], show that incorporation of a spatially dependent diffusion coefficient D(x) produces images that are in good agreement with real images of gliomas produced by magnetic resonance imaging (MRI).

Other spatial models, like that of equation (27), describe the movement of certain chemical signals, such as growth factors (GFs), which dictate directions for tumor growth [74, 75].

$$\frac{dC}{dt} = D\nabla^2 C - \gamma C + \sigma S(\boldsymbol{r}) \tag{27}$$

In equation (27), the first term on the right-hand side describes the diffusion of the concentration of GFs denoted by C, while the second term describes the degradation of C with rate γ. The third term describes production of C at a rate σ by a source $S(\boldsymbol{r})$. The source is a function of the tumor radius $\boldsymbol{r}$. Some models incorporate integro-differential equations that describe the radius of tumors over time [75], providing a type of spatial information for each cell type within the tumor. For example, the nutrient concentration often dictates where each type of cell is most likely located within the tumor. Thus, in a radially-symmetric cell, there are typically zero nutrients located at the center of the tumor, and so cells there are dead, defining a necrotic core, whereas the outermost layer is generally nutrient rich, and so it is composed of proliferating cells. Other models, such as those incorporating systems of PDEs, are used to describe the spatial movement over time of proliferating and quiescent cancer cells in the presence of GFs [76].

Typically, solid tumors initially grow to about 2 mm in size. To grow larger, tumors require nutrients. Such nutrients are additionally acquired through the process of angiogenesis, the formation of blood vessels that connect the tumor to the circulatory system [68]. The switch to this angiogenic stage can occur due to multiple factors, many of which are outlined in Semenza [77]. Tumor-induced angiogenesis, caused by the release of GFs or tumor angiogenic factors (TAFs) from the tumor, promotes the growth of endothelial cells (EC), which make up the linings of capillaries and other vessels. ECs then migrate towards the TAF source by chemotaxis. In other words, blood vessels grow towards the tumor.

The first model (given by equation (28)) to describe chemotaxis in a biological system using PDEs was developed by Keller and Segel [78]. Such an equation was initially developed to describe the chemotactic movement of slime molds towards an attracting source called cyclic adenosine monophosphate (cAMP):

$$\frac{du}{dt} = \nabla D \nabla u - \chi u \nabla v \tag{28}$$

In the context of cancer cells, the first term on the right-hand side of equation (28) describes random motion (diffusion) D of a cell population u, while the second term corresponds to chemotaxis of the cells u towards a GF. The chemotactic effect is given by χ and υ is the concentration of the GF. The dynamics of the growth factor is typically modeled by reaction-diffusion equations similar to that given above in equation (27).

Many other models, typically PDE-type models, have extensions similar to those described above, incorporating not only the key interactions of the endothelial cells with angiogenic factors, but also the macromolecules of the extracellular matrix [79, 80]. For an excellent review on mathematical approaches to studying tumor-induced angiogenesis we refer the reader to Mantzaris, Webb, and Othmer [81].

After a tumor has progressed to the stage of angiogenesis, by successfully recruiting blood vessels, the tumor eventually becomes vascular, thus connecting itself to the circulatory system [68]. Once this happens, tumor cells have the ability to enter the circulatory system, depositing themselves at secondary locations within the body and possibly promoting the growth of a secondary tumor, a process referred to as metastasis [68]. As soon as this occurs the cancer has progressed to the point where it is nearly impossible to cure the patient. One of the key steps in the progression to metastasis is the degradation of the extracellular matrix (ECM) by matrix-degrading proteins deposited by cancer cells, and the movement of cancer cells through the ECM. Newer mathematical approaches, using integro-PDEs, have been able to capture the qualitative movement of such cells through the ECM [82–84]. One such model was proposed by Hillen [82].

$$\frac{dp(\boldsymbol{x},t,v)}{dt} + v\cdot\nabla p(\boldsymbol{x},t,v) = -\mu \int_V \frac{q(\boldsymbol{x},t,\tilde{v})}{w} p(\boldsymbol{x},t,\tilde{v})d\tilde{v} \tag{29}$$

Here, $p(\boldsymbol{x}, t, v)$ describes the density of tumor cells at location $\boldsymbol{x}$, time $t > 0$, and velocity v in $V = [v_{min}, v_{max}]$. The advection term on the left-hand side of equation (29) describes directed movement of cells along collagen fibers in the ECM with speed v. The right-hand side describes changes in the cell's velocity due to changes in the matrix orientation $q(\boldsymbol{x}, t, v)$, where w is an appropriate weighting function. This matrix can change over time due to cuts made by matrix degrading proteins. For ease in reading we do not give the evolution equation for the matrix distribution q.

Many different cancer treatment modalities are available, including the administration of chemotherapeutic drugs and their combinations [85] and radiation treatment [86]. Also, as is often the case with avascular tumors, the solid tumor may be surgically removed, if discovered early enough. With the advancement in imaging techniques, much work has been done to extend earlier modelling of glioma to better model tumor growth as it applies to treatment (removal and radiation). In particular, better models for predicting survival rates have been developed [87], as well as models that predict efficacy of radiotherapy for individual patients [88]. Typically, MRI is used to detect tumors, and provides the imaging information needed for validating mathematical models. Other imaging techniques, such as Diffusion Tensor Imaging (DTI) (a technique which measures the anisotropic diffusion of water molecules in a tissue), have been used to better establish diffusion parameters required for models to predict the appropriate boundary of a tumor. Such information can be used to describe the appropriate boundary required for either surgical removal of a tumor or radiation treatment [89, 90]. Even though many treatment protocols exist, the probability of individuals surviving advanced stages of cancer is very low. As a result, many questions still remain as to how chemotherapeutic drugs work at a molecular level (e.g., which proteins they target by design and which by accident), and how radiation treatments should be delivered in order to maximize the death of cancer cells, while minimizing the harm to normal healthy tissue. Mathematical and computational modeling plays an important role in understanding radiation treatment protocols. For example, many models have been developed to define a tumor control probability (TCP) [91], where TCP is defined as the probability that no clonogenic cells survive radiation treatment [92]. TCP models have been incorporated into existing cancer models that describe cancer growth without treatment, such as those described above, to better understand the effects of radiation treatment on normal tumor dynamics [93].

The models discussed above describe the spatial and temporal changes of certain quantities that are of interest in various biological systems. In particular, the differential equations described above give temporally dependent solutions (and spatially dependent solutions in the case of the PDEs described) for various quantities, including the total populations of individuals, the total number/density of cells, or the total molecular concentration of a certain compound. Many of the successes and limitations of a differential equation modeling approach are highlighted above. One limitation, not highlighted in the above sections, is that such methods (those that use only a handful of differential equations) are not appropriate for describing the smaller scale movements of molecules. The movement and structure of an individual molecule is based on the many complex interactions between the individual atoms within a molecule, as well as its interactions with surrounding molecules. In order to follow the motions of every atom and molecule over extremely small timescales, computational methods such as molecular dynamic simulations (designed to solve extremely large systems of differential equations over very small timescales) can be applied. This technique is described in the following section.

MOLECULAR DYNAMICS

The sophistication of the model used to study a given system depends on the property of interest. Often, a 3D model of a molecule or complex that shows the spatial relationships between atoms is the best way to understand a system. Such computational models provide a means of observing the structure and motion of individual atoms within complex biomolecular systems. Although a physical model of a small molecule with less than 20 atoms can be easily made from plastic or wire in a few minutes, a similar model of a protein or an enzyme involves hundreds or thousands of atoms. Over the last decade improvements in a combination of computer graphics programs, and molecular modeling techniques and hardware have resulted in an unprecedented power to create and manipulate 3D models of molecules.

Molecular dynamics (MD) simulations follow the motions of atoms and molecules, and provide a means of investigating biological problems at a molecular level. This is achieved by solving Newton's equations of motion (EOM) for interacting atoms and evolving a system through time and space. Changes in atomic positions and velocities over time, usually ranging from nano- to milliseconds, result in a trajectory. In simple cases with few atoms, analytic solutions to Newton's EOM can be obtained, giving a trajectory that is a continuous function of time (Figure 1). However, in a computer simulation with many atoms, the EOM are solved numerically. Forces are evaluated for discrete intervals, or time steps (Δt, Figure 1) on the order of femtoseconds, where the forces are considered constant over a given time step. The goal is to follow the continuous function of time as closely as possible, which requires small time steps to ensure the motions of all atoms are resolved.

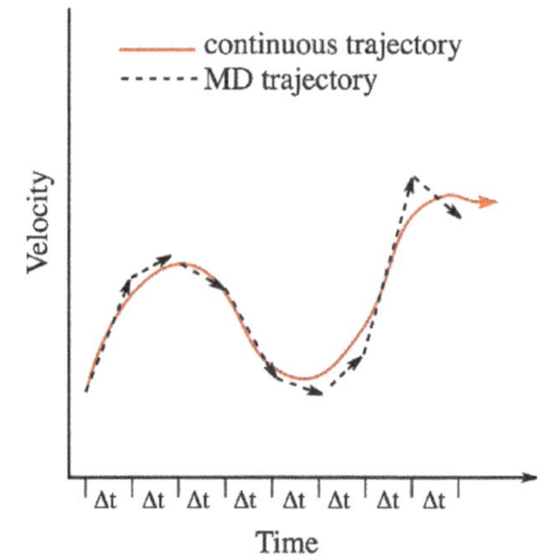

Figure 1: The evolution of a trajectory, showing the continuous, true trajectory (red), being closely followed by the MD trajectory (black).

The forces on each atom are derived from the potential energy of the system, which can be described with quantum or classical mechanics. Since quantal descriptions are generally limited to small systems, classical descriptions are commonly used when studying biological systems and will be discussed in this section. It is worth noting that MD is a deterministic approach for exploring the potential energy surface of a system, while a stochastic approach can be obtained using Monte Carlo methods.

MD trajectories are analyzed to obtain information about the system, including structural changes as measured by atomic root-mean-square deviation (RMSD), non-covalent interactions, binding free energies [94], structural stability, short-lived reaction intermediates [95], conformational changes, flexibility, ligand binding modes [96], as well as ionic conductivity and mobility [97]. Numerous and diverse applications include the investigation of clinically important proteins such as HIV-1 gp120 [98], protein binding sites [99], drug resistance mechanisms of HIV-1 protease [100], protein folding [101, 102] and the role of crystal water molecules in ligand-protein binding [103].

Force fields for protein simulations

Newton's second law, $m\ddot{x} = -\nabla V$, establishes the relation between mass (m) and acceleration ($\ddot{x}$), as well as force ($-\nabla V$), which is the negative of the gradient of the potential energy function (V). During an MD simulation, the forces acting on each atom of the system are calculated and atoms are moved according to those forces. In a classical MD simulation, the potential is calculated with a force field. The potential terms in a force field are a sum of contributions due to covalently-bonded interactions (V_{bond}, V_{angle}, $V_{torsion}$) and non-bonded interactions.

Non-bonded interactions are calculated pairwise between two atoms, denoted i and j, and commonly include van der Waals ($V_{ij,\, vdW}$) and electrostatic ($V_{ij,electrostatic}$) contributions.

$$V = \sum_{bonds} V_{bond} + \sum_{angles} V_{angle} + \sum_{torsions} V_{torsion} + \sum_{i<j} V_{ij,vdW} + \sum_{i<j} V_{ij,electrostatic} \quad (30)$$

Due to the pairwise calculation of non-bonded interactions, such force fields scale as N^2, whereN is the number of atoms in the system. Each potential term contains parameters, where fitting to experiment is required to calculate these interactions [104]. Parameter fitting is done to reproduce the behavior of real molecules. This includes determining the van der Waals' radii, partial charges on atoms, bond lengths, bond angles and force constants. These parameters, along with the functional form of each potential term, collectively define a force field. Today, several types of force fields are available: (a) all-atoms force fields (parameters are considered for every atom), (b) united-atoms force fields (aliphatic hydrogen atoms are represented implicitly) and (c) coarse-grained force fields (groups of atoms are treated as super atoms). For a list of all force fields discussed, see Table 1.

Table 1: List of force fields

Name	Family	Type	Reference
ff94	Amber	All-atoms	[104]
ff99	Amber	All-atoms	[107]
ff99SB	Amber	All-atoms	[106]
ff03	Amber	All-atoms	[112,113]
ff03.r1	Amber	All-atoms	
ff03ua	Amber	United-atoms	[114]
ff12SB	Amber	All-atoms	[110]
ff14SB	Amber	All-atoms	
GAFF	Amber	All-atoms	[115]
CHARMM	CHARMM	United-atoms	[117]
CHARMM19	CHARMM	United-atoms	[118]
CHARMM22	CHARMM	United-atoms	[119]
CHARMM22/CMAP	CHARMM	United-atoms	[120]
CHARMM27	CHARMM	United-atoms	[116]
CGenFF	CHARMM	United-atoms	[121]
OPLS-UA	OPLS	United-atoms	[117]
OPLS-AA	OPLS	All-atoms	[122]
GROMOS (A-version)	GROMOS	United-atoms	[125]
GROMOS (B-version)	GROMOS	United-atoms	[125]
43A1	GROMOS	United-atoms	[126]
45A3	GROMOS	United-atoms	[127]
53A5	GROMOS	United-atoms	[128]
53A6	GROMOS	United-atoms	[128]
54A7	GROMOS	United-atoms	[129,130]
54B7	GROMOS	United-atoms	[129,130]
54A8	GROMOS	United-atoms	[129,130]
MARTINI	MARTINI	Coarse-grained	[131-134]

Most all-atom force fields for proteins use relatively simple functions for modeling the potential energy surface [105], which correspond to the terms in Equation (30):

$$V = \sum_{bonds} K_b(b-b_0)^2 + \sum_{angles} K_\alpha(\alpha-\alpha_0)^2 + \sum_{torsions} K_\phi[\cos(n\phi+\varphi_0)+1] + \sum_{i<j}^{N}\left(\frac{A_{ij}}{r_{ij}^{12}} - \frac{B_{ij}}{r_{ij}^{6}}\right) + \sum_{i<j}^{N} \frac{q_i q_j}{4\pi\varepsilon_0 r_{ij}} \quad (31)$$

The variables are indicated in Figure 2. The first term includes the stretching of bonds, where the potential is calculated using parameters K_b (related to force constant) and b_0 (related to equilibrium bond length), as well as the variable, b, which is the distance between bonded atoms. Similarly, the second term includes angle bending, where parameters K_α and α_0, represent force constants and equilibrium angles, and the variable (α) is the angle obtained from the structure of interest. The third term, representing the potential from dihedral angles, involves parameters K_Φ, (barrier for rotation), n (number of maxima) and φ_0 (angular offset), as well as the variable φ, obtained from dihedral angles in the structure. The final two terms in Equation (31) represent non-bonded interactions between atoms i and j, which are summed over N atoms of the system. In this potential function, the van der Waals term is represented by the Lennard-Jones potential, where A_{ij} and B_{ij} are atom specific parameters related to atom size, and r_{ij} is a variable representing the distance between atoms i and j. The electrostatic term in Equation (31) is calculated by a Coulomb potential, where parameters q_i and q_j are (fixed) charges on atoms i and j, respectively, and the constant ϵ_0 is the permittivity of free space. The force constants are obtained by empirical methods or by quantum mechanics calculations, depending on the force field. Here we present a summary of the most commonly used force fields in MD simulations, namely Amber, CHARMM, OPLS, GROMOS and MARTINI.

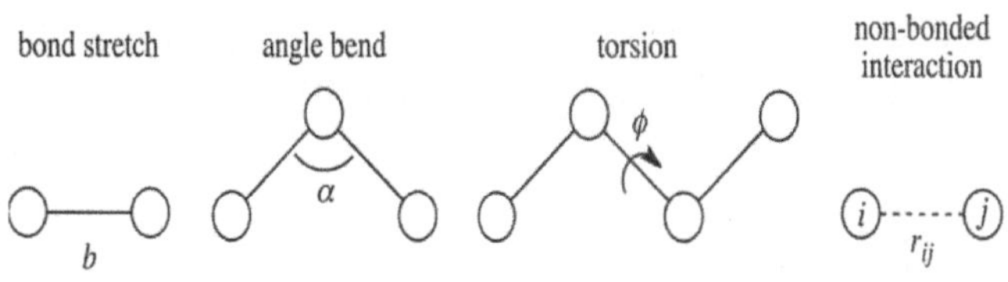

Figure 2: An illustration of the variables involved in a basic all-atom force field, corresponding to Equation (31).

Amber force fields

The functional form from which most of the Amber (Assisted Model Building with Energy Refinement) force fields come uses Equation (31) and was developed by Cornell and co-workers (denoted ff94) [104]. However, various revisions have since been developed that vary in their parameterization, with the goal of improving results. One notable revision of Amber for proteins and nucleic acids is ff99SB [106], which was developed at Stony Brook University as a modification of the old ff99 force field [107] and improves on ff99 in its description of the φ and ψ dihedral angles of the protein backbone, resulting in a better balance between secondary structures, and improved treatment of glycine [108, 109]. The ff12SB revision [110] reparameterizes backbone torsion angles, side chain torsions in select amino acids, and incorporates improved backbone torsions in DNA and RNA, with recent studies finding ff12SB performs better than ff99SB [111]. The current and most recent version, ff14SB, is recommended by Amber developers and minimizes dependencies of protein side chain conformations on backbone conformations by including side-chain corrections, and improves upon dihedrals in DNA and RNA, particularly χ. Another extensively-used Amber force field is ff03 [112, 113] (the latest version is ff03.r1), which improves upon the charges calculated by ff99 by using charges derived from quantum calculations with a continuum dielectric to simulate the solvent polarization. A united-atom version of ff03, ff03ua, is also available [114]. Expanding the utility of the Amber force fields beyond peptides and nucleotides is the general Amber force field (GAFF) [115], which includes a complete set of parameters for a large number of small molecules while still remaining fully compatible with the other versions of the Amber force fields discussed above.

This allows for Amber and GAFF to be combined to examine protein-ligand complexes, as well as modified proteins, DNA or RNA.

CHARMM force fields

The CHARMM (Chemistry at Harvard Macromolecular Mechanics) force fields are also a prominent set of force fields for studying biological systems. The functional form of the force field used by CHARMM is based on Equation (31), but also includes additional terms to treat improper torsions and atoms separated by two bonds (Urey-Bradley term) [116]. The CHARMM force fields use classical (empirical or semi-empirical) and quantum mechanical (semi-empirical or ab initio) energy functions for different types of molecular systems. They include parameters for proteins, nucleic acids, lipids and carbohydrates, allowing simulations on many common biomolecules. The initial version of the CHARMM force field was developed in the early 1980s, and used an atom force field with no explicit hydrogens [117]. However, in 1985, CHARMM19 parameters were developed in which hydrogen atoms bonded to nitrogen and oxygen were explicitly represented; hydrogens bonded to carbon or sulfur were still treated as extended atoms [118]. CHARMM19 parameters aimed to provide a balanced interaction between solute-water and water-water energies. Although this force field was tested primarily on gas-phase simulations, it is now used for peptide and protein simulation with implicit solvent models. Newer versions of CHARMM, such as CHARMM22, include atomic partial charges that are derived from quantum chemical calculations of the interactions between model compounds and water [119]. Although CHARMM22 is parameterized for the TIP3P explicit water model, it is frequently used with implicit solvents. CHARMM27 parameters were developed for nucleic acids (RNA, DNA) and lipid simulations [116]. Since both CHARMM27 and CHARMM22 are compatible, it is recommended CHARMM27 be used for DNA, RNA and lipids, while CHARMM22 should be applied to protein components [116]. A more recent, dihedral-corrected version of CHARMM22 was developed, denoted CHARMM22/CMAP, which improves the parameters describing the φ and ψ dihedral angles of the protein backbone [120]. A general version of the CHARMM force field (CGenFF) also exists which allows to the treatment of drug-like while maintaining compatibility with other the CHARMM force fields [121].

OPLS force fields

For the OPLS (Optimized Potentials for Liquid Simulations) family of force fields the form of the potential energy function differs from Equation (28) in the dihedral term, and was parameterized by Jorgensen and co-workers [117]. OPLS force fields were parameterized simulate the properties of the liquid states of water and organic liquids [118]. For proteins, a united-atom version was first developed (OPLS-UA), followed by an all-atoms version (OPLS-AA) [122]. Charges and van der Waals terms were extracted from liquid simulations. The OPLS-AA force field uses the same parameters as the Amber force fields for bond stretching and angles. The torsional parameters were obtained by using data from ab initio molecular orbital calculations for 50 organic molecules and ions [123]. Several improvements and re-parameterizations were proposed in later years for this set of force fields [122, 124], including for simulations of phospholipid molecules [125].

GROMOS force fields

The GROMOS (Groningen Molecular Simulation) force fields are united-atom force fields that that were developed in conjunction with the software package of the same name to facilitate research efforts in the field of biomolecular simulation in a university environment [125]. Its functional form varies from Equation (31) in the dihedral term, and differs from other force fields in its goal of reproducing enthalpies of hydration and solvation. The initial GROMOS force field (A-version) was developed for applications to aqueous or apolar solutions of proteins, nucleotides and sugars. However, a gas phase version (B-version) for the simulation of isolated molecules is also available [125]. Important versions of the GROMOS force fields include GROMOS 43A1 [126] (improved treatment of lipid bilayers), GROMOS 45A3 [127] (relevant to lipid membranes and micelles) and GROMOS 53A5 and 53A6 [128] (recommended for the simulation of biomolecules in explicit water). Recent releases such as GROMOS 54A7 and 54B7 involve modifications to φ and ψ protein dihedral angles, new lipid atoms types, new ion parameters and additional improper dihedral types, while GROMOS 54A8 accurately models the structural of lipid bilayers, proteins and electrolyte solutions [129, 130].

MARTINI force fields

The semi-empirical MARTINI force field is the most commonly used coarse-grained force field for biomolecular system simulations; it was originally developed for lipid simulations [131]. In this potential energy surface, four heavy atoms in a molecule are considered as a single interaction site, and only four types of interaction have been considered, namely polar, non polar, apolar and charged; moreover, each particle has different subtypes for taking in account the underlying atomic structure; a special particle types has been introduced for the ring conformations, as the four-to-one mapping is not appropriate to represent small ring molecules. Initially developed for lipid simulations, this force field includes now extensions for the parameterization of proteins [132], carbohydrates [133] and more recently glycolipids [134]. With the approximations introduced by the coarse grain approach it is possible to increase both the timescale and the size of systems compared with all-atoms or united-atoms force field simulations. The list of fields where MARTINI was employed includes characterization of lipid membranes [135], protein-protein interactions [136], self assembly of peptides and proteins [137] and interactions between nanoparticles and biological molecules, as for example the study of the mechanism by which fullerene can penetrate the cellular lipid membrane performed by Wong-Ekkabut et al. [138] and other applications [139]. However, MARTINI cannot be used for protein-folding studies [139], as the secondary structure is a required input parameter.

Calculation of solvation free energies

The calculation of solvation free energies is a challenging problems in MD simulations. Determining solvation free energy is especially difficult in aqueous bio-systems due to the size of the system [140]. Solvation free energy, ΔG_{solv}, is a thermodynamic property defined as the net energy change upon transferring a molecule from the gas phase into a solvent with which it equilibrates [141]. Solvation effects can change the physical and chemical properties of biomolecules including charge distribution, geometry, vibrational frequencies, electronic transition energies, NMR constants and chemical reactivity [142].

Several methods have been developed for modeling solvation and one can select the most advantageous choice among them based on the required accuracy and computational cost. To simulate effects of solvent on biomolecules, one can use explicit or implicit solvent models. While explicit solvent models include solvent molecules in the system, implicit models use a mean field approach [143, 144]. Although explicit solvent simulations are computationally expensive because of the enormous numbers of atoms involved, they provide a more realistic picture of solute-solvent interactions, reflecting the molecular complexity of the biomolecule and its environment. In comparison, implicit solvent models increase the speed of the simulation since the Newtonian equations of motion are not solved for additional solvent molecules. Table 2 lists the solvation models discussed.

Table 2: List of solvation models

Name	Type	Reference
Poisson–Boltzmann (PB)	Implicit	[145,146]
Generalized Born (GB)	Implicit	[140,141,147,149-151]
3D-RISM	RISM	[140,152-158]
3D-RISM-KH	RISM	[156,164]
MTS-MD/OIN/ASFE/3D-RISM-KH	RISM	[149]
3D-RISM-KH-NgB	RISM	[168]
SPC	Explicit	[174]
SPC/E	Explicit	[174]
POL3	Explicit	[175]
TIP3P	Explicit	[176]
TIP3P/F (TIP3P-PME/LRC)	Explicit	[177]
TIP4P	Explicit	[176,178]
TIP4P/Ew	Explicit	[179]
TIP5P	Explicit	[180]

Implicit Water Models

The simplest approach to solvation is to treat the effects (e.g. electrostatic interactions, cavitation, dispersion attraction and exchange repulsion) of solvent on the solute with an implicit model. These methods represent the solvent as a continuum environment, where the quality of the results is most affected by the electrostatic and cavitation (the size and shape of a cavity that the solute occupies) contributions. The solvation free energy of a molecule, ΔG_{solv}, can be divided into two parts: electrostatic (ΔG_{el}) and non-electrostatic (ΔG_{nonel}). The electrostatic energy is defined as the free energy required to remove all the charges in vacuum and add them back to the solute in the presence of continuum solvent [140, 141]. The origin of the non-electrostatic energy is a combination of favorable solute-solvent van der Waals interactions and the unfavorable disruption of the water structure by solute molecules (cavitation), and corresponds to solvating the neutral solute. There are several different implicit solvent models discussed below: the Poisson–Boltzmann model and the generalized Born model, which differ in how ΔG_{el} is obtained.

Poisson–Boltzmann model

Solving Poisson's equation, which is valid under conditions where ions are absent, gives a second order differential equation describing the electrostatic environment that is modeled with a dielectric continuum model [145],

$$\nabla \cdot \varepsilon(r) \nabla \phi(r) = -4\pi\rho(r), \tag{32}$$

where $\varphi(r)$ is the electrostatic potential, $\epsilon(r)$ is the dielectric constant and $\rho(r)$ is the charge density.

Poisson's equation cannot be solved analytically for most systems, and must be solved using computers and adopting numerical methods. The Boltzmann contribution, along with the assumptions of the Debye–Hückel theory, describes the charge density due to ions in solution. This results in the (non-linearized) Poisson–Boltzmann (PB) equation [146]:

$$\nabla \cdot \varepsilon(r) \nabla \phi(r) - \frac{\varepsilon_s kT}{e} S(r) \kappa^2 \sinh[e\phi(r)/kT] = -4\pi\rho(r), \tag{33}$$

where κ denotes the Debye–Hückel parameter, ϵ_s is the solvent dielectric constant, S(r) is a "masking" function with value 1 in the region accessible to the ions in the solvent and value 0 elsewhere; e is the protonic charge; k is Boltzmann's constant; T is the absolute temperature. Here, the charge density on the right represents the partial charges in the cavity. When the ionic strength of the solution or the potential is low, Equation (33) can be linearized by expanding the second term on the left into a Taylor series and retaining only the first term:

$$\nabla \cdot \varepsilon(r) \nabla \phi(r) - \varepsilon_s \kappa^2 S(r) \phi(r) = -4\pi\rho(r), \tag{34}$$

The non-electrostatic contribution (ΔG_{nonel}) to the solvation free energy is calculated by empirical methods and is proportional to the solvent accessible surface area. This is added to the electrostatic part to yield the solvation free energy. Although the PB approach is mathematically rigorous, it is computationally expensive to calculate without approximations [141, 147, 148]. The generalized Born model provides a more efficient means of including solvent in biomolecular simulations.

Generalized born model

The generalized Born (GB) model is based on the Born approximation of point charges, modeling solute atoms as charged spheres with an internal dielectric (generally equal to 1) that differs from the solvent (external) dielectric. The polarization effects of the solvent are represented by a dielectric continuum represents the polarization effects of the solvent. Numerical methods are used to determine the charges on the solute spheres that result in the same electrostatic potential on the cavity surface that mimics that of the solute in a vacuum.

By making approximations to the linear Poisson-Boltzmann equation (Equation 31), the electrostatic contribution of the generalized Born model is obtained:

$$\Delta G_{el} = -\frac{1}{2}\left(\frac{1}{\varepsilon_{int}} - \frac{1}{\varepsilon_{ext}}\right) \sum_{ij} \frac{q_i q_j}{\left[r_{ij}^2 + \alpha_i \alpha_j \exp\left(-r_{ij}^2 / 4\alpha i \alpha j\right)\right]^{1/2}}, \tag{35}$$

where α_i is effective Born radius of particle i, r_{ij} is the distance between atoms i and j, ϵ_{int} and ϵ_{ext} the internal and external dielectric constants, respectively, and q_i is the electrostatic charge on particle i. Like the PB method, ΔG_{nonel} is calculated from the solvent-accessible surface area [140, 141, 147, 149–151].

Reference interaction site model

Another type of solvation model is a probabilistic method known as the 3D reference interaction site model (3D-RISM) [140, 152–158]. This molecular theory of solvation simulates the solvent distributions rather than the individual solvent molecules. However, the solvation structure and the associated thermodynamics are obtained from the first principles of statistical mechanics.

In this method, the 3D site density distributions of the solvent are obtained, which accounts for different chemical properties of the solvent and solute. These properties include hydrogen bonding, hydrophobic forces and solvation thermodynamics, such as the partial molar compressibility and volume. In addition, the solvation free energy potential and its energetic and entropic components can be calculated. The solvation free energy is calculated from the RISM equation as well as the closure relation [159–163].

Several additional advances have been made in formulating improved versions of the 3D-RISM theory including the hypernetted chain (HNC) closure approximation [152, 153]. Another derivation came from the molecular Ornstein–Zernike integral equation [163] for the solute-solvent correlation functions [155, 156, 164]. Sometimes the calculated solvation free energy for ionic and polar macromolecules involves large errors due to the loss of long-range asymptotics of the correlation functions. Work has been done to account for the analytical corrections of the electrostatic long-range asymptotics for the 3D site direct correlation functions as well as the total correlation functions [157, 158, 164]. Other developments include the closure approximation, 3D-RISM-KH closure, for solid-liquid interfaces, fluid systems near structural and phase transitions, as well as poly-ionic macromolecules [156, 164].

Two methods have been developed to couple 3D-RISM with MD. The first method makes use of a multiple time step (MTS) algorithm [165, 166] wherein the 3D-RISM equations are solved for a snapshot of the solute conformation, then solved again after a few MD steps. This method is limited by the requirement to re-solve the 3D-RISM equations every few MD steps, which is computationally expensive for large biomolecular systems. The second of these methods involves the contraction of the solvent degrees of freedom and the extrapolation of the solvent-induced forces. These methods are aimed at speeding up the calculations, which is useful for larger systems [149].

Other work has involved the development of the multi-scale method of multiple time steps molecular dynamics (MTS-MD) in a method referred to as MTS-MD/OIN/ASFE/3D-RISM-KH [149]. Specifically, this method converges the 3D-RISM-KH equations at large outer time steps and uses advanced solvation force extrapolation to calculate the effective solvation forces acting on the biomolecule at inner time steps. The integration between the inner and outer time steps is stabilized by the optimized isokinetic Nosé–Hoover chain (OIN) ensemble, which enables an increase of the outer time step. Furthermore, effort was expended on MTS-MD aimed at converging the 3D-RISM-KH integral once every few OIN outer time steps, and the solvation forces in between were obtained by using solvation force-coordinate extrapolation (SFCE) in the subspace of previous 3D-RISM-KH solutions [167]. Another developed model is the 3D-RISM-KH-NgB [168]. In this model the non-polar component of the hydration free energy obtained from 3D-RISM-KH is corrected using a modified Ng bridge function [169]. Calibration of this model is based on the experimental hydration free energy values of a set of organic molecules.

Lastly, work has been done to improve the performance of 3D-RISM calculations by running them on graphical processing units (GPUs). To overcome memory issues, a modification of the Anderson Method [170] that accelerates convergence was introduced [171]. This method was reported to be eight times faster on an NVIDIA Tesla C2070 GPU as compared to the time taken on an eight-core Intel Xeon machine running at 3.33 GHz.

Although 3D-RISM calculations are more computationally expensive than GB and PB based solvation methods [172], they overcome some of the inherent shortcomings of these empirical methods [173].

Explicit water models

Explicit solvation is characterized by modeling individual water molecules around a solute. Several explicit water models are available in the Amber, NAMD and Gromacs MD simulation packages. These include simple explicit solvent models (SPC [174], SPC/E [174]), polarizable models (such as POL3 [175]), and fixed-charge explicit solvent models (TIP3P [176], TIP3P/F [177], TIP4P [176, 178], TIP4P/Ew [179] and TIP5P [180]).

Examples of explicit water models are the simple point charge (SPC) model and the extended simple point charge (SPC/E) model [174]. In both of these models the water molecules are rigid. A derivative of SPC with flexible water molecules has been developed [181]. Another simple explicit model is the POL3 water model, which is a polarizable model [175].

More complex explicit water models include the transferable intermolecular potential n point (TIPnP), where n represents the number of interaction sites on each model. These are the most common classes of explicit solvent models in use [147]. In the case of TIP3P, the most simple TIPnP model, the interaction sites includes the oxygen and two hydrogen atoms [176]. A re-parameterized model of TIP3P is the TIP3P-PME/LRC, also referred to as TIP3P/F [177], which calculates electrostatic contributions by particle mesh Ewald (PME) summation and includes a long-range van der Waals correction (LRC). TIP4P [176, 178] introduced a fourth dummy atom bonded to the oxygen to improve the electron distribution in the water molecule. This model has been re-parameterized for use with Ewald sums: TIP4P/Ew [176, 178]. The five interaction points in the TIP5P [180] model include two dummy atoms near oxygen, which further improves the charge distribution around the water molecule.

Molecular dynamics methods

Examining the dynamics of a system at an atomic scale requires beginning with a model having atomic-level resolution. For biological macromolecules, this may be experimentally obtained from nuclear magnetic resonance (NMR) spectroscopy or X-ray crystallographic data. Although electron microscopy data does not provide structures with atomic resolution, this data can be combined with structural data from NMR or crystallography to obtain a high-resolution structure. NMR, crystallographic and electron microscopy structures of bio-macromolecules can be downloaded from the Protein Data Bank (http://www.pdb.org). In the absence of experimental data, homology modeling may be used to generate the 3D structure of a protein (the target) using its amino acid sequence and an experimentally-available 3D structure of a homologous protein (the template). Homology modeling can produce high-quality structures of a target protein if the sequence identity of the target and template is sufficiently high (typically > 40%).

As with any computational modeling, performing MD simulations requires a balance between accuracy and efficiency. Ideally, a system should be allowed to evolve in time indefinitely so that all states of a system may be sampled. However, this is not possible in practice. Several techniques have been developed so improve simulation efficiency, while still maintaining accuracy. During a simulation, costly force calculations are performed in discrete time increments, known as time steps, typically on the order of femtoseconds. Although it is necessary to have small time steps to properly resolve the motion of atoms, this results in many force calculations and therefore large computational costs. To address this, one can restrain the fastest vibrations, which involve hydrogen, giving the SHAKE algorithm [182]. This allows a larger time step to be used during simulations [183]. Additionally, the number of time-consuming non-bonded potentials calculated can be limited by using a cut-off based method [184]. Here, interactions are calculated between pairs of atoms within the cutoff distance, but neglected for atom pairs that are far apart. Although this is generally appropriate for the short-range nature of van der Waals interactions, using cutoffs to calculate long-range Coulomb interactions leads to instabilities [185]. Alternatively, the particle mesh Ewald (PME) approach [185–188] can be used to calculate electrostatic potentials, which involves the calculation of the short range electrostatic component in real space and the long-range electrostatic component in Fourier space.

The simplest way to enhance sampling during an MD simulation is to increase the time duration of a simulation, which is usually on the order of nano- to milli-seconds. Researchers may also conduct multiple simulations that begin with the same initial structures [189], providing denser sampling of the conformational space by utilizing multiple trajectories. Advanced techniques also exist to increase the number of states of a system that are sampled during an MD run [190]. Enhanced sampling can be achieved by metadynamics [191], which allows for the high-energy regions between minima to be explored. This helps the system to escape local free energy minima and explore metastable states separated by large free energy barriers [192]. Metadynamics, which is used to calculate static properties, can also be used to calculate dynamic properties by introducing a history-dependent biasing potential as a function of a few collective variables. Selection of collective variables is an essential part of a metadynamics run as these variables help sample different energy basins. The use of metadynamics, which is a combination of ideas involving coarse-grained dynamics in space and the introduction of a history-dependent bias, can overcome the problem of limited time scale exploration by existing sampling algorithms and computational resources. Replica-exchange molecular dynamics (REMD) has also been preferred over standard MD to enhance sampling by allowing systems of similar potential energies to sample conformations at different temperatures [193, 194]. This overcomes the energy barriers on potential energy surfaces and helps explore more conformational space. REMD can effectively sample energy landscapes, including both high- and low-energy structures, which is especially important in the case of protein folding and unfolding processes [195].

In addition to obtaining structural information about a system, molecular dynamics provides information about energetics, particularly binding. Given the inherent flexibility of the these biomolecules over time, and throughout an MD simulation, properties such as the free energy are calculated as the time-average of an ensemble of snapshots obtained from MD trajectories [196–198]. Binding calculations have applications in drug design, protein-protein interactions and DNA stability. For example, the binding free energy of a ligand to a protein is calculated as the difference in free energy between the complex, and the receptor and ligand ($\Delta G_{bind} = G_{complex} - G_{receptor} - G_{ligand}$) [199]. The free energy is calculated as a sum of the energy contributions from the force field, free energy of solvation and entropy ($G = E_{MM} + G_{sol} + T\,\Delta S$). Commonly in MD, the solvation free energy is obtained from an implicit solvent model (Poisson-Boltzmann, generalized Born) and the solvent accessible surface area (denoted MM-PBSA and MM-GBSA, respectively) [199]. Decomposition of the binding free energies provides a means of obtaining information about the residues that significantly contribute to the binding affinity of a ligand. Pairwise decomposition may also provide insight into changes in binding that result from mutations, especially single point mutations [200].

Molecular dynamics is increasingly being used to solve a host of problems [201–207]. The simulation of bio-macromolecules, especially in conjunction with solvent, is very computationally demanding. This demand is being met by the increasing power and speed of modern computers, including special purpose computers such as Graphical Processing Units (GPUs) [208]. Novel methods have been developed to enhance the exploration of conformational space. The accuracy of force fields continues to improve with recent reparameterizations. A variety of solvation models exist, which can also be used to calculate binding free energies, which are immensely important in drug discovery applications, identifying enzyme inhibitors or compounds that block protein-protein interactions. Although classical MD simulations have been proven useful in studying a variety of biomolecular systems, they are limited in their application to stable structures. The examination of systems involving chemical reactions or quantum phenomena requires treatment using quantum mechanics methods, or the combination of quantum mechanics and molecular mechanics methods (denoted QM/MM). The application of quantum methods to biomolecules is discussed in the next section.

QUANTUM MECHANICS IN BIOPHYSICAL MODELING

Quantum mechanics (QM) calculations, being highly accurate and rigorous, are an essential tool in computational chemistry studies. Unfortunately, the prohibitive size of many biological systems has limited theoretical and computational studies of them to the realm of classical mechanics, largely utilizing non-polarizable force fields in MD simulations. This has necessarily reduced the scope of studies to conformational or structural aspects of these bio-systems, rather than more complex problems such as chemical reactions or quantum phenomena (excited states and charge-transfer, for example). Although hybrid quantum mechanics/molecular mechanics (QM/MM) approaches have increased accuracy, recent improvements in software, hardware and theory have allowed for full quantum mechanical studies of biochemical systems. The QM methods and functionals discussed are listed in Table 3.

Table 3: List of QM methods and functionals

Name	Type	Reference
Hartree–Fock (HF)	*ab initio* method	[209]
Møller–Plesset (MP2)	*ab initio* method	[209]
Coupled-cluster (CC)	*ab initio* method	[209]
CCSD(T)	*ab initio* method	[209]
CCSD(T)/CBS	*ab initio* method	[220]
MNDO	Semi-empirical method	[210-212]
AM1	Semi-empirical method	[210-212]
PM3	Semi-empirical method	[210-212]
PM6	Semi-empirical method	[210-212]
OMx	Semi-empirical method	[210-212]
B3LYP	DFT functional	[214]
PBE	DFT functional	[214]
TPSS	DFT functional	[214]
DFT-D2	DFT method	[215,218]
DFT-D3	DFT method	[215,218]
Effective fragment potential (EFP)	Fragmentation method	[241-243]
Fragment molecular orbital (FMO)	Fragmentation method	[244]
Elongation (ELG)	Fragmentation method	[245]
Divide and conquer (DC)	Fragmentation method	[246]

QUANTUM MECHANICS METHODS

The wavefunction, which contains all the information describing a quantum system, is obtained by solving the Schrödinger equation, usually in its time-independent, non-relativistic form, invoking the Born–Oppenheimer approximation.

$$\hat{H}\psi\left(\zeta_1,\zeta_2,...\zeta_N;\vec{R}_A,\vec{R}_B,...\vec{R}_M\right)=E\psi\left(\zeta_1,\zeta_2,...\zeta_N;\vec{R}_A,\vec{R}_B,...\vec{R}_M\right), \quad (36)$$

This wavefunction, ψ, is a function of N electronic coordinates and depends parametrically on M nuclear coordinates. The coordinates of an electron are denoted $\zeta_N = (x_N, y_N, z_N, \sigma_N)$, where x_N, y_N, z_N are the spatial coordinates of electron N, and σ_N is the spin of this electron. The spatial coordinates of nucleus M are indicated by $\vec{R}_M$.

The Hartree–Fock (HF) method is the simplest ab initio QM method for obtaining the wavefunction, where the N-electron wavefunction is approximated as a product of N one-electron wavefunctions, expressed in a Slater determinant. However, the HF theory is insufficient in describing electron correlation. Electron correlation is included in post-Hartree–Fock ab initio methods such as the Møller–Plesset perturbation theory (MP2) or coupled-cluster (CC) calculations, although at increased computational cost. While HF scales formally asK^4, where K is the number of basis functions, MP2 scales as K^5, and the gold-standard CCSD(T) scales as K^7. A detailed description of these methods, and many others, is available in theEncyclopedia of Computational Chemistry[209].

An alternative approach to including electron correlation is using semi-empirical quantum methods, where expensive two-electron integrals in the HF method are eliminated and the method is supplemented with parameters derived from experimental data. Although these methods are computationally very efficient, their applicability is generally limited to systems similar to the parameterization set. Many variants of semi-empirical methods exist, with applicability to biological systems [210–212], including MNDO, AM1, PM3, PM6, and the OMx methods.

The electron density, a physical observable, also determines all properties of a system, and it is this fact that is utilized in the density functional theory (DFT) [213].

$$\rho\left(\vec{r}_1\right)=N\int\left|\psi\left(\vec{r}_1,\vec{r}_2,\vec{r}_3,...\vec{r}_N\right)\right|^2 dr_2...dr_N \quad (37)$$

Since DFT methods scale as K^3, they are more efficient than the HF method while also containing electron correlation. Therefore, DFT has been successful in the study of a variety of biological systems [214]. Many functionals exist (for example, B3LYP, PBE and TPSS [214]), which vary based on the form of the exchange-correlation functional, and the performance of each functional is highly dependent on the system and properties of interest [213]. Even with the formal inclusion of electron correlation, many DFT functionals are inadequate in modeling dispersion forces, a dominant source of stabilization in bio-macromolecules. A common approach for improving the performance of DFT functionals is to add a dispersion correction, which can be derived either empirically or from high-level ab initio calculations [215–217]. For example, the DFT-D2 and DFT-D3 methods are popular [215, 218]. It is worth noting that the dispersion terms have also been applied to improve semi-empirical methods [210, 212, 217,219].

The electronic structure methods mentioned above, particularly the ab initio methods, have been largely limited to studying small molecules or truncated systems of biological and medicinal/pharmaceutical interest. For example, extensive benchmarking studies have examined non-covalent interactions to identify efficient methods of comparable accuracy to CCSD(T)/CBS [220]. QM methods are also routinely used to develop MD force field parameters and charges using small models, as well as to parameterize docking scoring functions [219]. Additionally, ligand strain may be evaluated with QM methods [219]. Quantitative structure–activity relationship (QSAR) models also utilize QM methods to calculate the predictor variables such as electrostatic potentials, orbital energies, charges, and dipoles in small molecules [219,221]. The QM methods, specifically the semi-empirical and DFT approaches, were combined with truncated protein models to examine protein-ligand interactions [211]. Nevertheless, the use of QM methods to study very large systems remains limited. In order for the extensive applications of QM methods to macromolecular systems of biological and medicinal interest to be feasible, significant advances in software, hardware, and theory must be achieved [222].

Acceleration of quantum mechanics with graphical processing units

Graphics processing units (GPUs) have led to substantial accelerations in high-performance computing, with significant advancements for costly QM calculations. GPU-accelerated code has been developed for the HF method [223], correlated ab initio methods [223–225], semi-empirical methods [226, 227], and DFT [228]. In fact, GPU acceleration is now implemented in many quantum chemistry programs. NVIDIA reports the following programs as having GPU support: ABINIT, BigDFT, CP2K, GAMESS-US, GAMESS-UK, GPAW, LATTE, MOLCAS, MOPAC2012, NWChem, OCtopus, PEtot, Q-Chem, QMCPACK, Quantum Espresso, TeraChem, and VASP [208]. However, the most extensive use of GPUs in a QM code [229–234] has been implemented in TeraChem, an electronic-structure package specifically designed for use with GPUs. This has enabled a QM description of a protein to examine charge-transfer and polarization in a solvated environment [235], full QM optimizations of protein structures [233], and the examination of excitations in the green fluorescent protein (GFP) chromophore [236].

Fragment-based quantum mechanics methods

One approach to making QM methods more tractable to biological and medicinal applications is the modification of methods so that they scale linearly with system size [237, 238]. One such approach is based on fragmentation methods, which have been developed to facilitate the application of wavefunction and density functional methods to macromolecular structures. These methods partition a macromolecular system and perform QM calculations on each fragment to obtain their wavefunctions and properties, which are then combined to arrive at properties of the macromolecular system as a whole. Fragmentation methods also benefit from their ability to be massively parallelized. A comprehensive summary of the many fragmentation methods available, with many applications to biological systems, can be found in a recent review [239].

Of the electronic structure software available, the GAMESS program includes the greatest variety of fragmentation methods [239, 240], which include the effective fragment potential (EFP) methods [241–243], the fragment molecular orbital (FMO) method [244], the elongation (ELG) method [245], and divide and conquer (DC) approaches [246]. Although the EFP methods utilize intermolecular potentials, these are distinct from those in force fields since they are rigorously derived from ab initio calculations rather than empirical parameters. Of these fragmentation methods, the FMO approach is arguably the most robust and has been widely applied to biological systems [239, 240, 247–249], which include drug discovery, protein-ligand binding, protein-protein interactions, enzymatic catalysis, and DNA.

Application of quantum mechanics/molecular mechanics to computational enzymology

Enzymology investigates, among other topics, enzyme kinetics and mechanisms of inhibition in steady-state turnover. Advances in technology and methods have led to more detailed information about enzyme structures and mechanisms. With an explosion in the number of novel and uncharacterized enzymes identified from the vast number of genome sequences, it has become evident that the structural and functional properties of these enzymes need to be elucidated to establish precisely their mechanisms of action and how the enzymes fit into the complex webs of metabolic reactions found in even the simplest of organisms [250].

Vast changes have occurred in the science of enzymology since molecular simulations and modeling were first developed. Calculations can provide detailed, atomic-level insights into the fundamental mechanisms of biological catalysis. Computational enzymology was launched in the 1970s [251]. The pioneering studies of Warshel are particularly notable [252, 253]. By the early 1990s the number of computational mechanistic studies of enzymes was still relatively small [254, 255], but recently there have been a great number of computational studies of enzymatic reaction mechanisms published [253, 256–258]. Currently, computational enzymology is a rapidly developing area, focused on testing theories of catalysis, challenging "textbook" mechanisms, and identifying novel catalytic mechanisms [259].

The choice of an appropriate method for the particular enzyme being modeled is vital. Quantitative predictions of reaction rates or the effects of mutations remain very challenging, but with appropriate methods, useful predictions can be made with some confidence. Careful testing and experimental validation are important. For example, a comparison of calculated barriers for a series of alternative substrates with experimentally determined activation energies demonstrated good correlation validating mechanistic calculations [260, 261]. Some enzymes have become important model systems in the development and testing of computational methods and protocols; these include chorismate mutase [259, 262–266], citrate synthase [267–269], P450 [257, 259], para-hydroxybenzoate hydroxylase [257, 260, 265] and triosephosphate isomerase [255, 270, 271].

Modeling enzyme-catalyzed reactions

The usual starting point for modeling an enzyme-catalyzed reaction is an enzyme structure from X-ray crystallography. When this is not available, sometimes a model may be constructed based on homology to other structures that have been solved [272], though such models should be treated with much more caution. The first step in studying an enzyme-catalyzed reaction is to establish its chemical mechanism. Its goal is to determine the functions of catalytic residues, which are often not obvious. Even the identities of many important groups may not be certain. Any specific interactions that stabilize transition states or reactive intermediates should also be identified and analyzed.

Enzymes, representing usually large molecules, need sophisticated modeling steps as the reactions that they catalyze are complex. This can be complicated further by the need to include a part of a particular enzyme's molecular environment, such as the surrounding solvent, cofactors, other proteins, a lipid membrane, or DNA. There are many practical considerations in simulating such complex systems, such as the proper interpretation of crystal structures and the choice of protonation states for ionizable amino acids [273]. Here, we illustrate these challenges with recent examples of modeling enzyme-catalyzed reactions.

The empirical valence bond method

The empirical valence bond (EVB) is considered to be one of the main methods used for modeling enzyme-catalyzed reactions [274]. In the EVB method, a few resonance structures are chosen to represent the reaction. The energy of each resonance form is given by a simple empirical force field, with the potential energy given by solving the related secular equation. The EVB Hamiltonian can be calibrated to reproduce experimental data for a reaction in solution, orab initio results can be used [275]. The surrounding protein and solvent are modeled by an empirical force field, using proper long-range electrostatics. The free energy of activation is calculated from free energy perturbation simulations [276]. The free energy surfaces can be calibrated by comparison with experimental data for reactions in solution. The EVB method allows the use of a non-geometrical reaction coordinate, which helps to evaluate non-equilibrium solvation effects [274]. A mapping procedure gradually moves the system from the reactants to products. The simplicity of the EVB potential function allows extensive molecular dynamics (MD) simulations, giving good sampling [277]. The EVB method has been widely used for studying reactions in condensed phases, particularly in enzymes [278–284].

Quantum chemical methods

Another approach to the modeling of enzyme-catalyzed reactions is to study only the active site using quantum chemical methods. This methodology is usually named the cluster approach or the supermolecule approach. The active site model should contain molecules representing the substrate(s), any cofactors, and enzyme residues involved in the chemical reaction or in binding substrates. Important functional groups are represented by small molecules (e.g., acetate can represent an aspartate side chain). The initial positions of these groups are usually coordinates taken from a crystal structure, or from an MD simulation of an enzyme complex.

Quantum chemical methods can give excellent results for reactions of small molecules. Semi-empirical techniques, such as AM1 and PM3, can model larger systems that contain hundreds of atoms. However, semi-empirical methods must be applied with caution due to their sensitivity to parameterization, because typical errors may be over 10 kcal/mol for barriers and reaction energies [285, 286]. DFT methods are considerably more accurate, while also allowing calculations on relatively large systems (e.g., active site models on the order of 100 atoms), larger than is feasible with correlated ab initio calculations. Many DFT methods, however, do not properly account for dispersion forces, which are important in the binding of ligands to proteins and can also be important in the calculation of energy barriers [287]. Dispersion corrections may be required in such cases [215–217]. Calculations on active site models can provide models of transition states and intermediates, which has proved particularly useful for studying metalloenzymes using DFT methods [288, 289].

Combined quantum mechanics/molecular mechanics methods

Combined quantum mechanics/molecular mechanics (QM/MM) methods are widely applied to accurately model enzymes. This has been made possible by increased computer power, and improved software packages. QM/MM approaches treat a small part of the system quantum mechanically (describing the electronic structure of molecules) and the rest of the system with a molecular mechanics (MM) method (using a classical potential energy function [290]). The QM treatment accounts for the electronic rearrangements involved in bond breaking and bond making, while the MM treatment allows to include the effects of the environment on the reaction energetics.

There are two general types of QM/MM methods. The first is additive,

$$E_{\text{total}} = E_{\text{QM}}(\text{QM}) + E_{\text{MM}}(\text{MM}) + E_{\text{QM-MM,interaction}} \tag{38}$$

where $E_{QM}(QM)$ is the energy of the QM region according to the QM method, $E_{MM}(MM)$ is the energy of the MM region according to the MM method, and $E_{QM\text{-}MM,interaction}$ is the interaction energy between the two regions.

The second type of QM/MM method is subtractive,

$$E_{\text{total}} = E_{\text{MM}} + E_{\text{QM}}(\text{QM}) - E_{\text{MM}}(\text{QM}) \tag{39}$$

where E_{MM} is the energy of the total system as calculated by the MM method, $E_{QM}(QM)$ is the energy of the QM region as calculated by the QM method, and $E_{MM}(QM)$ is the energy of the QM region as calculated by the MM method. This is used, for example, in the ONIOM method [291].

In the subtractive approach, the active site region is modeled at the MM level, and the choice of suitable MM parameters (e.g., atomic charges) for all states of the reaction is an important and delicate consideration. Until recently, calculations using the subtractive approach typically used the more approximate "mechanical embedding" scheme, whereas currently most implementations of both approaches allow "electrostatic embedding" [291] which takes into account the electrostatic influence of the MM region on the QM region, i.e., polarization of the QM region by the atomic charges of the MM region. Five general aspects are important in a QM/MM calculation on an enzyme:

- Choice of the QM method.
- Choice of the MM force field (including the MM parameters for the QM region).
- Partitioning of the system into QM and MM regions with due attention paid to any chemical bonds that straddle the two regions.
- Type of simulation (e.g., an MD simulation, or calculation of potential energy profiles); whether extensive conformational sampling will be performed.
- Construction (and testing) of an accurate molecular model of the enzyme complex.

The MM force field employed in a QM/MM study should be chosen to describe the part of the system outside the QM region and its interactions with the QM region. For proteins, standard all-atom force fields such as CHARMM27, AMBER ff99 or ff99SB, and OPLS-AA are commonly used. Apart from selecting suitable QM and MM methods and a QM/MM approach, modeling an enzyme reaction with a QM/MM method requires other important choices, such as deciding which atoms to include in the QM region and how to treat covalent bonds that cross the QM/MM boundary [292]. Another important choice is determining the protonation states of residues, and how (long-range) electrostatic interactions are treated. The influence of such choices on the results should be tested [267, 293–296] in order to be able to draw reliable conclusions. Recent improvements allow relocating the QM-MM boundary on-the-fly (adaptive partitioning) [297].

Modeling enzyme reactions by calculating potential energy surfaces

With QM or QM/MM methods, potential energy surfaces of enzyme reaction mechanisms can be explored accurately enough to enable discrimination between different mechanisms, e.g., if the barrier for a proposed mechanism is significantly larger than that derived from experiment (using transition state theory), within the limits of accuracy of the computational method and experimental error, then that mechanism can be considered to be unlikely. A mechanism with a calculated barrier comparable to the apparent experimental barrier (for that step, or failing that for the overall reaction) is more likely. However, to calculate rate constants also requires reliable estimates of enthalpies, internal energies, and free energies of a given reaction and activation, given the potential energy surface. Traditional approaches to modeling reactions rely on the identification of stationary points (reactants, products, intermediates, transition states) via geometry optimization, followed by the computation of second derivatives to enable relatively simplistic evaluation of zero-point corrections, thermal and entropy terms. Algorithms developed for small molecules are often not suitable for large systems, storage and manipulation of Hessian matrices become extremely difficult. A basic means of modeling approximate reaction paths is the "adiabatic mapping" or "coordinate driving" approach. The energy of the system is found by minimizing the energy at a series of fixed (or restrained) values of a reaction coordinate, e.g., the distance between two atoms. This approach has been successfully applied to many enzymes [257], but it is only valid if one conformation of the protein can represent the state of the system at a particular value of the reaction coordinate.

Calculations of a potential energy surface may not consider significant conformational fluctuations of the enzyme. Conformational changes, even on a small scale, may cause significant chemical changes. Conformational changes of the active site can greatly affect the energy barrier. To take the fatty acid amide hydrolase as an example, conformational fluctuations do not affect the general shape of the potential energy surfaces, but consistency between experimental and calculated barriers is observed only with a specific infrequent arrangement of the enzyme-substrate complex [298]. These findings indicate that investigation of different protein conformations is essential for a meaningful determination of the energetics of enzymatic reactions for calculations of potential energy profiles or surfaces.

Calculating free energy profiles for enzyme-catalyzed reactions

According to transition state theory, the rate constant of a reaction is related to the free energy barrier. The techniques described previously calculate potential energy barriers for a particular conformation. Techniques that illustrate configurations along a reaction coordinate give a more sophisticated and extensive description by taking account of multiple conformations and estimating entropic effects, and can be essential for modeling enzyme reactions. Simulations of this type provide estimates of the free energy profile along a specific reaction coordinate, which is often referred to as the potential of mean force. MD and Monte Carlo methods allow such illustration, but do not provide a sufficiently detailed view of high energy regions, such as those in the vicinity of transition states. Conformational illustration of processes of chemical change requires specialized techniques, e.g., to bias the simulation to sample the transition state region. Umbrella sampling, which is widely used in MD simulations, when combined with QM/MM techniques, can be used to model enzymatic reactions [262]. QM/MM umbrella sampling simulations are possible with semi-empirical molecular orbital methods (e.g., AM1 or PM3). Often, such methods are highly inaccurate for reaction barriers and energies but their accuracy can be improved significantly by re-parameterization for a specific reaction.

CONCLUSIONS

This review paper has aimed to provide a comprehensive guide to a plethora of mathematical and computational methods developed in the past few decades to tackle key quantitative problems in the life sciences. Our critical overview covers methods used across the life sciences, starting from macroscopic systems such as those in evolutionary biology, and ending with atomic level descriptions of biomolecules including quantum mechanical or hybrid classical/quantum approaches. Particular attention was given to large-scale computational methods, such as molecular dynamics, which play pivotal roles in the development of our understanding of molecular mechanisms at the level of molecular, structural, and cell biology. Important applications in medicine and pharmaceutical sciences have been discussed, in particular in the context of extracting crucial conclusions about complex system behavior with information limitations. We hope the reader will be encouraged to explore particular topics at a deeper level using the information and references provided in this review.

ACKNOWLEDGEMENTS

The authors are grateful to the Natural Sciences and Engineering Research Council of Canada, the Li Ka Shing Institute of Applied Virology and the Alberta Cancer Foundation for funding support.

REFERENCES

1. Caticha A: Entropic Inference and the Foundations of Physics (monograph Commissioned by the 11th Brazilian Meeting on Bayesian Statistics–EBEB-2012. 2012, Sao Paulo: USP Press
2. Cox RT: Probability, Frequency, and Reasonable Expectation. 1946
3. Caticha A, Giffin A: Updating probabilities. ArXiv Prepr Physics. 2006, 0608185
4. Chang C-H, Hsieh L-C, Chen T-Y, Chen H-D, Luo L, Lee H-C: Shannon information in complete genomes. J Bioinform Comput Biol. 2005, 3: 587-608.
5. Dover Y: A short account of a connection of power laws to the information entropy. Phys Stat Mech Its Appl. 2004, 334: 591-599.
6. Chang DT-H, Oyang Y-J, Lin J-H: MEDock: a web server for efficient prediction of ligand binding sites based on a novel optimization algorithm. Nucleic Acids Res. 2005, 33 (suppl 2): W233-W238.
7. Tseng C-Y, Ashrafuzzaman M, Mane JY, Kapty J, Mercer JR, Tuszynski JA: Entropic fragment-based approach to aptamer design. Chem Biol Drug Des. 2011, 78: 1-13.
8. Shannon CE: A mathematical theory of communication. Bell Syst Tech J. 1948, 27: 379-423.
9. Stelling J, Sauer U, Szallasi Z, Doyle FJ, Doyle J: Robustness of cellular functions. Cell. 2004, 118: 675-685.
10. Li F, Long T, Lu Y, Ouyang Q, Tang C: The yeast cell-cycle network is robustly designed. Proc Natl Acad Sci U S A. 2004, 101: 4781-4786.
11. Bak P, Tang C, Wiesenfeld K: Self-organized criticality. Phys Rev A. 1988, 38: 364-374.
12. Honma T: Recent advances in de novo design strategy for practical lead identification. Med Res Rev. 2003, 23: 606-632.
13. Schneider G, Fechner U: Computer-based de novo design of drug-like molecules. Nat Rev Drug Discov. 2005, 4: 649-663.
14. Morris GM, Huey R, Lindstrom W, Sanner MF, Belew RK, Goodsell DS, Olson AJ: AutoDock4 and AutoDockTools4: Automated docking with selective receptor flexibility. J Comput Chem. 2009, 30: 2785-2791.
15. Nimjee SM, Rusconi CP, Sullenger BA: Aptamers: an emerging class of therapeutics. Annu Rev Med. 2005, 56: 555-583.
16. James W: Aptamer. Encycl Anal Chem. 2000, Hoboken, N.J.: Wiley & Sons Inc, 4848-4871.
17. Hamula CLA, Guthrie JW, Zhang H, Li X-F, Le XC: Selection and analytical applications of aptamers. TrAC Trends Anal Chem. 2006, 25: 681-691. On-Site Instrumentation and Analysis
18. James W: Aptamers in the virologists' toolkit. J Gen Virol. 2007, 88 (Pt 2): 351-364.
19. Ashrafuzzaman M, Tseng C-Y, Kapty J, Mercer JR, Tuszynski JA: A computationally designed DNA aptamer template with specific binding to phosphatidylserine. Nucleic Acid Ther. 2013, 23: 418-426.
20. Brauer F, Castillo-Chávez C: Mathematical Models in Population Biology and Epidemiology. 2001, New York: Springer
21. Brauer F, Van den Driessche P, Wu J, Allen LJS: Mathematical Epidemiology. 2008, Berlin: Springer
22. Keener JP, Sneyd J: Mathematical Physiology. 1998, New York: Springer
23. Preziosi L: Cancer Modelling and Simulation. 2003, Boca Raton, Fla: Chapman & Hall/CRC
24. Edelstein-Keshet L: Mathematical Models in Biology. 1988, New York: Random House
25. Murray JD: Mathematical Biology I: An Introduction. 2003, New York: Springer Verlag, 3
26. Murray JD: Mathematical Biology II: Spatial Models and Biomedical Applications. 2003, New York: Springer Verlag, 3
27. Britton NF: Essential Mathematical Biology. 2003, London: Springer Verlag, 2
28. Merrill RM: Introduction to Epidemiology. 2013
29. Kermack MD, Mckendrick AG: Contributions to the mathematical theory of epidemics. Part I. In Proc R Soc A. 1927, 115: 700-721.
30. Hethcote HW: Qualitative analyses of communicable disease models. Math Biosci. 1976, 28: 335-356.
31. Soper HE: The interpretation of periodicity in disease prevalence. J R Stat Soc. 1929, 92: 34-73.
32. Kermack WO, McKendrick AG: Contributions to the mathematical theory of epidemics. II. The problem of endemicity. Proc R Soc Lond Ser A. 1932, 138: 55-83.
33. Smith HL, Wang L, Li MY: Global Dynamics of an SEIR Epidemic Model with Vertical Transmission. SIAM J Appl Math. 2001, 62: 58-69.
34. Dunham JB: An agent-based spatially explicit epidemiological model in MASON. J Artif Soc Soc Simul. 2005, 9: 3-
35. Tian C, Ding W, Cao R, Jiang S: Extensive Epidemic Spreading Model Based on Multi-agent System Framework. Lect Notes Comput Sci. 2007, 4490: 129-

133.

36. Newman MEJ: Spread of epidemic disease on networks. Phys Rev E Stat Nonlin Soft Matter Phys. 2002, 66 (1 Pt 2): 016128-
37. Webb GF: Theory of Nonlinear Age-Dependent Population Dynamics. 1985, New York: M. Dekker
38. Murray JD, Stanley EA, Brown DL: On the spatial spread of rabies among foxes. Proc R Soc Lond Ser B Contain Pap Biol Character R Soc G B. 1986, 229: 111-150.
39. Wonham MJ, de-Camino-Beck T, Lewis MA: An epidemiological model for West Nile virus: invasion analysis and control applications. Proc Biol Sci. 2004, 271: 501-507.
40. Lloyd AL, May RM: Spatial heterogeneity in epidemic models. J Theor Biol. 1996, 179: 1-11.
41. Fraser C: Epidemiology and Control of Infectious Diseases. 2014, London: Department of Infectious Disease Epidemiology Imperial College London
42. Lehninger AL, Nelson DL, Cox MM: Principles of Biochemistry. 1993, New York, NY: Worth Publishers
43. Köhler J, Baumbach J, Taubert J, Specht M, Skusa A, Rüegg A, Rawlings C, Verrier P, Philippi S: Graph-based analysis and visualization of experimental results with ONDEX. Bioinforma Oxf Engl. 2006, 22: 1383-1390.
44. Michaelis L, Menten ML, Johnson KA, Goody RS: The original Michaelis constant: translation of the 1913 Michaelis-Menten paper. Biochemistry (Mosc). 2011, 50: 8264-8269.
45. Briggs GE, Haldane JB: A Note on the Kinetics of Enzyme Action. Biochem J. 1925, 19: 338-339.
46. Monod J, Wyman J, Changeux JP: On the nature of allosteric transitions: a plausible model. J Mol Biol. 1965, 12: 88-118.
47. Hübner CA, Jentsch TJ: Ion channel diseases. Hum Mol Genet. 2002, 11: 2435-2445.
48. Mendes P, Messiha H, Malys N, Hoops S: Enzyme kinetics and computational modeling for systems biology. Methods Enzymol. 2009, 467: 583-599.
49. Chance B, Garfinkel D, Higgins J, Hess B: Metabolic control mechanisms. 5. A solution for the equations representing interaction between glycolysis and respiration in ascites tumor cells. J Biol Chem. 1960, 235: 2426-2439.
50. Cornish-Bowden A: Why is uncompetitive inhibition so rare? A possible explanation, with implications for the design of drugs and pesticides. FEBS Lett. 1986, 203: 3-6.
51. Abou El-Magd RM, Park HK, Kawazoe T, Iwana S, Ono K, Chung SP, Miyano M, Yorita K, Sakai T, Fukui K: The effect of risperidone on D-amino acid oxidase activity as a hypothesis for a novel mechanism of action in the treatment of schizophrenia. J Psychopharmacol Oxf Engl. 2010, 24: 1055-1067.
52. Westley AM, Westley J: Enzyme inhibition in open systems. Superiority of uncompetitive agents. J Biol Chem. 1996, 271: 5347-5352.
53. Hucka M, Finney A, Sauro HM, Bolouri H, Doyle JC, Kitano H, Arkin AP, Bornstein BJ, Bray D, Cornish-Bowden A, Cuellar AA, Dronov S, Gilles ED, Ginkel M, Gor V, Goryanin II, Hedley WJ, Hodgman TC, Hofmeyr J-H, Hunter PJ, Juty NS, Kasberger JL, Kremling A, Kummer U, Le Novère N, Loew LM, Lucio D, Mendes P, Minch E, Mjolsness ED, et al: The systems biology markup language (SBML): a medium for representation and exchange of biochemical network models. Bioinforma Oxf Engl. 2003, 19: 524-531.
54. Hoops S, Sahle S, Gauges R, Lee C, Pahle J, Simus N, Singhal M, Xu L, Mendes P, Kummer U: COPASI–a COmplex PAthway SImulator. Bioinforma Oxf Engl. 2006, 22: 3067-3074.
55. Funahashi A, Morohashi M, Kitano H, Tanimura N: Cell Designer: a process diagram editor for gene-regulatory and biochemical networks. Biosilico. 2003, 1: 159-162.
56. Shannon P, Markiel A, Ozier O, Baliga NS, Wang JT, Ramage D, Amin N, Schwikowski B, Ideker T: Cytoscape: a software environment for integrated models of biomolecular interaction networks. Genome Res. 2003, 13: 2498-2504.
57. Rojas I, Golebiewski M, Kania R, Krebs O, Mir S, Weidemann A, Wittig U: Storing and annotating of kinetic data. In Silico Biol. 2007, 7 (2 Suppl): S37-S44.
58. Sauro HM, Hucka M, Finney A, Wellock C, Bolouri H, Doyle J, Kitano H: Next generation simulation tools: the Systems Biology Workbench and BioSPICE integration. Omics J Integr Biol. 2003, 7: 355-372.
59. Frankenhaeuser B: A quantitative description of potassium currents in myelinated nerve fibres of Xenopus laevis. J Physiol. 1963, 169: 424-430.
60. Hodgkin AL, Huxley AF, Katz B: Measurement of current-voltage relations in the membrane of the giant axon of Loligo. J Physiol. 1952, 116: 424-448.
61. Hodgkin AL, Huxley AF: Currents carried by sodium and potassium ions through the membrane of the giant axon of Loligo. J Physiol. 1952, 116: 449-472.
62. Hodgkin AL, Huxley AF: The components of membrane conductance in the giant axon of Loligo. J Physiol. 1952, 116: 473-496.
63. Hodgkin AL, Huxley AF: The dual effect of membrane potential on sodium conductance in the giant axon of Loligo. J Physiol. 1952, 116: 497-506.
64. Hodgkin AL, Huxley AF: A quantitative description of membrane current and its application to conduction and excitation in nerve. J Physiol. 1952, 117: 500-544.
65. Moreno JD, Zhu ZI, Yang P-C, Bankston JR, Jeng M-T, Kang C, Wang L, Bayer JD, Christini DJ, Trayanova NA, Ripplinger CM, Kass RS, Clancy CE: A computational model to predict the effects of class I anti-arrhythmic drugs on ventricular rhythms. Sci Transl Med. 2011, 3: 98ra83-
66. Welsh MJ: Abnormal regulation of ion channels in cystic fibrosis epithelia. FASEB J Off Publ Fed Am Soc Exp Biol. 1990, 4: 2718-2725.
67. Keiser NW, Engelhardt JF: New animal models of cystic fibrosis: what are they teaching us?. Curr Opin Pulm Med. 2011, 17: 478-483.

68. Knowles MA, Selby P: Introduction to the Cellular and Molecular Biology of Cancer. 2005, New York: Oxford University Press
69. Vaidya VG, Alexandro FJ: Evaluation of some mathematical models for tumor growth. Int J Biomed Comput. 1982, 13: 19-36.
70. Casey AE: The experimental alteration of malignancy with an homologous mammalian tumor material: I. results with intratesticular inoculation. Am J Cancer. 1934, 21: 760-775.
71. Cruywagen GC, Woodward DE, Tracqui P, Bartoo GT, Murray JD, Alvord EC: The modelling of diffusive tumours. J Biol Syst. 1995, 03: 937-945.
72. Tracqui P, Cruywagen GC, Woodward DE, Bartoo GT, Murray JD, Alvord EC: A mathematical model of glioma growth: the effect of chemotherapy on spatio-temporal growth. Cell Prolif. 1995, 28: 17-31.
73. Swanson KR, Alvord EC, Murray JD: A quantitative model for differential motility of gliomas in grey and white matter. Cell Prolif. 2000, 33: 317-329.
74. Chaplain MAJ: Avascular growth, angiogenesis and vascular growth in solid tumours: The mathematical modelling of the stages of tumour development. Math Comput Model. 1996, 23: 47-87.
75. Chaplain MA, Ganesh M, Graham IG: Spatio-temporal pattern formation on spherical surfaces: numerical simulation and application to solid tumour growth. J Math Biol. 2001, 42: 387-423.
76. Sherratt JA, Chaplain MA: A new mathematical model for avascular tumour growth. J Math Biol. 2001, 43: 291-312.
77. Semenza GL: Angiogenesis in ischemic and neoplastic disorders. Annu Rev Med. 2003, 54: 17-28.
78. Keller EF, Segel LA: Model for chemotaxis. J Theor Biol. 1971, 30: 225-234.
79. Balding D, McElwain DL: A mathematical model of tumour-induced capillary growth. J Theor Biol. 1985, 114: 53-73.
80. Anderson AR, Chaplain MA: Continuous and discrete mathematical models of tumor-induced angiogenesis. Bull Math Biol. 1998, 60: 857-899.
81. Mantzaris NV, Webb S, Othmer HG: Mathematical modeling of tumor-induced angiogenesis. J Math Biol. 2004, 49: 111-187.
82. Hillen T: M5 mesoscopic and macroscopic models for mesenchymal motion. J Math Biol. 2006, 53: 585-616.
83. Preziosi L: Modeling cell movement in anisotropic and heterogeneous network tissues. Netw Heterog Media. 2007, 2: 333-357.
84. Preziosi L: Modelling the motion of a cell population in the extracellular matrix. DISCRETE Contin Dyn Syst Ser B. 2007, Supplement: 250-259.
85. Collins I, Workman P: New approaches to molecular cancer therapeutics. Nat Chem Biol. 2006, 2: 689-700.
86. Baskar R, Lee KA, Yeo R, Yeoh K-W: Cancer and radiation therapy: current advances and future directions. Int J Med Sci. 2012, 9: 193-199.
87. Swanson KR, Rostomily RC, Alvord EC: A mathematical modelling tool for predicting survival of individual patients following resection of glioblastoma: a proof of principle. Br J Cancer. 2008, 98: 113-119.
88. Rockne R, Rockhill JK, Mrugala M, Spence AM, Kalet I, Hendrickson K, Lai A, Cloughesy T, Alvord EC, Swanson KR: Predicting the efficacy of radiotherapy in individual glioblastoma patients in vivo: a mathematical modeling approach. Phys Med Biol. 2010, 55: 3271-3285.
89. Clatz O, Sermesant M, Bondiau P-Y, Delingette H, Warfield SK, Malandain G, Ayache N: Realistic simulation of the 3-D growth of brain tumors in MR images coupling diffusion with biomechanical deformation. IEEE Trans Med Imaging. 2005, 24: 1334-1346.
90. Painter KJ, Hillen T: Mathematical modelling of glioma growth: the use of Diffusion Tensor Imaging (DTI) data to predict the anisotropic pathways of cancer invasion. J Theor Biol. 2013, 323: 25-39.
91. Hillen T, de Vries G, Gong J, Finlay C: From cell population models to tumor control probability: including cell cycle effects. Acta Oncol Stockh Swed. 2010, 49: 1315-1323.
92. Munro TR, Gilbert CW: The relation between tumour lethal doses and the radiosensitivity of tumour cells. Br J Radiol. 1961, 34: 246-251.
93. Zaider M, Minerbo GN: Tumour control probability: a formulation applicable to any temporal protocol of dose delivery. Phys Med Biol. 2000, 45: 279-293.
94. Zhu T, Lee H, Lei H, Jones C, Patel K, Johnson ME, Hevener KE: Fragment-based drug discovery using a multidomain, parallel MD-MM/PBSA screening protocol. J Chem Inf Model. 2013, 53: 560-572.
95. Fujihashi M, Ishida T, Kuroda S, Kotra LP, Pai EF, Miki K: Substrate distortion contributes to the catalysis of orotidine 5′-monophosphate decarboxylase. J Am Chem Soc. 2013, 135: 17432-17443.
96. Tiwari G, Mohanty D: An in silico analysis of the binding modes and binding affinities of small molecule modulators of PDZ-peptide interactions. PLoS One. 2013, 8: e71340-
97. Zahn D: Molecular dynamics simulation of ionic conductors: perspectives and limitations. J Mol Model. 2011, 17: 1531-1535.
98. Wood NT, Fadda E, Davis R, Grant OC, Martin JC, Woods RJ, Travers SA: The influence of N-linked glycans on the molecular dynamics of the HIV-1 gp120 V3 loop. PLoS One. 2013, 8: e80301-
99. Chaudhuri A, Sarkar I, Chakraborty S: Comparative analysis of binding sites of human meprins with hydroxamic acid derivative by molecular dynamics simulation study. J Biomol Struct Dyn. 2013
100. Leonis G, Steinbrecher T, Papadopoulos MG: A contribution to the drug resistance mechanism of darunavir, amprenavir, indinavir, and saquinavir complexes with HIV-1 protease due to flap mutation I50V: a systematic MM-PBSA and thermodynamic

integration study. J Chem Inf Model. 2013, 53: 2141-2153.

101. Yoda T, Sugita Y, Okamoto Y: Hydrophobic core formation and dehydration in protein folding studied by generalized-ensemble simulations. Biophys J. 2010, 99: 1637-1644.

102. Yoda T, Sugita Y, Okamoto Y: Salt effects on hydrophobic-core formation in folding of a helical miniprotein studied by molecular dynamics simulations. Proteins. 2014, 82: 933-943.

103. Fukunishi Y, Nakamura H: Improved estimation of protein-ligand binding free energy by using the ligand-entropy and mobility of water molecules. Pharm Basel Switz. 2013, 6: 604-622.

104. Cornell WD, Cieplak P, Bayly CI, Gould IR, Merz KM, Ferguson DM, Spellmeyer DC, Fox T, Caldwell JW, Kollman PA: A second generation force field for the simulation of proteins, nucleic acids, and organic molecules. J Am Chem Soc. 1995, 117: 5179-5197.

105. Ponder JW, Case DA: Force fields for protein simulations. Adv Protein Chem. 2003, 66: 27-85.

106. Hornak V, Abel R, Okur A, Strockbine B, Roitberg A, Simmerling C: Comparison of multiple Amber force fields and development of improved protein backbone parameters. Proteins. 2006, 65: 712-725.

107. Wang J, Cieplak P, Kollman PA: How well does a restrained electrostatic potential (RESP) model perform in calculating conformational energies of organic and biological molecules?. J Comput Chem. 2000, 21: 1049-1074.

108. Wickstrom L, Okur A, Simmerling C: Evaluating the performance of the ff99SB force field based on NMR scalar coupling data. Biophys J. 2009, 97: 853-856.

109. Cino EA, Choy W-Y, Karttunen M: Comparison of secondary structure formation using 10 different force fields in microsecond molecular dynamics simulations. J Chem Theory Comput. 2012, 8: 2725-2740.

110. Case DA, Babin V, Berryman JT, Betz RM, Cerutti T, Cheatham I, Darden TA, Duke RE, Gohlke H, Goetz AW, Gusarov S, Homeyer N, Janowski P, Anderson J, Kolossváry I, Kaus J, Kovalenko A, Lee TS, LeGrand S, Luchko T, Luo R, Madej B, Merz KM, Paesani F, Roe DR, Roitberg A, Sagui C, Salomon-Ferrer R, Seabra G, Simmerling CL, et al: AMBER 14. 2014

111. Zhang Y, Sagui C: The gp41(659-671) HIV-1 antibody epitope: a structurally challenging small peptide. J Phys Chem B. 2014, 118: 69-80.

112. Duan Y, Wu C, Chowdhury S, Lee MC, Xiong G, Zhang W, Yang R, Cieplak P, Luo R, Lee T, Caldwell J, Wang J, Kollman P: A point-charge force field for molecular mechanics simulations of proteins based on condensed-phase quantum mechanical calculations. J Comput Chem. 2003, 24: 1999-2012.

113. Lee MC, Duan Y: Distinguish protein decoys by Using a scoring function based on a new AMBER force field, short molecular dynamics simulations, and the generalized born solvent model. Proteins Struct Funct Bioinforma. 2004, 55: 620-634.

114. Yang L, Tan C-H, Hsieh M-J, Wang J, Duan Y, Cieplak P, Caldwell J, Kollman PA, Luo R: New-generation amber united-atom force field. J Phys Chem B. 2006, 110: 13166-13176.

115. Wang J, Wolf RM, Caldwell JW, Kollman PA, Case DA: Development and testing of a general amber force field. J Comput Chem. 2004, 25: 1157-1174.

116. MacKerell AD, Banavali N, Foloppe N: Development and current status of the CHARMM force field for nucleic acids. Biopolymers. 2000, 56: 257-265.

117. Jorgensen WL, Tirado-Rives J: The OPLS [optimized potentials for liquid simulations] potential functions for proteins, energy minimizations for crystals of cyclic peptides and crambin. J Am Chem Soc. 1988, 110: 1657-1666.

118. Kaminski G, Jorgensen WL: Performance of the AMBER94, MMFF94, and OPLS-AA force fields for modeling organic liquids. J Phys Chem. 1996, 100: 18010-18013.

119. MacKerell AD, Bashford D, Bellott M, Dunbrack RL, Evanseck JD, Field MJ, Fischer S, Gao J, Guo H, Ha S, Joseph-McCarthy D, Kuchnir L, Kuczera K, Lau FTK, Mattos C, Michnick S, Ngo T, Nguyen DT, Prodhom B, Reiher WE, Roux B, Schlenkrich M, Smith JC, Stote R, Straub J, Watanabe M, Wiórkiewicz-Kuczera J, Yin D, Karplus M: All-atom empirical potential for molecular modeling and dynamics studies of proteins. J Phys Chem B. 1998, 102 (18): 3586-3616.

120. Feig M, MacKerell AD, Brooks CL: Force field influence on the observation of π-helical protein structures in molecular dynamics simulations. J Phys Chem B. 2003, 107: 2831-2836.

121. Vanommeslaeghe K, Hatcher E, Acharya C, Kundu S, Zhong S, Shim J, Darian E, Guvench O, Lopes P, Vorobyov I, Mackerell AD: CHARMM general force field: A force field for drug-like molecules compatible with the CHARMM all-atom additive biological force fields. J Comput Chem. 2010, 31: 671-690.

122. Jorgensen WL, Maxwell DS, Tirado-Rives J: Development and testing of the OPLS all-atom force field on conformational energetics and properties of organic liquids. J Am Chem Soc. 1996, 118: 11225-11236.

123. Maxwell DS, Tirado-Rives J, Jorgensen WL: A comprehensive study of the rotational energy profiles of organic systems by ab initio MO theory, forming a basis for peptide torsional parameters. J Comput Chem. 1995, 16: 984-1010.

124. Kahn K, Bruice TC: Parameterization of OPLS-AA force field for the conformational analysis of macrocyclic polyketides. J Comput Chem. 2002, 23: 977-996.

125. Siu SWI, Pluhackova K, Böckmann RA: Optimization of the OPLS-AA force field for long hydrocarbons. J Chem Theory Comput. 2012, 8: 1459-1470.

126. Daura X, Mark AE, Van Gunsteren WF: Parametrization of aliphatic CHn united atoms of GROMOS96 force field. J Comput Chem. 1998, 19: 535-547.

127. Schuler LD, Daura X, van Gunsteren WF: An improved

GROMOS96 force field for aliphatic hydrocarbons in the condensed phase. J Comput Chem. 2001, 22: 1205-1218.

128. Oostenbrink C, Villa A, Mark AE, Van Gunsteren WF: A biomolecular force field based on the free enthalpy of hydration and solvation: The GROMOS force-field parameter sets 53A5 and 53A6. J Comput Chem. 2004, 25: 1656-1676.
129. Reif MM, Winger M, Oostenbrink C: Testing of the GROMOS force-field parameter set 54A8: structural properties of electrolyte solutions, lipid bilayers, and proteins. J Chem Theory Comput. 2013, 9: 1247-1264.
130. Schmid N, Eichenberger AP, Choutko A, Riniker S, Winger M, Mark AE, van Gunsteren WF: Definition and testing of the GROMOS force-field versions 54A7 and 54B7. Eur Biophys J EBJ. 2011, 40: 843-856.
131. Marrink SJ, Risselada HJ, Yefimov S, Tieleman DP, de Vries AH: The MARTINI Force Field: Coarse Grained Model for Biomolecular Simulations. J Phys Chem B. 2007, 111: 7812-7824.
132. Monticelli L, Kandasamy SK, Periole X, Larson RG, Tieleman DP, Marrink S-J: The MARTINI Coarse-Grained Force Field: Extension to Proteins. J Chem Theory Comput. 2008, 4: 819-834.
133. López CA, Rzepiela AJ, de Vries AH, Dijkhuizen L, Hünenberger PH, Marrink SJ: Martini Coarse-Grained Force Field: Extension to Carbohydrates. J Chem Theory Comput. 2009, 5: 3195-3210.
134. López CA, Sovova Z, van Eerden FJ, de Vries AH, Marrink SJ: Martini Force Field Parameters for Glycolipids. J Chem Theory Comput. 2013, 9: 1694-1708.
135. Marrink SJ, de Vries AH, Tieleman DP: Lipids on the move: Simulations of membrane pores, domains, stalks and curves. Biochim Biophys Acta BBA - Biomembr. 2009, 1788: 149-168. Lipid Interactions, Domain Formation, and Lateral Structure of Membranes
136. Yoo J, Cui Q: Membrane-mediated protein-protein interactions and connection to elastic models: a coarse-grained simulation analysis of gramicidin A association. Biophys J. 2013, 104: 128-138.
137. Gautieri A, Russo A, Vesentini S, Redaelli A, Buehler MJ: Coarse-Grained Model of Collagen Molecules Using an Extended MARTINI Force Field. J Chem Theory Comput. 2010, 6: 1210-1218.
138. Wong-Ekkabut J, Baoukina S, Triampo W, Tang I-M, Tieleman DP, Monticelli L: Computer simulation study of fullerene translocation through lipid membranes. Nat Nanotechnol. 2008, 3: 363-368.
139. Marrink SJ, Tieleman DP: Perspective on the Martini model. Chem Soc Rev. 2013, 42: 6801-6822.
140. Freedman H: Solvation Free Energies from a Coupled Reference Interaction Site Model/simulation Approach (Thesis). 2005, Department of Chemistry: University of Utah
141. Chipman DM: Vertical electronic excitation with a dielectric continuum model of solvation including volume polarization. I. Theory. J Chem Phys. 2009, 131: 014103-
142. Canuto S: Solvation Effects on Molecules and Biomolecules: Computational Methods and Applications. 2008, Sao Paulo, Brazil: Springer
143. Arakawa T, Kamiya N, Nakamura H, Fukuda I: Molecular dynamics simulations of double-stranded DNA in an explicit solvent model with the zero-dipole summation method. PLoS One. 2013, 8: e76606-
144. Liu Y, Haddadian E, Sosnick TR, Freed KF, Gong H: A novel implicit solvent model for simulating the molecular dynamics of RNA. Biophys J. 2013, 105: 1248-1257.
145. Yu Z, Jacobson MP, Friesner RA: What Role Do Surfaces Play in GB Models? A New-Generation of Surface-Generalized Born Model Based on a Novel Gaussian Surface for Biomolecules. J Comput Chem. 2006, 27: 72-89.
146. Pang X, Zhou H-X: Poisson-Boltzmann Calculations: van der Waals or Molecular Surface?. Commun Comput Phys. 2013, 13: 1-12.
147. Salomon-Ferrer R, Case DA, Walker RC: An overview of the Amber biomolecular simulation package. Wiley Interdiscip Rev Comput Mol Sci. 2013, 3: 198-210.
148. Höfinger S: Solving the Poisson-Boltzmann equation with the specialized computer chip MD-GRAPE-2. J Comput Chem. 2005, 26: 1148-1154.
149. Luchko T, Gusarov S, Roe DR, Simmerling C, Case DA, Tuszynski J, Kovalenko A: Three-dimensional molecular theory of solvation coupled with molecular dynamics in Amber. J Chem Theory Comput. 2010, 6: 607-624.
150. Orozco M, Luque FJ: Theoretical methods for the description of the solvent effect in biomolecular systems. Chem Rev. 2000, 100: 4187-4226.
151. Simkin BI: Quantum Chemical and Statistical Theory of Solutions: A Computational Approach. 1995, London; New York: Ellis Horwood, Ellis Horwood Series in Physical Chemistry
152. Chandler D, McCoy JD, Singer SJ: Density functional theory of nonuniform polyatomic systems. I. General formulation. J Chem Phys. 1986, 85: 5971-
153. Chandler D, McCoy JD, Singer SJ: Density functional theory of nonuniform polyatomic systems. II. Rational closures for integral equations. J Chem Phys. 1986, 85: 5977-
154. Beglov D, Roux B: An integral equation to describe the solvation of polar molecules in liquid water. J Phys Chem B. 1997, 101: 7821-7826.
155. Kovalenko A, Hirata F: Potential of mean force between two molecular ions in a polar molecular solvent: a study by the three-dimensional Reference Interaction Site Model. J Phys Chem B. 1999, 103: 7942-7957.
156. Kovalenko A, Hirata F: Self-consistent description of a metal–water interface by the Kohn–Sham density functional theory and the three-dimensional reference interaction site model. J Chem Phys. 1999, 110: 10095-
157. Kovalenko A, Hirata F: Potentials of mean force of simple ions in ambient aqueous solution. I. Three-dimensional reference interaction site model approach. J Chem Phys. 2000, 112: 10391-

158. Kovalenko A, Hirata F: Potentials of mean force of simple ions in ambient aqueous solution. II. Solvation structure from the three-dimensional reference interaction site model approach, and comparison with simulations. J Chem Phys. 2000, 112: 10403-
159. Andersen HC: Optimized cluster expansions for classical fluids. I. General theory and variational formulation of the mean spherical model and hard sphere Percus-Yevick equations. J Chem Phys. 1972, 57: 1918-
160. Rossky PJ, Friedman HL: Accurate solutions to integral equations describing weakly screened ionic systems. J Chem Phys. 1980, 72: 5694-
161. Hirata F, Rossky PJ: An extended rism equation for molecular polar fluids. Chem Phys Lett. 1981, 83: 329-334.
162. Hirata F: Application of an extended RISM equation to dipolar and quadrupolar fluids. J Chem Phys. 1982, 77: 509-
163. Hansen J-P: McDonald IR: Theory of Simple Liquids with Applications to Soft Matter. 2013, Burlington: Elsevier Science
164. Hirata F: Molecular Theory of Solvation. 2003, Dordrecht; Boston: Kluwer Academic Publishers, Understanding Chemical Reactivity, v. 24
165. Tuckerman ME, Berne BJ, Martyna GJ: Molecular dynamics algorithm for multiple time scales: Systems with long range forces. J Chem Phys. 1991, 94: 6811-
166. Tuckerman M, Berne BJ, Martyna GJ: Reversible multiple time scale molecular dynamics. J Chem Phys. 1992, 97: 1990-
167. Omelyan I, Kovalenko A: Generalised canonical–isokinetic ensemble: speeding up multiscale molecular dynamics and coupling with 3D molecular theory of solvation. Mol Simul. 2013, 39: 25-48.
168. Truchon J-F, Pettitt BM, Labute P: A cavity corrected 3D-RISM functional for accurate solvation free energies. J Chem Theory Comput. 2014, 10: 934-941.
169. Ng K-C: Hypernetted chain solutions for the classical one-component plasma up to $\Gamma = 7000$. J Chem Phys. 1974, 61: 2680-
170. Anderson DG: Iterative procedures for nonlinear integral equations. J ACM. 1965, 12: 547-560.
171. Maruyama Y, Hirata F: Modified Anderson method for accelerating 3D-RISM calculations using graphics processing unit. J Chem Theory Comput. 2012, 8: 3015-3021.
172. Miller BR, McGee TD, Swails JM, Homeyer N, Gohlke H, Roitberg AE: MMPBSA.py : An Efficient Program for End-State Free Energy Calculations. J Chem Theory Comput. 2012, 8: 3314-3321.
173. Genheden S, Luchko T, Gusarov S, Kovalenko A, Ryde U: An MM/3D-RISM Approach for Ligand Binding Affinities. J Phys Chem B. 2010, 114: 8505-8516.
174. Berendsen HJC, Grigera JR, Straatsma TP: The missing term in effective pair potentials. J Phys Chem. 1987, 91: 6269-6271.
175. Caldwell JW, Kollman PA: Structure and properties of neat liquids using nonadditive molecular dynamics: water, methanol, and N-methylacetamide. J Phys Chem. 1995, 99: 6208-6219.
176. Jorgensen WL, Chandrasekhar J, Madura JD, Impey RW, Klein ML: Comparison of simple potential functions for simulating liquid water. J Chem Phys. 1983, 79: 926-
177. Price DJ, Brooks CL: A modified TIP3P water potential for simulation with Ewald summation. J Chem Phys. 2004, 121: 10096-
178. Jorgensen WL, Madura JD: Temperature and size dependence for Monte Carlo simulations of TIP4P water. Mol Phys. 1985, 56: 1381-1392.
179. Horn HW, Swope WC, Pitera JW, Madura JD, Dick TJ, Hura GL, Head-Gordon T: Development of an improved four-site water model for biomolecular simulations: TIP4P-Ew. J Chem Phys. 2004, 120: 9665-
180. Mahoney MW, Jorgensen WL: A five-site model for liquid water and the reproduction of the density anomaly by rigid, nonpolarizable potential functions. J Chem Phys. 2000, 112: 8910-
181. Toukan K, Rahman A: Molecular-dynamics study of atomic motions in water. Phys Rev B. 1985, 31: 2643-2648.
182. Ryckaert J-P, Ciccotti G, Berendsen HJC: Numerical integration of the cartesian equations of motion of a system with constraints: molecular dynamics of n-alkanes. J Comput Phys. 1977, 23: 327-341.
183. Elber R, Ruymgaart AP, Hess B: SHAKE parallelization. Eur Phys J Spec Top. 2011, 200: 211-223.
184. Ye X, Cai Q, Yang W, Luo R: Roles of boundary conditions in DNA simulations: analysis of ion distributions with the finite-difference Poisson-Boltzmann method. Biophys J. 2009, 97: 554-562.
185. Crowley M, Darden T, Iii TC, Ii DD: Adventures in Improving the Scaling and Accuracy of a Parallel Molecular Dynamics Program. J Supercomput. 1997, 11: 255-278.
186. Darden T, York D, Pedersen L: Particle mesh Ewald: An N log(N) method for Ewald sums in large systems. J Chem Phys. 1993, 98: 10089-10092.
187. Essmann U, Perera L, Berkowitz ML, Darden T, Lee H, Pedersen LG: A smooth particle mesh Ewald method. J Chem Phys. 1995, 103: 8577-8593.
188. Darden T, Perera L, Li L, Pedersen L: New tricks for modelers from the crystallography toolkit: the particle mesh Ewald algorithm and its use in nucleic acid simulations. Struct Lond Engl. 1993 1999, 7: R55-R60.
189. Adler M, Beroza P: Improved ligand binding energies derived from molecular dynamics: replicate sampling enhances the search of conformational space. J Chem Inf Model. 2013, 53: 2065-2072.
190. Mitsutake A, Sugita Y, Okamoto Y: Generalized-ensemble algorithms for molecular simulations of biopolymers. Biopolymers. 2001, 60: 96-123.
191. Laio A, Parrinello M: Escaping free-energy minima. Proc Natl Acad Sci U S A. 2002, 99: 12562-12566.
192. Tiwary P, Parrinello M: From metadynamics to dynamics.

Phys Rev Lett. 2013, 111: 230602-

193. Hukushima K, Nemoto K: Exchange Monte Carlo Method and Application to Spin Glass Simulations. J Phys Soc Jpn. 1996, 65: 1604-1608.

194. Hukushima K, Takayama H, Nemoto K: Application of an extended ensemble method to spin glasses. Int J Mod Phys C. 1996, 07: 337-344.

195. Chu W-T, Zhang J-L, Zheng Q-C, Chen L, Zhang H-X: Insights into the folding and unfolding processes of wild-type and mutated SH3 domain by molecular dynamics and replica exchange molecular dynamics simulations. PLoS One. 2013, 8: e64886-

196. Patra MC, Rath SN, Pradhan SK, Maharana J, De S: Molecular dynamics simulation of human serum paraoxonase 1 in DPPC bilayer reveals a critical role of transmembrane helix H1 for HDL association. Eur Biophys J EBJ. 2014, 43: 35-51.

197. Harris RC, Boschitsch AH, Fenley MO: Influence of grid spacing in Poisson-Boltzmann equation binding energy estimation. J Chem Theory Comput. 2013, 9: 3677-3685.

198. Reddy MR, Reddy CR, Rathore RS, Erion MD, Aparoy P, Reddy RN, Reddanna P: Free Energy Calculations to Estimate Ligand-Binding Affinities in Structure-Based Drug Design. Curr Pharm Des. 2014, 20: 3323-3337.

199. Hou T, Wang J, Li Y, Wang W: Assessing the performance of the MM/PBSA and MM/GBSA methods. 1. The accuracy of binding free energy calculations based on molecular dynamics simulations. J Chem Inf Model. 2011, 51: 69-82.

200. Kar P, Lipowsky R, Knecht V: Importance of polar solvation and configurational entropy for design of antiretroviral drugs targeting HIV-1 protease. J Phys Chem B. 2013, 117: 5793-5805.

201. Osterberg F, Aqvist J: Exploring blocker binding to a homology model of the open hERG K+ channel using docking and molecular dynamics methods. FEBS Lett. 2005, 579: 2939-2944.

202. Jenwitheesuk E, Samudrala R: Virtual screening of HIV-1 protease inhibitors against human cytomegalovirus protease using docking and molecular dynamics. AIDS Lond Engl. 2005, 19: 529-531.

203. Tatsumi R, Fukunishi Y, Nakamura H: A hybrid method of molecular dynamics and harmonic dynamics for docking of flexible ligand to flexible receptor. J Comput Chem. 2004, 25: 1995-2005.

204. Hu S, Yu H, Liu Y, Xue T, Zhang H: Insight into the binding model of new antagonists of kappa receptor using docking and molecular dynamics simulation. J Mol Model. 2013, 19: 3087-3094.

205. Hu W, Deng S, Huang J, Lu Y, Le X, Zheng W: Intercalative interaction of asymmetric copper(II) complex with DNA: experimental, molecular docking, molecular dynamics and TDDFT studies. J Inorg Biochem. 2013, 127: 90-98.

206. Huang X, Zheng G, Zhan C-G: Microscopic binding of M5 muscarinic acetylcholine receptor with antagonists by homology modeling, molecular docking, and molecular dynamics simulation. J Phys Chem B. 2012, 116: 532-541.

207. Moitessier N, Henry C, Maigret B, Chapleur Y: Combining pharmacophore search, automated docking, and molecular dynamics simulations as a novel strategy for flexible docking. Proof of concept: docking of arginine - glycine - aspartic acid-like compounds into the αvβ3 binding site. J Med Chem. 2004, 47: 4178-4187.

208. NVIDIA: Computational Chemistry & Biology Benchmark Report. 2013, NVIDIA: Santa Clara, CA

209. von R SP: Encyclopedia of Computational Chemistry. 1998, Chichester: New York: J. Wiley

210. Tuttle T, Thiel W: OMx-D: semiempirical methods with orthogonalization and dispersion corrections. Implementation and biochemical application. Phys Chem Chem Phys PCCP. 2008, 10: 2159-2166.

211. Yilmazer ND, Korth M: Comparison of molecular mechanics, semi-empirical quantum mechanical, and density functional theory methods for scoring protein-ligand interactions. J Phys Chem B. 2013, 117: 8075-8084.

212. Voityuk AA: Semi-empirical methods: current status and future directions. Drug Des Strateg Comput Tech Appl. Edited by: Banting L, Clark T. 2012, Cambridge: Royal Society of Chemistry, 107-119.

213. Koch W, Holthausen MC: A Chemist's Guide to Density Functional Theory. 2001, Weinheim, New York: Wiley-VCH

214. Putz MV, Mingos DMP: Applications of Density Functional Theory to Biological and Bioinorganic Chemistry. Berlin. 2013, New York: Springer

215. Grimme S: Density functional theory with London dispersion corrections. Wiley Interdiscip Rev Comput Mol Sci. 2011, 1: 211-228.

216. Klimeš J, Michaelides A: Perspective: Advances and challenges in treating van der Waals dispersion forces in density functional theory. J Chem Phys. 2012, 137: 120901-

217. Korth M: Error estimates for (semi-)empirical dispersion terms and large biomacromolecules. Org Biomol Chem. 2013, 11: 6515-6519.

218. Antony J, Grimme S: Fully ab initio protein-ligand interaction energies with dispersion corrected density functional theory. J Comput Chem. 2012, 33: 1730-1739.

219. Mucs D, Bryce RA: The application of quantum mechanics in structure-based drug design. Expert Opin Drug Discov. 2013, 8: 263-276.

220. Hobza P: Calculations on noncovalent interactions and databases of benchmark interaction energies. Acc Chem Res. 2012, 45: 663-672.

221. De Benedetti PG, Fanelli F: Computational quantum chemistry and adaptive ligand modeling in mechanistic QSAR. Drug Discov Today. 2010, 15: 859-866.

222. Haag MP, Reiher M: Real-time quantum chemistry. Int J Quantum Chem. 2013, 113: 8-20.

223. Asadchev A, Gordon MS: New multithreaded hybrid CPU/GPU approach to Hartree–Fock. J Chem Theory Comput. 2012, 8: 4166-4176.

224. Olivares-Amaya R, Watson MA, Edgar RG, Vogt L, Shao Y, Aspuru-Guzik A: Accelerating correlated quantum chemistry calculations using graphical processing units and a mixed precision matrix multiplication library. J Chem Theory Comput. 2010, 6: 135-144.

225. Asadchev A, Gordon MS: Fast and flexible coupled cluster implementation. J Chem Theory Comput. 2013, 9: 3385-3392.

226. Maia JDC, Urquiza Carvalho GA, Mangueira CP, Santana SR, Cabral LAF, Rocha GB: GPU linear algebra libraries and GPGPU programming for accelerating MOPAC semiempirical quantum chemistry calculations. J Chem Theory Comput. 2012, 8: 3072-3081.

227. Wu X, Koslowski A, Thiel W: Semiempirical quantum chemical calculations accelerated on a hybrid multicore CPU–GPU computing platform. J Chem Theory Comput. 2012, 8: 2272-2281.

228. Yasuda K: Accelerating density functional calculations with graphics processing unit. J Chem Theory Comput. 2008, 4: 1230-1236.

229. Ufimtsev IS, Martínez TJ: Quantum chemistry on graphical processing units. 1. Strategies for two-electron integral evaluation. J Chem Theory Comput. 2008, 4: 222-231.

230. Ufimtsev IS, Martinez TJ: Quantum chemistry on graphical processing units. 2. Direct self-consistent-field implementation. J Chem Theory Comput. 2009, 5: 1004-1015.

231. Ufimtsev IS, Martinez TJ: Quantum chemistry on graphical processing units. 3. Analytical energy gradients, geometry optimization, and first principles molecular dynamics. J Chem Theory Comput. 2009, 5: 2619-2628.

232. Luehr N, Ufimtsev IS, Martínez TJ: Dynamic precision for electron repulsion integral evaluation on graphical processing units (GPUs). J Chem Theory Comput. 2011, 7: 949-954.

233. Kulik HJ, Luehr N, Ufimtsev IS, Martinez TJ: Ab initio quantum chemistry for protein structures. J Phys Chem B. 2012, 116: 12501-12509.

234. Titov AV, Ufimtsev IS, Luehr N, Martinez TJ: Generating efficient quantum chemistry codes for novel architectures. J Chem Theory Comput. 2013, 9: 213-221.

235. Ufimtsev IS, Luehr N, Martinez TJ: Charge transfer and polarization in solvated proteins from ab Initio molecular dynamics. J Phys Chem Lett. 2011, 2: 1789-1793.

236. Mori T, Martínez TJ: Exploring the conical intersection seam: the seam space nudged elastic band method. J Chem Theory Comput. 2013, 9: 1155-1163.

237. Kussmann J, Beer M, Ochsenfeld C: Linear-scaling self-consistent field methods for large molecules. Wiley Interdiscip Rev Comput Mol Sci. 2013, 3: 614-636.

238. Leszczynski J: Linear-Scaling Techniques in Computational Chemistry and Physics: Methods and Applications. 2011, New York: Springer

239. Gordon MS, Fedorov DG, Pruitt SR, Slipchenko LV: Fragmentation methods: a route to accurate calculations on large systems. Chem Rev. 2012, 112: 632-672.

240. Alexeev Y, Mazanetz MP, Ichihara O, Fedorov DG: GAMESS as a free quantum-mechanical platform for drug research. Curr Top Med Chem. 2012, 12: 2013-2033.

241. Day PN, Jensen JH, Gordon MS, Webb SP, Stevens WJ, Krauss M, Garmer D, Basch H, Cohen D: An effective fragment method for modeling solvent effects in quantum mechanical calculations. J Chem Phys. 1996, 105: 1968-1986.

242. Gordon MS, Freitag MA, Bandyopadhyay P, Jensen JH, Kairys V, Stevens WJ: The effective fragment potential method: a QM-based MM approach to modeling environmental effects in chemistry. J Phys Chem A. 2001, 105: 293-307.

243. Gordon MS, Slipchenko L, Li H, Jensen JH: The effective fragment potential: a general method for predicting intermolecular interactions. Annu Rep Comput Chem. Edited by: Spellmeyer DC, Wheeler R. 2007, Oxford: Elsevier Science, 3: 177-193.

244. Fedorov DG, Kitaura K: The importance of three-body terms in the fragment molecular orbital method. J Chem Phys. 2004, 120: 6832-6840.

245. Imamura A, Aoki Y, Maekawa K: A theoretical synthesis of polymers by using uniform localization of molecular orbitals: Proposal of an elongation method. J Chem Phys. 1991, 95: 5419-5431.

246. Kobayashi M, Nakai H: Divide-and-Conquer Approaches to Quantum Chemistry: Theory and Implementation. 2011

247. Fedorov DG, Kitaura K: Extending the power of quantum chemistry to large systems with the fragment molecular orbital method. J Phys Chem A. 2007, 111: 6904-6914.

248. Fedorov D, Kitaura K: The Fragment Molecular Orbital Method: Practical Applications to Large Molecular Systems. 2009, Boca Raton: CRC Press/Taylor & Francis

249. Fedorov DG, Nagata T, Kitaura K: Exploring chemistry with the fragment molecular orbital method. Phys Chem Chem Phys PCCP. 2012, 14: 7562-7577.

250. Munro AW, Scrutton NS: Enzyme mechanisms: fast reaction and computational approaches. Biochem Soc Trans. 2009, 37 (Pt 2): 333-335.

251. Scheiner S, Lipscomb WN: Molecular orbital studies of enzyme activity: catalytic mechanism of serine proteinases. Proc Natl Acad Sci U S A. 1976, 73: 432-436.

252. Warshel A, Levitt M: Theoretical studies of enzymic reactions: dielectric, electrostatic and steric stabilization of the carbonium ion in the reaction of lysozyme. J Mol Biol. 1976, 103: 227-249.

253. Kamerlin SCL, Haranczyk M, Warshel A: Progress in ab initio QM/MM free-energy simulations of electrostatic energies in proteins: accelerated QM/MM studies of pKa, redox reactions and solvation free energies. J Phys Chem B. 2009, 113: 1253-1272.

254. Mulholland AJ, Grant GH, Richards WG: Computer modelling of enzyme catalysed reaction mechanisms. Protein Eng. 1993, 6: 133-147.

255. Mulholland AJ, Karplus M: Simulations of enzymic

reactions. Biochem Soc Trans. 1996, 24: 247-254.

256. Friesner RA, Guallar V: Ab initio quantum chemical and mixed quantum mechanics/molecular mechanics (QM/MM) methods for studying enzymatic catalysis. Annu Rev Phys Chem. 2005, 56: 389-427.
257. Senn HM, Thiel W: QM/MM methods for biomolecular systems. Angew Chem Int Ed Engl. 2009, 48: 1198-1229.
258. Lin H, Truhlar DG: QM/MM: what have we learned, where are we, and where do we go from here?. Theor Chem Acc. 2007, 117: 185-199.
259. Lonsdale R, Ranaghan KE, Mulholland AJ: Computational enzymology. Chem Commun Camb Engl. 2010, 46: 2354-2372.
260. Ridder L, Harvey JN, Rietjens IMCM, Vervoort J, Mulholland AJ: Ab Initio QM/MM Modeling of the Hydroxylation Step in p-Hydroxybenzoate Hydroxylase. J Phys Chem B. 2003, 107: 2118-2126.
261. Ridder L, Mulholland AJ: Modeling biotransformation reactions by combined quantum mechanical/molecular mechanical approaches: from structure to activity. Curr Top Med Chem. 2003, 3: 1241-1256.
262. Ranaghan KE, Mulholland AJ: Conformational effects in enzyme catalysis: QM/MM free energy calculation of the 'NAC' contribution in chorismate mutase. Chem Commun. 2004, 1238-1239. 10
263. Martí S, Roca M, Andrés J, Moliner V, Silla E, Tuñón I, Bertrán J: Theoretical insights in enzyme catalysis. Chem Soc Rev. 2004, 33: 98-107.
264. Warshel A, Sharma PK, Kato M, Xiang Y, Liu H, Olsson MHM: Electrostatic basis for enzyme catalysis. Chem Rev. 2006, 106: 3210-3235.
265. Claeyssens F, Harvey JN, Manby FR, Mata RA, Mulholland AJ, Ranaghan KE, Schütz M, Thiel S, Thiel W, Werner H-J: High-accuracy computation of reaction barriers in enzymes. Angew Chem Int Ed Engl. 2006, 45: 6856-6859.
266. Woodcock HL, Hodošček M, Sherwood P, Lee YS, Iii HFS, Brooks BR: Exploring the quantum mechanical/molecular mechanical replica path method: a pathway optimization of the chorismate to prephenate Claisen rearrangement catalyzed by chorismate mutase. Theor Chem Acc. 2003, 109: 140-148.
267. Van der Kamp MW, Zurek J, Manby FR, Harvey JN, Mulholland AJ: Testing high-level QM/MM methods for modeling enzyme reactions: acetyl-CoA deprotonation in citrate synthase. J Phys Chem B. 2010, 114: 11303-11314.
268. Van der Kamp MW, Mulholland AJ: Computational enzymology: insight into biological catalysts from modelling. Nat Prod Rep. 2008, 25: 1001-1014.
269. Mulholland AJ, Lyne PD, Karplus M: Ab initio QM/MM study of the citrate synthase mechanism. A low-barrier hydrogen bond is not involved. J Am Chem Soc. 2000, 122: 534-535.
270. Garcia-Viloca M, Gao J, Karplus M, Truhlar DG: How enzymes work: analysis by modern rate theory and computer simulations. Science. 2004, 303: 186-195.
271. Cui Q, Karplus M: Quantum mechanics/molecular mechanics studies of triosephosphate isomerase-catalyzed reactions: effect of geometry and tunneling on proton-transfer rate constants. J Am Chem Soc. 2002, 124: 3093-3124.
272. Bjelic S, Aqvist J: Computational prediction of structure, substrate binding mode, mechanism, and rate for a malaria protease with a novel type of active site. Biochemistry (Mosc). 2004, 43: 14521-14528.
273. Van der Kamp MW, Shaw KE, Woods CJ, Mulholland AJ: Biomolecular simulation and modelling: status, progress and prospects. J R Soc Interface R Soc. 2008, 5 (Suppl 3): S173-S190.
274. Warshel A: Computer simulations of enzyme catalysis: methods, progress, and insights. Annu Rev Biophys Biomol Struct. 2003, 32: 425-443.
275. Bentzien J, Muller RP, Florián J, Warshel A: Hybrid ab initio quantum mechanics/molecular mechanics calculations of free energy surfaces for enzymatic reactions: the nucleophilic attack in subtilisin. J Phys Chem B. 1998, 102: 2293-2301.
276. Warshel A: Computer Modeling of Chemical Reactions in Enzymes and Solutions. 1997, New York: Wiley
277. Villà J, Warshel A: Energetics and dynamics of enzymatic reactions. J Phys Chem B. 2001, 105: 7887-7907.
278. Warshel A, Sharma PK, Chu ZT, Aqvist J: Electrostatic contributions to binding of transition state analogues can be very different from the corresponding contributions to catalysis: phenolates binding to the oxyanion hole of ketosteroid isomerase. Biochemistry (Mosc). 2007, 46: 1466-1476.
279. Bjelic S, Aqvist J: Catalysis and linear free energy relationships in aspartic proteases. Biochemistry (Mosc). 2006, 45: 7709-7723.
280. Trobro S, Aqvist J: Mechanism of peptide bond synthesis on the ribosome. Proc Natl Acad Sci U S A. 2005, 102: 12395-12400.
281. Sharma PK, Xiang Y, Kato M, Warshel A: What are the roles of substrate-assisted catalysis and proximity effects in peptide bond formation by the ribosome?. Biochemistry (Mosc). 2005, 44: 11307-11314.
282. Hammes-Schiffer S: Quantum-classical simulation methods for hydrogen transfer in enzymes: a case study of dihydrofolate reductase. Curr Opin Struct Biol. 2004, 14: 192-201.
283. Liu H, Warshel A: Origin of the temperature dependence of isotope effects in enzymatic reactions: the case of dihydrofolate reductase. J Phys Chem B. 2007, 111: 7852-7861.
284. Sharma PK, Chu ZT, Olsson MHM, Warshel A: A new paradigm for electrostatic catalysis of radical reactions in vitamin B12 enzymes. Proc Natl Acad Sci U S A. 2007, 104: 9661-9666.
285. Gao J, Truhlar DG: Quantum mechanical methods for enzyme kinetics. Annu Rev Phys Chem. 2002, 53: 467-505.
286. Ranaghan KE, Mulholland AJ: Computer simulations of quantum tunnelling in enzyme-catalysed hydrogen

transfer reactions. Interdiscip Sci Comput Life Sci. 2010, 2: 78-97.

287. Lonsdale R, Harvey JN, Mulholland AJ: Inclusion of dispersion effects significantly improves accuracy of calculated reaction barriers for cytochrome P450 catalyzed reactions. J Phys Chem Lett. 2010, 1: 3232-3237.

288. Himo F, Siegbahn PEM: Quantum chemical studies of radical-containing enzymes. Chem Rev. 2003, 103: 2421-2456.

289. Siegbahn PEM, Himo F: Recent developments of the quantum chemical cluster approach for modeling enzyme reactions. J Biol Inorg Chem JBIC Publ Soc Biol Inorg Chem. 2009, 14: 643-651.

290. Mackerell AD: Empirical force fields for biological macromolecules: Overview and issues. J Comput Chem. 2004, 25: 1584-1604.

291. Vreven T, Byun KS, Komáromi I, Dapprich S, Montgomery JA, Morokuma K, Frisch MJ: Combining quantum mechanics methods with molecular mechanics methods in ONIOM. J Chem Theory Comput. 2006, 2: 815-826.

292. Zhang Y: Pseudobond ab initio QM/MM approach and its applications to enzyme reactions. Theor Chem Acc. 2006, 116: 43-50.

293. Meier K, Thiel W, van Gunsteren WF: On the effect of a variation of the force field, spatial boundary condition and size of the QM region in QM/MM MD simulations. J Comput Chem. 2012, 33: 363-378.

294. Riccardi D, Schaefer P, Cui Q: pKa calculations in solution and proteins with QM/MM free energy perturbation simulations: a quantitative test of QM/MM protocols. J Phys Chem B. 2005, 109: 17715-17733.

295. Rodríguez A, Oliva C, González M, van der Kamp M, Mulholland AJ: Comparison of different quantum mechanical/molecular mechanics boundary treatments in the reaction of the hepatitis C virus NS3 protease with the NS5A/5B substrate. J Phys Chem B. 2007, 111: 12909-12915.

296. Sumowski CV, Ochsenfeld C: A convergence study of QM/MM isomerization energies with the selected size of the QM region for peptidic systems. J Phys Chem A. 2009, 113: 11734-11741.

297. Pezeshki S, Lin H: Adaptive-partitioning redistributed charge and dipole schemes for QM/MM dynamics simulations: on-the-fly relocation of boundaries that pass through covalent bonds. J Chem Theory Comput. 2011, 7: 3625-3634.

298. Lodola A, Mor M, Zurek J, Tarzia G, Piomelli D, Harvey JN, Mulholland AJ: Conformational effects in enzyme catalysis: reaction via a high energy conformation in fatty acid amide hydrolase. Biophys J. 2007, 92: L20-L22.

CHAPTER 14 PHYSICS-BASED RNA STRUCTURE PREDICTION

Xiaojun Xu, and Shi-Jie Chen

Keywords: RNAVfold2D structure prediction,3D structure prediction, Tertiary motif

INTRODUCTION

The increasing discoveries of noncoding RNAs demand more than ever the information about RNA structure (Bachellerie et al. 2002; Kertesz et al. 2007; He et al. 2008; Bartel 2009; Gong and Maquat 2011; Wang et al. 2013). For example, the 3D structure of a microRNA–target complex is crucial for understanding microRNA's binding affinity and efficacy in gene regulation (Kertesz et al. 2007; Bartel 2009). However, the time-consuming, laborious, expensive experimental determination, such as X-ray crystallographic and NMR spectroscopic measurements, alone cannot catch up the pace with the rapidly increasing number of biologically significant RNAs such as noncoding regulatory RNAs. This problem highlights the need for computational prediction of RNA folding.

The structure of an RNA is determined by the complex pattern of base–base interactions, including base-paired secondary structures and long-range tertiary interactions. Existing RNA folding theories mainly focus on the secondary structures. However, knowing the secondary structure information alone is not sufficient to determine the 3D structure, because a 3D structure often involves long-range tertiary interactions such as kissing interactions between the different loops. Therefore, for a physics-based approach, accurate evaluation of the energetic parameters for tertiary interactions is critical for 3D structure prediction. Moreover, RNA function is correlated not only to the minimum free energy state of an RNA, but also to the potentially large conformational changes it can undergo. Understanding RNA function requires the understanding of the full energy landscape.

Statistical mechanics-based modeling has led to significant success in RNA structure prediction, folding stabilities, and folding kinetics for structures with different structural complexities (Liu et al. 2006; Chen 2008). For example, a recently developed statistical mechanics-based RNA folding model, "Vfold" model, has provided a wide range of quantitative predictions and novel insights for a variety of experiments and RNA functions, such as the pseudoknot-involved conformational switch between bistable secondary structures (Xu and Chen 2012), microRNA gene regulation through microRNA/mRNA-binding interactions (Cao and Chen 2012), and RNA/RNA dimerization critical for viral replication (Cao and Chen 2011; Cao et al. 2014). However, despite the success of this approach, several key issues remain. Estimation of the entropies for RNA tertiary folds and extraction of the energy parameters for noncanonical tertiary interactions from thermodynamic data or known structures present major challenges hampering the structural modeling for large and complex RNAs. The primary focus of this article is on the statistical mechanics-based methods for predicting RNA 3D structures and folding energy landscapes, and the related quantitative insights into RNA functions.

AN OVERVIEW OF COMPUTATIONAL METHODS FOR RNA FOLDING

RNA folding process is believed to be partly hierarchical, whereby secondary structural motifs fold first followed by the tertiary contacts formation. The secondary structure is a set of helices containing canonical base pairs (A–U, G–C, and G–U) and contributes to the major part of the free energy of an RNA system. Canonical base pairing and base stacking within helices are generally stronger than the non-canonical interactions in loop parts of an RNA system. Therefore, many computational models dissect the RNA folding problem into

Citation: Xiaojun Xu and Shi-Jie Chen, Physics-based RNA structure prediction, DOI 10.1007/s41048-015-0001-4.

two steps: from sequence to two-dimensional (2D) structure and from 2D structure to three-dimensional (3D) structure, where a 2D structure is defined by base pairs including tertiary cross-linked base pairs such as kissing base pairs. With the 2D structure as constraint, the accuracy of 3D structure prediction can be significantly improved.

2D structure predictions

Computational models for RNA 2D structure prediction fall into two general categories: free energy minimization (Ding and Lawrence 2003; Hofacker 2003; Zuker 2003; Xayaphoummine et al. 2005; Mathews and Turner 2006; Parisien and Major 2008; Bellaousov et al. 2013; Xu et al. 2014), and sequence comparison (Gutell et al. 2002; Hofacker et al. 2002; Mathews and Turner 2002; Havgaard et al. 2005; Bindewald and Shapiro 2006; Bernhart et al. 2008; Sato et al. 2009). Most free energy minimization methods employ the empirical thermodynamic parameters [the Turner parameters (Turner and Mathews 2010)] for the different secondary structural elements. Other models, such as MC-Fold (Parisien and Major 2008), use knowledge-based energy functions extracted from the known PDB structures. However, not all the interactions (such as long-range tertiary contacts) can be captured by these parameters. As a result, the accuracy of prediction falls off rapidly with the length of the sequence, because larger RNAs are more prone to the formation of long-range tertiary contacts. The accuracy of computational predictions is usually better for methods that consider "fold recognition": structure is usually more conserved than sequence, and the functional core regions are usually more conserved at all levels. In general, sequence comparison-based methods can give more improved predictions than free-energy-based predictions if the homologous sequences are available. However, these methods depend strongly on the availability of the sequence database. To overcome the above limitations, several hybrid algorithms that combine free energy minimization and sequence comparison have been developed (Mathews and Turner 2002; Havgaard et al. 2005; Bernhart et al. 2008). For example, Dynalign (Mathews and Turner 2002) combines free energy minimization and comparative sequence analysis to find a low free energy structure common to two sequences without requiring any sequence identity. On average, Dynalign predicted 86.1 % of known base-pairs in the tRNAs, compared to 59.7 % by free energy minimization alone. For the 5S rRNAs, the average accuracy improves from 47.8 to 86.4 %.

Another way to improve the accuracy of structure prediction is to incorporate experimental data to the secondary structure prediction modeling (Mathews et al. 2004; Deigan et al. 2009; Low and Weeks 2010; Kladwang et al. 2011; Hajdin et al. 2013; Leonard et al. 2013). Selective 2′-hydroxyl acylation analyzed by primer extension (SHAPE) probing data has proved useful for RNA secondary structure modeling (Deigan et al. 2009; Low and Weeks 2010; Kladwang et al.2011; Hajdin et al. 2013; Leonard et al. 2013). The SHAPE information provides refinements for the experimental determined thermodynamic parameters (Turner and Mathews 2010) for RNA folding. Benchmark test for a set of 21 RNAs of size from 34 to 530 nt shows that 93 % on average of known base pairs can be predicted. and all pseudoknots in well-folded RNAs can be identified (Hajdin et al. 2013).

Tertiary motifs: free energy models

Unlike the entropy (free-energy) parameters for simple loops (hairpin, bulge, and internal loops), which have been determined from thermodynamic experiments (Turner and Mathews2010). Quantitative understanding of many other interactions remains very limited. Moreover, because of the possible conformational coupling between the different loops and between loops and helices, the loop entropies are not additive for tertiary motifs such as loop–loop kissing contacts (Fig. 1). Previous studies on the kissing complexes and other RNA folding systems such as pseudoknots suggested that a reliable estimation for the entropy is indispensable for folding predictions (Cao and Chen 2006, 2009; Andronescu et al. 2010a, b). Accurate treatment for the entropy and free energy for tertiary structure formation is a bottleneck.

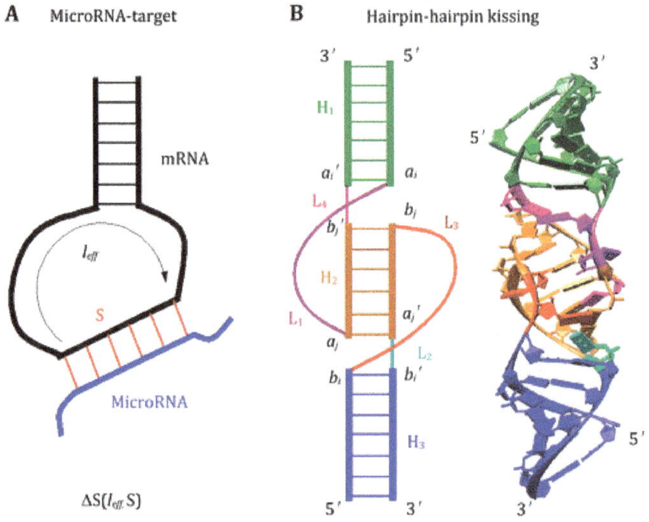

Figure 1: **A** A schematic figure for the microRNA–target-binding complex. The entropic change upon the binding between microRNA and the mRNA ΔS(l_{eff}, S) depends on the length of the binding site S and the effect loop length l_{eff}. **B** A schematic diagram and all-atom structure of the hairpin–hairpin kissing complex

Dirks and Pierce (2003) introduced a simplified energy model for H-type pseudoknots:

$$G_{pseudo} = \beta_1 + \beta_2 B^P + \beta_3 U^P,$$

where β_1 is the penalty for introducing a pseudoknot, B^P is the number of base pairs that border the interior of the pseudoknot, and U^P is the number of unpaired bases inside the pseudoknot. Later, this energy model was extended (Sperschneider et al. 2011) to parameterize hairpin–hairpin kissing motifs, as shown in Fig. 1. In essence, by decoupling the interplay between helices and loops in the tertiary motifs, this energy model approximates the non-additive energy with an additive model.

Based on the polymer physics theory (de Cloizeaux 1974; Grosberg and Khokhlov 1994), Aalberts et al. (2013) proposed the following expressions (Meng and Aalberts 2013) for the free energy cost of stretching mRNA hairpin loops upon small RNA binding (Fig. 1):

$$\beta G(N,M,z) = \frac{3}{2}\ln\left(N^{\frac{6}{5}}a^2 + M^{\frac{6}{5}}b^2\right) + \beta C_0 - \frac{5}{18}\ln\left(\frac{z-2.4}{R_F-2.4}\right)$$
$$-(N+M-2)\ln\left(\frac{Na+Mb-z}{Na+Mb-R_F}\right)$$
$$+\left(\frac{5}{2R_F} + \frac{5}{18(R_F-2.4)} - \frac{N+M-2}{Na+Mb-R_F}\right)$$
$$(z-R_F),$$

where N is the number of single-stranded backbone segments (of length a = 6.2 Å), and M is the number of helix crossing segments (of length b = 15 Å). The Flory radius $R_F = \left(N^{\frac{6}{5}}a^2 + M^{\frac{6}{5}}b^2\right)^{1/2}$ represents the characteristic end-to-end separation of a self-avoiding chain (Aalberts and Hodas 2010). $\beta = k_B T$ and the constant C_0 can be set on the basis of experiment. The parameter z is the end-to-end separation of a helix, which can be calculated as

$$z(s) = \left\{(hs)^2 + r^2\left[1-\cos\left(\frac{2\pi s}{11}\right)\right]^2 + r^2\sin^2\left(\frac{2\pi s}{11}\right)\right\}^{1/2}$$
$$= \left[(hs)^2 + 4r^2\sin^2\left(\frac{2\pi s}{11}\right)\right]^{1/2}$$

for an A-form RNA helix of (s + 1) base pairs, with h = 2.7 Å and r = 9.9 Å. We note that the freely jointed chain (FJC) model used to derive the above free energy upon small RNA binding to mRNA hairpin loops does not consider the excluded volume effect between the A-form helix and the single-stranded chain; moreover, the FJC model can only provide an estimation for long chains.

To compute entropy of hairpin, bulge/internal and multibranch loops of long length (up to 50 nt), with an efficient sampling method based on the sequential Monte Carlo principle, Zhang et al. (2008) developed optimized discrete k-state models based on RNA backbone conformations in known RNA structures. The method is general and can be applied to calculating entropy of loops with high complexity.

3D structure predictions

RNA 3D structure prediction is still at its early stage (Shapiro et al. 2007; Andersen 2010; Laing and Schlick 2011; Rother et al. 2011; Sim et al. 2012). Current RNA folding algorithms are generally limited to simple (short) structures, hampered by the challenges including adequate treatment of the conformational sampling and the evaluation of the energies for the tertiary contacts. Table 1 describes some of the most recently developed algorithms, ranging from coarse-grained modeling to various structure–assembly, and other conformational sampling approaches.

Table 1: A partial list for the computational models for RNA tertiary structure prediction and interactive manipulation

Model	Simulation method	References
YAMMP/YUP	Coarse-grained (1-bead/nt)	Wang et al. (1999) and Tan et al. (2006)
NAST	Coarse-grained (1-bead/nt)	Jonikas et al. (2009)
iFoldRNA	Coarse-grained (3-bead/nt)	Sharma et al. (2008)
CG	Coarse-grained (3-bead/nt)	Shi et al. (2014)
CG	Coarse-grained (5-bead/nt)	Xia et al. (2010, 2013)
HiRE-RNA	Coarse-grained (6- or 7-bead/nt)	Pasquali and Derreumaux (2010)
DMD	Discrete molecular dynamics	Ding et al. (2008)
RAG	Graph theory	Izzo et al. (2011) and Kim et al. (2014)
FARNA/FARFAR	Fragment assembly	Das and Baker (2007) and Das et al. (2010)
MC-Sym	Fragment assembly	Parisien and Major (2008)
Vfold	Coarse-grained (3-bead/nt) and motif-based template assembly	Cao and Chen (2011)
3dRNA	Secondary-elements assembly	Zhao et al. (2012)
BARNACLE	Probabilistic model for sampling	Frellsen et al. (2009)
RNA2D3D	Interactive manipulation	Martinez et al. (2008)
Assemble	Interactive manipulation	Jossinet et al. (2010)

Coarse-grained approaches

Coarse-grained representation can largely reduce the degrees of freedom and thus enhance the conformational sampling. YAMMP/YUP (Wang et al. 1999; Tan et al. 2006) and NAST (Jonikas et al. 2009) represent RNA with just one pseudo-atom per nucleotide residue: P and C3′, respectively. iFoldRNA (Sharma et al. 2008) and Vfold (Cao and Chen 2005; Shi et al. 2014) represent RNA by three pseudo-atoms per residue. Ren (Xia et al. 2010, 2013) uses 5-bead to represent each nucleotide, and HiRE-RNA (Pasquali and Derreumaux 2010) uses six or seven pseudo-atoms for purine and pyrimidine residues, respectively. Coarse-grained systems are usually modeled with knowledge-based potentials that are derived from known structures. Combined with discrete molecular dynamics (DMD) (Ding et al. 2008) or other similar methods, this approach has the potential to predict structures and folding mechanism for large RNAs. For example, a recently developed 3-bead model (Shi et al. 2014) can achieve 3.5 Å RMSD on average for 46 small RNAs including pseudoknots. Combined with Monte Carlo-simulated annealing algorithm and a coarse-grained force field with implicit salt, the model may provide reliable predictions for the stability and salt effect with the mean deviation ~1.0 °C of melting temperatures, compared with the extensive experimental data for 30 RNA hairpins.

Another coarse-grained approach is the graph theory-based tool (RAG) (Izzo et al. 2011; Kim et al. 2014) for sampling RNA global helical topologies. RAG represents RNA 2D structure as planar tree or dual graphs to assist the cataloging, analyzing, and designing of RNA structures. With the knowledge-based potential for internal loop orientations, such as bending and torsion of internal loops, the combination of graph theory and Monte Carlo-simulated annealing sampling shows great promise for assembling global features of RNA architecture: graph RMSDs range from 2.52 to 28.24 Å for RNAs of the size 25–158 nucleotides.

Structure–assembly approaches

Based on the assumption that the 3D fold is more conserved and can be recognized by the alignment of sequences and secondary structure patterns, the template-based modeling (Das and Baker 2007; Parisien and Major 2008; Das et al. 2010; Cao and Chen 2011; Zhao et al.2012) has shown promising achievements in RNA 3D structure predictions. In structure-assembly approaches, RNA 3D structures are built based on the known structures modules ranging from fragments of 1–3 nucleotides to 2D structural motifs.

FARNA/FARFAR (Das and Baker 2007; Das et al. 2010) models RNA 3D structures by assembling of short fragments (1–3 nucleotides) from a single crystal structure via a Monte Carlo procedure guided by a knowledge-based energy function that encodes base-stacking and base-pairing potentials. It can reach atomic resolution (<3.0 Å) for most short RNAs (<30 nt). MC-Sym (Parisien and Major 2008) builds all-atom structures using the 3D version of the nucleotide cyclic motif (NCM) fragments. The 3D NCM library was built from a list of 531 known RNA 3D structures. Due to the limited NCM fragments for large, complex NCM motifs, such as 6-way junctions and kissing loops, current MC-Sym is limited to short RNAs requiring 2D structures as input. 3dRNA (Zhao et al. 2012) builds the whole RNA structure from the smallest secondary elements (SSEs) by a two-step assembling procedure. The SSEs are defined as base-pair hairpin, internal/bulge loop, pseudoknot loop and junction, which are extracted from known structures.

One of the common limitations to the structure–assembly approaches is the degree of divergence of the fragment library. Given the limited number of known RNA structures, structural motif templates with the required high sequence identity are difficult to attain. The lack of reliable structural motifs for loops and junctions greatly hampers accurate 3D RNA structure prediction. Moreover, the template-based structure prediction models cannot predict structures with "new" motifs.

As for the coarse-grained models, incorporating experimental data can dramatically improve the accuracy for the structure–assembly approaches. For instance, constraints using structural inference of native RNAs by high throughput contact mapping, such as the multiplexed hydroxyl radical (–OH) cleavage analysis (MOHCA), improve the FARNA's prediction (Das et al. 2008). For the 158-nt P4–P6 domain of the group I intron, MOHCA leads to an improvement of RMSD from 35 Å with FARNA to 13 Å.

Sampling algorithms

One of the challenges for current RNA structure prediction is the problem of conformational sampling. Even for the DMD with knowledge-based energy functions at different coarse-grained levels, a major issue is that sampled conformations often remain close to the initial starting model (Sim et al. 2012). The molecular system is trapped in its local energy minima for the most part of the computational time, and the barriers between local minima on the energy landscape hinder transitions between different low-energy states. To overcome this difficulty requires the use of special simulation techniques (Li and Scheraga 1987; Rahman and Tully 2002; Minary et al. 2004; Curuksu and Zacharias 2009) to achieve effective sampling of conformational space.

A probabilistic model, called BARNACLE (Frellsen et al. 2009), allows for efficient sampling of RNA conformations in continuous space and with related probabilities. Using coarse-grained base-pairing information, BARNACLE generates reasonable RNA-like structures for small RNAs (<50 nt). However, the method is mostly limited to short RNAs because of the rapid increase in complexity of the probabilistic model.

Interactive manipulation

Many RNA structure design algorithms, such as RNA2D3D (Martinez et al. 2008) and Assemble (Jossinet et al. 2010), are quite efficient. This interactive graphical tools are useful to analyze and build RNA architectures, but have less ability for RNA structure predictions, since they rely on manual application of expert knowledge.

VFOLD: FROM SEQUENCE TO 3D ALL-ATOM STRUCTURES

Vfold (web server (Xu et al. 2014): http://rna.physics.missouri.edu) is a model used to predict RNA 2D and 3D structures and the folding stability from the sequence. The model distinguishes itself from other models by two unique features: physics-based modeling of conformational entropy for 2D structure prediction, and template-based multiscale modeling for 3D structure prediction.

Entropy parameters for tertiary motifs

Using the P-C4′ and C4′-P virtual bonds to represent the backbone conformations, the Vfold model (Cao and Chen 2005) samples loops/junction conformations in the 3D space through conformational enumeration (Xu et al. 2014). By calculating the probability of loop formation, the model estimates the conformational entropy parameters for the formation of the different types of loops such as pseudoknot loops and hairpin–hairpin kissing motifs. The model has the advantage of accounting for chain connectivity, excluded volume, and the completeness of conformational ensemble. Studies by us and other groups show that an accurate entropy parameter improves the prediction of RNA secondary structures and thermodynamic stabilities (Andronescu et al. 2010). Here, we use the hairpin–hairpin kissing motif to illustrate the Vfold calculation for the entropy of an RNA/RNA kissing complex.

The hairpin–hairpin kissing complex, shown in Fig. 1B, consists of three stems and four loops. We assume loop l_2 and l_4 are short, with ≤ 1 nucleotide, which favors the formation of coaxial stacking interaction between stem H_1 and H_2 and between H_2 and H_3. Therefore, the entropic cost upon the formation of loop–loop kissing $S(H_2, l_1, l_3)$

depends on the length of the stem H_2 and the single-stranded loops of l_1 and l_3. The computation involves three steps:

- Due to the nature of the coaxial stacking between stems of H_1, H_2, and H_3, the relative orientation between stems of H_1 and H_3 is determined by the length of stem H_2. The coordinates of the 8 nt $(a_i, a_i', a_j, a_j', b_i, b_i', b_j, b_j',$, shown in Fig. 1B) are adopted from the known NMR structure as the template. The final coordinates of the 8 nt for different length of H_2 are generated according to the A-formed H_2 and the template.
- For each helix orientation, with well-defined (a_i, a_j) of the starting and ending nucleotides for the loop l_1 and (b_i, b_j) of the starting and ending nucleotides for the loop l_3, we model loop conformations as self-avoiding walks of the virtual bonds on diamond lattice (Cao and Chen 2005) to sample loops/junctions 3D conformations. The connection between the A-form helix and the discrete loop conformations is realized through an iterative optimized algorithm (Ferro and Hermans 1971).
- A key issue in the conformational count is the excluded volume interaction between loop and helix and between the different loops. In the Vfold model, this can be explicitly taken into account by disallowing overlapping virtual bonds when the loop conformations are generated in the virtual bond diamond lattice. Assuming the interactions in the loops are weaker than the base stacking interactions that stabilize a 2D structure, we can estimate the loop entropy parameter as the logarithm of the conformational count.

The Vfold-predicted loop entropies (Table 2) enable folding free energy calculations for RNA/RNA complex such as for microRNA–target-binding (Cao and Chen 2012) and hairpin–hairpin kissing complex systems (Cao and Chen 2011; Cao et al. 2014). For example, for the kissing complex shown in Fig. 2, the free energy is computed as

$$\begin{aligned}\Delta G &= \Delta G_{\text{stem1}} + \Delta G_{\text{stem2}} + \Delta G_{\text{kissing-stem}} \\ &\quad - T\Delta S(S, l_1, l_2, l_3, l_4) \\ &= -15.7 - 15.7 - 14.2 + 8.62(k_B T) \\ &= -40.2 \text{ kcal/mol},\end{aligned}$$

where S is the number of base pairs in the kissing stem, and l_1, l_2, l_3, l_4 are the length of loops 1, 2, 3, and 4, respectively. The entropic energy of kissing loop is estimated by $k_B \ln \frac{\omega_{6,2,2}}{\omega_{\text{coil}}}$, with $\ln\omega_{6,2,2}$ is read from Table 1, and $\ln\omega_{\text{coil}} = 2.05\, l + 0.1$ (l is the chain length of loop l_1 or l_3) from the polymer physics theory.

Table 2: The Vfold-derived conformational entropies $\ln\omega_{H_2,l_1,l_3}$ for the kissing complex for the different stem lengths and different loop lengths

l_3	$H_2 = 3$							$H_2 = 4$						
	1	2	3	4	5	6	7	1	2	3	4	5	6	7
$l1 = 2$	–	0	0	1.8	2.6	4.2	5.8	–	1.1	0.7	1.4	3.4	5.0	6.7
$l_1 = 3$	–	0	–	1.6	1.1	1.4	2.5	–	0.7	1.4	0.7	3.4	4.9	6.6
$l_1 = 4$	–	1.8	1.6	3.8	4.2	5.8	7.4	–	1.4	0.7	–	2.7	4.1	5.7
$l_1 = 5$	–	2.6	1.1	4.2	4.1	5.4	7.0	–	3.4	3.4	2.7	5.3	6.7	8.4
$l_1 = 6$	–	4.2	1.4	5.8	5.4	6.3	7.8	–	5.0	4.9	4.1	6.7	7.9	9.5
$l_1 = 7$	–	5.8	2.5	7.4	7.0	7.8	9.3	–	6.7	6.6	5.7	8.4	9.5	11.2
l_3	$H_2 = 5$							$H_2 = 6$						
	1	2	3	4	5	6	7	1	2	3	4	5	6	7
$l_1 = 1$	0	1.4	1.4	2.8	3.7	5.2	6.7	–	–	–	–	–	–	–
$l_1 = 2$	1.4	2.8	2.4	4.1	4.8	6.3	7.8	–	0	0.7	1.1	2.2	3.3	4.7
$l_1 = 3$	1.4	2.4	2.1	3.7	4.4	5.8	7.3	–	0.7	1.8	2.3	3.7	5.1	6.7
$l_1 = 4$	2.8	4.1	3.7	5.4	6.1	7.6	9.0	–	1.1	2.3	2.7	4.0	5.2	6.8
$l_1 = 5$	3.7	4.8	4.4	6.1	6.8	8.3	9.7	–	2.2	3.7	4.0	5.5	6.6	8.2
$l_1 = 6$	5.2	6.3	5.8	7.6	8.3	9.7	11.2	–	3.3	5.1	5.2	6.6	7.6	9.2
$l_1 = 7$	6.7	7.8	7.3	9.0	9.7	11.2	12.6	–	4.7	6.7	6.8	8.2	9.2	10.9
l_3	$H_2 = 7$							$H_2 = 8$						
	1	2	3	4	5	6	7	1	2	3	4	5	6	7
$l1 = 2$	–	–	–	–	–	–	–	–	–	–	–	–	–	–
$l_1 = 3$	–	–	–	–	–	–	–	–	–	–	0	2.2	4.1	6.1
$l_1 = 4$	–	–	–	2.2	3.3	5.0	6.7	–	–	0	0.7	2.4	4.2	6.1
$l_1 = 5$	–	–	–	3.3	4.2	6.0	7.6	–	–	2.2	2.4	3.9	5.5	7.3
$l_1 = 6$	–	–	–	5.0	6.0	7.8	9.5	–	–	4.1	4.2	5.5	7.0	8.7
$l_1 = 7$	–	–	–	6.7	7.6	9.5	11.2	–	–	6.1	6.1	7.3	8.7	10.4

The unit of the entropies is k_B

The unit of the entropies is k_B

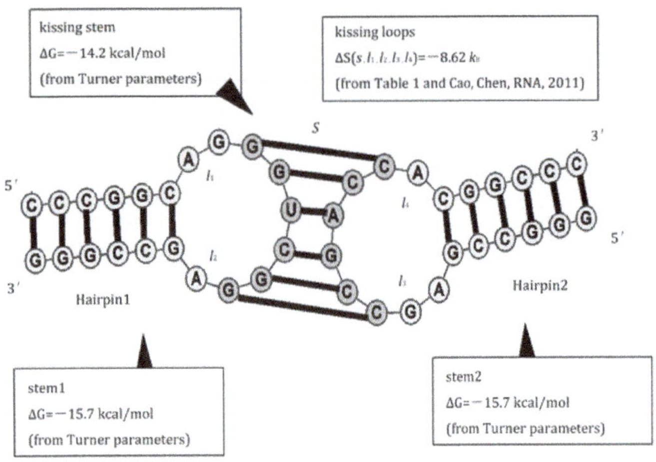

Figure 2: The evaluation of the free energy for a hairpin–hairpin kissing complex using the loop entropy parameters in Table 1 in Cao and Chen (2011) and the Turner parameters (Turner and Mathews 2010)

To treat more complicated (general) loop–loop kissing complexes, as shown in the Fig. 2 of Cao et al. (2014), the stem-loop substructure's impact on loop entropy is approximated by replacing the terminal base pairs (of the stems) by single nucleotides. Then, the effective loop lengths of l_1, l_2, l_3, and l_4 are equal to the sum of the number of unpaired nucleotides and the number of stem-loop substructures. This approximation ignores the (weak) excluded volume interference between the stem-loop substructure and the loop, thus enabling us to treat general kissing motifs.

Template-based RNA 3D structure prediction

Predicting RNA 3D structure is not a solved problem (Shapiro et al. 2007; Andersen 2010; Laing and Schlick 2011; Rother et al. 2011; Sim et al. 2012). Extensive efforts have been made to enhance the conformational sampling (Das and Baker 2007; Frellsen et al. 2009) and to establish accurate scoring functions for the ranking of the different structures (Ding et al. 2008; Parisien and Major 2008). The Vfold model can predict the 2D structures of RNA/RNA kissing complex including inter- and intramolecular base pairings. In general, a 2D structure can correspond to a large number of 3D structures due to the multiplicity of flexible loop conformations. The Vfold model-predicted virtual bond structure provides a scaffold for the construction of all-atom models of the 3D structure. The prediction of the all-atom 3D structures from a given sequence and 2D structure (base pairs) involves the following three steps (see Fig. 3):

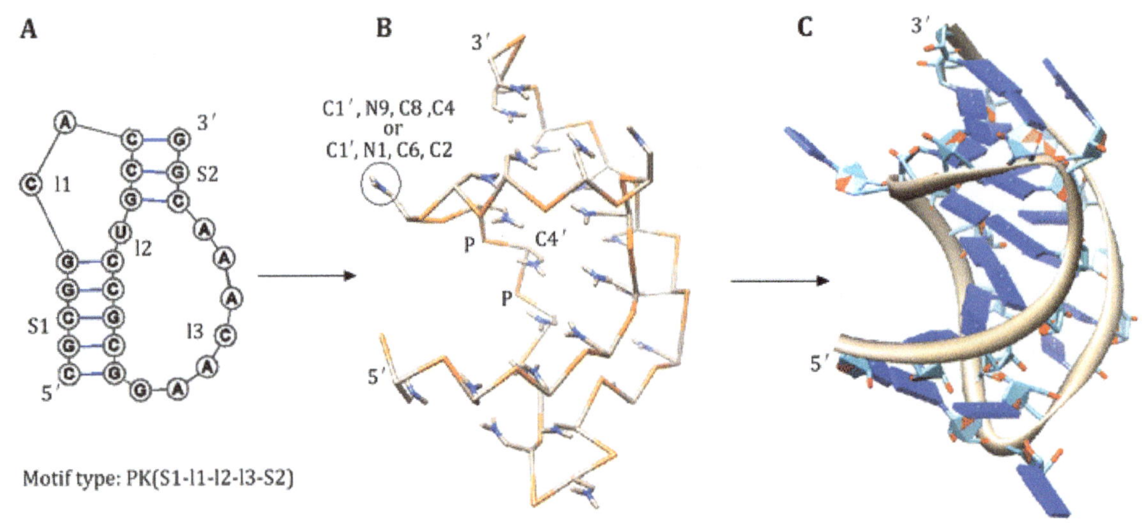

Figure 3: **A** The 2D structure of the BWYV pseudoknot. Vfold identifies it as a motif of "PK(5-2-1-7-3)". **B** The virtual-bond (low-resolution) structure built from the motif-based template library. **C** The all-atom 3D structure refined by Amber energy minimization

- To build the 3D virtual bond structure. Helices are modeled as A-form virtual-bonded helix structures. The loop/junction structures are built from the virtual bond fragments of the template structures. To identify the optimal template structure for the loops/junctions, the model screens the pretabulated template library according to the loop size (first) and the sequence (second) matches. If necessary, this step may involve sequence replacement in order to match the (same size) sequences in the template library. The model assembles the helix and loop 3D virtual-bonded structures to construct the 3D scaffold of the whole RNA.
- To add all atoms to the virtual-bonded structure. For nucleotides in each helix, atoms are added according to the A-form helix atomic structure. The 3D conformation of the nucleotides in loops are generated by adding atoms according to the templates for base configurations, by aligning the C1′, N9, C4, and C8 for purine (A or G) or C1′, N1, C2, and C6 for pyrimidine (C or U) with those of a nucleotide in a helix. This step results in an "atomistic version" of the Vfold structure. Using three atoms instead of one atom per base, the current Vfold can better capture the base orientation from templates and also can easily replace base type accordingly.
- Energy minimization of the whole atomistic structure using AMBER molecular dynamics simulations. The above pre-refinement structure may contain atoms/groups that clash sterically with each other. Such steric clashes can be readily resolved by the all-atom molecular dynamics simulations. With the above pre-refinement structure as the initial state, the AMBER energy minimization (Case et al.2005) yields reliable

predictions for all-atom 3D structures. In the energy minimization, the negative charges on phosphates are neutralized by Na+ cations added to the solution. The nonbonded interactions are truncated at 12 Å. Water molecules are treated by the standard TIP3P model included in AMBER software. For the most predicted structures, we found that the minimization causes only small change in the RMSD of the structure. The main advantage of the multiscale approach in the Vfold model is that the virtual-bonded tertiary structures as the initial state may already lie in the free energy basin, so the structure refinement can avoid large structural rearrangements and can thus lead to the native structure effectively.

The Vfold model predicts the 3D structure for a 2D structure based on the structural templates. To construct the template library, the model classifies the structure into different motifs, such as helices, hairpin loops, internal/bulge loops, pseudoknots, and N-way junctions ($N \geq 3$). The motif-based template library was built from 2621 PDB structures. With the increasing number of known RNA structures, the larger and more divergent pools of the known loop/junction structures with the different types and different lengths would lead to better predictions of the 3D structure.

As shown in Fig. 3, the above strategy gives reliable predictions for the all-atom 3D structures for simple tertiary folds, such as pseudoknots and hairpin–hairpin kissing complexes. The predicted structures as a 3D scaffold will provide highly needed guidance for experiments. For example, the sequential resonance assignments from the Nuclear Overhauser Effect (NOE) data may become efficient and more accurate if the information on the nucleotide spatial proximity from the predicted (low-resolution) structure is combined with for the NMR structural determination of RNA.

QUANTITATIVE PREDICTION FOR THE FOLDING OF HIV-1 DIS COMPLEX

Intermolecular loop–loop base pairing is a widespread and functionally important tertiary structure motif in RNA. Loop–loop interactions often facilitate dimerization reactions between RNA molecules. For example, in HIV-1 virus, the loop–loop kissing interaction is critical for one form of HIV-1 dimerization (Laughrea and Jette 1994; Muriaux et al. 1996; Paillart et al. 2004). In bacteria, loop–loop interaction can regulate gene expression and affect replication and translation of the bacteria (Schmidt et al. 1995; Argaman and Altuvia 2000; Repoila et al. 2003; Bossi and Figueroa-Bossi 2007; Vogel and Wagner 2007). A well-documented case is OxyS RNA repression of fhlA translation in Escherichia Coli through the formation of a stable loop-kissing interaction (Argaman and Altuvia 2000).

The dimerization process is essential for the HIV-1 replication. Muriaux et al. proposed a two-step dimerization process (Muriaux et al. 1996a, b). The kissing loop–loop complex is formed followed by a conversion to form the extended-duplex dimer due to temperature increase or protein binding. Both the kissing-loop dimer and the extended-duplex have been found in the structural measurement (Ennifar et al. 1999, 2001). Due to the lack of the thermodynamic parameters for the kissing-loop dimer, it has been difficult to determine the relative population of each dimer at the different temperature. Also, it would be biologically important to understand if the kissing-loop dimer is a kinetic intermediate or a thermodynamic stable state at room temperature. Vfold model provides a useful tool to quantitatively predict the thermodynamic stabilities for the different dimes by computing the free energy landscape of the two-stranded system (Cao and Chen 2011, 2012; Cao et al. 2014). Recently developed RNA structure prediction models are good at predicting some structures at high-accuracy resolution. For example, de novo predictive models can accurately predict the simple and short hairpin and internal loop structures (Das and Baker 2007; Ding et al. 2008; Parisien and Major 2008). However, the models cannot predict the kissing complex. The Vfold model enables the prediction of kissing complexes (Cao and Chen 2011).

Quantitative prediction of HIV-1 DIS complex requires modeling of the folding energy landscape and the structures of dimers. The partition function for the two-stranded system $Q(T) = Q_1 \cdot Q_2 + e^{-\Delta G_{\text{associate}}/k_B T} \cdot Q12$ is the sum over the unbound and bound systems. Here, Q_1, Q_2, and Q_{12} are the partition functions of the (unbound) strand 1 RNA, of the (unbound) strand 2 RNA and of the kissing (bound) system, respectively. $\Delta G_{\text{associate}}$ is dependent on the RNA concentration C_T: $\Delta G_{\text{associate}} = \Delta G_{\text{init}} - k_B T \ln(C_T/4)$. We choose ΔG_{init} to be 4.1 kcal/mol according to the experimental result (Serra and Turner 1995; Zuker 2003). The calculation of Q_1 and Q_2 for single-stranded RNA can be achieved by many RNA secondary structure prediction models; however, the computation of Q_{12} requires a statistical mechanical model such as the Vfold model.

The predicted free energy landscape for HIV-1 Mal shows two free energy minima, indicating two coexisting structures at room temperature, shown in Fig. 4A. The structural (base-pairing probability) calculations show that the free energy minima correspond to the kissing-loop dimer and the extended-duplex dimer, respectively. The extended-duplex dimer is slightly more stable than the kissing-loop dimer, with the free energy difference $\Delta G < 1.0$ kcal/mol. The result suggests that the two modes of dimerization of HIV-1 Mal can coexist in thermodynamic equilibrium and can possibly interconvert with the change of the temperature and solution condition.

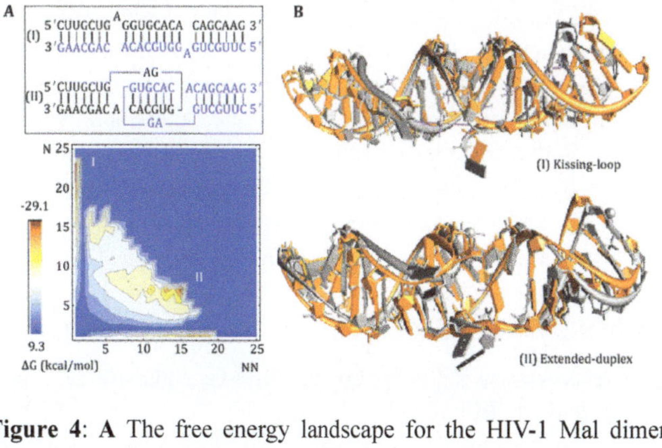

Figure 4: **A** The free energy landscape for the HIV-1 Mal dimer at T = 20 °C. The Vfold model predicts two coexisting structure (I, II), corresponding to the extended-duplex and kissing-loop dimers; respectively. In the energy landscape, N and NN are the numbers of the native and non-native base pairs, respectively. **B** The Vfold predicts 3D structures (inorange) for the kissing-loop and extended-duplex dimers for HIV-1 Mal dimer. The all-atom RMSDs are 3.1 and 2.9 Å

with respect to the experimental structures (in gray) with PDB IDs 1xpe and 462d, respectively

Moreover, based on the Vfold study, we find that the kissing-loop dimer of HIV-1 Mal is stabilized by the coaxial stacking. We built the 3D structure of the kissing-loop dimer according to the above multiscale strategy. The all-atom RMSD between predicted structure and the experiment solved NMR structure (PDB id: 1xpe) is 3.1 Å, as shown in Fig. 4B. For the extended-duplex dimer (structure I on the energy landscape), Vfold predicts the 3D structure with an RMSD of 2.9 Å (PDB id: 462d).

CONCLUSIONS

The bottleneck for RNA tertiary structure prediction is the inability to treat the free energy, especially the entropy, for structures with long-range tertiary interactions. The virtual-bond based low-resolution conformational model (Vfold) allows us to estimate the entropy and the full free energy landscape for RNA tertiary global folds. The predicted 2D structures provide scaffolds for the construction of all-atom 3D models through molecular dynamics calculations. Validation by the experimental data for the RNA 2D and 3D structures and the folding thermodynamics as well as kinetics suggests that the statistical mechanics-based approaches can be quite reliable.

By considering the non-canonical interactions at both secondary structure and tertiary structure levels, Vfold model improves the accuracy of the secondary structure prediction and introduces more detailed constraints, besides the canonical base-pairing information, to the 3D structure modeling. However, a fast sampling algorithm balancing the completeness of the conformational space including non-canonical base-pairing information and the computational time is needed to treat large RNAs. Moreover, further development of the model should go beyond the simple hairpin–hairpin kissing complexes by estimating the entropic parameters for global folds with more complicate tertiary interactions, such as the SAM riboswitch.

With the rapidly growing size of the database of the experimentally measured RNA structures, motif template-based methods shows increasingly promising results, especially when the homologous conformations can be identified from the known structures. However, a backup plan is always needed for a good model if no known homologous conformations can be found in the PDB database. For instance, there is only a few of hairpin–hairpin kissing motifs in the current motif library. Further development of the model should address the motifs involving tertiary interactions with the ability of de novo construction.

ACKNOWLEDGMENTS

This research was supported by the NIH Grant GM063732 and NSF Grant MCB0920411.

REFERENCES

1. Aalberts DP, Hodas NO (2010) A two-length-scale polymer theory for RNA loop free energies and helix stacking. RNA 16:1350–1355
2. Andersen ES (2010) Prediction and design of DNA and RNA structures. New Biotechnol 27:184–192
3. Andronescu MS, Pop C, Condon A (2010a) Improved free energy parameters for RNA pseudoknotted secondary structure prediction. RNA 16:26–42
4. Andronescu MS, Condon A, Hoos HH, Mathews DH, Murphy KP (2010b) Computational approaches for RNA energy parameter estimation. RNA 16:2304–2318
5. Argaman L, Altuvia S (2000) fhlA repression by OxyS RNA: kissing complex formation at two sites results in a stable antisense-target RNA complex. J Mol Biol 300:1101–1112
6. Bachellerie JP, Cavaille J, Huttenhofer A (2002) The expanding snoRNA world. Biochimie 84:774–790
7. Bartel DP (2009) MicroRNAs: target recognition and regulatory functions. Cell 136:215–233
8. Bellaousov S, Reuter JS, Seetin MG, Methews DH (2013) RNAstructure: web servers for RNA secondary structure prediction and analysis. Nucleic Acids Res 41:W471–W474
9. Bernhart SH, Hofacker IL, Will S, Gruber AR, Stadler PF (2008) RNAalifold: improved consensus structure prediction for RNA alignments. BMC Bioinformatics 11:474
10. Bindewald E, Shapiro BA (2006) RNA secondary structure prediction from sequence alignments using a network of k-nearest neighbor classifiers. RNA 12:342–352
11. Bossi L, Figueroa-Bossi N (2007) A small RNA downregulates LamB maltoporin in Salmonella. Mol Microbiol 65:799–810
12. Cao S, Chen S-J (2005) Predicting RNA folding thermodynamics with a reduced chain representation model. RNA 11:1884–1897
13. Cao S, Chen S-J (2006) Predicting RNA pseudoknot folding thermodynamics. Nucleic Acids Res 34:2634–2652
14. Cao S, Chen S-J (2009) Predicting structures and stabilities for H-type pseudoknots with interhelix loops. RNA 15:696–706
15. Cao S, Chen S-J (2011a) Structure and stability of RNA/RNA kissing complex: with application of HIV dimerization initiation signal. RNA 17:2130–2143
16. Cao S, Chen S-J (2011b) Physics-based de novo prediction of RNA 3D structures. J Phys Chem B 115:4216–4226
17. Cao S, Chen S-J (2012a) Predicting kissing interactions in microRNA-target complex and assessment of microRNA activity. Nucleic Acids Res 40:4681–4690
18. Cao S, Chen S-J (2012b) A domain-based model for predicting large and complex pseudoknotted structures. RNA Biol 9:201–212
19. Cao S, Xu X, Chen S-J (2014) Predicting structure and stability for RNA complexes with intermolecular loop–loop base pairing. RNA 20:835–845
20. Case DA, Cheatham TE, Darden T, Gohlke H, Luo R, Merz KM, Onufriev A, Simmerling C, Wang B,

Woods RJ (2005) The Amber biomolecular simulation programs. J Comput Chem 26:1668–1688

21. Chen SJ (2008) RNA folding: conformational statistics, folding kinetics, and ion electrostatics. Annu Rev Biophys 37:197–214
22. Curuksu J, Zacharias M (2009) Enhanced conformational sampling of nucleic acids by a new Hamiltonian replica exchange molecular dynamics approach. J Chem Phys 130:104110
23. Das R, Baker D (2007) Automated de novo prediction of native-like RNA tertiary structures. Proc Natl Acad Sci USA 104:14664–14669
24. Das R, Kudaravalli M, Jonikas M, Laederach A, Fong R, Schwans JP, Baker D, Piccirilli JA, Altman RB, Herschlag D (2008) Structural inference of native and partially folded RNA by high-throughput contact mapping. Proc Natl Acad Sci USA 105:4144–4149
25. Das R, Karanicolas J, Baker D (2010) Atomic accuracy in predicting and designing noncanonical RNA structure. Nat Methods 7:291–294
26. Deigan KE, Li TW, Mathews DH, Weeks KM (2009) Accurate SHAPE-directed RNA structure determination. Proc Natl Acad Sci USA 106:97–102
27. des Cloizeaux J (1974) Lagrangian theory for a self-avoiding random chain. Phys Rev A 10:1665–1669
28. Ding Y, Lawrence CE (2003) A statistical sampling algorithm for RNA secondary structure prediction. Nucleic Acids Res 31:7280–7301
29. Ding F, Sharma S, Chalasani P, Demidov VV, Broude NE, Dokholyan NV (2008) Ab initio RNA folding by discrete molecular dynamics: from structure prediction to folding mechanisms. RNA 14:1164–1173
30. Dirks RM, Pierce NA (2003) A partition function algorithm for nucleic acid secondary structure including pseudoknots. J Comput Chem 24:1664–1677
31. Ennifar E, Yusupov M, Walter P, Marquet R, Ehresmann B, Ehresmann C, Dumas P (1999) The crystal structure of the dimerization initiation site of genomic HIV-1 RNA reveals an extended duplex with two adenine bulges. Structure 7:1439–1449
32. Ennifar E, Walter P, Ehresmann B, Ehresmann C, Dumas P (2001) Crystal structures of coaxially stacked kissing complexes of the HIV-1 RNA dimerization initiation site. Nat Struct Biol 8:1064–1068
33. Ferro DR, Hermans J (1971) A different best rigid-body molecular fit routine. Acta Crystallogr A 33:345–347
34. Frellsen J, Moltke I, Thiim M, Mardia KV, Ferkinghoff-Borg J, Hamelryck T (2009) A probabilistic model of RNA conformational space. PLoS Comput Biol 5:e1000406
35. Gong C, Maquat LE (2011) lncRNAs transactivate STAU1-mediated mRNA decay by duplexing with 3 UTRs via Alu elements. Nature 470:284–288
36. Grosberg AY, Khokhlov A (1994) Statistical physics of macromolecules. AIP Press, New York
37. Gutell RR, Lee JC, Connone JJ (2002) The accuracy of ribosomal RNA comparative structure models. Curr Opin Struct Biol 12:301–310
38. Hajdin CE, Bellaousov S, Huggins W, Leonard CW, Mathews DH, Weeks KM (2013) Accurate SHAPE directed RNA secondary structure modeling, including pseudoknots. Proc Natl Acad Sci USA 110:5498–5503
39. Havgaard JH, Lyngso RB, Gorodkin J (2005) The FOLDALIGN web server for pairwise structural RNA alignment and mutual motif search. Nucleic Acids Res 33:W650–W653
40. He S, Su H, Liu C, Skogerbo G, He H, He D, Zhu X, Liu T, Zhao Y, Chen R (2008) MicroRNA-encoding long non-coding RNAs. BMC Genom 9:236
41. Hofacker IL (2003) Vienna RNA secondary structure server. Nucleic Acids Res 31:3429–3431
42. Hofacker IL, Fekete M, Stadler PF (2002) Secondary structure prediction for aligned RNA sequences. J Mol Biol 319:1059–1066
43. Izzo JA, Kim N, Elmetwaly S, Schlick T (2011) RAG: an update to the RNA-As-Graphs resource. BMC Bioinformatics 31:219
44. Jonikas MA, Radmer RJ, Laederach A, Das R, Pearlman S, Herschlag D, Altman RB (2009) Coarse-grained modeling of large RNA molecules with knowledge-based potentials and structural filters. RNA 15:189–199
45. Jossinet F, Ludwig TE, Westhof E (2010) Assemble: an interactive graphical tool to analyze and build RNA architectures at the 2D and 3D levels. Bioinformatics 26:2057–2059
46. Kertesz M, Iovino N, Unnerstall U, Gaul U, Segal E (2007) The role of site accessibility in microRNA target recognition. Nat Genet 39:1278–1284
47. Kim N, Laing C, Elmetwaly S, Jung S, Curuksu J, Schlick T (2014) Graph-based sampling for approximating global helical topologies of RNA. Proc Natl Acad Sci USA 111:4079–4084
48. Kladwang W, VanLang CC, Cordero P, Das R (2011) Understanding the errors of SHAPE-directed RNA structure modeling. Biochemistry 50:8049–8056
49. Laing C, Schlick T (2011) Computational approaches to RNA structure prediction, analysis, and design. Curr Opin Struct Biol 21:306–318
50. Laughrea M, Jette L (1994) A 19-Nucleotide sequence upstream of the 5′ major splice donor is part of the dimerization domain of human immunodeficiency virus 1 genomic RNA. Biochemistry 33:13464–13474
51. Leonard CW, Hajdin CE, Karabiber F, Mathews DH, Favorov OV, Dokholyan NV, Weeks KM (2013) Principles for understanding the accuracy of SHAPE-directed RNA structure modeling. Biochemistry 52:588–595
52. Li Z, Scheraga HA (1987) Monte Carlo-minimization approach to the multiple-minima problem in protein folding. Proc Natl Acad Sci USA 84:6611–6615
53. Liu F, Tong H, Ou-Yang Z (2006) Force unfolding single RNAs. Biophys J 90:1895–1902
54. Low JT, Weeks KM (2010) SHAPE-directed RNA secondary structure prediction. Methods 52:150–158

55. Martinez HM, Maizel JV Jr, Shapiro BA (2008) RNA2D3D: a program for generating, viewing, and comparing 3-dimensional models of RNA. J Biomol Struct Dyn 25:669–683
56. Mathews DH, Turner DH (2002) Dynalign: an algorithm for finding the secondary structure common to two RNA sequences. J Mol Biol 317:191–203
57. Mathews DH, Turner DH (2006) Prediction of RNA secondary structure by free energy minimization. Curr Opin Struct Biol 16:270–278
58. Mathews DH, Disney MD, Childs JL, Schroeder SJ, Zuker M, Turner DH (2004) Incorporating chemical modification constraints into a dynamic programming algorithm for prediction of RNA secondary structure. Proc Natl Acad Sci USA 101:7287–7292
59. Meng Y, Aalberts DP (2013) Free energy cost of stretching mRNA hairpin loops inhibits small RNA binding. Biophys J 104:482–487
60. Minary P, Tuckerman ME, Martyna GJ (2004) Long time molecular dynamics for enhanced conformational sampling in biomolecular systems. Phys Rev Lett 93:150201
61. Muriaux D, De Rocquigny H, Roques BP, Paoletti J (1996a) NCp7 activates HIV-1Lai RNA dimerization by converting a transient loop–loop complex into a stable dimer. J Biol Chem 271:33686–33692
62. Muriaux D, Fosse P, Paoletti J (1996b) A kissing complex together with a stable dimer is involved in the HIV-1 Lai RNA dimerization process in vitro. Biochemistry 35:5075–5082
63. Paillart JC, Shehu-Xhilaga M, Marquet R, Mak J (2004) Dimerization of retroviral RNA genomes: an inseparable pair. Nat Rev Microbiol 2:461–472
64. Parisien M, Major F (2008) The MC-Fold and MC-Sym pipeline infers RNA structure from sequence data. Nature 452:51–55
65. Pasquali S, Derreumaux P (2010) HiRE-RNA: a high resolution coarse-grained energy model for RNA. J Phys Chem B 114:11957–11966
66. Rahman JA, Tully JC (2002) Puddle-skimming: an efficient sampling of multidimensional configuration space. J Chem Phys 116:8750–8760
67. Repoila F, Majdalani N, Gottesman S (2003) Small non-coding RNAs, co-ordinators of adaptation processes in Escherichia coli: the RpoS paradigm. Mol Microbiol 48:855–861
68. Rother K, Rother M, Boniecki M, Puton T, Bujnicki JM (2011) RNA and protein 3D structure modeling: similarities and differences. J Mol Model 17:2325–2336
69. Sato K, Hamada M, Asai K, Mituyama T (2009) CENTROIDFOLD: a web server for RNA secondary structure prediction. Nucleic Acids Res 37:W277–W280
70. Schmidt M, Zheng P, Delihas N (1995) Secondary structures of Escherichia coli antisense micF RNA, the 5′-end of the target ompF mRNA, and the RNA/RNA duplex. Biochemistry 34:3621–3631
71. Serra MJ, Turner DH (1995) Predicting thermodynamic properties of RNA. Methods Enzymol 259:242–261
72. Shapiro BA, Yingling YG, Kasprzak W, Bindewald E (2007) Bridging the gap in RNA structure prediction. Curr Opin Struct Biol 17:157–165
73. Sharma S, Ding F, Dokholyan NV (2008) iFoldRNA: three-dimensional RNA structure prediction and folding. Bioinformatics 24:1951–1952
74. Shi YZ, Wang FH, Wu YY, Tan ZJ (2014) A coarse-grained model with implicit salt for RNAs: predicting 3D structure, stability and salt effect. J Chem Phys 141:105102
75. Sim AY, Minary P, Levitt M (2012) Modeling nucleic acids. Curr Opin Struct Biol 22:1–6
76. Sperschneider J, Datta A, Wise MJ (2011) Heuristic RNA pseudoknot prediction including intramolecular kissing hairpins. RNA 17:27–38
77. Tan RK, Petrov AS, Harvey SC (2006) YUP: a molecular simulation program for coarse-grained and multiscaled models. J Chem Theory Comput 2:529–540
78. Turner DH, Mathews DH (2010) NNDB: the nearest neighbor parameter database for predicting stability of nucleic acid secondary structure. Nucleic Acids Res 38:D280–D282
79. Vogel J, Wagner EG (2007) Target identification of small noncoding RNAs in bacteria. Curr Opin Microbiol 10:262–270
80. Wang R, Alexander RW, VanLoock M, Vladimirov S, Bukhtiyarov Y, Harvey SC, Cooperman BS (1999) Three-dimensional placement of the conserved 530 loop of 16 S rRNA and of its neighboring components in the 30 S subunit. J Mol Biol 286:521–540
81. Wang W, Wang L, Wu J, Gong Q, Shi Y (2013) Hfq-bridged ternary complex is important for translation activation of rpoS by DsrA. Nucleic Acids Res 41:5938–5948
82. Xayaphoummine A, Bucher T, Isambert H (2005) Kinefold web server for RNA/DNA folding path and structure prediction including pseudoknots and knots. Nucleic Acids Res 33:W605–W610
83. Xia Z, Gardner DP, Gutell RR, Ren P (2010) Coarse-grained model for simulation of RNA three-dimensional structures. J Phys Chem B 114:13497–13506
84. Xia Z, Bell DR, Shi Y, Ren P (2013) RNA 3D structure prediction by using a coarse-grained model and experimental data. J Phys Chem B 117:3135–3144
85. Xu X, Chen S-J (2012) Kinetic mechanism of conformational switch between bistable RNA hairpins. J Am Chem Soc 134:12499–12507
86. Xu X, Zhao P, Chen S-J (2014) Vfold: a web server for RNA structure and folding thermodynamics prediction. PLoS ONE 9:e107504
87. Zhang J, Lin M, Chen R, Wang W, Liang J (2008) Discrete state model and accurate estimation of loop entropy of RNA secondary structures. J Chem Phys 128:125107

88. Zhao Y, Huang Y, Gong Z, Wang Y, Man J, Xiao Y (2012) Automated and fast building of three-dimensional RNA structures. Sci Rep 2:734

89. Zuker M (2003) Mfold web server for nucleic acid folding and hybridization prediction. Nucleic Acids Res 31:3406–3415

CHAPTER 15 RANDOM MATRIX THEORY

Feng Luo, Yunfeng Yang, Jianxin Zhong, Haichun Gao, Latifur Khan, Dorothea K Thompson, and Jizhong Zhou

BACKGROUND

The cellular system, similar to engineering systems, is modular [1]. Hartwell et al. defined a module in biological system as "a discrete unit whose function is separable from those of other modules" and suggested that the functional modules are a "critical level of biological organization" [1]. One of important characteristics of modular system is collectivity. Namely, the similarities of behaviour or properties between elements in the same module are significantly higher than similarities between elements from different modules. Moreover, the cell is a complex system with many functionally diverse elements, including proteins, DNA, RNA and small molecules. Cellular functionalities involve groups of molecules interacting to each other. Modelling cellular systems as networks with connected elements allows us to understand the properties of cellular systems [2, 3]. Thus, a module in a biological network can be defined as a sub-network that structurally has more insider links than outsider links and functionally is enriched with genes (proteins) in the same functional module.

The microarray technology, which enables massive parallel measurement of expressions of thousands of genes simultaneously, has opened up great opportunities for the systems-level understanding and elucidating of gene networks [4–6]. Various methods have been developed for inferring gene networks, such as differential equation-based network methods [7–10], Bayesian network methods [11, 12] and relevance/co-expression network methods [13–15]. Nevertheless, the inference of genome-wide gene networks currently is still constrained by the dimensionality problem, namely, number of genes is far greater than the number of experiments in microarray data.

Because of its computational simplicity and the nature of microarray data (typically noisy, highly dimensional and significantly under-sampled) [16], co-expression network methods are most commonly used for identifying cellular networks [14, 15, 17–19]. As the expressions of genes in the same function modules generally are highly correlated, gene functional modules can be revealed from gene co-expression network as network modules. The co-expression network methods first construct a correlation matrix of gene expressions, in which the Pearson correlation and the mutual information are often used. Then, the co-expression network methods assign a link to a pair of genes when the correlation between their expressions exceeds a threshold [14, 15, 17–19]. Consequently, the network structure and topology, e.g. the number, size, content and connections of modules, are subjective, depending on the thresholds chosen. Thus, it is critical to appropriately define the threshold of correlation. Currently, thresholds are usually determined by either known biological information [17–19], or by statistical comparison to randomized expression data [20, 21]. New approaches are urgently needed to determine gene networks in an automatic and objective fashion [3]. To tackle this, we developed a novel random matrix theory (RMT)-based approach to determine the threshold in this report.

Initially proposed by Wigner and Dyson in the 1960s for studying the spectrum of complex nuclei [22], RMT is a powerful approach for identifying and modelling phase transitions associated with disorder and noise in statistical physics and materials science. It has been successfully used for studying the behaviour of complex systems, such as spectra of large atoms [23], metal insulator transitions in disorder systems [24, 25], spectra of quasiperiodic systems [23, 26, 27], chaotic systems [28], brain response [29] and the stock market

Citation: Feng Luo, Yunfeng Yang, Jianxin Zhong, Haichun Gao, Latifur Khan, Dorothea K Thompson, and Jizhong Zhou, Random Matrix Theory, 8:299 doi:10.1186/1471-2105-8-299.

[30]. However, its suitability for biological systems remains largely unexplored.

RMT makes two universal predictions for real symmetric matrices: the nearest neighbour spacing distribution (NNSD) of eigenvalues (i.e., the distribution of the difference of two nearest neighbour eigenvalues) follows Gaussian orthogonal ensemble (GOE) statistics if there exists correlation between nearest-neighbour eigenvalues, while it follows Poisson statistics if there is no correlation [23]. Deviations from GOE universal prediction can be used to distinguish system-specific, non-random properties of complex systems from random noise [30]. It has been well recognized that only a portion of genes change their expressions under different experimental conditions. Thus, the correlation matrix of gene expressions is the combination of the high correlation part Mc, which signifies the correlation of gene expressions specified to changes in biological systems, and the weak correlation part Mr or so-called noise, which signifies random relations between gene expressions: M = Mc + Mr. The modularity of the cellular systems indicates that Mc is non-random and will emerge collectivity property. Based on RMT, we hypothesized that the two universal predictions are applicable to biological systems. The NNSD of M will follow GOE and the NNSD of Mc will follow Poisson distribution. The transition of NNSD between GOE and Poisson distributions can serve as a reference point to distinguish system-specific, non-random relationship embedded in correlation matrix of gene microarray data from random noise. This reference point is mathematically defined and can be used as a threshold to identify gene co-expression networks in an automatic and objective fashion.

In this report, we describe the development and application of an RMT-based approach to determine the correlation threshold for identifying co-expression networks based on the microarray data from such simple-to-complex organisms as *S. oneidensis*, *E. coli*, yeast, *A. thaliana*, *Drosophila*, mouse and human. Moreover, the resulting co-expression networks are useful for predicting the function of unknown genes, which is supported by existing information and our experimental verification.

RESULTS

In silico evaluation of RMT approach to determine correlation threshold

To test the effectiveness of the RMT based criterion to determine correlation threshold for constructing co-expression network, we constructed an *in silico* model to simulate a simplified gene co-expression network. A correlation matrix of 2,000 genes with a designated correlation threshold of 0.7 was constructed (See Methods and Materials for details). The NNSD of the *in silico* correlation matrix showed a dramatic transition at the designated threshold (Fig. 1). NNSD followed Poisson distribution at the cutoff 0.7, whereas it obeyed GOE at cutoff 0.69. The RMT approach reliably identified the designated threshold.

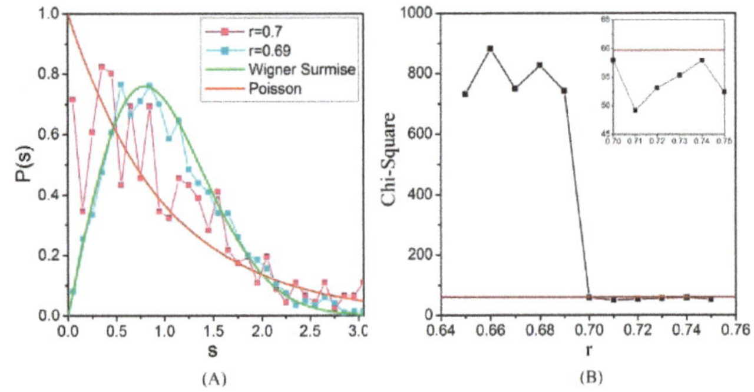

Figure 1: *In silico*evaluation of RMT approach. An *in silico* modular system was constructed at cutoff of 0.7 to simulated a simplified gene co-expression network. (A) The normalized NNSDs of correlation matrices of the system at cutoff 0.69 (cyan) and 0.7 (pink) were compared to the curves of Wigner surmise (green) and Poisson distribution (red). The x-axis is the level spacing s and the y-axis is probability of NNSDs. (B) Chi-Square Test at different cutoff values. The red line in the inset indicates the critical value of Chi-Square test of $p = 0.001$. X and Y axes represent the cutoff and the chi-square test values, respectively.

Sharp transitions of NNSD from GOE to Poisson distributions in correlation matrices from yeast microarray data

We then proceeded to apply RMT-based approach to real biological data. Yeast cell cycling microarray data [31] was selected because it has been extensively studied, making it easy to evaluate whether the results from RMT-based method are consistent with existing biological knowledge. A total of 5,293 genes with 70 time points available in the dataset were used. A correlation matrix based on pair-wise Pearson correlation coefficient in the range of (-1, 1) was calculated (See Methods for details). To simplify the analysis, the absolute cutoff values were set to be the same for both positive and negative correlations, though different cutoffs for positive and negative correlations were tested separately and similar results were obtained (data not shown).

A clear sharp transition of NNSD from GOE to Poisson distribution was observed (Fig. 2A). Based on χ^2 test ($p = 0.001$), NNSD started to deviate from GOE at the correlation coefficient $r_l = 0.62$ and completely transformed into Poisson distribution at the correlation coefficient $r_h = 0.77$. Similarly, after the missing values is estimated using the nearest neighbour based method [32], sharp transition (see additional file 1, Figure 1) also observed and the threshold remains the same, which was possible due to the small number of missing values allowed in our study (only 7 missing values allowed in total 77 experiments). Furthermore, sharp transition from GOE to Poisson was also observed in correlation matrix using mutual information (data not shown). In addition, we applied this method to another yeast microarray dataset generated from environmental stress responses of yeast [33]. A clear transition from GOE to Poisson distributions was observed likewise (r = 0.60–0.89) (data not shown).

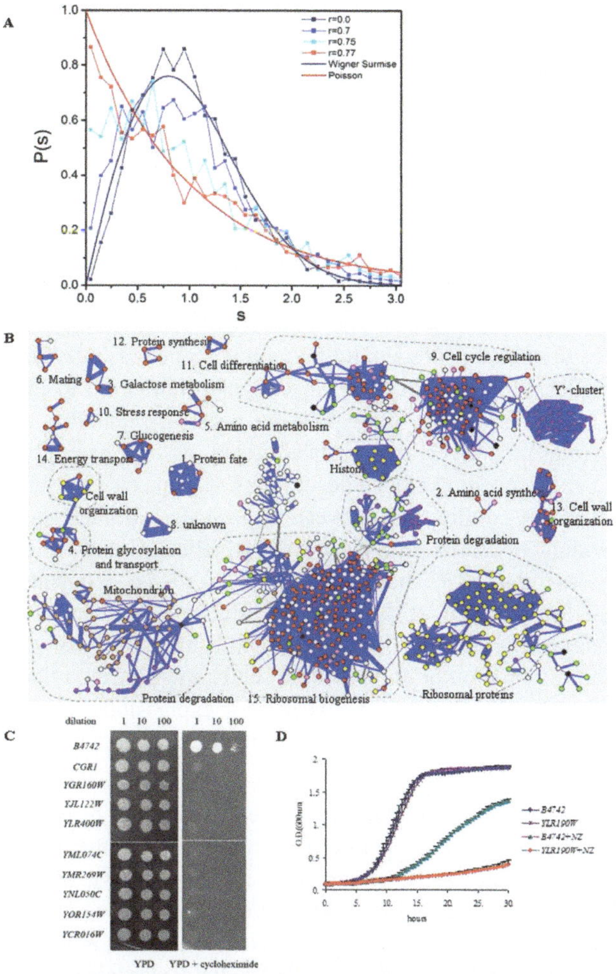

Figure 2: Transition of nearest neighbour spacing distribution and gene co-expression network from yeast cell cycle microarray profiles. (A) The normalized NNSDs of correlation matrices of yeast cell cycle gene expressions at different cutoff values. They were plotted against the curves of Wigner surmise (navy) and Poisson distribution (red). The x-axis is the level spacing s and the y-axis is the probability of NNSDs. (B) Fifteen significant gene co-expression subnetworks (modules) of the yeast cell cycling dataset were revealed at cutoff 0.77. All modules that have more than 4 genes are shown. For modules that have 3 genes, only those modules that form a cycle are shown, because only these kinds of modules are statistically significant [61]. Each node represents a gene and the width of line represents the Pearson correlation coefficient of two linked genes. Blue and gray lines indicate positive and negative correlation coefficients, respectively. Colors were assigned to nodes according to their functional categories: Red represents the major functional category of each module while purple, yellow and tan represent other functional categories, which are often clustered into sub-modules. Genes in lavender participate in processes closely related to genes in red. White nodes are unknown genes while black nodes are genes whose functional links to other genes are not currently understood. Green nodes are genes in metabolic processes, which are influenced by many biological processes. LightCyan nodes in Module 15 are genes involved in cell cycling regulation and related processes. Text in the map indicates the major functional category of each module, as represented by red. Dashed circles separate modules into sub-modules, which form independent modules at higher cutoffs. Deletion mutants and the wild-type strain *B4742*were grown in YPD overnight to saturation. Then cells were diluted 1:10 and 1:100 in water prior to spotting onto YPD or YPD containing 1 μg/ml cycloheximide plates. Images were obtained after incubation at 30°C for 4 or 7 days, respectively. (D) Growth curves of deletion mutant *YLR190W* and the parental strain *B4742*. Triplicates of *B4742* and the*YLR190W* mutant were grown in YPD or YPD containing nocodazole (NZ) at 30°C with constant agitation for 30 hrs.

Comparison of threshold obtained by RMT to that obtained by randomization

To evaluate the effectiveness of the threshold determined by RMT method, we compared it to the widely used method of determining the threshold by randomizing gene expression profiles [14, 34], using the yeast cell cycle data [31] and environmental stress responses data [33]. Gene Ontology Slim category from *Saccharomyces* Genome Database (SGD) database [35] was used to classify links. To simplify the comparison, a link connecting two genes in the some Gene Ontology Slim category is deemed to be true. As summarized in Table 1, more than half of links in all networks obtained by RMT method are true links. However, for randomized method, the networks constructed from yeast environmental stress responses data contain very low percentage of true links. This comparison indicates that randomization is poor for certain microarrays and RMT method has an advantage over randomization in identifying system-specific information embedded inside microarray profiles.

Table 1: Comparison of thresholds obtained by RMT approach and randomization method and their corresponding co-expression networks. The thresholds determined by RMT approach and randomization method on two yeast microarray expression profiles and their corresponding co-expression networks are compared. Abbreviations: MI – mutual information, and Pearson – Pearson correlation coefficient.

Microarray data	Yeast cell cycle				Yeast environmental stress responses			
	RMT		Randomization		RMT		Randomization	
Correlation measure	MI	Pearson	MI	Pearson	MI	Pearson	MI	Pearson
r_h	1.17	0.77	1.253065	0.68836	1.361	0.91	0.666697	0.533674
Number of genes	861	966	342	2398	402	523	3883	5266
Number of links	1764	3346	591	15853	4991	7160	625161	1585805
"true" links (%)	57.88	64.94	64.13	52.27	77.28	75.53	32.30	29.86

Gene co-expression network based on yeast cell cycle microarray data

From correlation matrix of yeast cell cycle microarray data, we have constructed a co-expression network at the cutoff value $r_h = 0.77$, where NNSD is completely transformed into Poisson distribution. The resulting network contains a total of 804 genes that are partitioned into 15 sub-networks (modules). Both positive and negative correlations are present in the network, as depicted in Figure 2B. To achieve an accurate evaluation based on current biological knowledge, we manually analyzed the biological coherence of modules according to gene annotations from the *Saccharomyces* Genome Database (SGD) and Munich Information centre for protein sequences (MIPS). Remarkably, all modules contain functionally coherent set of genes (Fig. 2B), demonstrating that RMT analysis faithfully reveals biologically meaningful networks in yeast. Indeed, among links of known genes, 85.4% of the links are between genes in the same or related functional pathways, whereas the rest 14.6% links are not supported by current experimental results, which might reflect the existence of systematic errors in microarray data or alternatively, the insufficiency of current biological knowledge of yeast.

Many large modules can be visually divided into smaller sub-modules, as indicated by the dashed circle in Figure 2B. For instance, Module 15 can be divided into four sub-modules: (1) ribosome proteins; (2) genes involved in ribosomal biogenesis; (3) mitochondrion proteins and (4) genes involved in protein degradation, while Module 9 contains distinct sub-modules of Y›-cluster genes, cell cycle regulators and histones. These have suggested the co-regulation at the gene expression level between different sub-modules, which often display evident functional association. For example, in Module 15, genes involved in ribosomal biogenesis are surely related to ribosome proteins, while a large portion of sub-module of mitochondrion proteins are indeed mitochondrion ribosome proteins. These results signify the presence of modular hierarchy in the network. Furthermore, sub-modules can be separated from each other by raising the cutoff. For instance, sub-modules of Module 15 are separated at cutoff of 0.79 (Additional File 1, Figure 2). Likewise, submodules of Module 9 are separated at cutoff of 0.79 and 0.81 (Additional File 1, Figure 2). Therefore, different levels of modularity of the yeast co-expression networks can be identified by further raising the cutoff values above the mathematically defined threshold.

Different types of modules are observed in the yeast co-expression network. Many (sub) modules are mainly comprised of components of protein complexes. Remarkable examples include the ribosomal protein sub-module in Module 15, in which about 90% of the known genes encode ribosomal or ribosome-associated proteins. Also, out of 10 genes of histone sub-module of Module 9, eight are histone subunits. In the case of mitochondrion, genes encoding proteins located to this small subcellular organelle are clustered into a sub-module of Module 15. In contrast, other modules cannot be classified by co-presence in protein complexes or subcellular organelles; instead, they are comprised of components in the same cellular processes. For instance, module 3 is composed of five genes participating in galactose metabolic pathway: *GAL1*, *2*, *3*, *7* and *10*. Gal1p, Gal2p, Gap7p and Gal10p function in consecutive steps of glycolysis, whereas Gal3p is a regulatory protein exerting tight transcriptional control over the galactose metabolism pathway. These genes are co-regulated at expression level but might not interact with each other directly [36]. Similarly, all five known genes in Module 7 participate in gluconeogenesis, despite the lack of physical interaction between their protein counterparts.

It has been noted that genes with similar functions do not always have similar expression profiles [17]. Although two genes are not strongly correlated in gene expression, they could both be strongly correlated with the same set of other genes, a characteristic named as "transitivity". All of these transitive genes should be grouped together in the same modules. However, major clustering algorithms fail to do so [17]. In contrast, the co-expression network method is able to detect transitively co-regulated genes, as best exemplified by genes encoding ribosome proteins. The pairwise Pearson correlation coefficients between expressions of three genes encoding ribosome proteins, RPL19B, RPL26B and RPL1A, are less than 0.3 and hence unlikely to be grouped by clustering, whereas RMT analysis correctly links them within Module 15. Similarly, both Smc1p and Bim1p are involved in spindle formation or chromosome segregation during mitosis. Although the correlation between their expressions is as low as 0.2, they are grouped together by their linkages to other cell cycle regulators such as RAD53 and KCC4.

We also constructed the gene co-expression network from the yeast cell cycle microarray profiles with the missing values estimated [32]. As shown in Table 5, the two gene co-expression networks obtained from yeast cell cycle microarray profiles with or without missing value estimation are almost the same with 95.5% genes and 93.8% links overlapped.

Functional predictions of unknown genes and experimental validation

The fact that functionally related genes are connected together in the co-expression networks sheds the light for predicting the cellular roles of hypothetical genes by "guilt-by-association" [37]. Although confidence level of the predictions has not been quantified at this moment, it can be inferred by functional uniformity among the associated genes. We have tentatively predicted the functions of 136 genes based on yeast cell cycling datasets. A few selected examples are listed in Table 2. For example, yeast*YCR072C* is associated with many genes of ribosomal biogenesis and protein synthesis in Module 15. Accordingly, the protein product of this gene has been reported recently to participate in several complexes involved in protein synthesis and RNA turnover metabolism [38]. It was also co-purified with the 60S ribosomal subunit [39]. Notably, several predictions were consistent with experimental results but were not made by other network identification methods (Table 2).

To experimentally evaluate the predictive power of co-expression network, we examined the functional association of six unknown proteins (*YJL122W*, *YML074C*, *YMR269W*, *YNL050C*,*YOR154W*, and *YCR016W*) predicted to be involved in ribosomal biogenesis. While *YJL122W*and *YCR016W* were suggested to be involved in ribosomal biogenesis by other methods (Table2), the other genes were not previously predicted to be associated with this functional process. Since we hypothesized that these genes participate in ribosomal biogenesis, we predicted that their deletion mutants might have defective ribosomes and deficiency in protein synthesis. Consequently, they should be sensitive to the protein synthesis inhibitor, cycloheximide. Indeed, a deletion mutant of *CGR1*, which is known to be involved in ribosomal biogenesis, failed to grow on YPD plates containing cycloheximide (Fig. 2D). Similarly, the deletion mutants of these six unknown genes, but not their parental strain *B4742*, failed to grow on YPD plates containing cycloheximide, indicating that protein synthesis in these mutants is defective. A possible function of ribosomal biogenesis of these unknown genes is also supported by recent high-throughput findings [40, 41]. For example, *YML074C* and *YMR269W* localize to the nucleolus [40], the cellular organelle for ribosome biogenesis.

We also examined another unknown gene, *YLR190W*, for its role in cytokinesis. The deletion mutant and its parental strain *B4742* grew similarly in YPD medium. However, the mutant showed a severe growth defect compared to *B4742* in the presence of the cytokinesis inhibitor nocodazole (Fig. 2E). This has suggested that *YLR190W* is involved in cytokinesis. In conclusion, our experiments demonstrated the prediction power of the co-expression networks.

RMT-based approach is applicable to microarray data of other tested organisms

The yeast co-expression network above was validated using existing gene annotations and experiments with deletion mutants for genes of unknown function. However, it is more desirable to apply the RMT-based method to determine correlation threshold for constructing gene co-expression networks from microarray data of little understood organisms. Since not much existing biological knowledge of these organisms could be employed to determine the threshold, it is appealing to use an automatic method to defined confident threshold. To test whether the NNSD transitions are present in correlation matrices of gene expression data from other organisms, the RMT method has been used to analyze correlation matrices of microarray data from a variety of organisms, e.g. *Shewanella oneidensis* [42], *Escherichia coli* [43],*Arabidopsis thaliana* [44], *Drosophila*[45], mouse [46] and human [47]. A clear NNSD transition from GOE to Poisson distribution has been observed in the range of 0.73–0.9 for the little understood bacterium *S. oneidensis* (Fig. 3A). Likewise, transitions have been revealed in other organisms: *E. coli* (r = 0.72–0.86), *A. thaliana* (r = 0.86–0.94), *Drosophila* (r = 0.76–0.93), mouse (r = 0.67–0.89) and human (r = 0.67–0.87) (data not shown). These results have demonstrated that RMT is applicable to gene expression data from all of the tested organisms.

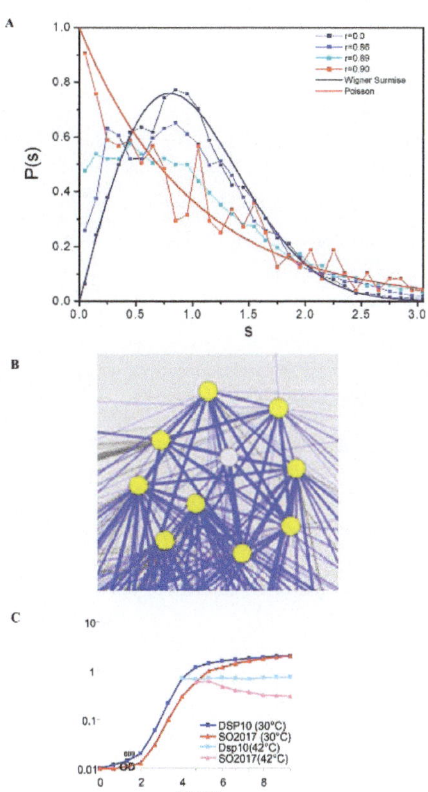

Figure 3: Transition of nearest neighbour spacing distribution and gene co-expression network from*S. oneidensis*microarray profiles. (A) The normalized NNSDs of correlation matrices of *S. oneidensis* heat/cold shock gene expressions at different cutoff values. They were plotted against the curves of Wigner surmise (navy) and Poisson distribution (red). The x-axis is the level spacing s and the y-axis is the probability of NNSDs. (B) A node representing a hypothetic protein SO2017 is interconnected to many heat shock proteins, suggesting a possible role in heat shock. The heat shock proteins are grpE, lon, SO3681, dnaJ, dnaK, groES, groEL, prlC and hslV. (C) Growth curves of SO2017 deletion mutant and its parental strain DSP10. Both strains were initially grown in LB media and shifted to 42°C when OD_{600} reached 0.69 and 0.62, respectively. In an independent experiment, the viability of the ΔSO2017 strain was reduced by 46% at 10 min after exposure to 42°C (data not shown).

The power of RMT method for determining the correlation threshold has been further evaluated in details based on genome-wide expression data from *S. oneidensis* and *E. coli*. A co-expression network of 7 modules was constructed for *S. oneidensis* heat/cold shock microarray data at cutoff of 0.90 (Additional File 1, Figure 3), and a co-expression network containing 30 modules was identified in *E. coli* dataset at a cutoff of 0.86 (Additional File 1, Figure 4). Similar to yeast data, functional modules was identified in the networks [62]. For example, modules of energy transport (#1, 2 and 4) were isolated in *S. oneidensis* co-expression network, while the big module #25 in *E. coli* dataset was dedicated to display sugar to the outer structure of bacterial surface. Taken together, the RMT-approach is useful to determine the correlation threshold for identifying gene co-expression networks in different species.

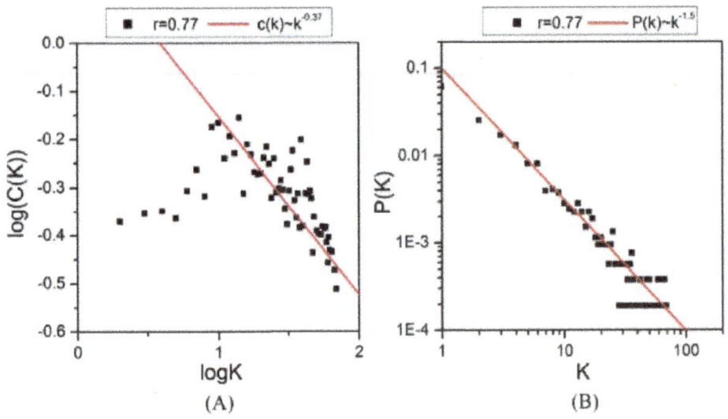

Figure 4: Structural properties of gene co-expression network from yeast cell cycling data at threshold of 0.77. (A) The dependence of the clustering coefficient on the node's degree $C(K)$. (B) The connectivity distribution $P(K)$.

Table 2: Representative functional predictions of hypothetical proteins with high confidence.

Species	Module	Gene Designation	Predicted biological pathway/ localization	Identified by other methods?	Experimental verification in this paper	Experimental supporting evidences from annotation databases
Yeast	Module 9	*YLR183C*	cell cycle	yes [19, 63]	ND	transcription factor regulating several promoters of genes involved in pheromone response and cell cycle;
	Module 9	*YOL007C*	cell cycle	yes [19, 63]	ND	structural component of the chitin synthase 3 complex
	Module 9	*YLR190W*	cytokinesis	yes [19]	Consistent with prediction	localized to small buds, bud neck, and incipient bud sites; mRNA is targeted to the bud via the mRNA transport system involving She2p
	Module 9	*YNL058C*	cell cycle	yes [18, 19]	ND	potential Cdc28p substrate
	Module 15	*YHL021C*	Stress response	yes [19, 63]	ND	
	Module 15	*YGR160W*	ribosome biogenesis	yes [19, 63]	Consistent with prediction	
	Module 15	*YIL127C*	ribosome biogenesis	yes [63]	ND	
	Module 15	*YJL122W*	ribosome biogenesis	yes [63]	Consistent with prediction	
	Module 15	*YLR196W*	ribosome biogenesis	yes [19, 63]	ND	nucleolar protein
	Module 15	*YLR400W*	ribosome biogenesis	no	Consistent with prediction	
	Module 15	*YML074C*	ribosome biogenesis	yes [19]	Consistent with prediction	nucleolar peptidyl-prolyl cis-trans isomerase (PPIase); FK506 binding protein; phosphorylated by casein kinase II (Cka1p-Cka2p-Ckb1p-Ckb2p) and dephosphorylated by Ptp1p
	Module 15	*YMR269W*	ribosome biogenesis	no	Consistent with prediction	nucleolar protein [40]; protein possibly involved in protein synthesis [64]
	Module 15	*YNL050C*	ribosome biogenesis	no	Consistent with prediction	
	Module 15	*YOR146W*	ribosome biogenesis	no	ND	
	Module 15	*YOR154W*	ribosome biogenesis	no	Consistent with prediction	
	Module 15	*YCR016W*	ribosome biogenesis	yes [63]	Consistent with prediction	nucleolar protein
	Module 15	*YPR169W*	ribosome biogenesis	no	ND	nucleolar protein [40]
	Module 15	YCR072C	ribosome biogenesis	yes [19, 63]	ND	present in several complexes involved in protein synthesis and RNA turnover metabolism [38]; co-purified with the 60S ribosomal subunit [39].
	Module 15	*YDL063C*	ribosome biogenesis	yes [19]	ND	
S. oneidensis	Module 4	SO3725	central intermediate metabolism; protein modification	no	ND	
	Module 5	SO2017	heat shcok response	yes [42].	Consistent with prediction	
	Module 5	SO2042	heat shcok response			
	Module 5	SO2375	metabolism			
	Module 5	SO3298	metabolism	no	ND	
	Module 5	DsbD	metabolism	no	ND	thiol:disulfide interchange protein

E. c	Module 143	yaeC	metabolism, cell surface transporter	no	ND
	Module 143	ybgF	protein synthesis	no	ND
	Module 141	envR	function related to cell surface structure	no	ND
	Module 2531	b1505	transport protein	no	ND
	Module 2531	yqcB	energy metabolism	no	ND

The function of the listed hypothetical proteins was not clearly understood. However, many of our predictions were supported by existing experimental evidence from other laboratories (see last column), which was cited and summarized in the *Saccharomyces* Genome Database or TIGR's *S. oneidensis* or *E. coli* annotation database [59].
ND: not determined.

Table 3: Effect of different *Ct* value on sampling to recover 99% of "true" links for a system of 2000 genes

Ct	0.3	0.5	0.7
Number of sampled expressions	600	200	50

From the co-expression networks, we predicted functions of 32 unknown genes for *E. coli* and 47 unknown genes for *S. oneidensis* . Some representative examples are listed in Table 2. For example, *E. coli* gene *yaeC* is predicted to function in metabolite transport. This prediction is supported by a recent report that *yaeC* is a component of the methionine uptake system [48], though this information is missing in the annotation database from the institute for genomic research (TIGR). The predictive power of co-expression network was further experimentally tested on a hypothetical protein SO2107 from *S. oneidensis*. It formed a compact sub-network with other known heat shock proteins, indicating a role of this gene in heat shock response (Fig. 2B). An in-frame deletion mutant was generated, and it was indeed sensitive to heat shock (Fig. 3C). In addition, an RpoH (σ^{32}) binding site was identified in the upstream region of SO2017 [42]. These results clearly supported the reliability of the gene function predictions made using co-expression network.

Topological properties of gene co-expression networks

Biological networks are considered to be small world, modular, hierarchical and scale-free [3,49]. To determine whether the obtained co-expression networks are consistent with general network theory, the topological properties of the co-expression network from yeast cell cycle microarray data have been examined. The average path length of this network is 7.81, which is quite small compared to the size of the network (804 genes). This result suggests that the network is a small world. The average clustering coefficient of this network is 0.323, implying a high degree of modularity. Also, the average clustering coefficient (C(k)) of all genes with k links follows the scaling law: $C(k) \sim k^{-0.37}$ as shown in Figure 4A. This signifies high hierarchical modularity although the scaling exponent of 0.37 differs from the values obtained from metabolic modular networks [50]. Analysis of connectivity properties of this network revealed a power-law distribution with a degree exponent of 1.5 (Fig. 4B), which is in accordance with the previous results on microarray expression profiles [51]. Taken together, the properties of RMT networks are consistent with general network theory.

Computational evaluations of RMT approach

To determine the sensitivity of RMT-based approach to determine correlation threshold for identifying gene co-expression networks, we randomly have rewired a small percentage of the links in the network from the yeast cell cycling data. As low as 0.4% random rewiring is able to make the NNSD deviating from Poisson distribution (Fig. 5). Therefore, the RMT approach is sensitive to detect even small topological changes in the networks. In addition, since microarray data typically contain high inherent variability, we have examined whether the networks are stable when additional noise is added. Different levels (1–50%) of Gaussian noise have been added to the entire dataset; new correlation thresholds have been determined for the perturbed data and corresponding networks have been constructed. When 30% noise has been added, 79.4% of the original links and 86.5% of the original genes are still preserved (Fig. 6), indicating that the RMT approach is robust in tolerating noises. Together, these statistical evaluations have indicated that the RMT approach is sensitive and robust to noise for determining correlation threshold.

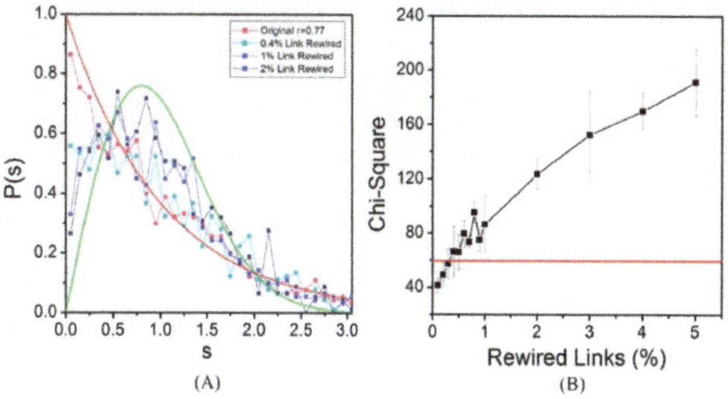

Figure 5: Sensitivity of RMT approach. (A) Wigner surmise (green) and Poisson distribution (red) were compared to the normalized NNSDs of the yeast dataset at cutoff 0.77 and its derived correlation matrices in which 1% (cyan), 2% (blue), 3% (navy) links were rewired. (B) Chi-Square test. Poisson distribution is plotted against NNSDs of correlation matrix of yeast cell cycling dataset at cutoff 0.77 and its derived correlation matrices in which 0.1%, 0.2%, 0.3%, 0.4%, 0.5%, 0.6%, 0.7%, 0.8%, 0.9%, 1%, 2%, 3%, 4%, 5% links are rewired. The red line indicates the critical value of Chi-Square test of $p = 0.001$.

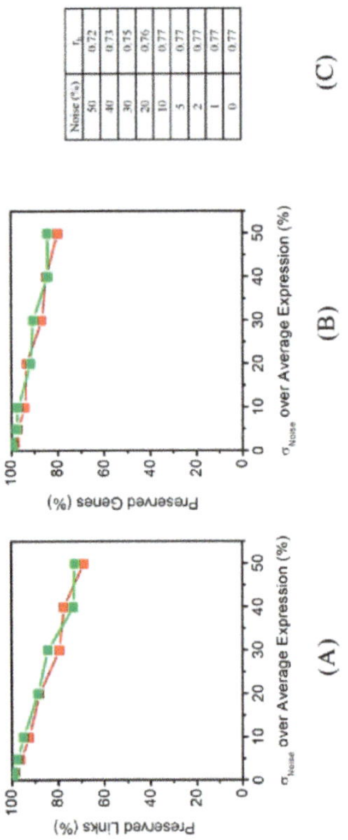

Figure 6: Analysis of RMT approach for robustness to noise. Increasing levels of Gaussian noise are added to the yeast cell cycling microarray expression profiles. The mean of noise is zero and its standard deviation (σ_{Noise}) is set to 1, 2, 5, 10, 20, 30, 40, and 50% of the average of absolute expression value of whole dataset. (A) Percentage of preserved links over total links in the modules perturbed by noise (red), and over total links in original modules (green) at different levels of noise. (B) Percentage of preserved Genes over total genes in the modules perturbed by noise (red), and over total genes in original modules (green) at different levels of noise. (C) Increased noise decreases the cutoff to separate modules.

In silico evaluation of the effect of sampling complexity on RMT approach

The microarray data is under-sampled, that is, the number of experiments is fewer than the number of genes. To evaluate the effect of under-sampling to RMT based method, we developed an *in silico* modular gene system (see Methods and Materials for details), in which correlations between expressions of genes inside modules ("true" links) have been designed to be a value $Ct(0 < Ct < 1)$; and other expression correlations have been designed to be zero. Then, RMT based method has applied to construct gene co-expression network from sampled expression data of the modular system. First, effect of different *Ct* value on the number of samplings to recover 99% of "true" links is tested. As shown in Table 3, when the *Ct* value is decreased (from 0.7 to 0.3), the number of expressions needed to separate the "true" links from random is increased (from 50 to 600). However, the number of experiments is still very small comparing to the total genes in the system. In addition, the effect of the size of system on the number of samplings needed is limited (Table 4). We also examined the percentage of «true» links over total links in the network. For $Ct = 0.7$ and a system size of 2000 genes, the percentage of "true" links over total links is 98.4% for only 20 expressions sampled (data not shown). In summary, even with the under-sampled expressions, the RMT approach is still able to recover the original co-expression networks in the designed model.

Table 4: Effect of the size of gene systems on sampling to recover 99% of "true" links for certain $Ct= 0.7$.

Size of gene system	2000	3000	5000	7000
Number of sampled expressions	50	70	100	100

Table 5: Comparison of gene co-expression networks obtained from yeast cell cycle microarray profiles with and without missing values estimation. The gene co-expression network obtained from original yeast cell cycle data is compared with the gene co-expression network obtained from a derived yeast cell cycle microarray data with missing value of original file is estimated by nearest neighbour method [32].

	Number of Genes	Number of Links
W/O	966	3346
With	941	3318
Overlap	922	3139

DISCUSSION

RMT has been used in characterizing the non-random phenomena in physical, material and social systems, including heavy nuclei, metal insulator transitions and the stock markets. It has been well recognized in these systems that RMT analyses are efficient for distinguishing system-specific, nonrandom properties from random noise. In this study, our observations with the microarray data from various organisms support our central hypothesis that the two universal predictions are applicable to biological systems. RMT might be particularly suitable for microarray data, which usually have high inherent variations. Based on yeast gene annotation, 85% of the functionally known genes are correctly linked. In addition, we demonstrated that the structure of network obtained from original microarray profiles differs from the networks obtained from randomized expression data [52]. Furthermore, computational analysis showed that all modules and links in an *in silico* network were correctly identified at the expected cutoff value. Together, these results suggest that the RMT-based approach can reliably identify gene co-expression networks.

Previously, we have applied the random matrix theory to study the properties of microarray data [50] and biological networks [54]. In our Physics Letters A paper [54], we have demonstrated that the NNSDs of the adjacent matrix of protein interaction network and metabolic network follow universal predictions of RMT. The current manuscript is a follow-on study of our Physical Review E paper [50]. The Physical Review E paper has just shown that NNSDs of correlation matrices from microarray data follow the universal description in random matrix theory. However, it did not provide a solid and complete algorithm for inferring gene co-expression networks from microarray data. Moreover, no rigorous biological tests of the predicted networks were performed. In the current manuscript, we systematically proposed a method for inferring gene co-expression network by utilizing the transition of NNSDs of correlation matrices from microarray data. And we provided computational analyses to show that this approach is reliable, sensitive and robust to noise. Furthermore, we demonstrated that the resulting co-expression networks are biologically meaningful. We provided evidences that the gene grouped in the network participate in the same biological pathway; the function of unknown genes can be accurately predicted, as shown by the experimentally validation; and the network are hierarchical, modular, small-world and scale-free, which are typical properties of biological networks. Together, the current manuscript is dedicated to a more practical method for inferring biological meaningful gene networks from microarray data, which is certainly not tackled at all in the previous publication. The current manuscript is a necessary follow-on study and also the first manuscript to provide a useful RMT based method for systems biology to identify biological pathways that are regulated by the given condition, to annotate function of unknown genes, and to dissect the global network properties.

There are two kinds of properties of eigenvalues of a real symmetric matrix: global properties and local properties. For example, eigenvalue distribution that changes based on large scale of eigenvalues is a global property of eigenvalues [23]. On the other hand, NNSD is a local property of eigenvalues. The global and local properties usually are unrelated. Numerical experiments showed that local properties of eigenvalues of a real symmetric matrix become independent of the probability distribution of matrix elements and the global properties of matrix when $N \rightarrow \infty$ [23]. The local properties, like NNSD, only dependent on over-all symmetries of the system, like real symmetric, or Hermitian.

As a matter of fact, the correlation matrices of microarray data are not general matrices with random elements, or even normal correlation matrices in statistic due to the number of microarray experiments is much less than the number of genes analyzed. Hence, current predictions of RMT about the global properties, such as eigenvalue distribution, may be invalid for these correlation matrices from microarray data. Additional File 1, Figure 5 has showed that the distribution of elements in correlation matrix of yeast cell cycle microarray data follows the Gaussian distribution and Additional File 1, Figure 6 has showed that the eigenvalue distribution of this matrix follows Cauchy distribution. However, the overall symmetry of these correlation matrices is still a real symmetry. The independent character of NNSD makes it possible to compare the NNSD of correlation matrices from microarray profiles with the theoretical predictions of RMT. Noted that the correlation matrix used in the report actually even is not a normal semi-positively defined correlation matrix as there are missing values in the data set and we have only used the experiments both genes having values to calculate the correlation. However, the same NNSD transitions have also been observed on normal semi-positively defined correlation matrices that calculated from same microarray data after estimating the missing value [32].

The co-expression networks identified based on RMT criterion can serve as a useful tool for predicting functions of hypothetical proteins. Genome sequencing projects indicate that substantial portions of open reading frames in a variety of organisms are functionally unknown. Defining the functions of such genes is a formidable task [53]. In this study, the cellular roles of 215 functionally unknown genes from yeast, *S. oneidensis* and *E. coli* were predicted, many of which were supported by existing information and our experiments. Such predictions will provide directions and guidance for future experimental design and verification, and hence facilitate the studies of functionally poorly characterized genes. In addition, the co-expression networks appear to be fairly sensitive in identifying functional modules, as shown by our computational analyses and comparison to relevance network. Indeed, the functions of several unknown genes identified and experimentally verified in this report were not reported before by other commonly used methods including relevance network.

We selected the threshold at which the NNSD finished the transformation into Poisson distribution from GOE distribution with the probability of $p = 0.001$. Based on the microarray data examined, it appears that the sharpest changes of χ^2 values were observed when NNSD is changed to a Poisson distribution at the probability level of $p = 0.001$. Thus $p = 0.001$ appears to be a good choice. However, due to the nature of microarray data (e.g. high noise), insufficient datasets to resolve the interactive relationships among different genes and/or the complexity of biological processes, complete removal of noises is unlikely. It is expected that some false links could still exist above the threshold. To further remove false links, one could select the threshold at other correlation values at which the NNSD is changed into a Poisson distribution with different probabilities such as p = 0.01, 0.05, or 0.1, which correspond to higher threshold correlation values. This should enable the isolation of network connections with higher confidence and further division of larger modules into smaller modules.

Our results have indicated that a transition zone can be defined based on the two critical points (r_l and r_h). A transition zone of r = 0.62–0.77 has been identified for yeast cell cycling data. In order to achieve high confidence on the co-expression network, we have chosen the upper bound of the NNSD transition as the threshold, with trade-off of possible loss of some correct information. For instance, at the threshold value of r = 0.77, a module involved in galactose metabolism has been identified (Module 3). But a key regulatory protein in this pathway, Gal80, is missing. This protein has been identified after lowering threshold value to 0.70 (data not shown). Therefore, lowering the threshold in the transition zone can enable us to identify additional correct links. However, this could also lead to much more false links (data not shown). One solution to solve this problem is called soft thresholding [15]. A further study that combines RMT based criterion with soft threshold to identify more true links in the transition zone will be worthy of investigation in the future.

When Mc can not be easily distinguished from Mr in correlation matrix, the RMT approach could not be able to generate a meaningful threshold. The examples include occasions that the experiment points are very limited and similarities between expressions of all genes are high; or very messy microarrays that have been poorly carried out.

CONCLUSION

Although high throughput genomics technologies such as microarray are powerful tools for studying gene functions and global regulations, identifying cellular networks in an automatic and objective fashion from genome-wide gene expression data remains challenging [6, 19, 54]. The RMT-based approach has been presented here provides a reliable, sensitive and robust method for determining correlation threshold; and then dissecting gene co-expression networks. The automatic and objective fashion of RMT-based approach makes it more advantageous in studying little understood organisms. The RMT-based approach could also be applied to other high throughput data for proteomes and metabolomes or combinations of these datasets. Similar characteristics have been also observed with the matrices from yeast protein interaction and metabolic pathway data [55]. Moreover, we expect that RMT is applicable to complex biological systems such as communities, and ecosystems. Further exploration of this method should provide valuable insights into the modular networks across different levels of biological organization.

METHODS

Determining correlation threshold by RMT based approach

First, a gene expression correlation matrix M, whose elements are Pair-wise Pearson correlation coefficients (r) in the range of (-1.0, 1.0), was constructed. If there are missing values in the expression files, only the experiments that both genes have values are used to calculate Pearson correlation. Then, a series of correlation matrices were constructed using different cutoff values. If the absolute value of an element in the original correlation matrix is less than the selected cutoff, it is set to 0. Eigenvalues of each correlation matrix were obtained by direct diagonalization of the matrix. Standard spectral unfolding techniques [26] were applied to have a constant density of eigenvalues and subsequently the nearest neighbour spacing distribution P(s), which is employed to describe the fluctuation of eigenvalues of the correlation matrix. We used the Chi square test to determine two critical threshold values, r_l at which P(s) start to deviate from GOE at a confidence level of $p= 0.001$, and r_h at which P(s) follows the Poisson distribution at a confidence level of $p = 0.001$. The critical point r_h is chosen to be the threshold used for constructing the gene co-expression network. Same procedure is used for analyzing the correlation matrix based on mutual information to determine the threshold.

Constructing gene co-expression network

Based on RMT, the complete transition from GOE to Poisson distributions can serve as a reference point to distinguish system-specific, nonrandom properties embedded in gene expression data from random noise. Thus, we used the r_h at which NNSD is completely transformed into Poisson distribution at the significance level of $p = 0.001$ as the threshold to define gene co-expression networks. From the correlation matrix, we can easily construct a co-expression network, in which each gene is a node and there is a link between two genes if the correlation measure between their expressions is greater than the threshold. All co-expression networks with cutoff beyond threshold will also provide system-specific relationship. However, the co-expression network at the threshold will provide the most relationship.

Construction of an *in silico* model

An *in silico* gene co-expression network of 2,000 genes with 10 modules of the size ranging from 5 to 200 genes (200, 100, 50, 50, 20, 20, 10, 10, 5, and 5 genes) was constructed using the threshold value of $r_h = 0.7$. The corresponding *in silico* correlation matrix was constructed as following: First, for each module whose size is greater than 5, we assigned the correlation coefficient of a random value in the range of +/-(0.7, 1.0) to each pair of element inside the module with a probability of 0.35. This probability is chosen to make the modules have the similar number of "true" links as the real gene co-expression modules. Second, for two modules with a size of 5, we linked the 5 elements to form some topology structure similar in real gene co-expression networks by assigning certain correlation coefficient with a random value in the range of +/-(0.7, 1.0). Finally, we assigned the rest correlation coefficient in the correlation matrix of these 2000 elements to a random value in the range of (-0.695, 0.695). The links with weight between 0.7 and 1.0 correspond to the "true" correlation inside modules and the links with weight between 0.0 and 0.7 are noise. The RMT method was applied to this system as described above.

Sampling expressions of an in silico modular gene system

A modular gene system has been developed to examine the effect of under-sampling on RMT based method. In this gene system (with size > 1000), there are 1000 genes divided into ten modules with equal size. Correlations between expressions of genes inside modules ("true" links) have been designed to a value *Ct* ($0 < Ct < 1$); and other expression correlations have been designed to be zero. Expressions of genes are sampled from Gaussian distribution to make their correlation matrix following previous designed pattern. Note that the real correlations from the sampled expression will fluctuate around the original designed values. Then, RMT based method was applied to sampled expressions to construct gene co-expression network. Multiple sampling experiments have been conducted. First, for a system of 2000 genes and different *Ct*value (0.3, 0.5, and 0.7), we started from sampling 20 expressions, and then sampled 10 more expressions every next time until 99% of "true" links recovered. Second, for certain *Ct* value (0.7) and different sizes of system (2000, 3000, 5000, 7000), we started from sampling 20 expressions, and then 10 more expressions every next time until 99% of "true" links recovered.

Evaluations of RMT-based network identification approach

(A) Sensitivity. The Maslov-Sneppen procedure [56] was used to rewire a small percentage (0.1%, 0.2%, 0.3%, 0.4%, 0.5%, 0.6%, 0.7%, 0.8%, 0.9%, 1%, 2%, 3%, 4%, 5%) of links in the gene co-expression network derived from yeast cell cycling data. The Maslov-Sneppen procedure randomizes the links in the network while keeping the degree of each vertex unchanged. For each rewiring, two links, A-B and C-D, are randomly selected. If there is no link between A and D and no link between C and B, the original two links A-B and C-D are disconnected and two new links A-D and B-C are constructed. After finishing rewiring the network, we constructed the corresponding correlation matrices, from which NNSDs were obtained.

(B) Robustness to noise. Increasing levels of Gaussian noise (1–50%) were added to the entire yeast cell cycling data. Then NNSDs from perturbed data were calculated as described above.

Network visualization and annotation

The software program Pajek [57] was used to visualize modular networks. Annotation databases for yeast, *S. oneidensis* and *E. coli* were from the *Saccharomyces* Genome Database (SGD) [58] and The Institute for Genomic Research [59], respectively.

Strains and physiological studies

All yeast strains were purchased from Open Biosystems (Huntsville, AL). For dilution assays, cultures were grown in YPD media (1% yeast extract, 2% bactopeptone, and 2% dextrose) overnight to stationary phase, suspended in water, and then spotted onto chemical-containing YPD plates at 10-fold serial dilutions followed by incubation at 30°C. Growth assay of cultures in liquid medium was performed in YPD with a BioScreen C (MTX Lab Systems, Inc., Vienna, VA).

To generate an in-frame *S. oneidensis* mutant of SO2017, the majority of the ORF was removed by an in-frame deletion mutagenesis approach [60]. The desired gene region was removed using PCR amplification with the primers A1 (5›AGC CTG TGA GCT CAC GGG), A2 (5›TGT TTA AAC TTA GTG GAT GGG GGT TAG ATC GAG GAT ATT), B1 (5›CCC ATC CAC TAA GTT TAA ACA GTT TGG CAA ACC AAT ATC) and B2 (5›ACA ATC GAG CTC TGC GAT), and a second cross-over PCR amplification with A1 and B2 using the mixed amplified fragments as templates. The resulting product was cloned into the suicide plasmid pDS3.0 and transformed into *E. coli* S17-1/λ_{pir} prior to conjugal transfer into *S. oneidensis* strain DSP10. Correct in-frame deletion was verified by DNA sequencing.

Growth of the *SO2017* mutant was examined in Luria-Bertani (LB) medium under both optimal temperature (30°C) and temperature shifting (30 to 42°C at OD_{600} = 0.62). The survival of mutant cells was compared to the parental strain DSP10 at 5, 10, 15, 25, 50 min after temperature shift.

ACKNOWLEDGEMENTS

The authors thank Dr. Vladimir Valtchinov for providing the RelNet package and thank T. Tschaplinski and C. Schadt for comments on the manuscript. The authors thank Dr. Xiaoqian Sun for valuable discussion on matrix analysis. The authors also thank four reviewers for their helpful suggestions. This work was supported by The United States Department of Energy under the Genomics:GTL through Shewanella Federation, the Virtual Institute of Microbial Stress and Survival (VIMSS [65]), Microbial Genome Program and Natural and Accelerated Bioremediation Research Programs of the Office of Biological and Environmental Research, Office of Science. Oak Ridge National Laboratory is managed by University of Tennessee-Battelle LLC for the Department of Energy under contract DE-AC05-00OR22725.

REFERENCES

1. Hartwell LH, Hopfield JJ, Leibler S, Murray AW: From molecular to modular cell biology. Nature 1999, 402(6761 Suppl):C47–52. 10.1038/35011540
2. Alon U: Biological networks: the tinkerer as an engineer. Science 2003, 301(5641):1866–1867. 10.1126/science.1089072
3. Barabasi AL, Oltvai ZN: Network biology: understanding the cell's functional organization. Nat Rev Genet 2004, 5(2):101–113. 10.1038/nrg1272
4. Alm E, Arkin AP: Biological networks. Curr Opin Struct Biol 2003, 13(2):193–202. 10.1016/S0959-440X(03)00031-9
5. Liao JC, Boscolo R, Yang YL, Tran LM, Sabatti C, Roychowdhury VP: Network component analysis: reconstruction of regulatory signals in biological systems. Proc Natl Acad Sci USA 2003, 100(26):15522–15527. 10.1073/pnas.2136632100
6. Basso K, Margolin AA, Stolovitzky G, Klein U, Dalla-Favera R, Califano A: Reverse engineering of regulatory networks in human B cells. Nat Genet 2005, 37(4):382–390. 10.1038/ng1532
7. Gardner TS, di Bernardo D, Lorenz D, Collins JJ: Inferring genetic networks and identifying compound mode of action via expression profiling. Science 2003, 301(5629):102–105. 10.1126/science.1081900
8. Yeung MK, Tegner J, Collins JJ: Reverse engineering gene networks using singular value decomposition and robust regression. Proc Natl Acad Sci USA 2002, 99(9):6163–6168. 10.1073/pnas.092576199
9. Liang S, Fuhrman S, Somogyi R: Reveal, a general reverse engineering algorithm for inference of genetic network architectures. Pac Symp Biocomput 1998, 18–29.
10. Akutsu T, Miyano S, Kuhara S: Identification of geneticnetworks from a small number of gene expression patterns under theBoolean network model. Pac Symp Biocomput 1999, 17–28.
11. van Berlo RJP, van Someren EP, Reinders MJT: Studying the conditions for learning dynamic Bayesian networks to discover genetic regulatory networks. Simulation 2003, 79: 689. [http://sim.sagepub.com/cgi/content/abstract/79/12/689] 10.1177/003754903040942
12. Friedman N, Linial M, Nachman I, Pe'er D: Using Bayesian networks to analyze expression data. J Comput Biol

2000, 7(3–4):601–620. 10.1089/10665270075005096l

13. Schmitt WA Jr, Raab RM, Stephanopoulos G: Elucidation of gene interaction networks through time-lagged correlation analysisof transcriptional data. Genome Res 2004, 14(8):1654–1663. 10.1101/gr.2439804
14. Butte AJ, Kohane IS: Mutual information relevance networks: functional genomic clustering using pairwise entropy measurements. Pac Symp Biocomput 2000, 418–429.
15. Zhang B, Horvath S: A General Framework for Weighted Gene Co-Expression Network Analysis. Stat Appl Genet Mol Biol. 2005;4:Article17 2005, 4: Article17.
16. Gardner TS, Faith J: Reverse-engineering transcription control networks. Physics of Life Reviews 2005, 2(1):65–88. 10.1016/j.plrev.2005.01.001
17. Zhou X, Kao MC, Wong WH: Transitive functional annotation by shortest-path analysis of gene expression data. Proc Natl Acad Sci USA 2002, 99(20):12783–12788. 10.1073/pnas.192159399
18. Bar-Joseph Z, Gerber GK, Lee TI, Rinaldi NJ, Yoo JY, Robert F, Gordon DB, Fraenkel E, Jaakkola TS, Young RA, et al.: Computational discovery of gene modules and regulatory networks. Nat Biotechnol 2003, 21(11):1337–1342. 10.1038/nbt890
19. Ihmels J, Friedlander G, Bergmann S, Sarig O, Ziv Y, Barkai N: Revealing modular organization in the yeast transcriptional network. Nat Genet 2002, 31(4):370–377.
20. Provero P: Gene networks from DNA microarray data: centrality and lethality.2002. [http://arxivorg/abs/cond-mat/0207345]
21. De Smet F, Mathys J, Marchal K, Thijs G, De Moor B, Moreau Y: Adaptive quality-based clustering of gene expression profiles. Bioinformatics 2002, 18(5):735–746. 10.1093/bioinformatics/18.5.735
22. Wigner EP: Random Matrices in Physics. SIAM Review 1967, 9: 1–23. 10.1137/1009001
23. Mehta ML: Random Matrices. 2nd edition. Academic Press; 1990.
24. Alt'shuler BL, Shklovskii BI: Repulsion of energy-levels and the conductance of small metallic samples. Sov Phys JETP 1986, 64: 127.
25. Hofstetter E, Schreiber M: Statistical properties of the eigenvalue spectrum of the three-dimensional Anderson Hamiltonian. Physical Review B Condensed Matter 1993, 48(23):16979–16985.
26. Zhong JX, Geisel T: Level fluctuations in quantum systems with multifractal eigenstates. Physical Review 1999, E 59: 4071–4074.
27. Zhong JX, Grimm U, Romer RA, Schreiber M: Level-spacing distribution of planar quasiperiodic tight-bindind models. Physical Review Letter 1998, 80: 3996–3999. 10.1103/PhysRevLett.80.3996
28. Bohigas O, Giannoni MJ, Schmit C: Spectral properties of the laplactian and random matrix theories. Physical Review Letter 1984, 52: 1. 10.1103/PhysRevLett.52.1
29. Seba P: Random matrix analysis of human EEG data. Phys Rev Lett 2003, 91(19):198104. 10.1103/PhysRevLett.91.198101
30. Plerou V, Gopikrishnan P, Rosenow B, Luis A, Amaral N, Stanley E: Universal and nonuniversal properties of cross correlation in financial time series. Physical Review Letters 1999, 83: 1471–1474. 10.1103/PhysRevLett.83.1471
31. Spellman PT, Sherlock G, Zhang MQ, Iyer VR, Anders K, Eisen MB, Brown PO, Botstein D, Futcher B: Comprehensive identification of cell cycle-regulated genes of the yeast Saccharomyces cerevisiae by microarray hybridization. Mol Biol Cell 1998, 9(12):3273–3297.
32. Troyanskaya O, Cantor M, Sherlock G, Brown P, Hastie T, Tibshirani R, Botstein D, Altman RB: Missing value estimation methods for DNA microarrays. Bioinformatics 2001, 17(6):520–525. 10.1093/bioinformatics/17.6.520
33. Gasch AP, Spellman PT, Kao CM, Carmel-Harel O, Eisen MB, Storz G, Botstein D, Brown PO: Genomic expression programs in the response of yeast cells to environmental changes. Mol Biol Cell 2000, 11(12):4241–4257.
34. Moriyama M, Hoshida Y, Otsuka M, Nishimura S, Kato N, Goto T, Taniguchi H, Shiratori Y, Seki N, Omata M: Relevance network between chemosensitivity and transcriptome in human hepatoma cells. Mol Cancer Ther 2003, 2(2):199–205.
35. Christie KR, Weng S, Balakrishnan R, Costanzo MC, Dolinski K, Dwight SS, Engel SR, Feierbach B, Fisk DG, Hirschman JE, et al.: Saccharomyces Genome Database (SGD) provides tools to identify and analyze sequences from Saccharomyces cerevisiae and related sequences from other organisms. Nucleic Acids Res 2004, (32 Database):D311–314. 10.1093/nar/gkh033
36. Ideker T, Thorsson V, Ranish JA, Christmas R, Buhler J, Eng JK, Bumgarner R, Goodlett DR, Aebersold R, Hood L: Integrated genomic and proteomic analyses of a systematically perturbed metabolic network. Science 2001, 292(5518):929–934. 10.1126/science.292.5518.929
37. Quackenbush J: Genomics. Microarrays – guilt by association. Science 2003, 302(5643):240–241. 10.1126/science.1090887
38. Gavin AC, Bosche M, Krause R, Grandi P, Marzioch M, Bauer A, Schultz J, Rick JM, Michon AM, Cruciat CM, et al.: Functional organization of the yeast proteome by systematic analysis of protein complexes. Nature 2002, 415(6868):141–147. 10.1038/415141a
39. Nissan TA, Bassler J, Petfalski E, Tollervey D, Hurt E: 60S pre-ribosome formation viewed from assembly in the nucleolus until export to the cytoplasm. Embo J 2002, 21(20):5539–5547. 10.1093/emboj/cdf547
40. Huh WK, Falvo JV, Gerke LC, Carroll AS, Howson RW, Weissman JS, O'Shea EK: Global analysis of protein localization in budding yeast. Nature 2003, 425(6959):686–691. 10.1038/nature02026
41. Krogan NJ, Cagney G, Yu H, Zhong G, Guo X, Ignatchenko A, Li J, Pu S, Datta N, Tikuisis AP, et al.: Global landscape of protein complexes in the yeast Saccharomyces cerevisiae. Nature 2006, 440(7084):637–

643. 10.1038/nature04670

42. Gao H, Wang Y, Liu X, Yan T, Wu L, Alm E, Arkin A, Thompson DK, Zhou J: Global transcriptome analysis of the heat shockresponse of Shewanella oneidensis. J Bacteriol 2004, 186(22):7796–7803. 10.1128/JB.186.22.7796-7803.2004
43. [http://chase.ou.edu/oubcf/]
44. Birnbaum K, Shasha DE, Wang JY, Jung JW, Lambert GM, Galbraith DW, Benfey PN: A gene expression map of the Arabidopsis root. Science 2003, 302(5652):1956–1960. 10.1126/science.1090022
45. Arbeitman MN, Furlong EE, Imam F, Johnson E, Null BH, Baker BS, Krasnow MA, Scott MP, Davis RW, White KP: Gene expression during the life cycle of Drosophila melanogaster . Science 2002, 297(5590):2270–2275. 10.1126/science.1072152
46. Schadt EE, Monks SA, Drake TA, Lusis AJ, Che N, Colinayo V, Ruff TG, Milligan SB, Lamb JR, Cavet G, et al.: Genetics of gene expression surveyed in maize, mouse and man. Nature 2003, 422(6929):297–302. 10.1038/nature01434
47. Welsh JB, Zarrinkar PP, Sapinoso LM, Kern SG, Behling CA, Monk BJ, Lockhart DJ, Burger RA, Hampton GM: Analysis of gene expression profiles in normal and neoplastic ovarian tissue samples identifies candidate molecular markers of epithelial ovarian cancer. Proc Natl Acad Sci USA 2001, 98(3):1176–1181. 10.1073/pnas.98.3.1176
48. Merlin C, Gardiner G, Durand S, Masters M: The Escherichia coli metD locus encodes an ABC transporter which includes Abc (MetN), YaeE (MetI), and YaeC (MetQ). J Bacteriol 2002, 184(19):5513–5517. 10.1128/JB.184.19.5513-5517.2002
49. Stuart JM, Segal E, Koller D, Kim SK: A gene-coexpression network for global discovery of conserved genetic modules. Science 2003, 302(5643):249–255. 10.1126/science.1087447
50. Ravasz E, Somera AL, Mongru DA, Oltvai ZN, Barabasi AL: Hierarchical organization of modularity in metabolic networks. Science 2002, 297(5586):1551–1555. 10.1126/science.1073374
51. Bhan A, Galas DJ, Dewey TG: A duplication growth model of gene expression networks. Bioinformatics 2002, 18(11):1486–1493. 10.1093/bioinformatics/18.11.1486
52. Luo F, Zhong J, Yang Y, Zhou J: Application of random matrix theory to microarray data for discovering functional gene modules. Phys Rev E Stat Nonlin Soft Matter Phys 2006, 73(3 Pt 1):031924.
53. Zhou J, Thompson DK, Xu Y, Tiedje JM: Microbial functional genomics. Wiley-Liss; 2004.
54. Segal E, Shapira M, Regev A, Pe'er D, Botstein D, Koller D, Friedman N: Module networks: identifying regulatory modules and their condition-specific regulators from gene expression data. Nat Genet 2003, 34(2):166–176.
55. Luo F, Zhong J, Yang Y, Scheuermann R, Zhou J: Application of Random Matrix Theory to Biological Networks. Phys Letter A 2006, 357(6):420–423. 10.1016/j.physleta.2006.04.076
56. Maslov S, Sneppen K: Specificity and stability intopology of protein networks. Science 2002, 296(5569):910–913. 10.1126/science.1065103
57. Batagelj V, Mrvar A: Pajek – Analysis and Visualization of Large Networks. In Graph Drawing Software. Edited by: Jünger M, Mutzel P. Springer, Berlin; 2003:77–103.
58. [http://www.yeastgenome.org/]
59. [http://www.tigr.org/]
60. Wan XF, Verberkmoes NC, McCue LA, Stanek D, Connelly H, Hauser LJ, Wu L, Liu X, Yan T, Leaphart A, et al.: Transcriptomic and proteomic characterization of the Fur modulon in the metal-reducing bacterium Shewanella oneidensis. J Bacteriol 2004, 186(24):8385–8400. 10.1128/JB.186.24.8385-8400.2004
61. Shen-Orr SS, Milo R, Mangan S, Alon U: Network motifs in the transcriptional regulation network of Escherichia coli. Nat Genet 2002, 31(1):64–68. 10.1038/ng881
62. [http://www.cs.clemson.edu/~luofeng/RMT.html]
63. Bergmann S, Ihmels J, Barkai N: Iterative signature algorithm for the analysis of large-scale gene expression data. Phys Rev E Stat Nonlin Soft Matter Phys 2003, 67(3 Pt 1):031902.
64. Schlitt T, Palin K, Rung J, Dietmann S, Lappe M, Ukkonen E, Brazma A: From gene networks to gene function. Genome Res 2003, 13(12):2568–2576. 10.1101/gr.1111403
65. PubMedCentralCrossRefPubMed
66. [http://www.vimss.lbl.gov]

CHAPTER 16

A COMPUTATIONAL METHOD FOR THE COUPLED SOLUTION OF REACTION–DIFFUSION EQUATIONS ON EVOLVING DOMAINS AND MANIFOLDS: APPLICATION TO A MODEL OF CELL MIGRATION AND CHEMOTAXIS

G. MacDonald, J.A. Mackenzie, M. Nolan, and R.H. Insall

Keywords: Reaction–diffusion, Bulk–surface equations, Cell migration, Chemotaxis, Evolving finite elements, ALE methods, Moving mesh methods

INTRODUCTION

Coupled bulk–surface problems arise in many areas of engineering and the applied and natural sciences. Examples include crystal growth [29], interfacial fluid flows using soluble and insoluble surfactants [7], [27], proton diffusion along biological membranes [34] and cell signalling [47].

Our motivation for the development of a method for coupled problems on evolving domains comes from the study of eukaryotic cell migration and chemotaxis. Chemotaxis, in particular, is essential during embryonic development, immune cell function, and cancer metastasis. Eukaryotic cells typically crawl by protruding pseudopods, which are dynamic structures based on actin fibres, at the front of the cell [26]. Actin is a globular protein which spontaneously polymerises into linear filaments that make up a large fraction of the cell›s cytoskeleton. The key step limiting actin polymerisation is the slow initiation of new filaments. Actin assembly can therefore be stimulated by "nucleating factors" which generate new actin filaments. In most cells, actin filaments are concentrated just beneath the cellular membrane and crosslinked into a relatively rigid cortex. Motor proteins such as myosin II bind to actin filaments in the cortex, crosslinking and contracting them, causing cortical tension and mechanical resistance, which together are key determinants of cells› overall behaviour. New actin polymerisation occurs between the cortex and the membrane, giving rise to a pressure pushing the cell membrane outward at the pseudopod; in other regions, cortical tension pulls along the remainder of the cell body. Together these processes lead to cell movement.

In [40] we developed a preliminary model of cell migration and chemotaxis where we only considered processes taking place on or close to the cell membrane. The model was based on a system of three reaction–diffusion equations posed on an evolving curve in two dimensions. An evolving surface finite element method was used to approximate the solution of the PDEs and a level set method was used to move the cell boundary. The method was later made more robust and more efficient by replacing the level set method by a parameterised finite element approach [39]. The model was shown to predict many aspects of real cell behaviour such as cell polarisation, persistent random walk migration in the absence of external signals and directed migration in the presence of a gradient in an external chemotactic field [41].

Citation: Feng Luo, Yunfeng Yang, Jianxin Zhong, Haichun Gao, Latifur Khan, Dorothea K Thompson, and Jizhong Zhou, Random Matrix Theory, 8:299 doi:10.1186/1471-2105-8-299.

The modelling of a number of important biological processes however also requires the ability to solve model equations in the extracellular and intracellular domains which are then coupled to equations posed on the cell membrane. It is also well appreciated that migrating cells have the ability to shape external chemotactic fields through the use of membrane-bound enzymes that degrade the ligand field and by the self-secretion of chemoattractants that allow signal relay to neighbouring cells [50]. The modelling of these important effects require the computational ability of solve partial differential equations on evolving bulk–surface domains.

Some previous computational studies looking at biological signalling have been carried out assuming fixed cell boundaries [36], [30], [42]. Methods have also been proposed to treat the deforming cell shape; for a recent review see [18]. For example, a phase-field approach has been used to model the combined effect of intracellular actin-flow, cell adhesions and morphology on cell motility [48]. Normally phase-field methods use isotropic stationary meshes which makes their implementation relatively straightforward. However, as the phase-field is smoothed over a small width of size ε, the computational mesh needs to be fine enough to resolve this spatial scale which can make the method computationally expensive unless some form of local mesh adaptation is used. Another well known embedding method is the level set method (LSM) [44]. This approach uses an evolution equation to move a signed distance function; the zero level set of this function is then identified as the moving cell boundary. LSMs have been used successfully in cell motility models [49],[54]. The disadvantage of LSMs is the requirement to maintain a high quality signed distance function which can become difficult if the evolving cell morphology becomes complicated. Bulk–surface reaction–diffusion systems on stationary domains have also been addressed using a simple point cloud method in [31]. An analysis of a finite element method for a steady state bulk–surface problem is presented in [13]. Analysis and computations of the diffusion-driven instability properties of coupled bulk–surface reaction–diffusion systems on stationary domains is presented in [32], [33].

To approximate the solution of bulk–surface systems of equations we will use an Arbitrary Lagrangian–Eulerian (ALE) approach. Traditionally, ALE methods have been used to solve problems on moving domains using a reformulation of the original problem with respect to an alternative reference frame rather than the standard fixed Eulerian frame. For fluid dynamics problems one could decide to use a Lagrangian transformation to follow the fluid flow. More generally however, there may be no obvious or preferred reference frame, and if the domain moves in time one may simply be satisfied with a transformation from a fixed stationary domain Ωc onto the physically evolving domain $\Omega(t)$. The ALE formulation was introduced for this purpose and it has been used successfully to tackle a number of physical applications such as fluid–structure interaction systems (see [24], [11]). In this paper we will use a conservative finite element ALE scheme. An advantage of the finite element framework is it allows the natural incorporation of flux boundary conditions linking the solution components in the bulk and surface domains. Another potential advantage of the ALE approach is the ability to accommodate arbitrary mesh movements which are not necessarily Lagrangian. This may help the robustness, accuracy and efficiency of the method with fewer instances of remeshing being required.

Crucial to the success of the front-tracking approach used here is the robust generation of cell body-fitted meshes. For this purpose we use an adaptive moving mesh procedure based on the solution of a moving mesh partial differential equations (MMPDEs) [9], [19], [4], [8], [23]. Although adaptive moving meshes have been used successfully to solve a range of physical problems [5], [51], [6], to our knowledge they have not been applied to the generation of meshes for curves evolving according to a geometric evolution law. An important contribution of this paper therefore is the development of an MMPDE method for the tangential motion of grid nodes along a curve which is moving with a specified normal velocity. The proposed method is motivated by the approach of Pan and Wetton [45], who devised a finite difference scheme for the generation of meshes which equidistribute the arc-length between grid nodes. The new method allows a degree of control over the generation of the surface mesh which is then used as known boundary conditions for the generation of the bulk mesh. The additional control of the boundary mesh also adds considerable robustness to the grid generation algorithm leading to far fewer instances for the need of global remeshing due to poor mesh quality.

The layout of this paper is as follows. In the next section we introduce the necessary notation for the description of coupled bulk–surface conservation laws on evolving domains which are then reformulated using a conservative weak ALE formulation. In section 3 we present a moving finite element discretisation of the coupled equation system. We then describe the generation of computational meshes covering the evolving bulk and surface domains in section 4. The complete algorithm is presented in section 5 and applied in section 6 to the simulation of cell migration in a chemotactic field. Finally, we draw some conclusions and directions for further research in section 7.

MODEL SYSTEM EQUATIONS

The physical layout for the cell migration simulations considered later is shown in Fig. 1. The cell membrane will be denoted by the evolving curve $\Gamma(t)$. The cell will be assumed to be moving through fixed lab frame of reference Λ. To improve computational efficiency, the governing equations for the extracellular region will only be computed over the time-dependent two-dimensional domain $\Omega(t)$, which is centred on the centroid of the cell. We will assume that a material particle P located at $X_p(t)$ on $\Gamma(t)$ has velocity $\dot{\boldsymbol{X}}_p(t)$. Therefore, we assume that there exists a velocity field $\boldsymbol{u}$ so that points on $\Gamma(t)$ evolve with a velocity field $\dot{\boldsymbol{X}}_p(t) = \boldsymbol{u}(\boldsymbol{X}_p(t), t)$. Let $\boldsymbol{n}=(n_1,n_2)$ denote the unit outward normal to $\Gamma(t)$ and let N(t) be any open subset of R^2 containing $\Gamma(t)$. For any function ζ which is differentiable in N(t), we define the tangential gradient on $\Gamma(t)$ by $\nabla_\Gamma\zeta=\nabla\zeta-(\nabla\zeta\cdot\boldsymbol{n})\boldsymbol{n}$, where $\cdot$ denotes the usual scalar product and $\nabla\zeta$ denotes the usual gradient on R^2. For a vector function $\zeta=(\zeta_1,\zeta_2)\in R^2$, the tangential divergence is defined by

$$\nabla_\Gamma \cdot \zeta = \nabla \cdot \zeta - \sum_{i=1}^{2}(\nabla\zeta_i \cdot \boldsymbol{n})n_i.$$

The Laplace–Beltrami operator on $\Gamma(t)$ is defined as the tangential divergence of the tangential gradient $\Delta_\Gamma\zeta=\nabla_\Gamma\cdot(\nabla_\Gamma\zeta)$.

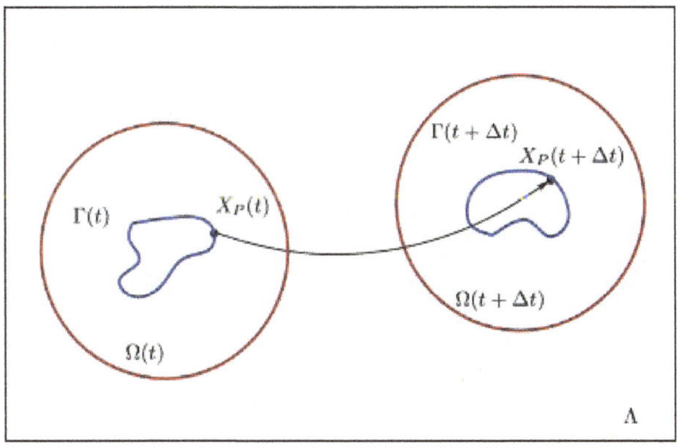

Figure 1: We consider the simulation of a motile cell through a fixed lab frame of reference Λ. The cell membrane is denoted by $\Gamma(t)$ and the extracellular region close to the cell is denoted by $\Omega(t)$. After a time interval of size Δt, the material point located at $\boldsymbol{X}p(t)$ on the cell membrane $\Gamma(t)$ evolves to the new location $\boldsymbol{X}p(t + \Delta t)$.

Reaction–diffusion systems of the type studied in pattern formation generally exclude cross-diffusion and are only coupled by the reaction kinetics terms. Therefore, we consider the behaviour of a single chemical species with a straightforward generalisation to a system of interacting chemicals. Let us define

$$Q_T = \{(\boldsymbol{x},t) \in \mathbb{R}^3 : \boldsymbol{x} \in \Omega(t), t \in (0,T]\}. \tag{1}$$

Application of Reynolds transport theorem to the equation for mass conservation for a chemical C, which diffuses with constant D , undergoing reaction at a rate f(c), gives

$$\frac{\partial c}{\partial t} = D\Delta c + f(c), \quad (\boldsymbol{x},t) \in Q_T, \tag{2}$$

where $c(\boldsymbol{x},t)$ is the concentration at position $\boldsymbol{x}\in\Omega(t)$ at time t.

We also consider the evolution of a chemical species C_s, that resides on the boundary $\Gamma(t)$. The bulk species C will be coupled to C_s through the generally nonlinear flux boundary condition

$$\overbrace{-D\frac{\partial c}{\partial \boldsymbol{n}}\bigg|_{\Gamma(t)}}^{\text{Diffusive flux}} + \overbrace{(\boldsymbol{u}\cdot\boldsymbol{n})c|_{\Gamma(t)}}^{\text{Advective flux}} = \overbrace{g(c|_{\Gamma(t)}, c_s)}^{\text{Rate of surface reaction}}, \tag{3}$$

where $c_s(\boldsymbol{x},t)$ denotes the concentration of C_s at the point $\boldsymbol{x}\in\Gamma(t)$ and $\boldsymbol{n}$ is the outward unit normal to $\Gamma(t)$. At the front of a motile cell $(\boldsymbol{u}\cdot\boldsymbol{n})>0$, leading to an advective flux onto $\Gamma(t)$, and at the back of the cell $(\boldsymbol{u}\cdot\boldsymbol{n})<0$ leading to a flux off of$\Gamma(t)$. This so-called windshield effect can potentially have a bearing on the ability of highly motile cells to chemotax efficiently in shallow gradients unless some additional mechanisms are employed by the cell such as surface ligand degradation or receptor internalisation [16].

In an analogous manner to the equation for the bulk species, we will assume that the boundary species evolves such that

$$\frac{\partial c_s}{\partial t} + \nabla_\Gamma\cdot(\boldsymbol{u}c_s) = D_s\Delta_\Gamma c_s + g(c|_{\Gamma(t)}, c_s) + h(c_s), \quad (\boldsymbol{x},t) \in \Gamma(t)\times(0,T), \tag{4}$$

where D_s is the boundary diffusion coefficient, $h(c_s)$ is a surface reaction term.

ALE reformulation

For the reasons mentioned earlier, when the domain is moving a common frame of reference adopted for computational purposes is the Arbitrary Lagrangian Eulerian (ALE) frame. Let A_t be a family of bijective mappings, which at each $t\in I=[0,T]$, map points in a reference or computational configuration Ω_c with coordinates $\xi=(\xi,\eta)$, to points in the current physical configuration $\Omega(t)$ with coordinates $\boldsymbol{x}=(x,y)$, so that

$$\mathcal{A}_t : \Omega_c \subset \mathbb{R}^2 \to \Omega(t) \subset \mathbb{R}^2, \qquad \boldsymbol{x}(\boldsymbol{\xi},t) = \mathcal{A}_t(\boldsymbol{\xi}). \tag{5}$$

The computational configuration could simply be the initial physical configuration $\Omega(0)$. We leave the discussion on how to construct the mapping A_t to section 4.

For an arbitrary function $g:Q_T\to R$, defined on the fixed Eulerian frame, its temporal derivative in the ALE frame is defined as

$$\frac{\partial g}{\partial t}\bigg|_\xi : Q_T \to \mathbb{R}, \quad \frac{\partial g}{\partial t}\bigg|_\xi(\boldsymbol{x},t) = \frac{\partial \hat{g}}{\partial t}(\xi,t), \quad \xi = \mathcal{A}_t^{-1}(\boldsymbol{x}), \tag{6}$$

where $\hat{g} : \Omega_c \times I \to \mathbb{R}$ is the corresponding function in the ALE frame; that is $\hat{g}(\xi,t) = g(\boldsymbol{x}(\xi,t),t) = g(\mathcal{A}_t(\xi),t)$. Taking the time derivative of the ALE mapping defines the ALE velocity $\boldsymbol{w}$ as

$$\boldsymbol{w}(\boldsymbol{x},t) = \frac{\partial \boldsymbol{x}}{\partial t}\bigg|_\xi (\mathcal{A}_t^{-1}(\boldsymbol{x}),t). \tag{7}$$

It is important to note that the ALE velocity $\boldsymbol{w}$ on $\Gamma(t)$ will, in general, be different from the material velocity $\boldsymbol{u}$. When $\boldsymbol{w}=\boldsymbol{u}$ the ALE transformation will be purely Lagrangian in nature. To relate the time derivatives with respect to the ALE transformation to the material derivative, a standard application of the chain rule gives

$$\frac{\partial c}{\partial t}\bigg|_\xi = \frac{\partial c}{\partial t}\bigg|_{\boldsymbol{x}} + \boldsymbol{w}\cdot\nabla c. \tag{8}$$

The reformulation of (2) in terms of the ALE reference frame therefore takes the form

$$\frac{\partial c}{\partial t}\bigg|_\xi - \boldsymbol{w}\cdot\nabla c = D\Delta c + f(c), \qquad \xi \in \Omega_c. \tag{9}$$

Similarly, on the boundary the ALE formulation of (4) takes the form

$$\frac{\partial c_s}{\partial t}\bigg|_\xi + \nabla_\Gamma\cdot(\boldsymbol{u}c_s) - \boldsymbol{w}\cdot\nabla_\Gamma c = D_s\Delta_\Gamma c + g(c|_{\Gamma(t)}, c_s) + h(c_s), \qquad \xi \in \partial\Omega_c. \tag{10}$$

The ALE reformulated equations (9) and (10) remain coupled through the flux boundary condition (3).

A conservative weak ALE formulation

To construct a weak formulation of (9), we consider a space of admissible test functions defined on the reference domain made of functions $\hat{v} \in H^1(\Omega_c)$. The ALE mapping then defines a set $H(\Omega(t))$ of test functions on the domain $\Omega(t)$, as follows:

$$\mathcal{H}(\Omega(t)) = \left\{ v : \Omega(t) \to \mathbb{R} : v = \hat{v}\circ\mathcal{A}_t^{-1},\ \hat{v} \in H^1(\Omega_c) \right\}, \quad t \in I.$$

A weak formulation of (9) can be obtained using Reynolds

transport formula which states that if $\psi(x,t)$ is a function defined on $\Omega(t)$, and $V_t \subseteq \Omega(t)$ such that $V_t = A_t(V_c)$ with $V_c \subseteq \Omega_c$, then

$$\frac{d}{dt}\int_{V_t} \psi(\boldsymbol{x},t)\,dx = \int_{V_t}\left(\left.\frac{\partial\psi}{\partial t}\right|_{\xi} + \psi\nabla\cdot\boldsymbol{w}\right)dx = \int_{V_t}\left(\left.\frac{\partial\psi}{\partial t}\right|_{x} + \nabla\psi\cdot\boldsymbol{w} + \psi\nabla\cdot\boldsymbol{w}\right)dx. \quad (11)$$

As functions $\hat{v} \in H^1(\Omega_c)$ do not depend on time, then for any $v \in H(\Omega(t))$ we can establish from (11) that

$$\frac{d}{dt}\int_{\Omega(t)} v\,dx = \int_{\Omega(t)} v\nabla\cdot\boldsymbol{w}\,dx \quad (12)$$

and

$$\frac{d}{dt}\int_{\Omega(t)} v\psi\,dx = \int_{\Omega(t)} v\left(\left.\frac{\partial\psi}{\partial t}\right|_{\xi} + \psi\nabla\cdot\boldsymbol{w}\right)dx. \quad (13)$$

Multiplying (9) by a test function $v \in H(\Omega(t))$, integrating over $\Omega(t)$ and the use of(3), (12) and (13) gives the conservative weak form: find c such that

$$\frac{d}{dt}\int_{\Omega(t)} cv\,dx - \int_{\Omega(t)}(\nabla\cdot(\boldsymbol{w}c))v\,dx + D\int_{\Omega(t)}\nabla c\cdot\nabla v\,dx + \int_{\Gamma(t)}(g - (\boldsymbol{u}\cdot\boldsymbol{n})c)v\,ds$$
$$= \int_{\Omega(t)} fv\,dx, \quad \forall v \in \mathcal{H}(\Omega(t)). \quad (14)$$

Similarly, on the boundary we have the conservative weak formulation: find c_s such that

$$\frac{d}{dt}\int_{\Gamma(t)} c_s v_s\,ds + \int_{\Gamma(t)}(\nabla_\Gamma\cdot((\boldsymbol{u}-\boldsymbol{w})c_s))\,v_s\,ds + D_s\int_{\Gamma(t)}\nabla_\Gamma c_s\cdot\nabla_\Gamma v_s\,ds$$
$$= \int_{\Gamma(t)}(g+h)v_s\,ds, \quad \forall v_s \in \mathcal{H}_s(\Gamma(t)), \quad (15)$$

where

$$\mathcal{H}_s(\Gamma(t)) = \left\{v_s : \Gamma(t) \to \mathbb{R} : v_s = \hat{v}_s \circ \mathcal{A}_t^{-1},\ \hat{v}_s \in H^1(\Gamma_c)\right\}, \quad t \in I,$$

is the space of test function on $\Gamma(t)$.

MOVING FINITE ELEMENT DISCRETISATION

Spatial semi-discretisation

We will assume that, for each $t \in [0,T]$, the physical and reference domains $\Omega(t)$ and Ω_c are approximated by polygonal domains $\Omega_h(t)$ and $\Omega_{c,h}$, respectively. We will assume that $\Omega_{c,h}$ is covered by a fixed triangulation $T_{h,c}$ with straight edges, so that$\Omega_{c,h} = \cup_{K \in T_{h,c}} K$. The approximation of the boundary domain $\Gamma_h(t)$ is chosen to be simply the boundary of $\Omega_h(t)$. The total number of elements of $T_{h,c}$ will be denoted by N . The total number of vertices of $T_{h,c}$ will be denoted by N and the number of vertices on the boundary as N_s. We define the Lagrangian finite element spaces on $T_{h,c}$ as

$$\mathcal{L}^1(\Omega_{c,h}) = \{\hat{v}_h \in H^1(\Omega_{c,h}) : \hat{v}_h|_K \in \mathbb{P}_1(K),\ \forall K \in \mathcal{T}_{h,c}\},$$
$$\mathcal{L}_0^1(\Omega_{c,h}) = \{\hat{v}_h \in H^1(\Omega_{c,h}) : \hat{v}_h|_K \in \mathcal{L}^1(\Omega_{c,h}) : \hat{v}_h = 0,\ \xi \in \Gamma_{c,h}\}. \quad (16)$$

where $\mathbb{P}_1(K)$ is the space of linear polynomials on K.

In section 4 we will describe a procedure for evolving the nodal positions of the triangulation covering $\Omega_h(t)$. Given the location of the mesh nodes, the ALE mapping will be interpolated using piecewise linear elements giving rise to a discrete mapping$A_{h,t} \in L^1(\Omega_{c,h})^2$ of the form

$$\boldsymbol{x}_h(\boldsymbol{\xi},t) = \mathcal{A}_{h,t}(\boldsymbol{\xi}) = \sum_{i=1}^{\mathcal{N}} \boldsymbol{x}_i(t)\hat{\phi}_i(\boldsymbol{\xi}), \quad (17)$$

where $\boldsymbol{x}_i(t) = A_{h,t}(\hat{\xi}_i)$ denotes the position of node i at time t , and $\hat{\phi}_i$ is the associated nodal basis function in $L^1(\Omega_{c,h})$. The discretised ALE velocity therefore takes the form

$$\boldsymbol{w}_h(\boldsymbol{\xi},t) = \sum_{i=1}^{\mathcal{N}} \dot{\boldsymbol{x}}_i(t)\hat{\phi}_i(\boldsymbol{\xi}).$$

Let $T_{h,t}$ be the image of the reference triangulation $T_{h,c}$ under the discrete ALE mapping $A_{h,t}$. Since the mapping is linear, each K_t which is the image of a triangle$K \in T_{h,c}$, is also a triangle with straight edges. Using the ALE mapping, the finite element test space on $\Omega_h(t)$ is therefore defined as

$$\mathcal{H}_h(\Omega_h(t)) = \{v_h : \Omega_h(t) \to \mathbb{R} : v_h = \hat{v}_h \circ \mathcal{A}_{h,t}^{-1},\ \hat{v} \in \mathcal{L}^1(\Omega_{c,h})\}.$$

The finite element spatial discretisation of the conservative ALE formulation (14)therefore takes the form: find $c_h(t) \in H_h(\Omega_h(t))$ such that

$$\frac{d}{dt}\int_{\Omega_h(t)} c_h v_h\,dx - \int_{\Omega_h(t)}(\nabla\cdot(\boldsymbol{w}_h c_h))v_h\,dx + D\int_{\Omega_h(t)}\nabla c_h\cdot\nabla v_h\,dx + \int_{\Gamma_h(t)}(g - (\boldsymbol{u}\cdot\boldsymbol{n})c_h)v_h\,ds$$
$$= \int_{\Omega_h(t)} fv_h\,dx, \quad \forall v_h \in \mathcal{H}_h(\Omega_h(t)). \quad (18)$$

Similarly, on the boundary, we have the weak formulation: find $c_{s,h} \in H_{s,h}(\Gamma_h(t))$such that

$$\frac{d}{dt}\int_{\Gamma_h(t)} c_{s,h}v_{s,h}\,ds + \int_{\Gamma_h(t)}(\nabla_\Gamma\cdot((\boldsymbol{u}-\boldsymbol{w}_h)c_{s,h}))\,v_{s,h}\,ds + D_s\int_{\Gamma_h(t)}\nabla_\Gamma c_{s,h}\cdot\nabla_{\Gamma_h} v_{s,h}\,ds$$
$$= \int_{\Gamma_h(t)}(g+h)v_{s,h}\,ds, \quad \forall v_{s,h} \in \mathcal{H}_{s,h}(\Gamma_h(t)). \quad (19)$$

The finite element approximation of the bulk and surface species can be expressed as

$$c_h(\boldsymbol{x},t) = \sum_{j=1}^{\mathcal{N}} c_j(t)\phi_j(\boldsymbol{x},t), \quad \text{and} \quad c_{s,h}(\boldsymbol{x},t) = \sum_{j=1}^{\mathcal{N}_s} c_{s,j}(t)\phi_{s,j}(\boldsymbol{x},t),$$

where $\{\phi_j(\boldsymbol{x},t)\}_{j=1}^{\mathcal{N}}$ and $\{\phi_{s,j}(\boldsymbol{x},t)\}_{j=1}^{\mathcal{N}_s}$ are the time-dependent bulk and surface nodal basis functions. If $\boldsymbol{C}(t) = \{c_j(t)\}_{j=1}^{\mathcal{N}}$ and $\boldsymbol{C}_s(t) = \{c_{s,j}(t)\}_{j=1}^{\mathcal{N}_s}$, we can express (18) as the system of ordinary differential equations

$$\frac{d}{dt}(M(t)\boldsymbol{C}(t)) + [K(t) + A(t,\boldsymbol{w}_h(t)) - B(t,\boldsymbol{w}_h(t))]\boldsymbol{C}(t) + \boldsymbol{D}(\boldsymbol{C}(t),\boldsymbol{C}_s(t)) = \boldsymbol{F}(\boldsymbol{C}(t)), \quad (20)$$

where

$$[M(t)]_{ij} = \int_{\Omega_h(t)} \phi_i(t)\phi_j(t)\,d\boldsymbol{x}$$

is the (time-dependent) mass matrix, while

$$[K(t)]_{ij} = D \int\limits_{\Omega_h(t)} (\nabla\phi_j(t) \cdot \nabla\phi_i(t))\, d\boldsymbol{x},$$

$$[A(t, \boldsymbol{w}_h(t))]_{ij} = \int\limits_{\Gamma_h(t)} [(\boldsymbol{u} - \boldsymbol{w}_h) \cdot \boldsymbol{n}]\phi_i(t)\phi_j(t)\, ds,$$

$$[B(t, \boldsymbol{w}_h(t))]_{ij} = -\int\limits_{\Omega_h(t)} [\boldsymbol{w}_h \cdot \nabla\phi_i(t)]\phi_j(t)\, d\boldsymbol{x},$$

$$[\boldsymbol{D}(\boldsymbol{C}(t), \boldsymbol{C}_s(t))]_i = \int\limits_{\Gamma_h(t)} [g(c_h(t), c_{s,h}(t)) - (\boldsymbol{u} \cdot \boldsymbol{n})c_h(t)]\phi_i(t)\, ds,$$

and the load vector

$$[\boldsymbol{F}(\boldsymbol{C}(t))]_i = \int\limits_{\Omega_h(t)} f(c_h(t))\phi_i(t)\, d\boldsymbol{x}.$$

Note that the vector $\boldsymbol{D}$ will be sparse as only those values of i corresponding to boundary vertices will be non-zero. The spatial discretisation of the boundary equation(19) results in a system of ODEs

$$\frac{d}{dt}(M_s(t)\boldsymbol{C}_s(t)) + [K_s(t) + A_s(t, \boldsymbol{w}_h(t))]\boldsymbol{C}_s(t) = \boldsymbol{D}_s(\boldsymbol{C}(t), \boldsymbol{C}_s(t)) + \boldsymbol{H}(\boldsymbol{C}_s(t)), \tag{21}$$

where

$$[M_s(t)]_{ij} = \int\limits_{\Gamma_h(t)} \phi_{s,i}(t)\phi_{s,j}(t)\, ds,$$

$$[K_s(t)]_{ij} = D_s \int\limits_{\Gamma(t)} (\nabla_\Gamma\phi_{s,j}(t) \cdot \nabla_\Gamma\phi_{s,i}(t))\, ds,$$

$$[A_s(t, \boldsymbol{w}_h(t))]_{ij} = \int\limits_{\Gamma_h(t)} \big([\nabla_\Gamma \cdot (\boldsymbol{u} - \boldsymbol{w}_h)]\phi_{s,i}(t)\phi_{s,j}(t) + [(\boldsymbol{u} - \boldsymbol{w}_h) \cdot \nabla_\Gamma\phi_{s,j}(t)]\phi_{s,i}(t)\big)\, ds,$$

$$[\boldsymbol{H}(\boldsymbol{C}_s(t))]_i = \int\limits_{\Gamma_h(t)} h(c_{s,h}(t))\phi_{s,i}(t)\, ds,$$

and $\boldsymbol{D}_s$ are the appropriately reordered non-zero elements of $\boldsymbol{D}$.

Temporal integration

To obtain a temporal discretisation of (20) and (21) we subdivide [0,T] into N_t equal time intervals of size $\Delta t = T/N_t$ and denote $t^n = n\Delta t$, $n = 0,1,\ldots,N_t$. We will discretise the ALE mapping using linear interpolation between time levels. That is we will define

$$\mathcal{A}_{h,\Delta t}(\boldsymbol{\xi}, t) = \frac{t - t^n}{\Delta t}\mathcal{A}_{h,t^{n+1}}(\boldsymbol{\xi}) + \frac{t^{n+1} - t}{\Delta t}\mathcal{A}_{h,t^n}(\boldsymbol{\xi}), \quad t \in [t_n, t_{n+1}), \tag{22}$$

where $A_{h,t}$ is the piecewise linear map at time t. The mesh velocity is therefore piecewise constant in time and is given by

$$\boldsymbol{w}^{n+1}_{h,\Delta t}(\boldsymbol{\xi}) = \frac{\mathcal{A}_{h,t^{n+1}} - \mathcal{A}_{h,t^n}}{\Delta t}, \quad t \in [t^n, t^{n+1}),$$
$$\boldsymbol{w}^{n+1}_{h,\Delta t}(\boldsymbol{x}, t) = \boldsymbol{w}^{n+1}_{h,\Delta t}(\boldsymbol{\xi}) \circ \mathcal{A}^{-1}_{h,\Delta t}(\boldsymbol{x}). \tag{23}$$

The temporal discretisation of the coupled systems (20) and (21) is obtained using a modified Crank–Nicolson semi-implicit approach. We first predict the boundary solution $\tilde{\boldsymbol{C}}^{n+1}_s$ using a semi-implicit backward Euler method, where the linear diffusion and mesh movement terms are treated implicitly, and the nonlinear reaction and coupling terms are treated explicitly. The predicted boundary solution therefore satisfies the linear system

$$[M^{n+1}_s + \Delta t(K^{n+1}_s + A^{n+1}_s)]\tilde{\boldsymbol{C}}^{n+1}_s = M^n_s\boldsymbol{C}^n_s + \Delta t[\boldsymbol{D}_s(\boldsymbol{C}^n, \boldsymbol{C}^n_s) + \boldsymbol{H}(\boldsymbol{C}^n_s)]. \tag{24}$$

The bulk approximation is then updated using a Crank–Nicolson step

$$[M^{n+1}+\tfrac{1}{2}\Delta t(K^{n+1}+A^{n+1}+B^{n+1})]\boldsymbol{C}^{n+1}$$
$$=[M^{n}-\tfrac{1}{2}\Delta t(K^{n}+A^{n}+B^{n})]\boldsymbol{C}^{n}+\tfrac{1}{2}\Delta t[\boldsymbol{F}(\boldsymbol{C}^{n+1})+\boldsymbol{F}(\boldsymbol{C}^{n})-\boldsymbol{D}(\boldsymbol{C}^{n+1},\tilde{\boldsymbol{C}}_s^{n+1})-\boldsymbol{D}(\boldsymbol{C}^{n},\boldsymbol{C}_s^{n})]. \quad (25)$$

The nonlinear system (25) is solved using Newton iteration. Finally, to ensure that the boundary solution is second-order in time, we perform a Crank–Nicolson correction step

$$[M_s^{n+1}+\tfrac{1}{2}\Delta t(K_s^{n+1}+A_s^{n+1})]\boldsymbol{C}_s^{n+1}$$
$$=[M_s^{n}-\tfrac{1}{2}\Delta t(K_s^{n}+A_s^{n})]\boldsymbol{C}_s^{n}+\tfrac{1}{2}\Delta t[\boldsymbol{D}_s(\boldsymbol{C}^{n+1},\tilde{\boldsymbol{C}}_s^{n+1})+\boldsymbol{D}_s(\boldsymbol{C}^{n},\boldsymbol{C}_s^{n})+\boldsymbol{H}(\tilde{\boldsymbol{C}}_s^{n+1})+\boldsymbol{H}(\boldsymbol{C}_s^{n})]. \quad (26)$$

Note that this correction step only requires the solution of a linear system of equations even if the reaction terms on the surface, g and h are nonlinear. The linear systems arising above are solved using the iterative method BiCGSTAB [52] and an incomplete LU (ILU) factorisation as a preconditioner.

For the cell migration application considered later, we note that the diffusive time scales in the extracellular region are often much shorter than the time scale associated with cell migration. As the time integration scheme above is fully implicit in the diffusive terms it is therefore robust to the choice of the time step for these applications.

A model bulk–surface problem on a stationary domain

To get an indication of the spatial and temporal convergence rate of the coupled bulk–surface finite element discretisation, we apply it to the solution of the following model problem:

$$\frac{\partial c}{\partial t}=\Delta c, \qquad \boldsymbol{x}\in\Omega \quad (27)$$

$$\frac{\partial c_s}{\partial t}=\Delta_\Gamma c_s+c-c_s, \quad \boldsymbol{x}\in\Gamma, \quad (28)$$

$$-\frac{\partial c}{\partial n}=c-c_s, \quad \boldsymbol{x}\in\Gamma, \quad (29)$$

where Ω is the unit circle.

This problem can be tackled analytically using polar coordinates $x=r\cos\theta,\ y=r\sin\theta$ so that

$$\frac{\partial c}{\partial t}=\frac{\partial^2 c}{\partial r^2}+\frac{1}{r}\frac{\partial c}{\partial r}+\frac{1}{r^2}\frac{\partial^2 c}{\partial \theta^2}, \qquad \boldsymbol{x}\in\Omega$$
$$\frac{\partial c_s}{\partial t}=\frac{\partial^2 c_s}{\partial \theta^2}+c|_{r=1}-c_s, \quad \boldsymbol{x}\in\Gamma,$$

and

$$-\frac{\partial c}{\partial r}\bigg|_{r=1}=c|_{r=1}-c_s.$$

Similar to Novak et al. [43], we look for a solution in the form

$$c_s(\theta,t)=Ae^{-k^2t}\cos\theta,$$

and

$$c(r,\theta,t)=\rho(r)c_s(\theta,t).$$

The following exact solution has been used to test the convergence of the finite element discretisation:

$$c(r,\theta,t)=J_1(rk)e^{-k^2t}\cos\theta$$

and

$$c_s(\theta,t)=\frac{J_1(k)}{2-k^2}e^{-k^2t}\cos\theta,$$

where J_1 is the first-order Bessel function of the first kind and k=1.177706027.

Fig. 2 shows the computed approximate solutions in the bulk domain and on the domain boundary using an isotropic mesh with maximal cell diameter h=0.1 and a time step$\Delta t=2\times10^{-4}$. We can see that the method performs well for these values of the discretisation parameters. To test the spatial rate of convergence of the algorithm, simulations were performed on a sequence of increasingly refined isotropic meshes. To ensure that the error was dominated by its spatial component, a sufficiently small time step of $\Delta t=10^{-3}$ was used. Fig. 3(a) shows the maximum error over all grid nodes for both the bulk and surface numerical solutions and we can see that both converge at the rate of $O(h^2)$, as expected. To investigate the temporal rate of converge, simulations were performed using a fine mesh with N=150,000 elements and various time steps. We can see from Fig. 3(b), that the three-step solution procedure (24), (25) and (26)furnishes approximations which are second-order accurate in time. Note that if the surface solution correction step (26) were omitted then, as expected, the resulting approximations were only temporally first-order accurate.

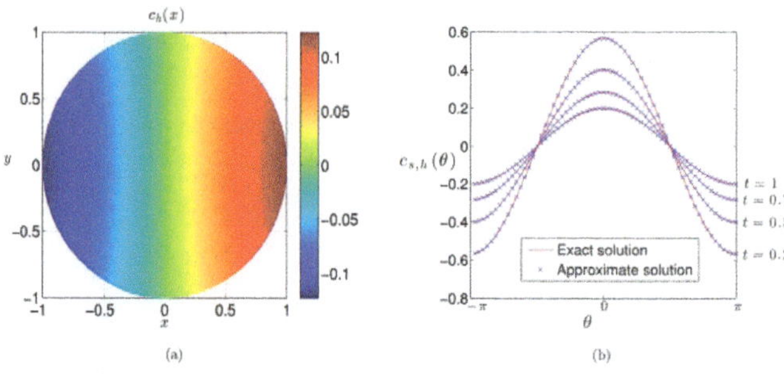

Figure 2: Numerical solution of coupled model problem (27), (28) and (29) on a stationary unit circle. (a) Approximate bulk solution ch($\boldsymbol{x}$) at t = 1. (b) Approximate solution csh(θ) on the boundary compared to the exact solution.

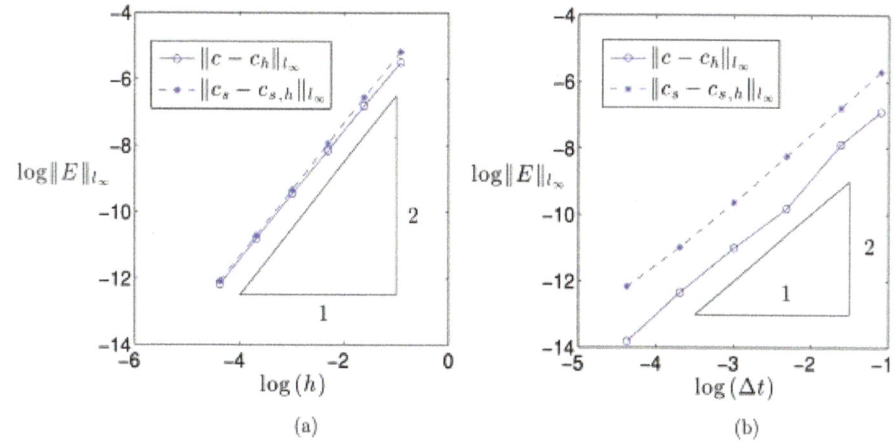

Figure 3: Second-order convergence of the maximum nodal error as (a) h → 0 and (b) Δt → 0 for a coupled model problem on a unit circle.

PRACTICAL EVALUATION OF ALE MAPPING

For the ALE mapping to be useful, it must satisfy a number of properties. First, it must be inexpensive to construct, relative to the cost of solving the physical problem. Second, the resulting meshes covering Ω(t) must be of good quality in the sense that mesh elements are adapted to salient features such at steep boundary or interior layers in the physical solution. Finally, the meshes must evolve smoothly in time to avoid the need to use small time steps to maintain numerical stability.

Bulk domain mesh generation

To avoid potential mesh crossings or foldings, we derive a suitable evolution equation for the inverse ALE mapping $\mathcal{A}_t^{-1}(x) = \xi(x,t)$ rather than $A_t(\xi)=x(\xi,t)$ (see, for example, the discussion in [12]). As shown in Fig. 4, a mesh $T_{h,t}$ on $\Omega_h(t)$ can then be generated as the preimage of a fixed mesh $T_{h,c}$ on $\Omega_{c,h}$. As introduced in [22], we choose the mapping $\xi(x)$ corresponding to a fixed value of t in order to minimise the functional

$$I[\xi] = \frac{1}{2}\int_{\Omega_t} [(\nabla\xi)^T G^{-1}(\nabla\xi) + (\nabla\eta)^T G^{-1}(\nabla\eta)]\,d\mathbf{x}, \tag{30}$$

where G is a 2×2 symmetric positive definite matrix referred to as a monitor matrix and ∇ is the gradient operator with respect to $\boldsymbol{x}$. It can be shown that the functional(30) is coercive and uniformly convex and hence has a unique minimiser [23].

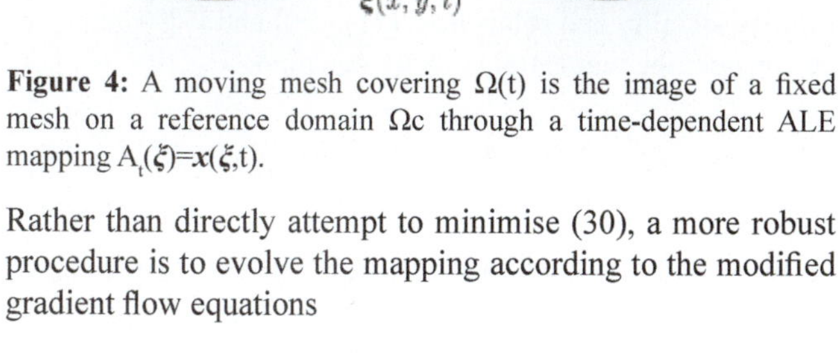

Figure 4: A moving mesh covering Ω(t) is the image of a fixed mesh on a reference domain Ωc through a time-dependent ALE mapping $A_t(\xi)=\boldsymbol{x}(\xi,t)$.

Rather than directly attempt to minimise (30), a more robust procedure is to evolve the mapping according to the modified gradient flow equations

$$\frac{\partial\xi}{\partial t} = \frac{P}{\tau}\nabla\cdot(G^{-1}\nabla\xi), \quad \text{and} \quad \frac{\partial\eta}{\partial t} = \frac{P}{\tau}\nabla\cdot(G^{-1}\nabla\eta). \tag{31}$$

Here, τ>0 is a user-specified temporal smoothing parameter which affects the temporal scale over which the mesh adapts and P is a positive function of ($\boldsymbol{x}$,t), chosen such that the mesh movement has a spatially uniform time scale [19]. The gradient flow structure of (31) ensures that the mesh evolves smoothly in time and improves robustness in the choice of the initial mesh.

The selection of an appropriate monitor matrix is crucial to the success of mesh adaptation. In this paper, we will consider the monitor matrix proposed by Winslow [53]

$$G = \begin{bmatrix} M & 0 \\ 0 & M \end{bmatrix}, \tag{32}$$

where M($\boldsymbol{x}$,t) is a positive weight function called a monitor function. A suitable monitor function is essential to the success of adaptive moving mesh methods. The monitor function should be taylored to the type of PDE being solved and the numerical method being used. Monitor functions based on interpolation estimates, a posteriori error estimates and adaptation to solution features have all been proposed [3],[8] and [23]. If no such estimate exists then the monitor function could be any smooth function designed to adapt the mesh towards important solution features such moving interfaces or boundaries (see for example [5]).

In practice, we interchange the roles of the dependent and independent variables in (31), since it›s the location of the physical mesh points $\{x_i(t)\}_{i=1}^{N}$ that defines the ALE map. With a Winslow-type monitor matrix (32) the resulting MMPDEs take the form

$$\tau\frac{\partial\boldsymbol{x}}{\partial t} = P(a\boldsymbol{x}_{\xi\xi} + b\boldsymbol{x}_{\xi\eta} + c\boldsymbol{x}_{\eta\eta} + d\boldsymbol{x}_{\xi} + e\boldsymbol{x}_{\eta}), \quad (\xi,\eta)\in\Omega_c, \tag{33}$$

where

$$a = \frac{1}{M}\frac{x_\eta^2 + y_\eta^2}{J^2}, \quad b = -\frac{2}{M}\frac{(x_\xi x_\eta + y_\xi y_\eta)}{J^2}, \quad c = \frac{1}{M}\frac{x_\xi^2 + y_\xi^2}{J^2},$$

$$d = \frac{1}{(MJ)^2}[M_\xi(x_\eta^2 + y_\eta^2) - M_\eta(x_\xi x_\eta + y_\xi y_\eta)], \quad e = \frac{1}{(MJ)^2}\left[-M_\xi(x_\xi x_\eta + y_\xi y_\eta) + M_\eta(x_\xi^2 + y_\xi^2)\right],$$

and $J=x_\xi y_\eta - x_\eta y_\xi$ is the Jacobian of the ALE mapping. To complete the specification of the coordinate transformation, the MMPDE must be supplemented by initial and boundary conditions. In all the applications described later, the computational domain $\Omega_{c,h}=\Omega_h(0)$, and the initial mesh over the physical domain is obtained using the Matlab toolbox Distmesh [46]. To allow the mesh points to move along the boundary $\Gamma_h(t)$, they are obtained using a one-dimensional moving mesh approach outlined in section 4.2. Once the boundary nodes have been evolved over one time step, they then become Dirichlet conditions completing the specification for (33).

The numerical solution of (33) requires spatial and temporal discretisation. In space, we discretise using standard linear Galerkin finite elements. In the time direction, we use a backward Euler integration scheme to update the solution at $t=t^{n+1}$, and to avoid solving nonlinear algebraic systems, we evaluate the coefficients a,c,…,e at the time $t=t^n$. We therefore seek $\mathbf{x}_h^{n+1} \in (\mathcal{L}^1(\Omega_{c,h}))^2$ such that

$$\tau \int_{\Omega_{c,h}} \left(\frac{\mathbf{x}_h^{n+1}-\mathbf{x}_h^n}{\Delta t}\right)\cdot \hat{\mathbf{v}}_h \, d\xi + \int_{\Omega_{c,h}} \Big[(\mathbf{x}_h^{n+1})_\xi \cdot (a^n \hat{\mathbf{v}}_h)_\xi + (\mathbf{x}_h^{n+1})_\eta \cdot (c^n \hat{\mathbf{v}}_h)_\eta$$
$$+\frac{1}{2}[(\mathbf{x}_h^{n+1})_\xi \cdot (b^n \hat{\mathbf{v}}_h)_\eta + (\mathbf{x}_h^{n+1})_\eta \cdot (b^n \hat{\mathbf{v}}_h)_\xi] - [d^n (\mathbf{x}_h^{n+1})_\xi + e^n (\mathbf{x}_h^{n+1})_\eta]\cdot \hat{\mathbf{v}}_h\Big] \, d\xi = 0, \quad (34)$$

for all $\hat{\mathbf{v}}_h \in (\mathcal{L}_0^1(\Omega_{c,h}))^2$. The resulting linear systems are again solved using the iterative method BiCGSTAB and an incomplete LU (ILU) factorisation as a preconditioner. An analysis of the performance of this iterative solver for the discretised MMPDE equations can be found in [4].

Boundary mesh generation

We consider the class of boundary movements given by the following evolution law for the normal velocity

$$\mathcal{V}(\mathbf{x},t) = \alpha\kappa + \beta, \quad \mathbf{x}\in\Gamma(t), \quad (35)$$

where κ is the curvature. Within the context of biological cell applications, the functions α and β could model the physical forces exerted on the membrane by cortical tension, and protrusions caused by actin polymerisation, respectively.

We consider the solution of (35) using a Lagrangian approach, where the main idea is to represent the flow of curves by the position vector $\mathbf{x}$ which is the solution of the geometric evolution equation

$$\dot{\mathbf{x}} = \mathcal{V}\mathbf{n} + \mathcal{B}\mathbf{t}, \quad (36)$$

where $\mathbf{n}$ and $\mathbf{t}$ are unit normal and tangent vectors, respectively. Note that the presence of the tangential velocity B has no effect on the shape of the evolving curve but it is well known that the incorporation of a suitably chosen non-zero value of B can avoid the major drawback of the Lagrangian approach which is the possibility that grid points merge, resulting in a loss of stability and accuracy [2], [1] and [37].

An embedded regular plane curve Γ can be parameterised by the smooth function $x: S^1 \to \mathbb{R}^2$, i.e. $\Gamma = \text{Image}(\mathbf{x}) := \{\mathbf{x}(\sigma),\ \sigma\in S^1\}$. Taking into account the periodic boundary conditions at σ=0,1, we shall hereafter identify S^1 with the interval [0,1]. The unit arc-length parameterisation will be denoted by s and it can be shown that $\mathbf{x}_s=\mathbf{t}$ and $\mathbf{x}_{ss}=\kappa\mathbf{n}$. The arc-length of the curve from the point $\mathbf{x}(\sigma_0)$ to $\mathbf{x}(\sigma)$ is given by

$$s(\sigma) = \int_{\sigma_0}^{\sigma} \sqrt{x_\sigma^2 + y_\sigma^2}\, d\sigma = \int_{\sigma_0}^{\sigma} \|\mathbf{x}_\sigma\|\, d\sigma,$$

where $\|\cdot\|$ denotes the l_2-norm. Using the chain rule

$$\mathbf{x}_\sigma = \mathbf{x}_s \frac{ds}{d\sigma} = \mathbf{x}_s \|\mathbf{x}_\sigma\| \quad (37)$$

and differentiation of (37) leads to the relation

$$\mathbf{x}_{\sigma\sigma} = \mathbf{x}_{ss}\|\mathbf{x}_\sigma\|^2 + \mathbf{x}_s \|\mathbf{x}_\sigma\|_s \|\mathbf{x}_\sigma\|. \quad (38)$$

Multiplying through (38) by $\mathbf{n}$, we get an expression for κ and hence in terms of the parameterisation σ, we can express the normal velocity as

$$\mathcal{V} = \dot{\mathbf{x}}\cdot\mathbf{n} = \alpha\left(\frac{\mathbf{x}_{\sigma\sigma}\cdot\mathbf{n}}{\|\mathbf{x}_\sigma\|^2}\right) + \beta. \quad (39)$$

Control of the mesh spacing in the tangential direction can be achieved by ensuring that a weighted arc-length between grid nodes is constant. That is, if $M(\mathbf{x},t)>0$ is a positive monitor function, then we set

$$M\frac{ds}{d\sigma} = \text{constant}$$

and hence

$$\left(M\frac{ds}{d\sigma}\right)_\sigma = (M\|\mathbf{x}_\sigma\|)_\sigma = 0. \quad (40)$$

The differential-algebraic system (39) and (40) could, in theory, be used to evolve the boundary Γ(t); the case α=1, β=0 and M=1, which aims to construct a uniform arc-length parameterisation, is exactly the equation system used by Pan and Wetton [45]. In [39] we used the parameterised finite element method (PFEM) [2] to evolve the approximation of a curve describing an evolving cell membrane. This method introduces an intrinsic tangential velocity B such that the arc-length between grid nodes is equidistributed. When using the PFEM to generate boundary node displacements, which are then used as boundary data for meshes covering bulk regions, we have found that the lack of control of the PFEM intrinsic tangential velocity can lead to poor quality bulk meshes which require frequent remeshing. Experience with the generation of adaptive moving meshes suggests however that the introduction of an evolution equation for the tangential velocity, driven by the mesh equidistribution condition (40), can significantly improve stability and robustness [21], [20] and [3]. Therefore, we consider the tangential velocity equation

$$\mathcal{B} = \dot{\mathbf{x}}\cdot\mathbf{t} = \frac{P}{\tau}(M\|\mathbf{x}_\sigma\|)_\sigma, \quad (41)$$

where, as before, τ and P are a temporal smoothing parameter and spatial balancing operator, respectively. The proposed procedure therefore involves the simultaneous solution of (39) and (41).

For simplicity, we discretise (39) and (41) using a finite difference method. The evolving curve at time t^n is represented by the discrete plane points x_i^n, i=1,…,N_s. The linear approximation of the curve is therefore given by the polygon with vertices x_i^n, i=1,…,N_s. Due to the periodicity conditions, we also use the additional values $x_{-1}^n = x_{N_s-1}^n$, $x_0^n = x_{N_s}^n$ and $x_{N_s+1}^n = x_1^n$. We use a uniform discretisation of the parameterisation interval $\sigma \in [0,1]$ with step size $\Delta\sigma = 1/N_s$. We use central finite differences to approximate the spatial derivatives in (39) and a first-order fully implicit temporal discretisation. This leads to a nonlinear system which is solved using Picard iteration. Let $x^{[n,m]}$ denote the approximation of x at time level t^n and iteration level m, then for the normal velocity equations we have the linear system

$$[-\mu_i^{[n+1,m]} x_{i-1}^{[n+1,m+1]} + (1 + 2\mu_i^{[n+1,m]}) x_i^{[n+1,m+1]} - \mu_i^{[n+1,m]} x_{i+1}^{[n+1,m+1]}] \cdot n_i^{[n+1,m]}$$
$$= x_i^{[n+1,m]} \cdot n_i^{[n+1,m]} + \Delta t \beta_i, \tag{42}$$

for i=1,…,N_s, where $\mu_i^{[n,m]} = 4\Delta t \alpha_i / (\|x_{i+1}^{[n,m]} - x_{i-1}^{[n,m]}\|^2)$. We approximate the unit tangent vector

$$t_i^n = \frac{x_{i+1}^n - x_{i-1}^n}{\|x_{i+1}^n - x_{i-1}^n\|} = (t_1^n, t_2^n),$$

and hence we set $n_i^n = (t_2^n, -t_1^n)$. Using a similar discretisation of the tangential velocity equation (41) we have

$$[-\nu_i^n x_{i-1}^{[n+1,m+1]} + (1 + 2\nu_i^n) x_i^{[n+1,m+1]} - \nu_i^n x_{i+1}^{[n+1,m+1]}] \cdot t_i^{[n+1,m]}$$
$$= x_i^n \cdot t_i^{[n+1,m]} + \frac{\Delta t P_i}{4\tau(\Delta\sigma)^2}(M_{i+1}^n - M_{i-1}^n)(\|x_{i+1}^{[n+1,m]} - x_{i-1}^{[n+1,m]}\|), \tag{43}$$

for i=1,…,N_s, where $\nu_i^n = \Delta t M_i^n P_i / (\tau(\Delta\sigma)^2)$. The coupled set of $2N_s$ equations (42)and (43) are solved for $x^{[n+1,m+1]}$ and the Picard iteration is stopped when

$$\|x^{[n+1,m+1]} - x^{[n+1,m]}\| < \mathrm{TOL}.$$

The last iteration is then used as the approximation x^{n+1}.

Example

To demonstrate the capability of the proposed adaptive grid generation procedures, we consider the evolution of an ellipse under mean curvature flow where $V = -\alpha\kappa$, and simultaneously we require the bulk and surface meshes evolve to resolve the travelling wave profile

$$u(x, y, t) = \frac{1}{2}\left[1 + \tanh\left(\frac{x + t - 0.7}{0.6}\right)\right].$$

There are many possible choices for the monitor function but for illustrative purposes we will set

$$M(x, y, t) = 1 + \mathrm{sech}^2\left(\frac{x + t - 0.7}{0.6}\right).$$

The monitor function takes its maximum value at the centre of the travelling wave and decreases smoothly to a non-zero constant value far from the front. The resulting adapted mesh should therefore be clustered around the wave but almost uniform in the flat regions. The initial physical domain consists of the region exterior to the ellipse $4x^2+16y^2=1$ and interior to a unit circular far-field boundary; the fixed computational domain Ω_c is chosen to be the initial physical domain. The computational mesh is obtained using the Matlab toolbox Distmesh [46] and has N=9832 elements and N_s=98 vertices on the boundary of the ellipse. The simulations are performed over the time interval $t \in [0, 7\times10^{-2}]$ using a constant time step $\Delta t = 10^{-3}$. The coefficient of the normal velocity of the evolving curve is α=0.75, and the temporal smoothing parameters in the moving mesh PDEs are both $\tau = 10^{-4}$. The computed meshes at four representative times are shown in Fig. 5. We can see that the initial mesh is adapted towards the wave front which is centred to the right of the ellipse at x=0.7. Initially the boundary mesh points are distributed to equidistribute the arc-length as the monitor function $M \approx 1$ along the ellipse. At t=0.02 and t=0.04, we can see that the bulk meshes have moved to follow the wave front; the boundary meshes have also adapted correctly in the tangential direction and evolved in the normal direction according to the curvature of the boundary. Finally, at t=0.05, the wave front has passed by the inner curve and the boundary mesh relaxes back to equidistribute the arc-length between mesh points. Fig. 6 shows the adapted boundary meshes; for clarity, at the two times when the wave front intersects the boundary, we have indicated its location by a vertical dashed line. We can see that at these times the boundary mesh has adapted well to the wave front and at the other two times the mesh equidistributes the arc-length of the evolving curve. To test the accuracy of the computed moving boundary, Fig. 7 shows the evolution of the exact and approximate area enclosed by the closed curve; the exact area for mean curvature flow takes the form $A(t)=A(0)-2\pi\alpha t$. We can see that the approximate area decreases linearly in time, and at the times indicated it agrees well with the exact area.

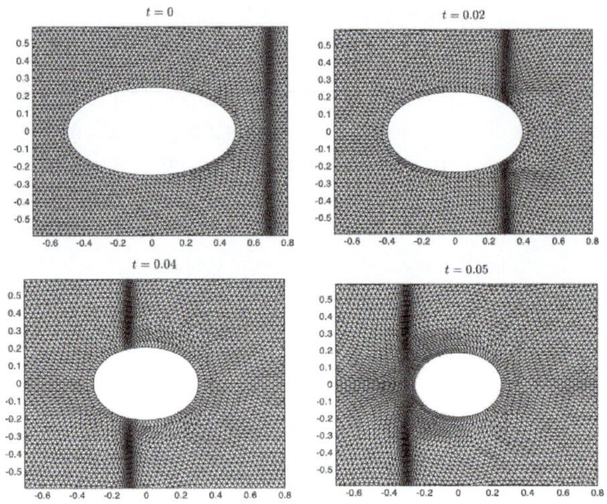

Figure 5: Adaptive bulk meshes for a time-dependent domain where the inner boundary evolves by mean curvature flow. The meshes are adapted to a travelling wave profile moving across the domain from right to left.

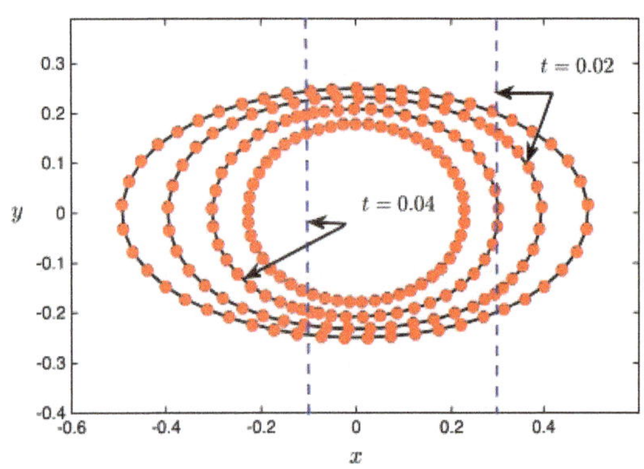

Figure 6: Computed meshes for a curve evolving in the normal direction by mean curvature flow; the meshes are also adapted in the tangential direction to a travelling wave profile moving across the domain from right to left. Meshes are shown at t = 0, 0.02, 0.04 and t = 0.05. The dashed vertical lines correspond to the location of the wave front at the times indicated.

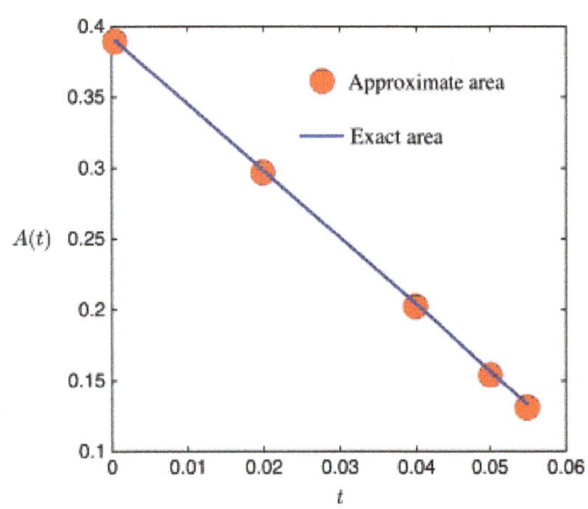

Figure 7: Comparison of the evolution of the exact and computed area enclosed by a closed curve under mean curvature flow at t = 0, 0.02, 0.04, 0.05 and t = 0.055.

THE COMPLETE ALGORITHM

We now describe the complete algorithm implementing the ALE–FEM scheme and the generation of the surface and bulk meshes. At time $t=t^n$ we assume we have a mesh$T_{h,t}^{\,n}$ and the finite element approximations c_h^n and $c_{s,h}^n$ of the bulk and surface species, respectively. The following steps are then carried out to advance the mesh and the finite element approximations.

1. Update the physical mesh

(a) Using $c_{s,h}^n$, calculate the normal velocity of Γ given by (35).

(b) Update the location of the manifold Γ_h^{n+1} by solving the simultaneous systems(42) and (43).

(c) Using the bulk approximation c_h^n, determine the mesh adaptation monitor function M.

(d) Using the updated boundary points as fixed Dirichlet data, update the interior mesh points by solving (34).

(e) Test for mesh quality. If the mesh is fine then goto 2. If not then re-grid and interpolate the solutions onto the new mesh and re-do the time step.

2. Update the finite element solutions in the bulk and the surface

(a) Use the meshes $T_{h,t}^{\,n+1}$ and $T_{h,t}^{\,n}$ to define the discrete ALE velocity $\boldsymbol{w}_h$.

(b) Update the solutions $c_{s,h}^{n+1}$ and c_h^{n+1} by solving (24), (25) and (26).

To determine the quality of the bulk meshes we measure the minimum angle over all of the triangles and decide to remesh when this is below a given tolerance. The new mesh covering the physical domain is then used as the fixed computational mesh. The bulk and boundary solutions are then linearly interpolated onto the new mesh to allow further time integration.

CELL MIGRATION AND CHEMOTAXIS

We now consider the application of the developed algorithm to the computational modelling of eukaryotic cell migration and chemotaxis. In [40] and [41] we developed a "pseudopod-centered" [25] model based on a system of reaction–diffusion equations that gives rise to a suitable spatiotemporal activator profile that can be used for the generation of pseudopods without the need for a driving external signal. The following set of equations was derived from a well-established discrete model developed by Meinhardt[35] (M model). The model describes the dynamic interaction between a membrane-bound local autocatalytic activator a, a rapidly distributed global inhibitor b and a local inhibitor c . Assuming that the cell boundary $\Gamma(t)$ moves with velocity $\boldsymbol{u}$, then for$\boldsymbol{x}\in\Gamma(t)$ the equations take the form

$$\frac{\partial a}{\partial t}+\nabla_\Gamma\cdot(\boldsymbol{u}a)=D_a\Delta_\Gamma a+\frac{s(a^2/b+b_a)}{(s_c+c)(1+s_a a^2)}-r_a a, \tag{44}$$

$$\frac{\partial b}{\partial t}+\nabla_\Gamma\cdot(\boldsymbol{u}b)=D_b\Delta_\Gamma b-r_b b+\frac{r_b}{|\Gamma(t)|}\oint_{\Gamma(t)} a\,dx, \tag{45}$$

$$\frac{\partial c}{\partial t}+\nabla_\Gamma\cdot(\boldsymbol{u}c)=D_c\Delta_\Gamma c+b_c a-r_c c. \tag{46}$$

Here, r_a,r_b and r_c denote decay rates of the local activator, global inhibitor and local inhibitor, respectively. The corresponding diffusion coefficients are D_a,D_b and D_c. In the nonlinear reaction term in the activator equation, s_a is a saturation coefficient, s_c is a Michaelis–Menten constant and b_a is a basal production rate of the activator. The constant b_c determines the growth of the local inhibitor c in the presence of the activator a. The effect of any external chemotactic field is incorporated in the signal term s. All of the variables appearing in this model are assumed to be non-dimensional. The activators and inhibitors in this model are not intended to represent molecular species. Although some molecular-based models have been successful in describing individual processes, the global morphology of real cells has proven too complex for such a description at present. We therefore have used a top-down approach, where each parameter can represent several molecular species. The pseudopod activator could read out at the level of actin nucleation, for example, through SCAR/WAVE proteins.

We assume that actin polymerisation creates a protrusive pressure that pushes the cell membrane outward in the normal direction. Recent detailed investigation of pseudopod formation suggests that this assumption is valid for cells migrating in the absence of external cues and in the presence of chemotactic gradients [5] and [4]. We will assume that the rate of polymerisation is proportional to the concentration of the local activator. At rest, the cell experiences pressure from cortical tension, which maintains the spherical shape of the cell. Using a cortical shell-liquid drop model [7], the pressure generated by the cortical tension is assumed to act normally on the cell membrane and depends on the local surface curvature. The cell boundary is therefore assumed to evolve with the normal velocity

$$\mathcal{V}(\boldsymbol{x},t) = K_{prot}a(\boldsymbol{x},t) - \lambda(t)\kappa, \quad \boldsymbol{x} \in \Gamma(t). \tag{47}$$

To control the area enclosed by Γ(t), we have used a spatially constant but time dependent cortical tension factor λ(t). Larger values of λ will increase the cortical tension, which will eventually result in a decrease in the cell area. Conversely, by decreasing λ the cortical tension is weakened, leading to an increase in cell area. There are many possible forms that a dynamic equation for λ could take, but we have found through numerical experimentation that the following works well:

$$\frac{d\lambda}{dt} = \frac{\lambda_0\lambda(A - A_0 + dA/dt)}{A_0(\lambda + \lambda_0)} - \beta\lambda. \tag{48}$$

Here, λ_0 and β are positive parameters and A_0 is the initial prescribed area of the cell. Equation (48) is solved numerically using an explicit Euler method.

The simulations of cell chemotaxis presented in [40] and [41] were obtained using an approximation of the local fractional receptor occupancy, i.e. the ratio of the local number of ligand-bound receptors to the total number of receptors. The receptor occupancy was estimated using a ligand–receptor binding model which assumed the extracellular ligand field was unaffected by the binding process and the fact that the cell is moving and constantly changing morphology. While the predictions in [40] and [41] displayed many attributes of real cell chemotaxis, we now consider a model to account for the interaction of the cell on the external chemotactic field. In the extracellular region, Ω(t), we assume that the concentration of ligand molecules evolves according to a linear diffusion equation. At the cell membrane, Γ(t), we assume that a chemoattractant ligand L binds reversibly to a receptor R to form a receptor–ligand complex LR. The coupled bulk–surface system for the evolution of the ligand concentration, l , and the concentration of bound receptors, l_s, therefore takes the form

$$\frac{\partial l}{\partial t} = D\Delta l, \qquad \boldsymbol{x} \in \Omega(t) \tag{49}$$

$$\frac{\partial l_s}{\partial t} + \nabla_\Gamma \cdot (\boldsymbol{u}l_s) = D_s\Delta_\Gamma l_s + k_1(R_{tot} - l_s)l - k_{-1}l_s, \qquad \boldsymbol{x} \in \Gamma(t). \tag{50}$$

Here, D and D_s are the diffusion coefficients for the ligand and receptor–ligand complex, respectively. We assume that the total concentration of bound and unbound receptors is constant and takes the value R tot. The constant k_1 is the rate of ligand association and k_{-1} the rate of disassociation. The normal flux boundary condition between the extracellular region and the cell membrane takes the form

$$-D\frac{\partial l}{\partial n} - [(\boldsymbol{u} \cdot \boldsymbol{n})]l = k_1(R_{tot} - l_s)l - k_{-1}l_s, \qquad \boldsymbol{x} \in \Gamma(t). \tag{51}$$

Determining the concentration of bound receptors, l_s allows the estimation of the local fractional receptor occupancy

$$R_o(\boldsymbol{x},t) = \frac{l_s(\boldsymbol{x},t)}{R_{tot}}.$$

It has been observed that some cells move randomly in the absence of any external cues. We have therefore included an intrinsic noise component to our system that is independent of the external chemotactic signal. For this purpose we will assume that the intrinsic noise η^t satisfies a stochastic differential equation of mean reverting type[40]. The combined effect of the response to the external signal and random intrinsic noise is modelled by the term

$$s(\boldsymbol{x},t) = r_a(\eta^t + R_o(\boldsymbol{x},t)),$$

which feeds in multiplicatively to the autocatalytic activator equation (44).

Migration in a linear gradient

The following numerical experiments were performed using the parameters found inTable 1 and a uniform time step Δt=0.1. We first simulate a cell moving in a linear gradient of the chemoattractant field given initially by

$$l(x,y) = 5.3 + 8.5(x + 0.1). \tag{52}$$

With the receptor–ligand disassociation constant $K_d=k_{-1}/k_1=30$, the receptor occupancy initially at the back of the cell $R_o^b = 0.15$, and at the front of the cell $R_o^f = 0.19$. To improve computational efficiency when there is significant cell movement, the diffusion equation is solved over a time-dependent annular region Ω(t), where the internal boundary represents the cell membrane, and the outer far-field boundary is a circle centred on the moving cell. The radius of the far-field boundary $r_f=3r_0$, where the initial radius of the cell $r_0=0.1$. At the far-field boundary the ligand concentration is given by (52). The initial mesh has N=6250 elements and is adapted isotropically towards the cell membrane; this mesh is then kept fixed as the initial computational mesh covering Ω_c. Fig. 8 shows four snap-shots of a simulated cell moving in the correct direction to the right. The cell migrates by the generation and splitting of pseudopods at the front of the cell while retracting the back. The generated mesh at one particular time is shown in Fig. 9. We can see that the bulk and boundary meshes are of high quality and follow the changing morphology of the migrating cell. Of the5×10^5 time steps used in this simulation, only 42 remeshing steps were required demonstrating the robustness of the grid generation algorithms. The computational method has been implemented in MATLAB to make use of the Distmesh algorithm for the generation of the initial mesh and meshes when re-gridding is needed. The simulations in this section took approximately 3 hours of computing time on a desktop machine using an Intel i7 2600 quad core processor running at 3.4 GHz. In this time the simulated cell had migrated a distance of approximately 10 cell diameters. No major attempt was made to optimise the code; it can be expected that run times would be reduced dramatically by computing in an appropriate low-level language.

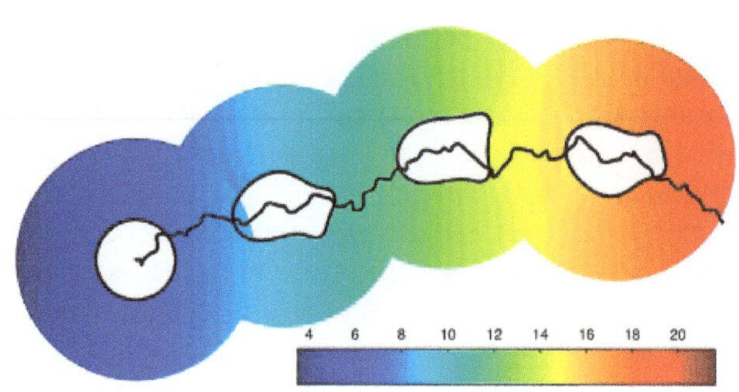

Figure 8: Simulation of single-cell chemotaxis. Although the cell's morphology is constantly changing by pseudopod extension and retraction, the cell's pathway exhibits a strong degree of directional persistence up the chemoattractant gradient to the right.

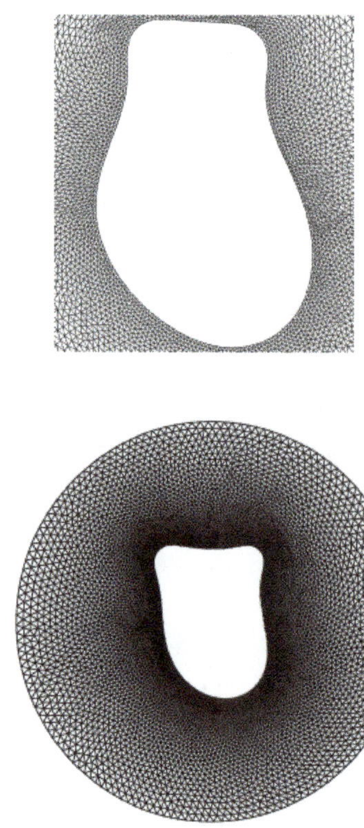

Figure 9: Global and close up view of a typical mesh used for the simulation of cell chemotaxis. The mesh follows faithfully the migrating cell and is of good quality close to the cell membrane.

Table 1: Non-dimensional default parameter values for cell migration simulations. Parameters for the pseudopod-centred cell migration model are taken from [40].

Quantity	Symbol	Value
Decay rate of activator	r_a	2×10^{-2}
Basic production rate of activator	b_a	1×10^{-1}
Saturation of activator autocatalysis	s_a	5×10^{-4}
Diffusion coefficient of activator	D_a	4×10^{-7}
Production & decay rate of global inhibitor	r_b	3×10^{-2}
Diffusion coefficient of global inhibitor	D_b	4×10^{-5}
Production rate of local inhibitor	b_c	7×10^{-3}
Decay rate of local inhibitor	r_c	1.3×10^{-2}
Diffusion coefficient of local inhibitor	D_c	2.8×10^{-6}
Michaelis-Menten constant	s_c	2×10^{-1}
Scaling of protrusive velocity	K_{prot}	1×10^{-5}
Cortical tension factor	λ_0	2×10^{-6}
Cortical tension factor	β	2×10^{-2}
Bulk ligand diffusion coefficient	D	10
Ligand association rate	k_1	$\frac{1}{30}$
Ligand disassociation rate	k_{-1}	1
Ligand-receptor complex diffusion coefficient	D_s	1×10^{-6}
Total number of receptors on cell	$2\pi r_0 R_{tot}$	7×10^{4}

Migration in an initially homogeneous field

We now consider the possibility of a cell to shape an initially homogeneous ligand field and investigate the feedback of this interaction on the cell's migratory pattern. The biological interest in this problem comes from recent experimental observations that some migrating cells locally self-generate chemotactic gradients thus leading to increased persistence of migration or sustained directed migration [10] and [38]. Fig. 10shows a simulation of a typical cell, where the initial ligand concentration is constant throughout the extracellular domain. We can see that movement of the cell, and changes to the cell morphology affect the ligand field (a). In particular, when a new pseudopod is created (b), the cell pushes out its membrane and this results in the local dilution of the concentration of the receptor–ligand complex (c). As the total receptor concentration is assumed constant, this means that additional unbound receptors become available to bind ligand molecules at the cell membrane (d). The bulk concentration at the cell surface is then depleted resulting in a local gradient in the bulk ligand field towards the cell (e). Conversely, in areas where the cell membrane is retracted such as the back of the cell and at retracting pseudopods, the local concentration of receptor–ligand complex increases slightly and this creates a gradient in the bulk ligand field away from the cell surface which drives ligand molecules away from the cell surface in these areas. The overall effect of these interactions between the extracellular ligand field and the surface receptor binding kinetics is to enhance the polarity and the persistence of the migrating cell even though the unperturbed background field is homogeneous. Additional reactions such as the degradation of ligand molecules by membrane bound enzymes and receptor internalisation could also play a major role in the generation of self-generated gradients and we intend to consider these possibilities in the future.

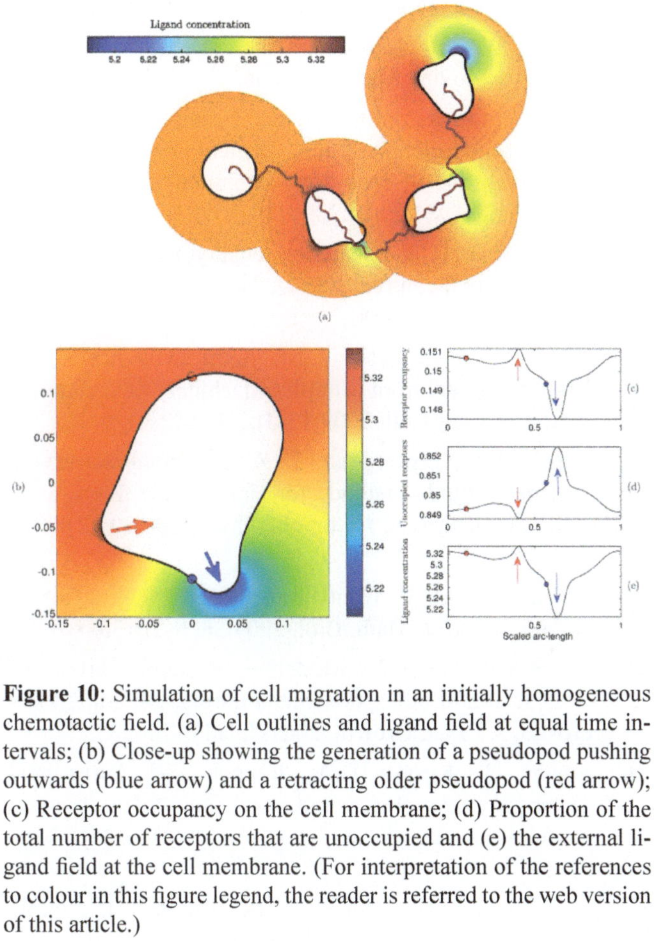

Figure 10: Simulation of cell migration in an initially homogeneous chemotactic field. (a) Cell outlines and ligand field at equal time intervals; (b) Close-up showing the generation of a pseudopod pushing outwards (blue arrow) and a retracting older pseudopod (red arrow); (c) Receptor occupancy on the cell membrane; (d) Proportion of the total number of receptors that are unoccupied and (e) the external ligand field at the cell membrane. (For interpretation of the references to colour in this figure legend, the reader is referred to the web version of this article.)

CONCLUSIONS AND FURTHER WORK

In this paper, we have developed a computational framework for the solution of coupled bulk–surface reaction–diffusion equations in two dimensions. The proposed algorithm is based on a conservative finite element ALE scheme to approximate the solution of the PDEs, and an adaptive moving mesh method for grid generation. A novel MMPDE approach has been developed to simulate a curve moving in the normal direction by a geometric evolution law that also allows control of the tangential distribution of mesh points. The overall algorithm has been shown to work well when applied to model problems with known analytical solutions. The method has also been applied to a model of cell migration and chemotaxis and shown to predict a possible novel mechanism for cells to increase their persistence and polarity by generating their own local gradients of chemoattractants. The numerical algorithm has proved to be very robust and requires few global remeshing steps even though the simulated cell morphology is constantly changing while it is migrating. The methodology is simple to implement using standard finite element procedures and freely available routines for initial grid generation.

In future, we aim to use the developed computational framework to tackle intracellular cell processes. The main issue there will be the proper description of the deformation of material points within the moving cell. This is of course a highly non-trivial issue although there have been attempts to model the cell mechanics as an active viscoelastic medium[28] and [17].

The computational approach presented here has been applied to a model of single cell migration and chemotaxis. It is of course of great interest how populations of cells migrate and interact with their micro-environment and each other. To apply the approach used here would be a major challenge if many cells were to be simulated given the fitted nature of the bulk mesh covering the highly dynamic extracellular region. We think the approach presented here is best suited to obtaining detailed information from simulations of single cells or the interaction of a small number of cells. This information can then be used to properly inform population models based on partial differential equations for the density of cells using upscaling techniques.

Ultimately, we would like to extend our approach to model cell migration and chemotaxis in three dimensions. Using a suitable variational formulation, the adaptive moving mesh method used here to evolve grid nodes in the tangential direction should extend to two-dimensional surfaces. This equation system could then be coupled to a geometric evolution law for the surface normal velocity. The MMPDE approach could then be used to generate evolving meshes for three-dimensional bulk domains. The ALE surface finite element method has already been utilised for solving PDEs on evolving surfaces[14] and [15] and the bulk ALE–FEM method proposed here also extends naturally to three dimensions. The main challenge in three dimensions however is not likely to be the development of efficient computational techniques but the increased complexity of modelling the interaction of migrating cells with their extracellular environment. The insights gained from modelling attempts in two dimensions would therefore be an essential first step in this direction.

ACKNOWLEDGEMENTS

GM and MN gratefully acknowledge the financial support from the UK Engineering and Physical Sciences Research Council (EPSRC) and Cancer Research UK for postgraduate studentship awards. Core support for RHI was provided by Cancer Research UK, grant number 15672. This work was completed while JAM was a Leverhulme Trust Research Fellow, supported by the award RF-2014-522.

REFERENCES

1. Balazovjech M., Mikula K. A higher order scheme for a tangentially stabilized plane curve shortening flow with a driving force. SIAM J. Sci. Comput. 2011;33(5):2277–2294.
2. Barrett J.W., Garcke H., Nürnberg R. On the variational approximation of combined second and fourth order geometric evolution equations. SIAM J. Sci. Comput. 2007;29:1006–1041.
3. Beckett G., Mackenzie J.A., Ramage A., Sloan D.M. On the numerical solution of one-dimensional PDEs using adaptive methods based on equidistribution. J. Comput. Phys. 2001;167(2):372–392.
4. Beckett G., Mackenzie J.A., Ramage A., Sloan D.M. Computational solution of two-dimensional unsteady PDEs using moving mesh methods. J. Comput. Phys. 2002;182(2):478–495.
5. Beckett G., Mackenzie J.A., Robertson M.L. A moving mesh finite element method for the solution of two-dimensional Stefan problems. J. Comput. Phys. 2001;168(2):500–518.
6. Beckett G., Mackenzie J.A., Robertson M.L. An r-adaptive finite element method for the solution of the two-dimensional phase-field equations. Commun. Comput. Phys. 2006;1(5):805–826.
7. Booty M., Siegel M. A hybrid numerical method for interfacial fluid flow with soluble surfactant. J. Comput. Phys. 2010;229(10):3864–3883.
8. Budd W., Huang C.J., Russell R.D. Adaptivity with moving grids. Acta Numer. 2009;18:111–241.
9. Cao W., Huang W., Russell R.D. An r-adaptive finite element method based upon moving mesh PDEs. J. Comput. Phys. 1999;149(2):221–244.
10. Donà E., Barry J.D., Valentin G., Quirin C., Khmelinskii A., Kunze A., Durdu S., Newton L.R., Fernandez-Minan A., Huber W., Knopp M., Gilmour D. Directional tissue migration through a self-generated chemokine gradient. Nature. 2013;503:285–289.
11. Donea J., Giuliani S., Halleux J.P. An arbitrary Lagrangian–Eulerian finite element method for transient dynamic fluid–structure interactions. Comput. Methods Appl. Mech. Eng. 1982;33(1–3):689–723.
12. Dvinsky A.S. Adaptive grid generation from harmonic maps on Riemannian manifolds. J. Comput. Phys.1991;95(2):450–476.
13. Elliott C.M., Ranner T. Finite element analysis for a coupled bulk–surface partial differential equation.IMA J. Numer. Anal. 2013;33:377–402.
14. Elliott C.M., Stinner B., Venkataraman C. Modelling cell motility and chemotaxis with evolving surface finite elements. J. R. Soc. Interface. 2012;9(76):3027–3044.
15. Elliott C.M., Styles V. An ALE ESFEM for solving PDEs on evolving surfaces. Milan J. Math.2012;80(2):469–501.
16. Endres R.G., Wingreen N.S. Accuracy of direct gradient sensing by single cells. Proc. Natl. Acad. Sci. USA. 2008;105(41):15749–15754.
17. George U.Z., Stéphanou A., Madzvamuse A. Mathematical modelling and numerical simulations of actin dynamics in the eukaryotic cell. J. Math. Biol. 2013;66(3):547–593.
18. Holmes W.R., Edelstein-Keshet L. A comparison of computational models for eukaryotic cell shape and motility. PLoS Comput. Biol. 2012;8
19. Huang W. Practical aspects of formulation and solution of moving mesh partial differential equations. J. Comput. Phys. 2001;171(2):753–775.
20. Huang W., Ren Y., Russell R.D. Moving mesh methods based on moving mesh partial differential equations. J. Comput. Phys. 1994;113(2):279–290.
21. Huang W., Ren Y., Russell R.D. Moving mesh PDEs based on the equidistribution principle. SIAM J. Numer. Anal. 1994;31(3):709–730.
22. Huang W., Russell R.D. Moving mesh strategy based on a gradient flow equation for two-dimensional problems. SIAM J. Sci. Comput. 1999;20(3):998–1015.
23. Huang W., Russell R.D. vol. 174. Springer; 2010. Adaptive Moving Mesh Methods.
24. Hughes T.J.R., Liu W.K., Zimmermann T.K. Lagrangian–Eulerian finite element formulation for incompressible viscous flows. Comput. Methods Appl. Mech. Eng. 1981;29(3):329–349.
25. Insall R.H. Understanding eukaryotic chemotaxis: a pseudopod-centred view. Nat. Rev. Mol. Cell Biol.2010;11:453–458.
26. Insall R.H., Machesky L.M. Actin dynamics at the leading edge: from simple machinery to complex networks. Dev. Cell. 2009;17(3):310–322.
27. James A.J., Lowengrub J. A surfactant-conserving volume-of-fluid method for interfacial flows with insoluble surfactant. J. Comput. Phys. 2004;201(2):685–722.
28. Kuusela E., Alt W. Continuum model of cell adhesion and migration. J. Math. Biol. 2009;58(1–2):135–161.
29. Kwon Y.-I., Derby J.J. Modeling the coupled effects of interfacial and bulk phenomena during solution crystal growth. J. Cryst. Growth. 2001;230(1):328–335.
30. Levine H., Kessler D.A., Rappel W.-J. Directional sensing in eukaryotic chemotaxis: a balanced inactivation model. Proc. Natl. Acad. Sci. USA. 2006;103(26):9761–9766.
31. Macdonald C.B., Merriman B., Ruuth S.J. Simple computation of reaction–diffusion processes on point clouds. Proc. Natl. Acad. Sci. USA. 2013;110(23):9209–9214.
32. Madzvamuse A., Chung A.H.W., Venkataraman C. Stability analysis and simulations of coupled bulk–surface reaction–diffusion systems. Proc. R. Soc., Math. Phys. Eng. Sci. 2015;471(2175)
33. Madzvamuse A., Chung A.H.W. The bulk–surface finite element method for reaction–diffusion systems on stationary volumes. Finite Elem. Anal. Des. 2016;108:9–21.
34. Medvedev E.S., Stuchebrukhov A.A. Proton diffusion

along biological membranes. J. Phys. Condens. Matter. 2011;23(23)

35. Meinhardt H. Orientation of chemotactic cells and growth cones: models and mechanisms. J. Cell Sci.1999;112:2867–2874.

36. Meyers J., Craig J., Odde D.J. Potential for control of signaling pathways via cell size and shape. Curr. Biol. 2006;16(17):1685–1693.

37. Mikula K., Remešíková M., Sarkoci P., Ševčovič D. Manifold evolution with tangential redistribution of points. SIAM J. Sci. Comput. 2014;36(4):A1384–A1414.

38. Muinonen-Martin A.J., Susanto O., Zhang Q., Smethurst E., Faller W.J., Veltman D.M., Kalna G., Lindsay C., Bennett D.C., Sansom O.J. Melanoma cells break down LPA to establish local gradients that drive chemotactic dispersal. PLoS Biol. 2014;12(10)

39. Neilson M.P., Mackenzie J.A., Webb S.D., Insall R.H. Use of the parameterised finite element method to robustly and efficiently evolve the edge of a moving cell. Integr. Biol. 2010;2(11–12):687–695.

40. Neilson M.P., Mackenzie J.A., Webb S.D., Insall R.H. Modelling cell movement and chemotaxis using pseudopod-based feedback. SIAM J. Sci. Comput. 2011;33(3):1035–1057.

41. Neilson M.P., Veltman D.M., van Haastert P.J.M., Webb S.D., Mackenzie J.A., Insall R.H. Chemotaxis: a feedback-based computational model robustly predicts multiple aspects of real cell behaviour. PLoS Biol.2011;9(5)

42. Neves S.R., Tsokas P., Sarkar A., Grace E.A., Rangamani P., Taubenfeld S.M., Alberini C.M., Schaff J.C., Blitzer R.D., Moraru I.I., Iyengar R. Cell shape and negative links in regulatory motifs together control spatial information flow in signaling networks. Cell. 2008;133(4):666–680.

43. Novak I.L., Gao F., Choi Y.S., Resasco D., Schaff J.C., Slepchenko B.M. Diffusion on a curved surface coupled to diffusion in the volume: application to cell biology. J. Comput. Phys. 2007;226(2):1271–1290.

44. Osher S., Sethian J.A. Fronts propagating with curvature-dependent speed: algorithms based on Hamilton–Jacobi formulations. J. Comput. Phys. 1988;79(1):12–49.

45. Pan Z., Wetton B. A numerical method for coupled surface and grain boundary motion. Eur. J. Appl. Math. 2008;19(3):311–327.

46. Persson P.-O., Strang G. A simple mesh generator in MATLAB. SIAM Rev. 2004;46(2):329–345.

47. Rätz A., Röger M. Turing instabilities in a mathematical model for signaling networks. J. Math. Biol.2012;65(6):1215–1244.

48. Shao D., Levine H., Rappel W. Coupling actin flow, adhesion, and morphology in a computational cell motility model. Proc. Natl. Acad. Sci. USA. 2012;109(18):6851–6856.

49. Strychalski W., Adalsteinsson D., Elston T.C. Simulating biochemical signaling networks in complex moving boundaries. SIAM J. Sci. Comput. 2010;32(5):3039–3070.

50. Sucgang R.H., Weijer C.J., Siegert F., Franke J., Kessin R.H. Null mutations of the Dictyostelium cyclic nucleotide phosphodiesterase gene block chemotactic cell movement in developing aggregates. Dev. Biol.1997;192(1):181–192.

51. Tenchev R.T., Mackenzie J.A., Scanlon T.J., Stickland M.T. Finite element moving mesh analysis of phase change problems with natural convection. Int. J. Heat Fluid Flow. 2005;26(4):597–612.

52. van der Vorst H.A. Bi-CGSTAB: a fast and smoothly converging variant of Bi-CG for the solution of nonsymmetric linear systems. SIAM J. Sci. Stat. Comput. 1992;13(2):631–644.

53. Winslow A. Numerical solution of the quasilinear Poisson equation in a nonuniform triangular mesh. J. Comput. Phys. 1966;1(2):149–172.

54. Wolgemuth C.W., Zajac M. The moving boundary node method: a level set-based, finite volume algorithm with applications to cell motility. J. Comput. Phys. 2010;229(19):7287–7308.

CHAPTER 17

STABILITY ANALYSIS AND SIMULATIONS OF COUPLED BULK-SURFACE REACTION–DIFFUSION SYSTEMS

Anotida Madzvamuse, Andy H. W. Chung, and Chandrasekhar Venkataraman

Keywords: Bulk-surface reaction–diffusion equations, bulk-surface finite-elements, Turingdiffusively driveninstability, linear stability, pattern formation, Robin-type boundary conditions

INTRODUCTION

In many fluid dynamics applications and biological processes, coupled bulk-surface partial differential equations naturally arise in (2D+3D) [1–3]. In most of these applications and processes, morphological instabilities occur through symmetry breaking resulting in the formation of heterogeneous distributions of chemical substances [4]. In developmental biology, it is essential for the emergence and maintenance of polarized states in the form of heterogeneous distributions of chemical substances such as proteins and lipids. Examples of such processes include (but are not limited to) the formation of buds in yeast cells, and cell polarization in biological cells owing to responses to external signals through the outer cell membrane [5,6]. In the context of reaction–diffusion processes, such symmetry breaking arises when a uniform steady state, stable in the absence of diffusion, is driven unstable when diffusion is present thereby giving rise to the formation of spatially inhomogeneous solutions in a process now well known as the Turing diffusion-driven instability [7]. Classical Turing theory requires that one of the chemical species, typically the inhibitor, diffuses much faster than the other, theactivator resulting in what is known as the long-range inhibition and short-range activation [8,9].

Recently, there has been a surge in studies on models that couple bulk dynamics to surface dynamics. For example, Rätz & Röger [6] study symmetry breaking in a bulk-surface reaction–diffusion model for signalling networks. In this work, a single diffusion partial differential equation (the heat equation) is formulated inside the bulk of a cell, whereas on the cell surface, a system of two membrane reaction–diffusion equations is formulated. The bulk and cell-surface membrane are coupled through Robin-type boundary conditions and a flux term for the membrane system [6]. Elliott & Ranner [10] study a finite-element approach to a sample elliptic problem: a single elliptic partial differential equation is posed in the bulk, and another is posed on the surface. These are then coupled through Robin-type boundary conditions. Novak et al.[11] present an algorithm for solving a diffusion equation on a curved surface coupled to a diffusion model in the volume. Chechkin et al. [12] study bulk-mediated diffusion on planar surfaces. Again, diffusion models are posed in the bulk and on the surface coupling them through boundary conditions. In the area of tissue engineering and regenerative medicine, electrospun membrane are useful in applications such as filtration systems and sensors for chemical detection. Understanding of the fibres' surface, bulk and architectural properties is crucial to the successful development of integrative technology. Nisbet et al. [13] present a detailed review on surface and bulk characterization of electrospun membranes of porous and fibrous polymer materials. To explain the long-range proton translocation along biological mombranes, Medvedev & Stuchebrukhov [14] propose a model that takes into account coupled bulk diffusion that accompanies the migration of protons on the surface. More recently, Rozadaet

Citation: Anotida Madzvamuse, Andy H. W. Chung, and Chandrasekhar Venkataraman, Stability analysis and simulations of coupled bulk-surface reaction–diffusion systems, http://dx.doi.org/10.1098/rspa.2014.0546.

al. [15] present singular perturbation theory for the stability of localized spot patterns for the Brusselator model on the sphere.

In most of the work above, either elliptic or diffusion models in the bulk have been coupled to surface-elliptic or surface-diffusion or surface-reaction–diffusion models posed on the surface through Robin-type boundary conditions [5,6,11–14,16]. Here, our focus is to couple systems of reaction–diffusion equations posed both in the bulk and on the surface, setting a mathematical and computational framework to study more complex interactions such as those observed in cell biology, tissue engineering and regenerative medicine, developmental biology and biopharmaceuticals [5,6,11–14,16,17]. We employ the bulk-surface finite-element method as introduced by Elliott & Ranner in [10] to numerically solve the coupled system of bulk-surface reaction–diffusion equations (BSRDEs). Details of the surface-finite-element can be found in reference [18]. The bulk and surface reaction–diffusion systems are coupled through Robin-type boundary conditions. Details of the coupled bulk-surface finite-element method can be found in [19]; the finite element algorithm is implemented in deall II [20].

The key contributions of our work to the theory of pattern formation are:

- we derive and prove Turing diffusion-driven instability conditions for a coupled system of BSRDEs;
- using the bulk-surface finite-element method, we approximate the solution to the model system within the bulk and on the boundary surface of a sphere of radius one;
- our results show that if the surface-reaction–diffusion system has the long-range inhibition, short-range activation form and the bulk-reaction–diffusion system has equal diffusion coefficients, then the surface-reaction–diffusion system can induce patterns in the bulk close to the surface and no patterns form in the interior, far away from the surface;
- on the other hand, if the bulk-reaction–diffusion system has the long-range inhibition, short-range activation form and the surface-reaction–diffusion system has equal diffusion coefficients, then the bulk-reaction–diffusion system can induce pattern formation on the surface;
- furthermore, we prove that if the bulk and surface reaction–diffusion systems have equal diffusion coefficients, no patterns form; and
- these theoretical predictions are supported by numerical simulations.

Hence, this article is outlined as follows. In §2, we present the coupled bulk-surface reaction–diffusion system on stationary volumes with appropriate boundary conditions coupling the bulk and surface partial differential equations. The main results of this article are presented in §2b where we derive Turing diffusion-driven instability conditions for the coupled system of BSRDEs. To validate our theoretical findings, we present bulk-surface finite-element numerical solutions in §3. In §4, we conclude and discuss the implications of our findings.

COUPLED BULK-SURFACE REACTION–DIFFUSION SYSTEMS ON STATIONARY VOLUMES

Here, we present a coupled system of BSRDEs posed in a three-dimensional volume as well as on the boundary surface enclosing the volume. We impose Robin-type boundary conditions on the bulk reaction–diffusion system, whereas no boundary conditions are imposed on the surface reaction–diffusion system since the surface is closed.

(a) A coupled system of bulk-surface reaction–diffusion equations

Let Ω be a stationary volume (whose interior is denoted the bulk) enclosed by a compact hypersurface $\Gamma:=\partial\Omega$ which is C^2. In addition, let $I=[0,T](T>0)$ be some time interval. Moreover, let $\boldsymbol{v}$ denote the unit outer normal to Γ, and let U be any open subset of $\mathbb{R}^{N+1}$ containing Γ, then for any function u which is differentiable in U, we define the tangential gradient on Γ by, $\nabla_\Gamma u=\nabla u-(\nabla u\cdot\boldsymbol{v})\,\boldsymbol{v}$, where $\cdot$ denotes the regular dot product and ∇ denotes the regular gradient in $\mathbb{R}^{N+1}$. The tangential gradient is the projection of the regular gradient onto the tangent plane, thus $\nabla_\Gamma u\cdot\boldsymbol{v}=0$. The Laplace–Beltrami operator on the surface Γ is then defined to be the tangential divergence of the tangential gradient $\Delta_\Gamma u=\nabla_\Gamma\cdot\nabla_\Gamma u$. For a vector function $\boldsymbol{u}=(u_1,u_2,\ldots,u_{N+1})\in\mathbb{R}^{N+1}$, the tangential divergence is defined by

$$\nabla_\Gamma\cdot\boldsymbol{u}=\nabla\cdot\boldsymbol{u}-\sum_{i=1}^{N+1}(\nabla u_i\cdot\boldsymbol{v})v_i.$$

To proceed, we denote by $u:\Omega\times I\to\mathbb{R}$ and $v:\Omega\times I\to\mathbb{R}$ two chemical concentrations (species) that react and diffuse in Ω and $r:\Gamma\times I\to\mathbb{R}$ and $s:\Gamma\times I\to\mathbb{R}$ be two chemical species residing only on the surface Γ which react and diffuse on the surface. In the absence of cross-diffusion and assuming that coupling is only through the reaction kinetics, we propose to study the following non-dimensionalized coupled system of BSRDEs

$$\left.\begin{array}{l}\begin{cases}u_t=\nabla^2u+\gamma_\Omega f(u,v),\\ v_t=d_\Omega\nabla^2v+\gamma_\Omega g(u,v),\end{cases}\quad \text{in } \Omega\times(0,T],\\ \begin{cases}r_t=\nabla_\Gamma^2r+\gamma_\Gamma(f(r,s)-h_1(u,v,r,s)),\\ s_t=d_\Gamma\nabla_\Gamma^2s+\gamma_\Gamma(g(r,s)-h_2(u,v,r,s)),\end{cases}\quad \text{on } \Gamma\times(0,T],\end{array}\right\}\tag{2.1}$$

with coupling boundary conditions

$$\begin{cases}\dfrac{\partial u}{\partial\boldsymbol{v}}=\gamma_\Gamma h_1(u,v,r,s),\\ d_\Omega\dfrac{\partial v}{\partial\boldsymbol{v}}=\gamma_\Gamma h_2(u,v,r,s),\end{cases}\quad \text{on } \Gamma\times(0,T].\tag{2.2}$$

In the above, $\nabla^2=\partial^2/\partial x^2+\partial^2/\partial y^2+\partial^2/\partial z^2$ represents the Laplacian operator. d_Ω and d_Γ are a positive diffusion coefficients in the bulk and on the surface, respectively, representing the ratio between u and v, and r and s, respectively. γ_Ω and γ_Γ represent the length-scale parameters in the bulk and on the surface, respectively. In this formulation, we assume that f(·,·) and g(·,·) are nonlinear reaction kinetics in the bulk and on the surface. $h_1(u,v,r,s)$ and $h_2(u,v,r,s)$ are reactions representing the coupling of the internal dynamics in the bulk Ω to the surface dynamics on the surface Γ. As a first attempt, we consider a more generalized form of linear coupling of the following nature [**21**]

$$h_1(u,v,r,s) = \alpha_1 r - \beta_1 u - \kappa_1 v \tag{2.3}$$

and

$$h_2(u,v,r,s) = \alpha_2 s - \beta_2 u - \kappa_2 v, \tag{2.4}$$

where α_1, α_2, β_1, β_2, κ_1 and κ_2 are constant non-dimensionalized parameters. Initial conditions are given by the positive-bounded functions $u_0(\boldsymbol{x})$, $v_0(\boldsymbol{x})$, $r_0(\boldsymbol{x})$ and $s_0(\boldsymbol{x})$.

(i) Activator-depleted reaction kinetics: an illustrative example

From now onwards, we restrict our analysis and simulations to the well-known activator-depletedsubstrate reaction model [**8,22–25**] also known as the Brusselator given by

$$f(u,v) = a - u + u^2 v \quad \text{and} \quad g(u,v) = b - u^2 v, \tag{2.5}$$

where a and b are positive parameters. For analytical simplicity, we postulate the model system (2.1) in a more compact form given by

$$\left.\begin{aligned} &\begin{cases} u_t = \nabla^2 u + f_1(u,v,r,s), \\ v_t = d_\Omega \nabla^2 v + f_2(u,v,r,s), \end{cases} && x \text{ on } \Omega,\ t>0, \\ &\begin{cases} r_t = \nabla_\Gamma^2 r + f_3(u,v,r,s), \\ s_t = d_\Gamma \nabla_\Gamma^2 s + f_4(u,v,r,s), \end{cases} && x \text{ on } \Gamma,\ t>0, \end{aligned}\right\} \tag{2.6}$$

with coupling boundary conditions (2.2)–(2.4). In the above, we have defined appropriately

$$f_1(u,v,r,s) = \gamma_\Omega (a - u + u^2 v), \tag{2.7}$$

$$f_2(u,v,r,s) = \gamma_\Omega (b - u^2 v), \tag{2.8}$$

$$f_3(u,v,r,s) = \gamma_\Gamma (a - r + r^2 s - \alpha_1 r + \beta_1 u + \kappa_1 v) \tag{2.9}$$

and

$$f_4(u,v,r,s) = \gamma_\Gamma (b - r^2 s - \alpha_2 s + \beta_2 u + \kappa_2 v). \tag{2.10}$$

(b) Linear stability analysis of the coupled system of BSRDEs

Definition 2.1 (Uniform steady state).

A point (u*,v*,r*,s*) is a uniform steady state of the coupled system of BSRDEs (2.6) with reaction kinetics (2.5) if it solves the nonlinear algebraic system given by fi(u*,v*,r*,s*)=0, for alli=1,2,3,4, and satisfies the boundary conditions given by (2.2)–(2.4).

Proposition 2.2 (Existence and uniqueness of the uniform steady state).

The coupled system of BSRDEs (2.6) with boundary conditions (2.2) admits a unique steady state given by

$$(u^*,v^*,r^*,s^*) = \left(a+b, \frac{b}{(a+b)^2}, a+b, \frac{b}{(a+b)^2}\right), \tag{2.11}$$

provided the following compatibility condition on the coefficients of the coupling is satisfied

$$(\beta_1 - \alpha_1)(\kappa_2 - \alpha_2) - \kappa_1 \beta_2 = 0. \tag{2.12}$$

Proof.

The proof follows immediately from the definition of the uniform steady state satisfying reaction kinetics (2.7)–(2.10). It must be noted that in deriving this unique uniform steady state the compatibility condition (2.12) coupling bulk and surface dynamics must be satisfied.

Remark 2.3: The constraint condition (2.12) on the parameter values α_i, β_i and κ_i, i=1,2 is a general case of the specific parameter values given in reference [**21**] where the following parameter values were selected $\alpha_1 = \beta_1 = \frac{5}{12}$, $\alpha_2 = \kappa_2 = 5$, $\kappa_1 = 0$ and $\beta_2 = 0$.

Remark 2.4: Note that there exists an infinite number of solutions to problem (2.12).

(i) Linear stability analysis in the absence of diffusion

Next, we study Turing diffusion-driven instability for the coupled system of BSRDEs (2.1)–(2.4) with reaction kinetics (2.5). To proceed, we first consider the linear stability of the spatially uniform steady state. For the sake of convenience, let us denote by $\boldsymbol{w}=(u,v,r,s)^T$, the vector of the species u, v, r and s. Furthermore, defining the vector ξ such that $|\xi_i|<1$ for all i=1,2,3 and 4, it follows that writing $\boldsymbol{w}=\boldsymbol{w}^*+\xi$, the linearized system of coupled BSRDEs can be posed as

$$w_t = \xi_t = J_F \xi, \tag{2.13}$$

where J_F represents the Jacobian matrix representing the first linear terms of the linearization process. Its entries are defined by

$$J_F = \begin{pmatrix} \frac{\partial f_1}{\partial u} & \frac{\partial f_1}{\partial v} & \frac{\partial f_1}{\partial r} & \frac{\partial f_1}{\partial s} \\ \frac{\partial f_2}{\partial u} & \frac{\partial f_2}{\partial v} & \frac{\partial f_2}{\partial r} & \frac{\partial f_2}{\partial s} \\ \frac{\partial f_3}{\partial u} & \frac{\partial f_3}{\partial v} & \frac{\partial f_3}{\partial r} & \frac{\partial f_3}{\partial s} \\ \frac{\partial f_4}{\partial u} & \frac{\partial f_4}{\partial v} & \frac{\partial f_4}{\partial r} & \frac{\partial f_4}{\partial s} \end{pmatrix} = \begin{pmatrix} f_{1_u} & f_{1_v} & 0 & 0 \\ f_{2_u} & f_{2_v} & 0 & 0 \\ f_{3_u} & f_{3_v} & f_{3_r} & f_{3_s} \\ f_{4_u} & f_{4_v} & f_{4_r} & f_{4_s} \end{pmatrix}$$

$$:= \begin{pmatrix} f_u & f_v & 0 & 0 \\ g_u & g_v & 0 & 0 \\ -h_{1u} & -h_{1v} & f_r - h_{1r} & f_s - h_{1s} \\ -h_{2u} & -h_{2v} & g_r - h_{2r} & g_s - h_{2s} \end{pmatrix}, \tag{2.14}$$

where by definition $f_{1_u} := \partial f_1/\partial u$ represents a partial derivative of $f_1(u,v)$ with respect to u. We are looking for solutions to the system of linear ordinary differential equations (2.13) which are of the form $\xi \propto e^{\lambda t}$. Substituting into (2.13), results in the following classical eigenvalue problem

$$|\lambda I - J_F| = 0, \tag{2.15}$$

where $\boldsymbol{I}$ is the identity matrix. Making appropriate substitutions and carrying out standard calculations, we obtain the following dispersion relation for λ

$$|\lambda I - J_F| = \begin{vmatrix} \lambda - f_{1_u} & f_{1_v} & 0 & 0 \\ f_{2_u} & \lambda - f_{2_v} & 0 & 0 \\ f_{3_u} & f_{3_v} & \lambda - f_{3_r} & f_{3_s} \\ f_{4_u} & f_{4_v} & f_{4_r} & \lambda - f_{4_s} \end{vmatrix} = 0,$$

$$\Longleftrightarrow \quad p_4(\lambda) = \lambda^4 + a_1\lambda^3 + a_2\lambda^2 + a_3\lambda + a_4 = 0,$$

where

$$a_1 = -(f_{1_u} + f_{2_v} + f_{3_r} + f_{4_s}), \tag{2.16}$$

$$a_2 = (f_{1_u}f_{2_v} - f_{1_v}f_{2_u}) + (f_{3_r}f_{4_s} - f_{3_s}f_{4_r}) + (f_{1_u} + f_{2_v})(f_{3_r} + f_{4_s}), \tag{2.17}$$

$$a_3 = -[(f_{1_u}f_{2_v} - f_{1_v}f_{2_u})(f_{3_r} + f_{4_s}) + (f_{3_r}f_{4_s} - f_{3_s}f_{4_r})(f_{1_u} + f_{2_v})] \tag{2.18}$$

and

$$a_4 = (f_{1_u}f_{2_v} - f_{1_v}f_{2_u})(f_{3_r}f_{4_s} - f_{3_s}f_{4_r}). \tag{19}$$

For the sake of convenience, let us denote by

$$(J_F)_\Omega := \begin{pmatrix} f_{1_u} & f_{1_v} \\ f_{2_u} & f_{2_v} \end{pmatrix} \quad \text{and} \quad (J_F)_\Gamma := \begin{pmatrix} f_{3_r} & f_{3_s} \\ f_{4_r} & f_{4_s} \end{pmatrix} \tag{2.20}$$

the submatrices of J_F corresponding to the bulk reaction kinetics and the surface reaction kinetics, respectively. We can now define

$$\mathrm{Tr}(J_F) := f_{1_u} + f_{2_v} + f_{3_r} + f_{4_s}, \quad \mathrm{Tr}(J_F)_\Omega := f_{1_u} + f_{2_v}, \quad \mathrm{Tr}(J_F)_\Gamma := f_{3_r} + f_{4_s},$$
$$\mathrm{Det}(J_F)_\Omega := f_{1_u}f_{2_v} - f_{1_v}f_{2_u}, \quad \text{and} \quad \mathrm{Det}(J_F)_\Gamma := f_{3_r}f_{4_s} - f_{3_s}f_{4_r}.$$

Theorem 2.5 Necessary and sufficient conditions for Re(λ) < 0).

(The necessary and sufficient conditions such that the zeros of the polynomial $p_4(\lambda)$ have $\mathrm{Re}(\lambda)<0$ are given by the following conditions

$$\mathrm{Tr}(J_F) < 0, \tag{2.21}$$

$$\mathrm{Det}(J_F)_\Omega + \mathrm{Det}(J_F)_\Gamma + \mathrm{Tr}(J_F)_\Omega \mathrm{Tr}(J_F)_\Gamma > 0, \tag{2.22}$$

$$\mathrm{Det}(J_F)_\Omega \mathrm{Tr}(J_F)_\Gamma + \mathrm{Det}(J_F)_\Gamma \mathrm{Tr}(J_F)_\Omega < 0, \tag{2.23}$$

$$\mathrm{Det}(J_F)_\Omega \mathrm{Det}(J_F)_\Gamma > 0, \tag{2.24}$$

$$[\mathrm{Tr}(J_F)_\Gamma \mathrm{Tr}(J_F) - 2\mathrm{Det}(J_F)_\Omega]\,\mathrm{Tr}(J_F)_\Omega + [\mathrm{Tr}(J_F)_\Omega \mathrm{Tr}(J_F) - 2\,\mathrm{Det}(J_F)_\Gamma]\,\mathrm{Tr}(J_F)_\Gamma > 0 \tag{2.25}$$

and

$$[(\mathrm{Det}(J_F)_\Omega + \mathrm{Det}(J_F)_\Gamma)^2 - (\mathrm{Det}(J_F)_\Omega \mathrm{Tr}(J_F)_\Gamma + \mathrm{Det}(J_F)_\Gamma \mathrm{Tr}(J_F)_\Omega)\,\mathrm{Tr}(J_F)]\,\mathrm{Tr}(J_F)_\Omega \mathrm{Tr}(J_F)_\Gamma > 0. \tag{2.26}$$

Proof.

The proof enforces that $p_4(\lambda)$ is a Hurwitz polynomial and therefore satisfies the Routh–Hurwitz criterion in order for $\mathrm{Re}(\lambda)<0$. The first condition to be satisfied is that $a_4 \neq 0$ otherwise $\lambda=0$ is a trivial root, thereby reducing the fourth-order polynomial to a cubic polynomial. The first four conditions are a result of requiring that each coefficient a_i with i=1,2,3 and 4 of the polynomialp$_4(\lambda)$ are all positive. The rest of the conditions are derived as shown below.

We require that the determinant of the matrix

$$\begin{vmatrix} a_1 & a_3 \\ 1 & a_2 \end{vmatrix} = a_1a_2 - a_3 > 0.$$

Substituting a_1, a_2 and a_3 appropriately, we obtain

$$[\mathrm{Det}(J_F)_\Omega + \mathrm{Det}(J_F)_\Gamma + \mathrm{Tr}(J_F)_\Omega \mathrm{Tr}(J_F)_\Gamma][-\mathrm{Tr}(J_F)] + [\mathrm{Det}(J_F)_\Omega \mathrm{Tr}(J_F)_\Gamma + \mathrm{Det}(J_F)_\Gamma \mathrm{Tr}(J_F)_\Omega] > 0. \tag{2.27}$$

Exploiting the fact that

$$\mathrm{Tr}(J_F) = \mathrm{Tr}(J_F)_\Omega + \mathrm{Tr}(J_F)_\Gamma,$$

it then follows that

$$a_1a_2 - a_3 = \mathrm{Tr}(J_F)_\Omega \mathrm{Tr}(J_F)_\Gamma \mathrm{Tr}(J_F) - [\mathrm{Det}(J_F)_\Omega \mathrm{Tr}(J_F)_\Omega + \mathrm{Det}(J_F)_\Gamma \mathrm{Tr}(J_F)_\Gamma] > 0$$

if and only if

$$\tfrac{1}{2}[\mathrm{Tr}(J_F)_\Gamma \mathrm{Tr}(J_F) - 2\,\mathrm{Det}(J_F)_\Omega]\,\mathrm{Tr}(J_F)_\Omega + \tfrac{1}{2}[\mathrm{Tr}(J_F)_\Omega \mathrm{Tr}(J_F) - 2\,\mathrm{Det}(J_F)_\Gamma]\,\mathrm{Tr}(J_F)_\Gamma > 0.$$

Multiplying throughout by 2, we obtain condition (2.25) in theorem 2.5.

The last condition results from imposing the condition that

$$\begin{vmatrix} a_1 & a_3 & 0 \\ 1 & a_2 & a_4 \\ 0 & a_1 & a_3 \end{vmatrix} = a_3(a_1a_2 - a_3) - a_1^2a_4 > 0.$$

It can be shown that

$$\begin{aligned} a_3(a_1a_2 - a_3) = &-[\mathrm{Det}(J_F)_\Omega \mathrm{Tr}(J_F)_\Omega \mathrm{Tr}^2(J_F)_\Gamma + \mathrm{Det}(J_F)_\Gamma \mathrm{Tr}^2(J_F)_\Omega \mathrm{Tr}(J_F)_\Gamma]\,\mathrm{Tr}(J_F) \\ &+ \mathrm{Det}^2(J_F)_\Omega \mathrm{Tr}(J_F)_\Omega \mathrm{Tr}(J_F)_\Gamma + \mathrm{Det}(J_F)_\Omega \mathrm{Det}(J_F)_\Gamma \mathrm{Tr}^2(J_F)_\Omega \\ &+ \mathrm{Det}(J_F)_\Omega \mathrm{Det}(J_F)_\Gamma \mathrm{Tr}^2(J_F)_\Omega + \mathrm{Det}^2(J_F)_\Gamma \mathrm{Tr}(J_F)_\Omega \mathrm{Tr}(J_F)_\Gamma. \end{aligned} \tag{2.28}$$

On the other hand,

$$
\begin{aligned}
a_1^2 a_4 &= \mathrm{Tr}^2(J_F)\,\mathrm{Det}(J_F)_\Omega\,\mathrm{Det}(J_F)_\Gamma \\
&= (\mathrm{Tr}(J_F)_\Omega + \mathrm{Tr}(J_F)_\Gamma)^2\,\mathrm{Det}(J_F)_\Omega\,\mathrm{Det}(J_F)_\Gamma \\
&= \mathrm{Det}(J_F)_\Omega\,\mathrm{Det}(J_F)_\Gamma\,\mathrm{Tr}^2(J_F)_\Omega + 2\,\mathrm{Det}(J_F)_\Omega\,\mathrm{Det}(J_F)_\Gamma\,\mathrm{Tr}(J_F)_\Omega\,\mathrm{Tr}(J_F)_\Gamma \\
&\quad + \mathrm{Det}(J_F)_\Omega\,\mathrm{Det}(J_F)_\Gamma\,\mathrm{Tr}^2(J_F)_\Gamma.
\end{aligned}
\tag{2.29}
$$

Hence, combining (2.28) and (2.29) and simplifying conveniently, we have

$$
\begin{aligned}
a_3(a_1a_2 - a_3) - a_1^2 a_4 = [(\mathrm{Det}(J_F)_\Omega + \mathrm{Det}(J_F)_\Gamma)^2 - (\mathrm{Det}(J_F)_\Omega\,\mathrm{Tr}(J_F)_\Gamma \\
+ \mathrm{Det}(J_F)_\Gamma\,\mathrm{Tr}(J_F)_\Omega)\,\mathrm{Tr}(J_F)] \times \mathrm{Tr}(J_F)_\Omega\,\mathrm{Tr}(J_F)_\Gamma > 0,
\end{aligned}
$$

resulting in condition (2.26). ▪

Remark 2.6

The characteristic polynomial,

$$p_4(\lambda) = \lambda^4 + a_1\lambda^3 + a_2\lambda^2 + a_3\lambda + a_4,$$

can also be written more compactly in the form of

$$p_4(\lambda) = (\lambda^2 + \lambda(f_{1_u} + f_{2_v}) + f_{1_u}f_{2_v} - f_{1_v}f_{2_u})(\lambda^2 + \lambda(f_{3_r} + f_{4_s}) + f_{3_r}f_{4_s} - f_{3_s}f_{4_r}),$$

thereby coupling the bulk and surface dispersion relations in the absence of spatial variations.

(ii) Linear stability analysis in the presence of diffusion

Next, we introduce spatial variations and study under what conditions the uniform steady state is linearly unstable. We linearize around the uniform steady state by taking small spatially varying perturbations of the form

$$w(x,t) = w^* + \epsilon\xi(x,t), \quad \text{with } \epsilon \ll 1. \tag{2.30}$$

Substituting (2.30) into the coupled system of BSRDEs (2.1)–(2.4) with reaction kinetics (2.5), we obtain a linearized system of partial differential equations

$$\xi_{1t} = \nabla^2\xi_1 + \gamma_\Omega(f_u\xi_1 + f_v\xi_2), \tag{2.31}$$

$$\xi_{2t} = d_\Omega\nabla^2\xi_2 + \gamma_\Omega(g_u\xi_1 + g_v\xi_2), \tag{2.32}$$

$$\xi_{3t} = \nabla_\Gamma^2\xi_3 + \gamma_\Gamma(f_r\xi_3 + f_s\xi_4 - h_{1_u}\xi_1 - h_{1_v}\xi_2 - h_{1_r}\xi_3 - h_{1_s}\xi_4) \tag{2.33}$$

and

$$\xi_{4t} = d_\Gamma\nabla_\Gamma^2\xi_4 + \gamma_\Gamma(g_r\xi_3 + g_s\xi_4 - h_{2_u}\xi_1 - h_{2_v}\xi_2 - h_{2_r}\xi_3 - h_{2_s}\xi_4), \tag{2.34}$$

with linearized boundary conditions

$$\frac{\partial\xi_1}{\partial\nu} = \gamma_\Gamma(h_{1_u}\xi_1 + h_{1_v}\xi_2 + h_{1_r}\xi_3 + h_{1_s}\xi_4) \tag{2.35}$$

and

$$d_\Gamma\frac{\partial\xi_2}{\partial\nu} = \gamma_\Gamma(h_{2_u}\xi_1 + h_{2_v}\xi_2 + h_{2_r}\xi_3 + h_{2_s}\xi_4). \tag{2.36}$$

In the above, we have used the original reaction kinetics for the purpose of clarity.

In order to proceed, we restrict our analysis to circular and spherical domains where we can transform the cartesian coordinates into polar coordinates and be able to exploit the method of separation of variables. Without loss of generality, we write the following eigenvalue problem in the bulk

$$\nabla^2\psi_{k_{l,m}}(r) = -k_{l,m}^2\psi_{k_{l,m}}(r), \quad 0 < r < 1, \tag{2.37}$$

where each ψk satisfies the boundary conditions (2.35) and (2.36). On the surface, the eigenvalue problem is posed as

$$\nabla^2_\Gamma \phi(y) = -l(l+1)\phi(y), \quad y \in \Gamma. \tag{2.38}$$

Remark 2.7

For the case of circular and spherical domains, if $r=1$, then $k^2_{l,m} = l(l+1)$.

Taking $x \in \mathbb{B},\ y \in \Gamma$, then writing in polar coordinates $x = ry,\ r \in (0,1)$ we can define, for all $l \in \mathbb{N}_0,\ m \in \mathbb{Z},\ |m| \le l$, the following power-series solutions [5,6]

$$\xi_1(ry,t) = \sum u_{l,m}(t)\psi_{k_{l,m}}(r)\phi_{l,m}(y), \quad \xi_2(ry,t) = \sum v_{l,m}(t)\psi_{k_{l,m}}(r)\phi_{l,m}(y), \tag{39}$$

and

$$\xi_3(y,t) = \sum r_{l,m}(t)\phi_{l,m}(y), \quad \text{and} \quad \xi_4(y,t) = \sum s_{l,m}(t)\phi_{l,m}(y). \tag{2.40}$$

On the surface, substituting the power series solutions (2.40) into (2.33) and (2.34), we have

$$\begin{aligned}\frac{dr_{l,m}}{dt} &= -l(l+1)r_{l,m} + \gamma_\Gamma(f_r r_{l,m} + f_s s_{l,m}) \\ &\quad - \gamma_\Gamma(h_{1_u} u_{l,m}\psi_{k_{l,m}}(1) + h_{1_v} v_{l,m}\psi_{k_{l,m}}(1) + h_{1_r} r_{l,m} + h_{1_s} s_{l,m}),\end{aligned} \tag{2.41}$$

and

$$\begin{aligned}\frac{ds_{l,m}}{dt} &= -d_\Gamma l(l+1)s_{l,m} + \gamma_\Gamma(g_r r_{l,m} + g_s s_{l,m}) \\ &\quad - \gamma_\Gamma(h_{2_u} u_{l,m}\psi_{k_{l,m}}(1) + h_{2_v} v_{l,m}\psi_{k_{l,m}}(1) + h_{2_r} r_{l,m} + h_{2_s} s_{l,m}).\end{aligned} \tag{2.42}$$

Similarly, substituting the power-series solutions (2.39) into the bulk equations (2.31) and (2.32), we obtain the following system of ordinary differential equations

$$\frac{du_{l,m}}{dt} = -k^2_{l,m} u_{l,m} + \gamma_\Omega(f_u u_{l,m} + f_v v_{l,m}) \tag{2.43}$$

and

$$\frac{dv_{l,m}}{dt} = -d_\Omega k^2_{l,m} v_{l,m} + \gamma_\Omega(g_u u_{l,m} + g_v v_{l,m}). \tag{2.44}$$

Equations (2.43) and (2.44) are supplemented with boundary conditions

$$u_{l,m}\psi'_{k_{l,m}}(1) = \gamma_\Gamma(h_{1_u} u_{l,m}\psi_{k_{l,m}}(1) + h_{1_v} v_{l,m}\psi_{k_{l,m}}(1) + h_{1_r} r_{l,m} + h_{1_s} s_{l,m}), \tag{2.45}$$

and

$$d_\Omega v_{l,m}\psi'_{k_{l,m}}(1) = \gamma_\Gamma(h_{2_u} u_{l,m}\psi_{k_{l,m}}(1) + h_{2_v} v_{l,m}\psi_{k_{l,m}}(1) + h_{2_r} r_{l,m} + h_{2_s} s_{l,m}), \tag{2.46}$$

where $\psi'_{k_{l,m}} := d\psi_{k_{l,m}}(r)/dr|_{r=1}$. Writing

$$(u_{l,m},\ v_{l,m},\ r_{l,m},\ s_{l,m},)^T = (u^0_{l,m},\ v^0_{l,m},\ r^0_{l,m},\ s^0_{l,m})^T e^{\lambda_{l,m} t},$$

and substituting into the system of ordinary differential equations (2.41)–(2.44), we obtain the following eigenvalue problem

$$(\lambda_{l,m}\boldsymbol{I} + \boldsymbol{M})\boldsymbol{\xi}^0_{l,m} = \boldsymbol{0}, \tag{2.47}$$

where

$$M = \begin{pmatrix} k_{l,m}^2 - \gamma_\Omega f_u & -\gamma_\Omega f_v & 0 & 0 \\ -\gamma_\Omega g_u & d_\Omega k_{l,m}^2 - \gamma_\Omega g_v & 0 & 0 \\ \psi'_{k_{l,m}}(1) & 0 & l(l+1) - \gamma_\Gamma f_r & -\gamma_\Gamma f_s \\ 0 & d_\Omega \psi'_{k_{l,m}}(1) & -\gamma_\Gamma g_r & d_\Gamma l(l+1) - \gamma_\Gamma g_s \end{pmatrix},$$

and

$$\boldsymbol{\xi}_{l,m}^0 = (u_{l,m}^0,\ v_{l,m}^0,\ r_{l,m}^0,\ s_{l,m}^0)^{\mathrm{T}}.$$

Note that the boundary conditions (2.45) and (2.46) have been applied appropriately to the surface-linearized reaction–diffusion equations. Because

$$(u_{l,m}^0,\ v_{l,m}^0,\ r_{l,m}^0,\ s_{l,m}^0)^{\mathrm{T}} \neq (0,\ 0,\ 0,\ 0)^{\mathrm{T}},$$

it follows that the coefficient matrix must be singular, hence we require that

$$|\lambda_{l,m}\boldsymbol{I} + \boldsymbol{M}| = 0.$$

Straightforward calculations show that the eigenvalue λl m solves the following dispersion relation written in compact form as

$$(\lambda_{l,m}^2 + \mathrm{Tr}(\boldsymbol{M})_\Omega \lambda_{l,m} + \mathrm{Det}(\boldsymbol{M})_\Omega)(\lambda_{l,m}^2 + \mathrm{Tr}(\boldsymbol{M})_\Gamma \lambda_{l,m} + \mathrm{Det}(\boldsymbol{M})_\Gamma) = 0, \tag{2.48}$$

where we have defined conveniently

$$\begin{aligned} \mathrm{Tr}(\boldsymbol{M})_\Omega &:= (d_\Omega + 1)k_{l,m}^2 - \gamma_\Omega(f_u + g_v), \\ \mathrm{Tr}(\boldsymbol{M})_\Gamma &:= (d_\Gamma + 1)l(l+1) - \gamma_\Gamma(f_r + g_s), \\ \mathrm{Det}(\boldsymbol{M})_\Omega &:= d_\Omega k_{l,m}^4 - \gamma_\Omega(d_\Omega f_u + g_v)k_{l,m}^2 + \gamma_\Omega^2(f_u g_v - f_v g_u), \\ \mathrm{Det}(\boldsymbol{M})_\Gamma &:= d_\Gamma l^2(l+1)^2 - \gamma_\Gamma(d_\Gamma f_r + g_s)l(l+1) + \gamma_\Gamma^2(f_r g_s - f_s g_r). \end{aligned}$$

The above holds true if and only if either

$$\lambda_{l,m}^2 + \mathrm{Tr}(\boldsymbol{M})_\Omega \lambda_{l,m} + \mathrm{Det}(\boldsymbol{M})_\Omega = 0 \tag{2.49}$$

or

$$\lambda_{l,m}^2 + \mathrm{Tr}(\boldsymbol{M})_\Gamma \lambda_{l,m} + \mathrm{Det}(\boldsymbol{M})_\Gamma = 0. \tag{2.50}$$

In the presence of diffusion, we require the emergence of spatial growth. In order for the uniform steady state $\boldsymbol{w}^*$ to be unstable, we require that either

1. $\mathrm{Re}(\lambda_{l,m}(k_{l,m}^2)) > 0$ for some $k_{l,m}^2 > 0$,
   or
2. $\mathrm{Re}(\lambda_{l,m}(l(l+1))) > 0$ for some $l(l+1) > 0$,
   or
3. Both.

Solving (2.49) (and similarly (2.50)), we obtain the eigenvalues

$$2\,\mathrm{Re}(\lambda_{l,m}(k_{l,m}^2)) = -\mathrm{Tr}(\boldsymbol{M})_\Omega \pm \sqrt{\mathrm{Tr}^2(\boldsymbol{M})_\Omega - 4\,\mathrm{Det}(\boldsymbol{M})_\Omega}. \tag{2.51}$$

It follows then that $\mathrm{Re}(\lambda_{l,m}(k_{l,m}^2)) > 0$ for some $k_{l,m}^2 > 0$ if and only if the following conditions hold and

$$\left.\begin{array}{l}\mathrm{Tr}(M)_{\Omega}<0 \iff (d_{\Omega}+1)k_{l,m}^{2}-\gamma_{\Omega}(f_u+g_v)<0\\ \mathrm{Det}(M)_{\Omega}>0 \iff d_{\Omega}k_{l,m}^{4}-\gamma_{\Omega}(d_{\Omega}f_u+g_v)k_{l,m}^{2}+\gamma_{\Omega}^{2}(f_ug_v-f_vg_u)>0,\end{array}\right\} \tag{2.52}$$

or

$$\left.\begin{array}{l}\mathrm{Tr}(M)_{\Omega}>0 \iff (d_{\Omega}+1)k_{l,m}^{2}-\gamma_{\Omega}(f_u+g_v)>0\\ \mathrm{Det}(M)_{\Omega}<0 \iff d_{\Omega}k_{l,m}^{4}-\gamma_{\Omega}(d_{\Omega}f_u+g_v)k_{l,m}^{2}+\gamma_{\Omega}^{2}(f_ug_v-f_vg_u)<0.\end{array}\right\} \tag{2.53}$$

and

Similarly, on the surface, $\mathrm{Re}(\lambda_{l,m}(l(l+1)))>0$ for some $l(l+1)>0$ if and only the following conditions hold

and

$$\left.\begin{array}{l}\mathrm{Tr}(M)_{\Gamma}<0 \iff (d_{\Gamma}+1)l(l+1)-\gamma_{\Gamma}(f_r+g_s)<0\\ \mathrm{Det}(M)_{\Gamma}>0 \iff d_{\Gamma}l^2(l+1)^2-\gamma_{\Gamma}(d_{\Gamma}f_r+g_s)l(l+1)+\gamma_{\Gamma}^{2}(f_rg_s-f_sg_r)>0,\end{array}\right\} \tag{2.54}$$

or

$$\left.\begin{array}{l}\mathrm{Tr}(M)_{\Gamma}>0 \iff (d_{\Gamma}+1)l(l+1)-\gamma_{\Gamma}(f_r+g_s)>0,\\ \mathrm{Det}(M)_{\Gamma}<0 \iff d_{\Gamma}l^2(l+1)^2-\gamma_{\Gamma}(d_{\Gamma}f_r+g_s)l(l+1)+\gamma_{\Gamma}^{2}(f_rg_s-f_sg_r)<0.\end{array}\right\} \tag{2.55}$$

and

We are in a position to state the theorem 2.8.

Theorem 2.8

Assuming that

$$\mathrm{Tr}(J_F)_{\Omega}=f_u+g_v<0 \quad and \quad \mathrm{Det}(J_F)_{\Omega}=f_ug_v-f_vg_u>0, \tag{2.56}$$

then the necessary conditions for $\mathrm{Re}(\lambda_{l,m}(k_{l,m}^2))>0\,for\ some\ k_{l,m}^2>0$ are given by

$$d_{\Omega}f_u+g_v>0, \quad and \quad (d_{\Omega}f_u+g_v)^2-4d_{\Omega}(f_ug_v-f_vg_u)>0. \tag{2.57}$$

Similarly, assuming that

$$\mathrm{Tr}(J_F)_{\Gamma}=f_r+g_s<0 \quad and \quad \mathrm{Det}(J_F)_{\Gamma}=f_rg_s-f_sg_r>0, \tag{2.58}$$

then the necessary conditions for $\mathrm{Re}(\lambda_{l,m}(l(l+1)))>0\,for\ some\ l(l+1)>0$ are given by

$$d_{\Gamma}f_r+g_s>0 \quad and \quad (d_{\Gamma}f_r+g_s)^2-4d_{\Gamma}(f_rg_s-f_sg_r)>0. \tag{2.59}$$

Proof.

The proof is a direct consequence of conditions (2.52)–(2.55). Assuming that conditions (2.56) and (2.58) hold, then one of the conditions in (2.52) and (2.54) is violated, which implies that $\mathrm{Re}(\lambda_{l,m}(k_{l,m}^2))<0$ for all $k_{l,m}^2>0$ and similarly $\mathrm{Re}(\lambda_{l,m}(l(l+1)))<0$ for all $l(l+1)>0$. This entails that the system can no longer exhibit spatially inhomogeneous solutions.

The only two conditions left to hold true are (2.53) and (2.55). This case corresponds to the classical standard two-component reaction–diffusion system which requires that (for details, see for example [9])

$$d_\Omega f_u + g_v > 0, \quad \text{and} \quad (d_\Omega f_u + g_v)^2 - 4d_\Omega(f_u g_v - f_v g_u) > 0, \tag{2.60}$$

and similarly

$$d_\Gamma f_r + g_s > 0 \quad \text{and} \quad (d_\Gamma f_r + g_s)^2 - 4d_\Gamma(f_r g_s - f_s g_r) > 0. \tag{2.61}$$

This completes the proof ▪

Remark 2.9

Assuming that conditions (2.56) and (2.58) both hold, then conditions (2.57) and (2.59) imply that $d\Omega \neq 1$ and $d\Gamma \neq 1$.

Remark 2.10

If condition (2.56) or (2.58) holds only, then either $d\Omega \neq 1$ or $d\Gamma \neq 1$ but not necessarily both.

Remark 2.11

If conditions (2.56) and (2.58) are both violated, then diffusion-driven instability cannot occur.

Remark 2.12

Similar to classical reaction–diffusion systems, conditions (2.57) and (2.59) imply the existence of critical diffusion coefficients in the bulk and on the surface whereby the uniform states lose stability. In order for diffusion-driven instability to occur, the bulk and surface diffusion coefficients must be greater than the values of the critical diffusion coefficients.

Next, we investigate under what assumptions on the reaction-kinetics do conditions (2.52) and (2.54) hold true.

– First, let us consider the case when

$$\mathrm{Tr}(J_F)_\Omega = f_u + g_v > 0 \quad \text{and} \quad \mathrm{Det}(J_F)_\Omega = f_u g_v - f_v g_u > 0,$$

and

$$\mathrm{Tr}(J_F)_\Gamma = f_r + g_s > 0 \quad \text{and} \quad \mathrm{Det}(J_F)_\Gamma = f_r g_s - f_s g_r > 0.$$

Then, $\mathrm{Tr}(J_F) = \mathrm{Tr}(J_F)_\Omega + \mathrm{Tr}(J_F)_\Gamma > 0$ which violates condition (2.21).

– Similarly, the case when

$$\mathrm{Tr}(J_F)_\Omega = f_u + g_v > 0 \quad \text{and} \quad \mathrm{Det}(J_F)_\Omega = f_u g_v - f_v g_u < 0,$$

and

$$\mathrm{Tr}(J_F)_\Gamma = f_r + g_s > 0 \quad \text{and} \quad \mathrm{Det}(J_F)_\Gamma = f_r g_s - f_s g_r < 0$$

violates condition (2.21).

– Let us consider the case when

$$\mathrm{Tr}(J_F)_\Omega = f_u + g_v < 0 \quad \text{and} \quad \mathrm{Det}(J_F)_\Omega = f_u g_v - f_v g_u < 0$$

and

$$\mathrm{Tr}(J_F)_\Gamma = f_r + g_s < 0 \quad \text{and} \quad \mathrm{Det}(J_F)_\Gamma = f_r g_s - f_s g_r < 0.$$

Then, it follows that condition (2.25) given by

$$[\mathrm{Tr}(J_F)_\Gamma\,\mathrm{Tr}(J_F) - 2\,\mathrm{Det}(J_F)_\Omega]\,\mathrm{Tr}(J_F)_\Omega + [\mathrm{Tr}(J_F)_\Omega\,\mathrm{Tr}(J_F) - 2\,\mathrm{Det}(J_F)_\Gamma]\,\mathrm{Tr}(J_F)_\Gamma < 0$$

is violated.

- – Next, we consider the case when

$$\mathrm{Tr}(J_F)_\Omega = f_u + g_v < 0 \quad \text{and} \quad \mathrm{Det}(J_F)_\Omega = f_u g_v - f_v g_u < 0,$$

and

$$\mathrm{Tr}(J_F)_\Gamma = f_r + g_s > 0 \quad \text{and} \quad \mathrm{Det}(J_F)_\Gamma = f_r g_s - f_s g_r < 0.$$

It follows then that none of the conditions (2.21)–(2.26) are violated. However, condition (2.52) does not hold.

- – Similarly, the case when

$$\mathrm{Tr}(J_F)_\Omega = f_u + g_v > 0 \quad \text{and} \quad \mathrm{Det}(J_F)_\Omega = f_u g_v - f_v g_u < 0,$$

and

$$\mathrm{Tr}(J_F)_\Gamma = f_r + g_s < 0 \quad \text{and} \quad \mathrm{Det}(J_F)_\Gamma = f_r g_s - f_s g_r < 0.$$

This implies that none of the conditions (2.21)–(2.26) are violated, while condition (2.54) fails to hold.

- – Finally, the cases when

$$\left.\begin{array}{l}\mathrm{Tr}(J_F)_\Omega = f_u + g_v > 0 \quad \text{and} \quad \mathrm{Det}(J_F)_\Omega = f_u g_v - f_v g_u > 0\\ \mathrm{Tr}(J_F)_\Gamma = f_r + g_s < 0 \quad \text{and} \quad \mathrm{Det}(J_F)_\Gamma = f_r g_s - f_s g_r > 0,\end{array}\right\} \tag{2.62}$$

and

$$\left.\begin{array}{l}\mathrm{Tr}(J_F)_\Omega = f_u + g_v < 0 \quad \text{and} \quad \mathrm{Det}(J_F)_\Omega = f_u g_v - f_v g_u > 0,\\ \mathrm{Tr}(J_F)_\Gamma = f_r + g_s > 0 \quad \text{and} \quad \mathrm{Det}(J_F)_\Gamma = f_r g_s - f_s g_r > 0,\end{array}\right\} \tag{2.63}$$

result in remark 2.10.

The above cases clearly eliminate conditions (2.52) and (2.54) as necessary for uniform steady state to be driven unstable in the presence of diffusion. We are now in a position to state our main result.

Theorem 2.13 (Necessary conditions for diffusion-driven instability)

The necessary conditions for diffusion-driven instability The necessary conditions for diffusion-driven instability for the coupled system of BSRDEs (2.1)–(2.4) are given by

$$\mathrm{Tr}(J_F) < 0, \tag{2.64}$$

$$\mathrm{Det}(J_F)_\Omega + \mathrm{Det}(J_F)_\Gamma + \mathrm{Tr}(J_F)_\Omega \, \mathrm{Tr}(J_F)_\Gamma > 0, \tag{2.65}$$

$$\mathrm{Det}(J_F)_\Omega \, \mathrm{Tr}(J_F)_\Gamma + \mathrm{Det}(J_F)_\Gamma \, \mathrm{Tr}(J_F)_\Omega < 0, \tag{2.66}$$

$$\mathrm{Det}(J_F)_\Omega \, \mathrm{Det}(J_F)_\Gamma > 0, \tag{2.67}$$

$$[\mathrm{Tr}(J_F)_\Gamma \, \mathrm{Tr}(J_F) - 2\,\mathrm{Det}(J_F)_\Omega]\,\mathrm{Tr}(J_F)_\Omega + [\mathrm{Tr}(J_F)_\Omega \, \mathrm{Tr}(J_F) - 2\,\mathrm{Det}(J_F)_\Gamma]\,\mathrm{Tr}(J_F)_\Gamma > 0, \tag{2.68}$$

$$[(\mathrm{Det}(J_F)_\Omega + \mathrm{Det}(J_F)_\Gamma)^2 - (\mathrm{Det}(J_F)_\Omega \, \mathrm{Tr}(J_F)_\Gamma + \mathrm{Det}(J_F)_\Gamma \, \mathrm{Tr}(J_F)_\Omega)\,\mathrm{Tr}(J_F)]\,\mathrm{Tr}(J_F)_\Omega \, \mathrm{Tr}(J_F)_\Gamma > 0 \tag{2.69}$$

and

$$d_\Omega f_u + g_v > 0, \quad \text{and} \quad (d_\Omega f_u + g_v)^2 - 4 d_\Omega \, \mathrm{Det}(J_F)_\Omega > 0, \tag{2.70}$$

or/and

$$d_\Gamma f_r + g_s > 0, \quad \text{and} \quad (d_\Gamma f_r + g_s)^2 - 4 d_\Gamma \, \mathrm{Det}(J_F)_\Gamma > 0. \tag{2.71}$$

(iii) Theoretical predictions

From the analytical results, we state the following theoretical predictions to be validated through the use of numerical simulations.

1. The bulk and surface diffusion coefficients d_Ω and d_Γ must be greater than one in order for diffusion-driven instability to occur. Taking $d_\Omega=d_\Gamma=1$ results in a contradiction between conditions (2.64), (2.70) and (2.71). As a result, the BSRDEs does not give rise to the formation of spatial structure. For this case, the uniform steady state is the only stable solution for the coupled system of BSRDEs (2.1)–(2.4).
2. The above imply that taking $d_\Omega>1$ and $d\Gamma=1$, the bulk-reaction–diffusion system has the potential to induce patterning in the bulk for appropriate diffusion-driven instability parameter values, whereas the surface-reaction–diffusion system is not able to generate patterns. Here, all the conditions (2.64)–(2.70) hold except (2.71).
3. Alternatively, taking $d_\Omega=1$ and $d_\Gamma>1$, the bulk-reaction–diffusion system fails to induce patterning in the bulk while the surface-reaction–diffusion system has the potential to induce patterning on the surface. Similarly, all the conditions (2.64)–(2.71) hold except (2.70).
4. On the other hand, taking $d_\Omega>1$ and $d_\Gamma>1$ appropriately, then the coupled system of BSRDEs exhibits patterning both in the bulk and on the surface. All the conditions (2.64)–(2.71) hold both in the bulk and on the surface.

NUMERICAL SIMULATIONS OF THE COUPLED SYSTEM OF BULK-SURFACE REACTION–DIFFUSION EQUATIONS

Here, we present bulk-surface finite-element numerical solutions corresponding to the coupled system of BSRDEs (2.1)–(2.5). Here, we omit the details of the bulk-surface finite-element method as these are given elsewhere (see **[19]** for details). Our method is inspired by the work of Elliott & Ranner **[10]**. We use the bulk-surface finite-element method to discretize in space with piecewise bilinear elements and an implicit second-order fractional-step θ-scheme to discretize in time using the Newton›s method for the linerization **[19,26]**. For details on the convergence and stability of the fully implicit time-stepping fractional-step θ-scheme, the reader is referred to Madzvamuse et al. **[19,26]**. In all our numerical experiments, we fix the kinetic model parameter values a=0.1 and b=0.9 since these satisfy the Turing diffusion-driven instability conditions (2.64)–(2.71). For these model parameter values, the equilibrium values are (u*,v*,r*,s*)=(1,0.9,1,0.9). Initial conditions are prescribed as small random perturbations around the equilibrium values. For illustrative purposes, let us take the parameter values describing the boundary conditions as shown in **table 1**; these are selected such that they satisfy the compatibility condition (2.12).

Table 1: Model parameter values for the coupled system of BSRDEs (2.1)–(2.4).

a	b	γ_Ω	γ_Γ	α_1	α_2	β_1	β_2	κ_1	κ_2
0.1	0.9	500	500	$\frac{5}{12}$	5	$\frac{5}{12}$	0	0	5

(a) A note on the bulk-surface triangulation

We briefly outline how the bulk-surface triangulation is generated. For further specific details, please see reference **[19]**. Let Ω_h be a triangulation of the bulk geometry Ω with vertices xi, $i=1,\ldots,N_h$, where N_h is the number of vertices in the triangulation. From Ω_h, we then construct Γ_h to be the triangulation of the surface geometry Γ by defining $\Gamma_h=\Omega_h|_{\partial}\Omega_h$, i.e. the vertices of Γ_h are the same as those lying on the surface of Ω_h. In particular, then, we have $\partial\Omega_h=\Gamma_h$. An example mesh is shown in **figure 1**. The bulk triangulation induces the surface triangulation as illustrated.

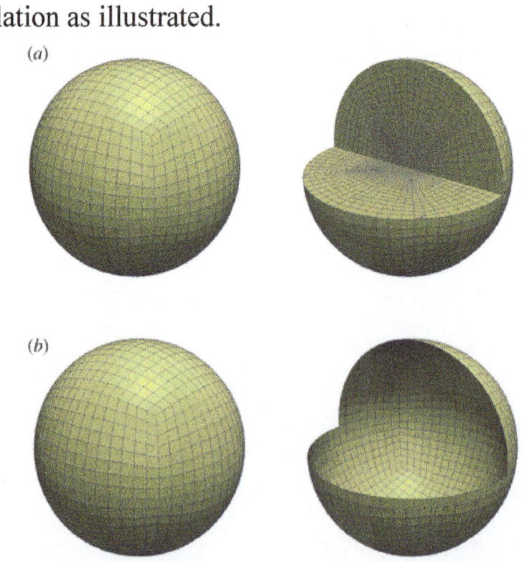

Figure 1: Example meshes for the bulk (a) and surface system (b). Part of the domain has been cut away and shown on the

right to reveal some internal mesh structure. (Online version in colour.)

(b) Numerical experiments

Here, we will present only four cases to validate our theoretical predictions outlined in §2b(iii). In most of our simulations, parameter values are fixed as shown in **table 1**, except for d_Ω and d_Γ whose values are varied to demonstrate the patterning mechanism of the coupled system of BSRDEs. We present patterns corresponding only to the chemical species u and r in the bulk and on the surface, respectively. Those corresponding to v and s are 180 degrees out of phase to those of u and r and are therefore omitted. It must be noted however that this is not always the case in general, Robin-type boundary conditions may alter the structure of the solution profiles depending on the model parameter values and the coupling compatibility boundary parameters.

(i) Simulations of the coupled system of BSRDEs with $(d\Omega, d\Gamma)=(1,1)$

The bulk-surface finite-element numerical simulations of the coupled system of BSRDEs with $d_\Omega=1$ in the bulk, $d_\Gamma=1$ on the surface are shown in **figure 2**. We observe that no patterns form in complete agreement with theoretical predictions. Similar to classical reaction–diffusion systems, diffusion coefficients must be greater than one. In particular, the diffusion coefficients must be greater than their corresponding respective critical diffusion coefficients in the bulk and on the surface. An example is shown next.

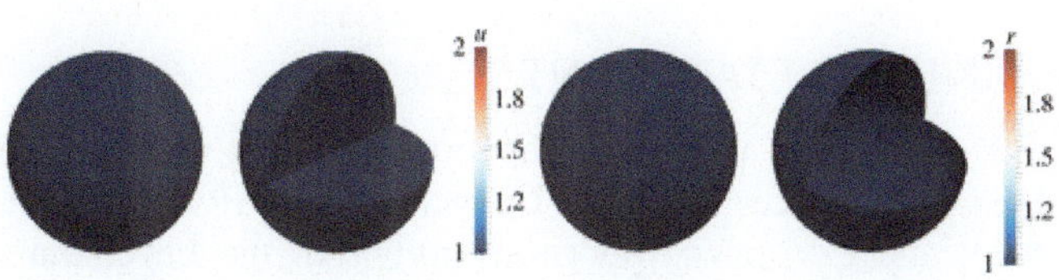

Figure 2: Numerical solutions corresponding to the coupled system of BSRDEs (2.1)–(2.5) with $d_\Omega=1$ in the bulk and $d_\Gamma=1$ on the surface. The uniform steady-state solutions are converged to and no patterns form. Columns 1 and 2: solutions in the bulk representing u. Columns 3 and 4: solutions on the surface representing r. Second and fourth columns represent cross sections of the bulk and the surface, respectively. (Online version in colour.)

(ii) Simulations of the coupled system of BSRDEs with $(d_\Omega, d_\Gamma)=(1,20)$

For illustrative purposes, let us take $d_\Omega=1$ in the bulk, $d_\Gamma = 20 > d_\Gamma^{\text{crit}} = 8.5$ on the surface.**Figure 3** illustrates pattern formation on the surface as well as within a small region in the vicinity of the surface membrane. Spots are observed to form on the surface, whereas in the bulk, small balls form inside. Far away from the surface, no patterns form, because the necessary conditions for diffusion-driven instability are not fulfilled in the bulk. These results confirm our theoretical predictions. We note that this particular example describes realistically pattern formation in biological systems. We expect skin patterning to manifest in the epidermis layer as well as on the surface.

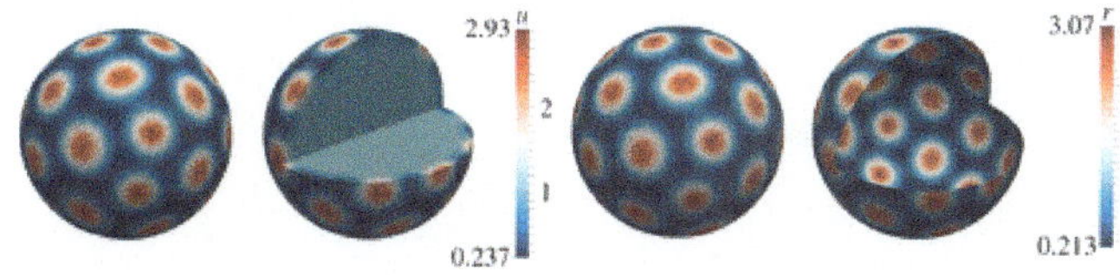

Figure 3: Numerical solutions corresponding to the coupled system of BSRDEs (2.1)–(2.5) with $d_\Omega=1$ in the bulk and $d_\Gamma=20$ on the surface. Columns 1 and 2: solutions in the bulk representing u. Columns 3 and 4: solutions on the surface representing r. Second and fourth columns represent cross sections of the bulk and the surface, respectively. Spot patterns form on the surface, whereas small balls form in the vicinity of the surface inside the bulk. (Online version in colour.)

(iii) Simulations of the coupled system of BSRDEs with $(d_\Omega, d_\Gamma)=(20,1)$

To generate patterns in the bulk, we take $d_\Omega = 20 > d_\Gamma^{\text{crit}} = 8.5$ and $d_\Gamma=1$ on the surface.**Figure 4** exhibits stripe, circular and spot patterns in the bulk as illustrated by the cross sections. On the surface, small amplitude patterns occur consistent with theoretical predictions. Although the patterns for the u species (columns one and two) appear uniform on the surface this is simply owing to the colour scale, with the amplitude of the patterns in the bulk larger than those on the surface. This difference in the amplitude of the pattern of the bulk solution in the bulk and on the surface is due to the Robin-type boundary conditions. Unlike zero-flux (also known as homogeneous Neumann), boundary conditions for standard reaction–diffusion systems which imply that no species enter or leave the domain, here, there is deposition or removal of chemical species through the flux on the surface, resulting in differences in amplitude between the bulk and surface solutions.

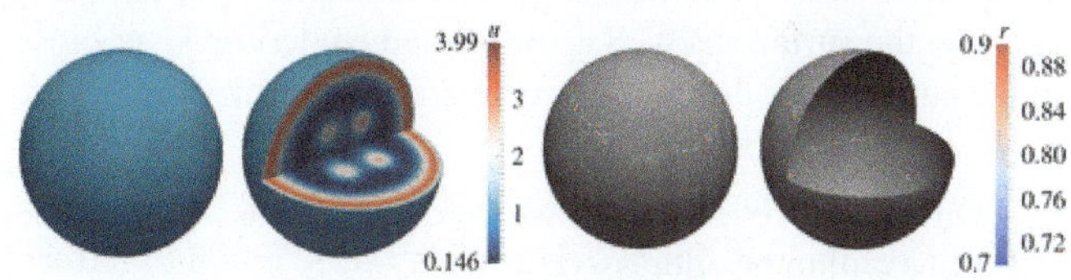

Figure 4: Numerical solutions corresponding to the coupled system of BSRDEs (2.1)–(2.5) with $d_\Omega=20$ in the bulk and $d_\Gamma=1$ on the surface. Columns 1 and 2: solutions in the bulk representing u. Columns 3 and 4: solutions on the surface representing r. Second and fourth columns represent cross sections of the bulk and the surface, respectively. Spectacular patterning occurs in the bulk exhibiting spots, stripes and circular patterns. The surface dynamics produce uniform patterning. (Online version in colour.)

(iv) Simulations of the coupled system of BSRDEs with $(d_\Omega, d_\Gamma)=(20,20)$

In this example, we illustrate how both bulk and surface dynamics induce patterning by taking $d_\Omega=20$ in the bulk, $d_\Gamma=20$ on the surface. **Figure 5** shows pattern formation in the bulk and on the surface. In the bulk, we observe the formation of balls (which can be seen as spots through cross sections) and these translate to spots on the surface. The surface dynamics themselves induce spot pattern formation.

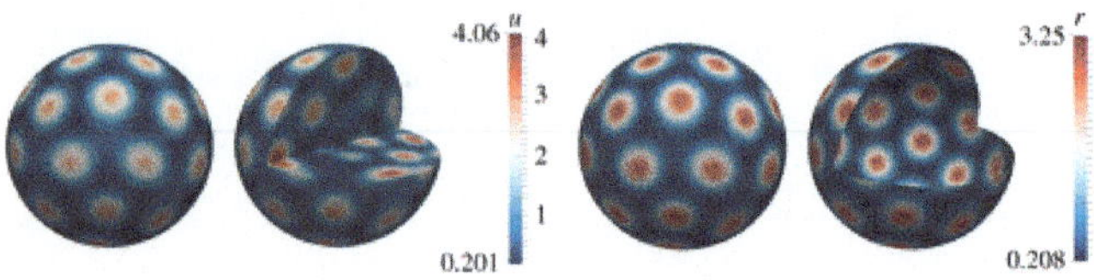

Figure 5: Numerical solutions corresponding to the coupled system of BSRDEs (2.1)–(2.5) with d_Ω=20 in the bulk and d_Γ=20 on the surface. Columns 1 and 2: solutions in the bulk representing u. Columns 3 and 4: solutions on the surface representing r. Second and fourth columns represent cross sections of the bulk and the surface, respectively. We observe spot pattern formation both in the bulk and on the surface. (Online version in colour.)

CONCLUSION, DISCUSSION AND FUTURE RESEARCH CHALLENGES

We have presented a coupled system of BSRDEs whereby the bulk and surface reaction–diffusion systems are coupled through Robin-type boundary conditions. Nonlinear reaction-kinetics are considered in the bulk and on the surface and for illustrative purposes, the activator-depleted model was selected because it has a unique positive steady state. By using linear stability theory close to the bifurcation point, we state and prove a generalization of the necessary conditions for Turing diffusion-driven instability for the coupled system of BSRDEs. Our most revealing result is that the bulk reaction–diffusion system has the capability of inducing patterning (under appropriate model and compatibility parameter values) for the surface reaction–diffusion model. On the other hand, the surface reaction–diffusion is not capable of inducing patterning everywhere in the bulk; patterns can be induced in only regions close to the surface membrane. For skin pattern formation, this example is consistent with the observation that patterns will form on the surface as well as within the epidermis layer close to the surface. We do not expect patterning to form everywhere in the body of the animals.

Our studies reveal the following observations and research questions still to be addressed:

- – our numerical experiments reveal that the Robin-type boundary conditions seem to introduce a boundary layer coupling the bulk and surface dynamics. However, these boundary conditions do not appear explicitly in the conditions for diffusion-driven instability and this makes it difficult to theoretically analyse their role and implications to pattern formation. Further studies are required to understand the role of these boundary conditions as well as the size of the boundary layer;
- – the compatibility condition (2.12) implies that the uniform steady state in the bulk is identical to the uniform state on the surface. We are currently studying the implications of relaxing the compatibility condition;
- – finally, in this manuscript, we have not carried out detailed parameter search and estimation to deduce the necessary and sufficient conditions for pattern generation as well as isolating excitable wavenumbers in the bulk and on the surface. Such studies might reveal more interesting properties of the coupled bulk-surface model and this forms part of our current studies.

We have presented a framework that couples bulk dynamics (three-dimension) to surface dynamics (two-dimension) with the potential of numerous applications in cell motility, developmental biology, tissue engineering and regenerative medicine and biopharmaceutical where reaction–diffusion-type models are routinely used [**5**,**6**,**11**–**14**,**16**,**17**].

We have restricted our studies to stationary volumes. In most cases, biological surfaces are known to evolve continuously with time. This introduces extra complexities to the modelling, analysis and simulation of coupled systems of BSRDEs. In order to consider evolving bulk-surface partial differential equations, evolution laws (geometrical) should be formulated describing how the bulk and surface evolve. Here, it is important to consider specific experimental settings that allow for detailed knowledge of properties (biomechanical) and processes (biochemical) involved in the bulk-surface evolution. Such a framework will allows us to study three-dimensional cell migration in the area of cell motility [**16**,**27**–**29**]. In future studies, we propose to develop a three-dimensional integrative model that couples bulk and surface dynamics during growth development or movement.

DATA ACCESSIBILITY

This manuscript does not contain primary data and as a result has no supporting material associated with the results presented.

FUNDING STATEMENT

This work (A.M. and C.V.) is supported by the Engineering and Physical Sciences Research Council grant: (EP/J016780/1). A.M. and C.V. acknowledge support from the Leverhulme Trust Research Project Grant (RPG-2014-149). A.H.C. was supported partly by the University of Sussex and partly by the Medical Research Council.

ACKNOWLEDGEMENTS

The authors thank anonymous reviewers for their constructive comments.

REFERENCES

1. Bianco S, Tewes F, Tajber L, Caron V, Corrigan OI, Healy AM. 2013 Bulk, surface properties and water uptake mechanisms of salt/acid amorphous composite systems. Int. J. Pharm. 456, 143–152. (doi:10.1016/j.ijpharm.2013.07.076)
2. Kwon Y, Derby JJ. 2001 Modeling the coupled effects of interfacial and bulk phenomena during solution crystal growth. J. Crystallogr. Growth 230, 328–335. (doi:10.1016/S0022-0248(01)01345-8)
3. Medvedev ES, Stuchebrukhov AA. 2011 Proton diffusion along biological membranes. J. Phys. Condensed Matter 23, 234103. (doi:10.1088/0953-8984/23/23/234103)
4. Levine H, Rappel WJ. 2005 Membrane-bound Turing patterns. Phys. Rev. E 72, 061912.
5. Rätz A, Röger M. 2012 Turing instabilities in a

mathematical model for signaling networks. J. Math. Biol. 65, 1215–1244. (doi:10.1007/s00285-011-0495-4) Medline

6. Rätz A, Röger M. 2013 Symmetry breaking in a bulk-surface reaction–diffusion model for signaling networks. (http://arxiv.org/abs/1305.6172v1)
7. Turing AM. 1952 The chemical basis of morphogenesis. Phil. Trans. R. Soc. Lond. B 237, 37–72. (doi:10.1098/rstb.1952.0012)
8. Gierer A, Meinhardt H. 1972 A theory of biological pattern formation. Kybernetik 12, 30–39. (doi:10.1007/BF00289234)
9. Murray JD 2003 Mathematical biology II: spatial models and biomedical applications, 3rd edn. Berlin, Germany: Springer.
10. Elliott CM, Ranner T. 2012 Finite element analysis for a coupled bulk-surface partial differential equation. IMA J. Numer. Anal. 33, 377–402. (doi:10.1093/imanum/drs022)
11. Novak IL, Gao F, Choi Y-S, Resasco D, Schaff JC, Slepchenko BM. 2007 Diffusion on a curved surface coupled to diffusion in the volume: application to cell biology. J. Comp. Phys. 226, 1271–1290. (doi:10.1016/j.jcp.2007.05.025)
12. Chechkin AV, Zaid IM, Lomholt MA, Sokolov IM, Metzler R. 2012 Bulk-mediated diffusion on a planar surface: full solution. Phys. Rev. E 86, 041101.
13. Nisbet DR, Rodda AE, Finkelstein DI, Horne MK, Forsythe JS, Shen W. 2009 Surface and bulk characterisation of electrospun membranes: problems and improvements. Colloids Surf. B, Biointerfaces 71, 1–12. (doi:10.1016/j.colsurfb.2009.01.022)
14. Medvedev ES, Stuchebrukhov AA. 2013 Mechanism of long-range proton translocation along biological membranes. FEBS Lett. 587, 345–349. (doi:10.1016/j.febslet.2012.12.010)
15. Rozada I, Ruuth S, Ward MJ. 2014 The stability of localized spot patterns for the Brusselator on the sphere. SIADS 13, 564–627. (doi:10.1137/130934696)
16. Elliott CM, Stinner B, Venkataraman C. 2013 Modelling cell motility and chemotaxis with evolving surface finite elements. J. R. Soc. Interface 9, 3027–3044. (doi:10.1098/rsif.2012.0276)
17. Venkataraman C, Sekimura T, Gaffney EA, Maini PK, Madzvamuse A. 2011 Modeling parr-mark pattern formation during the early development of Amago trout. Phys. Rev. E 84, 041923. (doi:10.1103/PhysRevE.84.041923)
18. Dziuk G, Elliott CM. 2007 Surface finite elements for parabolic equations. J. Comp. Math. 25, 385–407.
19. Madzvamuse A, Chung AHW. Submitted The bulk-surface finite element method for reaction–diffusion systems on stationary volumes. SIAM SISC.
20. Bangerth W, Heister T, Heltai L, Kanschat G, Kronbichler M, Maier M, Turcksin B, Young TD. 2013 The deal.II library, version 8.1 arXiv preprint.
21. Macdonald CB, Merrimanb B, Ruuth SJ. 2013 Simple computation of reaction–diffusion processes on point clouds. Proc. Natl Acad. Sci. USA 110, 9209–9214. (doi:10.1073/pnas.1221408110)
22. Lakkis O, Madzvamuse A, Venkataraman C. 2013 Implicit–explicit timestepping with finite element approximation of reaction–diffusion systems on evolving domains. SINUM 51, 2309–2330. (doi:10.1137/120880112)
23. Prigogine I, Lefever R. 1968 Symmetry breaking instabilities in dissipative systems. II. J. Chem. Phys. 48, 1695–1700. (doi:10.1063/1.1668896)
24. Schnakenberg J. 1979 Simple chemical reaction systems with limit cycle behaviour. J. Theor. Biol. 81, 389–400. (doi:10.1016/0022-5193(79)90042-0)
25. Venkataraman C, Lakkis O, Madzvamuse A. 2012 Global existence for semilinear reaction–diffusion systems on evolving domains. J. Math. Biol. 64, 41–67. (doi:10.1007/s00285-011-0404-x)
26. Madzvamuse A, Chung AHW. 2014 Fully implicit time-stepping schemes and non-linear solvers for systems of reaction–diffusion equations. Appl. Math. Comp. 244, 361–374. (doi:10.1016/j.amc.2014.07.004)
27. George UZ, Stephanou A, Madzvamuse A. 2013 Mathematical modelling and numerical simulations of actin dynamics in the eukaryotic cell. J. Math. Biol. 66, 547–593. (doi:10.1007/s00285-012-0521-1)
28. Madzvamuse A, George UZ. 2013 The moving grid finite element method applied to cell movement and deformation. Finite Element Anal. Des. 74, 76–92. (doi:10.1016/j.finel.2013.06.002)
29. Neilson MP, Mackenzie J, Webb S, Insall RH. 2011 Modelling cell movement and chemotaxis using pseudopod-based feedback. SIAM J. Sci. Comp. 33, 1035–1057. (doi:10.1137/100788938)

INDEX